Lecture Notes in Networks and Systems

Volume 202

The series “Lecture Notes in Networks and Systems” publishes the latest developments in Networks and Systems—quickly, informally and with high quality. Original research reported in proceedings and post-proceedings represents the core of LNNS.

Volumes published in LNNS embrace all aspects and subfields of, as well as new challenges in, Networks and Systems.

The series contains proceedings and edited volumes in systems and networks, spanning the areas of Cyber-Physical Systems, Autonomous Systems, Sensor Networks, Control Systems, Energy Systems, Automotive Systems, Biological Systems, Vehicular Networking and Connected Vehicles, Aerospace Systems, Automation, Manufacturing, Smart Grids, Nonlinear Systems, Power Systems, Robotics, Social Systems, Economic Systems and other. Of particular value to both the contributors and the readership are the short publication timeframe and the world-wide distribution and exposure which enable both a wide and rapid dissemination of research output.

The series covers the theory, applications, and perspectives on the state of the art and future developments relevant to systems and networks, decision making, control, complex processes and related areas, as embedded in the fields of interdisciplinary and applied sciences, engineering, computer science, physics, economics, social, and life sciences, as well as the paradigms and methodologies behind them.

Indexed by SCOPUS, INSPEC, WTI Frankfurt eG, zbMATH, SCImago.

All books published in the series are submitted for consideration in Web of Science.

More information about this series at http://www.springer.com/series/15179

Swagatam Das · Mihir Narayan Mohanty
Editors

Advances in Intelligent Computing and Communication

Proceedings of ICAC 2020

Springer

Editors
Swagatam Das
Electronics and Communication
Science Unit
Indian Statistical Institute
Kolkata, West Bengal, India

Mihir Narayan Mohanty
Department of Electronics
and Communication Engineering
Institute of Technical Education
and Research (ITER)
Siksha 'O' Anusandhan
(Deemed to be University)
Bhubaneswar, Odisha, India

ISSN 2367-3370 ISSN 2367-3389 (electronic)
Lecture Notes in Networks and Systems
ISBN 978-981-16-0694-6 ISBN 978-981-16-0695-3 (eBook)
https://doi.org/10.1007/978-981-16-0695-3

This Springer imprint is published by the registered company Springer Nature Singapore Pte Ltd.
The registered company address is: 152 Beach Road, #21-01/04 Gateway East, Singapore 189721, Singapore

ICAC-2020 Committee

Chief Patron

Prof. Manoj Ranjan Nayak, President, Siksha 'O' Anusandhan (Deemed to be University), Bhubaneswar, Odisha, India

Patron

Prof. Ashok Kumar Mohapatra, Vice-Chancellor, Siksha 'O' Anusandhan (Deemed to be University), Bhubaneswar, Odisha, India

General Chair

Prof. Mihir Narayan Mohanty, Professor, Department of ECE, ITER, Siksha 'O' Anusandhan (Deemed to be University), Bhubaneswar, Odisha, India

General Co-chair

Prof. Swagatam Das, ISI, Kolkata

Program Chair

Prof. Bibhuprasad Mohanty, Professor and HoD, Department of ECE, ITER, Siksha 'O' Anusandhan (Deemed to be University), Bhubaneswar, Odisha, India

Reviewers

Dr. Rashmirekha Ram, Siksha 'O' Anusandhan (Deemed to be University), Bhubaneswar
Dr. J. Naga Vishnu Vardhan, BVRIT Hyderabad College of Engineering for Women
Dr. Jayashree Ratnam, Siksha 'O' Anusandhan (Deemed to be University), Bhubaneswar
Dr. Monalisa Mohanty, Siksha 'O' Anusandhan (Deemed to be University), Bhubaneswar
Dr. Lambodar Jena, Siksha 'O' Anusandhan (Deemed to be) University, Bhubaneswar
Dr. Manoj Kumar Naik, Siksha 'O' Anusandhan (Deemed to be University), Bhubaneswar
Dr. Laxmi Prasad Mishra, Siksha 'O' Anusandhan (Deemed to be University), Bhubaneswar
Dr. Basanta Kumar Panigrahi, Siksha 'O' Anusandhan (Deemed to be University), Bhubaneswar
Dr. Hemanta Kumar Palo, ITER, Siksha 'O' Anusandhan (Deemed to be University), Bhubaneswar
Gyana Ranjan Patra, ITER, Siksha 'O' Anusandhan (Deemed to be University), Bhubaneswar
Dr. Smita Prava Mishra, Siksha 'O' Anusandhan (Deemed to be University), Bhubaneswar
Dr. Waheed Ahmad Lone, Yildiz Technical University
Dr. Debabrata Singh, ITER, Siksha 'O' Anusandhan (Deemed to be University), Bhubaneswar
Dr. Ashish Kumar Mourya, Gautam Buddha University
Mr. Amir Hassan, GDC Anantnag
Dr. Gaurav Tripathi, Central Research Lab, Bharat Electronics Limited
Dr. Sivkumar Mishra, CAPGS, BPUT, Odisha
Dr. Asit Kumar Subudhi, Siksha 'O' Anusandhan (Deemed to be University), Bhubaneswar

Dr. Akash Kumar Bhoi, Sikkim Manipal University
Dr. Rakesh Ranjan Kumar, GGSESTC Bokaro, Jharkhand
Dr. Sujata Chakravarty, Centurion University of Technology and Management, Odisha
Dr. Niva Das, Professor, ECE Department, ITER, Siksha 'O' Anusandhan (Deemed to be University), Bhubaneswar
Dr. Kaberi Das, Associate Professor, Department of CSE, ITER, Siksha 'O' Anusandhan (Deemed to be University), Bhubaneswar
Dr. Rudra Kalyan Nayak, Associate Professor, Department of Computer Science and Engineering, Koneru Lakshmaiah Education Foundation (Deemed to be University), Vaddeswaram, Guntur District, Andhra Pradesh, India
Dr. P. V. Praveen Sundar, Assistant Professor, Adhiparasakthi College of Arts and Science, Kalavai
Dr. Deepak Sharma, Professor and HOD, KJSCE Mumbai
Dr. Dhaval Pujara, Nirma University
Dr. Saroja Kumar Rout, Associate Professor, Department of CSE, Gandhi Institute for Technology, Bhubaneswar
Dr. Sukant Kishoro Bisoy, C. V. Raman Global University, Bhubaneswar
Dr. Dinesh Garg, Sri Sai College of Engineering & Technology
Dr. D. H. Manjaiah, Mangalore University
Dr. Kandarpa Kumar Sarma, Gauhati University
Dr. Sanjeeb Kumar Kar, Siksha 'O' Anusandhan (Deemed to be University), Bhubaneswar
Dr. Binod Kumar Pattanayak, Siksha 'O' Anusandhan (Deemed to be University), Bhubaneswar
Dr. Binod Kumar Sahu, Siksha 'O' Anusandhan (Deemed to be University), Bhubaneswar
Dr. R. Sridaran, Marwadi University
Dr. Manoj Kumar Naik, ITER, Siksha 'O' Anusandhan (Deemed to be University), Bhubaneswar
Dr. Santanu Kumar Sahoo, Siksha 'O' Anusandhan (Deemed to be University), Bhubaneswar
Dr. Mitrabinda Ray, Siksha 'O' Anusandhan (Deemed to be University), Bhubaneswar

Preface

This issue of *Lecture Notes in Networks and Systems* is dedicated to the Third International Conference on Intelligent Computing and Communication (ICAC-2020) at the campus of Institute of Technical Education and Research (Faculty of Engineering and Technology), Siksha 'O' Anusandhan (Deemed to be University), Bhubaneswar, from November, 25 to 26, 2020. This conference was organized by the Department of Electronics and Communication Engineering of the Institute of Technical Education and Research (Faculty of Engineering and Technology). The conference had three tracks, namely advances in communication systems, intelligent systems and signal processing. From 130 papers received from different authors, 67 papers were selected for inclusion into the conference proceedings. Each paper was peer reviewed by at least two reviewers. Due to the COVID-19 pandemic, this time the conference was organized by virtual mode.

The objective of the conference was to bring together experts from academic institutions, industries, research organizations and professional engineers for sharing of knowledge, expertise and experience in emerging trends related to the computer, communication and electrical topics. The aim of this international conference is to coverage all the issues on a single platform and provide international forum for researchers to discuss the real-time problems and solutions to exchange their valuable ideas and showcase the ongoing works which may lead to path-breaking foundation of the futuristic engineering. This conference mainly aims at advanced communication protocol, database security and privacy, advanced computing system, saving energy, etc., on several updated techniques. The conference offers a platform to focus on the inventive information and computing toward the investigation of cognitive mechanisms and processes of human information processing and the development of the next-generation engineering and advanced technological systems.

Siksha 'O' Anusandhan is a Deemed to be University located in Bhubaneswar, Odisha, India. It was originally founded as the Institute of Technical Education and Research (ITER) in the year 1996. The university has been at the forefront nourishing a learning ambience and encouraging academic, research and innovations since its inception in 2007. The university is composed of nine degree-granting schools with 10,000 students. Many of SOAU's programs are nationally accredited for meeting high standards of academic quality, including engineering, medicine, pharmacy, business, nursing, biotechnology, humanities, environment, nanotechnology, agriculture and law. The university was ranked 20th by National Institutional Ranking Framework (NIRF) under the aegis of Ministry of Human Resource Development, Government of India, and has been awarded 'A' Grade by NAAC. It has established 10 research centers and 23 research laboratories to fulfill the need of faculties and students.

Faculty of Engineering and Technology is a constituent of SOAU having thirteen departments with more than four hundred faculty members. The department of ECE is continually working to provide quality research outputs in the areas of signal and image processing, communication engineering and microelectronics devices.

The editors thank the authors for extending their fullest cooperation in preparing the manuscripts to the Springer Lecture Notes guidelines, taking aboard the additional review comments.

The editors would also like to convey their heartfelt thanks to Prof. M. R. Nayak, President, Siksha 'O' Anusandhan (Deemed to be University); Prof. Ashok Kumar Mohapatra, Vice-Chancellor, Siksha 'O' Anusandhan (Deemed to be University); Prof. M. K. Mallick, Director, ITER (FET), Siksha 'O' Anusandhan (Deemed to be University); Prof. P. K. Nanda, Dean (R&D), Siksha 'O' Anusandhan (Deemed to be University); and Prof. P. K. Sahoo, Dean, Siksha 'O' Anusandhan (Deemed to be University), for their constant inspiration and motivation in all stages of the conference.

Sincerely,

Bhubaneswar, India — Dr. Mihir Narayan Mohanty
Kolkata, India — Dr. Swagatam Das

Contents

About the Editors

Dr. Swagatam Das received his B.E. Tel. E., M.E. Tel. E (Control Engineering specialization), and Ph.D. degrees, all from Jadavpur University, India, in 2003, 2005, and 2009, respectively. Currently serving as an Associate Professor at the Electronics and Communication Sciences Unit of the Indian Statistical Institute, Kolkata, India, his research interests include evolutionary computing, pattern recognition, multi-agent systems, and wireless communication. Dr. Das has published more than 300 research articles in peer-reviewed journals and international conference proceedings. He is the founding co-editor-in-chief of Swarm and Evolutionary Computation, an international journal from Elsevier. He has also served as or is serving as an associate editor of various journals, including Pattern Recognition, Neurocomputing, Information Sciences, IEEE Access, and so on. He is an editorial board member of numerous other journals, including Progress in Artificial Intelligence, Applied Soft Computing, and Artificial Intelligence Review. Dr. Das is the recipient of the 2012 Young Engineer Award from the Indian National Academy of Engineering (INAE), and of the 2015 Thomson Reuters Research Excellence India Citation Award for the highest cited researcher from India in the Engineering and Computer Science category for the period 2010 to 2014.

Dr. Mihir Narayan Mohanty received his M.Tech. in Communication System Engineering from Sambalpur University, Odisha, and his Ph.D. in Applied Signal Processing from Biju Patnaik University of Technology, Odisha. He is currently serving as a Professor at the Department of Electronics and Communication Engineering, Institute of Technical Education and Research (FET), Siksha 'O' Anusandhan (Deemed University), Bhubaneswar, Odisha. With 24 years of teaching experience, his research interests include intelligent signal and image processing, digital signal/image processing, biomedical signal processing, microwave communication engineering, and antennas. He has published over 300 research papers in national and international journals and conference proceedings, and has received various national and international awards for his contributions.

He is an active member of numerous professional societies, e.g., the IEEE, Senior Member, IET, IETE, ISTE, EMC & EMI Engineers India, etc. Dr. Mohanty is also an active reviewer for the IEEE, Elsevier, Springer, and several international conferences.

Classification and Detection of Leaves Using Different Image Processing Techniques

Akshay Raina, Shubham Mahajan, Ch. Vanipriya, Anil Bhardwaj, and Amit Kant Pandit

Abstract Diversity of plants are present in earth's nature where each type has its individual exceptional highlights. Because of their gigantic advantages to humankind, many plant species are utilized in everyday lifetime. Hence, precise plant leaf acknowledgment over computer vision techniques has cleared its approach to a few fields such as ayurvedic and analysis of wellbeing matters. Innovation has consistently assumed a fundamental job in all parts of human turn of events. Accomplishing exact acknowledgment and arrangement of plant leaf is consistently a test to specialists. In this work, advance different procedures that are received for pre-preparing include extraction furthermore, characterization of leaf, in view of shape and surface highlights of leaves test. The paper further presents test outcomes did on Flavia dataset in request to perceive, arrange leaves utilizing dim level co-event network and various leveled centroid-based strategies. In this work, about 300 leaves are tested by 30 unique modules with end goal of examination. Abstract should summarize the contents of the paper and should contain at least 70 and at most 150 words. It should be set in 9-point font size and should be inset 1.0 cm from the right and left margins. There should be two blank (10-point) lines before and after the abstract. This document is in the required format.

Keywords Image processing · Leaves · Plant classification · Gray level co-occurrence network

---

A. Raina · S. Mahajan (✉) · A. Bhardwaj · A. K. Pandit
Shri Mata Vaishno Devi University, Katra, Reasi, Jammu and Kashmir, India
e-mail: mahajanshubham2232579@gmail.com

A. Raina
e-mail: 18bee003@smvdu.ac.in

A. Bhardwaj
e-mail: anil.bhardwaj@smvdu.ac.in

A. K. Pandit
e-mail: amitkantpandit@gmail.com

Ch. Vanipriya
Sir M. Visvesvaraya Institute of Technology, Bangalore, India
e-mail: vanipriya_is@sirmvit.edu

S. Das and M. N. Mohanty (eds.), *Advances in Intelligent Computing and Communication*, Lecture Notes in Networks and Systems 202,
https://doi.org/10.1007/978-981-16-0695-3_1

1 Introduction

Plants are maximum noteworthy and basic individual from earth's condition. Earth's atmosphere is inconceivable without plants. All things are considered to misuse plants restorative properties and protection highlights, discovery, and order of plants which is vital undertaking. Plants type is additionally utilized as elective vitality origin as bio-fuel. There are different approaches for plant recognition over root part, leaves, organic product, blossom and so on. Due to headway in science and mechanical space, picture preparing containing acknowledgment and grouping plan has become an indispensable exploration work in practically all wildernesses of uses. In any case, computer vision and example acknowledgment techniques are embraced for plant acknowledgment through a robotized procedure. In picture preparing, invariant moments are trendier for discoveries of leaves and items of plants. It is essential to take note of a few strategies that shading minutes were read for plant classification, while a few methodologies utilized picture correlogram for picture recovery. In this manner, a few strategies are created to identify and group discriminative highlights of plant types over characterization calculations [1]. Order and acknowledgment of leaves pictures of plants dependent on development rate could be accomplished by investigating flora stature at normal time stretches. Figuring of tallness of flora were utilized in recognizing surface highlights, shape, and shade of flora leaves images [2]. In clinical arena, different sicknesses are restored through plant types. Ayurvedic natural field utilizes a few plant leaves to sedate a few maladies. In this manner, characterization and looking at of plant leaf pictures is a significant advance in clinical field. This paper gives structure of a robotized leaf acknowledgment and order model on standard Flavia dataset. The conversion of RGB to grayscale change strategy is being utilized for picture pre-handling stage. Discrete wavelet changes (DWT) and dim level co-event matrix (GLCM) calculations are measured for surface component extraction of leaves tests. Also, utilizing progressive centroid strategy, shape highlights are examined. Information tests are grouped utilizing guideline part investigation (PCA) calculation to support up the characterization exactness.

2 Literature Survey

Juheon expressed to recognize different species and identify different vegetation attributes. The built up conspire displays nonexclusive work process to plan tree types. The model received PCA and SVM classifier for include extraction. The calculations utilized, support in delivering ITC level order precision to 61%-pixel level characterization exactness up to 91% [3]. While creators, Jou has projected novel learning-based leaves picture acknowledgment. The framework model shows data about sack of words based distinguishing proof plan each leaf pictures utilized for examination. Capable leaf acknowledgment strategy has been accomplished utilizing open leaves tests dataset [4]. Hicham has received an exact surface acknowledgment framework. The procedure comprises of removal of locally and internationally invariant portrayals. Received plan is another multifractal descriptor that catches rich surface data furthermore and is scientifically to different complex changes. PCA classifier and SVM support framework. The trial results showed that model is computationally efficient [5]. Author has explored on technique dark level co-event matrix to determine different utilizations of model. Co-event matrix jam recurrence and data. The pictures are procured with a slight change in edges and separations to test heartiness of artificial neural organize (ANN) classifier. Utilizations of gray level co-occurrence network (GLCM) beat haralick includes because of loose lighting [6]. In any case, Diwakar has projected vision-based investigation, qualities of burning flares or charge-coupled gadget (CCD) identified with computerized picture preparing. By utilizing GLCM, different factual picture highlights, grayscale test pictures are determined, corresponded with edge boundaries. LBP in view of edge identification of the red shading segment at different temperatures is completed. Test results show the great aftereffect of the measurable highlights of the image [7].

3 Results and Discussion

The result of projected model is assessed in area. Pre-preparing of information test picture is executed by the RBG to dark scale change technique. Surface highlights of info leaves tests are removed via utilizing discrete wavelet change and dark level co-event framework calculations. Its shape highlights are taken out by utilizing various leveled centroid calculations. The yield highlights are grouped by utilizing PCA approach for best possible acknowledgment of leaf test as shown in Fig. 1.

The measure of exactness is given:

$$\mathrm{ACC} = \frac{T_P + T_N}{P + N} = \frac{T_P + T_N}{T_P + T_N + F_P + F_N} \tag{1}$$

Exactness is only logical or mechanical precision which is given:

$$\mathrm{PPV} = \frac{T_P}{T_P + F_P} \tag{2}$$

Fig. 1 Different types of leaves with accuracy

Sensitivity is recall rate and true positive rate:

$$\text{TPR} = \frac{T_P}{P} = \frac{T_P}{T_P = F_N} \quad (3)$$

4 Conclusion

The projected method gives standard leaf affirmation framework by utilizing diverse component extraction with plan systems. By utilizing GLCM, DWT, and hierarchical centroid estimations, we contract spatial association between features exactly. These strategies will support us with boosting up nature of outcome. To accomplish request exactness, PCA is utilized. Completely 300 leaf tests are measured with 30 unmistakable class. The projected technique utilized Flavia dataset and attained all things considered execution precision up to 85%.

References

1. Lee, S.H., Chan, C.S., Wilkin, P., Remagnino, P.: Plant identification with convolutional neural networks, pp. 452–456. IEEE (2015)
2. Patil, S., Soma, S., Nandyal, S.: Identification of growth rate of plant based on leaf features using digital image processing techniques. Int. J. Emerg. Technol. Adv. Eng. **3**(8), 266–275 (2013)
3. Lee, J., Cai, X., Lellmann, J., Dalponte, M., Malhi, Y., Butt, N., Morecroft, M., Schönlieb, C. B., Coomes, D.A.: Individual tree species classification from airborne multisensor imagery using robust PCA. IEEE J. Sel. Top. Appl. Earth Obs. Remote Sens. **9**(6), 2554–2567 (2016)
4. Hsiao, J.K, Kang, L.W, Chang, C.L., Lin, C.Y.: Comparative study of leaf image recognition with a novel learning-based approach, pp. 389–393. IEEE (2014)
5. Badri, H., Yahia, H., Daoudi, K.: Fast and accurate texture recognition with multilayer convolution and multifractal analysis, pp. 505–519. Springer (2014)
6. Wong, W.K., Ali, C., Brandon, K., Wee, C.C., Marriapan, M.: Co-occurrence matrix with neural network classifier for weed species classification: a comparison between direct application of co-occurrence matrix (GLCM) and haralick features as inputs. Int. J. Enhanced Res. Sci. Eng. **2**(2), 1–6 (2013)
7. Agarwal, D.: GLCM based analysis of combustion flame parameters uses edge detection of flame radiation images, pp. 1–5. IEEE (2015)

COVID-19 Detection: An Approach Using X-Ray Images and Deep Learning Techniques

Akshay Raina, Shubham Mahajan, Ch. Vanipriya, Anil Bhardwaj, and Amit Kant Pandit

Abstract In the recent history of human civilization, a pandemic affecting such an enormous population like COVID-19 was about 140 years ago-The Smallpox Worldwide Epidemic (1877–1977, Deaths-500 M). It can be easily inferred that the health management system over the globe in the nineteenth century was too underdeveloped than that of today, which also refers to the fact that the present epidemic must not be allowed to last much longer as the number of deaths is increasing nonlinearly (506 K, with 10.3 M affected). While the medical community around the globe is striving to find a permanent cure, it becomes evident responsibility of all professionals who can contribute in stabilizing the medical management systems of countries particularly underdeveloped/developing countries or those with highest rate of increase in COVID-19 cases like USA, Brazil. In this regard, this study introduces a fast, robust and practically effective method for detection of COVID-19 from chest x-ray images utilizing enhanced deep learning techniques. An object detection network is proposed to be trained with publicly existing datasets. In this model, SSD is used with ResNet101 as a base layer and some pre-processing, achieving a sensitivity of 0.9495 and a specificity of 0.9247. If practically implemented, this can prove very beneficial in aiding economies and health systems of the above-mentioned countries.

A. Raina · S. Mahajan (✉) · A. Bhardwaj · A. K. Pandit
Shri Mata Vaishno Devi University, Katra, Reasi, Jammu and Kashmir, India
e-mail: mahajanshubham2232579@gmail.com

A. Raina
e-mail: 18bee003@smvdu.ac.in

A. Bhardwaj
e-mail: anil.bhardwaj@smvdu.ac.in

A. K. Pandit
e-mail: amitkantpandit@gmail.com

Ch. Vanipriya
Sir M. Visvesvaraya Institute of Technology, Bangalore, India
e-mail: vanipriya_is@sirmvit.edu

S. Das and M. N. Mohanty (eds.), *Advances in Intelligent Computing and Communication*, Lecture Notes in Networks and Systems 202,
https://doi.org/10.1007/978-981-16-0695-3_2

Keywords COVID-19 · COVIDx-Net · Deep learning · Medical images · X-ray · Object detection

1 Introduction

The disease with SARS-CoV-2, infection that causes COVID-19, has created widespread instability in almost every domain and made most sectors (particularly) newly formed and small-scale organizations) to the brink of collapse. Various studies have predicted that this would affect economic and professional instability across numerous domains. Presently, this has provoked border-sealing by most of the countries and confined millions of households to any social activities. The USA stands on the top of the table with most confirmed cases of 2.64 M and 128 K deaths followed by Brazil, Russia and India. It is worth stating that these numbers are increasing very rapidly.

There are many studies done and being done which state that even the rate of increase in these numbers is increasing. As such, origin of outbreak is officially unknown, but symptoms were detected in first cases in early December in China. The most widespread health signs are fever, dry cough and tiredness, while fewer common symptoms are aches and irritation, vomiting, sore throat, conjunctivitis, skin rash, or discoloration of fingers or toes and extreme symptoms such as trouble breathing or shortness of breath, chest pain or pressure, lack of speech or movement. There are a few approved drug-based detection methods presently! The polymerase chain reaction (PCR) is one of them. One is reverse transcription polymerase chain reaction (RT-PCR) process, which first practices invert translation for DNA creation, trailed by PCR for DNA intensification, producing enough to be investigated. This encourages the RT-PCR to recognize SARS-CoV-2, which contains just an RNA. This typically takes several hours, be that as it may. Ongoing PCR (qPCR) gives preferences including precision, better and increasingly solid instrumentation. It has become a typical technique. Tests can likewise be acquired utilizing an assortment of methods, including nasopharyngeal swab, sputum, throat swabs, profound aviation route material assembled by attractions catheter or salivation. It merits referencing that Drosten [1] noticed that for 2003 SARS, from an indicative perspective, notice that nasal and throat swabs seem, by all accounts, to be less fitting for conclusion, as these materials contain generously less popular RNA than sputum, and the infection may get away from distinguishing proof if just these materials are analyzed.

In such conditions, AI, radiomics, medical imaging applied to C: X-rays can prove to be a great support to the detection system mentioned above. As stated in [2, 3], the signs of COVID-19 are reciprocal appropriation of inconsistent shadows and prominent ground glass opacity lesions (GGO) in fringe and back lungs which can assume a significant job in the conclusion of COVID-19 through methods for medical imaging. The COVID-19 cases have fundamentally the same highlights with different viral pneumonia like GGO in beginning phases and pulmonary combination later on. Consequently, it is impossible for radiologists to distinguish COVID-19 from other viral pneumonia without any diagnostic examinations.

AI algorithms and radiomics technologies extracted from chest X-rays will be of considerable benefit in conducting major screening programs in almost every region. Therefore, it is also encouraged to set up plans to bring this alternative to detection through drugs into testing phase to begin with.

There have been number of works in this regard previously. More specifically, other than traditional machine learning approaches, deep learning techniques are widely accepted and preferred for such targets of interest. An obvious neural network category in this regard is the convolutional neural network (CNN). These have proved to be "revolution in image processing" since the advent of the twenty-first century. In the context of our discussion, one of such works is done by Wang et al. [4] trained 217 computed tomography (CT) images using the inception migration-learning model with random selection of Regions of Interest (ROIs) and total accuracy of 83% was achieved with a specificity of 80.5% and a sensitivity of 84% for validation. Another study and one of the highly cited, by Wang and Wong [5], proposed the COVID-Net CNN architecture, one of first open-source network for COVID-19 detection. This network resulted in comparable values of accuracy of 92.4%, sensibility of 80% and specificity of 88.9%. In the same regard, another proposal has been made by Hemdan et al. [6]. The COVIDx-Net network consists of seven separate architectures of deep convolutional neural network models (VGG19, DenseNet121, ResNet-V2, Inception-V3, InceptionResNet-V2, Xception and MobileNet-V2) and findings of the proposed COVIDx-Net confirmed that the highest success ratings of deep learning classifiers are for the VGG19 and DenseNet201 versions. Their tests were tested with f1 scores of 0.89 and 0.91 for stable and COVID-19 identification. Ultimately, a very recent analysis by Saiz and Barandiaran [7] suggested a deep learning object detector network with VGG16 as their base model and SSD300 as a detector. Researchers were able to reach a sensitivity of 94.92%, 92.00% of the precision of COVID-19 identification and increased accuracy with some pre-processing.

2 Model Overview

This study uses a robust detection network, the Single Shot MultiBox Detector (SSD) [8], to be specific. The base network chosen in this model is the residual network (ResNet101) [9], unlike the conventional visual geometry group (VGG) [10]. The images in the training data have also been pre-processed by means of contrast limited adaptive histogram equalization (CLAHE) [11]. Augmentation techniques like random cropping (in area of interest), random flipping, random rotation were also applied prior to training for effective results.

3 Architecture

In this section, we will first exploit the decisions made in this model like the choice of ResNet101 [9] over VGG [10] for base network, SSD512 [8] for detector network and preprocessing methods implemented. In Sect. 3, we will analyze the structure of dataset used, training of the model and hyper-parameter selection. Lastly, we will quote the concluding remarks for the model.

3.1 SSD512 and ResNet101

The Single Shot MultiBox Detector (SSD) [8] is installed on top of the so-called main network and is docked with some convolution layers to the end, and a sequence of increasingly smaller convolution layers is attached (blue in Fig. 1). Certain anchors are generated by tiling regularly across the image, a collection of boxes at variable aspect ratios and scales. The benefit of using a traditional anchor grid is that predictions for these boxes are described and referred to as tiled predictors on a picture with specific parameters (i.e., convolutions) and are reminiscent of typical sliding window techniques. Class scores and offsets for the pre-defined bounding boxes (predefined anchor boxes) are determined by the earlier base network layers and the layers introduced. Non-maximum suppression (NMS) is utilized to test predictions and final effects of the detection are obtained. The SSD with its variants has always been on the top of the bar in object detection (OD) deep learning networks. The speed versus accuracy trade-off is far better than other ODs of the time. It should not be a matter of further concern of choosing SSD over other detection networks. In image classification tasks, the residual networks [9] have proven to be better than VGG [10] because of the skip connections as depicted in Fig. 2 between convolutional blocks, helping diminish the effects of vanishing gradient, allowing networks to go deeper. Previously, SSD was extensively based on the VGG network, but there have been a number of studies that have achieved better accuracy for tasks using Residual-101. Also, as the dataset of the C: X-ray images grows, deeper networks can be expected to yield slightly better results.

3.2 Prediction Module

This is worth noting that just by docking the SSD with the residual network does not improve results considerably. Studies by Fu et al. [12] have shown that adding a prediction module increases the performance significantly. This can also be verified from Table 1. Cai et al. [13] in their study on MS-CNN have pointed out that the accuracy can be improved by modifying the sub-network for each task. The prediction modules (PMs) would do the necessary processing of feature maps which

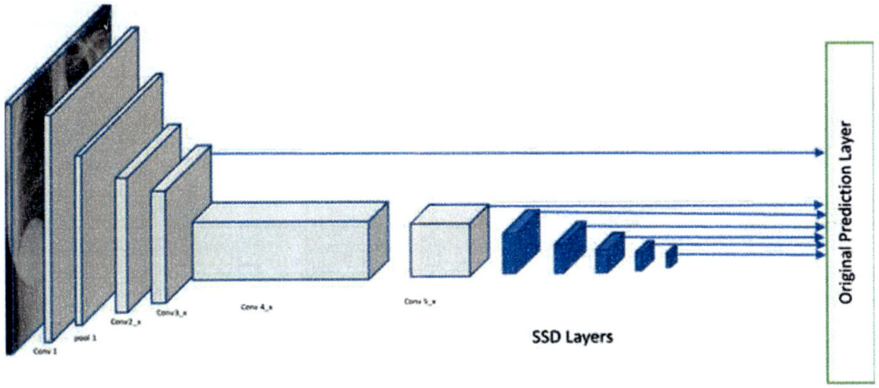

Fig. 1 Shows the SSD with base network

Fig. 2 Shows the residual learning

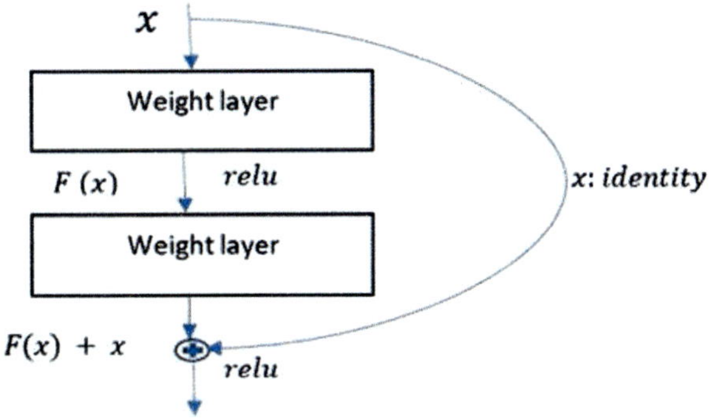

Table 1 Effects of various prediction modules on Pascal Voc 2007 test

Method	mAP
SSD 321	76.4
SSD 321 + PM (b)	76.9
SSD 321 + PM (c)	77.1
SSD 321 + PM (d)	77.0

Pm-Prediction module in Fig. 3

would lead to good classification because otherwise the layers in the feature extractor must learn to generate feature maps that represent spatial, semantic information from previous layers as well as the transformations for classification since in the original SSD architecture, feature maps are n-t well processed before application of the loss functions. Also, it has to undo previous transformations before selecting the best for a scale. Following this and as followed in [12], in this study, for each prediction layer, one residual block as in Fig. 3c is added. In [12], the researchers have also convinced well that the original SSD approach using

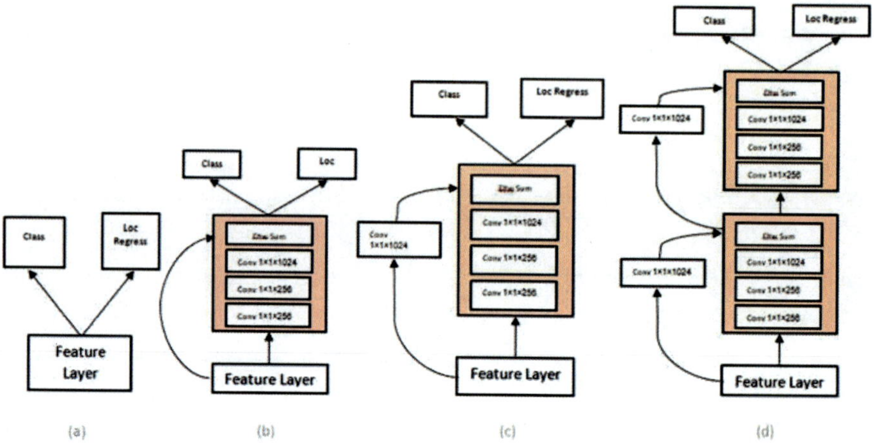

Fig. 3 Shows the variants of prediction modules (PM)

block in Fig. 3a, kind of residual block with skip connection as in Fig. 3b and two sequential residual blocks shown in Fig. 3d, all are out performed (Table 1) by the prediction module 3c. It should be well considered that with addition of PM to the original SSD architecture, the model evaluation measures (precision, recall) went up though by fractional amount, yet an improvement to previous works done.

3.3 *Preprocessing*

The authors of the article [14] presented that in an X-ray image formation, a tungsten filament is heated up with flow of an electric current. Some electrons are liberated as a result of thermionic emission from the filament and get electrically attracted toward the anode. Then, photons in X-ray spectrum are released as a result of the electrons hitting the target (tungsten) in a beam out of window of tube. This explains the basis for X-ray image formation. The dataset images come from various source locations/machines around world, and image processing settings and conditions are various on-platform. There is no realistic possibility that the parameters described above in the X-ray image formulation are consistent across the board. Exposure time, which refers to time period during which X-rays are made, is also a factor that influences the contrast of resulting image. Also, Reza [11] stated that in X-ray imaging a low-level exposure is maintained until the scanning process for the ROI is completed. Hence, the images so obtained are often with low signal-to-noise ratio.

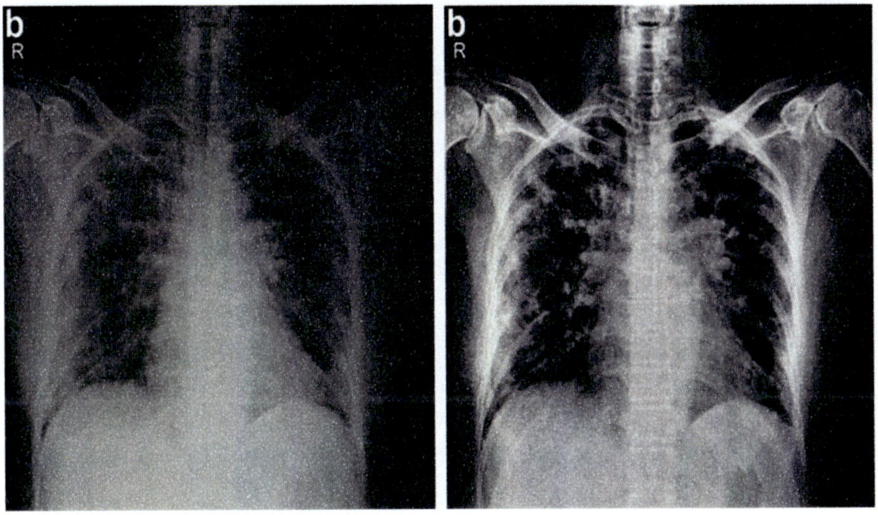

Fig. 4 Shows the original x-ray image on left, and CLAHE processed image to the right

Hence, it is evidently clear now that, all the X-ray images need to be unified in contrast. Therefore, the contrast limited adaptive histogram equalization (CLAHE) [11] is applied before training to input images' dataset. In contrast limited adaptive histogram equalization, contrast amplification is limited as a modification to adaptive histogram equalization. This is helpful in reducing the above-mentioned issue of amplification of noise. It has shown good results for medical pictures. The target image is divided into equally sized regions (non-overlapping). This algorithm applied to an X-ray image is shown in Fig. 4.

It has been observed that applying CLAHE to images before training improves the model feature measurements (accuracy, sensibility, specificity) by considerable amount. Extensive data augmentation techniques like randomly flipping, cropping, rotating and random photometric distortion are also used.

4 Data and Training

Based on previous studies on common pneumonia detection through C: X-rays and the assessment on chest X-ray scans of COVID-19 patients by Cellina et al. [15], it can be confidently concluded that the lung abnormalities in both types of infections are visually very similar to human perception.

Therefore, the dataset to be proposed must consists of both pneumonia and COVID-19 X-ray scans in order to obtain as less false positives as possible while inference. Hence, and similar to that proposed in [7], the dataset used in this model is composed of both COVID-19 and pneumonia images. The primary and direct source is the COVIDx dataset [16], proposed and introduced by Wang et al. [5]. This collection is further constructed of five different open-source chest radiography datasets.

Thus, the number for C: X-ray images of COVID-19 cases is 473 and for images of pneumonia cases consists of more than enough images required. Random selection was made from available pneumonia dataset keeping note of the limit of the required number of images for training as the COVID-19 X-ray images were limited and a balance between images in both the categories was required. The split of the dataset was done as follows-

- Balancing the image dataset for both categories with deviation roughly of not more than 10%.
- For the training and validation dataset, roughly (78 ± 2)% of available COVID-19 X-ray images and (73 ± 2)% of selected pneumonia X-ray images to be used.
- For testing dataset, roughly 20% of available COVID-19 X-ray images and (25 ± 1.5)% of available pneumonia X-ray images to be used.

It should be well noted that the number of pneumonia images selected is more than COVID images in testing dataset. This is because of the above-mentioned fact of visual similarities between both the infections on X-ray images, thereby attempting to make the model more robust, as in learning the difference between the two effectively.

The training policy is similar to that of SSD, i.e., firstly matching set of default boxes to target ground truth boxes as best overlapped and that with Jaccard overlap greater than a threshold. Certain "most-non matched" default boxes are chosen as negative samples such that their ratio with the matched boxes is at most 3: 1. Then after, minimizing the SoftMax (for confidence) and Smooth L1 (for joint localization) losses. The online version of the volume will be available in LNCS online. Members of institutes subscribing to the Lecture Notes in Computer Science series have access to all the pdfs of all the online publications. Non-subscribers can only read as far as the abstracts. If they try to go beyond this point, they are automatically asked, whether they would like to order the pdf and are given instructions as to how to do so.

With all of this, we achieved sensitivity of 94.95% and a specificity of 92.47% with considerable accuracy relative to previous works in the domain as shown in Table 2.

Table 2 Accuracy relative to previous works in the domain

Metric	Operation	Value
Sensibility	1015/1015 + 54	0.949485
Specificity	307/307 + 25	0.92469

5 Conclusion

This study demonstrates well the use and potential impact of deep convolution neural networks in the domain of medical imaging via detection networks. It must be considered that such models improve with time by a number of researchers around the globe, thereby making it a practical possibility if, to begin with, it is pushed to a testing phase. One of the most affecting pandemics in recent decades and even last century, this is growing at ever faster rate. This is also an appeal to all AI researchers to step and develop it to "beyond risk bar."

Moreover, this approach is by no means a production-ready solution; it is expected that researchers will accelerate the creation of highly accurate and realistic deep learning approaches to detect CXR images in COVID-19 cases and improve the treatment.

Lastly, this study also acknowledges [17, 18] for providing further insights about selection of hyper-parameters and regarding.

References

1. Xu, R., Cui, B., Duan, X., Zhang, P., Zhou, X., et al.: Saliva: potential diagnostic value and transmission of 2019-nCoV. Int. J. Oral Sci. **12**(1), 11 (2020). https://doi.org/10.1038/s41368-020-0080-z. PMC 7162686. PMID 32300101
2. Song, F., Shi, N., Shan, F., Zhang, Z., Shen, J., et al.: Emerging 2019 novel coronavirus (2019-nCoV) pneumonia. Radiology **295**, 210–217 (2020)
3. Wang, D., Hu, B., Hu, C., Zhu, F., Liu, X., et al.: Clinical characteristics of 138 hospitalized patients with 2019 novel coronavirus-infected pneumonia in Wuhan, China. Jama (2020)
4. Wang, S., Kang, B., Ma, et al.: A deep learning algorithm using CT images to screen for corona virus disease (COVID-19). Cold Spring Harbor Laboratory (2020). https://doi.org/10.1101/2020.02.14.20023028
5. Wang, L., Wong, A.: COVID-Net a tailored deep convolutional neural network design for detection of COVID-19 cases from chest radiography images. arXiv preprint: arXiv:2003.09871 (2020)
6. Hemdan, E.E.-D., Shouman, M.A., Karar, M.E.: COVIDX-Net a framework of deep learning classifiers to diagnose COVID-19 in x-ray images. arXiv preprint: arXiv:2003.11055 (2020)
7. Saiz, F.A., Barandiaran, I.: COVID-19 detection in chest x-ray images using a deep learning approach. Int. J. Interact. Multimedia Artif. Intell. (2020). https://doi.org/10.9781/ijimai.2020.04.003
8. Liu, W., Anguelov, D., Erhan, et al.: SSD: Single shot multibox detector. arXiv 2016, arXiv preprint: arXiv:1512.02325 (2020)
9. He, K., Zhang, X., Ren, et al.: Deep residual learning for image recognition. In: Proceedings of the IEEE Conference on Computer Vision and Pattern Recognition, pp. 770–778 (2016)
10. Simonyan, K., Zisserman, A.: Very deep convolutional networks for large-scale image recognition. arXiv preprint: arXiv:1409.1556 (2014)
11. Reza, A.M.: Realization of the contrast limited adaptive histogram equalization (CLAHE) for real-time image enhancement. J. VLSI Signal Process. Syst. Signal, Image, and Video Technol. **38**, 35–44 (2004). https://doi.org/10.1023/B:VLSI.0000028532.53893.82
12. Fu, C.Y., Liu, W., Ranga, A., et al.: DSSD: Deconvolutional single shot detector. arXiv 2017, arXiv preprint: arXiv:1701.06659 (2018)

13. Cai, Z., Fan, Q., Feris, et al.: A unified multi-scale deep convolutional neural network for fast object detection. In: European Conference on Computer Vision, pp. 354–370. Springer, Cham (2016)
14. https://www.radiologycafe.com/radiology-trainees/frcr-physics-notes/production-of-x-rays
15. Cellina, M., Orsi, M., Pittino, V., et al.: Chest computed tomography findings of COVID-19 pneumonia: pictorial essay with literature review. Japanese J. Radiol. (2020). https://doi.org/10.1007/s11604-020-01010-7
16. Gupta, R.: Generalized frequency domain formulation of the switching frequency for hysteresis current controlled VSI used for load compensation. IEEE Trans. Power Electron. **27**(5), 2526–2535 (2011)
17. Szegedy, C., Ioffe, S., Vanhoucke, et. al.: Inception-v4, inception-resnet and the impact of residual connections on learning. In: Thirty-first AAAI Conference on Artificial Intelligence (2017)
18. Huang, J., Rathod, V., Sun, et al.: Speed/accuracy trade-offs for modern convolutional object detectors. In: Proceedings of the IEEE Conference on Computer Vision and Pattern Recognition, pp. 7310–7311 (2017)

Smart Biometric Lock with Security Features

J. Naga Vishnu Vardhan, D. Antra, D. Eswari Ankitha, G. Akshitha, and B. Sumathi

Abstract Things are getting smarter day by day. With the present conventional locks, there is a possibility of keys being misplaced or lost or may fall into the hands of strangers. With the advances taking place in technology, people are also giving high preference for the technology that is safe, secure and handsfree. This made to the development of a biometric lock using Bluetooth technology working through Android app. To operate this, a person needs to enrol initially and later he can just operate the door using the fingerprint ensuring safety. When the user wants to lock/unlock the door, he needs to place the finger in Android app available in their mobile unit. This data via Bluetooth of mobile unit will be received by Bluetooth module attached to the Arduino. Then the Arduino unlocks the lock through relay if the fingerprint matches. This smart biometric lock avoids carrying keys and robbery. It enhances the security of their houses by sending an alert message, in case the fingerprint is not matched or trying to break the lock to the concerned person. A buzzer is also used which alerts the neighbourhood people. As the biometric system proposed here works on Bluetooth, it does not have any Internet issues. More importantly the system proposed is using the inbuilt app for fingerprint recognition in the mobile rather than physical device that prevents the mechanism of creating duplicate fingerprints and operates the lock. This makes the system more secure, cost effective and simple to implement.

Keywords Biometric lock · Bluetooth technology · Android app

J. Naga Vishnu Vardhan (✉) · D. Antra · D. Eswari Ankitha · G. Akshitha · B. Sumathi
Department of Electronics and Communication Engineering, BVRIT HYDERABAD College of Engineering for Women, Hyderabad, Telangana, India
e-mail: jnvvardhan@gmail.com

S. Das and M. N. Mohanty (eds.), *Advances in Intelligent Computing and Communication*, Lecture Notes in Networks and Systems 202,
https://doi.org/10.1007/978-981-16-0695-3_3

1 Introduction

In the present scenario, security is a major issue in our daily life, and usage of technology is becoming prominent. Digital locks are becoming necessitous for these security systems. Different security systems based on passive infrared sensor (PIR), radio frequency identification (RFID), digital lock system, bio-matrix systems, electronics code lock, etc., are into usage and have their own features. Fingerprint-based security systems are considered as one of the safest and secured way to lock or unlock any system as it recognizes any person uniquely and cannot replicate. This paper presents fingerprint-based biometric security system to lock/unlock the door.

Biometric locking system implemented using Arduino for shared access using centralized control system is discussed in [1]. Home automation, where the home appliances can be remotely operated using the Bluetooth technology via an android app, installed in the smartphone is into use. To provide robust security, fingerprint-based biometric system is detailed in [2]. Security system using fingerprint along with password authentication to operate the door and also communicating to the concerned person in case of misoperation using GSM with smart mobile is presented in [3]. For various applications like security for an exam control room and cash deposit rooms, a security system is discussed in [4], where the door will be opened automatically if the fingerprint matches. If not, the system camera captures the image of unauthorized person trying to operate the lock. A biometric-based security lock for bank locker is discussed in [5]. A smart door lock based on IoT that uses digital password for opening and closing the door is presented in [6] where Internet or public network is required. Security system using one-time password (OTP) is discussed in [7] where OTP will be sent to that authorized person mobile number. If the fingerprint does not match with the fingerprints stored, then OTP will not be sent and buzzer will turn on indicating that someone is trying to access door. In this case, due to some technical issues or signal strength if OTP is not received, it may cause inconvenience. To overcome the various limitations in the above mentioned, a simple effective method that does not require Internet connectivity and uses inbuilt fingerprint is discussed below.

Usually, in biometric applications, a biometric device is installed near the door where fingerprints of authenticated people are stored to access or allow into that room. But now, there is a big misusage of technology where strangers are smartly/intelligently arranging some mechanism to capture fingerprints of the authorized persons unknowingly and then developing similar artificial fingerprints to get access to open. This leads to big threat or damage to the system. The proposed system will come over such effect as the fingerprint is given through app in the mobile unit rather than the device.

2 Proposed System

The unique feature of the proposed system of smart biometric lock is that the user needs to enrol finger in that android app deployed in their mobile unit instead of device which therefore prevents possible threat of damage discussed above. This does not require any Internet facility and uses Bluetooth which is commonly available in all mobile phones. As mobile is a part of everyone's life and carries with them every second, this system works and even simple smartphone of low cost serves the purpose. This system is further integrated to GSM to send the alert message to the authorized person in case if any stranger detects the unlock notification using reed switch and also in case if any stranger tries to break the system.

Moreover, as discussed in [8], the proposed method comes under multiple protocols that have several security features that ensure more safety and security.

3 Description

The controlling device of the project is Arduino microcontroller. Bluetooth module, relay, reed switch and GSM are interfaced to Arduino microcontroller. When the user wants to lock/unlock the door, he/she need to place the finger in android app over Bluetooth. And this data is received by Bluetooth module which is attached to the Arduino. Then the Arduino locks/unlocks the door through relay if the fingerprint is matched. If any unauthorized person wants to open the door, the reed switch detects that and this data is sent to the Arduino and the GSM interfaced with the circuit sends the SMS to the predefined mobile number.

The controlling device used to develop biometric lock is Arduino microcontroller. Bluetooth module, relay and GSM are interfaced to it. GSM is used to send alert messages to specified people in case of emergency. As fingerprint of a person is unique, it is considered as one of the safest keys to operate any system as it can recognize any person uniquely and cannot be copied easily. Fingerprints of all the required members can be stored to provide access (Fig. 1).

4 Software Used for App

A MIT App Inventor is a web application integrated development environment originally developed by Google and presently operated by Massachusetts Institute of Technology (MIT). It allows creating application software (apps) for the operating systems (OS)—Android and iOS, which is a free and open-source software released under dual licensing.

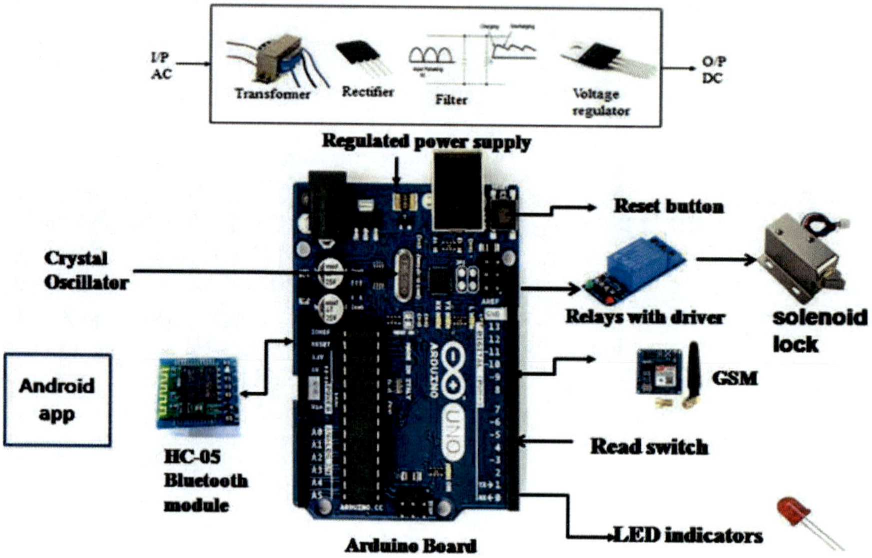

Fig. 1 Interface diagram

It uses a graphical user interface (GUI) and the Star Logo TNG user interface that make users to drag and drop visual objects for creating an application that can be operated on mobile units.

Projects based on App Inventor are informed by constructionist learning theories, that emphasize programming can be a vehicle for engaging powerful ideas. App Inventor also supports the use of cloud data via an experimental Firebase Real-time Database component (Fig. 2).

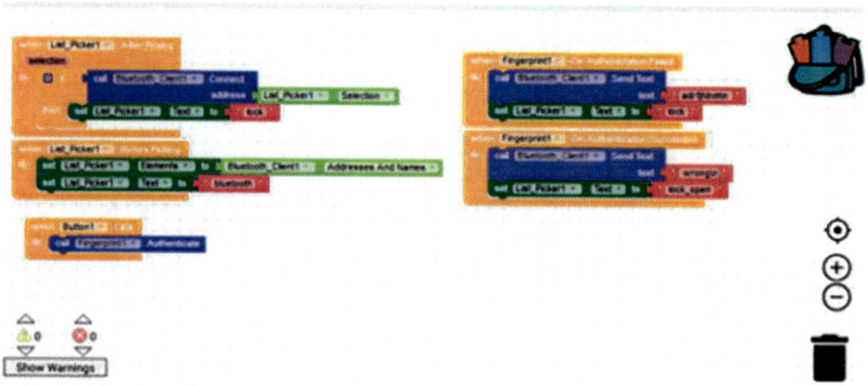

Fig. 2 Code blocks

5 Working

1. The controlling device of the project is Arduino Uno. Bluetooth module and relay are interfaced to Arduino Uno.
2. Instead of traditional locks, we have used solenoid lock in this project. This solenoid lock is connected to relay circuit to control the in and out function of the lock. The relay circuit is connected to a transistor and a resistor.
3. Relay circuit is given to the PIN 11 of Arduino board.
4. HC-05 is given to PIN 8 and PIN 9 of the Arduino board. (Txd. Rxd)
5. An adapter is given to the Arduino board for continuous power supply to the device.
6. Reed switch is connected to the door prototype as breaking of door needs to be sensed by the switch.
7. GSM module is connected to Arduino board along with a sim inserted into the module for sending message alert to the owner.
8. A continuous power supply is given to the entire device other than Arduino board as solenoid lock needs more than 12 V power supply to work.
9. Data is received by the Bluetooth module, the Arduino locks/unlocks through relay if the fingerprint matches.
10. If fingerprint does not match, the app will not get connected to the Arduino (Fig. 3).

Fig. 3 Working prototype

6 Testing

Initially, power the lock and servo with a battery of 5 V and then open the app installed. Later, turn on Bluetooth on phone and pair Bluetooth HC 05 of the circuit with Bluetooth of phone in operation. After successful connection with Bluetooth HC 05 with app, it can be noted that Bluetooth icon changes to lock icon (Fig. 4).

Now, touch the fingerprint icon. When one touches it, a message will be displayed in pop-up box asking to unlock with fingerprint. Now touch the fingerprint sensor on the phone. If it matches with the fingerprint set in the phone, it turns the lock to 'on position' and simultaneously the lock icon changes into unlocked icon (Fig. 5).

If wrong fingerprint is provided, then the lock moves back to its locking position (Fig. 6).

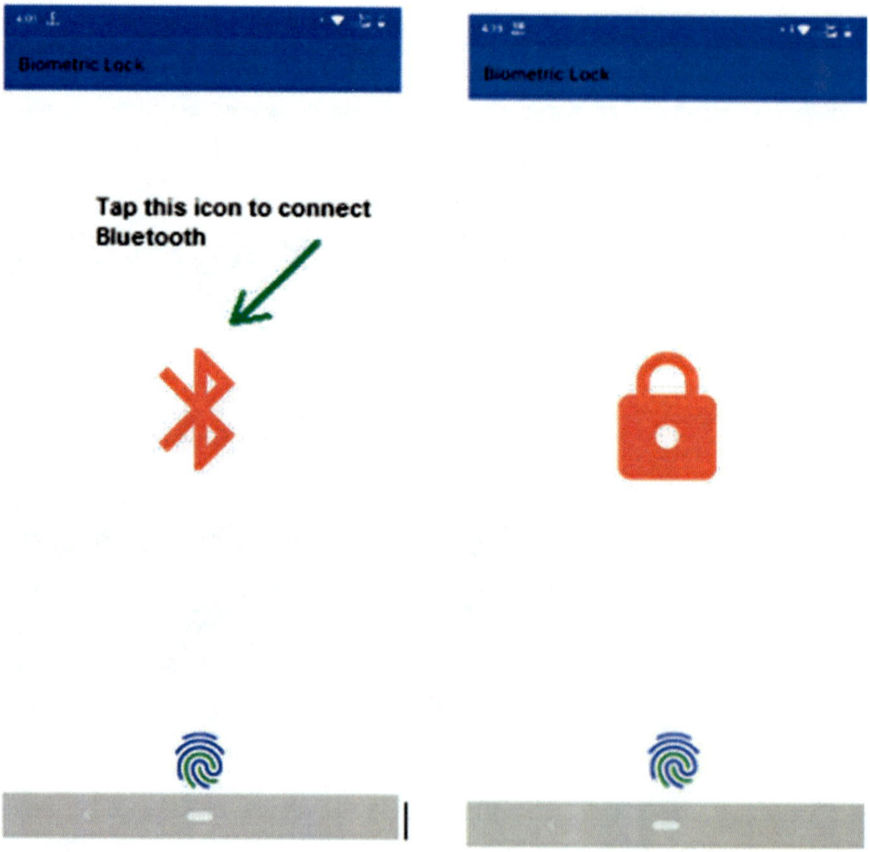

Fig. 4 App layout

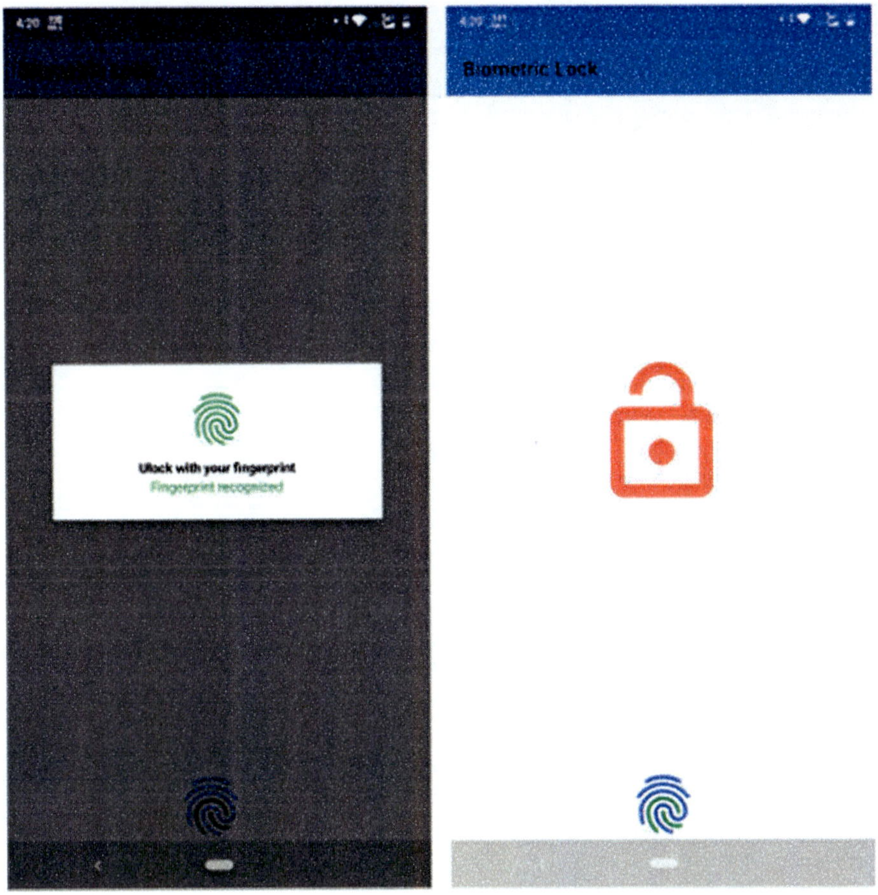

Fig. 5 App after asking for fingerprint

If someone forcibly tries to open or break the lock, an alert message is sent to the predefined number (Fig. 7).

7 Advantages

1. It uses inbuilt fingerprint sensor in smartphones instead of biometric sensors available in market. This prevents duplication of fingerprint by misusage of technology.
2. Real-time authentication system using fingerprint module.
3. Highly secured device because of multiple protocols.

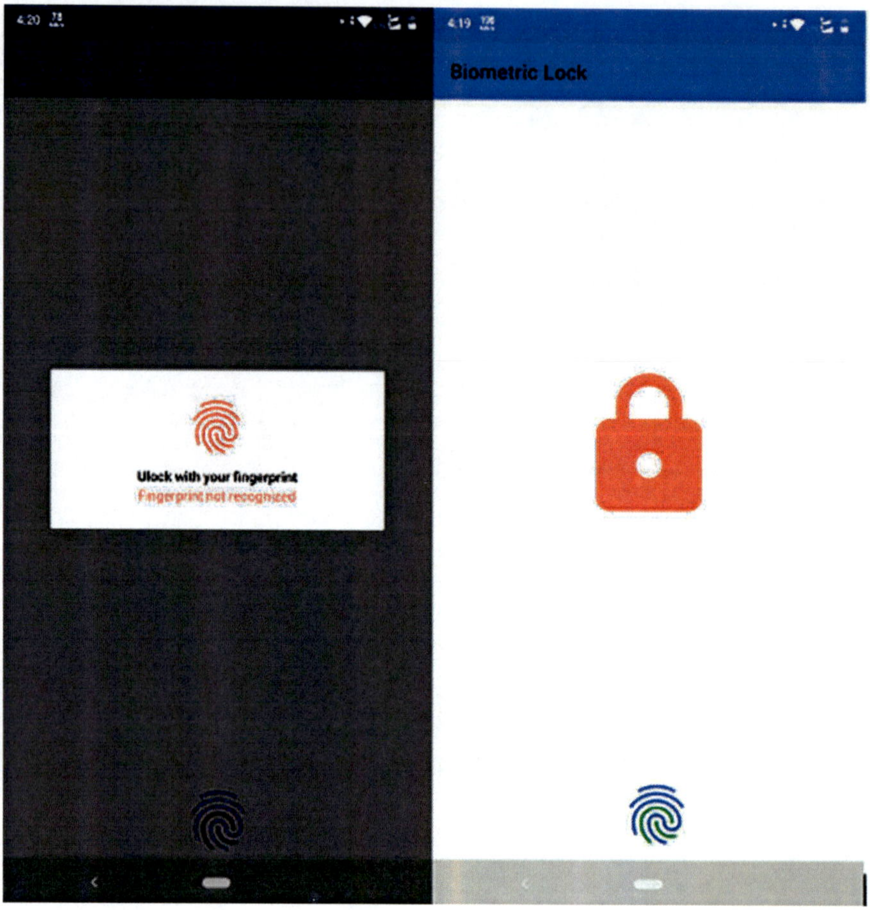

Fig. 6 Fingerprint does not match

4. No Internet connection is required as it works on wireless communication using HC-05 Bluetooth and does not suffer signal strength related issues.
5. Avoid discomfort situation in case of keys lost and also promotes handsfree of carrying keys.

8 Applications

This smart biometric lock can be used in real-time applications such as homes, industries, and offices. This is quite useful mainly to restrict entry to authenticated rooms/sections in various offices/industries.

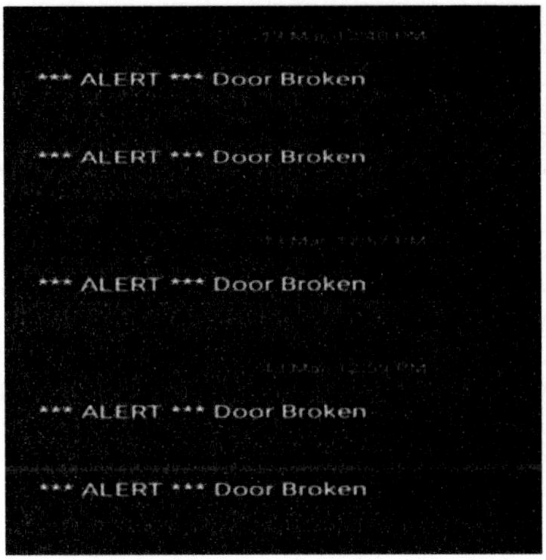

Fig. 7 Alert message from GSM module

9 Conclusion

The model proposed is very useful in present situation where conventional locks have become outdated with high risk of theft. The use of easily available components and fingerprint sensor in smartphones instead of biometric sensors available in market makes the project very cost effective. GSM module used in the project provides additional security. In case of theft or someone forcibly trying, reed switch present near the device detects unauthorized door opening and an alert message is sent to the owner's mobile phone. Moreover, the app of one device cannot be used to open lock of another house hence making the project highly secure and efficient. Purpose of every module is elucidated and placed accordingly, contributing to the best working of the unit.

10 Future Scope

We can extend this project to

- Interfacing camera through which the person's image can be captured and send the image to mail/mobile unit if any unauthenticated person tries to break the key. This helps in identifying the person.

References

1. Baidya, J., Saha, T., Moyashir, R., Palit, R.: Design and implementation of a fingerprint based lock system for shared access. In: 2017 IEEE 7th Annual Computing and Communication Workshop and Conference (CCWC), Las Vegas, NV, pp. 1–6 (2017). https://doi.org/10.1109/ccwc.2017.7868448
2. Swati, Gupta, R.P.: Implementation of biometric security in a smartphone based domotics. In: 2018 International Conference on Advances in Computing, Communication Control and Networking (ICACCCN), Greater Noida (UP), India, pp. 80–85 (2018). https://doi.org/10.1109/icacccn.2018.8748672
3. Rama Murthy, B., Jagadish, O., Tanveer Alam, K., Mahammad Dada, V., Priyanka Gandhi, K.: Development of GSM based advanced alert home locker safety security system using arduino UNO. IJSRST **4**(2) (2018). Print ISSN: 2395-6011, Online ISSN: 2395-602X. Themed Section: Science and Technology
4. Sonawane, V.M., Tarate, S. G., Vyavahare, V.A, Divekar, S.N.: Exam control room smart security system. IJSRSET **4**(4) (2018). Print ISSN: 2395-1990, Online ISSN: 2394-4099, Themed Section: Engineering and Technology
5. Chavan, S.P., Pagare, T.G., Sapkal, S.K., Shimpi, G.S.: Smart locker system using biometrics authentication. Int. J. Innovative Res. Technol. **4**(12), May (2018). ISSN: 2349-6002
6. Satoskar, R., Mishra, A.: Smart door lock and lighting system using internet of things. Int. J. Comput. Sci. Inf. Technol. **9**(5), 132–135 (2018). ISSN: 0975-9646
7. Nayana, R., Shashidhar, R.: Smart door lock system. Int. J. Mod. Trends Sci. Technol. **5**(2) (2019). ISSN: 2455-3778
8. Anaza, S.O., Jiya, J.D., Haruna, Y.S.: A review of intelligent lock system. Am. J. Eng. Res. (AJER) **6**(6), 9–15 (2017). e-ISSN: 2320-0847, p-ISSN: 2320-0936

Realization of a Vehicular Robotic System Using the Principle of Photonics

K. P. Swain, G. Palai, and M. N. Mohanty

Abstract Optical filter-based vehicular robotic system is envisaged in this paper where optical filters are explored at the infrared regime to control the movement of the robot. Here, the physics and mathematics of the research deal with the silicon grating structure and plane wave expansion method, respectively. Further, numerical computation using MATLAB–Simulink reveals that the controlling of the movement of the robot can be made with the help of the proposed photonic filter and Arduino development board. Finally, the outcomes of the optical filter of present research could be deployed with human robots for different purposes.

Keywords Photonic monochromatic filter · Vehicular robotic system · PWE · MATLAB–Simulink

1 Introduction

Nowadays, photonics research is not confined to the area of optical technology only. The phenomenon of photonics has been expanding toward the various arenas to understand the multidisciplinary applications. To envisage the same, several types of photonic devices including 1D structure have been designed to explore the photonic application. For example, one-dimensional photonic crystal is fabricated using Si and SiO_2 materials as in [1] to enhance the detectivity of a photodetector. Aside from this, recently a variety of works using 1D photonic crystal structures underwent through numerous investigations to design different kinds of optical MUX/DEMUX where optical filters are the backbone of the same [2–5]. In these references [2–5], SiO and Si are used in alternative fashion to form a total of 68 layers with defects in 18th and 52nd places.

K. P. Swain (✉) · G. Palai
Department of Electronics and Communication Engineering, GITA, Bhubaneswar, India
e-mail: kaleep.swain@gmail.com

M. N. Mohanty
Department of ECE, I. T. E. R, SoA Unversity, Bhubaneswar, India

S. Das and M. N. Mohanty (eds.), *Advances in Intelligent Computing and Communication*, Lecture Notes in Networks and Systems 202,
https://doi.org/10.1007/978-981-16-0695-3_4

In addition to the above-said work, various applications of 1D photonic crystal structure have been illustrated in [6–8]. In [6], sensing application is explored at Tera-Hertz region using BK-7 glass and air alternatively to fabricate the photonic structure wherein the reference [7], a gradient refractive index (GRIN) is sandwiched by two microfluidic channels for biosensing and chemical sensing application which is very expedient for real-time monitoring in medical science predominantly to analyze the amyloid beta's aggregation in Alzheimer's disease. Again in reference [8], one-dimensional photonic crystal is discussed using TiO_2 and SiO_2 for thermophotovoltaic optical filter application.

Apart from this, microwave photonic filter in the frequency range of 10 MHz–10 GHz is experimentally demonstrated in [9] for video transmission by using four numbers of bandpass windows positioned at 4.62, 6.86, 4.0, and 6.0 GHz. A near-infrared polarization filtering at 1210 and 1170 nm is proposed in [10] using photonic crystal fiber and the finite element method is implemented to study the optical characteristics whereas a tunable photonic filter is demonstrated in [11] for the application of tunable filter in the microwave range.

Nonetheless, different works pertaining to the 1D photonic structure have been disclosed, the present work is also divulging a new type of work with the help of 1D photonic structure, where photonic structure deals with SiO and Si-based 1D structure. Further, the operational mechanism deals with the lower range of infrared signal (800–900 nm). The reason for choosing such 1D structure is that the controlling of the vehicular robot is possible at this structure only. Again the reason for choosing such operational wavelength (800–900 nm) is that this range is particularly selected to operate the proposed system at night time which will be very useful for maneuvering a vehicular robot in any difficult or human unreached region to perform any type of surveillance work. As far as the segregation of this paper is concerned, the system overview along with the possible combination of filter outputs is discussed in Sect. 2 and the result of both PWE and MATLAB simulation is described in Sect. 3, where the conclusion is conferred in Sect. 4.

2 System Overview

Before going to discuss the intrinsic mechanism, let us concentrate on the overview of the proposed work which is disclosed in Fig. 1.

In Fig. 1, a light source with having signal ranges from 800 to 900 nm (wavelength) is incident on the photonic crystal system which is a combination of six optical filters and all the filters are designed in such a way that, they allow six optical signals of wavelength 833.1, 854.9, 857.4, 867.8, 888.2, and 893.7 nm only. The reason for choosing such wavelengths is due to the realization of controlling the robot at night time. As far as the internal elements of the filter are concerned, each filter is designed by 68 layers of alternate combination of SiO and Si with a defect at two positions, where the thickness of these layers controls the allowed wavelength as described in Fig. 2 where $t1$ and $t2$ represent the thickness of SiO

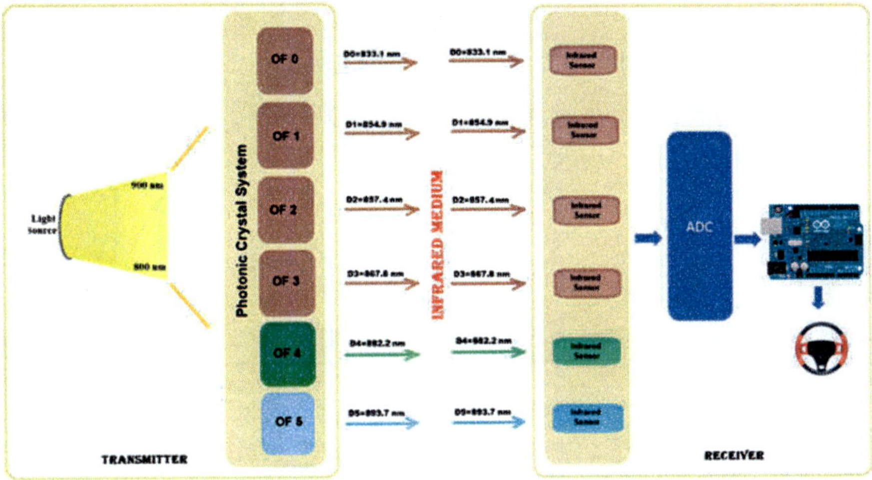

Fig. 1 Overview of the proposed application

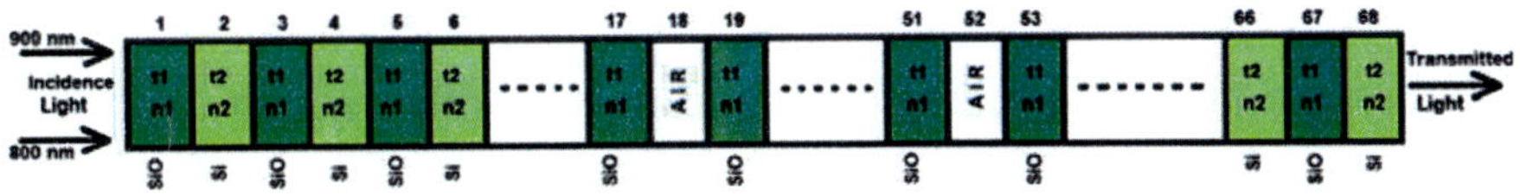

Fig. 2 Internal structure of the optical filter

and Si, respectively, and corresponding refractive indexes of SiO and Si are represented by $n1$ and $n2$ and regarding defect position, Si is replaced by air at 18th and 52nd layer.

A smart meter will be installed in every house registered on a given network, which sends the meter readings at regular intervals to a decentralized blockchain. Moreover, these signals are sensed by the infrared receivers which convert them into electrical equivalent voltages to process through an Arduino development board containing ATmega 328 microcontroller and ADC. Here, a program is written using Arduino IDE and included in the permanent memory of the microcontroller to control these incoming signals from infrared sensors. According to data given in Table 1, with the help of a different combination of output signals from the sensor, the movement of a vehicular robot can be accomplished. Here, the movement referred to as forward, backward, left, and right motion, and also the degree of rotation can also be controlled with the left and right movement. In Fig. 1, a motor steering represents the different movement with the degree of rotation of the vehicular robotic system.

As far as the working mechanism is concerned, the outcomes of six filters named D0, D1, D2, D3, D4, D5 (as shown in Fig. 4) are used in the proper sequence to carry out the different movement of the vehicular robot. The filter outputs are

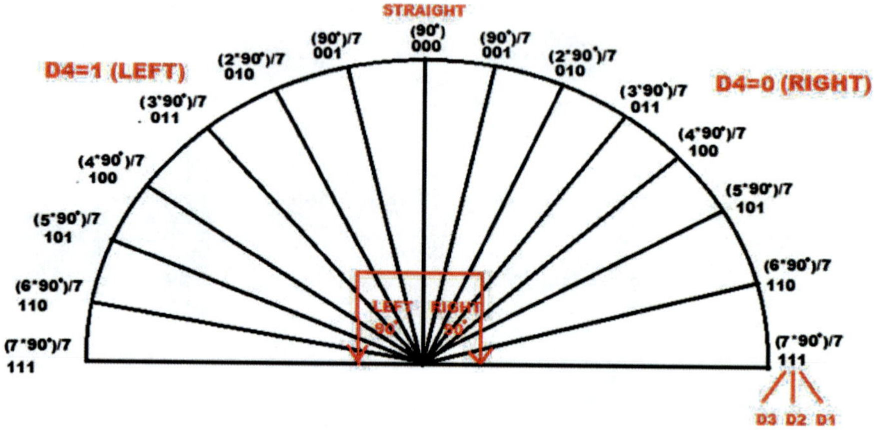

Fig. 3 Combination of D4, D3, D2, D1 for the degree of rotation in both left/right movement

Table 1 Thickness of SiO (*t*1) and Si (*t*2) layer w. r. t allowable signal wavelength

Thickness (*t*1) in μm	Thickness (*t*2) in μm	Allowable wavelength in nm
0.104	0.055	833.1
0.105	0.058	854.9
0.1055	0.058	857.4
0.1059	0.0595	867.8
0.1062	0.0629	888.2
0.1064	0.0637	893.7

controlled by ATmega 328 microcontroller using a program written in Arduino IDE. Physically, all the outputs of the filter are connected to the input pins of Arduino Uno through the internal ADC and the output pins are connected to the motor driver circuit of the vehicular robot. Among the filters, D0 is used for forward or backward motion, the combination of D1, D2, and D3 is used for the degree of rotation in left/right direction, D4 selects the left or right motion whereas D5 controls the breaking the system. To clarify further, eight combinations are used using three bits (D1, D2, D3) to control the degree of rotation where the first combination '000' is used for the straight drive for both left and right cases and the rest combinations like '001', '010', '011', '100', '101', '110', and '111' are assigned by (90/7) degree each. For example, '001' is assigned to an angle of (90/7) degree, '010' is assigned to an angle of (2 * 90/7) degree, and so on as given in Table 1. Here, each left and right motion is divided into seven equal divisions of (90/7) degree for different angles of movement, which is clearly understood from Fig. 3. The validation for the above-said angular movements (Fig. 3) can be verified through MATLAB–Simulink which is discussed in the subsequent section.

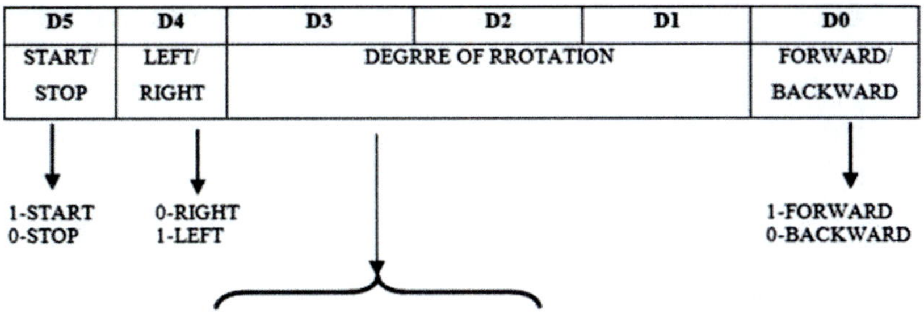

D5	D4	D3	D2	D1	D0	DEGREE OF ROTATION	
1	0	0	0	0	1	STRAIGHT (0 Degree)	
1	0	0	0	1	1	(1*90/7) Degree	
1	0	0	1	0	1	(2*90/7) Degree	
1	0	0	1	1	1	(3*90/7) Degree	LEFT
1	0	1	0	0	1	(4*90/7) Degree	
1	0	1	0	1	1	(5*90/7) Degree	
1	0	1	1	0	1	(6*90/7) Degree	
1	0	1	1	1	1	(7*90/7) Degree	
1	1	0	0	0	0	STRAIGHT (0 Degree)	
1	1	0	0	1	0	(1*90/7) Degree	
1	1	0	1	0	0	(2*90/7) Degree	
1	1	0	1	1	0	(3*90/7) Degree	RIGHT
1	1	1	0	0	0	(4*90/7) Degree	
1	1	1	0	1	0	(5*90/7) Degree	
1	1	1	1	0	0	(6*90/7) Degree	
1	1	1	1	1	0	(7*90/7) Degree	
0	X	X	X	X	X	STOP	

Fig. 4 Combination of filter outputs w. r. t the robotic movement

3 Result and Discussion

As far as the result and interpretation of this work are concerned, it deals with the two intrinsic mechanisms such as the filter outcomes, which can be obtained with the help of SiO and Si-based photonic structure, and the optical receiver which is made by using optical sensors and microcontroller with ADC. In the beginning, PWE simulation methods have been applied to find the reflectance characteristics of six optical filters that are operated in the infrared region. Each filter is realized by using the alternative combination of SiO and Si layers and a total of 68 layers is used by taking two defects into account where Si is replaced by air which is discussed in the previous section. Different thickness of these SiO and Si layers (in figure marked as *t*1 and *t*2) allows the desired signal to pass through the structure while rejecting others completely as indicated in Fig. 5 which is the reflectance

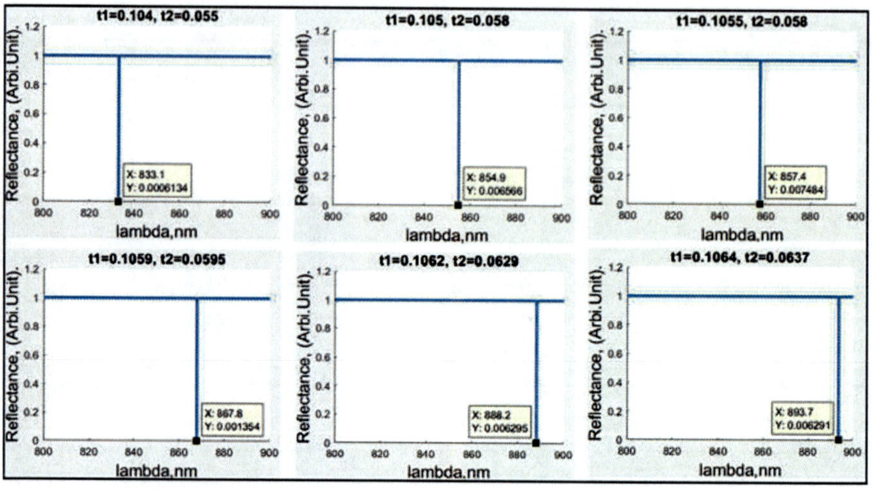

Fig. 5 Reflectance characteristics curve for six allowable signals

curve of each allowable signal. In this, reflectance (Arbi unit) and wavelength (nm) are chosen along vertical and horizontal axis, respectively. Since the proposed structure is manipulated with the signals of the wavelengths 833.1, 854.9, 857.4, 867.8, 888.2, and 893.7 nm to control the vehicular robots, the structure parameters such as the thickness of SiO and Si along with the nature of the structure and configuration determine the said operational mechanism of the robot. For example, thickness 0.104 and 0.055 μm of SiO and Si layer allow the signal of 833.1 nm. Similarly, the allowed signal would be changed with the variation of these thicknesses. The details of thickness of SiO (*t*1) and Si (*t*2) layers of each allowable signal are given in Table 1 and the validation of these parameters is exhibited by the simulation outcomes in Fig. 5.

In Fig. 5, it is seen that the reflectance is found at the wavelength 833.1, 854.9, 857.4, 867.8, 888.2, and 893.7 nm. As the absorption of structure pertaining to the operating signal is found zero, the transmittance corresponding to the aforementioned signals would be one (100%). Further MATLAB–Simulink is implemented to realize the entire structure for different combinations of filter outputs as given in Fig. 4. Though the simulations have been carried out for all combinations but only two cases are presented in Figs. 6 and 7 to indicate the left and right movement of the vehicular robot.

In the simulation diagram, two circular symbols are used; one to indicate START or STOP condition and the other for FORWARD or BACKWARD motion. The red color is used to indicate for STOP and BACKWARD condition and green color is used for START and FORWARD movement. A pointer is used to indicate the degree of rotation for each left and right movement.

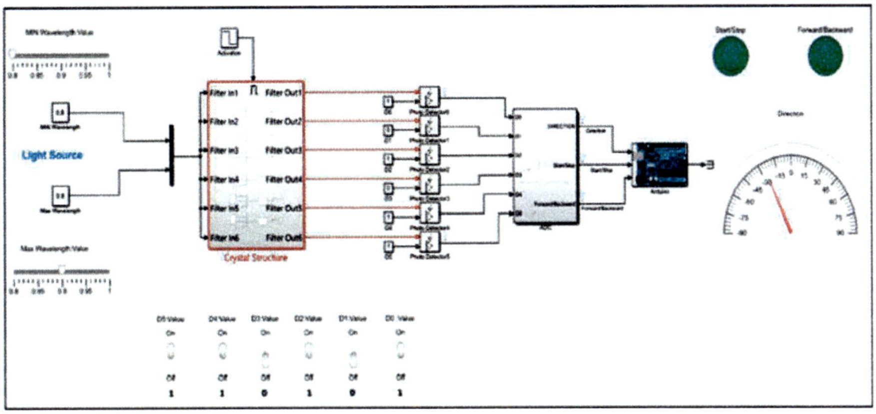

Fig. 6 Simulink output for START-FORWARD LEFT movement

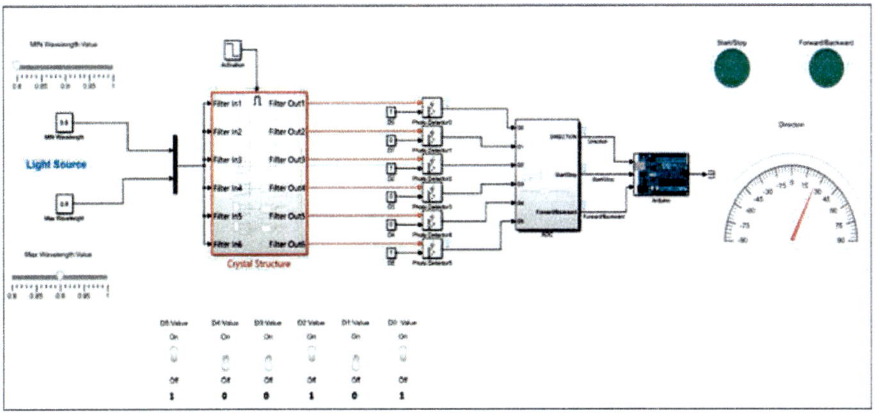

Fig. 7 Simulink output for START-FORWARD RIGHT movement

In Figs. 6 and 7, forward motions of the vehicular robot are indicated for the combination of '110101' and '100101' for 'D5 D4 D3 D2 D1 D0' filter outcomes. Here, both circular symbols are green in color which indicates the vehicle in forward motion and the only difference is marked by pointer which presents left movement for Fig. 6 and right movement for Fig. 7. In the above cases, the values of D3, D2, and D1 are taken as '010' indicating an angular movement of 2 * (90/7) which is equal to 25.7° in both left and right direction.

To sum up, the outcomes of the research work would be a new promising field of photonics because the optical filter could be a suitable candidate for controlling the movement of the robots including vehicular robot as well as humanoid.

4 Conclusion

An optical filter-based automatic vehicular robot which is operated at the near-infrared region is designed and implemented in this paper. Both PWE simulation and MATLAB–Simulink are employed to realize the filters and entire research, respectively. The outcomes of the proposed optical filters have successfully been realized to guide a vehicular robot in the night time with the near-infrared range. Finally, it is inferred that the controlling of any kind of robot could be possible with the help of one-dimensional photonics and shall be a new paragon to the silicon photonics.

References

1. Djuric, Z., Petrovic, R., Randelovic, D., Dankovic, T., Jaksic, Z., Ehrfeld, W., Feiertag, G., Freimuth, H.: One-dimensional Si–SiO_2 photonic crystal with defects intended for use in infrared spectral region. ICM **1**, 14–17 (1997)
2. Swain, K.P., Palai, G., Moharana, J.K.: Analysis for '101' channels of MUX/DEMUX using grating SOI structure at sub nanometer scale. Optik **129**, 78–82 (2017)
3. Dalai, P.K., Sarkar, P., Palai, G.: Analysis of silicon waveguide structure for realization of optical MUX/DEMUX circuit: an application of silicon photonics. Optik **127**, 10569–10574 (2016)
4. Palai, G., Nayak, B., Rout, S.R.: Realisation of optical demux vis-a-vis 850/1310/1550 nm using photonic crystal fiber. Optik **159**, 344–347 (2018)
5. Palai, G., Beura, S.K., Gupta, N., Sinha, R.: Optical MUX/DEMUX using 3D photonic crystal structure: a future application of silicon photonics. Optik **128**, 224–227 (2017)
6. Silveira, L.C.A., Hejase, J.A., Chahal, P.: A terahertz photonic crystal structure for sensing applications. Int. Symp. Ant. Propag, 1–2 (2012)
7. Bayat, F., Kandjani, S.A., Tajalli, H.: Designing real time biosensors and chemical sensors based on defective 1D photonic crystals. J. Latex Class Files **28**, 1843–1846 (2016)
8. Mbakop, F.K., Djongyang, N., Raidandi, D.: One–dimensional TiO_2/SiO_2 photonic crystal filter for thermo photovoltaic application. J. Eur. Opt. Soc. Rapid Publ. **12**, 1–9 (2016)
9. Huerta, I.E.Z., Montana, D.F.P., Nava, P.H., Juarez, A.G., Asomoza, J.R., Cruz, A.L.L.: Transmission system for distribution of video over long-haul optical point-to-point links using a microwave photonic filter in the frequency range of 0.01–10 GHz. Opt. Fiber Technol. **13**, 665–670 (2019)
10. Zhenkai, F., Yingying, W., Zhang, Z., Wang, B., Zhao, R., Zhang, X., Wu, R.: Near infrared polarization filtering characteristics of photonic crystal fibers filled with silver nanowire. Opt. Fiber Technol. **49**, 1–6 (2019)
11. Zhang, Q., Zhang, J., Li, Q., Wang, Y., Sun, X., Dong, W., Zhang, X.: Bandwidth tunable microwave photonic filter based on digital and analog modulation. Opt. Fiber Technol. **42**, 34–38 (2018)

A Modified Hybrid Planar Antenna for Cognitive Radio Application

Arjuna Muduli and Rabindra Kishore Mishra

Abstract This article approaches an integrated planar antenna configuration for CR communication. There is a rectangular microstrip antenna for narrowband operation and a monopole antenna along with CPW feed concept for wideband operation which is printed on the single side of substrate. This configuration reduces coupling effect. The narrowband microstrip antenna helps the CR for communication, and the sensing of spectrum is done by the ultra-wideband monopole structure.

Keywords Microstrip antenna · MSA · UWB · CR

1 Introduction

The facility augmentation in communication system establishes scarcity in the radio frequency spectrum. Nowadays, due to static frequency allocation, some of the spectrum remains unutilized. As a result, the researchers devised so many dynamic spectrum access techniques. In 1999, the CR concept was proposed by Mitola et al. [1] based upon the principle of the spectrum access technique dynamically. This cognitive radio is capable for operating itself from its past learned knowledge [2] and reconfigurable itself to match its characteristics with radio frequency environment. Nowadays, two types of CR concepts are available [3]: the spectrum interweaves and underlay. For spectrum underlay cognitive radio, the utilization of wideband antenna can efficiently receive the signal, because here, secondary users operate under the permitted noise floor fixed by the primary user. Here, the condition is as follows: both users (secondary and primary) have to utilize same

A. Muduli (✉)
Department of ECE, Koneru Lakshmaiah Education Foundation,
Vaddeswarm, Vijayawada, Andhra Pradesh, India
e-mail: arjunamuduli@gmail.com

R. K. Mishra
Department of Electronic Science, Berhampur University, Berhampur, Odisha, India

S. Das and M. N. Mohanty (eds.), *Advances in Intelligent Computing and Communication*, Lecture Notes in Networks and Systems 202,
https://doi.org/10.1007/978-981-16-0695-3_5

channel [4]. Hence, the interference is more compared to the interweave mode. In the interweave mode, the band is utilized by the secondary user where it is not occupied by the primary user and transmits the signal without any constraints. In this approach, the secondary (unlicensed) user avoids the primary (licensed) user using "spectrum sensing and adaptive allocation" techniques where, at any time, licensed users can utilize the channel, and the unlicensed user can access that spectrum if the primary (licensed) user does not access that channel [4]. So, this type of transmission does not create any interference between licensed (primary) and unlicensed (secondary) users; hence, the interference level in interweave CRs decreases. The interweave CR requires two types of antennas. The first type is one narrowband antenna (reconfigurable) used for communication and another ultra-wideband antenna [5–14] for monitoring and sensing the availability of spectrum [15]. Dual-feeding techniques by considering the antenna requirement for spectrum interweave (overlay) CR have been suggested in the literature [16, 17]. However, the problem arises for narrowband communication antenna. At the resonance frequency of NB antenna, the effect of coupling between the ultra-wideband antenna and the microstrip patch antenna is more (above—7 dB) because one antenna is placed just backside of the other antenna (Fig. 3) which creates coupling between feed line of patch antenna, the monopole structure (ground) and patch.

In this communication, a double-feed patch antenna structure is designed for the interweave CR. This structure consists of two patch antennas printed back to back on a substrate with the double-feed mechanism. A wideband about 3.25–12.75 GHz is sensed by the monopole patch. A narrowband rectangular microstrip antenna operates from 10.3 to 10.7 GHz. Here, we try to minimize coupling effect between these two antennas by placing both the antennas on one side of the substrate.

2 Design Consideration

Figure 1 depicts an UWB monopole antenna printed on a substrate (FR4) with a thickness of 1.6 mm and dielectric constant of 4.3. Table 1 shows the parameters of the monopole structure. In this arrangement, the monopole structure gives a bandwidth of 9.5 GHz in UWB frequency range. A rectangular microstrip antenna is designed above the monopole structure in Fig. 1. A ground plane is designed on the opposite side of the rectangular patch (Fig. 2). This rectangular microstrip patch is used for CR communication due to its narrow bandwidth (Fig. 3).

This rectangular MSA operates from 10.3 to 10.7 GHz (narrowband) band with resonant frequency (f_p) at 10.5 GHz. The dimension of the narrowband microstrip antenna is calculated from the given formula in [18].

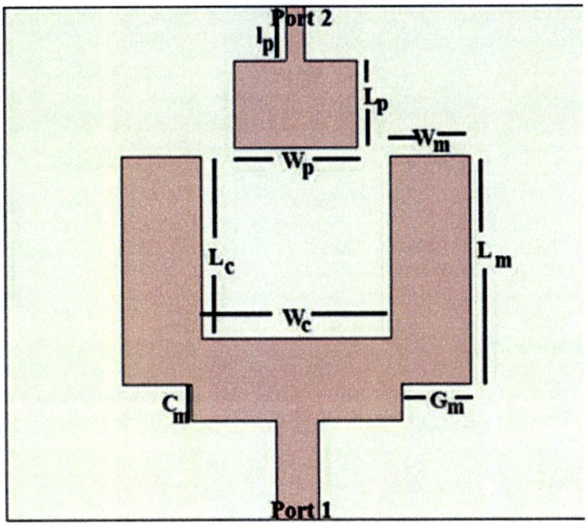

Fig. 1 Front view of the proposed structure

Table 1 Design parameter consideration

Sl. No.	Dimension	Parameter values in mm
1	W_m	5.5
2	L_m	15
3	G_m	4.75
4	C_m	2.5
5	L_c	12
6	W_c	13
7	L_g	12
8	W_g	12
9	L_G	6

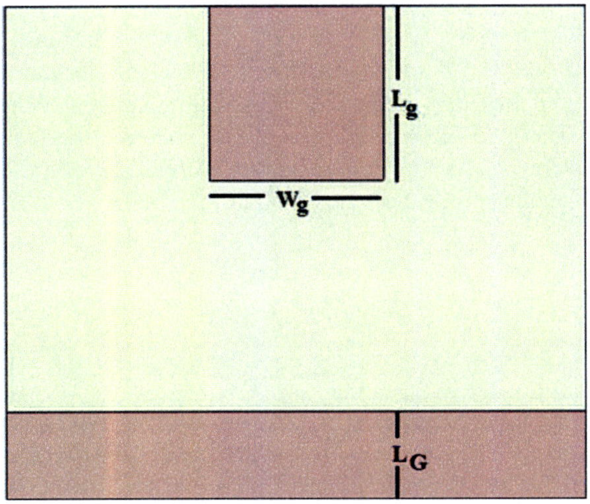

Fig. 2 Backside view

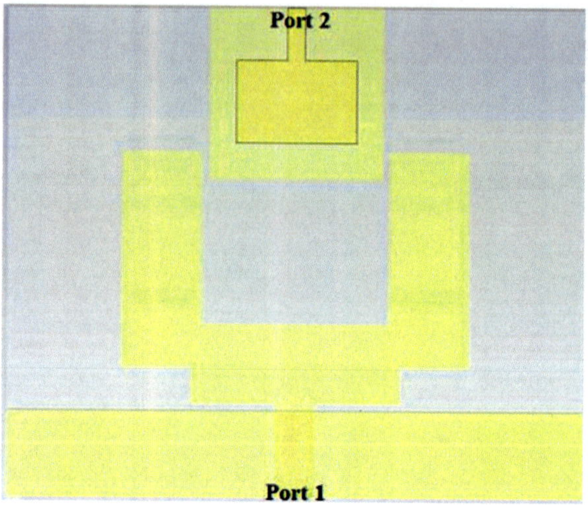

Fig. 3 Proposed antenna structure (transparent view)

$$W_p = \frac{c}{2f_p}\sqrt{\frac{2}{(\varepsilon_r + 1)}} \tag{1}$$

$$L_p = \frac{c}{2f_p\sqrt{\varepsilon_{\text{eff}}}} - 2\Delta L_p \tag{2}$$

3 Results and Discussion

The return loss parameter of the UWB antenna (Fig. 4) shows that the impedance bandwidth from 3.25 to 12.75 GHz with a maximum return loss of around −23 dB. In Fig. 5, the return loss of narrowband antenna is maximum at 10.5 GHz and has a bandwidth of 0.4 GHz. The S_{12} curve in Fig. 6 shows that the coupling between the NB and UWB structure is less than −17 dB for the whole band.

Figure 7 shows the E-field radiation patterns of the monopole structure at 4.51, 7.08, 11.85 GHz, and Fig. 8 shows pattern of radiation for narrowband microstrip antenna at 10.5 GHz. The pattern consistency of the NB and UWB hybrid antenna is evident from these Figs. 7 and 8.

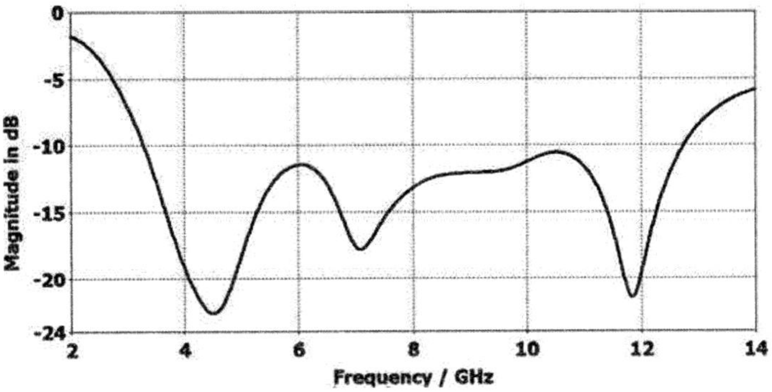

Fig. 4 S_{11} for UWB monopole antenna

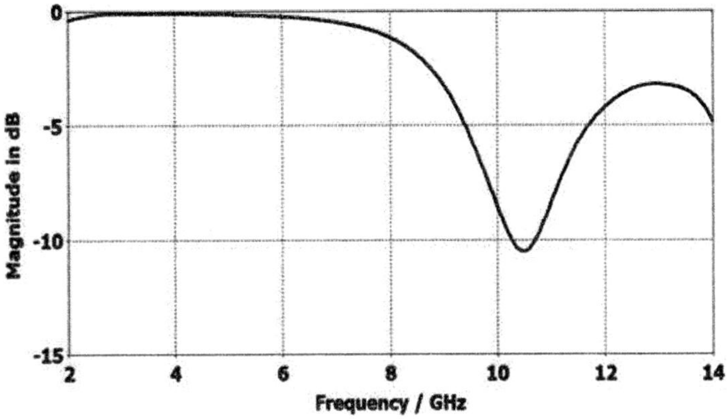

Fig. 5 S_{22} for narrowband patch antenna

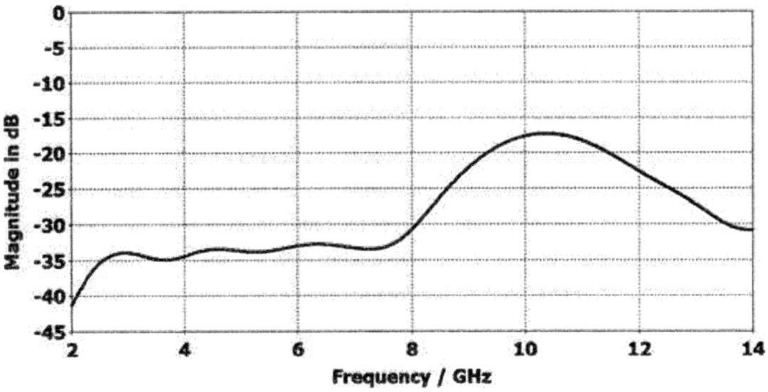

Fig. 6 S_{12} curve for integrated structure

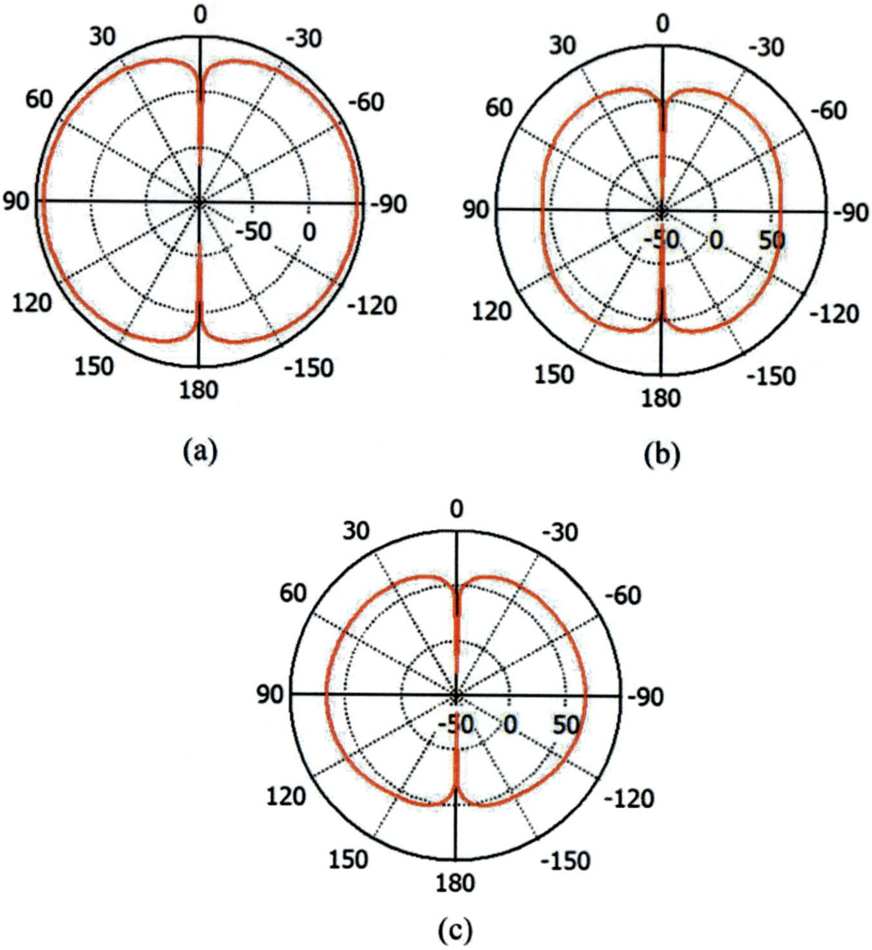

Fig. 7 Radiation pattern of monopole antenna at **a** 4.51 GHz, **b** 7.08 GHz and **c** 11.85 GHz

Fig. 8 Rectangular patch antenna radiation pattern at $f = 10.5$ GHz

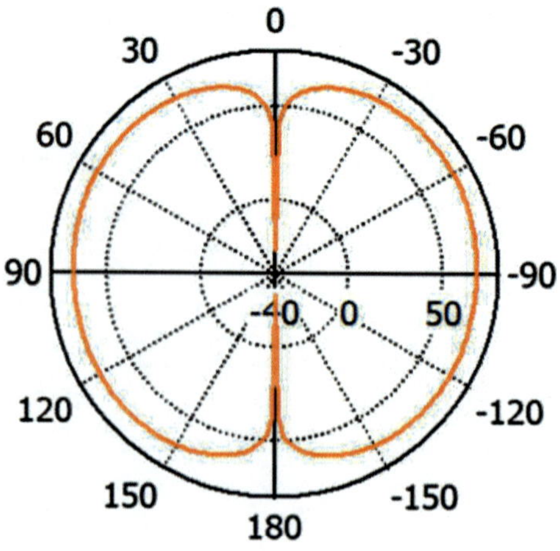

4 Conclusions

The article investigated the performance of broad- and narrowband hybrid planar antenna. Here, coupling problem is minimized by optimizing the position of the patch. The rectangular MSA is designed to achieve the narrowband communication after spectrum sensing by the wideband monopole antenna. By adding tuner externally to the MSA, our proposed structure can sense and communicate throughout the ultra-wideband of consideration. Hence, it is suitable for CR application.

References

1. Mitola, J., Maguire, G.Q.: Cognitive radio: making software radios more personal. IEEE Pers. Commun. **6**(4), 13–18 (1999)
2. Haykin, S.: Cognitive radio: brain-empowered wireless communications. IEEE J. Sel. Areas Commun. **23**, 201–220 (2005)
3. Al-Husseini, M., Kabalan, K., El-Hajj, A., Christodoulou, C.: Cognitive radio: UWB integration and related antenna design. In: Joo, E.M. (ed.) New Trends in Technologies: Control, Management, Computational Intelligence and Network Systems, ISBN 978-953307, 213-5, Sciyo, Nov. 2010
4. Jayaweera, S.K., Christodoulou, C.G.: Radiobots: architecture, algorithms and realtime reconfigurable antenna designs for autonomous, self-learning future cognitive radios. University of New Mexico, Technical Report EECE-TR-11-0001, Mar. 2011
5. Saleem Akram, P.: Investigations on metamaterial slot antenna for different wireless applications. Int. J. Sci. Technol. Res. **8**(11), 34–39 (2019)
6. Venu Madhav, P., Siva Ganga Prasad, M.: Fabrication of hybrid fractal microstrip antenna for mobile and wi-max applications. Int. J. Innovative Technol. Exploring Eng. **8**(11), 3373–3377 (2019)
7. Ketavath, K.N.: Enhancement of gain with coplanar concentric ring patch antenna. Wirel. Pers. Commun. **108**(3), 1447–1457 (2019)
8. Aradhyula, R., Rama Krishna, T.V., Madhav, B.T.P.: Gain enhancement modelling of coplanar waveguide fed circular monopole antenna with EBG placement. Int. J. Eng. Adv. Technol. **8**(6), 1986–1991 (2019)
9. Aradhyula, R., Rama Krishna, T.V., Madhav, B.T.P.: Electromagnetic bandgap structured CPW fed circular monopole antenna with bandwidth enhancement for wideband applications. Int. J. Innovative Technol. Exploring Eng. **8**(10), 879–882 (2019)
10. Subba Reddy, V., Siva Ganga Prasad, M., Madhav, B.T.P.: Design and implementation of dual notch band characteristics in UWB antenna for wireless personal communications. Int. J. Adv. Trends Comput. Sci. Eng. **8**(4), 1719–1725 (2019)
11. Tilak, G.B.G., Kotamraju, S.K., Madhav, B.T.P., Sri Kavya, C., Venkateswara Rao, M.: Broken heart shaped monopole antenna for WiMAX applications. Int. J. Recent Technol. Eng. **8**(2), 2333–2337 (2019)
12. Preetham Reddy, C.S., Sai Krishna, K., Parvathi, K., Eswar Pavan Kumar, M., Sai Raja, P.M. S., Rajesh Kumar, K.: Design and simulation of ultra-wide band mimo antenna with band notched function. Int. J. Eng. Adv. Technol. **8**(5), 304–307 (2019)
13. Yathiraju, R., Pardhasaradhi, P., Madhav, B.T.P.: Experimental investigations on DGS monopole antenna for LTE applications by few iterative techniques for achieving stable gain. Int. J. Innovative Technol. Exploring Eng. **8**(8), 1946–1950 (2019)

14. Madhav, B.T.P., Mayukha, K., Mahitha, M., Manisha, M., Somlal, J.: Circularly polarized dielectric resonator disc monopole antenna for mobile communication and IoI applications. Int. J. Innovative Technol. Exploring Eng. **8**(8), 166–169 (2019)
15. Reconfigurable filtennas and MIMO in cognitive radio applications. IEEE Trans. Antennas Propag. **62**(3), 1074, 1083 (2014)
16. Muduli, A., Mishra, R.K.: Modified UWB microstrip monopole antenna for cognitive radio application. In: IEEE AEMC-2015 Conference, IIT Guwahati, 18–21 Dec 2015, Guwahati, India
17. Ebrahimi, E., Hall, P.S.: A dual port wide-narrowband antenna for cognitive radio. In: 2009 3rd European Conference on Antennas and Propagation, Berlin, pp. 809–812 (2009)
18. Garg, R., Bhartia, P., Bahl, I., Itlipiboon, A.: Microstrip Antenna Design Handbook. Artech House Inc., Nonvood, USA (2001)

Detection of Broken and Good Medical Tablets Using Various Machine Learning Models

Suvajit Lodh, Shubham Mahajan, Ch. Vanipriya, Anil Bhardwaj, and Amit Kant Pandit

Abstract The computer vision has played a very important role in the fields of medical science. Not only computer vision but also machine learning has been an important aspect in medical applications. One such application of both of them together is explained in paragraphs given below. Detection medicines of different shapes and size have already been done before. But, in this paper, we have proposed various machine learning models to segregate damaged tablets and capsules from the good ones and also compared accuracy of each models separately.

Keywords Gaussian regression · Feature detection · Medical tablets · Machine learning

1 Introduction

Medicines play an important role in human's life. Medicines are manufactured by various industries and factories. They often face a problem of manufacturing damaged products in their factories. In order to counter with such problems, we have proposed various techniques for detection of broken tablets using various machine learning techniques.

S. Lodh · S. Mahajan (✉) · A. Bhardwaj · A. K. Pandit
Shri Mata Vaishno Devi University, Katra, Reasi, Jammu and Kashmir, India
e-mail: mahajanshubham2232579@gmail.com

S. Lodh
e-mail: 18bec062@smvdu.ac.in

A. Bhardwaj
e-mail: anil.bhardwaj@smvdu.ac.in

A. K. Pandit
e-mail: amitkantpandit@gmail.com

Ch. Vanipriya
Sir M. Visvesvaraya Institute of Technology, Bangalore, India
e-mail: vanipriya_is@sirmvit.edu

S. Das and M. N. Mohanty (eds.), *Advances in Intelligent Computing and Communication*, Lecture Notes in Networks and Systems 202,
https://doi.org/10.1007/978-981-16-0695-3_6

Fig. 1 Tablet detection

Since tablet detection is delicate to color and various other factors in the background, execution is for the most part influenced by tablet identification calculations. Every tablet detection calculation works dependent on a tablet indicator that finds best tablet bouncing box. Figure 1 shows distinction of jumping box from different tablets and results of tablet location.

This shows significance of beginning detection which is examined in future. In deep learning-based techniques for broken tablet discovery, one of the fundamental premiums is foreseeing probabilistic solid areas of tablets' spots for more. So as to concentrate on a solid area, we misuse introduction techniques for tablets' regression by utilizing Kazemi's decision tree learning (TREE) [1, 2] calculation which gives quick and exact execution utilizing little preparing datasets. Later, the tablet detection calculations aim to deal with limited varieties under particular classifications, and tablet images are usually compiled under "controlled" conditions. For example, under "regulated" circumstances, the postures of the tablet and the outward appearance must be in certain classifications. More recently, the research centers around challenging "in-wild" conditions in which image of tablets can encounter particular external appearances and so on. When this is said to be finished, there is still a lack of an efficient approach which can deal with each of these varieties.

Prior to artificial intelligence (AI)-based methods, two ways are used in recognition of tablet. One is algorithmic state machine (ASM) [3] which utilizes principal component analysis (PCA) on facial shapes and another is active appearance model (AAM) [4] which utilizes PCA on tablet structure. Starting with these techniques, a tablet-based parametric identification calculation has been developed.

Non-parametric methodologies, as Gaussian model, the Supervised Descent Method, and the Tree regression, to understand this nonlinearity. In Kazemi's Tree [1] is methodology that reduce regression. It has run season of 10 ms and is generally used.

2 Work Done

We have used histogram of oriented gradients (HOG) + support vector machines (SVM) identifier, best jumping box disclosure procedure, as reason for Tree computation (Fig. 1). As stated above, due to use of multi-task cascade convolutional neural network (MTCNN), which is beneficial in tablet identifier that learns that three systems conduct multiple learning strategies for different scales, off-base fitting problem will occur in underlying jumping box following the Tree calculation. The dataset comprises 2225 pictures which were extracted from strips of tablets. The test dataset comprises 225 pictures which were again extracted for testing purpose.

So as to apply regression method for Tree computation, preparation pictures found by HOG + SVM are used for the arranging. Thus, we looked at 1047 images of 2225 for preparation.

For regression network, we changed ResNet which is difficult to deal with, and because of interference of a deeper neural system structure, slope data vanishes as depth of system increases through the remaining interaction (Fig. 2).

Regression system is portrayed as shown in Fig. 3. Utilizing earlier data got from the Tree, the info picture size is made; then, Gaussian bit separating technique [5] is used. The information picture is linked in channel measurement to utilize as an 8-channel info picture of regression network.

Using the Gaussian highlight map, in Fig. 4, the information on region of interest area is accentuated and provided to allow the device to accomplish increasingly accurate tablet detection [6, 7]. The element guide is adjusted by the use of distinctive Gaussian portion sizes, which will be addressed in the trial area. If we apply regression from Tree:

$$L = \|x - (y_{\mathrm{TREE}} - y_{\mathrm{PRE}})\|^2 \tag{1}$$

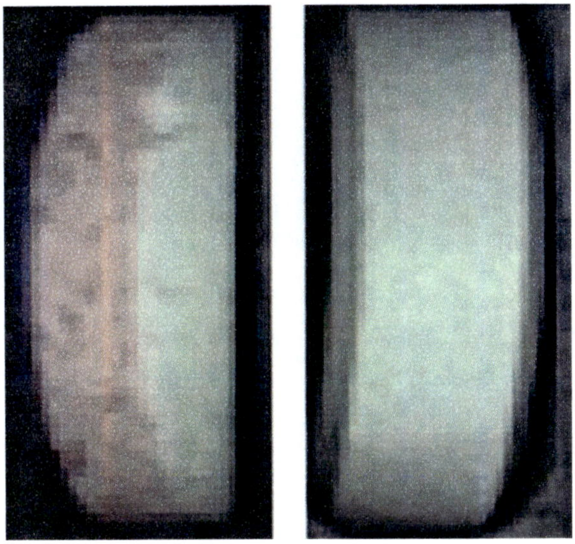

Fig. 2 Eextracted tablets from the strip

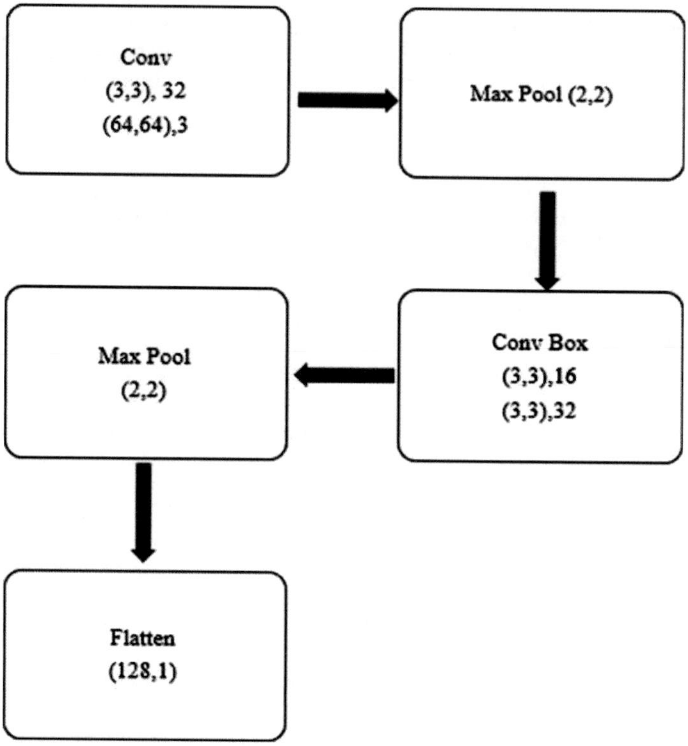

Fig. 3 Regression guided network

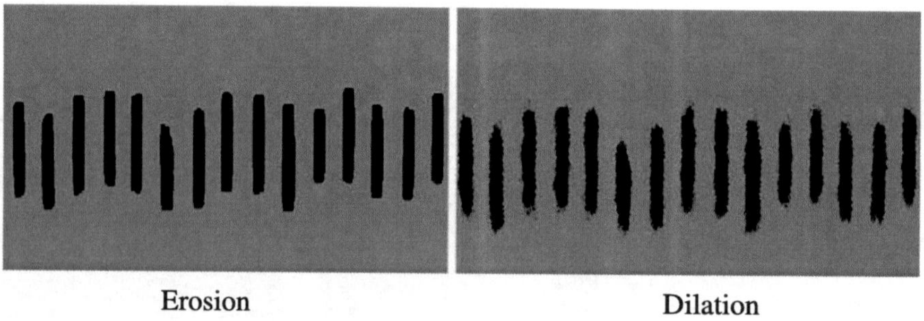

Fig. 4 Eroded image after dilation

where

x = Ground truth,
y = Vector information about how far to pass the previous region of interest to ground reality.

In particular, accuracy of tablet detection is measured as mean value of point-wise normalized Euclidean distance. We utilize normalized mean error:

$$\mathrm{NME} = \sum_{k}^{N} \frac{\|x_k - y_k\|^2}{d_{\mathrm{inter}}} \tag{2}$$

x = Ground truth,
y = Expectation of relapse arranges,
d = Intraocular separation,
N = Quantity tourist spots.

After detection of tablet erosion, dilation of the image is done. Based on the result obtained, sharpening of the image is done; then, morphological features are calculated.

Then, plots of both a broken and good tablet are plotted as shown (Fig. 5).

Quantitative findings related with HOG + SVM discovery with face alignment tool (TCDCN) and Tree bouncing box reveal terrible execution due to misalignment between place and tourist spots. The best implementation of our relapse method is refinement adaptation of the Tree. This illustrates the importance of the previous underlying data for relapse assignments.

The findings in Table 1 show the use of ResNet34, ResNet152, ResNet101, ResNet50, and ResNet18 in analysis of a deeper system does not necessarily indicate better execution and that larger Gaussian component size, data could be transmitted to system. However, as weight of former data builds up, weight of unsuitable position also increases.

Our experiments indicate that desired relapse is arranged, and Gaussian part size ResNet50 and is chosen to allow machine to forecast the precise path vectors.

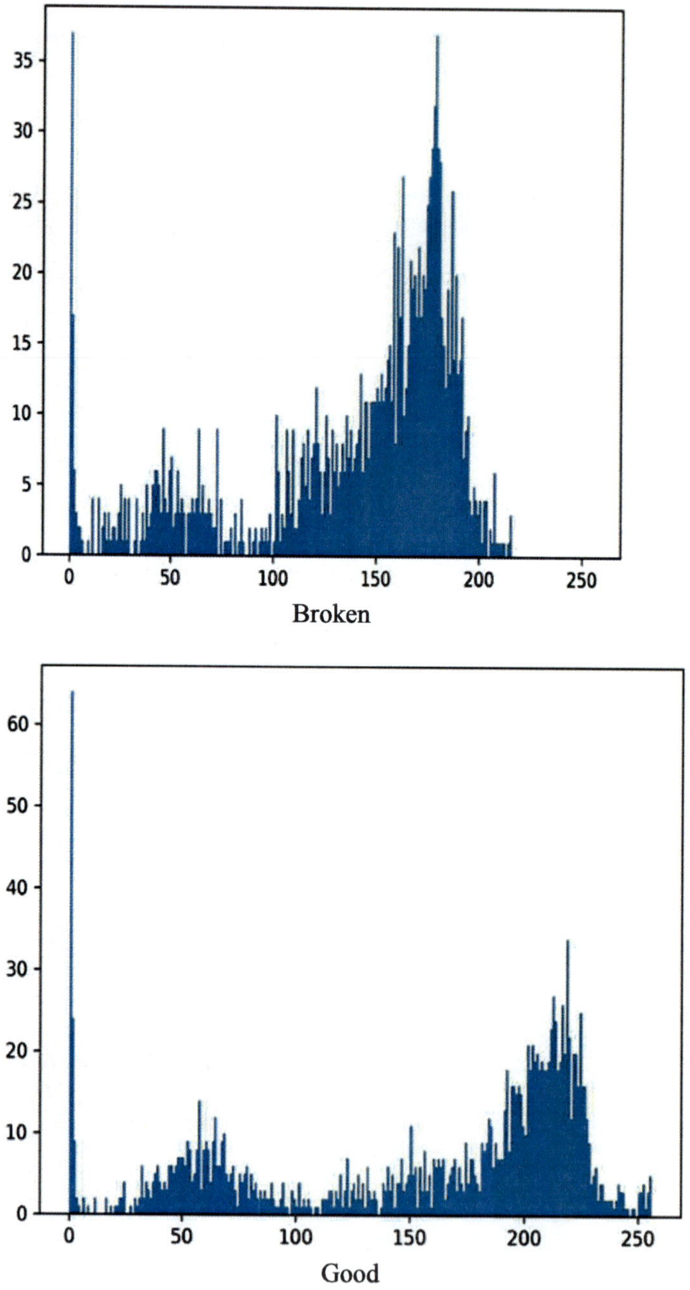

Fig. 5 Graph of a broken tablet and a good one. Our milestone identification uses 1800 images for testing, with the exception of 15 recognition disappointments that are not identified by the HOG + SVM model, which is one of the 1910 test datasets

Table 1 Normalized mean error (NME) result using ResNet

Kernel	Res18	Res34	Res50	Res101	Res152
(3, 3)	2.564	2.491	2.479	2.402	3.580
(5, 5)	6.496	6.585	6.461	7.441	7.455

3 Conclusion

In this work, the technique utilizes Tree techniques to form contours and then crops tablets out of strips. After cropping SVM + Hogg technique is used to prepare a model that detects broken tablets and enables the system to successfully improve the table discovery execution by utilizing less datasets. Trial results and removal results show viability of regression arrange as broken tablet detection refined calculation.

References

1. Kazemi, V., Sullivan, J.: One millisecond face alignment with an ensemble of regression trees. In: Proceedings of the IEEE Conference on Computer Vision and Pattern Recognition (2014)
2. Cootes, T.F., Edwards, G.J., Taylor, C.J.: Active appearance models. IEEE Trans. Pattern Anal. Mach. Intell. **6**, 681685 (2001)
3. Zhang, Z., et al.: Facial landmark detection by deep multi-task learning. In: European Conference on Computer Vision. Springer, Cham (2014)
4. Bulat, A., Tzimiropoulos, G.: How far are we from solving the 2d & 3d face alignment problem (and a dataset of 230,000 3d facial landmarks). In: Proceedings of the IEEE International Conference on Computer Vision (2017)
5. Wu, Z., et al.: A comprehensive survey on graph neural networks. IEEE Trans. Neural Netw. Learn. Syst. (2020)
6. Kingma, D.P., Welling, M.: An introduction to variational autoencoders. arXiv preprint arXiv:1906.02691 (2019)
7. Chalapathy, R., Chawla, S.: Deep learning for anomaly detection: a survey. arXiv preprint arXiv:1901.03407 (2019)

Lungs Nodule Prediction Using Convolutional Neural Network and K-Nearest Neighbor

Vijaya Patnaik and Chandrabhanu Mishra

Abstract Detection of nodules remains a very worrying problem because of its changeable shapes and sizes. Deep learning (CNN architecture) can detect the cancerous nodules at an early stage as it uses different approaches for solving problem, and are good enough to find the typical features of nodules is implemented, because deep learning methods are good to acquire characteristics from raw image data and provide proper answer to this kind of problem. Also by using KNN, better performance and accurate results are obtained. Confusion matrix for both methods is plotted to correlate how accurate both methods are. The performance measures like the classification rate , the FP rates, accuracy, and sensitivity are analyzed. Publicly available database having a wide range of data known as LIDC_IDRI database particularly for pulmonary lung nodule detection that focuses on an evaluation of automatic nodule detection.

Keywords Deep learning · Convolutional neural network · KNN · Lung image database consortium

1 Introduction

One of the leading deaths is lung cancer that causes a numerous deaths worldwide [1]. Patients on first stage of cancer do not show any clinical symptoms, so early detection is decisive for lung cancer survivability. In images of lungs CT, automatic recognition of nodules is a necessity in a computer-aided diagnosis (CAD) system [2] which finds out the probable value of a CT scan for cancerous nodules. A system that predicts the cancerous nodules normally consists of different stages

V. Patnaik (✉) · C. Mishra
Department of Instrumentation and Electronics Engineering,
College of Engineering and Technology, Bhubaneswar, India
e-mail: vijayapatnaik99@gmail.com

C. Mishra
e-mail: cmishra@cet.edu.in

S. Das and M. N. Mohanty (eds.), *Advances in Intelligent Computing and Communication*, Lecture Notes in Networks and Systems 202,
https://doi.org/10.1007/978-981-16-0695-3_7

like segmentation of lungs, prediction of nodules, determining features of nodules, and false positive cancelation which is known as CAD systems [3]. Recent clinical methods use many images per patient that makes a difficulty for radiologist's to properly analyze all the CT images in detail. Computers by certain programming features, within in a short time interval and with high precision, are advantageous. In short, the expertise of many expert radiologists can be put into action computationally; thus, computers are usually prepared in a particular area by a team of numerous human experts. In today's scenario, the progress of medicine toward CAD systems creates better CT image findings (which may contain cancerous nodules), for the disease, predicting a smallest hole in the nodules (which is almost difficult for a radiologist for detecting) and decreasing the duration diagnosis [4]. An automated image classification an approach that is usually used by CAD systems that help radiologists concerning both with accuracy and their speed [5]. Deep learning is a vastly growing approach that has an important role experience in CAD systems [6]. For the beginning stage, revelation of lung cancers CNN is a well-known technique of deeper learning as it has deeper networks and has sparse connectivity [7]. CNNs have much deeper layers with less networks and parameters so it is not hard to train with sufficient gathered information [8]. An essential characteristic of CAD is that it can determine the size, volume, and density of nodules quickly within minutes that make it an integral part in screening lung cancer and also in prediction [9]. A difference between sizes of nodules, i.e., cancerous nodules or non-cancerous nodules can get good results on 8–30 mm nodule sizes [10]. Here, a model is projected using neural networks to predict whether the nodules are benign or malignant. Also, employing KNN algorithm, to the data's provided, the average gap between all the testing and training specimens is computed at primary stage, and K-neighbors with most distance are taken for classification.

1.1 CAD Systems Using CNN for Detection of Pulmonary Nodules

A. CNN Flowchart for Detection of Nodules

CNNs are considered as well-known deep learning methods for its deeper networks and for its kernels and filters. Here, an advanced model is projected using deep learning's CNN approach to predict cancerous (malignant) nodules that can eliminate constants of the nodules and giving better sensitivity in detection. Figure 1 demonstrates a basic CNN step for predicting cancerous nodules. Convolutional neural networks (CNN) are fast ascendable neural network that also works with several fully connected layers that classify the input as one known category with few particulars and spatial characteristics of the image (Fig. 2).

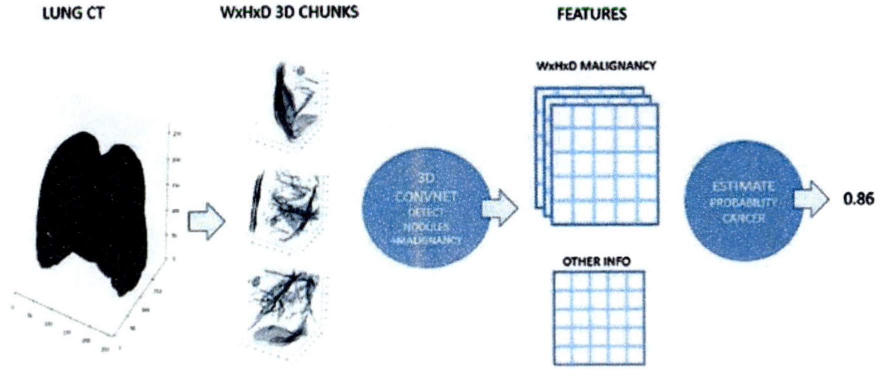

Fig. 1 CNN detection of pulmonary nodules. *Source* https://doi.org/10.1371/journal.pone.0219369.g001

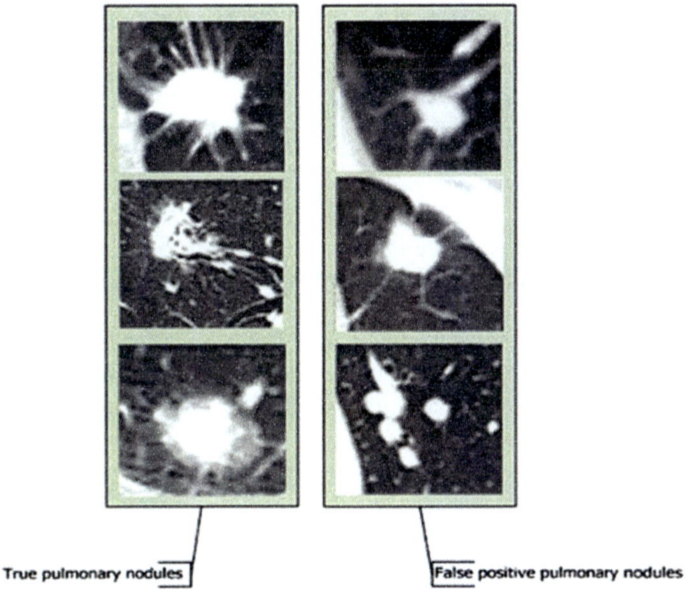

Fig. 2 Comparison of true positive nodules and false positive nodules. *Source* https://doi.org/10.1371/journal.pone.0219369.g002

B. Classification of Nodules

Nodules can be categorized as benign and malignant nodules, and they are even further categorized. These nodules are very crucial complications due to very close similarity during its initial stages. Benignes and malignancy of nodules can be

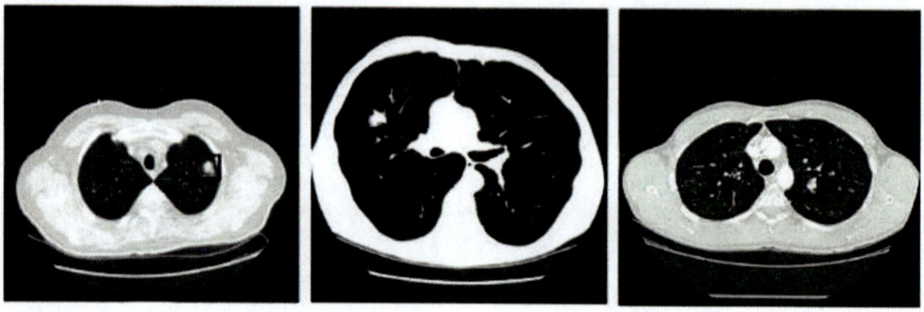

Fig. 3 Categories of lung nodules in a CT scan (benign, primary, malignant, and metastatic malignant), (left to right). *Source* https://www.ncbi.nlm.nih.gov

differentiated by structure, shape, and location at primary stages. The three different categories (benign, primary malignant, and metastatic malignant) of lung nodules are shown in Fig. 3.

2 Convolutional Neural Network

Convolutional neural networks employ neurons/activations and weights and an objective or loss function, considered as special type of neural networks that take images as inputs. Since images are its inputs, it allows modifications to the architecture so that there are sparse connections between layers and weights which are shared between output neurons in hidden layers. CNN stacks a sequence of layers followed by an output layer—Classification layer. Each layer performs a convolution or pooling operation. One of the important elements also parameter sharing as each member of convolutional kernel is used at every instance of the input dramatically reducing the network sizes.

A. *Convolutional Layers*

Convolutional layer is the core of the convolutional neural network. Highly important parameters are the kernels and the kernel size. Due to its sparse connectivity, every node in the convolutional layer receives input from a small number of nodes in the previous layer needing smaller specifications. Every output neuron is connected to a small neighborhood of intake through weight matrix filter or kernel. Each filter is moved around the input giving rise to one 2D output.

B. *Pooling*

Feature reduction is an important aspect of pooling as it provides translational invariance by sub-sampling. Commonly used pooling are max pooling and average pooling. Every location of output is replaced by an output node at certain locations with a summary static to nearby location. Here, max pooling is used.

C. *Activation Function*

Essential feature operation is performed that is used in each component of feature map that takes the place all negative pixel values in the feature map by replacing with zero. Activation function is also an important aspect in this model.

D. *Softmax Classifier*

A softmax classifier can be used for classification purpose which determines the probable output which is used to threshold or classify; it is a neural network transfer function that calculates the layers output from its net input.

E. *Classification Layer*

It is used for classification problems as it minimizes the gap of the probable distribution of the anticipated value and actual value, where the output is a probable value between '0' and '1.'

3 KNN

One of the easily understandable algorithms of machine learning is K-nearest neighbor (KNN). KNN is a user defined function, where KNN rule is used, and implementation of KNN is easy. Any distance can be used for KNN but mostly Euclidean distance is used. One of the demerits of KNN is that it takes more memory as everything has to be loaded.

4 Training and Simulation

A. *Dataset Preparation*

The data taken here is from the lung image database consortium and image database resource initiative [(LIDC/IDRI) database]. This dataset is specially made for the researchers to globally analyze "Lung Cancer Detection" problem that was initially released for the contest LUNA16 in the year 2016. A CT image scan having dimensions of each image is of size $512 \times 512 \times n$, where n represents axial scans which is considered. There occurs a huge imbalance in the class because of the data analysis. Therefore, to deal handle this kind of issue, the data is under sampled by the major class, and the minor class is augmented through rotating images.

B. *Data Augmentation*

Usually, deep learning approaches have their own built-in data augmentation techniques, but those are not working well or lacking some performances; therefore, by flipping vertically or horizontally, or by rotating, the data's in desired degree would help to get more data's to train the model. A considerable amount of true positive and true negative samples are required for getting accurate results. Translation, rotation, and flip are some image processing approaches that are applied before the image was input to model/system, which in case helped to upgrade the sampled data.

Generating patches: Using python, a script is used that generates 50 × 50 grayscale images that are further used for validation, testing , and training a CNN. However, the dataset is vastly imbalanced for training, so augmentation technique is used. Total 9796 images are obtained out of which 1351 are malignant and 8445 are benign which are divided into training and testing as per the below table (Table 1).

A. *Training*

The proposed system/model is designed keeping the earlier CNN architectures in mind. Deep learning approach is an important aspect for functioning of the model. This model helps to decrease the false positives by increased accuracy and sensitivity. Even if by using deep CNN techniques, we figure out nodules of <6 mm in diameter, still benign and malignant nodules are a difficult problem due to its lying over characteristics. Figure 4 shows the proposed CNN model for the lung nodules detection.

Table 2 briefly describes about the layers that has been used in the proposed CNN model.

5 Results

The model's training data's is from LIDC-IDRI database. The training includes data's from LUNA16 dataset which holds 9496 images after extracting features and augmenting the data's, and the prediction is further split into training and testing. To expand the training data and for class imbalance reduction among the type-ii error generated candidates and true nodules, several augmentation techniques such as interpretation, turning the pictures, twisting, and trimming-off are applied. Well-labeled data's are easily accessible from the database, as that is a need for the training of the new models. To train the system accurately and get better results from it, the data is separated into different categories like training, testing, and validation. The purpose of training is to reduce the dissimilarity between the training and testing data. Here, training data is 80%, and the testing data is 20%. The aim for stopping the training process is that: firstly, the model has attained its

Table 1 Classification of nodules in training and testing

	Total	Training	Testing
Malignant	1351	845	506
Benign	8445	6756	1689

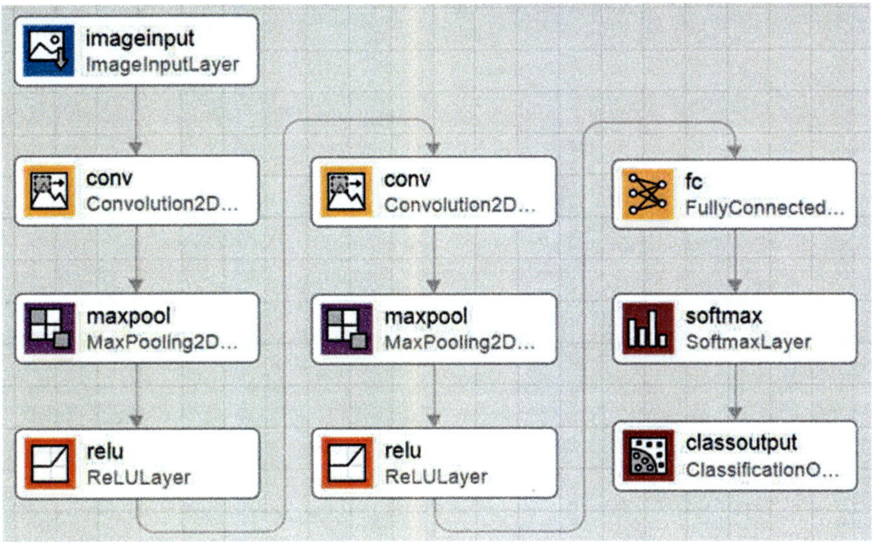

Fig. 4 Proposed CNN architecture

Table 2 Description of CNN layers

Layer	Name	Type	Description
1	Image input	Input as image	"50 × 50 × 1" CT scan formatted images
2	Conv layer_1	Convolution layer 1	20 images of size "9 × 9 × 1" convolutions with stride size [1 1] and padding [0 0 0 0] and window [1 1]
3	MaxPoollayer_1	Max pooling layer 1	"2 × 2" max pooling feature map reduction with stride, padding and window remaining same as above
4	relulayer_1	Activation layer 1	ReLU as considered as activation layer
5	Conv layer_2	Convolution layer 2	10 images of size "5 × 5 × 20" convolutions with stride size[1 1], padding [0 0 0 0], window [1 1]
6	MaxPoollayer_2	Max pooling layer 2	2 × 2 max pooling with stride, padding , and window same as above
7	Relulayer_2	Activation layer 2	ReLU replaces all negative pixel values by zero
8	Fc	Fully connected layer	Classifies the input as one of the known category that helps us to classify
9	Softmax	Softmax	Classifier that thresholds and gives the probable output
10	Class output	Classification output	Cross-entropy function used for classification problem

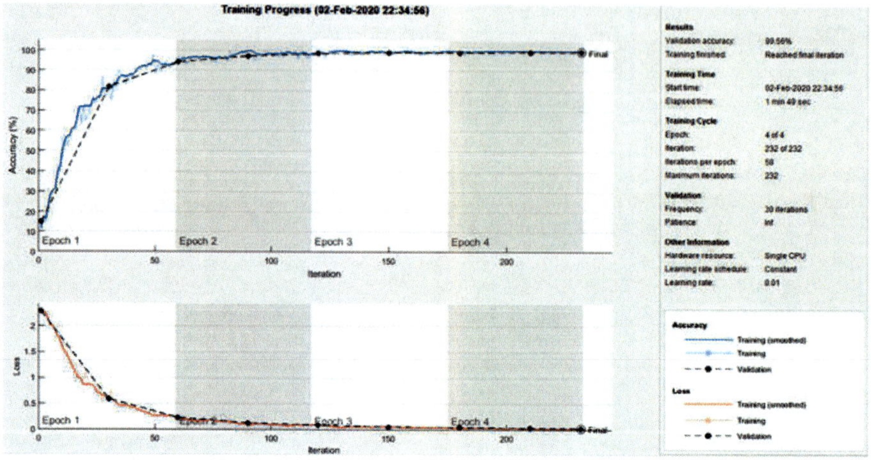

Fig. 5 Training progress

minimal error rate, along with the loss function on validation carry out no more changes. Figure 5 represents the training progress that has been conducted for the nodule detection.

5.1 Performance Metrics

Accuracy (understanding to characterize the nodule and non-nodule cases properly), sensitivity (capability to choose/decide the nodule cases precisely), and specificity (capability to choose/decide the non-nodule cases precisely) are used to calculate the exactness of the classification. For binary classification problems, extensively used metrics are mentioned above and in other words can be defined as:

$$\text{Accuracy} = \frac{\text{TP} + \text{TN}}{\text{TP} + \text{TN} + \text{FP} + \text{FN}}$$

$$\text{Sensitivity} = \frac{\text{TP}}{\text{TP} + \text{FN}}$$

$$\text{Specificity} = \frac{\text{TN}}{\text{TN} + \text{FP}}$$

where TP (true positive) represents rightly diagnosed nodules; FP (false positive) represents the incorrectly diagnosed nodules; TN (true negative) represents the correctly diagnosed as non nodules, and FN (false negative) represents incorrectly diagnosed non-nodules.

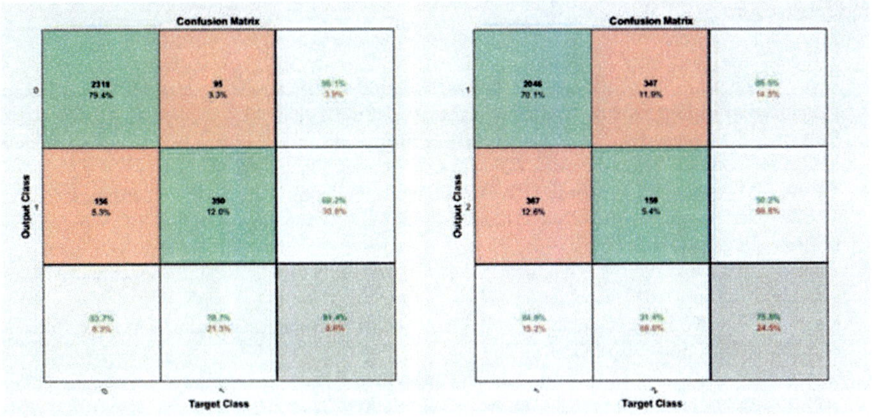

Fig. 6 Confusion matrix for CNN and KNN

Figure 6 above is a confusion matrix formed by training the system. Matrix can be defined as table that describes efficiency of a classification model (or “classifier”) for known true values. This table is a blend of four different combinations of predicted and actual values. KNN confusion matrix shows efficiency of the system taken from various distances with different data and data points.

6 Conclusion

After wide use of deep convolutional neural networks, here and after evaluating it below mentioned are some of the results. As the data’s were taken from the LIDC-IDRI database after the prediction of benign or malignant nodules they are compared with the database. From the results, the projected model results with minor estimation cost with an accuracy of 89%, sensitivity of 90% and specificity of 77.7%. The conclusion from the experiments suggests that the CNN model achieved an outstanding conclusion. At present, the trained CNN architecture resulted more than 90% of results, still the conclusion can be made-up with the inclusion of other feature extractors. As this could boost the feature volume, so that feature reduction is done. With other databases if the proposed model is tested possibly better accuracy can be achieved. Also KNN algorithm is trained and from comparison with the results of CNN we get a clearer image about the nodule prediction. Here, greater distances are taken for classification. Here, we achieve a good accuracy by plotting a confusion matrix of the KNN which is about 84%. Each time the top-most accuracy is recorded. However, it is promising that deep neural networks approach is in demand for computer-aided diagnosis system that help’s experts as an auxiliary tool in coming future.

References

1. Zuo, W., Zhou, F., Li, Z., Wang, L.: Multi-resolution CNN knowledge transfer for candidate classification in lung nodule detection. IEEE Acc. **7** (2019)
2. Hawkins, S., Wang, H., Liu, Y., Garcia, A., Stringfield, O., Krewer, H., et al.: Predicting malignant nodules from screening CT scans. J. Thorac. Oncol. (2016)
3. Firmino, M., Angelo, G., Morais, H., Dantas, M.R., Valentim, R.: Computer-aided detection (CADe) and diagnosis (CADx) system for lung cancer with likelihood of malignancy. Biomed. Eng. Online **2** (2016)
4. Shariaty, F., Mousavi, M.: Application of cad systems for automatic detection of lung nodules. Inform. Med. Unlocked **15** (2019)
5. Kang, G., Liu, K., Hou, B., Zhang, N.: 3D Multi-view convolutional neural networks for lung nodule classification. PLoS One (2017)
6. Gao, X.H.W., Hui, R., Tian, Z.M.: Classification of CT brain images based on deep learning networks. Comput. Meth. Prog. Biomed. **138**, 49–56 (2017)
7. Spinhoven, M.J., Desbuquoit, D., Snoeckx, A., Van Meerbeeck, J.P., Reyntiens, P., Parizel, P.M., Van Schil, P.E.: Evaluation of the solitary pulmonary nodule: size matters, but do not ignore the power of morphology. Insights Imaging (2018)
8. Krizhevsky, A., Sutskever, I., Hinton, G.E.: Imagenet classification with deep convolutional neural networks. Adv. Neural Inf. Process. Syst., 1097–1105 (2012)
9. Prakashini, K., Babu, S., Rajgopal, K.V., Kokila, K.R.: Role of computer aided diagnosis (CAD) in the detection of pulmonary nodules on 64 row multidetector computed-tomography. Lung India **33**, 391–397 (2016)
10. Silvestri, G.A., Tanner, N.T., Kearney, P., Vachani, A., Massion, P.P., Porter, A., Springmeyer, S.C., Fang, K.C., Midthun, D., Mazzone, P.J.: Assessment of plasma proteomics biomarker's ability to distinguish benign from malignant lung nodules: results of the PANOPTIC (Pulmonary Nodule Plasma Proteomic Classifier) trial. Chest **54**, 491–500 (2018)

Quantitative Structure–Activity Relationships (QSARs) Study for KCNQ Genes (Kv7) and Drug Discovery

Nilima Rani Das and P. Ganga Raju Achary

Abstract Neuronal Kv7 (KCNQ) genes are important group of voltage-gated K^+ channels, popularly known as the M-current or M-channel. These genes play a pivotal role in regulating neuronal excitability and access the risk associated with these neuronal excitability with reference to hypertension, diabetes and obesity. They have vital physiological functions in the nervous system, heart and inner ear. Failure in the functioning of Kv7 channels leads to severe genetic disorders. The drugs targeting Kv7 channels can be used in the treatment of neuronal channelopathies like chronic and neuropathic pain, deafness and mental illness. Owing to the prime importance of the KCNQ channels, novel quantitative structure–activity relationships (QSARs) models were built based on key descriptors using multiple linear regression (MLR) technique. A diverse set of 60 organic molecules were used to build the models which showed good experimental interaction with KCNQ. The models were validated by the external validation and standardization approach. The feed-forward neural network (FFNN) has also been used to establish the accuracy and efficiency of the models. When the FFNN is applied to the descriptors attained from the MLR, it exhibited a correlation coefficient of 0.839. A high correlation has been observed between predicted and experimental pEC50 (M), thereby confirming the high quality of the developed QSAR models. The obtained R^2 and R^2ext values for the models are significant and acceptable. The other statistical parameters are also considerable proving the robustness and effectiveness of the reported models. These models can provide mechanistic interpretations and can be the starting point to discover the best drugs for the diseases caused by the irregularities in the function of the KCNQ genes.

N. R. Das
Department of CA, Siksha 'O' Anusandhan (Deemed to be University), Bhubaneswar, Odisha, India
e-mail: nilimadas@soa.ac.in

P. G. R. Achary (✉)
Department of Chemistry, Siksha 'O' Anusandhan (Deemed to be University), Bhubaneswar, Odisha, India
e-mail: pgrachary@soa.ac.in

S. Das and M. N. Mohanty (eds.), *Advances in Intelligent Computing and Communication*, Lecture Notes in Networks and Systems 202,
https://doi.org/10.1007/978-981-16-0695-3_8

Keywords KCNQ genes · QSAR · SMILES · ANN

1 Introduction

Kv7 channels are a small family of voltage-gated potassium channel subunits (Kv7.1–Kv7.5) that are encoded by the KCNQ genes. These channels are considered important because they become active at the sub-threshold membrane potentials [1]. The Kv7 channels produce low voltage-activating M-current that can regulate excitability of the neurons and firing rate of action potential. KCNQ genes have a wide range of physiological correlates. An exceptional proportion (4 out of 5) of the KCNQ genes is subject to mutations that give rise to human genetic diseases. Initially, the M-current was characterized functionally as a 'braking current' that served to limit neuronal discharge frequency. Though confirmed by subsequent work, immunocytochemical localization by Kv7 antibodies, and the development and use of Kv7-specific drugs have highlighted important additional or consequential functions. Neural KCNQ channels are drug targets that can cure several diseases like epilepsy and other nervous system disorders like neuropathic pain, hearing loss and mental health disorders.

The foremost intension of drug design process is to formulate safe and efficient medications that can lessen pain and discomfort of the patient. However, the drug development procedure is extremely complicated, time-consuming and demands a lot of multidisciplinary proficiency and inventive methods. Current evaluations propose that it takes around 13.5 years and 1.8 billion US dollars to introduce a new drug in the market [2]. During recent years, the use of computational techniques has been remarkably increased in the process of finding new medications [3]. Realistic and sensible drug design techniques can decrease the time and cost required to design a particular drug. Quantitative structure–activity relationships (QSAR) are nowadays applied to prepare and find out new inhibitors with minimized toxicity level. QSAR techniques have a crucial function in the development of drug targets as they are low-cost substitute for in vitro and in vivo testing [4]. The basic theory on which the QSAR methods work is the correlation existing between biological activities of the different compounds having divergent molecular composition. The biological activities can be correlated with explicit molecular descriptors, and some regression methods can be employed to find out the amount of contribution of the descriptors in that biological process [5]. QSAR can discover chemical structures having excellent inhibitory properties on particular drug targets that have less toxicity level [6].

The importance of the KCNQ channels has lead to the construction of novel QSAR models using MLR and simplified molecular input line entry system (SMILES) based on single optimal descriptor. These models can provide the mechanistic interpretations and can be very helpful in discovering promising and effective Kv7 specific drug targets.

2 Molecules, Software Codes and Methods

2.1 Dataset Preparation and Data Reduction

The dataset of the 60 diverse molecules with their EC50 (nM) values for KCNQ molecules have been derived from the popular binding database [7]. The range of EC50 (nM) values for the molecules lies between 1 and 30,000, which is quite broad. Finding it difficult to build QSAR models directly from their EC50 scores, these EC50 values were converted into pEC50. pEC50 is the negative decimal logarithm of EC50. The software codes Spartan-10 (Wave function INC) and OpenBabel (ver. 2.3.2) were also used to generate the SMILES and MDL (.mol) chemical structures from the SDF structure file of these 60 molecules.

2.2 Descriptors

PaDEL_2.18 descriptor calculation code was used to obtain more than 500 molecular 2D and 3D descriptors. Similarly, CORAL software code was used to generate the optimal descriptor based on SMILES. The optimal descriptor derived by SMILES is DCW(1, 15).

2.3 Model Building

QSARINS (QSAR-Insubria) software [8] code developed at the University of Insubria was applied to build the desired robust QSAR models. Since more than 500 different descriptors were there, taking all descriptors simultaneously to build the model required heavy computational time. Preliminarily scans were performed using 100 descriptors in each case to identify potential descriptors for the KCNQ. Finally, a descriptor set of 60 was chosen to build the robust QSAR models employing a genetic algorithm (GA) technique, which is available in the QSARINS. GA was used to select suitable descriptors and ordinary least squares (OLS) method was used to achieve a linear regression model between the response variable pEC50 and the descriptors. Both internal and external validation methods, LOO validation and principle component analysis were explored to verify the robustness and stability of the constructed models. The model with minimum value of $R^2 - Q^2$ was considered as more stable.

2.4 Dataset Distribution

The 60 agonist dataset for the KCNQ was randomly distributed into different sets such as the training (75%), calibration (15%) and validation sets (15%). Different random splits were checked to build extrapolative capability of the optimal QSAR model. The robustness of a model basically depends on the quality of training set and during the training process the decrease in the performance of a model due to overtraining can be significantly removed by the calibration set [9, 10]. After proper distribution of the dataset into training, calibration and validation sets the models were built.

3 QSAR Model Validation Criteria

OECD principles have been used to establish the significance of the QSAR models, and in this work, QSARINS software was used to build QSAR models and validate the models using OECD principles. The OECD principles state that a model should congregate the following properties: (1) It should have a defined endpoint. (2) It should be represented using definite algorithm with no ambiguity. (3) It should have a definite domain of applicability. (4) It should follow appropriate measures of goodness of fit, robustness and predictability. (5) If possible, it should have a mechanistic interpretation. The first principle ensures lucidity in the measurement point calculated by a particular model. The endpoint based on which models are built is the KCNQ inhibition constant (EC50), which describes underlying inhibition by an efficient inhibitor expressed as pEC50. The second principle states that the QSAR models should be represented using unambiguous algorithms so that the relationship between the descriptors and the endpoint can be defined by mathematical models. The algorithm used in this work is a mathematical model of MLR that used OLS to obtain this relationship. According to the third principle, the model can make consistent predictions if the leverage values are less than the critical leverage value with ±3 standard deviations. Molecules not remaining within this range can possibly be treated as outliers. To evaluate the consistency of the predictions of experimental compounds William's graph has been used here that describes the applicability domain (AD) of the model. According to the fourth principle, the discrepancy between the experimental values and predicted values should be minimized. The goodness of fit of the models was estimated with the coefficient of determination (R^2) and adjusted R^2. Internal validation techniques, such as LOO (leave one out) and LMO (leave many out) techniques, were used to verify and confirm the strength of a model. To show that the models generated were not the outcome of a chance correlation the Y-scrambling method has been used here. External validation methods were used to prove the models' capability of predicting new compounds. The dataset was divided into two disjoint sets. One set was used for training the model, and another set was used for prediction. The performances of the models were evaluated by a variety of measures such as RMSEext, Q^2-$F1$, Q^2-$F2$, Q^2-$F3$, CCC and methods like Golbraikh and Tropsha methods [11]. The fifth principle is also fulfilled by presenting a common interpretation of the models using the descriptors selected to describe the KCNQ channel modulation.

4 Results and Discussion

The geometrical optimization for all the sixty 2D and 3D structures were carried out by the semi-empirical (AM1) quantum chemical calculation with the Spartan 10 (Wavefunction Inc). After that some models were developed having low multi-collinearity between the descriptors (2D, 3D and the optimal descriptor) and desirable correlation with the modelled response. Some of the QSAR models built by the QSARINS code, which showed interesting characteristics are:

1.
$$\mathrm{pEC50} = 4.7458 + 0.0356(\mathrm{DCW}(1, 15)) \tag{1}$$

2.
$$\begin{aligned}\mathrm{pEC50} =& 29.1166 + 0.0001(\mathrm{ATS6i}) - 0.0009(\mathrm{ATSC2m}) + 3.8829(\mathrm{MATS4i}) \\ & - 0.3731(\mathrm{GATS8e}) + 1.7047(\mathrm{VE1\_Dzs}) - 12.9266(\mathrm{SpMin2\_Bhs}) \\ & - 9.2893(\mathrm{ETA\_BetaP\_ns\_d})\end{aligned} \tag{2}$$

3.
$$\begin{aligned}\mathrm{pEC50} =& 26.7111 + 3.8971(\mathrm{MATS4i}) + 1.4222(\mathrm{VE1\_Dzs}) \\ & + 1.9152(\mathrm{SpMin5\_Bhm}) - 12.6648(\mathrm{SpMin2\_Bhs}) \\ & - 7.3340(\mathrm{ETA\_BetaP\_ns\_d})\end{aligned} \tag{3}$$

The statistical parameters for the training set such as R^2, R^2_{adj}, CCC_{tr}, LOF and s for the above models are given in Table 1 for a performance evaluation of the models.

s is standard error of estimate, F is the variance ratio, LOF is the Friedman lack of fit, K_{xx} denotes the correlation among the descriptors, $RMSE_{tr}$ is root mean square in fitting, MAE_{tr} is mean absolute error in fitting and CCC_{tr} is the concordance correlation coefficient over the training set. All the models are having good values for R^2 showing proper fit of the models. The values of R^2_{adj} show the convenience of adding new descriptors to the models. Low values of LOF show there is no possibility of overfitting. The value of K_{xx} is low showing minimum correlation between the descriptors. Values of RMSE, MAE and s are small showing small errors on training calculations and estimated parameters.

Table 1 Fitting parameters

Model	R^2	R^2_{adj}	s	F	$R^2 - R^2_{adj}$	LOF	K_{xx}	$RMSE_{tr}$	MAE_{tr}	CCC_{tr}
1 (Eq. 1)	0.699	0.693	0.283	116.404	0.006	0.083	0.000	0.278	0.182	0.823
2 (Eq. 2)	0.832	0.800	0.218	26.277	0.031	0.083	0.374	0.198	0.149	0.908
3 (Eq. 3)	0.742	0.709	0.264	22.444	0.033	0.100	0.350	0.246	0.195	0.852

5 Model Validation According OECD Principles

The OECD regulatory values were followed to ensure the validity of the models. The fitting and performance of the model were assessed by some internal validation techniques like LOO and LMO. The values of different parameters are described in Table 2.

The difference between R^2 and Q^2_{LOO} is very small that indicates the consistency and stability of the models. The error in the predictions is very low. The experimental versus predicted values for all training and test set compounds are presented in Fig. 1a. Figure 1b depicts a comparison between the values calculated by LOO and actual the experimental values of pEC50 for the first model. The values of R^2 and Q^2_{LMO} are nearly similar, confirming the robustness of the model. The values of Q^2_{LMO} are very alike authenticating the models as a good fit. The Y-scrambling technique has been used to demonstrate that the models are not the effect of casual correlations. The values of R^2 and Q^2 and their average (R^2_{Yscr}, Q^2_{Yscr}) show the goodness of the model. As the values of R^2_{Yscr} and Q^2_{Yscr} are very small, the models are believed to be good models. It confirms that the models are not product of a random correlation.

Table 2 Cross-validation parameters

Model	$R^2 - Q^2_{\mathrm{LOO}}$	$\mathrm{RMSE_{cv}}$	$\mathrm{MAE_{cv}}$	$\mathrm{CCC_{cv}}$	Q^2_{LMO}	R^2_{Yscr}	Q^2_{Yscr}
1	0.017	0.286	0.188	0.813	0.677	0.020	−0.061
2	0.066	0.234	0.180	0.873	0.743	0.161	−0.261
3	0.110	0.294	0.228	0.792	0.603	0.115	−0.193

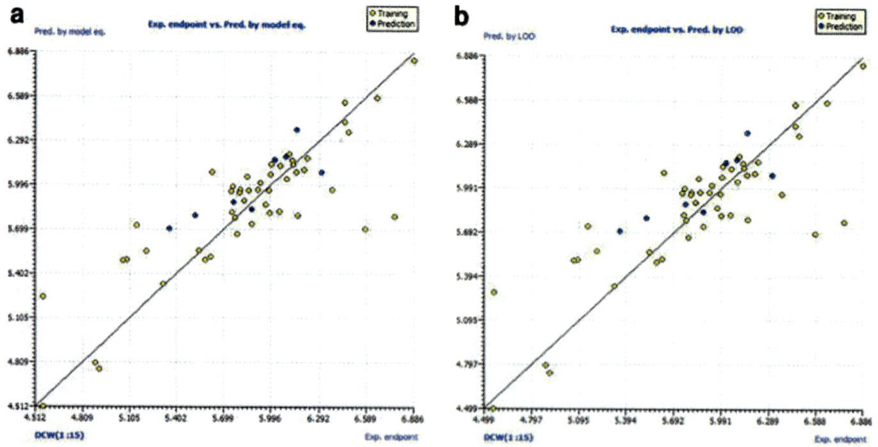

Fig. 1 **a** Exp. endpoint versus predicted by model Eq. 1. **b** Exp. endpoint versus predicted by LOO

The extrapolative capability of the model was evaluated using some external prediction tools such as R^2ext (external determination coefficient), RMSEext (RMSEext in external prediction), MAEext (MAE in external prediction), PRESSext (Predictive Residual Sum of Squares in external validation), Q^2-$F1$, Q^2-$F2$, Q^2-$F3$, CCCext (CCC in external prediction). Table 3 shows the values of external validation parameters. Q^2-$F1$, Q^2-$F2$ and Q^2-$F3$ are the variances in external prediction. In this work, the leverage (h) method was also used to check for the AD. In William's graph of the first model, most of the compounds are within the AD. Along with this, QSARINS provides a graph that helps to visualize the model's AD on the basis of leverage and model predictions called as graph of Insubria. This technique can find out the molecules that lack experimental response. The graph of Insubria (Fig. 2b) is similar to William's graph (Fig. 2a).

From the values presented in Table 3, it has become clear that external validation parameters are significant for model 1. As model 1 has satisfied all the internal and external validation criteria, it can be considered as effective and robust.

Table 3 External validation parameters

Model	R^2ext	RMSEext	MAEext	CCCext	PRESSext	Q^2-$F1$	Q^2-$F2$	Q^2-$F3$
1	0.700	0.206	0.184	0.723	0.340	0.549	0.538	0.835
2	0.005	0.448	0.343	0.046	2.217	0.102	−0.075	0.143
3	0.206	0.572	0.412	−0.339	3.603	−0.458	−0.747	−0.392

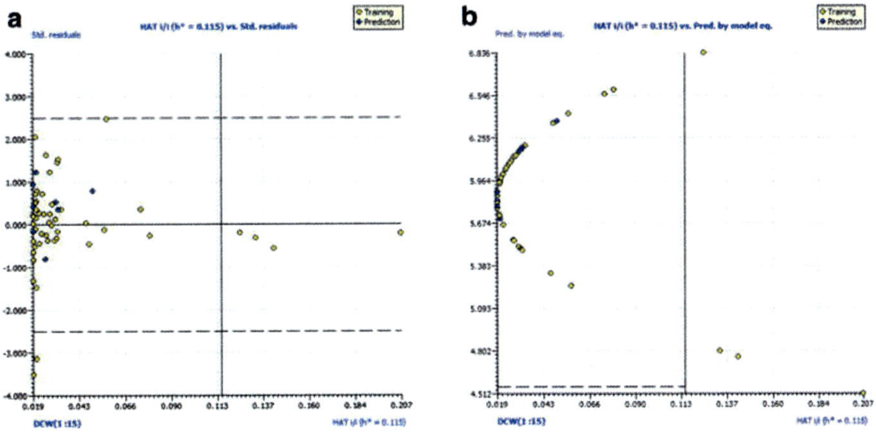

Fig. 2 **a** William's plot (Hat diagonal values versus residuals), **b** Insubria graph (Hat diagonal values versus predicted data)

6 Artificial Neural Networks

ANN has been used to increase the probability of good characterization of the compounds. To prove the robustness of the model generated by MLR, a two-layer feed-forward neural network (FFNN) trained with Levenberg–Marquardt was applied on the descriptors calculated by the model (model-1). The dataset was divided into three sets. 75% of the dataset was taken for training, 15% of the dataset formed the validation set and the rest 15% was taken for prediction. The correlation of the experiential values with those calculated using the neural network is presented in Fig. 3. The correlation coefficient ($R = 0.839$ for training set and 0.982 for the test set), the MSE (0.091) for the training set and the MSE (0.015) for the testing set attained with the neural network confirm that the descriptor selected by MLR is relevant and the model developed has good predictive capability.

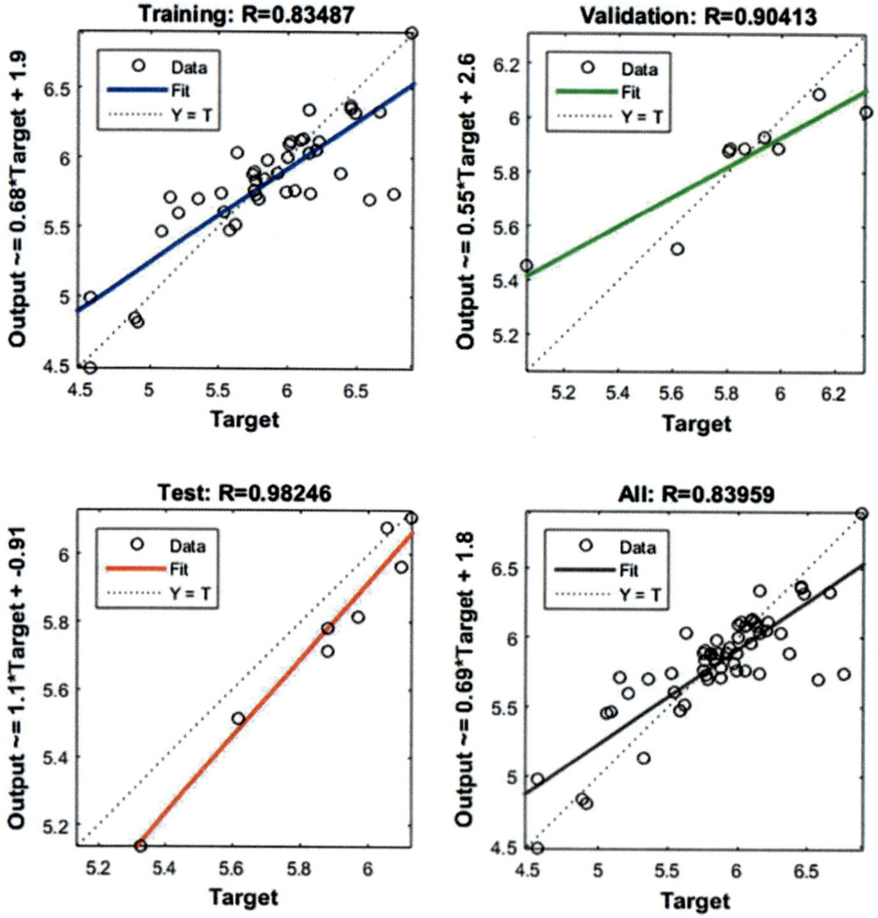

Fig. 3 Regression graph generated by the FFNN

7 Conclusion

With a belief that, computational methods can be used to predict the activities of KCNQ molecules to smoothen the progress of the drug discovery process by reducing cost and time, in the present communication, QSAR models were developed using 2D and 3D molecular descriptors which adequately predicted the activity of the KCNQ molecules. The proposed model satisfied all the required principles of the OECD. The internal validation (LOO, LMO and Y-scrambling) techniques helped to establish the robustness of the model and the external prediction set helped to validate the predictability of the model. The presented MLR-based QSAR model provides an effective method to test and develop better drug candidates. The incorporation of DCW descriptor in the model building decreased the number of descriptors and registered impressive statistical parameters, internal validation characteristics and external validation characteristics. The result of ANN proves that the model is really robust and stable. The proposed work creates a paradigm for future studies of the KCNQ genes for novel drug targets. The work on this interesting small family of K^+ channels can provide new insights into physiology, pharmacology, pathology and potential therapeutics.

References

1. Brown, D.A.: M currents. In: Narahashi, T. (ed.) Ion Channels, vol. 1, pp. 55–99. Plenum Press, New York (1988)
2. Kapetanovic, I.M.: Drug Discovery and Development—Present and Future. InTech (2016). https://doi.org/10.5772/1179
3. Badnjevic, A., Beganovic, E., Music, O.: Facts about solution based and cartridge based devices for blood gas analyses. In: IEEE 18th International Conference on System, Signals and Image Processing, pp. 1–5 (2011)
4. Santos-Filho, O.A., Hopfinger, A.J., Cherkasov, A., de Alencastro, R.B.: The receptor-dependent QSAR paradigm: an overview of the current state of the art. Med. Chem. **5**, 359–366 (2009)
5. Kubinyi, H.: QSAR: Hansch analysis and related approaches. In: Mannhold, R., Kroogsgard-Larsen, P., Timmerman, H. (eds) Methods and Principles in Medicinal Chemistry, vol. 1, pp. 240. Wiley-VCH, Weinheim, Germany (1993)
6. Doichinova, I.A., Natcheva, R.N., Mihailova, D.N.: QSAR studies of 8-substituted xanthines as adenosine receptor antagonists. Eur. J. Med. Chem. **29**(2), 133–138 (1994)
7. Gilson, M.K., Liu, T., Baitaluk, M., Nicola, G., Hwang, L., Chong, J.: BindingDB in 2015: a public database for medicinal chemistry, computational chemistry and systems pharmacology. Nucleic Acids Res. **44**(D1), D1045–D1053 (2016)
8. Gramatica, P., et al.: QSARINS: a new software for the development, analysis, and validation of QSAR MLR models. J. Comput. Chem. **34**(24), 2121–2132 (2013)
9. Achary, P.G.R.: Simplified molecular input line entry system-based optimal descriptors: QSAR modelling for voltage-gated potassium channel subunit Kv7.2. SAR QSAR Environ. Res. **25**(1), 73–90 (2014). https://doi.org/10.1080/1062936x.2013.842930

10. Toropov, A.A., Achary, P.G.R., Toropova, A.P.: Quasi-SMILES and nano-QFPR: the predictive model for zeta potentials of metal oxide nanoparticles. Chem. Phys. Lett. (2016). https://doi.org/10.1016/j.cplett.2016.08.018
11. Golbraikh, A., Tropsha, A.: Beware of q2! J. Mol. Graph. Model. **20**(4), 269–276 (2002)

Apple Fruit Disease Detection and Classification Using K-Means Clustering Method

Rishabh Tiwari and Manisha Chahande

Abstract Agriculture field has major role in the growth of Indian economy. Innovations that can detect fruit diseases at early stages would guarantee natural product quality for the organic product for agriculture industry. This proposed method was expected to explore the capability of utilizing K-Means segmentation, colour coherence vector, cluster selection and order shading surface highlights for identifying apple fruit infections. A shading imaging framework was created to obtain RGB pictures from apples with ordinary and four diseases affected conditions (i.e., apple blotch, apply scab, apple rot and apple rust). Selected image cluster with affected condition is examined for difference in features highlights based on the features extracted from the image histogram and LBR of each fruit disease test. Calculations for choosing affected image clusters were created dependent on a stepwise feature's examination of the cluster selected. Characterization models were built utilizing the diminished surface capabilities through a discriminant work dependent on a proportion of the summed up squared separation. In this work, we have analysed diseases that often appear on apple using K-Means clustering method. Out of these diseases, few may be cured at an early stage which can save the economic loss at later stage. Out of these diseases, few may be cured at an early stage which can save the economic loss at later stage.

Keywords Texture · Colour · Segmentation · K-Means clustering · Support vector machine (SVM)

R. Tiwari · M. Chahande (✉)
Amity School of Engineering and Technology, Amity University, Sector-125, Noida, Uttar Pradesh, India
e-mail: mmchahande@amity.edu

R. Tiwari
e-mail: rishabhtiwari0282@gmail.com

S. Das and M. N. Mohanty (eds.), *Advances in Intelligent Computing and Communication*, Lecture Notes in Networks and Systems 202,
https://doi.org/10.1007/978-981-16-0695-3_9

1 Introduction

As of late, apple fruit diseases have become a difficult issue for apple production in India. The diseases in plants are not only the problem in India but it exists in the whole world. Exploring for contaminated plants and creepy crawlies by human monitors has constrained value. In this manner, mechanized infection identification has a few valid statements, including bringing down creation costs and improving exploring productivity. Fruit sources (such as plant, bushes and trees) can display a large group of side effects reflecting different issues that can antagonistically affect their well-being, force and profitability to fluctuating degrees. Now and then, fruit diseases control activities or medicinal measures can be embraced if the side effects are recognized early. For more accuracy in fruit disease detection and control exists when correct farming methods are included, this will result in a better yield overall, alongside a worldwide situating framework to outline in the forest for future control activities. An automated disease detection framework may support ecological and horticultural assets. This automated vision-based method that can outwardly separate basic apple diseases affected images of apples using singular fruit image processing, segmentation feature extraction and classification. Fruit disease tests were done for only apple affected by certain conditions. Future investigations will extend the advancements to other fruit assessments [1, 2].

In the previous decade, different analysts have utilized image processing and example acknowledgement techniques in rural applications, for example, identification of weeds in the field and arranging of leafy foods. The hidden methodology for these techniques is the equivalent. To begin with, images are obtained from the earth utilizing simple, advanced, or camcorders. At that point, image processing techniques are applied to separate valuable highlights that are vital for additional examination of the images.

A while later, discriminant techniques, for example, parametric or nonparametric measurable classifiers and neural systems, are utilized to order the images. The determination of the image processing techniques and the arrangement procedures are significant for the effective execution of any machine vision framework. Item shape coordinating capacities, shading-based classifiers, reflectance-based classifiers and surface texture-based classifiers are a portion of the normal techniques that have been attempted before.

Various techniques have been concentrated to recognize deformities and diseases related to fruits and their production. The general goal of this exploration was to build up an automated vision-based technique for distinguishing different fruit diseases by utilizing features highlights processed by using various segmentation and clustering methods.

Explicit goals actualized to achieve the general goal were to: utilize a image processing framework to gather RGB images from apple with one healthy and four different disease affected apples (i.e., apple blotch, apply scab, apple rot and apple rust) finding cluster in the image which has the affected area under the disease dependent on the colour coherence vector (CCV) [3, 4].

2 Fruit Diseases Found in Apples

India is one of the largest producers of apples in the world. The fruit industry is very important to the Indian economy and hence improved, and accurate fruit disease detection techniques can really reduce fruit shortages and improve the quality of the fruits produced. These farming frameworks require more information on in-field knowledge and its relationship to trim yield and the impact of diseases on fruit and at last loss in the economy.

In the previous decade, different scientists have utilized image processing and design acknowledgement techniques in agricultural applications, for example, recognition of weeds in a field, arranging of foods grown from the ground, and rotting and disease discovery and so on. This section will present past research in farming-based machine vision, image processing and characterization techniques for agriculture applications, and furthermore present basic datasets for fruit diseases and rotting [5, 6].

2.1 *Apple Blotch*

Developing apples, be that as it may, isn't as simple as developing other tough perennials. In the event that you need your apples to flourish and produce heaps of fruits, you'll need to give close consideration to their consideration consistently. Apple blotch fungus disease is only one basic issue for both apple ranchers and mortgage holders the same.

Blotch fungus on apples is a typical disease brought about by an assortment of growths all through the fruiting season. Luckily, it is a difficult that is restricted to the skin of the apple. It is likewise sheltered to eat except if you have a form hypersensitivity, so for some property holders, apple blotch fungus disease may not represent a genuine enough danger to treat. For other people, some degree of treatment among none and plantation-level insurance may appear to be increasingly suitable. The shading might be overcast or dingy, regularly making the apple surface seem olive green. It is basic for littler territories to meet up to frame bigger, non-round about spots on the skin.

2.2 *Apple Scab*

Mainly seen on apple trees/leaves apple scab infections are started in late-winter on developing and youthful leaves; however, they also leave their symptoms on the apple fruit itself. Scab injuries that structure on youthful leaves may grow to more than 1 cm in breadth. Ontogenetic opposition of more established leaves, be that as it may, for the most part brings about littler injuries or no noticeable side effects.

Influenced tissues in the long run may get contorted and puckered, and leaf injuries become split and torn. Seriously contaminated leaves drop from trees.

A few sequential defoliation occasions can debilitate trees, bringing about a greater susceptibility to different burdens, for example, freeze harm, insect injury and different infections.

Organic product sores are by and large rankled and "scabby" in appearance, with a particular edge. The soonest perceptible manifestation on organic product is water-drenched regions that quickly form into smooth, green to olive-earthy coloured sores. Infections of young fruits cause natural product bending as solid tissue keeps on developing. Severely contaminated fruits regularly drop prematurely from trees.

2.3 Apple Rust

Cedar apple rust is brought about by a contagious pathogen called *Gymnosporangium Juniperi-Virginianae*. This fungus affects apples and eastern red cedar (Juniper) just as a couple of different junipers not normally developed. So as to endure, the fungus must move starting with switching of host from one sort of host then onto the next [7].

2.4 Apple Rot

Dark rot fruit infections, as a rule, create in the bloom end of the fruit. As rot extends, it is regularly portrayed by a progression of groups exchanging from dark to brown; contagious fruiting structures (pycnidia) show up as dark bits dissipated over rotted fruit surfaces. As rot advances, fruit injuries obscure, however, stay firm furthermore, not indented. In the end, rotted fruits evaporate and wither. Early infections may prompt preserved fruit, some of the time holding tight trees until the accompanying season. Foliar manifestations (frog eye leaf spot) [8] may likewise create if the parasite has gotten set up in plantations (Fig. 1).

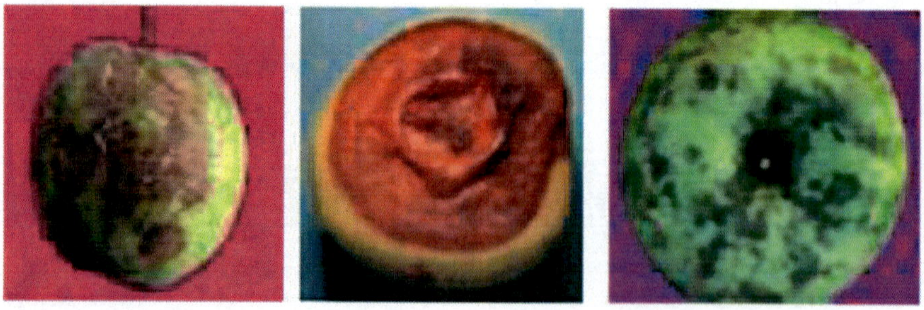

Fig. 1 Sample apple images with diseases

3 Methodology

The main aim of this work is to build up an automated vision-based technique for recognizing apple fruit diseases utilizing features highlights extraction from the image of the fruit. This algorithm follows the steps below

- Collecting image of the fruit affected by a disease find its informational indexes of different contaminated and find ordinary examples and using K-Means segmentation to gather RGB images histograms from the image of fruits with disease affected conditions.
- Determine image features highlights dependent on the colour coherence vector technique (CCV) and correct cluster selection.
- Create lots of datasets which reduced components information models through a stepwise classification end process and order for the classification of fruit diseases conditions and compare the order accuracy of utilizing different example acknowledgement techniques and select ideal classifier (Fig. 2).

4 Pre-processing of Datasets

This is one of the first most important steps as in this step, we remove the unwanted noise from the images which could produce unwanted pixels in the image if ignored. First ten coloured pictures were taken for every disease type found in apples and fixated on particular apples. The computerized pictures were stored in JPEG format for ease of access to MATLAB. The 500 images of diseases were partitioned into two datasets comprising of 250 examples for preparing and 250 examples for testing. The examples were first organized in ascending order for the time, and the pictures were obtained.

This methodology limits negative time-dependant inconstancy and diminishes the potential for information determination predisposition between the preparation and test datasets. All calculations for picture division and surface component age were created in MATLAB. In the underlying advance, the RGB pictures of all leaf tests were gotten. For diminishing the computational weight with negligible loss of surface element quality, the picture goals were decreased from their unique size to 256×256 pixels and the decreased pictures were then changed over from eight-piece to six-piece for each channel RGB group [9–11] (Fig. 3).

The resulting steps were rehashed for each picture in the dataset. After the pictures were decreased, edge recognition of the diseases influenced the group was finished on each picture of the leaf test utilizing MATLAB program record, which displays a detailed edge identification process. To start with, each RGB picture was changed over to a grey picture and afterwards a paired picture. Next, the edge of a binary image was identified by the order "impede" and "imitate" in MATLAB.

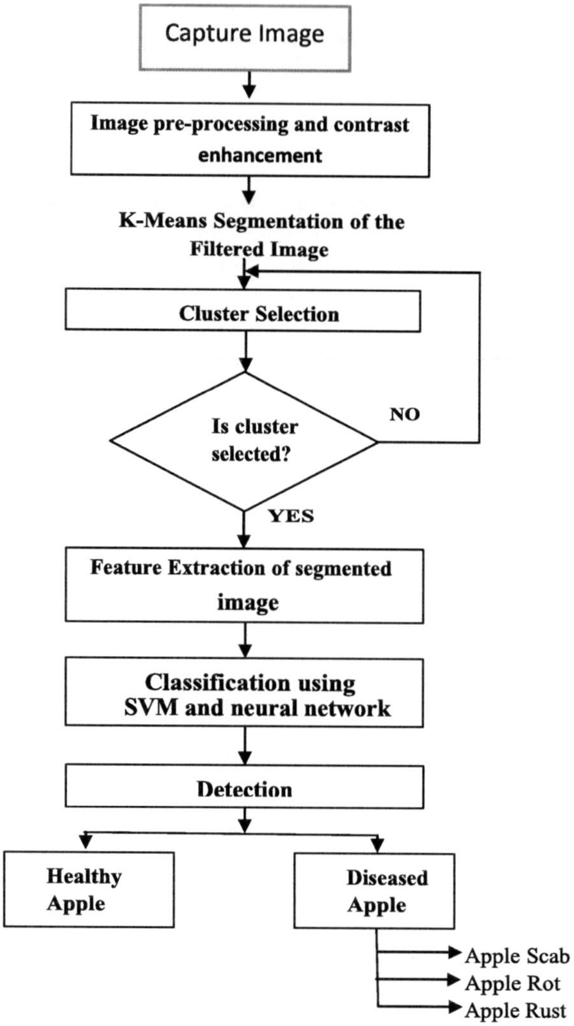

Fig. 2 Flowchart of the process in identifying the tomato disease

4.1 *Image Segmentation*

Image segmentation is a significant advance in image processing, and it appears to be all over the place in the event that we need to break down what is inside the image. For instance, on the off chance that we seek to discover if there is a seat or individual inside an indoor image, we may require image segmentation to isolate protests and break down each article exclusively to check what it is. Image segmentation normally fills in as the pre-processing before design acknowledgement, include extraction and pressure of the image. Image segmentation is the arrangement of an image into various gatherings. Numerous kinds of research have been done in the zone of image segmentation utilizing clustering [12] .

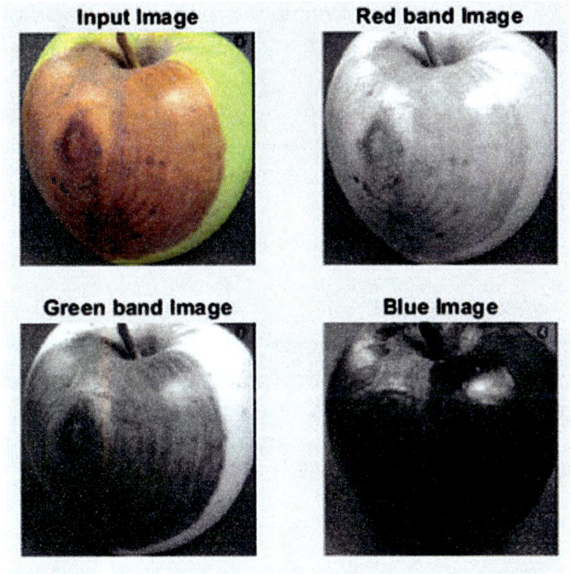

Fig. 3 RGB band of thc input image of the apple

Image segmentation includes changing over an image into an assortment of areas of pixels that are spoken to by a mask or a marked image. Separating an image into fragments, you can process just the significant portions of the image as opposed to processing the whole image.

A typical strategy is to look for unexpected discontinuities in pixel esteems, which commonly demonstrate edges that characterize an area. Another basic methodology is to recognize similarities in the regions of an image (Fig. 4).

A few methods that follow this methodology are locale developing, clustering and thresholding. An assortment of different ways to deal with perform image segmentation have been created throughout the years utilizing space explicit knowledge to adequately take care of segmentation issues in explicit application territories.

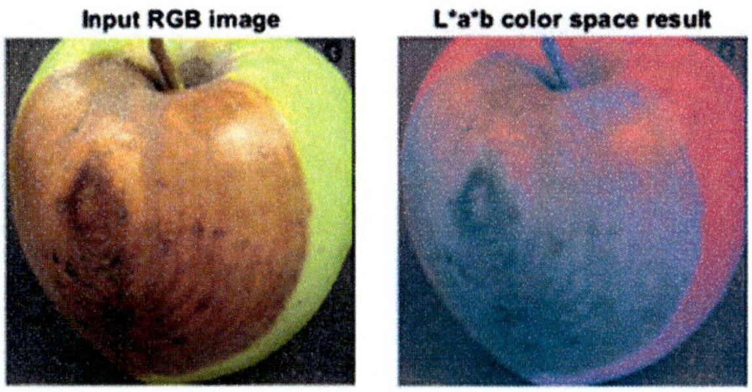

Fig. 4 Image segmentation and conversion of input RGB to LAB colour space

There are various techniques, and one of the most mainstream strategies is the K-Means clustering calculation. K-Means clustering calculation is a solo calculation, and it is utilized to portion the intrigue zone from the background. It groups, or segments the given information into K-bunches or parts dependent on the K-centroids.

$$J = \sum_{j=1}^{k} \sum_{i=1}^{n} \left\| x_i^{(j)} - c_j \right\|^2 \quad (1)$$

The calculation is utilized when you have unlabelled data (i.e. information without characterized classes or gatherings). The objective is to discover certain gatherings dependent on a comparability in the information with the quantity of gatherings spoke to by K.

The target of K-Means clustering is to limit the whole of squared separations between all focuses and the bunch place (Fig. 5).

4.2 Cluster Selection and Feature Extraction

After K-Means segmentation, highlighted features would be carefully separated by choosing the infected cluster as we need to choose one group out of "n" bunches acquired, on which the element extraction and arrangement will be handled. We alter the quantity of groups as per the need of the image and in this proposed methodology we have chosen three clusters (Fig. 6).

This is one of the most important steps as this plays a crucial role in the system. On the basis of the features extracted, the system classifies the segmented image and detects the diseases and the accuracy of the system is determined. The feature

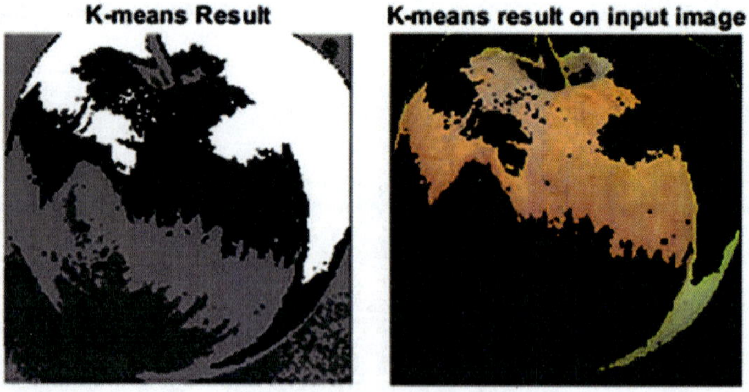

Fig. 5 K-Means segmentation results on input image

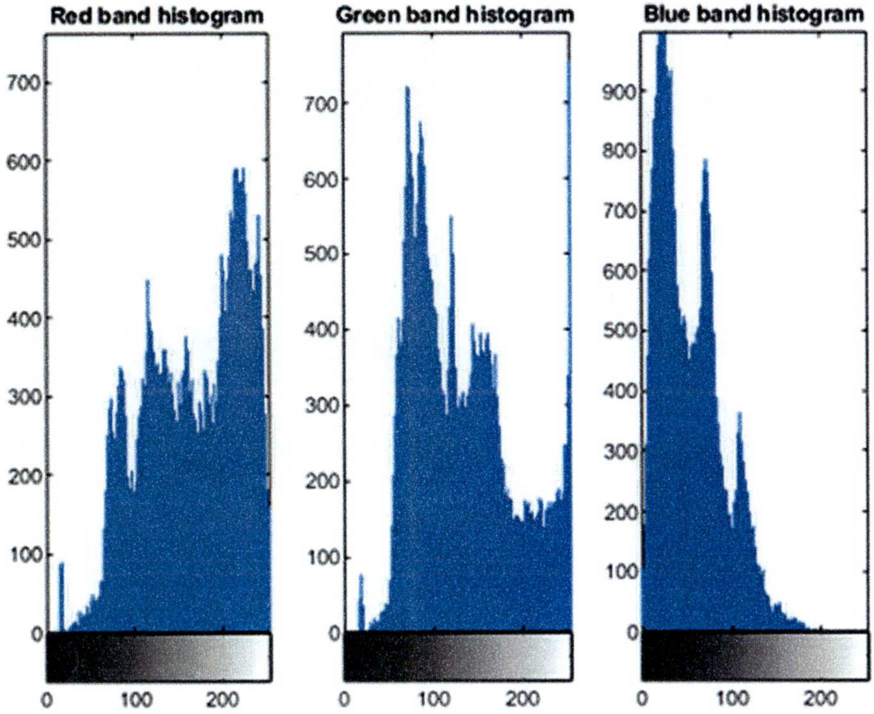

Fig. 6 RGB band histograms for image segmentation and cluster selection

extraction is done using the GLCM algorithm, which is the best method to classify on statistical basis for feature analysis. The GLCM algorithm is also used for the calculations of the texture's extractions intensity values.

4.3 Classification Using Neural Network

In the method of image segmentation, the final step is the classification of images based on its characteristics. This is done with the help of a reference picture which is a previously classified picture [5] (Fig. 7).

This picture can be made utilizing the fragment-based marks or some other methodology. The significant point here is that this reference picture is utilized to dole out the dominant part class inside each portion (Table 1).

In this method for the characterization and classification of images, SVM has been utilized on the grounds that it is among the broadly known ways for design acknowledgement and picture order (Fig. 8).

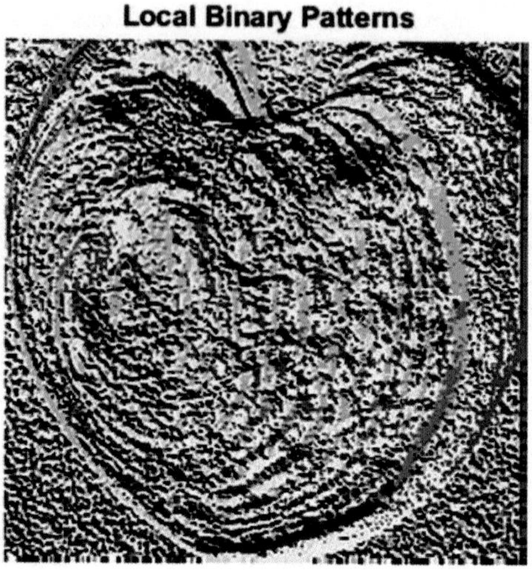

Fig. 7 LBP extract of the input image

It is a managed AI-based calculation. As we probably are aware in regulated learning, the information is named consequently bunching of information turns into the need therefore K-Means was used. SVM builds the ideal isolating hyperplanes. Unsupervised neural networks are trained by letting the neural network continually adjust itself to new inputs. They are used to draw inferences from data sets consisting of input data without labelled responses. You can use them to discover natural distributions, categories and category relationships within data.

Clustering is such an unsupervised learning approach in which artificial neural networks can be used for exploratory data analysis to find hidden patterns or groupings in data. This process involves grouping data by similarity. Applications for cluster analysis include gene sequence analysis, market research and object recognition.

In SVM grouping, portion stunt is furthermore implemented. The bit ways are fundamentally occurrence-based. Instead of learning some fixed arrangement of parameters as for the highlights of their information sources.

Table 1 Classification of different diseases of apples

Sr. No.	Input image	Pre-processing	Cluster a	Cluster b	Cluster c
Image 1					
Image 2			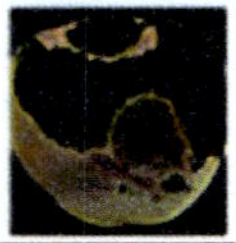		
Image 3					
Image 4					

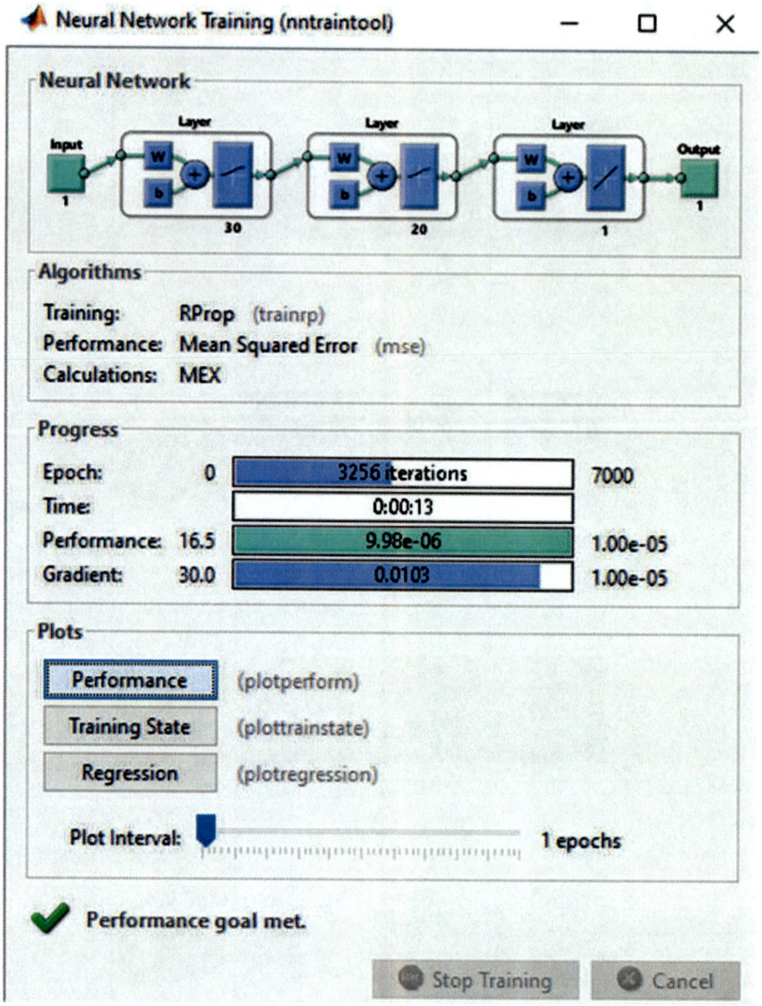

Fig. 8 Neural network training

5 Results

As there are numerous procedures by which we can distinguish the symptoms of a fruit disease. In this section, results and observations of the experimentation performed on apple is mentioned.

This approach is to make a framework that is less expensive, increasingly proficient and effectively available by anybody. Consequently, in this paper, a programmed fruit disease identification strategy is proposed. Using this method highest accuracy achieved is 98.31%. Table 1 shows the accuracy achieved with proposed method for apple data.

Table 2 Disease detected by classification

Image	Disease detected	Affected region	Accuracy of the classifier
	Healthy apple	None	98.3871
	Apple scab	23.4937	95.1613
	Black rot canker	43.8557	98.3871
	Core rot	26.2115	96.7742

6 Conclusion

There are fundamentally three stages—image segmentation, feature extraction and classification. Along these lines, making it increasingly precise as contrasted and the current classifiers, the total calculation runs truly quicker and does not hang in any event, when tried for a great many examples at the same time. Later on, it will be stretched out to the few different organic products like mangoes, pears and so forth, and furthermore on the inner inadequacies of apple and other fruit products. With very less computational efforts, the optimum results were obtained, which help for apple disease identification at early stage or the initial stage.

Experimental results showed the apple fruit disease recognition rate of the algorithm can reach to 98.3% in ideal conditions while generally it turns out to be 93.3%. The discussed methods of disease detection and classification show a good potential with an ability to detect fruit diseases accurately.

References

1. Sankarana, S., Mishra, A., Ehsania, R., Davis, C.: A review of advanced techniques for detecting plant diseases. Comput. Electron. Agric. **72**, 1–13 (2010)
2. Dubey, S.R., Dixit, P., Singh, N., Gupta, J.P.: Infected fruit part detection using k-means clustering segmentation technique. Int. J. Artif. Intell. Interact. Multi. **2**, 65–72 (2013)

3. Unnikrishnan, R., Pantofaru, C.E., Hebert, M.: Toward objective evaluation of image segmentation algorithms. IEEE Trans. Pattern Anal. Mach. Intell. **29**(6), 929–943 (2007)
4. Rocha, A., Hauagge, C., Wainer, J., Siome, D.: Automatic fruit and vegetable classification from images. Comput. Electron. Agric. **70**, 96–104 (2010)
5. Quan, J.J., Wen, X.B.: Multiscale probabilistic neural network method for SAR image segmentation. Appl. Math. Comput. **205**(2), 578–583 (2008)
6. Sankarana, S., Mishraa, A., Ehsania, R., Davisb, C.: A review of advanced techniques for detecting plant diseases. Comput. Electron. Agric. **72**, 1–13 (2010)
7. Lerner, B., Guterman, H., Aladjem, M., Dinstein, I.: A comparative study of neural network-based feature extraction paradigms. Pattern Recogn. Lett. **20**, 7–14 (1999)
8. Patil, S.B., Bodhe, S.K.: Greening the economy with climate smart agriculture food. Int. J. Eng. Technol. **3**(5), 297–301 (2011)
9. Meybeck, A., Gitz, V.: Leaf diseases severity measurement using image processing. In: Agriculture, Food Security and Climate Change Conference (2012)
10. Dubey, S.R., Jalal, A.S.: Species and variety detection of fruits and vegetables from images. Int. J. Appl. Pattern Recogn. (IJAPR) **1**(1), 108–126 (2013)
11. Teoh, C.C., Mohd Syaifudin, A.R.: Image processing and analysis techniques for estimating weight of Chokanan mangoes. J. Trop. Agric. Fd. Sc. **35**(1), 183–190 (2007)
12. Francisco, J.E., Jepson, A.D.: Benchmarking image segmentation algorithms. Int. J. Comput. Vis. **85**(2), 167–181 (2009)

A Detailed Review of the Optimal Distributed Generation Placement in Smart Power Distribution Systems

Binaya Kumar Malika, Vivekananda Pattanaik, Sarthak Mohanty, Binod Kumar Sahu, and Pravat Kumar Rout

Abstract The integration of distributed generators (DGS) attracts power engineers to circumvent many related issues like energy demand, congestion management, dependency on conventional energy sources, and overall system efficiency, etc. However, the interfacing of DGs with the power distribution grids for optimal operation with stability, control, and efficient energy management needs to solve many requirements and challenges during the establishment time. The optimal placement with the right size and power factor impacts the operation of the system. This paper presents a detailed review of the various proposals by many authors in recent times and also provides the pros and cons along with the future possibilities for further improvement. Various objectives, constraints, and variables along with methods are discussed as a primary knowledge for this problem to achieve line loss reduction, improvement of voltage stability, reliability improvement, economic benefits, security enhancement, and environmental pollution optimization, etc.

B. K. Malika
Department of Electrical Engineering, Einstein Academy of Technology and Management, Bhubaneswar 751015, India
e-mail: er.binaya@gmail.com

V. Pattanaik
Department of Electrical Engineering, Synergy Institute of Technology, Phulnakhara, Bhubaneswar 754001, India
e-mail: vivek7048@gmail.com

S. Mohanty (✉) · B. K. Sahu
Department of Electrical Engineering, Siksha 'O' Anusandhan University, Bhubaneswar 751030, India
e-mail: sarthak.mohanty1995@gmail.com

B. K. Sahu
e-mail: binoditer@gmail.com

P. K. Rout
Department of Electrical and Electronics Engineering, Siksha 'O' Anusandhan University, Bhubaneswar 751030, India
e-mail: pkrout_india@yahoo.com

S. Das and M. N. Mohanty (eds.), *Advances in Intelligent Computing and Communication*, Lecture Notes in Networks and Systems 202,
https://doi.org/10.1007/978-981-16-0695-3_10

Keywords Distributed generation · Optimal sizing and siting · Optimal DG allocation · Optimal distributed generation placement (ODGP) · Analytical methods · Heuristic methods

1 Introduction

The microgrid concepts bring a smart distribution system with local control facilities, balanced technology choice and operation, heterogeneous reliability and power quality, and resilience. This transition from the conventional distribution system to microgrid scenario integration bring for the overall distribution system is environmentally friendly, electric vehicle integration, lower energy utilization cost, energy savings (direct DC charging), renewable energy integration, enhanced system reliability, energy storage systems integration, and wireless charging, etc., [1, 2].

The major drivers for the attraction of microgrid implementation in the distribution section are as follows [3]: (1) The generation on the local load side can be made available, thereby deferring the energy cost and troubles related to regulatory norms of building utility-scale/central station generation [4]; (2) DGs in the distribution sector can be integrated with the existing distribution line, thereby reducing the installation cost related to upgrading and constructing new transmission lines [4]; (3) The placement of DG in the distribution sector enhances efficiency by lowering transmission losses [5]; (4) DG can lessen grid congestion; (5) DG often can make available reactive power; (6) The reliability of energy supply can be enhanced by the addition of DG [4]; (7) DG particularly building-mounted PV in remote areas requires less water and land than similar capacity totals provided by central station plants; (8) Enhance the stability of the distribution system with the grid; (9) Integrating in a coordinated way the electric and thermal storage capabilities, to provide local management of variable renewable generation; and (10) DGs can be used as a power backup in case of unexpected problems in the main distribution system.

The various significant benefits of DG planning are related to its placement, size, and type. The distribution system integrated with DG judiciously arrives at many benefits and has an immense impact on overall performance of the overall system in many angles as mentioned as [6]: (1) The real and reactive power loss related to distribution feeders is reduced [7]; (2) The power system oscillations with faster damping bring better stability to the system under disturbance conditions [5]; (3) The power system stability to voltage, frequency, and rotor angle is improved in wide margins than the conventional setup [5, 8, 9]; (4) The power system reliability and security are enhanced substantially than conventional distribution system [7, 8]; (5) The power transfer capacity of distribution system is enhanced [5]; (6) The power system loadability is improved [5]; (7) Power transfer capacity availability is improved [5]; (8) The power congestion of distribution line also reduces substantially [10]; and (9) The overall distribution system is more flexible operationally [5].

The placement and rating of the DGs play a vital role during the design and planning of the smart distribution system. The following factors justify the importance of this issue and should be carried out with further research in this direction to arrive at an optimal solution [10].

- The characteristics and dynamics of different types of DGs are different, and according to the nature of the DGs, the output response and support to the distribution system vary. That way the decision to use the type of DG plays a vital role in management to result in better system response in terms of benefits and performance.
- The sizing and location of DGs need to be chosen judiciously to enhance system performance and efficiency [4].
- The number and size also and issue during the planning stage from an economic point of view. Because of the issue like the large size of DG may give rise to bidirectional power flow and less impact on the overall performance [3].
- Placement of DG can be done to enhance various operating constraints and indices to bring index permissible standard limits like power loss index, voltage stability index, penetration level index, reliability index, and security index [8].

The installation of DG with optimal location, size, and rating is the major issue during the planning and design stage. This needs further devotion to research for an optimal solution satisfying all the operating constraints. The motivation of this review article is to make a brief presentation of various methods, formulations, and suggestions from various researchers on the optimal location and sizing of DG. In addition to all the major findings of the review and possible future areas to investigate these issues are presented. Detail of the various objectives, necessary operating constraints, and variables is discussed. The various methodologies applied for the optimal solution are systematically presented.

2 Mathematical Formulation

2.1 Methods

Different methods are applied for DGs placement and sizing in distribution planning. The algorithms aimed at DG allocation and sizing can be classified into two major classes as illustrated in Fig. 1 and described below.

2.1.1 Conventional/Classical Techniques

Classical approaches can be categorized as 1: Classic approaches, 2: Basic search methods. Analytical algorithms, linear programming, mixed-integer linear programming, nonlinear programming, mixed-integer nonlinear programming, and

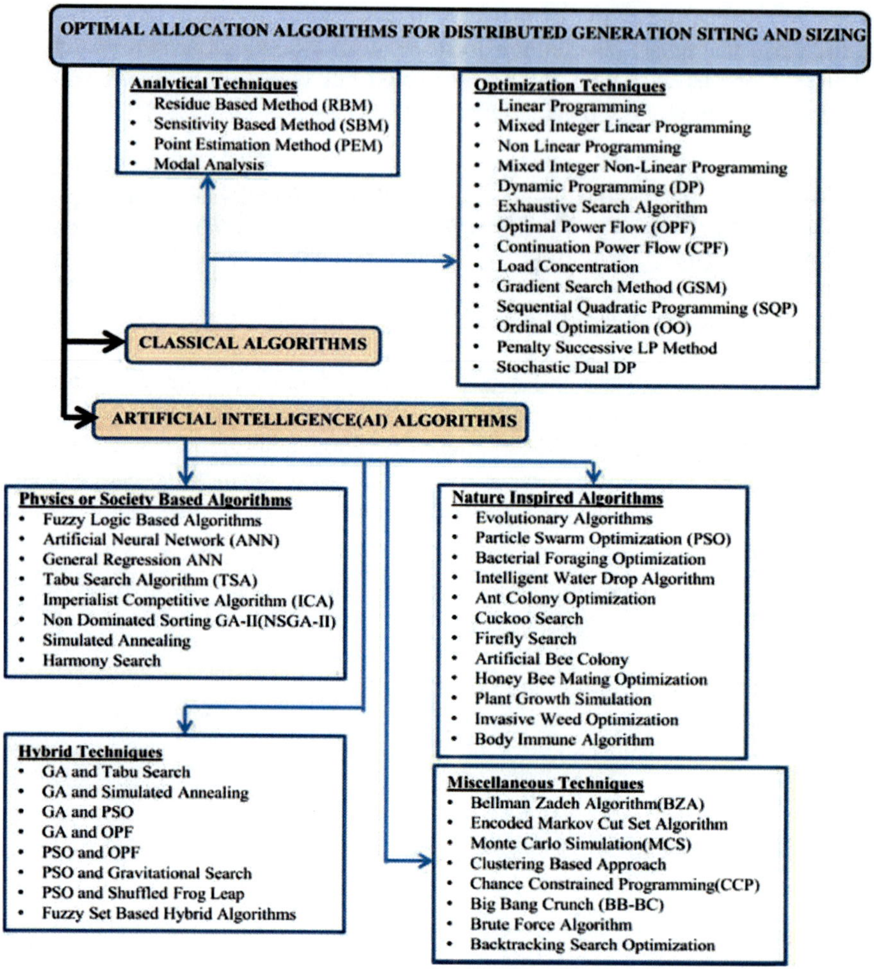

Fig. 1 Algorithms for optimal allocation of DG siting and sizing

dynamic programming-based approaches come under classic approaches. Simple algorithms, easy to implement, and high precision are the major advantages of classical techniques. On the other hand, large computation time and slow convergence are two major visible limitations particularly applying for complex problems [5, 8].

2.1.2 Artificial Intelligence Techniques

General artificial intelligence computational techniques: Fuzzy set of algorithms, artificial neural network, and Tabu search are the major techniques in this category and applied for DG placement and planning. The efficient performance, needs fewer iterations, handling capacity of complex problems, and easy in the formulation are the major positive aspects of these techniques and that' is why successfully applied in optimal placement and sizing problems in distribution planning. Difficult in coding, pre-mature convergence, unstable and inconsistency in answer running multiple times, the possibility of trapping at local minima, and many randomly setting parameters are the major limitations of these methods [6].

Natured inspired/evolutionary optimization techniques: There are many optimization techniques recently suggested by various authors, and many of these techniques are applied for distribution planning. Particle swarm optimization, ant colony, artificial bee colony, cuckoo search, invasive weed optimization technique, teaching–learning optimization, and differential evolution [6, 11, 12] are few prominent methods applied. However, less prone to local minima in comparison to conventional methods, non-differential-based approach, and arrive at a compromised solution are the major advantages of evolutionary-based approaches. However, inconsistencies in solution and reduction of efficiency in the case of large and complex methods are the major limitations in these methods.

Hybrid artificial intelligence computational techniques: Fuzzy set-based hybrid algorithms and hybrid techniques between two evolutionary algorithms are two major types applied in recent times for optimal location and sizing problems and also other optimization techniques in distribution planning and design [3]. Efficient results and performance, handling capacity in complex problems in comparison to other approaches, and faster convergence are the major advantages of these techniques. However, a large number of setting parameters and difficulty in coding are the drawbacks during applications of these techniques.

Miscellaneous techniques: Other AI techniques such as Bellman-Zadeh algorithm, Marko-cut set algorithm, Monte-Carlo optimization, Big-Bang Big-Crunch, and Brute force technique can be employed for more adaptable and multi-objective optimization oriented methods.

2.2 *Objective Function Formulation*

The placement of DGs and sizing optimization is generally formulated with the following objectives as illustrated in Fig. 2. However, many authors have formulated multi-objective approaches for optimal planning of DGs placement and sizing [11, 13].

- **Power and energy loss minimization**: Loss minimization indirectly increases the total energy utilization and energy transfer capability of the distribution system. Active power loss minimization is to provide more power to the loads reducing the loss in distribution lines. Reactive power loss minimization is for reactive power management in the distribution sector to bring better stability for the system [4].

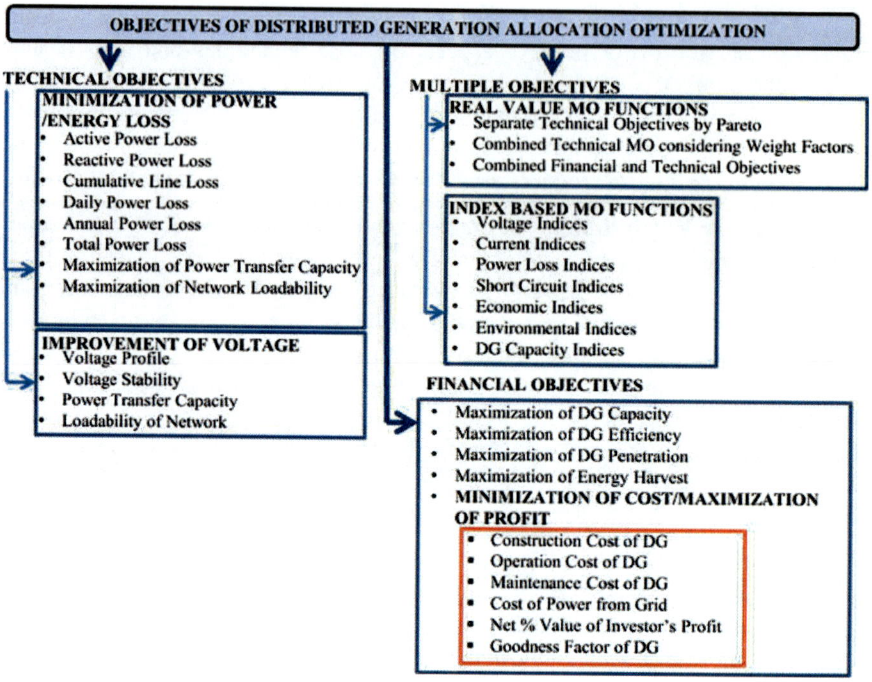

Fig. 2 Major objectives of DG allocation optimization

- **Improvement of reliability of the system**: The objective function can make the DG system more reliable by using reliability indices and other indices like power loss, short circuit current level, and penetration level indices [7, 9].
- **Maximization of power transfer capacity and network loadability**: Always there is a need to enhance the distribution and transmission line transfer capability to meet the exponential growth of load demand and consumers. However, due to the optimal placement of DGs in the distribution sector, power can be supplied locally reducing the long transmission of power. Similarly, this can be in other way taken into consideration in the form of network loadability enhancement. Network capabilities to handle more loads are maximized by locally installing the DGs [4].
- **Energy loss/Line loss minimization**: This objective function is formulated in terms of energy rather than power considering the entire system as static and formulating the objective function and solution. The study can be extended by formulating objective function in the form of daily energy loss, annual energy loss, and total energy loss. This objective function can also be used in the reduction of line losses and to ensure better control of reactive power flow within the power system.
- **Voltage profile and voltage stability improvement**: The objective function is formulated with an idea to improve the voltage profile to keep the bus voltages

always within the standard range. Related to the voltage, the objective function can be formulated based on voltage stability [14].

- **Enhancement of security**: The objective function can help with ensuring the dependability of critical loads, reduction of security risks in the grid, power distribution utility, protection from cyberattacks, and terrorist attacks.
- **Financial or economic factor improvement**: During the installation of DG integration to the grid or addition and restructure of microgrid with more DG penetration, it is necessary to look after the financial perspective against the achievement of the benefits. So, the major aim of the objective function is for the maximization of DG efficiency and energy harvest. Other aspects can be taken into consideration which is the minimization of cost and maximization of profit. Deferment in investments can be done as well as a reduction in general operation and management costs to maximize profits [4].
- **Environmental impact improvement**: The optimization of land use, reduced fuel usage, implementation of renewable energy sources (RES), and reduction of pollutants and emissions can be done by utilization of objective functions.
- **Multi-objective formulation**: This multi-objective approach is considered two or more approaches to simultaneously optimize. In the first method that is real-value multi-objective, each objective function considered is taken simultaneously and optimized. Separate technical objectives by Pareto set and combined technical approach by weighting factors are the most used approaches and applied extensively in distribution system planning and particularly in case of problems related to optimal placement and sizing. The second formulation in the form of multi-objective is an index-based function approach. Power loss indices, voltage indices, current indices, short circuit indices, DG capacity indices, financial indices, and environmental indices are few approaches suggested by various authors to formulate the objective functions in multi-objective form for finding out the optimal solution for the placement and sizing of DGs [4, 12].

2.3 Constraints

Constraints need to be satisfied for a feasible solution. However, consideration of all real-time operative and component constraints makes the problem more complex. The primary or most essential constraints are to be taken the most to justify the solution for its practical application. For the optimal placement and sizing of the DGs in the distribution system, *the power limitation constraints, voltage limitation constraints, current limitation constraints, transformer limitation constraints, and DG limitations constraints* [8] are most prominent. These constraints need to be satisfied during the study to make the solution more feasible. Figure 3 illustrates all the DG allocation optimization constraints as well as all the constraints related to the specific optimization factor.

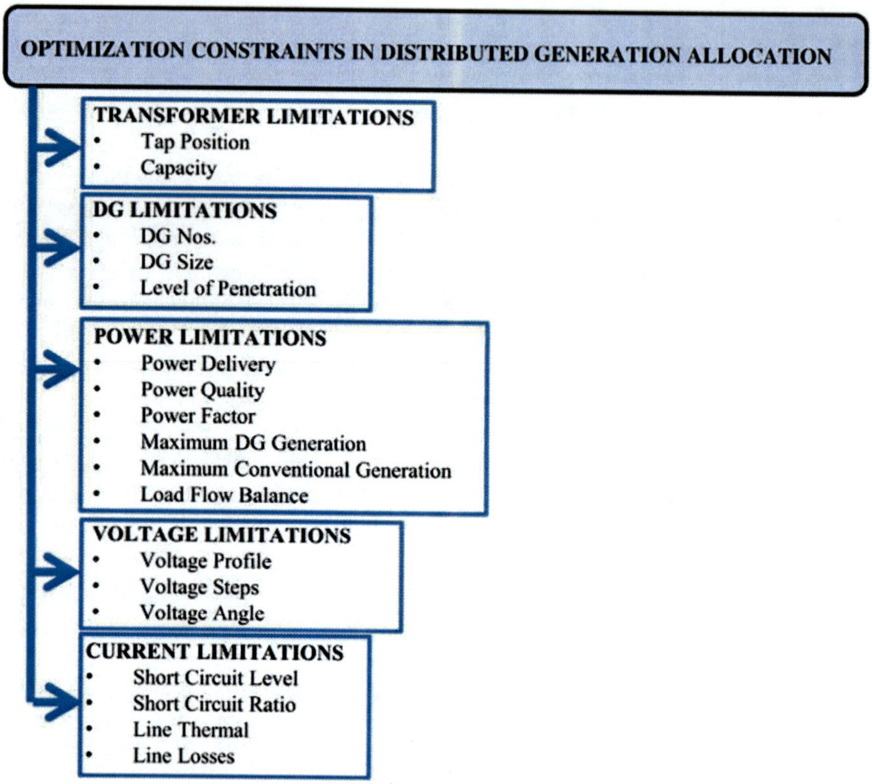

Fig. 3 DG allocation optimization constraints

2.4 *Set of Decision Variables*

Three major decision variables are considered during the problem formulation of the problem. These are 1: DG location considering the appropriate bus in the distribution system, 2: DG size considering the environmental factor, design factor, and economic factor constraints, 3: DG power factor signifying the response in terms of active and reactive power delivery [12, 15].

2.5 *Development of Techniques*

The development of techniques [11] is explained by the illustration in Fig. 4.

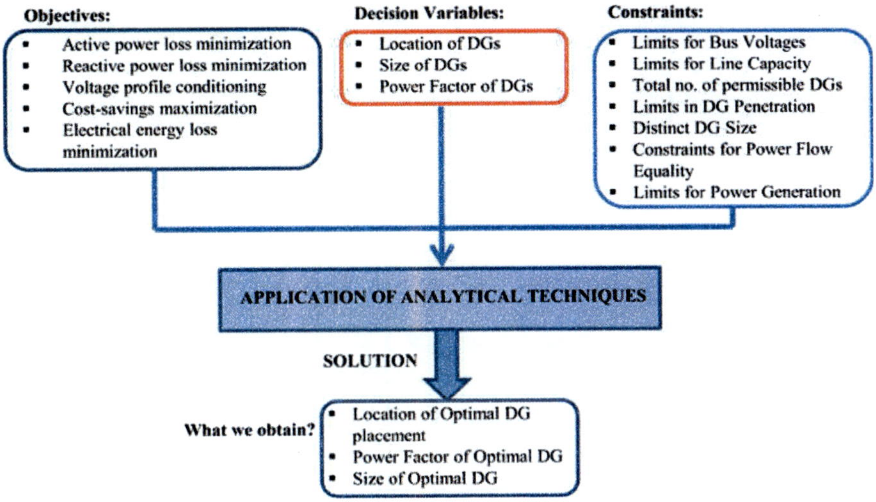

Fig. 4 Development of the problem-solution technique

2.6 General Algorithm for DG Placement and Size Allocation

The backward and forward sweep technique is very often applied for the load flow of a balanced radial distribution network. The appropriate flowchart of the demonstrated general algorithm for DG optimization is described in Fig. 5 [7, 15].

3 Discussions on Key Observations and Future Research

The following issues can be present as a shortcoming of the existing research and put as a direction for future research.

- Fewer investigations have been done on the optimal power factor of the DGs, unbalanced distribution systems, reactive power generation limits of the DGs, optimal selection criteria for the weighting factor values, uncertainties, and nonlinearities associated with load and renewable sources of DGs without violating the system constraints [4, 12, 16].
- To date, even though many studies are done on the optimal placement and sizing of DG, many considerations at standalone and islanding difficulties and its impact on smart distribution systems are not focused yet.
- There are many uncertainties related to DG output, load model, and electric prices. These factors need to be considered for long term and expansion

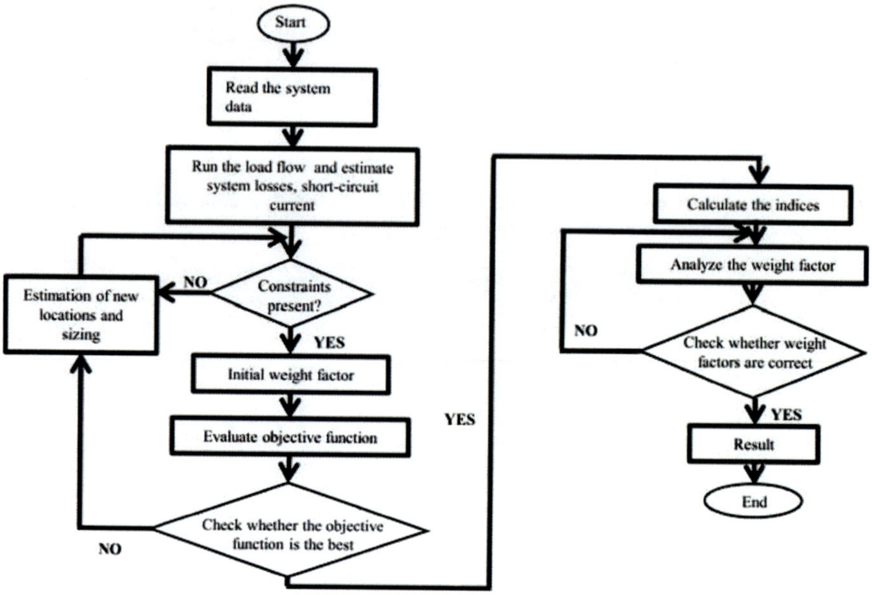

Fig. 5 General flowchart for the problem-solution approach

planning stage of the conventional distribution system transition toward smart grid development [11].

- The hybrid approaches are very effective computationally. Much work has not been exploited in this direction [5, 7].
- The major issues of DG integration like impact on voltage and frequency regulation, impact on protection coordination, impact on harmonics, and impact of reactive power management need to be studied parallel with the DG placement and sizing [4].
- Multi-objective formulation considering all objectives of DG integration like maximization of renewable DG penetration, maximization of system reliability, minimization of investment and operational cost, reduction in system losses, and improvement in voltage profile are to focus during the formulation of the methodology of DG placement and sizing [10].
- In the future, the placement of DGs with flexible AC transmission systems simultaneously can be studied [5].
- Multi-objective tasks are relatively less investigated in distribution planning and design problems [10].
- The studies were done in past mostly based on simple load modeling. Static and realistic load models, composite load models, and dynamically varying load models are not considered in many approaches to testing the feasibility of these approaches to apply in real-time conditions [5].

- Since DG optimization deals with placing DG for their protection, some DG applications can be further investigated to demonstrate their utilization as protection devices (to protect from events such as reverse power flow) [7].
- Aside from the fact that meta-heuristic algorithms can be considered as a suitable technique for optimization of DG, other analytical techniques can be further investigated for their accuracy and compatibility in numerical validation and power system modeling [11].
- For further research, the efficiency of meta-heuristic algorithms can be improved upon to develop more capable methods to discover global optimums [10].
- The dynamic and transient stability of DG systems during their allocation can be enhanced in further studies [10].

4 Conclusion

The following major conclusions can be highlighted as an outcome of this review:

- Artificial intelligence techniques arrive at better solutions comparatively than other conventional and optimization methods in terms of better operational performance viewpoint [17].
- Hybrid artificial intelligence techniques nullify the limitations of each technique due to a coordinated approach and perform better than the conventional and optimization techniques.
- The impact of different DGs in terms of their diversified characteristics is more beneficial to handle different issues than the single DG allocation [11].
- The objectives like accuracy, reliability, and computational time are mutually conflicting influence and sometimes highly difficult to achieve simultaneously in the solution obtained.
- The uncertainties arise due to system planning and operation, make the problem more complex, and need more investigation to arrive at a feasible solution to implement in real-time conditions [10].

References

1. Nayak, M.R., Kasturi, K., Rout, P.K.: Optimal placement and sizing of multiple capacitors in radial distribution systems using modified TLBO algorithm. Int. Rev. Autom. Control **6**(6), 701–713 (2013)
2. Nayak, M.R., Dash, S.K., Rout, P.K.: Optimal placement and sizing of distributed generation in radial distribution system using differential evolution algorithm. In: International Conference on Swarm, Evolutionary, and Memetic Computing. Springer, Berlin, Heidelberg (2012)
3. Viral, R., Khatod, D.K.: Optimal planning of distributed generation systems in the distribution system: a review. Renew. Sustain. Energy Rev. **16**(7), 5146–5165 (2012)

4. Paliwal, P., Patidar, N.P., Nema, R.K.: Planning of grid integrated distributed generators: a review of technology, objectives, and techniques. Renew. Sustain. Energy Rev. **40**, 557–570 (2014)
5. Singh, B., Sharma, J.: A review on distributed generation planning. Renew. Sustain. Energy Rev. **76**, 529–544 (2017)
6. Abdmouleh, Z., et al.: Review of optimization techniques applied for the integration of distributed generation from renewable energy sources. Renew. Energy **113**, 266–280 (2017)
7. Prakash, P., Khatod, D.K.: Optimal sizing and siting techniques for distributed generation in distribution systems: a review. Renew. Sustain. Energy Rev. **57**, 111–130 (2016)
8. Haji Abbas, M.P.: A review of the optimal allocation of distributed generation: objectives, constraints, methods, and algorithms. Renew. Sustain. Energy Rev. **75**, 293–312 (2016)
9. Farrokhabadi, M., et al.: Microgrid stability definitions, analysis, and examples. IEEE Trans. Power Syst. **35**(1), 13–29 (2020)
10. Jordehi, A.R.: Allocation of distributed generation units in electric power systems: a review. Renew. Sustain. Energy Rev. **56**, 893–905 (2016)
11. Ehsan, A., Yang, Q.: Optimal integration and planning of renewable distributed generation in the power distribution networks: a review of analytical techniques. Appl. Energy **210**, 44–59 (2018)
12. Georgilakis, P.S., Hatziargyriou, N.D.: Optimal distributed generation placement in power distribution networks: models, methods, and future research. IEEE Trans. Power Syst. **28**(3), 3420–3428 (2013)
13. Akorede, M.F., et al.: A review of strategies for optimal placement of distributed generation in power distribution systems. Res. J. Appl. Sci. **5**(2), 137–145 (2010)
14. Razavi, S.E., Rahimi, E., Javadi, M.S., et al.: Impact of distributed generation on protection and voltage regulation of distribution systems: a review. Renew. Sustain. Energy Rev. **105**, 157–167 (2019)
15. Badran, O., et al.: Optimal reconfiguration of distribution system connected with distributed generations: A review of different methodologies. Renew. Sustain. Energy Rev. **73**, 854–867 (2017)
16. Duong, M.Q., Pham, T.D., et al.: Determination of optimal location and sizing of solar photovoltaic distribution generation units in radial distribution systems. Energies **12**(1), 174 (2019)
17. Aldeferati, T., Bansal, R.C.: Integration of renewable distributed generators into the distribution system: a review. IET Renew. Power Gener. **10**(7), 873–884 (2016)

Virtual Inertia Control Strategy in Microgrid Stability Control: A Conceptual Synthesis and Discussion

Debabrata Routray, Vijaya Kumar Dunna, Pravat Kumar Rout, and Binod Kumar Sahu

Abstract The integration of distributed generations in a smart microgrid distribution system as focused in recent times and considered as a solution towards the exponential rise of energy demands. However, the inverter-based energy sources lead to a weak power system in the sense of having low inertia. This, in turn, make the overall system more prone to instability. The related issues and challenges along with its remedial misuse in terms of control prospective need to be discussed thoroughly. On the above-related problems, this paper presents all possible virtual inertia topologies in the implementation stage. Apart from that, this paper explores potential research directions, major challenges, and technical analysis on the state-of-the-art of virtual inertia. Comparative analysis also discussed mentioning the merits and limitations of the topologies. To understand the concept, the related mathematical expressions and schematic diagrams of different topologies and control strategies are illustrated.

Keywords Virtual inertia · Frequency stability · Renewable energy sources · Distributer generators · Microgrid control

D. Routray (✉) · V. K. Dunna · B. K. Sahu
Department of Electrical Engineering, ITER, Siksha 'O' Anusandhan (Deemed to be University), Bhubaneswar, Odisha 751030, India
e-mail: debabrata.r.india@gmail.com

V. K. Dunna
e-mail: vijayakumardunna@gmail.com

B. K. Sahu
e-mail: binoditer@gmail.com

P. K. Rout
Department of Electrical and Electronics Engineering, ITER, Siksha 'O' Anusandhan (Deemed to be University), Bhubaneswar, Odisha 751030, India
e-mail: pravatrout@soa.ac.in

S. Das and M. N. Mohanty (eds.), *Advances in Intelligent Computing and Communication*, Lecture Notes in Networks and Systems 202,
https://doi.org/10.1007/978-981-16-0695-3_11

1 Introduction

The operation of the power system needs to stabilize under any fluctuation or power discrepancy between the power generation and load demand. Synchronized operation with proper coordination and equilibrium is essential to maintain the desired frequency and voltage. For the above issues and challenges, the inertia of the power system and its components play a vital role, and its level indicates the system capability to handle the power deficiency and the subsequent effects.

The inertia of the rotating mass of synchronous generators (SGs) and turbines is a measure of their kinetic energy (KE) stored to regulate the power deviation and that in turn frequency fluctuations. This reserve K.E injects into the grid or absorbs from the grid to prevent any deviation in frequency. In other words, the frequency is a reflection of inequalities in the overall power balance, and a higher inertia system is more capable to stabilize fast due to the availability of K.E more in the rotating mass [1]. On the other hand, the higher the inertia, the higher is the damping and lesser fluctuation of grid frequency to sudden changes in load patterns as well as generation. The reverse dynamics play for the lower inertia systems. Figure 1 indicates the above dynamics related to inertia in microgrid system. This motivates to discuss further on inertia concept and its relevance in power systems.

In recent times to circumvent the deficit of power, distributed generators (DGs) are integrated at a smart distribution system. The microgrid system can operate in both grid-connected and islanded mode of operation. The dynamics, operational characteristics, and random intermittent power generation due to the environmental dependency of different types of DGs make the operation quite different in comparison to traditional centralized facilities. Due to the environmental dependency there is a chance of electrically decoupling of the DGs from the grid. So the DGs are unable to contribute inertia to the overall system. Moreover, the synchronous generator is providing inertia to the system that is very often missing

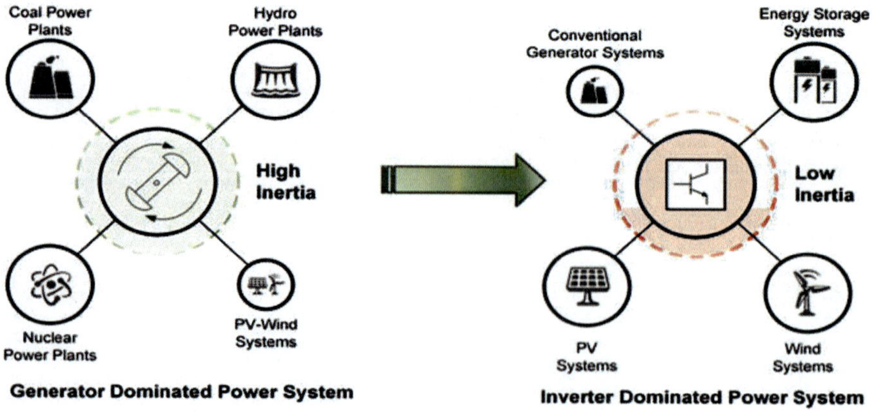

Fig. 1 Advancement towards an inverter-based power system

in these renewable generation units. So, the inertia of the modern smart power system decreases with the integration of a large number of converters connected generation units and loads. This in turn forced to behave a power system differently from before. This caused concerns for power engineers to stabilize the system. Therefore, a detailed study and revised analysis with possible futuristic methods to enhance the inertia through providing adequate control strategy is essential. In this paper, the relevance and possible inertia-based control strategy in power system sire discussed from a better operation, control, and stability of the smart microgrid system.

Specific to small and smart DGs integrated microgrid system, inertia study is very crucial due to the major factors like:

I The inertia of these types of systems is comparatively very low to large power systems or grid-connected power systems. This concept is illustrated in Fig. 2.

II Secondly, the ratio of the magnitude of a possible generation outage to the total spinning capacity is relatively large for these systems.

III The DGs are generally inverter based rather than rotational based like PV and wind turbines. The inverter-based generation does not give any mechanical inertial response, so that we compromised with frequency stability.

IV Frequency stability in recent times becomes a serious concern and needs a better control strategy due to a lack of inertia.

V There is a necessity to integrate DGs in the microgrid system with a larger percentage irrespective of the limitation. Inertial control brings stability to the system in a very short time interval (less than 10 s) automatically as illustrated in Fig. 3; so adequate inertia control is necessary to open up the high penetration level of the DGs [2].

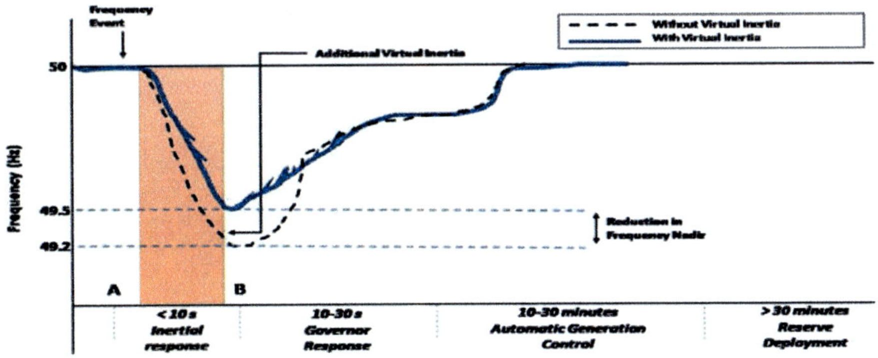

Fig. 2 Multiple time-frame frequency responses following a frequency event in a power system

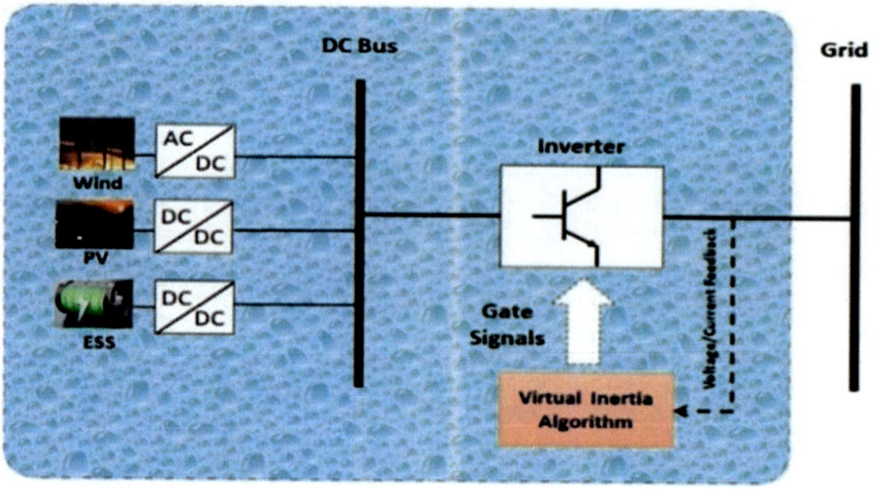

Fig. 3 Virtual inertia concept

2 Microgrid Instability with High Penetration of RES

This section presents inertia of the traditional power system (PS), frequency response of conventional power system, and the impact of DGs on inertia of the power system.

2.1 *Inertia of the Conventional Power System*

The frequency response of a PS on a frequency event/disturbance is regulated according to the swing equation.

$$P_g - P_l = \frac{d}{dt}(E_{K.E}) = \frac{d}{dt}\left(\frac{1}{2}J\omega_g^2\right) \tag{1}$$

$$P_g - P_l = \frac{J\omega_g d\omega_g}{dt} \tag{2}$$

where P_g, P_l, J, and ω_g are denoted for generated power, power demand (losses are included), total system inertia, and angular system, respectively. The normalized inertia constant of the PS 'H' can be defined in terms of apparent power S_g of the integrated generators in the system.

$$H = \frac{1}{2}\frac{J\omega_{\mathrm{g}}^{2}}{S_{\mathrm{g}}} \tag{3}$$

Substituting (3) in Eq. (2)

$$P_{\mathrm{g}} - P_{\mathrm{l}} = \left(\frac{2S_{\mathrm{g}}H}{\omega_{\mathrm{g}}^{2}}\right)\omega_{\mathrm{g}}\frac{\mathrm{d}}{\mathrm{d}t}(\omega_{\mathrm{g}}) \tag{4}$$

Simplifying

$$\frac{2H}{\omega_{\mathrm{g}}}\frac{\mathrm{d}\omega_{\mathrm{g}}}{\mathrm{d}t} = \frac{P_{\mathrm{g}} - P_{\mathrm{l}}}{S_{\mathrm{g}}} \tag{5}$$

Expressing Eq. (5) in terms of frequency (*f*) and substituting

$$\omega = 2\pi f \,\mathrm{rad/sec}$$

$$\frac{2H}{f}\frac{\mathrm{d}f}{\mathrm{d}t} = \frac{P_{\mathrm{g}} - P_{\mathrm{l}}}{S_{\mathrm{g}}} \tag{6}$$

where $\frac{\mathrm{d}f}{\mathrm{d}t}$ is the rate of change of frequency (ROCOF) of the system.

The following conclusions can be drawn from Eq. (6).

(a) $\frac{\mathrm{d}f}{\mathrm{d}t}$ is higher with less *H*. So, to integrate more DGs, there is a necessity to increase *H*.
(b) The value of *H* is very low in case of inverter-based DGs as shown in Fig. 1, as these sources are decoupled from the grid and not participate for inertia compensation.
(c) The K.E of generator is due to rotating mass that K.E contributes to the grid or absorbs from the grid under frequency variation. In other words, from any discrepancy between the demand and supply, frequency of the system will fluctuate proportional to the system inertia.
(d) The study interprets ‘inertia’ in conventional PS, as the opposition/resistance offered by the SG, to change in the rotational speed. In another aspect, any energy backup/support for an extremely short period of time during load fluctuations can be interpreted as inertia [3].

2.2 Frequency Response and Effect of Inertia on Frequency

In normal operation, under any abnormal event occurrence, the system frequency starts to decline due to generation-demand unbalance. However, the level of decline and rate of frequency both depend on total system inertia as illustrated in Fig. 2 and Eq. (6). Before any controller activate, due to the inbuilt inertia factor, the synchronous generator releases the K.E stored in its rotating mass lasting within 10 s duration from the time of inception as illustrated in Fig. 2. Despite of this, if the frequency declines beyond some threshold value, the primary frequency controllers are activated within 30 s. After that, if the frequency deviation will occur in the system (10–30 min), then the secondary control will take care of that and make the system frequency to a normal value. Finally, the tertiary control takes care of remaining power variation and subsequently frequency stabilization (>30 min). Reserve power availability plays a vital role of controlling in this stage [4].

The same thing can be analysed in case of inverter dominated power system; energy storage system (ESS) plays a prominent role for achieving normal frequency by providing 'inertia' to the microgrid.

2.3 DGs Impact on the Inertia of the PS

In the modern smart microgrid structure, the loads and renewable energy sources (RESs) are integrated with the grid by means of power electronics-based interfacing devices as shown in Fig. 3. As the inertia of the overall system reduces with RESs integration, the ROCOF and frequency variations would increase under any contingency events occurrence, and frequency instability that is why very likely to occur due to the low inertia, nonlinear dynamics, and damping influence of the grid [3]. The ROCOF is an indicator to the way how rapidly the frequency changes after a sudden imbalance between generator and load. The ROCOF can be defined as:

From Eq. (6),

$$\mathrm{ROCOF} = \frac{1}{2}\frac{f * \Delta P}{S_g H} \tag{7}$$

Due to DGs integration, there is a deviation in frequency (high ROCOF). Essential measures like emulated inertia controls for converts are required. Inertia can be increased by applying various topologies in the PS [5]. The application of various topologies in PS is another option for handling the issue which arises due to offering of less inertia from loads and sources.

3 Basic Concepts on Virtual Inertial Application Strategy

3.1 *Synchronverters-Based Virtual Inertia Control Topology*

Synchronverters are operated to behave exactly like the SG and try to reduce the issue of lack of inertia in DG operation in a microgrid. The dynamic equations for both are the same. The mechanical power exchanged with the prime mover (with the mechanical load in another case) is replaced by the power exchange with the inverter along with filter inductors and capacitors. As a result, the voltage and frequency in microgridare regulated by DG providing inertia and damping to the grid as like conventional SG performs [6]. The basic operating principle of synchronverter topology is illustrated by an overall schematic diagram in Fig. 4 showing operating principles. Even though the synchronverters successfully able to mimic the dynamics of SG, till the complexity of the differential equations applied can lead to numerical instability. Apart from that, the topology based on voltage source converter may fetch to protection problems.

3.2 *Virtual Synchronous Generator (A Swing Equation-Based Topology)*

The virtual inertia can be realized for any DGs by adding a short term and fast acting energy storage system (ESS) along with a optimal control strategy for the power electronics converter. In this way, the DG dynamics can act like a virtual synchronous generator (VSG). By using a ESS with DG, VSG can be realized having virtual inertia which acts equivalently to rotor's inertia of SGs [4]. A simple

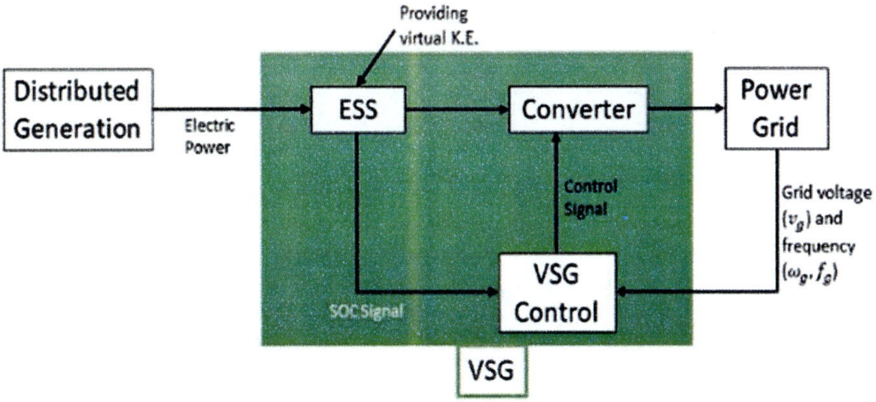

Fig. 4 Schematic diagram of synchronverter control

VSG model is illustrated in Fig. 5. A proper control signal is generated by VSG control unit based on the grid voltage/frequency and state of change (SOC) of ESS to the inverter to inject or absorb the requisite amount of power. This way the grid stability is maintained by using the virtual inertia of the VSG in the same way like the rotor inertia of SGs by using storages and keeping synchronization with other generators in microgrid [7, 8].

The K.E of rotor can be expressed as:

$$\text{K.E} = \frac{1}{2} J \omega_m^2 \tag{8}$$

where J and ω are the inertia coefficient and rotor speed (a virtual angular frequency), respectively.

The swing equation model for the VSG can be expressed as:

$$P_{\text{ref}} - P_{\text{act}} = J\omega_m \frac{d\omega_m}{dt} + D.\Delta\omega_m \tag{9}$$

where D and $\Delta\omega_m$ are the damping factor and $(\omega_m - \omega_g)$, respectively. Here, ω_g denotes the rotating speed of the grid voltage. Equation (9) is applied to drive the power converter. Due to that, the inverter performs like a SG. The grid frequency and power are measured to calculate $\Delta\omega_m$. Then the virtual rotor speed ω_m is computed by the VSG control block according to Eq. (9). Generally, Runge-Kutta method is used to obtain the ω_m from the differential Eq. (9). The virtual mechanical phase θ_m is computed from ω_m. This is used as the input (control signal) to the converter as a phase control command from controlling the converter output to regulate voltage and frequency as shown in Fig. 6.

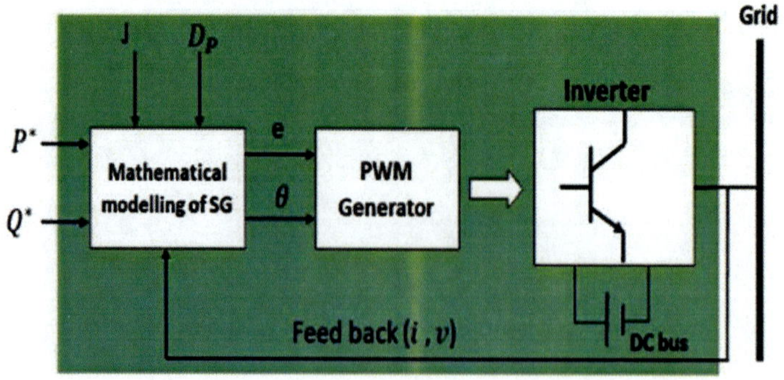

Fig. 5 VSG schematic diagram

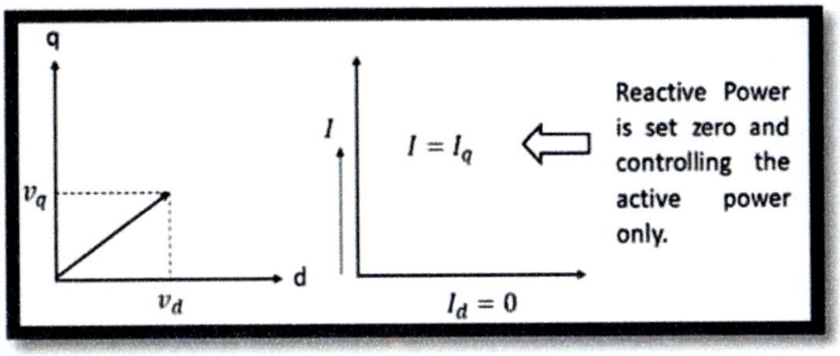

Fig. 6 *d-q* based approach

3.3 A Power-Frequency-Based Topology

The basic principle of VSG based on frequency-power response characteristics of an SG in a DG system is to emulate the inertia response characteristics with respect to frequency change; according to the frequency change response, the DG unit is operated such that it supplies or absorbed K.E exactly like the SG [5]. Under this condition, the DG acts as a dispatchable source and capable to contribute dynamic frequency control rather than only frequency regulation like conventional DG system. The basic strategy followed for this dynamic frequency control is to release inertial power by contribution of K.E or absorption of K.E same as the SG during power imbalance according to the derivative of the frequency measurement [9, 10]. The controlling output of VSG can be expressed as:

$$P_{\mathrm{VSG}} = K_{\mathrm{D}}\Delta\omega = \frac{K_{\mathrm{I}}\mathrm{d}\Delta\omega}{\mathrm{d}t} \tag{10}$$

Here, $\Delta\omega$ denote the change in angular frequency, $\frac{\mathrm{d}}{\mathrm{d}_t}\Delta\omega$ is the corresponding rate of change, K_{D} denotes the dropping constant, and K_{I} represents the inertial constant. The value of K_{D} factor regulates the frequency to decrease the frequency nadir and also achieve the steady-state value. The K_{I} seizes the ROCOF by providing a faster dynamic frequency response based on the derivative of frequency. On this approach, a detail control strategy to implement virtual inertia is illustrated in Fig. 6, based on *d-q* composition with deriving current reference as:

$$I_d^* = \frac{2}{3}\left(\frac{V_d P_{\mathrm{VSG}} - V_q Q}{V_d^2 + V_q^2}\right) \tag{11}$$

where V_d and V_q denote the *d*-axis and *q*-axis components of the grid voltage 'V'. As shown in Fig. 6, the inverter acts as a current controlled voltage source inverter, and its gate signal is generated by the current controller based on the grid current feedback [11]. Use of phase lock loop (PLL) from synchronization and high sensitiveness due to use of derivative term are two major limitations Fig. 7.

3.4 Droop-Based Virtual Inertia Control Concept

The frequency droop equation can be represented as:

$$\omega^* - \omega_g = m_p(P_{act} - P_{ref}) \tag{12}$$

where ω^*, ω_g, P_{ref}, and P_{act} denote the reference frequency, the local grid frequency, reference set active power, and active power output (measured) from the DG unit, respectively. m_P denotes the active power droop. Similarly, the voltage droop equation can be written as:

$$v^* - v_g = m_q(Q_{act} - Q_{ref}) \tag{13}$$

where v^*, v_g, Q_{ref}, and Q_{act} denote reference voltage, local grid voltage, reference set reactive power, and measured reactive power out from the DG unit [12]. The m_q represents the reactive power droop coefficient. Both frequency and voltage droop curves as expressed in Eqs. (12) and (13) are illustrated in Fig. 8.

The schematic diagram of implementation of droop-based virtual inertia control is illustrated in Fig. 9.

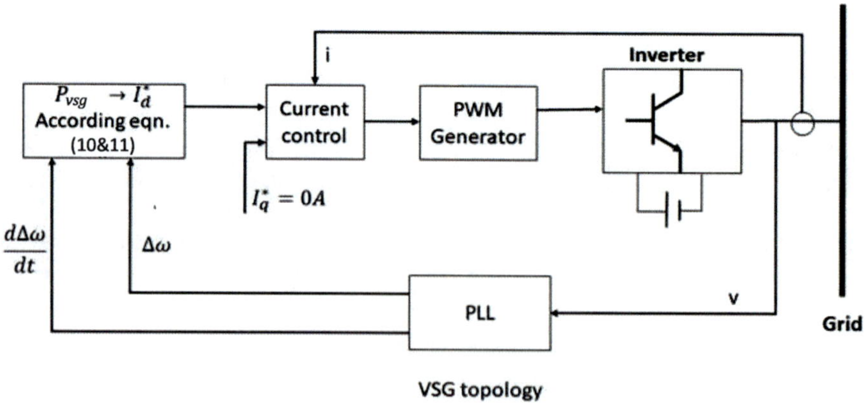

Fig. 7 Power-frequency-based topology

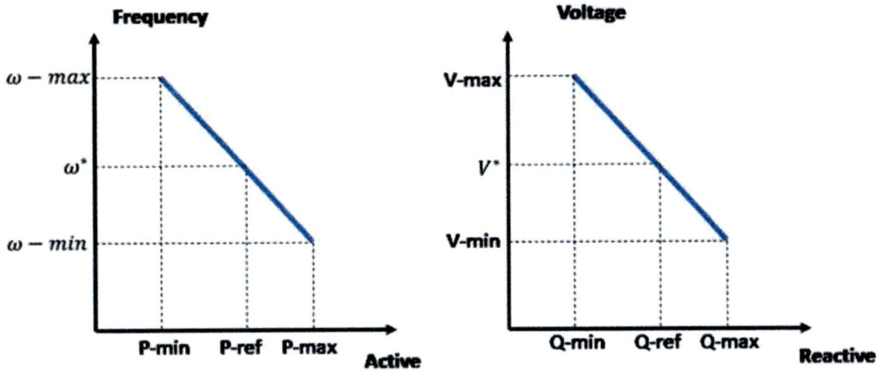

Fig. 8 Frequency and voltage droop curves

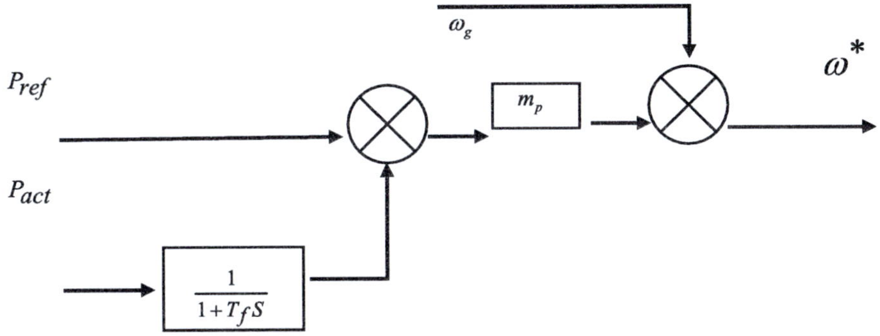

Fig. 9 Schematic diagram for frequency droop control

From Fig. 9,

$$P_{\text{act}} = \left(1 + T_f S\right)\left\{\frac{1}{M_P}\left(\omega_{\text{g}} - \omega^*\right) + P_{\text{ref}}\right\} \tag{14}$$

Rearranging

$$P_{\text{ref}} - P_{\text{act}} = \frac{1}{M_P}\left(\omega^* - \omega_{\text{g}}\right) + T_f \frac{1}{M_P} S\omega^* \tag{15}$$

Can be represented as:

$$P_{\text{ref}} - P_{\text{act}} = K_{\text{D}}\left(\omega^* - \omega_{\text{g}}\right) + K_{\text{I}} S\omega^* \tag{16}$$

where $K_{\text{I}} = T_f \cdot \frac{1}{M_P}$ and $K_{\text{D}} = \frac{1}{M_P}$

Equation (16) is similar to the virtual synchronous generator (VSG) as presented in Eq. (9). The insertion of filter mimics the dynamics of SG as represented in Eq. (16). The key factors for the filter integration are:

(a) The virtual inertia system is identical with the droop-based control approach dynamics and this is done by the filter.
(b) The filter put a delay in the measurement of power and mathematically, and it is equivalent to virtual inertia. The droop gain here is similar to damping. However, slow transient response and cannot perform satisfactorily beyond inductive grid systems are two major limitations [6, 13].

4 Discussions and Findings

Synchronous generator (SG) model-based virtual inertia control strategy based on the accurate replication of SG dynamics; in this approach, phase-locked loop (PLL) is essential for synchronization, and frequency derivation is not necessary for control; however, improvisation of this technique needs to be focused on over-current protection, numerical instability concerns, and voltage source implementation. The method based on swing equation is a very simple model in comparison to SG based model. However, the control strategy is the limitations like power and frequency oscillations. Later, to overcome the limitations of the over two methods, a novel approach based on frequency-power response has emerged and gain attraction due to straight forward implantation and inherent over current protection, However, it is found that instability due to PLL in week grid, the requirement of frequency derivative and system susceptibility to noise is the measure associated limitations; lastly as discussed, droop-based approach is communication less and having a similar concept to conventional droop control in SGs till so transient response and improper transient response active power sharing are the two associated measure limitations.

5 Conclusions

The major contribution of this paper can be summarized as follows:

i. An extensive literature review on virtual inertia systems and control in the smart distribution system with DGs is presented.
ii. Different topologies and principle of operation of various virtual inertia control are discussed and analysed.
iii. The major advantages and limitations of each approach are maintained to help the reader for further improvement.

iv. The basic concept and the relevance of inertia in power system for frequency stability are emphasized.

References

1. Ratnam, K.S., Palanisamy, K., Yang, G.: Future low-inertia power systems: requirements, issues, and solutions: a review. Renew. Sustain. Energy Rev. **124**, 109773 (2020)
2. Tamrakar, U., et al.: Virtual inertia: current trends and future directions. Appl. Sci. **7**(7), 654 (2017)
3. Yap, K.Y., Sarimuthu, C.R., Lim, J.M.Y.: Virtual inertia-based inverters for mitigating frequency instability in grid-connected renewable energy system: a review. Appl. Sci. **9**(24), 5300 (2019)
4. Zhong, Q.C., Weiss, G.: Synchronverters: inverters that mimic synchronous generators. IEEE Trans. Ind. Electron. **58**(4), 1259–1267 (2010)
5. Singh, D., Seethalekshmi, K.: A review on various virtual inertia techniques for distributed generation. In: 2020 International Conference on Electrical and Electronics Engineering (ICE3). IEEE (2020)
6. Meng, J., et al.: Adaptive virtual inertia control of distributed generator for dynamic frequency support in microgrid. In: 2016 IEEE Energy Conversion Congress and Exposition (ECCE). IEEE (2016)
7. Bhowmik, P., Rout, P.: Emulation of virtual inertia with the dynamic virtual damping in microgrids. In: 2019 International Conference on Applied Machine Learning (ICAML). IEEE (2019)
8. Samanta, S., Mishra, J.P., Roy, B.K.: Implementation of a virtual inertia control for inertia enhancement of a dc microgrid under both grid connected and isolated operation. Comput. Electr. Eng. **76**, 283–298 (2019)
9. Bhowmik, P., Rout, P.K.: Establishment of an auxiliary virtual damping loop for the superior inertial response in the microgrid. IET Smart Grid **3**(1), 42–50 (2020)
10. Shintai, T., Miura, Y., Ise, T.: Oscillation damping of a distributed generator using a virtual synchronous generator. IEEE Trans. Power Delivery **29**(2), 668–676 (2014)
11. Han, J., et al.: An autonomous power-frequency control strategy based on load virtual synchronous generator. Processes **8**(4), 433 (2020)
12. Ren, H.P., et al.: Inertia and feedback parameters adaptive control of virtual synchronous generator. arXiv preprint: arXiv:2009.05916 (2020)
13. Cheema, K.M.: A comprehensive review of virtual synchronous generator. Int. J. Electr. Power Energy Syst. **120**, 106006 (2020)

A Dynamic Load Scheduling Using Binary Self-adaptive JAYA (BSAJAYA) Algorithm in Cloud-Based Computing

Kaushik Mishra and Santosh Kumar Majhi

Abstract The load scheduling is a paramount concern in the cloud-based computing due to the involvement of conflicting parameters and fluctuating demands of users. Though the problem of load scheduling comes in the category of NP-hard problem, it is utmost essential to design a load scheduling approach to tackle the challenges in cloud computing. These challenges could be elevated by using metaheuristic approaches that offer an optimal solution to an NP-hard problem. In this research, authors propose a binary self-adaptive JAYA-based load scheduling algorithm to solve the dynamically independent load scheduling problem in cloud-based computing. The proposed algorithm is compared with the lately invented metaheuristic-based task scheduling algorithms such as bird swarm optimization (BSO), modified particle swarm optimization (MPSO), and the standard JAYA. The proposed algorithm is evaluated using a real-world dataset for the various QoS scheduling parameters to advocate the effectiveness of the algorithm. The simulated results show notable improvements over other compared algorithms for makespan, average resource utilization, and load-balancing.

Keywords Task scheduling · Load scheduling · Self-adaptive JAYA · Binary JAYA · Cloud computing

---

K. Mishra (✉) · S. K. Majhi
Department of Computer Science and Engineering, Veer Surendra Sai University of Technology, Burla 768018, India
e-mail: kaushikmishra1991@gmail.com

S. K. Majhi
e-mail: smajhi_cse@vssut.ac.in

S. Das and M. N. Mohanty (eds.), *Advances in Intelligent Computing and Communication*, Lecture Notes in Networks and Systems 202,
https://doi.org/10.1007/978-981-16-0695-3_12

1 Introduction

1.1 *Research Motivation*

The load scheduling problem is categorized under NP-hard problem [1, 2] due to the involvement of various conflicting scheduling parameters and nature of the problem. It is essential to assign the incoming requests of users to the underlying resources in a way that keeps the whole system balanced by utilizing the resources efficiently. This challenge coerces the broker to design an effective task scheduling algorithm that can cope up with the growing fluctuating demands.

The heuristic or the general scheduling algorithms are incapable of solving this deterministic problem within a stipulated time. Therefore, many researchers use the metaheuristic approaches to solve this problem deterministically that offers either an optimal solution or a near to optimal solution. In [3], authors have suggested that the metaheuristic algorithms are incorporated for escaping the local optima problem when there is diversity in independent requests and VMs. In this paper, authors implement a binary self-adaptive JAYA algorithm for scheduling of tasks (incoming requests) in cloud-based computing. The key reason for using the JAYA algorithm is its simplicity in implementation and use of no or significantly fewer tuning parameters. Besides, this algorithm does not get trapped in local optima yielding a high convergence rate. Selecting a population size for this algorithm is a cumbersome task for the researchers, so authors use a self-adaptive approach which determines the population size automatically. Therefore, the researchers need not focus on tuning the population size. This algorithm generates a series of continuous solutions which are not appropriate for the problem like task scheduling in cloud computing. Hence, a binary version of the self-adaptive JAYA algorithm is implemented to transform the generated continuous solutions into discrete solutions.

1.2 *A State-of-the-Art Literature*

In this section, authors discuss the works of different researchers in the context of task scheduling and load-balancing problem in cloud computing and summarized in Table 1.

1.3 *Structure of the Paper*

The rest of the paper is organized as follows. Section 2 explains the improved JAYA optimization algorithm. Section 3 presents the application of the BSAJAYA algorithm in solving the load scheduling in cloud-based computing by highlighting

Table 1 State-of-the-art of the related work

Reference	Algorithm	Advantages	Disadvantages
[4]	Honey bee-inspired load-balancing	• Reduced response time • Reduced degree of imbalance (DOI) • Improved throughput	• Suffers from starvation
[5]	RASA, MIN-MIN	• Low makespan	• Low scalability • Not implemented in real workload
[6]	DWOLB algorithm using GA	• Low-power consumption • Low total cost	• Low fault tolerance • Lack of considering other QoS metrics
[7]	PSO and load-balancing	• Low makespan • Improved resource utilization • High convergence	• Homogenous environment
[8]	JAYA	• Reduced response time • Low service time • Heterogenous resources	• Inefficient for dynamic independent workloads • Homogenous environment
[9]	BWM and VIKOR	• High throughput and VM utilization • Low makespan, response time • Heterogenous environment	• Tasks size with varying lengths could be better
[10]	MPSO with Q-learning (QMPSO)	• Low makespan, DOI, completion time, energy consumption • High resource utilization, throughput	• Inefficient for heterogeneous workloads

the load scheduling algorithm. Section 4 presents the experimental setup to conduct the simulations in cloud-based computing and an analysis of the results after that. Section 5 ends with the concluding remarks along with the future highlights.

2 The Improved JAYA Optimization Algorithm

2.1 The Standard JAYA

In 2016, a swarm-based optimization technique 'JAYA' is developed by Rao [11] for solving constrained and unconstrained optimization problems. In this algorithm, each searching candidate is called a particle, and each particle competes with each other to get the best position by evading through the worst solutions. The primary idea of this algorithm is that the particles always seek to move toward the best solution by avoiding the worst solutions in the search space [12]. Mathematically, the particles' position is modified as Eq. 1 [11, 12].

$$X_i^{k+1} = X_i^k + c_1\left(\text{best}_i - \left|X_i^k\right|\right) - c_2\left(\text{worst}_i - \left|X_i^k\right|\right) \quad (1)$$

Note X_i^{k+1} is the modified value of i at iteration $k + 1$, X_i^k is the current position of a particle i at iteration k, c_1 and c_2 are two random numbers varies in between (0, 1). The term $c_1\left(\text{best}_i - \left|X_i^k\right|\right)$ indicates the tendency of exploration and the term $-c_2\left(\text{worst}_i - \left|X_i^k\right|\right)$ indicates the tendency to escape the worst solution. The solution is accepted if the value X_i^{k+1} has a better fitness value than X_i^k. The two random numbers are used in the equation to maintain a balance between exploration and exploitation.

2.2 *The Self-adaptive JAYA*

The key idea behind the self-adaptive JAYA (SJAYA) is to set the population size automatically. Hence, the user need not focus on determining the population size. Rao and More [13], in the year of 2017, have proposed a self-adaptive-based JAYA algorithm for solving the unconstrained optimization problem. Let the random initial population is ($10 \times m$), where m is the number of design variables or VMs then the new population is formulated as,

$$n_{\text{new}} = \text{round}(n_{\text{old}} + r \times n_{\text{old}}) \quad (2)$$

where r is a random number ranges in [−0.5, 0.5].

Due to negative or positive random value of r, the population size may decrease or increase. Elitism technique is implemented when the size of the new population is greater than or less than or equal to the old population.

2.3 *The Binary Self-adaptive JAYA*

The generated continuous solutions by SAJAYA will not fit into the problem domain because the problem of load scheduling is dynamic in nature. Hence, it is required to convert the continuous solutions into discrete solutions. Therefore, this work implements a binary variant of the CJAYA algorithm using tangent hyperbolic logistic transfer function [14], which converts continuously generated solutions into discrete values using the following Eq. 3.

$$\tan h\left(\left|X_i^{k+1}\right|\right) = \frac{e^{\left(\left|2X_i^{k+1}\right|\right)} - 1}{e^{\left(\left|2X_i^{k+1}\right|\right)} + 1} \tag{3}$$

The updated value X_i^{k+1} is then represented in binary form using the following Eq. 4.

$$X_i^{k+1} = \begin{cases} 1, & \text{if } r\text{ and}() < \tanh\left(\left|X_i^{k+1}\right|\right) \\ 0, & \text{otherwise} \end{cases} \tag{4}$$

3 Application of the Proposed BSAJAYA Algorithm in Load Scheduling

3.1 Task-Resource Formulations

The load scheduling problem is elucidated as follows. Let us consider a datacenter or a cloud consists of a number of physical machines or hosts. An array of VMs $\text{VM} = \{\text{VM}_1, \text{VM}_2, \text{VM}_3, \ldots, \text{VM}_m\}$ are emulated under the series of hosts. Consider a task set $T = \{T_1, T_2, T_3, \ldots, T_n\}$ of non-preemptive and independent in nature, where $T_i, 1 \leq i \leq n$ is the ith task with a million sets of instructions (MI) in l_i. These tasks are scheduled to be executed under the array of VMs. Number of quality of service (QoS) parameters are considered to evaluate the performance of the proposed algorithm. Makespan, resource utilization, and load-balancing are considered as QoS parameters.

Task completion time is the time taken to execute a task T_i by VM_j. It is the difference time between the start time and the finish time that is denoted in Eq. 5.

$$\text{TCT}_j = \text{FT} - \text{ST} \tag{5}$$

Makespan refers to the maximum completion time of a task T_i among the VMs [15]. It is denoted in Eq. 6.

$$\text{MS} = \max\{\text{TCT}_j | j = 1, 2, 3, \ldots m\} \tag{6}$$

VM utilization is the notch of the utilization of VMs [16]. The utilization and the average utilization of all VMs are calculated using the Eq. 7 [15, 16], where m is the number of VMs.

$$\mathrm{AUtil}_{VM} = \frac{\sum_{i=1}^{n} \mathrm{TCT_j}}{\mathrm{MS} \times m} \tag{7}$$

Load-balancing is the degree of imbalances among VMs that is used to measure the imbalances [4]. Where $T_{\max}$ and $T_{\min}$ are the maximum and minimum completion time of tasks T_i among all VMs. Further, T_{avg} is the average of all tasks T_i of VMs. L is the length of total instruction, $\mathrm{PE_{num_i}}$ is the number of processing elements in the ith VM, and $\mathrm{PE_{MIPS_i}}$ are the million instructions per second of the ith VM.

$$\mathrm{LB} = \frac{T_{\max} - T_{\min}}{T_{\mathrm{avg}}} \tag{8}$$

with

$$T_i = \frac{L}{\mathrm{PE_{num_i}} \times \mathrm{PE_{MIPS_i}}} \tag{9}$$

Fitness function is used to evaluate the best and worst fitness values of each particle in the problem space. It is assumed to be problem-specific. The key objective of load scheduling is to diminish the makespan while maximizing resource utilization. These two parameters are associated with a reverse linear relationship. Therefore, the authors consider a single-objective function by keeping the above objective intact. The particle has a better position if the fitness value is less than the previous value. Hence, the fitness value f is defined as follows.

$$f = \frac{1}{\mathrm{MS}} \times \mathrm{AUtil_{VM}} \tag{10}$$

3.2 Load Scheduling Algorithm

The Algorithm 1 depicts the proposed load scheduling algorithm to schedule the loads among balanced VMs in a cloud-based computing environment.

Algorithm 1 Pseudocode for mapping of tasks onto VMs using BSAJAYA algorithm

```
Start
1. Set particles' dimension to the total number of tasks and VMs set
   and initialize population size (n = 10 × m);
2. Calculate Total completion time, Makespan, and Average resource
   utilization using Eqs. 5, 6, 7, and 8;
3. For all the particles, calculate fitness function using Eq. 10;
                         if(F < P_best)
      Set the new fitness value as the P_best;
4. For all the particles, create a new set of mapping by calculating
    the position using Eq. 1;
5. Update the position by converting continuous new solutionto dis-
    crete solution using Eq. 3 and Eq. 4;
6. Repeat until terminating condition;
     if (termination condition ≠ met)
7. Generate new set of population using Eq. 2;
if(n_new > n_old)
         assigns fit (best) solutions in the new population to the
         (n_new − n_old) solutions;
else if(n_new < n_old)
          assigns only the fit solutions in the new population to
         the n_new solutions;
else if(n_new < m)
          set n_new = m;
else
          Display the optimum solution;
8. Return best possible solution to the proposed mapping of tasks to VMs;
End
```

4 Experimental Setup and Discussions

4.1 Experimental Setup

Authors have extended the classes of CloudSim [17] for creating a virtualized simulation environment in order to process the incoming requests. For the simulation, we have created one data centre that consists of four hosts, each capable of emulating the physical machines and sharing the resources among VMs. Each host encompasses some processing cores, RAM, Storage, bandwidth (BW), and the speed of processing. The technical specifications of each host are shown in Table 2. We have taken 300 VMs that run on hosts with distinctive characteristics. The VM technical details are shown in Table 3. We have considered independent cloudlets up to 3000 with varying length of instructions ranges from 0 to 15,000 that gets

Table 2 Host technical details

Host ID	Processing cores	Speed, MIPS	RAM, GB	Storage, GB	BW, MIPS
1	1	3500	50	1024	102,400
2	2	4000	100	1024	102,400
3	3	4500	150	1024	102,400
4	4	5500	200	1024	102,400

Table 3 VM technical details

CPU	Number of cores	Speed, MIPS	RAM, GB	Storage, GB	BW, MIPS
Core_i5_Extreme_Edition	1–3	1000	2	20	1024

processed on the respective VMs. To validate the efficacy of the proposed algorithm, we evaluate the performance of the algorithm using a real-world dataset logged by NASA [18] against some conflicting scheduling parameters. The general structure of the instances of the dataset is denoted as $T \times VM$, where T is the set of tasks are to be scheduled on a set of virtual machines VM. The six instances are 500×50, 1000×100, 1500×150, 2000×200, 2500×250, and 3000×300. For this algorithm, the population size is set to 100, and the maximum termination criterion is set to 100.

4.2 *Results and Discussions*

In this section, we present the obtained values from the experiments. Simulation is carried out in a heterogeneous environment where both tasks and VMs vary. The influence of makespan, resource utilization, and load-balancing is recorded. To witness the effectiveness of the proposed algorithm, we have compared the proposed algorithm with (1) BSO, (2) MPSO, and (3) JAYA.

Figure 1a depicts the obtained simulation values of the proposed BSAJAYA algorithm over compared algorithms for makespan. The simulation graph shows a considerable reduction of makespan by the proposed algorithm over others. Figure 1b presents the simulated results of the proposed BSAJAYA algorithm over compared algorithms for resource utilization. It has been observed from the result that the proposed method outperforms over the compared algorithms.

Figure 1c showcases the experimental results of the proposed algorithm, along with the compared algorithms for load-balancing. It shows a significant improvement in terms of balancing out the loads in the overall system. The efficient load-balancing makes the VMs utilized extensively and hence, reduces in makespan.

Using of a self-adaptive variant of JAYA, the burden of setting population size is leveraged. The consideration of both the best and worst solutions in a single

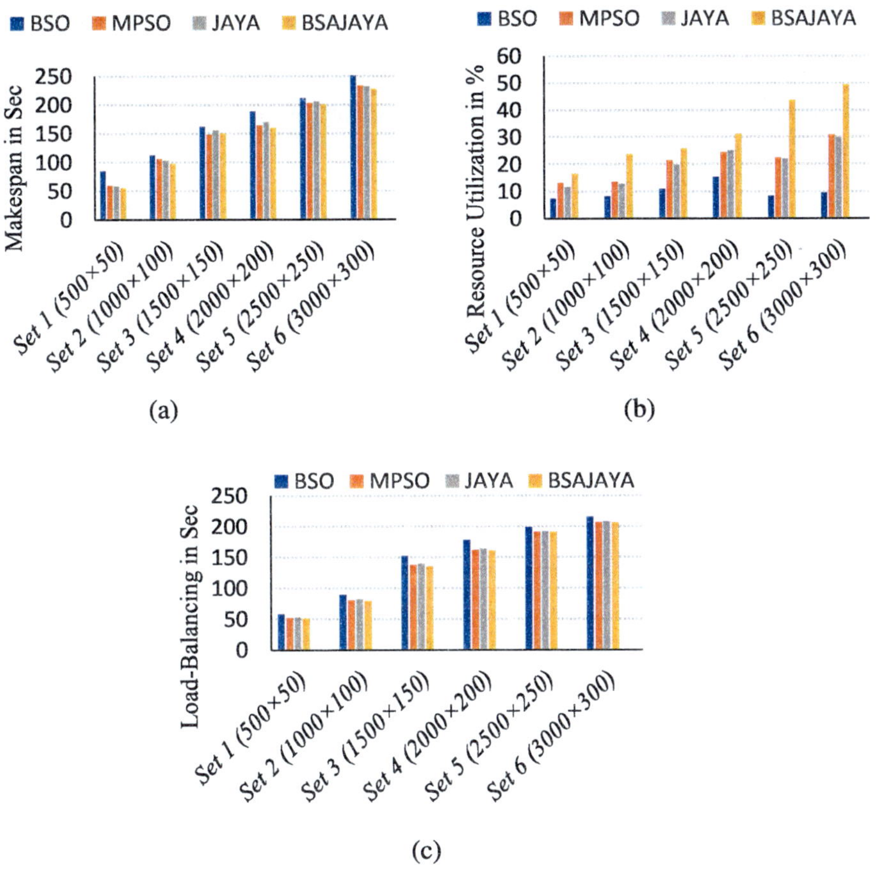

Fig. 1 Performance comparison of BSAJAYA over NASA dataset in terms of **a** Makespan, **b** resource utilization, and **c** load-balancing

equation strives the algorithm to reach an optimum point by evading through the local optima. This algorithm offers a low computational time as the solutions are modified in a single-phase equation. This very nature of self-adaptiveness of JAYA maintains the trade-off between the exploration and exploitation in the problem space.

5 Conclusion and Future Work

A swarm-based binary self-adaptive JAYA (BSAJAYA) algorithm is proposed to address the load scheduling problem in cloud-based computing. The proposed work is validated using a real-world dataset logged by NASA. The experimental results of the proposed algorithm show a significant amelioration in terms of makespan, resource utilization, and load-balancing with the considered objectives. The BSAJAYA algorithm outperforms while being compared with BSO, MPSO, and JAYA. The effectiveness of the proposed approach has experimented in a heterogeneous environment. In this work, the algorithm considers the varying length of independent tasks with increasing VMs for the simulation. But the tasks are not associated with priority and the dependency among tasks. In future, we would attempt to develop an energy-efficient scheduling algorithm by considering the above constraints.

References

1. Ullman, J.D.: NP-complete scheduling problems. J. Comput. Syst. Sci. **10**(3), 384–393 (1975)
2. Mishra, K., Majhi, S.: A state-of-art on cloud load balancing algorithms. Int. J. Comput. Digit. Syst. **9**(2), 201–220 (2020)
3. Kalra, M., Singh, S.: A review of metaheuristic scheduling techniques in cloud computing. Egypt. Inf. J. **16**(3), 275–295 (2015)
4. LD, D.B., Krishna, P.V.: Honey bee behavior inspired load balancing of tasks in cloud computing environments. Appl. Soft. Comput. **13**(5), 2292–2303 (2013)
5. Mittal, S., Katal, A.: An optimized task scheduling algorithm in cloud computing. In: IEEE 6th International Conference on Advanced Computing (IACC), pp. 197–202 (2016)
6. Vanitha, M., Marikkannu, P.: Effective resource utilization in cloud environment through a dynamic well-organized load balancing algorithm for virtual machines. Comput. Electr. Eng. **57**, 199–208 (2017)
7. Ebadifard, F., Babamir, S.M.: A PSO-based task scheduling algorithm improved using a load-balancing technique for the cloud computing environment. Concurrency Comput. Pract. Experience **30**(12), 1–16 (2018)
8. Mohanty, S., Patra, P.K., Ray, M., Mohapatra, S.: An approach for load balancing in cloud computing using JAYA algorithm. Int. J. Inf. Technol. Web Eng. (IJITWE) **14**(1), 27–41 (2019)
9. Rafieyan, E., Khorsand, R., Ramezanpour. M.: An adaptive scheduling approach based on integrated best-worst and VIKOR for cloud computing. Comput. Ind. Eng. **140**, 1062–1072 (2020)
10. Jena, U.K., Das, P.K., Kabat, M.R.: Hybridization of meta-heuristic algorithm for load balancing in cloud computing environment. J. King Saud Univ. Comput. Inf. Sci. (2020)
11. Rao, R.: Jaya: a simple and new optimization algorithm for solving constrained and unconstrained optimization problems. Int. J. Ind. Eng. Comput. **7**(1), 19–34 (2016)
12. Pradhan, C., Bhende, C.N.: Online load frequency control in wind integrated power systems using modified Jaya optimization. Eng. Appl. Artif. Intell. **77**, 212–228 (2019)
13. Rao, R.V., More, K.C.: Design optimization and analysis of selected thermal devices using self-adaptive Jaya algorithm. Energy Convers. Manag. **140**, 24–35 (2017)
14. Prakash, T., Singh, V.P., Singh, S., Mohanty, S.: Binary Jaya algorithm based optimal placement of phasor measurement units for power system observability. Energy Convers. Manag. **140**, 34–35 (2017)
15. Ebadifard, F., Babamir, S.M., Barani, S.: A dynamic task scheduling algorithm improved by load balancing in cloud computing, In: 6th IEEE International Conference on Web Research (ICWR), pp. 177–183 (2020)

16. Khorsand, R., Ghobaei-Arani, M., Ramezanpour, M.A.: Self-learning fuzzy approach for proactive resource provisioning in cloud environment. Softw. Pract. Experience **49**(11), 1618–1642 (2019)
17. Calheiros, R.N., Ranjan, R., Beloglazov, A., De Rose, C.A., Buyya, R.: CloudSim: a toolkit for modeling and simulation of cloud computing environments and evaluation of resource provisioning algorithms. Softw. Pract. Experience **41**(1), 23–50 (2011)
18. Feitelson, D.G., Nitzberg, B.: Job characteristics of a production parallel scientific workload on the NASA Ames iPSC/860. In: Workshop on Job Scheduling Strategies for Parallel Processing, pp. 337–360. Springer, Berlin, Heidelberg (1995)

Haze Identification and Classification Model for Haze Removal Techniques

Gaurav Saxena and Sarita Singh Bhadauria

Abstract In the modern age, several applications depend on the information collected from the images and its classification. However, due to the poor climatic conditions, the images captured by the camera are degraded in terms of contrast, brightness, color, and sharpness, which create hazy images. Therefore, several techniques have been suggested in the literature to restore the degraded image. Besides, poor climate conditions are not same all the time which leads the idea, that, if images are identified before applying haze removal technique, then it can save processing time which otherwise will be consumed for clear images as well. Therefore, in this article, the haze identification model is proposed to identify the hazy and non-hazy images which makes all haze removal systems more efficient. Moreover, proposed model also classifies the hazy images based on their densities which can be utilized for selection of configuration parameters of haze removal techniques. Finally, the performance of the proposed model is demonstrated and the valuable identification, as well as classification parameters, are computed and presented.

Keywords Image classification · Image formation model (IFM) · Haze removal · Principal component analysis (PCA) · Artificial neural network (ANN) · Convolutional neural network (CNN)

G. Saxena (✉)
Jaypee University of Engineering and Technology, A.B Road Raghogarh, Guna, Madhya Pradesh 473226, India
e-mail: gauravsagar311@gmail.com

S. S. Bhadauria
Madhav Institute of Technology and Science, Gole ka Mandir, Gwalior, Madhya Pradesh 474008, India
e-mail: saritamits61@yahoo.co.in

S. Das and M. N. Mohanty (eds.), *Advances in Intelligent Computing and Communication*, Lecture Notes in Networks and Systems 202,
https://doi.org/10.1007/978-981-16-0695-3_13

1 Introduction

Classification of an image in the group of images is not a difficult task but it is becoming a challenging issue whenever this task is performed with the help of a machine. The image classifier extracts various features from an image by applying various image processing tasks (start from image acquisition to classification) such as image sensing, feature extraction for the classification of an image. Image classification plays a vital role in a variety of image processing application fields like telemetry [1], remote sensing [2], satellite imaging [3], biomedical [4], submarine images [8], humanoid robotics, etc. In general, the image classification method is broadly divided into three categories: (i) Supervised classification; (ii) Unsupervised classification; and (iii) Hybrid classification. The selection of the classification methods is decided according to the requirement of the application. In recent years, the classification of images proves their importance in several fields. The applications of image classification in various fields are discussed below.

In the field of medical science, the image classification is to identify the portion of spread infection with high accuracy and reliability [4]. With the advancement of technology, the remote clinical computer-aided-diagnosis methodology is developed which is mainly based on image analysis and classification [5].

The utilization of the image classification in remote sensing applications is concentrated on spectra-based techniques whereas inherent spatial component information of the real world is overlooked [6]. In recent years, based on the noticeable performance of the CNN models, Peng et al. [7] introduce an efficient CNN two stages framework for remote sensing image scene classification which consists of search and evaluation stages. In the context of object classification through underwater images, Huo et al. [8] propose a deep CNN method to improve the accuracy of object classification through side-scan sonar images by effectively combined the semi-synthetic training data and deep transfer learning.

In recent years, the ground objects (such as trees, grass, lanes, and sidewalks) classification based on the multispectral images collected by unmanned aerial vehicles (UAV). Further, these images are used for the various feature extractions (like, vegetation indices, texture, gray mean value, etc.) and the three classification model is developed which is based on piece-wise regression, partial least squares, PCA - extreme machine learning, respectively [9]. Ichim and Popescu [10] collected the multispectral images through UAV and implemented a CNN-based system, classify the patches which are flooded or un-flooded type for the detection and control of flood damages.

From the literature survey, it is observed that image classification can also be applied to the atmospherically degraded images, but this type of classification is not available in the existing literature to the best of our knowledge. For the many automatic multimedia vision applications like the object or target recognition, video surveillance in the military, etc. [11]., get severely affected by the images captured during adverse climatic conditions. These images are generally degraded due to the availability of scattering particles (like dust, smog, water droplets, etc.) in the atmosphere. Further, it becomes the main cause of the accident at various positions such as road, sea, and air [5, 12, 13]. Therefore, the desired performance can only be achieved by using some good defogging techniques to get a pleasant view of the hazy image.

There are many fog/haze removal techniques are available with significant performance in this regard, but these methods create some processing leg in their respective application system. However, this processing leg collectively creates severe impact whenever a sequence of images is to be processed for the real-time systems. Therefore, there is a requirement of updating the system for the reduction of processing leg developed by dehazing techniques in real-time systems. To update the existing system, the classification stage is introduced so that image is classifying whether it is hazy or non-hazy before applying the haze removal process and thereafter the required parameters of haze removal techniques are updated according to the haze density. Moreover, it is also taken into consideration that the time required for the classification stage is very much less as compared to the processing time of the haze removal technique then only this approach becomes effective. Thus, in this article, very fast haze identification and classification technique are proposed. The rest of the paper is prepared as follows. Section 2 discusses the brief background. Section 3 describes the proposed haze identification model. Simulation results and discussion are carried out in Sect. 4. Finally, Sect. 5 presents the conclusion.

2 Brief Background

To understand the concept of the hazy image, it is important to know how the haze in the atmosphere affects the image acquisition of the image. The concept can be easily understood by the hazy image formation model, and it is also utilized for the restoration of the image by various methods.

Hazy image formation model: According to Koschmieder's optical image formation model [14], the formation of a foggy image can be expressed as

$$I_f(x) = R_d(x) \times t_r(x) + A(1 - t_r(x)) \quad (1)$$

where $I_f(x)$ is the foggy image, $R_d(x)$ is the clear image representing the scene radiance, $t_r(x)$ is known as the transmissivity function, and the term A denotes the global atmospheric light. When the light passes through an even (or homogenous) medium, $t_r(x)$, the function denoting transmission can be modeled as

$$t_r(x) = e^{-\beta . d(x)} \quad (2)$$

Here, β is the coefficient of scattering, and $d(x)$ is the depth of the underlying scene. The algorithm aims to store back the original image $R_d(x)$ from the captured image $I_f(x)$ via estimation of the transmission function $t_r(x)$ and A is the value of the atmospheric light.

3 Proposed Haze Identification Model

The proposed haze identification model is shown in Fig. 1. The proposed model can be applied before the haze removal algorithm is applied. In general, the hazy or non-hazy images contain three $n \times m$ matrices of pixels; each pixel value lies in the range of levels 0–255. The three matrices individually create a channel, namely red, green, and blue channels. These three channels create a 16 M color resolution for each pixel and responsible for the color quality of an image. In the proposed model, in the first stage all three channels are separated and stored individually as a function of $f^c(x,y); c \in \{r,g,b\}$. Now, collect the value of each pixel from a particular location in each plane and compare them to determine the minimum value.

The collection of these minimum values creates a plane (image) $J(x,y)$ of size $n \times m$ which can be mathematically expressed as:

$$J(x,y) = \min_{c \in \{r,g,b\}} [f^c(x,y)], \tag{3}$$

where f^c represents the function of individual channels of an input image and c denotes the color of channels red, green, and blue.

Further, the nonlinear filtering is applied to the image $J(x,y)$ to get the identifier image $K(x,y)$. The nonlinear filter is also known as order statistics filter because these filters are derived by the ordering of the pixels covered the image region by the filter kernel. The value of the central pixel is replaced by the smallest value of

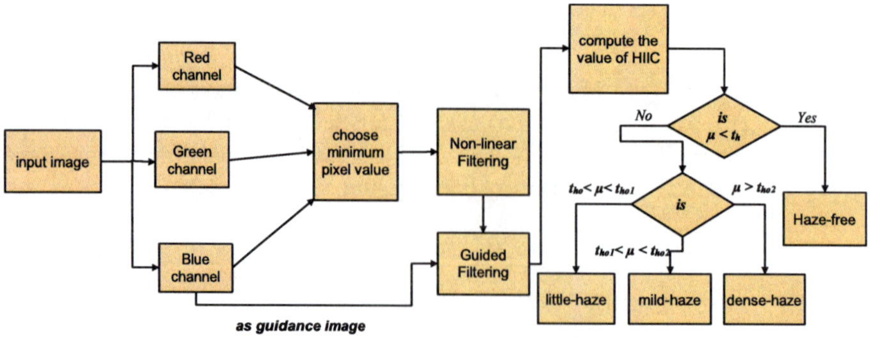

Fig. 1 Hazy image detection model

the neighboring pixel in the kernel window. This filter has good edge-preserving capability [15]. The mathematical operation of the order static filter can be expressed as:

$$K(n) = \min_{m \in \Omega(n)} [J(m)], \tag{4}$$

here, Ω denotes a kernel window focused on pixel n, and K represents the identifier image. Due to the minimum filtering used in the previous step to obtain the identifier image, the small square blocks are created on the output image as shown in Fig. 1. This problem of blocks raised on the identifier image can be removed by using guided filtering. The guided image filtering is utilized to perform smoothing of an image, and the filter also preserves the edges during the smoothening process [16]. In this context, the input image b-channel is selected as a guidance image for preserving the structure. The guidance filter works on the phenomenon that the same edges are present in the original as well as in the guidance image. The guided image filtering block receives the rough identifier image $K(x, y)$ and produces the refined identifier image $I(x, y)$ by suppressing the effect of blocking artifacts. After the suppression of the internal noise problem (blocking problem) of the model, the image obtained is now available for testing of two objectives: (i) input image is hazy or non-hazy, (ii) categorization of haze density.

3.1 *Test of the Input Image for Categorizing Hazy or Haze-Free*

In the case of a haze-free image, the output image of the model has been observed as a black/dark image. On the other side, if the input test image is hazy, it produces a brighter image from the output of the guided filter. To test the haziness of the image by the system itself as well as for the computation of haze density for the categorization of the input image, the image $I(x, y)$ obtained from the guided filtering block of the proposed model is taken into consideration. The pixel values of the image $I(x, y)$ are added one by one, and the total value of the summation is divided by $n \times m$ the size of an image so that the contribution of haze by each pixel can be called as hazy image identifier coefficient (HIIC) denoted by μ and it may be expressed as follows:

$$\mu = \frac{1}{n \times m} \sum_{i=1}^{n} \sum_{j=1}^{m} [I(i,j)], \tag{5}$$

The value of μ will be compared to the threshold value t_h determine the status of the image, i.e., image is hazy or non-hazy. For the computation of threshold value t_h, an experiment is done by selecting thousands of hazy and non-hazy images from the different dataset available on the internet such as the Waterloo Image and Vision Computing (IVC) Image Database [17], NITRE [18], and RESIDE [19]. The selected hazy and haze-free images are processed through the proposed model and generate the corresponding values of μ. After statistical analysis of the generated values of μ, it is observed that the value of μ corresponds to a haze-free image never comes more than 0.1, and thus, the value of the threshold $t_h = 0.1$ decided. After deciding the haziness of images, it becomes important to categorize them.

3.2 *Categorization of Haze Density*

It is a very tedious task to estimate the density of any input image due to the unavailability of the same image with different density of the fog. However, this type of set availability is practically unachievable. Therefore, the synthetic images are to be generated with different density of haze. To generate the hazy image from the clear image, the image formation model which is discussed in Sect. 2 is utilized by some prior assumptions. In the synthetic image generation, the value of the scattering coefficient β in Eq. 2 of the hazy image formation model is set to unity, i.e., with the assumption that there is a homogeneous haze/fog [20]. Further, the value of scene-depth is varied to generate the various values of transmissivity. After setting the parameters in the hazy image formation model Eq. 1 and applying the input clear image. The different density hazy image dataset of thousands of hazy images is generated according to the various values of transmissivity. By using these synthetic images, the experiment conducts on the proposed model and compute the statistical data for HIIC corresponding to the haziness of the synthetic images. After the comparative analysis among the synthetic haziness parameter and the value of μ, it is observed that the value of μ can be separated into three thresholds namely t_{ho}, t_{ho1}, and t_{ho2} and its estimated values are 0.1, 0.3, and 0.5, respectively. Finally, these decided threshold values corresponding to the haziness, i.e., dense hazy, mildly hazy, and little hazy, respectively, listed in Table 1.

Table 1 Threshold values for HIIC (μ) to decide the haziness

Threshold values of HIIC (μ)	Type
$\mu < t_{ho}$	Non-hazy
$t_{ho} < \mu < t_{ho1}$	Little-hazy
$t_{ho1} < \mu < t_{ho2}$	Mild-hazy
$\mu > t_{ho2}$	Dense-hazy

4 Simulation Results and Discussion

The dehazing method enhances the details of visibility, texture, and edges along with preserve the image structure and color. The experimental analysis for the proposed method has been carried out on MATLAB-19(a) with PC having 64-bit Intel I7.

Processor, 8 GB RAM, and NVIDIA GeForce GTX GPU. The testing of the proposed model is conducted on haze-free, synthetic, and natural hazy images. Further, the categorization of the input image is done based on HIIC values.

(i) *Testing of proposed Model*: To test the proposed model, nine different scene hazy, synthetic hazy, and haze-free images are collected from different sources on the Internet, and these selected images are applied to the model and the outcome of the model for the respective input image is shown in Fig. 2. From Fig. 2, it is clear that the output images obtained from the proposed model for the haze-free input image are completely dark is sufficient to shows that the input image is hazy-free and its value of HIIC which is less than the threshold $t_h = 0.1$, also support the perception. Besides, the refined identifier images generated from the hazy images are much brighter than the images obtained from haze-free images as shown in Fig. 2. Hence, it can be concluded that whenever the proposed model output is brighter, then the decision of identification is gone in favor of a hazy image and the value of HIIC is always greater. Moreover, if the identified image is hazy, then it is required to know the strength of haziness, which can be easily classified by utilizing the value of μ and compared them with the threshold values given in Table 1. This categorization supports the existing haze removal algorithms to decide the value of their different parameters to be set accordingly to achieve efficient restoration.

(ii) *Computation time complexity*: It is a very important aspect for the proposed model because the model is useful only when its computation time for the decision is very much less than the time required for the dehazing method; otherwise the purpose of the haze identifier model fails. The comparison of time consumption is computed on different size images and listed in Table 2. From Table 2, it is found that the proposed model takes very little time to provide satisfactory results and it is suitable for the implementation of real-time systems.

Input Image	Identifier image	Value of HIIC (μ)	Category
		0	Non-Hazy
		0.6308	Dense Haze
		3.0383e-06	Non-Hazy
		0.3184	Mild Haze
	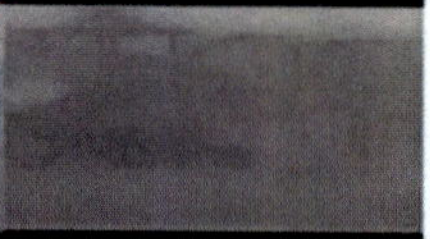	0.5416	Dense Haze
		0.0574	Non-Hazy
		0.6675	Dense Haze
	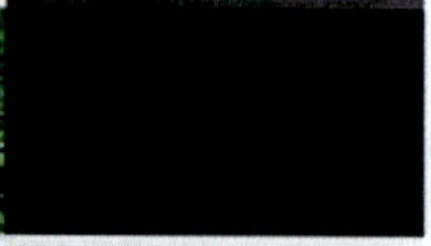	6.4714e-04	Non-Hazy
		0.2901	Little-Haze

Fig. 2 Columns (1–4) shows the haze-free images; refined identifier images of proposed model; value of HIIC (μ); and category of haze respectively

Table 2 Processing time (sec) of the proposed method

Size of image ($n \times m \times c$)	Processing time (second)
$341 \times 512 \times 3$	0.60
$384 \times 512 \times 3$	0.62
$450 \times 600 \times 3$	0.66
$512 \times 512 \times 3$	0.69

5 Conclusion

In this article, the identification and the classification of haze in image have been demonstrated. The decision of the identification and the classification of supporting parameters are taken based on experimental analysis conducted on proposed model using thousands of images related to different categories like dense hazy, mild hazy, little hazy, and haze-free. After critically examining the results, it can be observed that the proposed technique automatically detects the haze, and also categories the strength of haze which becomes helpful to decide the different configuring parameters (such as atmospheric light A, scattering coefficient β) of the existing haze removal technique. In the future, this model is clubbed with the existing haze removal algorithm, and then it enhances the performance efficiently.

References

1. Zheng, X., Yuan, Y., Lu, X.: A deep scene representation for aerial scene classification. IEEE Trans. Geosci. Remote Sens. **57**(7), 4799–4809 (2019)
2. Hu, F., Xia, G.-S., Hu, J., Zhang, L.: Transferring deep convolutional neural networks for the scene classification of high-resolution remote sensing imagery. Remote Sens. **7**(11), 14680–14707 (2015)
3. Hussein, I.I., Scheeres, D.J., Hyland, D.C.: Control of a satellite formation for imaging applications. In: Proceedings of the 2003 American Control Conference, 2003, vol. 1, pp. 308–313. IEEE (2003)
4. Lashari, T.A., Ibrahim, R.: A framework for medical image classification using soft set (2013)
5. Hosseini, M.S., Zekri, M.: Review of medical image classification using the adaptive neuro-fuzzy inference system. J. Med. Signals Sens. **2**(1), 49 (2012)
6. Moser, G., Serpico, S.B., Benediktsson, J.A.: Land-cover mapping by markov modeling of spatial contextual information in very-high-resolution remote sensing images. Proc. IEEE **101** (3), 631–651 (2012)
7. Peng, C., Li, Y., Jiao, L., Shang, R.: Efficient convolutional neural architecture search for remote sensing image scene classification. In: IEEE Transactions on Geoscience and Remote Sensing, pp. 1–14 (2020)
8. Huo, G., Wu, Z., Li, J.: Underwater object classification inside scan sonar images using deep transfer learning and semisynthetic training data. IEEE Access **8**, 47407–47447 418 (2020)
9. yao Wu, M., Cheng, S., Qin, L., Wu, G.: Ground object classification based on uav multispectral images. In 2019 Chinese Automation Cong. (CAC), pp. 4629–4634. IEEE (2019)

10. Ichim, L., Popescu, D.: Flooded areas evaluation from aerial images based on convolutional neural network. In: IGARSS 2019–2019 IEEE International Geoscience and Remote Sensing Symposium, pp. 9756–9759. IEEE (2019)
11. Wang, Z., Du, L., Zhang, P., Li, L., Wang, F., Xu, S., Su, H.: Visual attention based target detection and discrimination for high-resolution SAR images in complex scenes. IEEE Trans. Geosci. Remote Sens. **56**(4), 1855–1872 (2018)
12. Huang, D., Wang, Y., Song, W., Sequeira, J., Mavromatis, S.: Shallow-water image enhancement using relative global histogram stretching based on adaptive parameter acquisition. In: International Conference on Multimedia Modeling, pp. 453–465. Springer (2018)
13. Wu, D., Zhu, Q.: The latest research progress of image dehazing. Acta Automatica Sin. **41**(2), 221–239 (2015)
14. Narasimhan, S.G., Nayar, S.K.: Vision and the atmosphere. Int. J. Comput. Vision **48**(3), 233–254 (2002)
15. Van Herk, M.: A fast algorithm for local minimum and maximum filters on rectangular and octagonal kernels. Pattern Recogn. Lett. **13**(7), 517–521 (1992)
16. He, K., Sun, J., Tang, X.: Guided image _filtering. IEEE Trans. Pattern Anal. Mach. Intell. **35** (6), 1397–1409 (2013)
17. Ma, K., Liu, W., Wang, Z.: Perceptual evaluation of single image dehazing algorithms. In: 2015 IEEE International Conference on Image Processing (ICIP), pp. 3600–3604. IEEE (2015)
18. Arad, B., Ben-Shahar, O., Timofte, R., et al.: Ntire 2018 challenge on spectral reconstruction from rgb images. In 2018 IEEE/CVF Conference on Computer Vision and Pattern Recognition Workshops (CVPRW), pp. 1042–1042 09 (2018)
19. Li, B., Ren, W., Fu, D., Tao, D., Feng, D., Zeng, W., Wang, Z.: Benchmarking single-image dehazing and beyond. IEEE Trans. Image Process. **28**(1), 492–505 (2019)
20. Zhu, Q., Mai, J., Shao, L.: A fast single image haze removal algorithm using color attenuation prior. IEEE Trans. Image Process. **24**(11), 3522–3533 (2015)

Machine Learning Algorithms for Breast Cancer Detection and Prediction

Pawan Kumar, Ashutosh Bhatnagar, Roshan Jameel, and Ashish Kumar Mourya

Abstract Breast cancer has been recorded one of the most common disease among the females worldwide. It is the most common type of melanoma and responsible for the increasing mortality rate. The accurate and timely diagnosis of this fatal disease is particularly important to increase the chances of survival of the patients. There have been several implementations in this area, and numerous machine learning and soft computing algorithms have been proposed to analyze and detect the cancer and help the doctors to provide right medications on time. In this paper, a comparative analysis of prominent machine learning techniques is done, and their performance is evaluated. The techniques used for detecting and diagnosing the breast cancer are support vector machines (SVM), random forest, and k-nearest neighbor (k-NN). All experiments have been conducted with R programming software (data mining tool). The outcomes indicated that k-NN has achieved the highest accuracy (97.32%) in comparison with SVM, RF.

Keywords Breast cancer · k-nearest neighbor (k-NN) · Random forest · Support vector machines (SVM) · Random forest · Breast cancer wisconsin dataset (WDBC)

1 Introduction

Breast cancer is one of the predominant diseases worldwide and has caused the maximum number of cancer deaths in women. According to WHO, at least 2 million women suffer by this disease each year, and in year 2018, 15% of deaths

P. Kumar · A. Bhatnagar
Department of Computer Science, Pt. DDUMC, Meerut, India

R. Jameel
Department of Computer Science and Engineering, Jamia Hamdard, New Delhi, India

A. K. Mourya (✉)
Department of Computer Science and Engineering, Gautam Buddha University, Noida, India
e-mail: ashishkumarmourya@gmail.com

S. Das and M. N. Mohanty (eds.), *Advances in Intelligent Computing and Communication*, Lecture Notes in Networks and Systems 202,
https://doi.org/10.1007/978-981-16-0695-3_14

related to cancer in women are caused by breast cancer itself. In this context, Globocon reported that there were approximately 1.5 million new cases within the year 2018, and half of them died because of this disease in India [1]. It was ranked one in number of new cases of cancers reported by both male and females and number 1 in number of deaths caused by cancer. Therefore, accurate prediction of breast cancer is a puzzling task for the doctors, and advancements made in the technological area can help in developing new strategies for detecting it at early phases. However, the medical practitioners for diagnostics collect the data, but if the right tools will be amalgamated with the human intelligence, then the accuracy of the diagnostics would be facilitated. Several techniques are being advanced for early prediction and accurate diagnosis of this disease including machine learning algorithms [2].

Machine learning, which is sub-area of artificial intelligence, gives the machines the capability to improve with experience [3]. In the past few years, machine learning is developing and is being used for performing tasks accurately in almost every domain and transforming the entire face of IT industry. The factors on which the breast cancer depends can be multiple, including age, genetics, previous disease history, diet, etc. And to have a better treatment and chances of survival, early diagnosis is required. According to several studies in this area, it is suggested to analyze the various risk factors to find out the chances of developing cancer later in their lives [4]. If with certain risk factors, there were any chances that the person could have the disease, then the patient would be suggested to go for a screening test that would assure the presence or absence of the disease.

Several healthcare industries have started using machine learning techniques, to make the healthcare smarter and safer. The classification of breast cancer using machine learning is a three-step procedure: first to pre-process the data so that the missing attributes or any kind of ambiguity is removed, second step is to extract the desired and important features that contributes maximum in the accurate classification of the data, and third step is classification that tells whether the tumor in the data element is malignant or benign in nature. In this effort, the three most prominent learning techniques namely k-nearest neighbor (k-NN), RF, and support vector machines (SVM), are applied for classifying the tumor as benign or malignant, and their performance in measurement of accuracy is analyzed and assessed. The dataset used is breast cancer Wisconsin dataset (WDBC) [5], which is divided into the ratio of 80:20 as training and test datasets.

2 Related Works

The prediction of cancer requires the analysis of various generic features such as age, weight, smoking and drinking habits, family history, lifestyle, and numerous clinical factors such as type of tumor, size of tumor, pain induced or not, etc. Even with all this information available, it is still not easy for the doctors to accurately predict the cancer. Multiple studies have been conducted in predicting the breast melanoma, using learning algorithms, and yielded particularly good results. An SVM-based system was compared on two different breast cancer datasets, and the accuracy of 96.63 and 97.08 was achieved. Another comparative research focused on the chances of recurrence of the breast cancer, for which they used two breast cancer datasets and applied k-NN, logistic regression, naïve Bayes, and SVM using MATLAB platform, and the results were compared [6].

Artificial neural network (ANN) for classifying the breast cancer has been studied and applied in numerous works [7]. A comparison of four different types of ANNs, namely radial basis function, multilayer perceptron (MLP), probabilistic neural network (PNN), and self-organizing map (SOM), is performed in [8], on two datasets Namazi hospital breast cancer data (NHBCD) and WBC and found that the best performer among them is PNN. Another paper [9] combines the ANN with the association rule for reducing the number of features to only the best ones on Wisconsin breast cancer data (WDBC) dataset, and the performance after feature reduction was found to be more than without eliminating them.

For the forecast of breast cancer, another learning classifier called Bayesian is also extremely popular amongt the researchers. It has been compared to and applied in combination with various other techniques. In a work, it was implemented alone and in combination with ANN and executed on SEER dataset, and the results concluded that the hybrid approach with both the techniques performs better in terms of accuracy as well as speed that the standalones [10]. Moreover, fuzzy technique has also been applied in the context of predicting the breast cancer. In [11], fuzzy logic is applied in combination with evolutionary algorithms on WDBC dataset, while in [12], it was combined with decision tree called fuzzy decision tree to yield the best and accurate results in terms of recurrence of the disease on Surveillance, Epidemiology, and End Results (SEER) dataset.

Likewise, several researchers have also admired SVM for the purpose of classifying the datasets. A combination of SVM and rough set was proposed in [13], in which both of these algorithms works as a layer, the first layer being RS and the SVM being second, and the performance was evaluated on two datasets BRC1 and BRC2. Another work [14] has also used the same approach but on WDBC dataset. Apart from these works and approaches, several other techniques have also been applied in this area like decision tree [15], particle swarm optimization algorithm [16], mix integer programming [17], etc. After going through the literature related to the prediction of breast cancer using machine learning techniques, it was found that various algorithms work differently on different datasets. Also, the combination of more than one algorithm makes the system more effective and efficient.

3 Methodology

In this work, a comparison of three machine learning techniques, i.e., k-NN, SVM, and random forest is performed to predict the breast cancer as benign or malignant on the WDBC dataset.

3.1 Dataset

This research has used WDBC for analysis. This dataset has been pre-processed before used. Several ambiguities also have been removed from dataset. The features are extracted from the fine Needle Aspirate of breast masses in the form of images. The features found in the nuclei of the cell are shown in the images. And the result of the analysis of this data gives the type of tumor, i.e., benign or malignant. The dataset consists of thirty features out of which ten real valued features are given in Table 1 below.

3.2 Machine Learning Techniques

Machine learning, sub-branch of artificial intelligence which learns to recognize because of previous experience. The machine learning techniques can be classified into two classes, supervised and unsupervised. In supervised classification, the model is trained with the training dataset and is expected to learn from this training phase, so that when a new data comes, it recognizes and classifies it as desired. While in unsupervised learning, there is neither such predefined goal is given nor there is any training done. The machine is supposed to perform on its own. The machine learning has been successfully applied in various domains including health care, finance, speech recognition, etc. Supervised learning techniques are used in classification related problems. Following three prominent supervised machine learning classification techniques are applied in this work.

Table 1 Description of features of WDBC dataset

S. No.	Feature	Description
1	Radius	Mean of the distances between the points on perimeter and the center
2	Texture	Standard deviation of gray-scale values
3	Perimeter	Perimeter
4	Area	Area
5	Smoothness	Local variation in length of radius
6	Compactness	$(\text{Perimeter})^2/\text{area} - 1.0$
7	Concavity	Severity of concave areas of the contour
8	Concave points	Number of concave areas of the contour
9	Symmetry	Similarity with others
10	Fractal dimension	"Coastline approximation" – 1

3.2.1 k-Nearest Neighbor (k-NN)

It is one of the most used algorithms in the field of detection and prediction of breast cancer [18]. It does not need any kind of learning because it works based on instances. The technique works based on distance between the data point and the points belonging to several classes. After measuring the distance, the class that is found closely related to the data point is assigned to it. Firstly, a random value for 'k' is selected; then an input with N number of data elements and associated is given to the model. After that, a new sample arrives that is supposed to be classified, for which the Euclidean distance is calculated for finding the k-nearest neighbors, then the classes of these neighbors vote for the class of the new data element, and class is assigned accordingly. The Euclidean distance can be

$$\mathrm{d}(x,y) = \sqrt{\sum_{i=1}^{n}(x_r - y_r)^2} \quad (1)$$

Selecting the value of k is the most vital step in k-NN. It is not fixed; one may select the value of k as per the dataset. The value of k should neither be too small nor too large, because the stability and performance are totally dependent on this k.

3.2.2 Support Vector Machine (SVM)

SVM, a strong classifier, that has very vast area of applications and has also been applied successfully in the field of prediction of cancer [14]. It works the best when the number of features and instances is high and hence is suitable for various pattern recognition problems. In SVM, a hyperplane, i.e., separating region between the classes is selected to maximize the distance amongt them. The two core components of this approach are the kernel function and the regularization parameter that are to be defined before the training phase.

The main objective of SVM is found out the hyperplane with maximum margin. The implementation of SVM approach is dependent on the support vectors that are the data points that lie on the hyperplane boundary, because these are the points whose removal has no effect on the decision boundary or margin if they are removed unlike other data points. The hyperplane equation is $W^TX = 0$, where W and X are the vectors, W is perpendicular to the hyperplane, and W^TX gives the dot product of these vectors. The SVM model is created based on the training data elements, in which each item is allotted a categorical value. And the new data items are categorized based on the created hyperplane.

3.2.3 Random Forest

Random forest is a classification technique that works based on votes. In this technique, several decision trees ensemble to vote, and the class with maximum number of votes wins. It is more accurate and stable as compared to single decision tree, as more trees give better performance. The advantage of random forest is its insensitivity toward the noise and ambiguity of the data. It is a recursive approach that picks a random sample of dataset with replacement and one more dataset with no replacement. The acquired data is then divided, and the ambiguous data is removed from it, and this sequence of steps is repeated for n number of times [19].

where n is the number of decision trees in the forest. When a new data item arrives, it is classified by each of these trees, and the results generated by them play the role of the votes for every class. And finally, all the votes are counted, and the class with maximum number of votes is selected as the final class of the new data item.

4 Outcomes and Discussion

To assess these learning classifiers, UCI data repository has been chosen for experimental purposes. Table 1 confirms the outlines of predictor attributes. We apply the *k*-10 (cross-validation) trial, one of the most common method used in estimating analytical simulations that split up the original dataset into a training specimen to train the classifiers, and a test data to assess it. Later, applying the pre-processing technique and formulation methods, then try to analyze the data graphically and figure out the dissemination of values in terms of usefulness and effectiveness. Table 2 shows tuning parameters, area under ROC curve (AUC) and accuracy for various categorization techniques.

The proposed study has investigated the menace factors associated with foreseeing breast malignancy by using data mining technologies. Every approaches have its particular weaknesses and intensities, limited to the type of usage. The outcomes show that k-NN surpasses as compared to SVM and random forest with all aspects, such as AUC, Kappa, and accuracy. k-NN has maximum breast cancer relapse (Figs. 1, 2).

Table 2 Accuracy with tuning parameters

S. no	Classifier	Tuning parameter	Accuracy	Kappa	AUC
1	k-NN	K = 15	0.9732	0.9278	0.95
2	SVM (RBF)	C = 1	0.9632	0.9253	0.98
3	RF	Mtry = 2	0.9559	0.9101	0.97

Fig. 1 Combine accuracy of different models

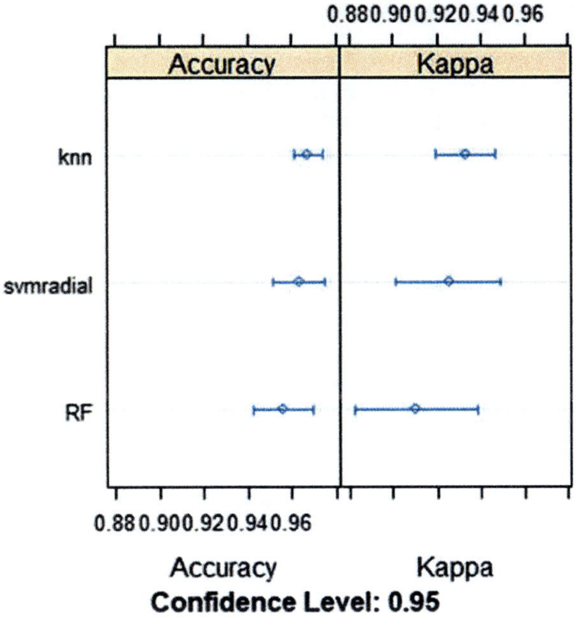

Fig. 2 k-NN accuracy

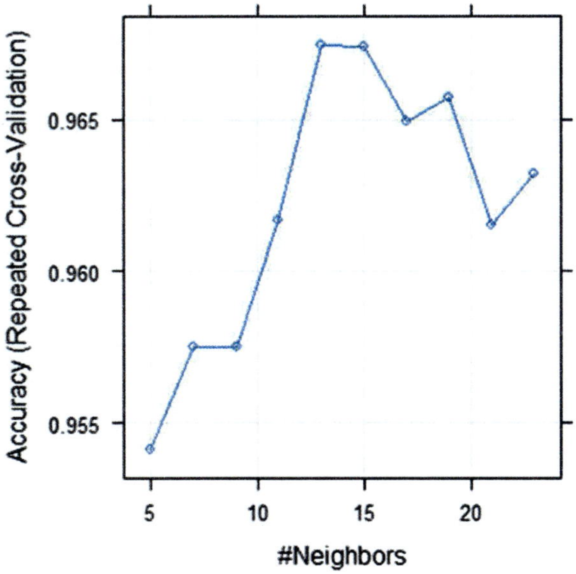

5 Conclusion

There are distinct learning classifiers which can be used for the estimating breast melanoma recurrence. In the proposed experiment, investigators analyzed breast melanoma disease using three methods for learning to anticipate the repetition of the melanoma and then contrasted the outcomes. The outcomes showed that the k-NN classifier represents the best learning performer on the test dataset, in comparison to rest classifiers. Additional research needs to be conducted to improve the efficiency and effectiveness of these learning methods by applying multiple attributes coupled with choice for an extended follow-up spell.

References

1. World Health Organization. World Health Statistics 2018: Monitoring Health for the SDGs Sustainable Development Goals. World Health Organization (2018)
2. Mao, N., Yin, P., Wang, Q., et al.: Added value of radiomics on mammography for breast cancer diagnosis: a feasibility study. J. Am. Coll. Radiol. **16**(4), 485–491 (2019)
3. Michie, E.D., Spiegelhalter, D.J., Taylor C.C.: Machine learning, neural and statistical classification. In: Proceeding (1994)
4. Han, J., Kamber, M.: Data mining: concepts and techniques, 3rd edn, pp. 5–6. Elsevier, USA (2012)
5. William, H., Wolberg, W., Street, N., Mangasarian, O.L.: Breast cancer Wisconsin (diagnostic) data set. UCI Machine Learning Repository. http://archive.ics.uci.edu/ml/ (1992)
6. Chaurasia, V., Tiwari, B.B., Pal, S.: Prediction of benign and malignant breast cancer using data mining techniques. J. Algorithms Comput. Technol.
7. Wang H, Yoon, S.W.: Breast cancer prediction using data mining method. In: Proceedings of the 2015 Industrial and Systems Engineering Research Conference
8. Sarvestan, A.S., Safavi, A.A., Parandeh, M.N., Salehi, M.: Predicting breast cancer survivability using data mining techniques. In: Software Technology and Engineering (ICSTE), 2nd International Conference, vol. 2, pp. 227–231 (2010)
9. Karabatak, M., Ince, M.C.: An expert system for detection of breast cancer based on association rules and neural network. Expert Syst. Appl. **36**(2), 3465–3469 (2009)
10. Delen, D., Walker, G., Kadam, A.: Predicting breast cancer survivability: a comparison of three data mining methods. Artif. Intell. Med. **34**(2), 113–127 (2005)
11. Pena-Reyes, C.A., Sipper, M.: A fuzzy-genetic approach to breast cancer diagnosis. Artif. Intell. Med. **17**(2), 131–155 (1999)
12. Khan, M.U., Choi, J.P., Shin, H., Kim, M.: Predicting breast cancer survivability using fuzzy decision trees for personalized healthcare. In: Proceedings of IEEE Engineering in Medicine and Biology Society, pp. 48–51 (2008)
13. Mourya, A.K.: Performance and evaluation of support vector machine and artificial neural network over heterogeneous data. In: Santosh, K., Hegadi, R. (eds.) Recent Trends in Image Processing and Pattern Recognition. RTIP2R 2018. Communications in Computer and Information Science, vol. 1035. Springer, Singapore. https://doi.org/10.1007/978-981-13-9181-1_51 (2019)
14. Chen, H.L., Yang, B., Liu, J., Liu, D.Y.: A support vector machine classifier with rough setbased feature selection for breast cancer diagnosis. Expert Syst. Appl. **38**(7), 9014–9022 (2011)

15. Lavanya, D.: Ensemble decision tree classifier for breast cancer data. Int. J. Inf. Technol. Convergence Serv. **2**(1), 17–24 (2012)
16. Gandhi Rajiv, K., Marcus, K., Kannan, S.: Classification rule construction using particle swarm optimization algorithm for breast cancer datasets. In: Signal Acquisition and Processing. ICSAP, International Conference, pp. 233–237 (2010)
17. Masmoudi, Y., Chabchoub, H., Hanafi, S., Rebaï, A.: A mathematical programming based procedure for breast cancer classification. J. Math. Model. Algorithms **9**(3), 247–255 (2010)
18. Pawlovsky, A.P., Nagahashi, M.: A method to select a good setting for the k-NN algorithm when using it for breast cancer prognosis. In: Proceedings of IEEE-EMBS International Conference on Biomedical and Health Informatics (BHI), Valencia (2014)
19. Speiser, J.L., Miller, M.E., Tooze, J., Ip, E.: A comparison of random forest variable selection methods for classification prediction modeling. Expert Syst. Appl. **134**, 93–101 (2019)

A Differential Squirrel Search Algorithm

Bibekananda Jena, Manoj Kumar Naik, Aneesh Wunnava, and Rutuparna Panda

Abstract This paper presents a new hybrid differential squirrel search algorithm optimization algorithm (DSSE) by combining the searching methods of a squirrel search algorithm and differential evolution optimization process for solving the global optimization problem. The main thrust is to include the crossover mechanism in differential evolution algorithm, to improve its exploration capability and modification of the updating rule of squirrel during foraging to improve its exploitation skill. To evaluate its performance, the proposed algorithm is benchmarked on a set of test functions defined in CEC2005. The results show that the proposed algorithm outperforms a well-known optimization algorithm including the squirrel search algorithm.

Keywords Nature-inspired algorithm · Benchmark function · Squirrel search optimization · Differential evolution

B. Jena
Dept. of Electronics and Communication Engineering, Anil Neerukonda Institute of Technology & Science, Sangivalasa, Visakhapatnam 531162, Andhra Pradesh, India
e-mail: bibekananda.jena@gmail.com

M. K. Naik (✉) · A. Wunnava
Faculty of Engineering and Technology, Siksha O Anusandhan, Bhubaneswar 751030, Odisha, India
e-mail: manojnaik@soa.ac.in

A. Wunnava
e-mail: aneeshwunnava@soa.ac.in

R. Panda
Dept. of Electronics and Communication Engineering, Veer Surendra Sai University of Technology, Burla 768018, Odisha, India
e-mail: r_ppanda@yahoo.co.in

S. Das and M. N. Mohanty (eds.), *Advances in Intelligent Computing and Communication*, Lecture Notes in Networks and Systems 202,
https://doi.org/10.1007/978-981-16-0695-3_15

1 Introduction

Optimization algorithms are generally used for purpose of obtaining the best solution to a problem under a given condition. In the past few years, several nature-inspired optimization algorithms have been proposed by many researchers with better performance than traditional optimization algorithms in solving a modern engineering problem. Three main categories of nature-inspired optimization algorithms are swarm intelligence-based algorithm (SI), evolutionary algorithms (EA), and physics-based (PB) algorithms. Optimization algorithms inspired by nature follow a mathematical model developed by imitating the biological behaviours of the creatures available in nature or the basics laws of physics that exist in the real world. Some of the standards swarm intelligence-based algorithms like particle swarm optimization (PSO) [1], ant colony optimization (ACO) [2], artificial bee colony optimization (ABC) [3], bacterial foraging optimization (BFO) [4], and grey wolf optimization (GWO) [5] follow the principle of the social behaviour of the animals during foraging to produce optimal results. Differential algorithm (DE) [6] and genetic algorithms [GA) [7] are two standard evolutionary-based algorithms follow the evolutionary behaviour of different types of the creature for solving optimization problems. Some of the well-known physics law-based algorithms are gravitational search algorithm (GSA) [8], charge system search (CSS) algorithm [9], and wind driven optimization (WDO) [10]. Though these algorithms are applied successfully in many engineering problems for getting an optimal solution, they are having the major drawback of a slow convergence rate.

Squirrel search algorithm (SSA) [11] is one of the recently proposed swarm intelligence-based algorithms that mimic the foraging behaviour of the flying squirrel found in the forest. Implementation of dynamic foraging behaviour for getting the optimal food source and season monitoring condition of squirrel in the algorithms makes it more powerful than other standard optimization algorithms. However, it has been observed that SSA struggles to get the desired solutions when tested with complex problems and sometimes trapped in the local minima.

This paper proposed a differential squirrel search algorithm (DSSA) by hybridizing the differential evolution algorithm with the squirrel search algorithm. The crossover property of the differential algorithm boosts the exploration ability of the squirrel during foraging which reduces the chance of getting trapped in local minima. Further, there is a modification to the updating rule of squirrel position in each iteration which makes the squirrel reach the optimal food source in less time before the arrival of the winter. The rest part of the paper is structured as follows: The details about the proposed method are presented in Sect. 2. Section 3 gives the experimental results and comparative studies with a well-known algorithm. Finally, the concluding remark is presented in Sect. 4.

2 Proposed Differential Squirrel Search Algorithm (DSSA)

This section presents a hybrid optimization, DSSA by combining the differential evolution algorithm (DE) with squirrel search algorithm (SSA). In SSA, the squirrels update their position by following the position of the squirrels either in the hickory tree or in the acorn tree. Therefore, the updating rules of these leading squirrels have been modified to improve its searching strategy. Exploration ability is further improved by including crossover operation inspired by DE.

The different foraging strategy included in DSSA is modelled mathematically and presented below.

2.1 Position Initialization and Fitness Evaluation

Initially, the positions of the squirrels are randomly paced in the search space. Once the positions of the squirrels are known, its fitness which resembles the quality of food source found by a squirrel in that position is evaluated substituting the position in the fitness function. Fitness values are then sorted to obtain the current best squirrel $\mathrm{PS}_{\mathrm{ht}}$ present in the hickory tree. The next three best function values represent the positions of the squirrels which are in the acorn tree $\mathrm{PS}_{\mathrm{at}}(1:3)$ and are assumed to be a move towards the optimal position in a subsequent iteration. The rest of the squirrels $\mathrm{PS}_{\mathrm{nt}}(1:\mathrm{NP}-4)$ have not yet found any food source, and their positions are in the normal tree.

2.2 Position Update

In the absence of a predator, the squirrels in the acorn tree update its position and move towards the optimal source by following the current best, $\mathrm{PS}_{\mathrm{ht}}$. The squirrels in the normal tree update its position by following the squirrels in the acorn tree or hickory tree. The squirrels change their path randomly during foraging in the presence of a predator. The mathematical models used for updating the position of the squirrel are as follows.

Squirrels in the acorn trees now update its position based on the positions of others which is expressed in Eq. (3):

$$\mathrm{PS}_{\mathrm{at}}^{\mathrm{new}} = \begin{cases} \mathrm{PS}_{\mathrm{at}}^{\mathrm{old}} + \mathrm{d}_g.G_c\left(\mathrm{PS}_{\mathrm{ht}}^{\mathrm{old}} - \mathrm{PS}_{\mathrm{at}}^{\mathrm{old}} - P_{\mathrm{avg}}\right), & r_1 \geq P_{\mathrm{dp}} \\ \text{random position}, & \text{otherwise}, \end{cases} \tag{1}$$

where P_{avg} is the average of all squirrels position in the current population.

Further, to avoid the chances of getting trapped in local minima, the crossover mechanism of DE is also used to increase the diversity of the squirrels. Crossover operation applied to a current position of the squirrel and its respective new position obtained from Eq. (3),

$$\mathrm{PS}^{\mathrm{cr}}_{\mathrm{at},i,j} = \begin{cases} \mathrm{PS}^{\mathrm{new}}_{\mathrm{at},i,j}, & \text{if } (\mathrm{rand}_j \leq \mathrm{Cr}) \text{ or } j = j_{\mathrm{rand}} \\ \mathrm{PS}^{\mathrm{old}}_{\mathrm{at},i,j}, & \text{if } (\mathrm{rand}_j > \mathrm{Cr}) \text{ or } j \neq j_{\mathrm{rand}} \end{cases}, \quad j = 1, 2, 3, \ldots\ldots D \tag{2}$$

where $i = 1, 2, 3, \ldots\ldots \mathrm{NP}$ and NP is the population size. $\mathrm{PS}^{\mathrm{cr}}_{\mathrm{at},i,j}$ is the positions of the squirrels after crossover operation, $\mathrm{PS}^{\mathrm{new}}_{\mathrm{at},i,j}$ and $\mathrm{PS}^{\mathrm{old}}_{\mathrm{at},i,j}$ are new and old positions of the squirrels present in acorn or normal trees, D is the dimension of the problem, and Cr is the crossover rate taken as 0.5. $j_{\mathrm{rand}} \in [1, D]$ is a randomly generated index, and $\mathrm{rand}_j \in [0, 1]$ represents the jth evaluation of random numbers uniformly generated in the given range.

Some of the squirrels present in the normal trees move to the new position by following the positions of the squirrel in acorn trees as

$$\mathrm{PS}^{\mathrm{new}}_{\mathrm{nt}} = \begin{cases} \mathrm{PS}^{\mathrm{old}}_{\mathrm{nt}} + \mathrm{d}_g.G_c\left(\mathrm{PS}^{\mathrm{old}}_{\mathrm{at}} - \mathrm{PS}^{\mathrm{old}}_{\mathrm{nt}}\right), & r_2 \geq P_{\mathrm{dp}} \\ \text{random position}, & \text{otherwise}, \end{cases} \tag{3}$$

where r_2 is a random number uniformly distributed in the range [0, 1]

The rest of the squirrels in normal trees follow the current best position, and their new locations are expressed as:

$$\mathrm{PS}^{\mathrm{new}}_{\mathrm{nt}} = \begin{cases} \mathrm{PS}^{\mathrm{old}}_{\mathrm{nt}} + \mathrm{d}_g.G_c\left(\mathrm{PS}^{\mathrm{old}}_{\mathrm{ht}} - \mathrm{PS}^{\mathrm{old}}_{\mathrm{nt}}\right), & r_3 \geq P_{\mathrm{dp}} \\ \text{random position}, & \text{otherwise}, \end{cases} \tag{4}$$

Crossover operation is also included for the squirrels in normal trees as follows:

$$\mathrm{PS}^{\mathrm{cr}}_{\mathrm{nt},i,j} = \begin{cases} \mathrm{PS}^{\mathrm{new}}_{\mathrm{nt},i,j}, & \text{if } (\mathrm{rand}_j \leq \mathrm{Cr}) \text{ or } j = j_{\mathrm{rand}} \\ \mathrm{PS}^{\mathrm{old}}_{\mathrm{nt},i,j}, & \text{if } (\mathrm{rand}_j > \mathrm{Cr}) \text{ or } j \neq j_{\mathrm{rand}} \end{cases}, \quad j = 1, 2, 3, \ldots\ldots D \tag{5}$$

The convergence speed is further increased by allowing the squirrel in the hickory tree to update its position concerning the average of the positions of the squirrels in the hickory tree as follows:

$$\mathrm{PS}^{\mathrm{new}}_{\mathrm{ht}} = \mathrm{PS}^{\mathrm{old}}_{\mathrm{ht}} + \mathrm{d}_g.G_c\left(\mathrm{PS}^{\mathrm{old}}_{\mathrm{ht}} - \mathrm{PS}^{\mathrm{avg}}_{\mathrm{at}}\right) \tag{6}$$

where $\mathrm{PS}^{\mathrm{avg}}_{\mathrm{at}}$ is the mean of all squirrel positions present in the acorn trees.

The best of the new position and its crossover positions are then compared with the old positions in the selection process for taking part in the next-generation population. The details of the steps involved in DSSA are presented in algorithm 1

Algorithm 1: Pseudocode of DSSA

Input $Itermax$, NP, Pdp, sf, Gc, ub, and lb

Initialize the flyingsquirrels' position randomly using Eq (1)

Calculate the fitness value using the given fitness function using Eq (2)

While $itr \leq itrmx$ do

  Sort the fitness function of all squirrels and determine the current best , PS_{ht}, positions of squirrel in acorn tree , PS_{at} (1: 3)and position of squirrel in normal tree , PS_{nt} (1: $NP - 4$)

  Generate the new location of squirrels in acorn trees by comparing the new positions obtained by using Eq (1) and Eq (2)

  Generate the new location of squirrels in normal trees by comparing the new positions obtained by using Eq (3), Eq (4), and Eq (5)

  Generate the new location of squirrels in hickory trees by comparing the new positions obtained by using Eq (6) and old position.

  Update the population with the best position obtained so far

End While

Return PS_{ht}

---

3 Results and Discussion

The performance of the proposed DSSA is compared with five standard and well-known optimization algorithms, including basic SSA, GWO, WDO, PSO, and DE. All these algorithms are tested with 23 benchmark functions [12]. Each algorithm is independently run 31 times, and the average and standard values obtained for each test function are used for comparison. The maximum number of iterations is set to 500 for all algorithms. The controlling parameters of the algorithms are chosen as those used in the original literature. The population size NP of for all the algorithms also fixed to 30. Table 1 shows the statical results obtained from scalable unimodal function (F1-F7) with the Friedman rank test result. It can be observed from the table that the proposed DSSA algorithm dominates the standard optimization algorithm with a high margin, whereas gives comparable result for F7 with SSA. In Friedman's average rank test, the proposed DSSA also achieve rank one. Statistical results for scalable multimodal benchmark function (F8-F13) are presented in Table 2. The result shows that the proposed DSSA performs significantly better than other algorithms used for comparison and achieve rank one in the Friedman test. Table 3 presents the statistical results obtained from the non-scalable test function (F14-F23). From the results, the proposed algorithms lag DE in most of the cases and achieved rank two on the Friedman test. It can also be observed from all tests that proposed DSSA struggles to reach an optimal solution for a low-dimensional problem, whereas performs better for the

Table 1 Statistical results obtained from unimodal benchmark function F1-F7

Function	Metric	DSSA	SSA	GWO	WDO	PSO	DE
F1	Ave	0	6.61E-25	3.04E-08	1.08E-01	5.45E-02	3.49E+00
	Std	0	2.50E-24	4.82E-08	1.08E-01	9.40E-03	1.07E+01
F2	Ave	8.41E-179	2.65E-14	2.01E-05	6.43E-01	1.08E+00	8.91E-02
	Std	0	1.36E-13	1.47E-05	8.61E-01	1.88E-01	3.10E-01
F3	Ave	1.11E-218	3.69E-13	4.20E-01	2.82E-01	5.17E-01	6.70E+02
	Std	0	2.05E-12	5.28E-01	1.91E-01	1.45E-01	3.23E+02
F4	Ave	4.86E-170	4.80E-15	1.57E-01	5.08E-02	1.45E-01	2.56E+01
	Std	0	2.49E-14	1.14E-01	8.69E-02	2.64E-02	6.89E+00
F5	Ave	0	1.40E-29	2.88E+01	3.57E+01	3.36E+01	1.71E+04
	Std	0	4.34E-29	2.56E-01	9.33E+00	1.58E+00	7.59E+04
F6	Ave	0	3.36E-19	3.49E+00	2.33E-01	5.97E-02	2.66E+01
	Std	0	1.87E-18	5.18E-01	1.10E-01	1.36E-02	6.59E+01
F7	Ave	2.52E-04	1.72E-04	5.43E-03	1.82E-01	1.22E-01	6.27E-02
	Std	2.80E-04	1.61E-04	3.11E-03	8.54E-02	3.72E-02	1.82E-02
Friedman's mean rank		1.143	1.857	3.714	4.429	4.429	5.429
Rank		1	2	3	4	4	5

Table 2 Statistical results obtained from multimodal benchmark function F8-F13

Function	Metric	DSSA	SSA	GWO	WDO	PSO	DE
F8	Ave	−1.49E+69	−3.19E+61	−4.02E+01	−9.28E+01	−1.54E+02	−7.11E+03
	Std	5.20E+69	5.71E+61	6.72E+00	1.11E+00	3.70E+01	1.14E+03
F9	Ave	0	3.43E+00	2.79E+01	6.07E+01	1.90E+01	1.65E+02
	Std	0	9.29E+00	2.24E+01	3.25E+01	4.89E+00	2.54E+01
F10	Ave	8.88E-16	8.57E-15	6.69E-05	1.90E-01	2.96E-01	2.51E+00
	Std	0	4.21E-14	6.95E-05	2.81E-01	9.79E-02	3.37E+00
F11	Ave	0	0	1.29E-09	2.52E-03	3.18E-03	1.69E-01
	Std	0	0	2.56E-09	4.05E-03	1.02E-03	1.51E-01
F12	Ave	1.57E-32	5.47E-32	2.64E-01	3.59E-04	8.38E-04	1.19E+04
	Std	5.56E-48	2.17E-31	6.23E-02	1.83E-04	1.40E-04	6.50E+04
F13	Ave	1.35E-32	1.29E-17	2.16E+00	8.30E-03	1.77E-02	4.51E+04
	Std	2.78E-48	7.18E-17	2.57E-01	1.40E-02	7.40E-03	1.81E+05
Friedman's mean rank		1.08	1.92	4.33	4.00	4.17	5.50
Rank		1	2	5	3	4	6

Table 3 Statistical results obtained from fixed dimensional multimodal benchmark function F14-F23

Function	Metric	DSSA	SSA	GWO	WDO	PSO	DE
F14	Ave	0.998	1.015	12.671	12.671	12.671	1.377
	Std	1.67E-16	6.63E-02	1.57E-10	1.41E-11	1.27E-12	1.78E+00
F15	Ave	5.70E-04	1.67E-03	6.81E-04	4.48E-04	4.16E-04	1.37E-03
	Std	1.23E-04	7.98E-19	4.49E-04	1.33E-04	1.43E-04	3.56E-03
F16	Ave	−1.028	−1.006	−1.026	−1.031	−1.032	−1.032
	Std	3.58E-03	2.04E-02	1.16E-02	3.18E-04	2.52E-06	6.71E-16
F17	Ave	0.402	0.418	0.759	0.398	0.548	0.398
	Std	3.35E-03	1.85E-02	1.19E+00	2.32E-04	8.34E-01	0.00E+00
F18	Ave	4.998	6.420	5.628	3.019	3.000	3.000
	Std	1.36E+00	3.91E+00	8.11E+00	1.86E-02	5.67E-05	2.33E-15
F19	Ave	−3.834	−3.847	−3.758	−3.855	−3.857	−3.863
	Std	2.20E-02	1.00E-02	5.36E-01	6.29E-03	3.62E-03	2.71E-15
F20	Ave	−3.022	−2.832	−2.493	−3.089	−3.034	−3.245
	Std	1.29E-01	1.13E-01	9.19E-01	2.04E-01	3.64E-01	5.78E-02
F21	Ave	−9.061	−9.012	−4.784	−5.031	−5.055	−9.990
	Std	2.07E+00	2.15E+00	1.04E+00	1.88E-02	1.83E-04	9.07E-01
F22	Ave	−10.045	−9.551	−4.681	−5.066	−5.087	−9.972
	Std	1.32E+00	1.97E+00	1.25E+00	1.66E-02	3.25E-04	1.67E+00
F23	Ave	−9.984	−8.980	−4.750	−5.101	−5.128	−9.884
	Std	1.49E+00	2.47E+00	1.17E+00	1.97E-02	2.29E-04	2.06E+00
Friedman’s mean rank		2.800	4.200	5.600	3.500	3.100	1.800
Rank		2	5	6	4	3	1

Table 4 Overall Friedman's mean rank

	DSSA	SSA	GWO	WDO	PSO	DE
Friedman's mean rank	1.85	2.89	4.70	3.91	3.78	3.87
Rank	1	2	6	5	3	4

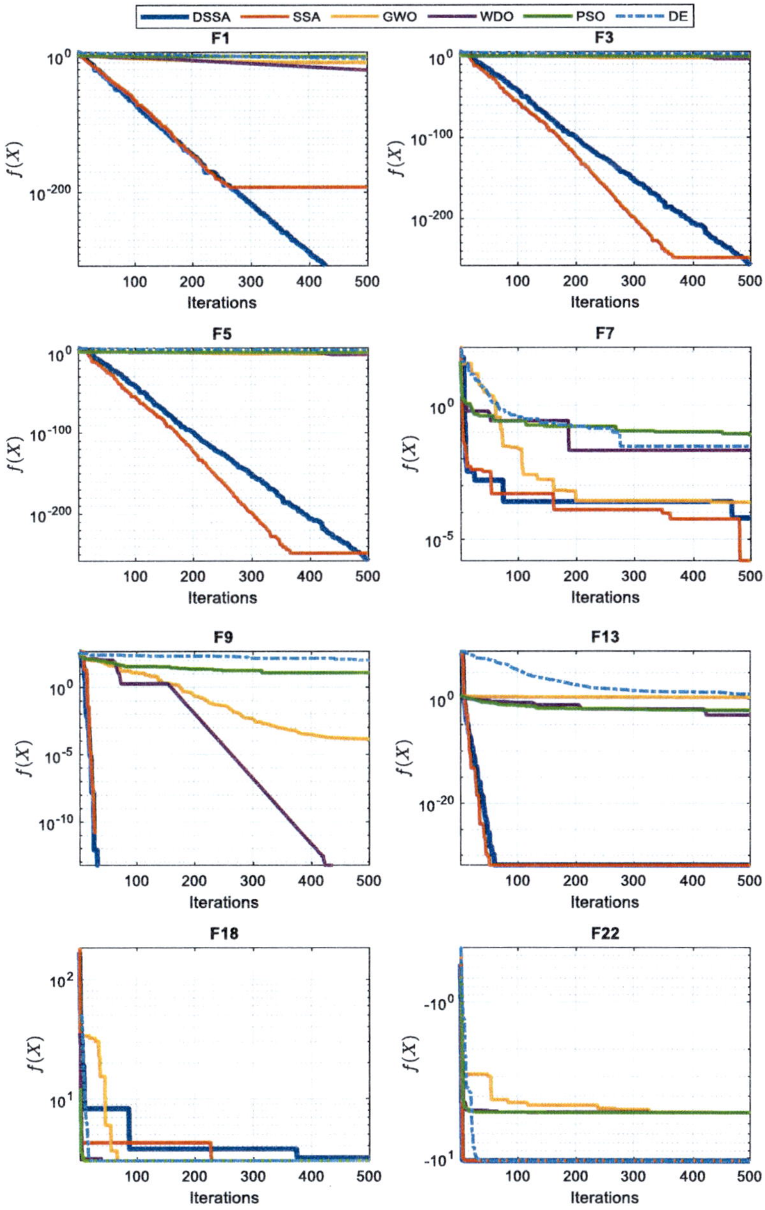

Fig. 1 Convergence plot of various optimization algorithms

higher-dimensional problem. The overall Friedman test shown in Table 4 proclaimed DSSA as superior to other algorithms in this study. The convergence plot of various optimization algorithms for the eight different test functions is presented in Fig. 1. It is clear from the figure that the DSSA converges faster than other standard algorithms used for comparison.

4 Conclusion

A new hybrid DSSA is proposed in this paper and is validated by testing with a set of benchmark functions. Statistical results obtained from various test functions show that the proposed DSSA outperformed SSA and other standard optimization algorithms. The DSSA may be used for solving the various optimization problem in the field of science and engineering.

References

1. Eberhart, R., Kennedy, J.: A new optimizer using particle swarm theory. In: Proceedings of the Sixth International Symposium on Micro Machine and Human Science, pp. 39–43. IEEE, (1995)
2. Dorigo, M., Birattari, M., Stutzle, T.: Ant colony optimization. IEEE Comput. Intell. Mag. **1** (4), 28–39 (2006)
3. Basturk, B., Karaboga, D.: An artificial bee colony (ABC) algorithm for numeric function optimization. In: IEEE Swarm Intelligence Symposium, vol. 8, pp. 687–697 (2006)
4. Das, S., Biswas, A., Dasgupta, S., Abraham, A.: Bacterial foraging optimization algorithm: theoretical foundations, analysis, and applications. In: Abraham, A., Hassanien, A.E., Siarry, P., Engelbrecht, A. (eds.) Foundations of Computational Intelligence Volume 3. Studies in Computational Intelligence, vol. 203. Springer, Berlin, Heidelberg (2009)
5. Mirjalili, S., Mirjalili, S.M., Lewis, A.: Grey wolf optimizer. Adv. Eng. Softw. **69**, 46–61 (2014)
6. Das, S., Suganthan, P.N.: Differential evolution: a survey of the state-of-the-art. IEEE Trans. Evol. Comput. **15**(1), 4–31 (2011)
7. Holland, J.H.: Adaptation in Natural and Artificial Systems. University of Michigan Press, Ann Arbor, MI (1975)
8. Rashedi, E., Nezamabadi-Pour, H., Saryazdi, S.: GSA: a gravitational search algorithm. Inf. Sci. **179**(13), 2232–2248 (2009)
9. Kaveh, A., Talatahari, S.: A novel heuristic optimization method: charged system search. Acta Mech. **213**(3), 267–289 (2010)
10. Bayraktar, Z., Komurcu, M., Werner, D.H.: Wind driven optimization (WDO): a novel nature-inspired optimization algorithm and its application to electromagnetics. In: 2010 IEEE Antennas and Propagation Society International Symposium, Toronto, ON (2010)
11. Jain, M., Singh, V., Rani, A.: A novel nature-inspired algorithm for optimization: squirrelsearch algorithm. Swarm. Evol. Comput. (2018)
12. Wunnava, A., Kumar Naik, M., Panda, R., Jena, B., Abraham.: An adaptive Harris hawks optimization technique for two dimensional grey gradient based multilevel image thresholding. Appl. Soft Comput. J. **95** (2020)

Dynamic Stability Improvement of Power System Utilizing Fuzzy Logic and Compare It with Conventional Stabilizer

Mohammad Istanbuly, M-Ramez Halloum, and Basanta K. Panigrahi

Abstract Control of excitation system is one of the methods used to dynamic stability for synchronous generator; for that, power system stabilizer (PSS) is used, but fuzzy logic controller (FIC) can deal better with nonlinear and complex systems; in this paper, we will design improved fuzzy stabilizer using deviation of speed-deviation and acceleration as input variables; we will use 49 if-then rules written using triangular shape membership functions of output and inputs, and we will apply the stabilizer on synchronous machine together with HTG and excitation system. Also, we will compare between fuzzy-based PSS and conventional one and analysis of the results. The simulation is achieved by MATLAB/Simulink program.

Keywords Stability · Power system stabilizer · Single synchronous generator · Simulink · Fuzzy logic

1 Introduction

The stability of power system allows the system to last in an equilibrium state during operating conditions. For damping the low-frequency oscillation, the power system stabilizer (PSS) is utilized to generate a control signal. Generally, in the existing power systems, there is a wide usage of conventional PSS which has made an enhancement of the power systems dynamic stability [1]. The conventional PSS parameters depend on making the power system model linear at a normal operating

M. Istanbuly · B. K. Panigrahi (✉)
Department of Electrical Engineering, ITER, S'O'A University, Bhubaneswar, India
e-mail: basanta1983@gmail.com

M. Istanbuly
e-mail: mohammadistanbuly@gmail.com

M.-R. Halloum
Department of Electrical Engineering, C. V. Raman College of Engineering, Bhubaneswar, India
e-mail: mramezhalloum@gmail.com

S. Das and M. N. Mohanty (eds.), *Advances in Intelligent Computing and Communication*, Lecture Notes in Networks and Systems 202,
https://doi.org/10.1007/978-981-16-0695-3_16

point. Generally, the nonlinearity in power systems is high, so the conventional stabilizer design of power system based on linear model of the system cannot guarantee good performance in real environment [2].

The conventional stabilizer, which uses lead–lag compensation, gives weak performance under different loading conditions [3]. Metwally and Malik [3] used the speed of generator and output power variation as input variables of the controller to describe fuzzy PSS. Hiyama [4] gets the deviation of speed, acceleration and the deviation of generator phase by measuring the real signal of power. Lau et al. in [5] modify the scheme of single-input FLC in [6]. They make a merge of the error E, one-level error difference E and the two-level error difference. This attitude gives a nonlinear scheme of PID-type fuzzy control. Lim and Elangovan [7] used a new method for designing decentralized stabilizers which determines the parameters of all stabilizers in the system, such that some or all of its mechanical mode eigenvalues in the complex plane have desired locations. Toliyat et al. [8] enhance power system stability by designing augmented fuzzy PSS.

Our paper submits the results of modeling a single synchronous machine infinite bus system and the design of fuzzy logic power system regulator using triangular shape membership function. In addition, we have used input variables for the deviation of speed (Δw) and the acceleration; after that, we study and compare between fuzzy-based PSS and conventional one.

2 Design of Fuzzy Logic Controller

The function of fuzzy logic controller (FIC) is receiving the crisp input values and converts it to fuzzy value performed by the use of the information in the knowledge base [2]. The FIC is two-input and single-output component. It is known as multiple inputs, single-output system (MISO) [9]. The following figure illustrates the principle of FIC (Fig. 1).

The steps of FIC design are along these lines:

1. Choose the outputs Y and inputs X.
2. Define the sets; universal and fuzzy of X and Y.

Fig. 1 Principle of FIC

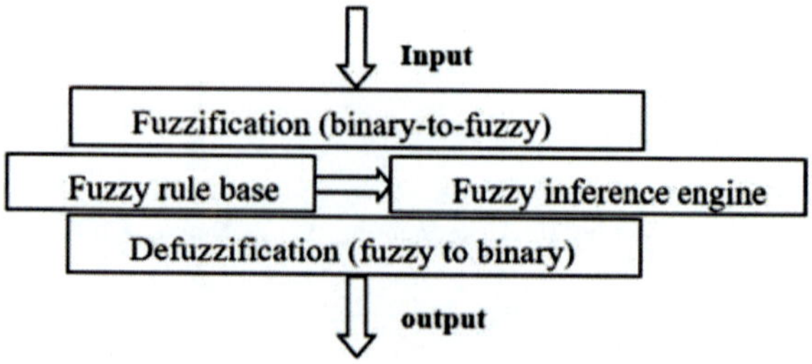

3. Define the variables using linguistic way and their correlated membership functions.
4. Choose mathematics relationships to calculate the control action [3].

In this paper, we will design two inputs to the fuzzy controller:

1. Error value which is the difference between system input and its output, which means the difference between output electrical power and mechanical power, and it is expressed physically by the term acceleration power.
2. Change in error value which is expressed physically by the term speed-deviation and is converted to linguistic variable.

The shape of linguistic groups for every input and output is chosen to achieve the goal needed from the control system, and in this paper, we choose the triangular shape. The number of linguistic groups for each input and output depends on the nature of the system and the time needed for calculations [4]. We consider seven linguistics groups for each input and output, and these groups represented by linguistic variables connected with each other by a number of rules equal to 7 * 7 = 49. The rules which describe the control action are if-then and are identified in this paper generally as follows: If the error (represented by a linguistic group) and the change in error (represented by a linguistic group) so the output will be (represented by a linguistic group). These rules may be represented in a matrix form between input linguistic group and output linguistic group [5], and it is called decision panel as follows (Table 1).

Fuzzy inference system editor:

The Fuzzy Logic Designer displays and opens a diagram of the fuzzy inference system (FIS)—on the left, the names of every input variable and on the right, those of every output variable, as illustrated in Fig. 2a. In this paper, we have speed-deviation and acceleration power as inputs and stabilizing voltage as output.

The Membership Function Editor shows several features to the Fuzzy Logic Designer, as illustrated in Fig. 2b, c and d; the membership function is triangular for the first input which is speed-deviation:

Table 1 Decision panel

	Acceleration							
Speed-deviation		**NB**	**NM**	**NS**	**Z**	**PS**	**PM**	**PB**
	NB	NB	NB	NB	NB	NM	NS	Z
	NM	NB	NB	NB	NM	NS	Z	PS
	NS	NB	NB	NM	NS	Z	PS	PM
	Z	NB	NM	NS	Z	PS	PM	PB
	PS	NM	NS	Z	PS	PM	PB	PB
	PM	NS	Z	PS	PM	PB	PB	PB
	PB	Z	PS	PM	PB	PB	PB	PB

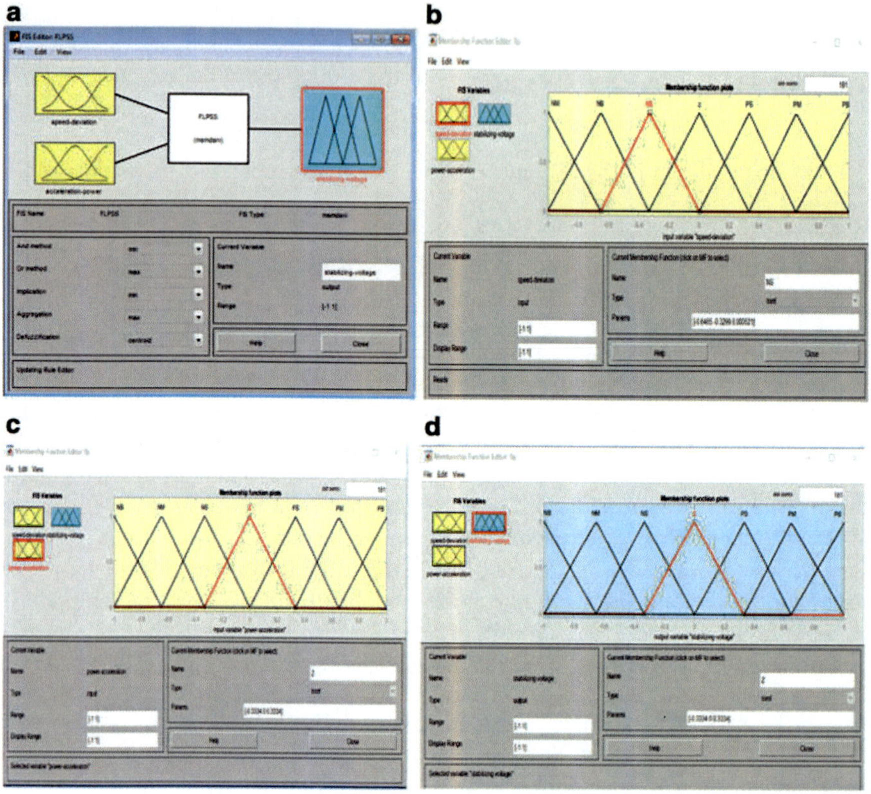

Fig. 2 **a** Fuzzy inference system editor, **b** membership function for the first input, **c** membership function for the second input which is acceleration power, **d** membership function for output

Rule Editor:

We utilize Rule Editor to modify or view the rules of our fuzzy logic system. In order to define rules, we have to specify the output and input variables of our FIS and their equivalent membership functions. In this paper, we use 49 fuzzy rules.

We use the Surface Viewer to show how one of the outputs depends on any one or more of the inputs. This tool plots and generates an output superficial map of the system. Figure 3 is the relationship between input and output which mean relationship between acceleration power, speed-deviation and stabilizing voltage. The membership function is triangular also for the second input which is acceleration power and for output which is stabilizing-voltage.

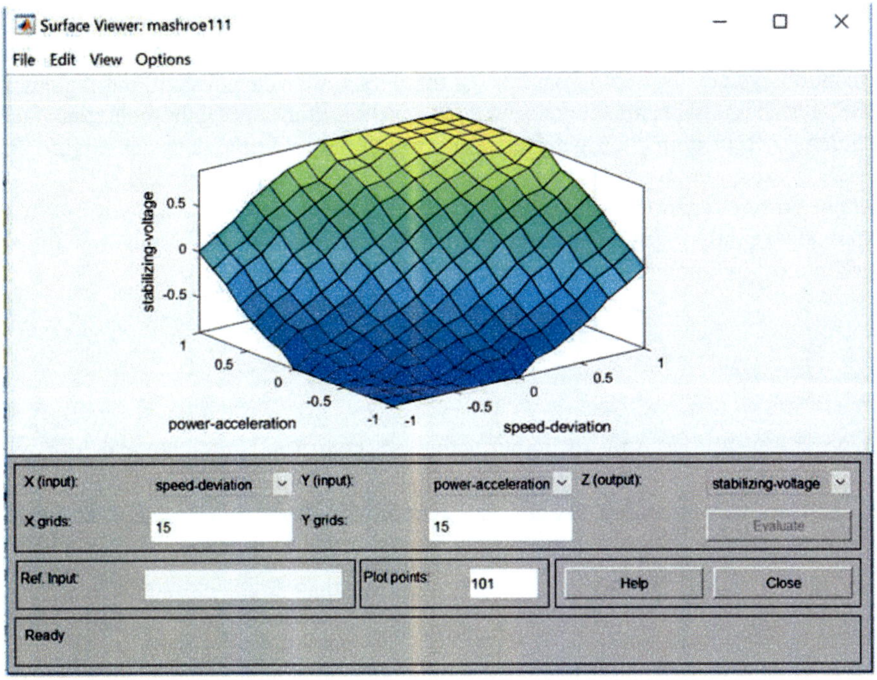

Fig. 3 Diagram of the relationship between output and input

3 System Modeling and Results

The system in this paper consists of synchronous generator connected with prime mover which is hydraulic turbine, and the excitation winding is fed from excitation system. The synchronous generator is a three-phase generator with the following characteristics: 200 MVA, 13.8 kV, 112.5 RBM, 64 poles connected with a grid which has frequency of 60 Hz and nominal voltage of 230 kV by using (star-delta) transformer rating 210 MVA; we will study this system in the steady state in three cases (without regulator, fuzzy logic power system regulator, power system regulator).

3.1 Modeling

We model a diagram of synchronous machine with HTG together with excitation system connected to IB. Here, we will get V_f from the excitation system and Pm (mechanical power) from HTG. The ground will be V_{stab} input to excitation system. The rotor type of the generator is salient pole. Figure 4a shows the diagram of single synchronous generator, P_m is connected to HTG, V_f is connected to excitation system, and PSS is connected to IB. The generic PSS will be V_{stab} input to excitation system. The configurations of this block are time constant and regulator gain, low-pass filter time constant Tr(s), exciter, transient gain reduction, and regulator output limits and gain, damping filter gain and time constant. Figure 4b shows the diagram of single synchronous generator, P_m is connected to HTG, V_f is connected to excitation system, and FLPSS is connected to IB. FIC will be V_{stab} input to excitation system. The configurations of HTG are gate opening limits, servomotor, hydraulic turbine and permanent droop and regulator.

3.2 Results

Figure 5a shows the result of output active power p_{eo}(pu) for single synchronous generator connected with HTG and excitation system connected to IB where ground is V_{stab} input to excitation system, and Fig. 5b shows the result of output active power p_{eo}(pu) for single synchronous generator connected with HTG, excitation system and PSS connected to IB where PSS is V_{stab} input to excitation system.

Figure 5c shows the result of output active power p_{eo}(pu) for single synchronous generator connected with HTG, excitation system and FLPSS connected to IB where FLPSS is V_{stab} input to excitation system.

4 Discussion and Analysis of the Results

From Fig. 5a, b and c, we can find generator stabilization time as illustrated in this Table 2.

We can see from Table 2 the difference between stabilization time without stabilizer state and with PSS is 6−2.5 = 3.5 s, and between without stabilizer state and with FLPSS state is 6−1.6 = 4.4 s and between PSS and FLPSS is 2.5−1.6 = 0.9 s. Our system reaches the steady state earlier with fuzzy logic stabilizer.

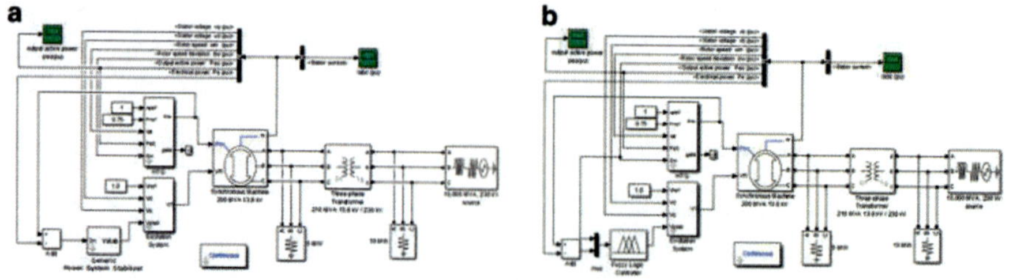

Fig. 4 **a** Synchronous machine together with HTG, excitation system using PSS, **b** synchronous machine together with HTG, excitation system using FLPSS

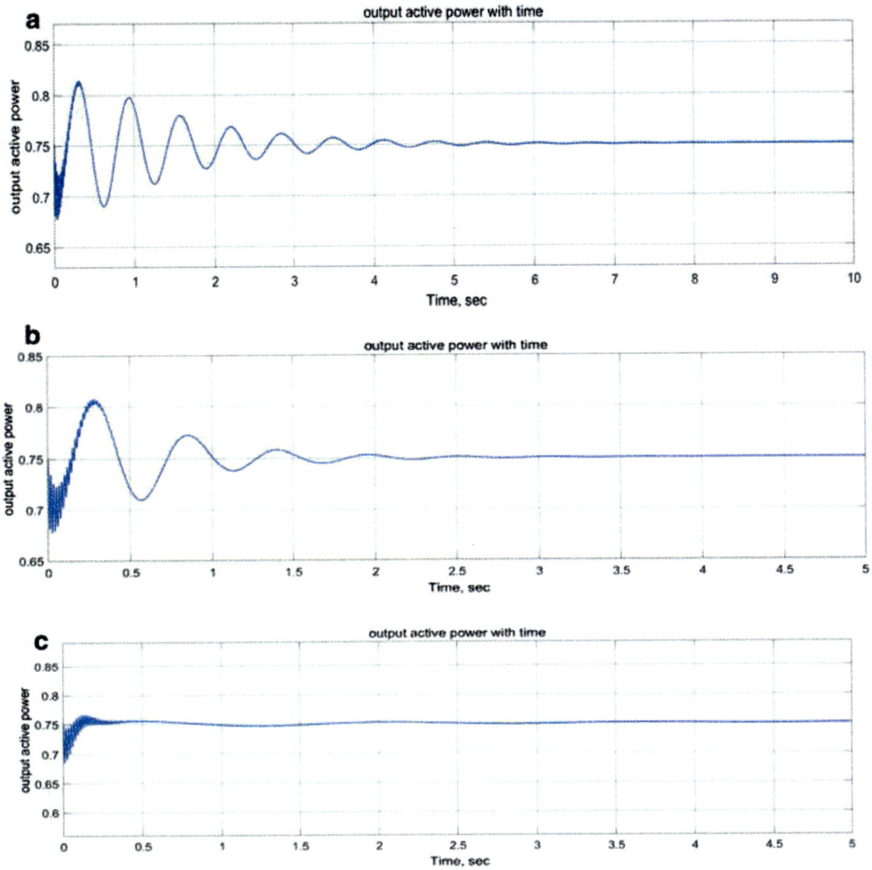

Fig. 5 **a** Generator response in case of no stabilizer, **b** response of generator when using PSS, **c** response of generator when using FLPSS

Table 2 Single generator connected to IB

Results of response for single generator connected to IB	
Without stabilizer (Fig. 5a)	6 s
With PSS (Fig. 5b)	2.5 s
With FLPSS (Fig. 5c)	1.6 s

This results are considered very good in comparison with recent papers using fuzzy logic to minimize settling time, for example, Shah [10] in 2013 used fuzzy PSS; the output reaches stability state after 8 s. Kumar et al. [11] in 2016 took 6.42 s to reach settling time, Kharadi et al. [12] in 2018 took 2.6 s to make the angular speed reach steady state, and Oraibi et al. [13] in 2020 took 2.1 s to make output power reach steady state.

Table 3 Comparison of results between PSS and FLPSS

Comparison of stability time results between PSS and FLPSS	
With PSS	140%
With FLPSS	275%
FLPSS better than PSS	[(275 − 140)/140] * 100 = 96.4%

We will calculate improvement percentage in case of (no stabilizer) with case of (PSS) and then calculate improvement percentage between the case (no stabilizer) and with case (FLPSS). Improvement percentage is calculated by the formula:

$$Y\% = [(X - X')/X'] * 100$$

Y% Improvement percentage
X The value of the studied factor before improvement (no stabilizer)
X′ The value of the studied factor after improvement (PSS or FLPSS)

We will calculate improvement percentage to the time of transient state ending and maximum amplitude to vibration and minimum amplitude to vibration.

Improvement percentage of stability time between (without stabilizer) and (PSS):

$$[(6 - 2.5)/2.5] * 100 = 140\%$$

Improvement percentage of stability time between (without stabilizer) and (PSS)

$$[(6 - 1.6)/1.6] * 100 = 275\%$$

5 Comparison of Results Between PSS and FLPSS

See the Table 3.

6 Conclusion

The simulation using MATLAB depicts that the fuzzy PSS has a very good response regarding small oscillation, while the response of conventional stabilizer has a ripple, and the steady state is reached after a lot of oscillations. Moreover, the fuzzy PSS performance is better because it is more effective, and the results show that the designed fuzzy stabilizer has the ability to maintain the system to operate at a better response regarding transient condition, steady-state error and the time of settling. By using this design of stabilizer, the response will be better 96.4% in comparison with CPSS.

References

1. Gupta, Neeraj, Jain, Sanjay K.: Comparative analysis of fuzzy power system stabilizer using different membership functions. Int. J. Comput. Electr. Eng. **2**(2), 262 (2010)
2. Corcau, J.I., Stoenescu, E.: A fuzzy logic controller as a power system stabilizer. Int. J. Circuits Syst. Signal Process. **1**(3), 10 (2007)
3. El-Metwally, K.A., Malik, O.P.: Fuzzy logic power system stabiliser. IEEE Proc.-Gener. Transm. Distrib. **142**(3), 277–281 (1995)
4. Hiyama, T.: Development of fuzzy logic power system stabilizer and further studies. In: IEEE SMC'99 Conference Proceedings. 1999 IEEE International Conference on Systems, Man, and Cybernetics (Cat. No. 99CH37028), vol. 6, pp. 545–550. IEEE (1999)
5. Lau, J.K., Low, W.Y., Lim, C.M.: A simplified fuzzy logic power system stabilizer. In: 2015 IEEE International Conference on Digital Signal Processing (DSP), pp. 996–1000. IEEE (2015)
6. Lim, C.M., Ouyang, Y.Z.: A simplified fuzzy logic control scheme. In: Proceedings of 6th Mechatronics Forum International Conference, pp. 229–233 (1998)
7. Xu, W., Lu, Q.: A new stabilizer design technique for multimachine power systems. Electr. Power Syst. Res. **15**(2), 89–97 (1988)
8. Toliyat, H.A., Sadeh, J., Ghazi, R.: Design of augmented fuzzy logic power system stabilizers to enhance power systems stability. IEEE Trans. Energy Convers. **11**(1), 97–103 (1996)
9. We, C.J., Hsu, Y.Y.: Design of self-tuning PID power system stabilizer for multimachine power system. IEEE Trans. on PWRS 3, no. 3 (1988)
10. Shah, B.: Comparative study of conventional and fuzzy based power system stabilizer. In: 2013 International Conference on Communication Systems and Network Technologies, pp. 547–551. IEEE (2013)
11. Kumar, P.K., Vivekanandan, S., Kumar, C.K., Chinnaiyan, V.K.: Neural network tuned fuzzy logic power system stabilizer design for SMIB. In: 2016 2nd International Conference on Contemporary Computing and Informatics (IC3I), pp. 446–451. IEEE (2016)
12. Kharadi, G.S., Gohil, G.D., Parmar, N.B.: Development of matlab simulink model of fuzzy based power system stabilizer for power quality
13. Oraibi, W.A., Hameed, M.J., Abbas, A.K.: An adaptive neuro-fuzzy based on reference model power system stabilizer

Facial Emotion Detection Using Deep Learning and Haar Cascade Face Identification Algorithm

Bhavya Alankar, Mohammad Sharay Ammar, and Harleen Kaur

Abstract Emotions are the determinant of what makes us human. They influence each and every aspect of our everyday life, including concentration, perception, memory, decision making and contact with society. Human beings are exceptionally skilled and competent at reading and interpreting faces through evolution. In addition, facial recognition is so essential to human survival that an area called the Fusiform Face Area (FFA) had been developed for it by the human brain. In this paper, the key proposal is implementing multilevel Haar wavelet-based technique also known as Haar cascade face identification algorithm, which pulls out several expression attributes from major face regions at different levels. The translation of the structure and the arrangement of 43 facial muscles into emotions can be a difficult task to execute. So, we will Train FER2013 Dataset in Google collab for 25 *epochs*, which will churn out training accuracy to more than 90%. Many biometric technologies are available in our day-to-day activities to identify people, such as eye recognition, fingerprint recognition, face recognition. Face is a prime portion of the human being and needs detection for various applications such as forensic investigation, protection. The accuracy of the given technique is trialed on popular FER2013 facial expression dataset, and it achieves 91.85% accuracy for the given dataset.

Keywords Convolutional neural network · Facial expression recognition (FER) · Edge detection · Haar cascade classifier · AdaBoost · Kernels · Face detection · Feature extraction

B. Alankar · M. S. Ammar · H. Kaur (✉)
Department of Computer Science and Engineering, School of Engineering Sciences and Technology, Jamia Hamdard, New Delhi, India
e-mail: harleen@jamiahamdard.ac.in

S. Das and M. N. Mohanty (eds.), *Advances in Intelligent Computing and Communication*, Lecture Notes in Networks and Systems 202,
https://doi.org/10.1007/978-981-16-0695-3_17

1 Introduction

Facial emotion recognition (FER) is a computer program's ability to accurately recognize and explain the facial and/or facial characteristics such as eyebrows, eyes, nose, shape of mouth, etc. Today, in the wide fields of face recognition, computer vision, machine learning and even artificial intelligence, it forms a major research and development area. Facial recognition is a major step in various implementations for the processing of documents, e-learning, entertainment, security and forensics, using digital image and video processing. This technology is becoming ever more reliable and will finally be capable to scrutinize emotions as well as our brains do.

Artificial Intelligence (AI) can detect emotions by learning the meaning of each face expression and applying that knowledge to the new information that is presented [1]. Emotional (AI) is a technology capable of understanding, imitating, interpreting and responding to gestures and emotions in human faces. Several of the examples used for emotion recognition include TSA airport screening, audience participation and much more.

Nowadays in the scope of artificial intelligence, there is a popular trend for a human–computer interaction (HCI) [2]. Real-time face detection and analysis of various facial emotions such as joyful, angry, sad, fear, surprise, etc., are based on facial features and their actions. The primary features of the face are known for "facial recognition" and "prediction of facial expressions or emotions." The variability in each facial characteristic is used to assess the various facial expressions. Machine learning algorithms are considered for the recognition and categorization of various types of facial expressions by training dataset images. The suggested algorithm uses open-source computer vision (OpenCV) and Python-based machine learning, such as Haar cascade image classification.

The paper aims to identify and describe as many as seven distinct facial emotions as possible. This runs on top of a convolutional neural network (CNN) developed with the aid of Keras, whose backend in Python is TensorFlow. The facial emotions which this device can detect and identify are Happy, Sad, Anger, Anxiety, Surprise and Neutral. We have used *OpenCV* for image filtering job where we identify a face from a computer web-camera feed that is processed into the instructed neural network for expression recognition.

1.1 Dataset Used

The name of the dataset is *Fer2013* [3] that consists of 32,298 different pictures which is further subdivided into training (28,709) and testing (3,589) pictures, respectively. Machine learning technique to FER all include a collection of image training examples, each labelled with a single category of emotions. A common collection of seven classifications of the emotions is sometimes used:

1. Anger
2. Sadness
3. Disgust
4. Surprise
5. Fear
6. Neutral
7. Happiness

These expression categorizations are demonstrated in the images below, presenting illustrative test images (Fig. 1).

2 Methodology/Experiment

2.1 What Is Convolutional Neural Network?

The ***convolutional neural network*** (CNN) is a class of neural profound learning network [4]. CNN marks a significant development in the identification of pictures. They are most widely used to examine visual imagery, and also function in image classification behind the scenes. They can be found at the heart of everything from photo-tagging on Facebook and Instagram to operation of self-driving cars. They are used extensively in everything from health care to defense, but from behind the scenes [5].

A convolutional neural network extracts features from 2D data and assigns weights to those features, leading to a prediction [4]. For instance, if we want to instruct a network to acknowledge handwritten numerical values, we will have a dataset of numbers 100×100 px. The CNN would identify "curves" and "straight" lines in 10×10 px sections, and after identifying these features, the

Fig. 1 Sample images from the (FER2013) dataset, representing variations in lighting, lifespan, posture and expressions, namely happy, anger, fear, sadness, disgust, surprise and neutral

model would understand that certain numbers are representative of combinations of certain curves and lineages. One especially curvy number such as 8 will be distinguishable from a straight number such as 1 or 7. In the case of identification of facial emotions, happiness will be linked to the upward curves of a smile.

CNN has an input layer, concealed layer and yield layer. The concealed layer consists typically of convolutional layers, pooling layers and connected linked layers.

- Convolutional layer applies a convolution action to the input. It passes on the information to the following level.
- Pooling merges the yield of collection of neurons into a solitary neuron in the upcoming layer.
- Fully linked layers join every neuron in single layer to every other neuron in the upcoming layer (Fig. 2).

Neurons only receive feedback from a sub-area of the preceding layer in a convolutional network. CNN, unlike other neural networks, absorbs and summons pictures as tensors, and tensors are numeral matrices with supplementary proportions. The machine interprets stimuli differently from humans. They basically see pixels or an array of matrix as pictures (Fig. 3).

A CNN [6] operates via withdrawing attributes from the input images. This removes the demand for manual attributes extraction. The attributes are not instructed, otherwise they are acknowledged while the neural network instructs on a bunch of pictures. They are trained while train on a collection of images by the network. This makes the deep learning model *highly effective* for tasks involving computer vision. CNNs learn to detect features across multiple layers in abstraction. Increasing layers increases the sophistication of the characteristics learned.

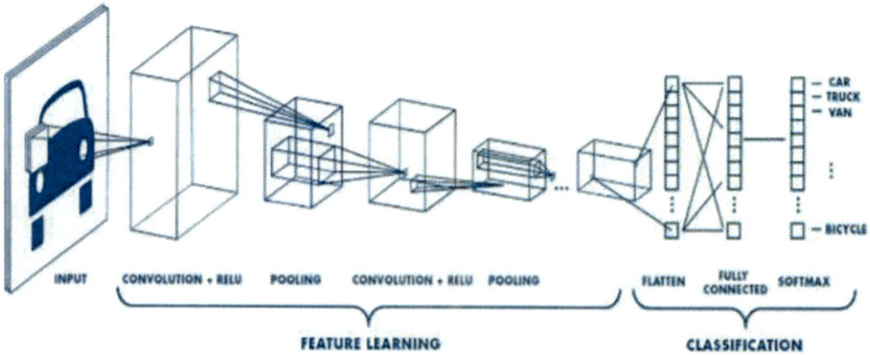

Fig. 2 Multiple convolutional layers in neural network

Fig. 3 Conversion of image into convolutional matrix

2.2 *Why Deep Learning Used in Face Recognition?*

A face is defined as the location of major face milestone such as the nose, eyes, mouth, cheeks and other regions, and then the distances are measured between these points. Then thresholds for detecting the emotion are set. If the face in the picture smiled, the locations of the jaw would be closer to the eyes, the mouth would be extended and the eyes squinted.

Deep learning has a number of uses, and the face recognition occupies a significant position [7]. First, deep learning gives the power to create biometric recognition software capable of recognizing or verifying an individual in a specific way. All this because deep learning approaches are capable of exploiting very large facial datasets and learning rich and compact facial representations, enabling modern models to perform as-well first, and later outperform human facial recognition capabilities (Fig. 4).

Deep learning systems in the human brain neocortex, where higher-level cognition exists, are prototyped after the neural network. A neuron in the human brain is a complex cell which transmits information via electrical/chemical signals. This forms a neural network, when attached to other neurons. The neuron in machines is implicit—basically a bit of cipher code sustaining statistical regression—compiling together adequate amount of virtual neuron(s) and you have a virtual (look alike) neural network.

To obtain a glance of deep learning [8], let us imagine a kid whose first word learned is a "dog." A child acknowledges what a dog is (and not) by indicating to pictures and beings and uttering the phrase "dog." The parent says, "Yes, it is a dog," or, "No, it is not a dog." As the kid begins to point to things, he becomes more profound of the characteristics that all dogs have. What the kid does, without understanding it, is to simplify a complex abstraction (the dog concept) by creating a ranking in which every abstraction level is generated with the information obtained from the previous hierarchy layer.

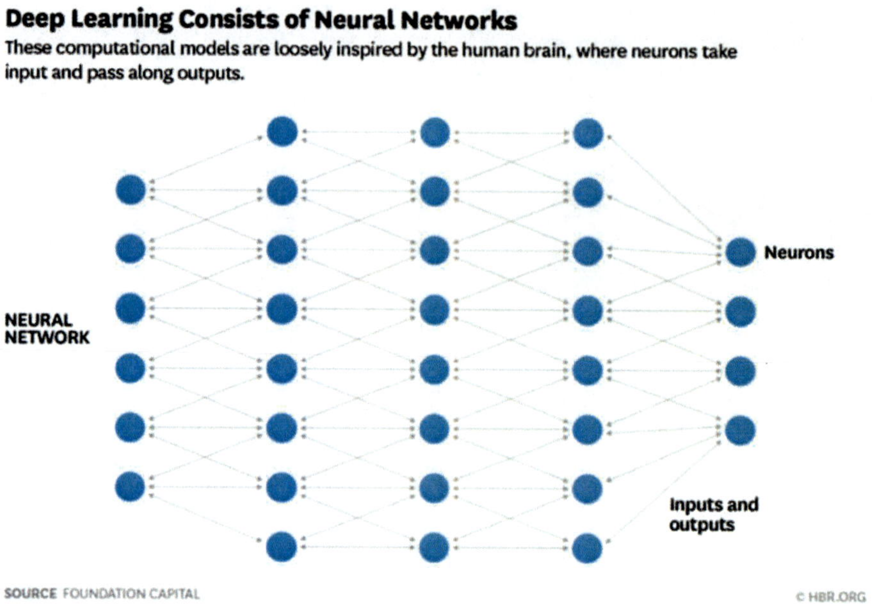

Fig. 4 A general model of a deep neural network

2.3 *Facial Recognition System Working*

In facial recognition software, the most important parameters are the space between the eyes, the spread of the nostrils, the stretch of the nose, the height and shape of the cheekbones, the breadth of the mouth, the height of the forehead and other parameters [9]. The facial detection system discovers the face in the picture and highlights this field. The system then uses classifier algorithms that detect a face (1) or not (0) in a dataset picture. Categorization has been instructed to use thousand (s) of sample pictures to identify faces to achieve precision in tests. OpenCV uses several classificatory forms, one of which is known as Haar cascade classifier.

2.4 *Haar Cascade Classifier Algorithm*

A *Haar cascade* is established on "Haar Wavelets" which is defined as: "**A series of rescaled "square-shaped" functions which together developed a wavelet family or basis.**"

Haar cascade [10] is a machine learning object detection algorithm proposed by *Michael Jones* and *Paul Viola* [3]. It is an approach based on machine learning, where a cascade function is instructed from multiple constructive and destructive test pictures (where constructive images are those where the object/faces to be identified are present, destructive are those where they are not). This process in the algorithm is feature extraction.

The classification of pixels [11] in the picture into squares by feature is built on the Haar wavelet technique. It uses machine learning approach to obtain high precision "training results." This uses concepts like "integral image" to measure the detected "features." Using the AdaBoost learning algorithm, Haar cascades sort out a fraction of key attributes from a wide collection to yield an effective classifier result, then utilize cascading approach to detect the face in an image.

The algorithm has four steps/stages:

(i) Haar feature selection
(ii) Creating integral images
(iii) AdaBoost training
(iv) Cascading classifiers (Fig. 5)

1. Collecting Haar features is the first step. A Haar feature takes an adjacent rectangular region in a detection window at a given location, sums up the intensities of pixels in each region, and measures the difference between those sums. Here are some Haar features. They are "edge features," used to detect edges in the image. The second is a "line feature," while the third is a "four rectangle features," most likely used to detected a slanted line (Fig. 6).
   On a target picture, a set of rectangular black and white masks in different shapes is used. The masks are superimposed on contrasting fragments of the picture, and the procedure applies the illumination of all the pixels of the picture

Fig. 5 In attribute extraction, the algorithm utilizes instructed dataset to identify best facial attributes that it can contemplate a face

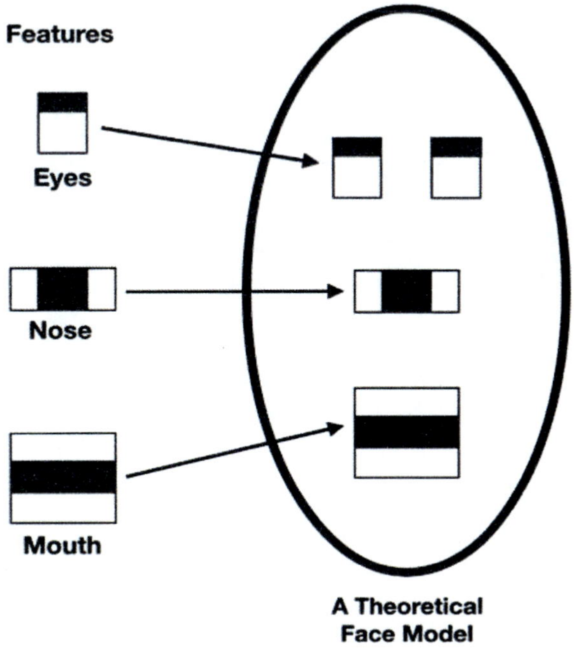

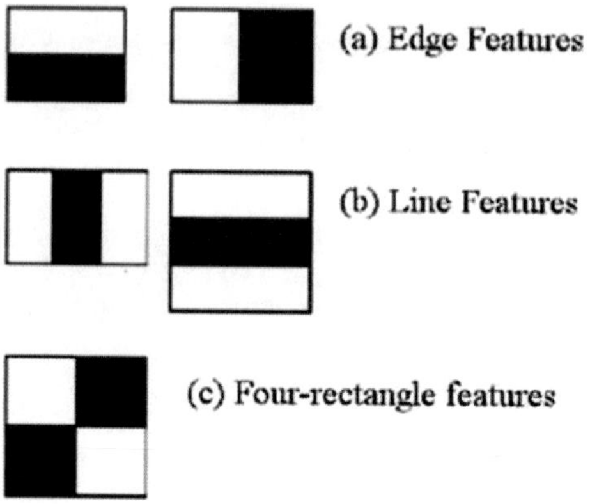

Fig. 6 Common Haar features

beneath the black and white fragments of the mask and then measures the difference between those values. The program then differentiates the output with the cumulative data and, after determining the face of the image, continues to monitor it in order to select the optimum angle and image quality [12].

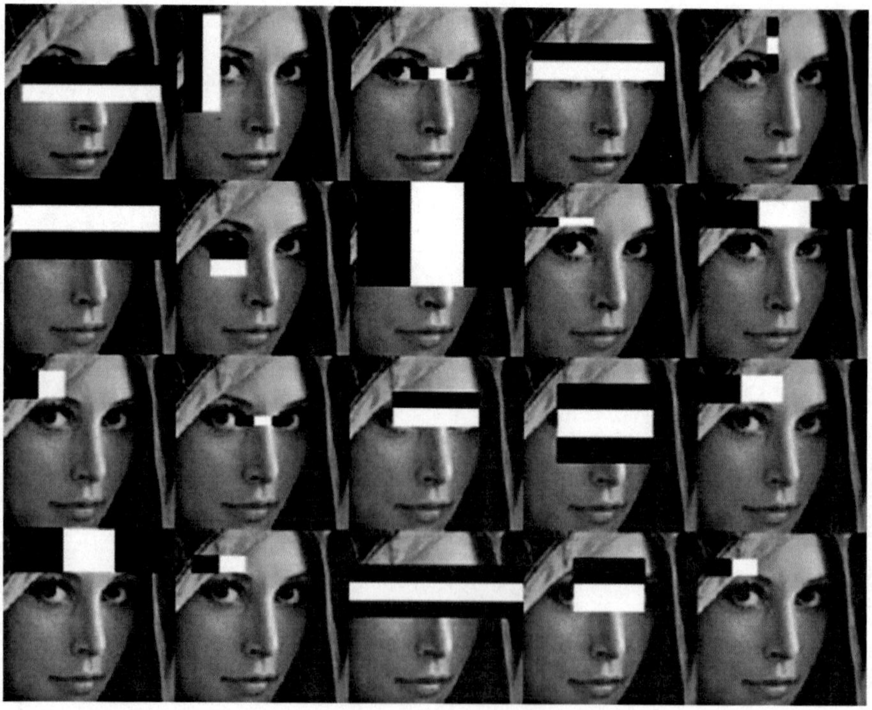

Fig. 7 Working of Haar features by parsing of rectangle from left to right, top to bottom

Here's a sample photo of Haar features working when it's added to a girl's face (Fig. 7).

2. Creating an integral image: It is an algorithm to generate the value sum in a rectangular grid subset. The aim is to reduce the number of computations required to achieve pixel intensity summations within a window.
The idea is to transform an input image into a combined area table where the value is the summation of all the pixels above and to the left of (x, y) at any point (x, y) in that table, including:

$$ii(x,y) = \sum_{x' < x, y' \leq y} i(x', y'),$$

$$h_j(x) = \begin{cases} 1 & \text{if } p_j f_j(x) < p_j \theta_j \\ 0 & \text{otherwise} \end{cases}$$

where $I(x, y)$ is the merit of the integrated photo pixels in the position (x, y), while $i(x, y)$ is the associative intensity in the original image.
Integral images are used to make the process more efficient thus making the computation super-fast. If we evaluate all viable variables of the Haar features like location, type and scale, we will calculate about 1,60,000+ features. So, we need to evaluate huge set of features for every 24 * 24 PX.

3. Efficiency and optimization are the next step. To pick the useful features and avoid redundant features out of 1,60,000+ features, we introduce an AdaBoost concept that sort out the foremost attributes from the image and also train the classifiers using it. This procedure builds a "strong" classifier as a linear union of simple weighted "weak" classifiers.
The pictorial diagram represents the working of edge detection in the process of facial recognition. The kernels, also known as Haar feature, determine the location of size (height and width) of human face in the form of digital image format. After the process of edge detection, face recognition comes into play to determine face and isolate it using crop and box method so that it can be recognized.

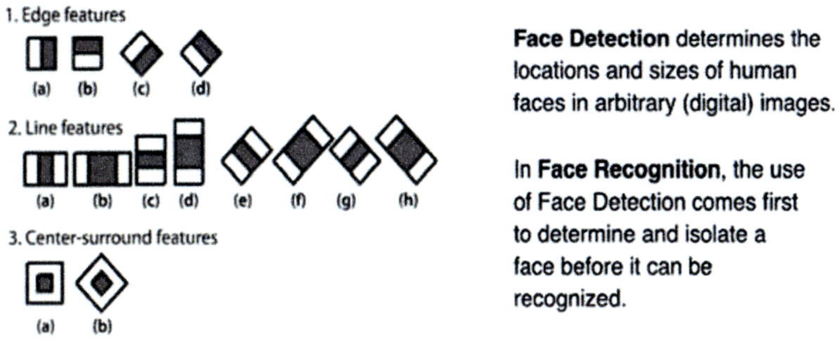

Most of the region on a picture [13] is non-face region. So, it is easier to have a clear way of testing if a window is not a face area. If the image does not have a region of the face, it will get rid of it in a single shot and will not compute it another time. It will rather concentrate on region where a face can be present. This allows us to find more time to examine a viable region of the face. To accomplish this, the introduction to the cascade of classifiers principle comes into play. Instead of estimating 2,500 elements, we use the concept of cascades for each window. We perform a sampling of 2,500 attributes into x different cascades, so we can linearly detect whether or not there is face in different cascades. If cascade discovers a face in a picture, then that picture will be passed to next cascade. If there is no face in a cascade, then we can move to the next window. That lowers the time complexity. The Eq. (1) for the Haar wavelet function $\psi(t)$ is given as:

$$\psi_{a,b}(t) = a^{-1/2}\psi\left(\frac{t-b}{a}\right) \tag{1}$$

where, a, $b \in$ R(a > 0), R is the wavelet space. Parameter 'a' is the scaling factor and 'b' is the shifting factor.

4. The cascade classifier [14] is made of an assembly of layers, in which each stage a simple classifier (also known as weak learners because of their low discriminative power) is built. Growing steps are trained using a boosting technique by merging of all the weak learners/classifiers that forms a strong classifier that can acknowledged any kind of object it was trained with. The problem is to look for a specific sized window over the full-sized image, applying the series of weak classifiers on each sub-frame of the image. Viola and Jones [3], cascade of classifiers was used (see Fig. 5) to tackle this problem: The first classifier (the most discriminatory) is extended to all the sub-windows of the image, even at various scales. The second classifier will be applied only to the sub-windows under which the first classifier proven successful. The cascade continues, applying all of the weak classifiers and discarding the negative sub-windows, concentrating the computation power only on the promising regions.

(i) A *true positive* occurs when a constructive specimen is correctly classified.
(ii) A *false positive* occurs when a destructive specimen is misinterpreted classified as positive.
(iii) A *false negative* occurs when constructive specimen is misinterpreted classified as negative (Fig. 8).

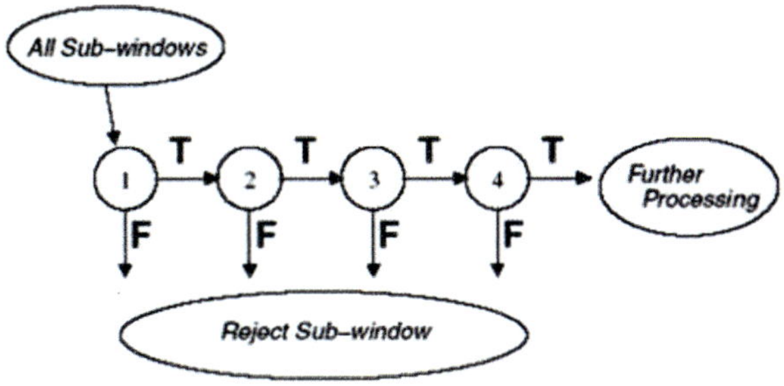

Fig. 8 Flowchart for 4 stages of cascade classifier

3 Procedure for Recognizing Emotions

3.1 Face Detection

A face placement is found in an image frame, which can be achieved by applying, for example, the cascade classifier algorithm. This may sound sophisticated, but this technology can actually be found in your iPhone or Android smartphone's camera as well. The effect is a face-box that frames the face detected.

Figure 9 illustrates how AdaBoost [15] (learning algorithm selecting a small no. of essential sighted attributes from a very large collection of potential features) selects those features.

Two instances of features are seen in the top row and then overlaid in the last row on a training mask. These features essentially measure the difference in the

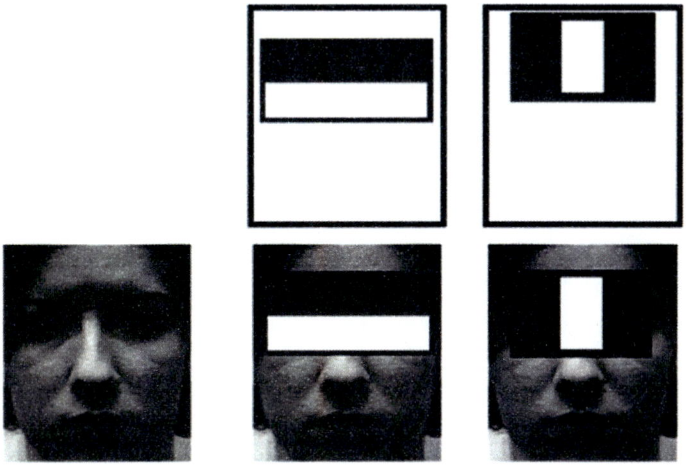

Fig. 9 Features selected by recurrence of AdaBoost training algorithm in a face from the FER database

magnitude between the black region and the white region. Typically, one region is comparatively shady than the other, and so are the features chosen to detect faces.

3.2 *Feature Extraction*

Facial attributes [16] such as eye corners, eyebrows, mouth corners, the nose tip, etc., are detected within the identified face. After that an internal face model is modified to suit the actual face of the respondent in position, height, and scale. The face model adapts and follows immediately whenever the respondent's face moves or changes expressions. Compared to the real face, it has much fewer specifics (so-called features), but it does include exactly certain facial features to get the job done. Single landmark points (eyebrow corners, mouth corners, nose tip) as well as feature groups (the entire mouth, the entire arch of the eyebrows, etc.) represent the entire "Gestalt" of an emotionally indicative face region are exemplary features.

3.3 *Feature Classification and Output*

Once the simplified facial model is available, position and orientation information of all the key features are given as input into classification algorithms which converts the features into Action Unit codes, emotional states, and other affective metrics (Fig. 10).

4 Experimental Observation

VGG-16 pre-trained neural network is used to result in bottleneck attributes. VGG-16 network consists of sixteen layers out of which thirteen levels are convolution layers/levels (CN). The neural network is trained well on sample image dataset which contains thousands of test images to train on.

We got final 91.85% accuracy on CV data of human images. We operated the model till for 25 epochs.

Below, we have the following figures from epoch 1 till epoch 20:

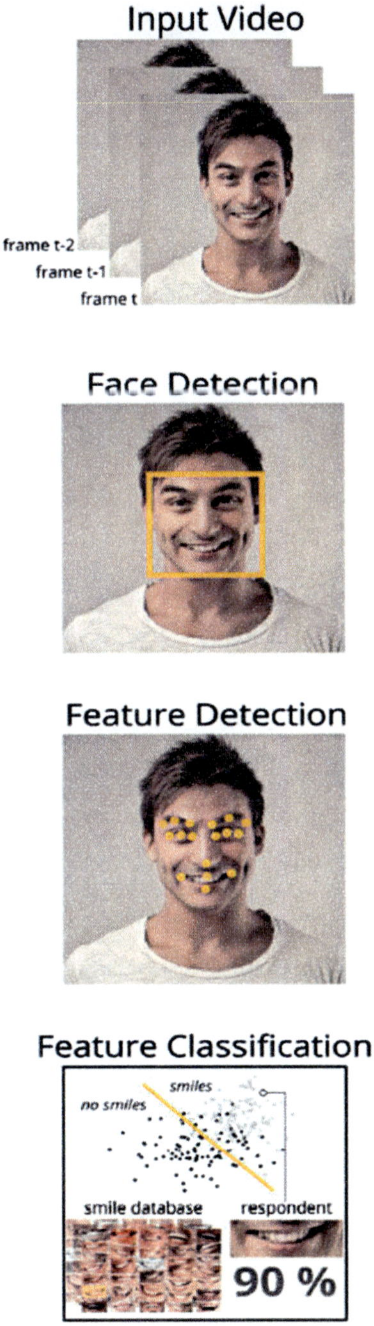

Fig. 10 Main three steps for facial emotion detection: face detection, feature detection and feature classification

	Epoch	Comb_Train_Loss	Comb_Train_Accuracy	CVHuman_Loss	CVHuman_Accuracy	CVAnime_Loss	CVAnime_Accuracy
0	1	2.45	0.17	2.89	0.15	2.21	0.19
1	2	2.19	0.21	2.25	0.21	1.55	0.34
2	3	1.98	0.25	1.97	0.25	1.09	0.51
3	4	1.78	0.31	1.77	0.29	0.70	0.71
4	5	1.63	0.35	1.65	0.35	0.40	0.85
5	6	1.55	0.41	1.44	0.48	0.22	0.91
6	7	1.32	0.51	1.34	0.50	0.11	0.93
7	8	1.11	0.57	1.22	0.54	0.07	0.97
8	9	0.95	0.61	1.20	0.55	0.04	0.98
9	10	0.77	0.69	1.10	0.57	0.01	0.99
10	11	0.65	0.75	1.09	0.60	0.01	0.99
11	12	0.44	0.81	1.03	0.61	0.01	0.99
12	13	0.21	0.89	0.89	0.69	0.01	1.00
13	14	0.11	0.93	0.77	0.75	0.00	1.00
14	15	0.07	0.97	0.65	0.83	0.00	1.00
15	16	0.04	0.99	0.50	0.85	0.00	1.00
16	17	0.03	0.99	0.54	0.84	0.00	1.00
17	18	0.04	0.99	0.49	0.86	0.00	1.00
18	19	0.03	0.99	0.49	0.87	0.00	1.00
19	20	0.04	0.99	0.46	0.87	0.00	1.00

Train Log

From the 1st epoch until the 20th epoch, we will reduce losses as follows:

Loss in training: 2.45–0.04
Human loss factor: 2.89–0.46
Animated loss factor: 2.21–0.00

At the same time, accuracy increased from 1st epoch till 20th epoch as follows:

Accuracy in training: 17–96.9%
Human accuracy percentage: 15–87%
Animated accuracy percentage: 19–100% (Fig. 11).

In the above given bar graph, Happy and Surprise expressions have the maximum visible facial deformation. Both expressions have the peak recognition rates among all seven classes of expression in the given plot.

In Fig. 10, confusion matrix is obtained from FER2013 dataset. The numbers along the diagonal in the matrix (grey boxes) refer to the times the sample images of various expression from the dataset were correctly classified, i.e., true label, whereas the white boxes, i.e., the off-diagonal elements, refer to the sample data that were incorrectly classified by the classifier algorithm. In short, the confusion matrix represents the right and wrong done by the algorithm by showing true and false positive and negative on the 7 × 7 confusion matrix (due to 7 expression elements to choose from). The confusion matrix was developed using Scikit learn using Python in Jupyter Notebook (Fig. 12).

When all false positives were removed, it gives a detection rate of around 91.85% for the eyes and nose. The mouth recognition rate is low due to the

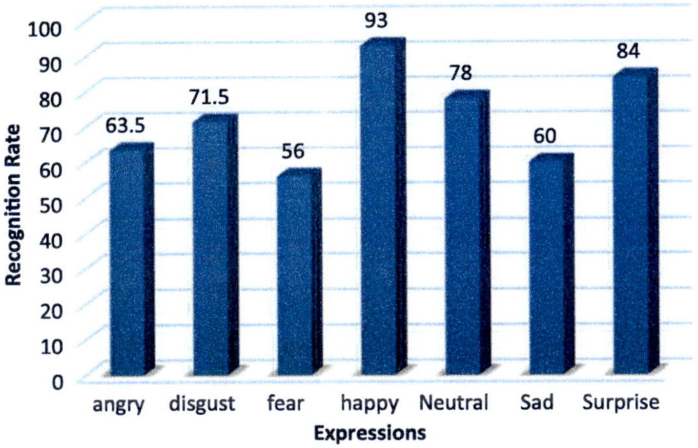

Fig. 11 Recognition rate (%) of individual expression

Fig. 12 Confusion matrix obtained on FER2013 dataset

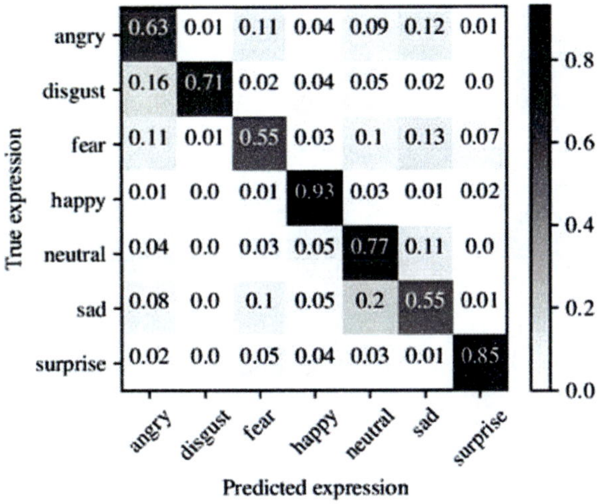

minimum size required for detection and due to facial hair of the subject face. By altering the length and width variables to more precise level, by re-training the classifier. The precision should increase the precision to that of the other features.

The line graph in Fig. 11 represents the performance parameter of the epochs (number of passes/iterations of the training dataset) that was executed on the machine for the accuracy. The first refers to epoch vs loss graph statistics and second graph refers to epoch vs accuracy statistics. Here, the loss is trained data is less than the loss in the validated (ingested) data, depicting that our model executed better in trained dataset as we can see in the graph due more downfall in the train line in the given graphical representation (Fig. 13).

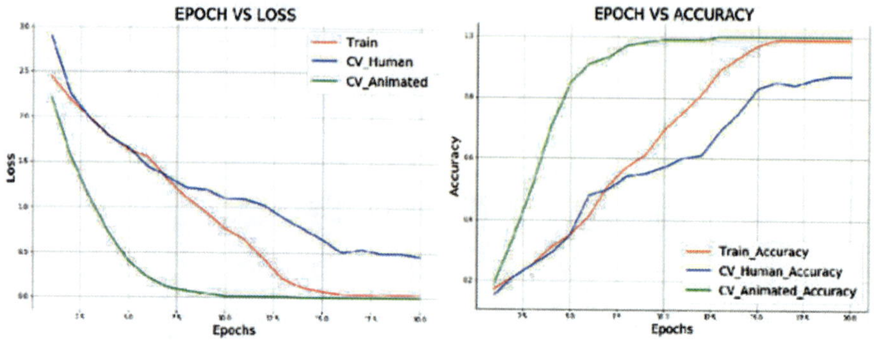

Fig. 13 Epoch loss accuracy

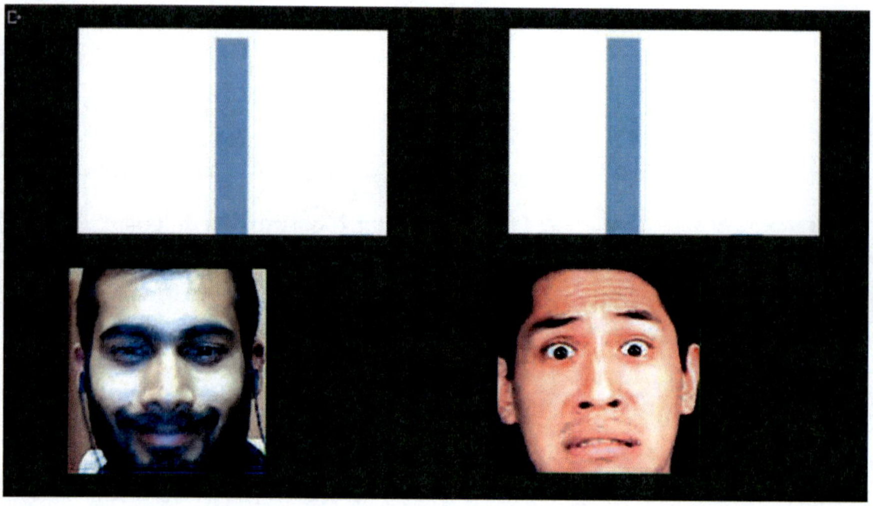

Fig. 14 Live testing module with input taken from computers webcam

Some of the real-world example are given in the given screenshot with magnitude of expression on subject's face in terms of bar graph representation (Fig. 14).

5 Conclusion

The recognition system of facial expression presented in this paperwork contributes to a robust model of facial detection focused on mapping behavioral characteristics with biometric physiological characteristics. The human face's physiological characteristics with respect to various expressions such as joy, surprise, sadness, fear, anger and disgust are correlated with geometric structures that restored the recognition mechanism as the foundation matching prototype. The behavioral dimension of this method is related to the attitude as the basis of property behind different words.

This research work promises a continued research path in the field of asymmetric biometric cryptosystems which is highly advisable in order to fully get rid of passcodes and smart cards. 140 face recognition system based on facial expression is made effective with facial surface genetic algorithm invariants resulting in a recognition figure of 91.85%.

Haar algorithm can effectively express the signal with comparatively less features. Functionality of Haar was to conserve the energy of the signal even with very few coefficients makes it a choice for pattern recognition. Haar features are computed only for these components rather than the full image. This reduces feature dimension without losing information.

In this research work, the conclusion is given to construct emotional depiction by using the principle of an embedding method based on hierarchy. For biometric authentication method, the facial emotion and representation model are employed on a laptop. This study makes it easier to identify practical concerns that should be taken into account when designing monitoring e-learning processes and interpreting the automatic emotion recognition results.

The facial expression recognition model, utilizing convolutional neural architecture, is 91.8% accurate, which is approximately the level of precision anticipated from an support vector machine (SVM) approach as well.

References

1. Cowie, R., Douglas-Cowie, E., Tsapatsoulis, N., Votsis, G., Kollias, S., Fellenz, W., Taylor, J.G.: Emotion recognition in human-computer interaction. IEEE Signal Process. Mag. **18**(1), 32–80 (2001)
2. Content derived from: https://www.researchgate.net/publication/264151703_Region-Based_Facial_Expression_
3. Viola, P., Jones, M.: Rapid object detection using boosted cascade of simple features. In: IEEE Conference on Computer Vision and Pattern Recognition (2001)
4. Dachapally, P.R.: Facial emotion detection using convolutional neural networks and representational autoencoder units. School of Informatics and Computing Indiana University, 5 Jun 2017
5. Baboota R., Kaur H.: Predictive analysis and modelling football results using machine learning approach for English Premier League. Int. J. Forecast. 35 (2):741-755 (2019)
6. Rzayeva, Z., Alasgarov, E.: Facial emotion recognition using convolutional neural networks. In: 2019 IEEE 13th International Conference on Application of Information and Communication Technologies (AICT)
7. Sang, D.V., Van Dat, N., Thuan, D.P.: Facial expression recognition using deep convolutional neural networks. In: 2017 9th International Conference on Knowledge and Systems Engineering (KSE), Hue, pp. 130–135 (2017)
8. Kharkovyna, O.: An intro to deep learning for facial recognition, 26 June 2019. www.medium.com

9. Sharma, H.: Facial Expressions Recognition, 21 April 2018. www.medium.com
10. Alto, V.: Face Recognition with OpenCV: Haar Cascade, 16 Jul 2019. www.medium.com
11. Wilson, P.I., Fernandez, J.D.: Facial Feature Detection Using Haar Classifier. J. Comput. Sci. (2006)
12. Inspiration derived from: https://medium.com/analytics-vidhya/haar-cascade-face-identification-aa4b8bc794
13. Minaee, S., Abdolrashidi, A.: Deep-Emotion: Facial Expression Recognition Using Attentional Convolutional Network. Expedia Group University of California, Riverside, USA
14. Sharma, N.: How to do Facial Emotion Recognition Using A CNN? 24 Dec 2018. www.medium.com
15. Michael Revina, I., Sam Emmanuel, W.R.: A Survey on Human Face Expression Recognition Techniques. Reg No. 12417, N.M. Christian College, Marthandam Affiliated to Manonmaniam Sunadaranar University
16. Blog on "Algorithmia" quoted. Introduction to Facial Emotion Recognition, 28 Feb 2018

A Novel Modified PSO Algorithm with Inertia Weight Parameter

Arabinda Pradhan and Sukant Kishoro Bisoy

Abstract Makespan is a big issue in cloud computing environment, when there is a large number of incoming tasks which is to be allocated among accessible virtual machines (VMs). To manage such an issue, we have proposed an effective task allocation scheduling algorithm followed on inertia weight named as global–local best modified PSO (GLMPSO) which avoids the problem of local optimum and slows convergence rate faced by standard PSO algorithm. Our proposed algorithm improves the searching capabilities of both local and global search according to the number of cycles. GLMPSO algorithm is implemented on CloudSim simulator with taking different datasets. Simulation results show that GLMPSO algorithm performs better in reducing almost 2% and 9% makespan in comparison with the existing algorithms such as MPSO and L-PSO, respectively.

Keywords Cloud computing · PSO · Task allocation · Makespan · Inertia weight

1 Introduction

Cloud computing provides various computing services to its user such as server, storage, databases, networking and software over an Internet [1]. These computing services are known as the resource unit that is assigned to the user as on-demand basis and charged some money for specific duration of time [2]. In cloud environment, all the needed services are stored in data center (DC) and controlled by data center controller. Each data center consists of number of hosts, and each physical host can stack at least one virtual machine with the goal that clients can run the applications freely [3]. When the number of requests increases at a particular time, then it transforms into a troublesome activity to manage the entire request in

A. Pradhan (✉) · S. K. Bisoy
Depertment of Computer Science, CGU, Bhubaneswar, Odisha, India
e-mail: arabindapradhan1@gmail.com

S. K. Bisoy
e-mail: sukantabisoyi@yahoo.com

S. Das and M. N. Mohanty (eds.), *Advances in Intelligent Computing and Communication*, Lecture Notes in Networks and Systems 202,
https://doi.org/10.1007/978-981-16-0695-3_18

the most restricted allocation time, execution time. Therefore, a better scheduling technique is required to assign all tasks into suitable VM that reduce execution time. Because of dynamic behavior of cloud, many researchers focused their research work on heuristic, meta-heuristic and hybrid scheduling strategies. According to [4], meta-heuristics accomplish the preferred outcomes over conventional heuristics. It handles huge computational and complex issues and gives faster and inexpensive response that can optimize the error in network [5]. PSO is one of the most famous and popular meta-heuristic algorithms that is widely used in swarm intelligence optimization technique to solve optimization issue. It was discovered by Kennedy and Eberhart [6]. PSO is appropriate for dynamic task scheduling and provides a best optimum result. But the major limitation is that PSO is effectively caught into local optimal solution. It is controlled by one of the best parameters of PSO algorithm which is inertia weight (ω) [7]. Commonly, its value lies in between 0 and 1. It controls and explores both local and global search capabilities of PSO [8]. Due to this reason, we focused the modified PSO algorithm [9] that achieves the targeted objective.

The rest of this paper is arranged as follows: Sect. 2 shows the related work. Section 3 shows the proposed approaches. Section 4 shows the experiment setup and results. Conclusion and future work of the paper are given in Sect. 5.

2 Related Work

Huang et al. [10] show L-PSO task scheduling algorithm, which is the focused logarithm decreasing inertia weight strategy to reduce the average makespan. Zhou et al. [11] proposed MPSO task scheduling approach to reduce the energy efficiency and enhance convergence speed according to the number of iterations by taking modified inertia weight method. Agarwal and Srivastava [12] present PSO-based task scheduling mechanism to reduce 30% in makespan and expand the asset use by 20% by using global–local best inertia weight. Ebadifard and Babamir [13] introduced PSO-TS method whose resource utilization is increased by 22% and a decrease of makespan by 33%. Jacob and Pradeep [14] proposed CPSO hybrid algorithm that decreases both cost and makespan value. Saleh et al. [15] introduced IPSO algorithm to reduce makespan and increase resource utilization by using clustering approach.

3 Proposed Approaches

In this section, the proposed approach is explained in detail. This section is comprised of GLMPSO model, problem formulation and proposed algorithm.

3.1 Model of GLMPSO

In this section, we propose GLMPSO model that minimizes the makespan time or execution time without compromising the allocation of all task to VM and shown in Fig. 1. Our GLMPSO algorithm is based on modified PSO and inertia weight parameter to achieve better result than PSO algorithm. Initially, our GLMPSO gathers all the required information about incoming tasks and VMs such as task length or size, bandwidth capacity, number of processor and processing speed that are used to get the correct result and managed in such a way that no task is left in buffer and total execution time as well as allocation time is optimized. In this model, numbers of tasks are going into CSP to get their services.

3.2 Problem Formulation

Suppose i number tasks are entering into the cloud such as: $\mathrm{T} = \{\mathrm{T}_1, \mathrm{T}_2, \ldots, \mathrm{T}_i\}$, and each task has task length $\mathrm{T}_{\mathrm{len}}$. Also, data center consists of j number of heterogeneous virtual machines such as: $\mathrm{VM} = \{\mathrm{VM}_1, \mathrm{VM}_2, \ldots, \mathrm{VM}_j\}$, and each

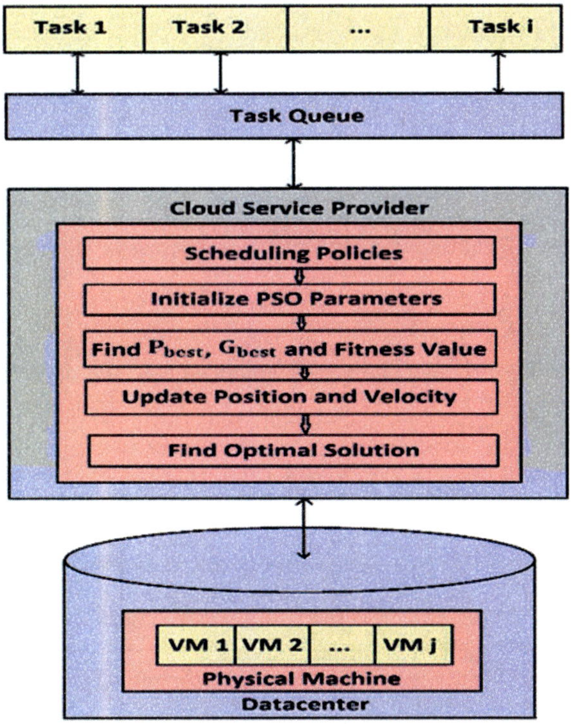

Fig. 1 Proposed model GLMPSO

VM having processing speed is showing as millions of instructions per second (MIPS). But the condition for allocating such tasks in VMs is $i > j$. Table 1 shows the term and meaning that are used in the proposed algorithm. To achieve our goal, we have calculated the capacity of VM using Eq. (1). At initial stage, all tasks are allocated to VM that takes some time. To reduce makespan, we need to reduce allocation time and increase the allocation speed. Task allocation time T_{at} depends upon task length with respect to speed. Task allocation time is shown in Eq. (2). Expected execution time of task T_i execute on virtual machine VM_j can be represented in Eq. (3). It is calculated by task length and capacity of VM. Finally, total execution time of all tasks at any VM_j is the summation of expected execution time and allocation time. It is shown in Eq. (4). Makespan time can be reduced when a greater number of tasks are executed successfully at data center. It also gives an impact of system throughput. The makespan (MS) is the highest execution time of a task and can be computed using Eq. (5). The fitness value is used to minimize the makespan time as shown in Eq. (6).

$$\mathrm{VM_{cap}} = \mathrm{PS} \times \mathrm{C} \tag{1}$$

$$\mathrm{T_{at}} = \mathrm{T_{len}}/\mathrm{T_s} \tag{2}$$

$$\mathrm{T_{eet}} = \mathrm{T_{len}} \times \mathrm{VM_{cap}} \tag{3}$$

$$\mathrm{T_{et}} = \mathrm{T_{eet}} + \mathrm{T_{at}}. \tag{4}$$

$$\mathrm{MS} = \max(\mathrm{T_{et}}). \tag{5}$$

$$\mathrm{F} = \min(\mathrm{MS}) \tag{6}$$

3.3 Proposed Algorithm

In our proposed algorithm, PSO works after the initialization of the task and virtual machine. It calculates the task execution time and minimizes the makespan by taking task allocation time, length of task and capacity of VM. The standard PSO, proposed by [6] has no inertia weight and shows the slow convergence rate as well

Table 1 Denote terms and meaning

Term	Meaning	Term	Meaning
VM_{cap}	Capacity of VM	T_s	Task allocation speed
PS	Processing speed of VM	T_{eet}	Expected execution time
C	Central processing unit	T_{et}	Execution time of task
T_{at}	Task allocation time	F	Fitness value
T_{len}	Length of task	MS	Makespan time

Table 2 Different inertia weights

Name	Value	Eq. No.	Ref.
Modified inertia weight	$\omega = \left(\frac{\alpha \times k_{\max}}{k^{\gamma} + k_{\max}}\right) + \beta$	(7)	[11]
Global–local best inertia weight	$\omega = \left(1.1 - \frac{\mathrm{G}_{\text{best}\,i}}{\mathrm{P}_{\text{best}\,i}}\right)$	(8)	[18]
Logarithm decreasing inertia weight	$\omega = w_{\max} + (\omega_{\min} - \omega_{\max}) \times \log_{10}\left(a + \frac{10k}{k_{\max}}\right)$	(9)	[19]

as local optimum problem. The final value is gradually reduced over the iteration. Inertia weight was first introduced by [16]. The author in [16, 17] proposed that an enormous inertia weight encourages a worldwide search while a little inertia weight encourages a local search. Later, different researchers have proposed a number of inertia weight values to overcome this problem in PSO. Similarly, in our proposed algorithm, we used the update position and velocity value of each particle, which is shown in Eqs. (10) and (11), respectively, and adding the inertia weight to investigate the search space in an ideal manner. We used global–local best inertia weight method. It is the better inertia weight strategy method from other inertia methods in our experiment section. Table 2 shows three important inertia weight setting methods, and Table 3 shows the notation. Figure 2 shows the pseudocode of the proposed algorithm.

$$X_i(k+1) = X_i(k) + \mathrm{V}_i(k+1) \tag{10}$$

$$\mathrm{V}_i(k+1) = \omega V_i(k) + c_1 r_1 (\mathrm{P}_{\text{best}} - X_i(k)) + c_2 r_2 (\mathrm{G}_{\text{best}} - X_i(k)) \tag{11}$$

4 Experiment Setup and Results

The proposed GLMPSO algorithm is executed on Eclipse Java Programming Environment and CloudSim toolkit. We have used our PC having configurations: Intel (R) Core (TM) i3-7100 Processor (2.40 GHz) and 4 GB of RAM. The

Table 3 Denote notation and meaning

Notation	Meaning	Notation	Meaning
ω	Inertia weight	$\mathrm{G}_{\text{best}\,i}$	Global best position of particle i
$\omega_{\max}$	Maximum inertia weight	$\mathrm{P}_{\text{best}\,i}$	Local best position of particle i
$\omega_{\min}$	Minimum inertia weight	a and β	Constant parameter, value is 1
$k_{\max}$	Total number of iterations	γ	Acceleration factor
k	Current iteration	c_1 and c_2	Cognitive factor
r_1 and r_2	Random numbers	X_i and V_i	Position and velocity

Input: VM set as $VM = \{VM_1, VM_2, \ldots, VM_j\}$, Task set as $T = \{t_1, t_2, \ldots, t_i\}$.
Output: Minimize makespan time
Parameter: $\omega,\ k,\ k_{max}, c_1,\ c_2,\ r_1 and\ r_2$

1. Start
2. Initializes number of tasks and VMs.
3. Initialize task position, allocation speed, capacity of VM and maximum iteration
4. Calculate P_{best} and G_{best} value of each task where P_{best} shows execution time is and G_{best} is shows optimal solution.
5. Calculate the fitness value (F) using the Eq. (5) and (6)
6. Compare F with P_{best} and G_{best}
7. Update the personal best P_{best}
8. Update the global best G_{best}
9. Update the iteration
10. Calculate the inertia weight by using Eq. (8)
11. Change the velocity and position using Eq. (10) and (11)
12. Repeat the step until we found most extreme point to reduce makespan
13. Stop

Fig. 2 Pseudocode of GLMPSO algorithm

Table 4 Parameter setting

Parameter name	Range	Parameter name	Range
Task number	100–1000	Population	30
Task length	1000–7000	Maximum iteration	50
VM number	3–5	c_1 and c_2	2
Processing speed	250–300	r_1 and r_2	0–1
Memory	512–1024	Stopping criterion	k_{max}

operating system was 64 sbits with Windows 10. The parameters used for the simulation are shown in Table 4.

The inspiration driving this work is to reduce the makespan value in dynamic environment by using suitable inertia weight parameter, i.e., global–local best inertia weight. The performance of GLMPSO is compared with two swarm optimization techniques, namely modified PSO (MPSO) and logarithm-PSO (L-PSO). GLMPSO takes two different testing scales, i.e., 3 and 5 VMs, and having 100–1000 numbers of tasks. Simulation results are shown in Figs. 3, 4, 5, 6. From Figs. 3 and 4, we obtained the result for scheduling the task with 3 and 5 VMs having 100–1000 tasks. From both the figures, it is clear that our proposed GLMPSO technique takes less makespan than other meta-heuristic PSO algorithms that are shown in Tables 5 and 6. In contrast to MPSO and L-PSO, the proposed algorithm GLMPSO reduces makespan by more than 2% and 8%, respectively, shown in Fig. 5. Similarly, GLMPSO can reduce makespan as 2% and 9%, respectively, in Fig. 6.

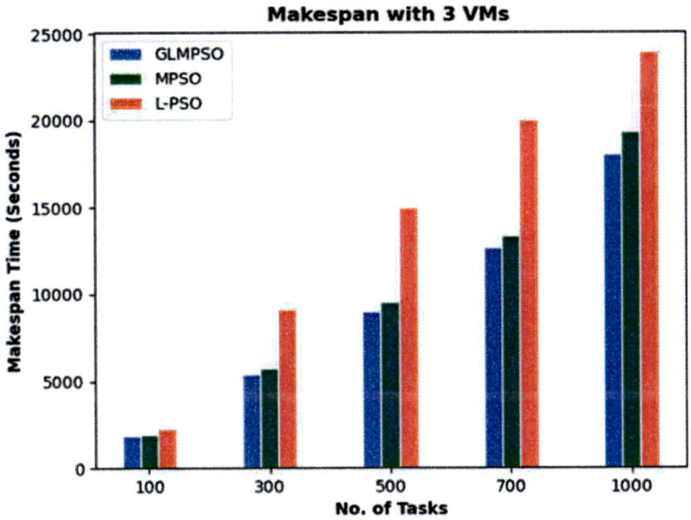

Fig. 3 Makespan value obtained for 3 VMs and 100–1000 task sets

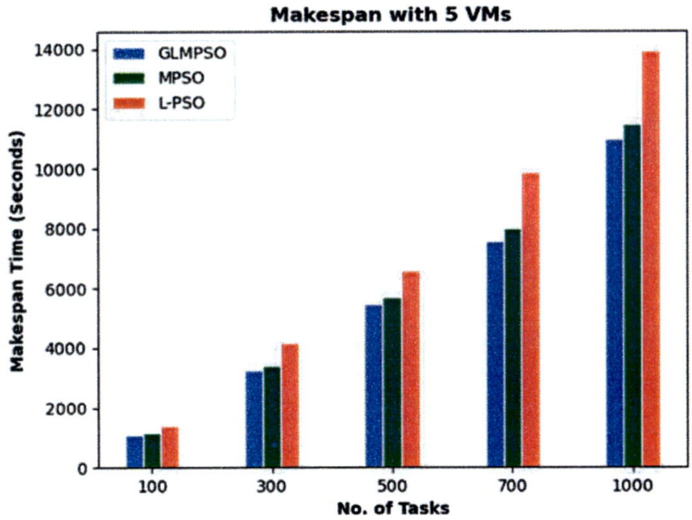

Fig. 4 Makespan value obtained for 5 VMs and 100–1000 task sets

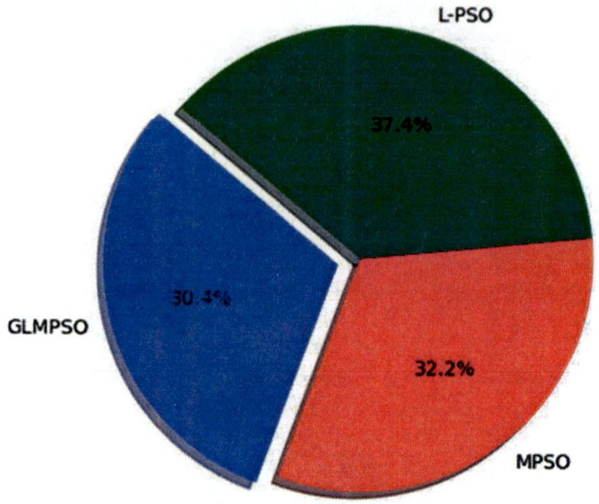

Fig. 5 Makespan over 3 VMs

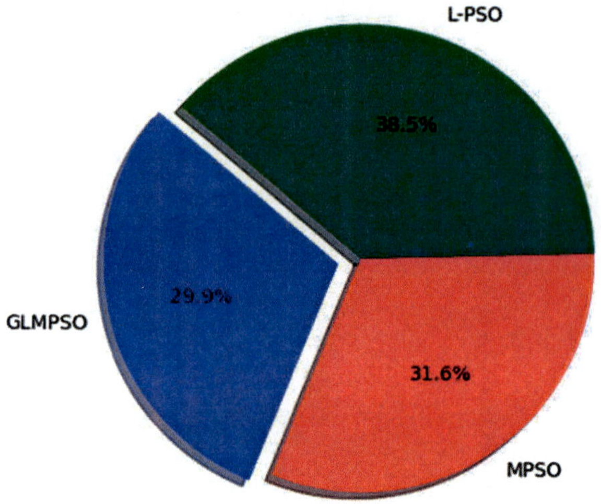

Fig. 6 Makespan over 5 VMs

Table 5 Minimum and maximum makespan for different scheduling algorithms for 3 VMs

Scheduling algorithm	100 tasks	300 tasks	500 tasks	700 tasks	1000 tasks
GLMPSO	1834.87	5394.74	9008.54	12,622.35	18,016.09
MPSO	1939	5701	9520	13,339	19,293
L-PSO	2253.53	9145.04	14.949.07	19,953.77	23,904.73

Table 6 Minimum and maximum makespan for different scheduling algorithms for 5 VMs

Scheduling algorithm	100 tasks	300 tasks	500 tasks	700 tasks	1000 tasks
GLMPSO	1079.74	3237.24	5448.54	7552.23	10,950.29
MPSO	1141	3421	5701	7981	11,458
L-PSO	1389.74	4170.32	6564.11	9851.26	13,907.98

5 Conclusion and Future Work

In this paper, we proposed a technique named GLMPSO which can decrease the overall makespan time and allocation time of task for allocating suitable VMs. As compared to standard PSO, GLMPSO algorithm has a faster convergence rate and better search capability during the period of iteration due to global-local best inertia weight method. In this, a huge amount of task can be handled by selecting suitable VM to minimize the allocation time, execution time and waiting time. Our simulation outcomes demonstrate that under each condition, GLMPSO diminishes the makespan time by 2% and 9% as compared to MPSO and L-PSO techniques, respectively. Further, we take some optimal value and compare our proposed algorithm with MPSO and L-PSO. Also, we take other parameters such as priority of task, deadline of task, energy consumption and other quality of service (QoS) which can be considered as future work.

References

1. Panda, S.K., Nanda, S.S., Bhoi, S.K.: A pair-based task scheduling algorithm for cloud computing environment. J King Saud University Comput. Inf. Sci. 1–10 (2018)
2. Xue, S., Shi, W., Xu, X.: A heuristic scheduling algorithm based on PSO in the cloud computing environment. Int. J. u- and e- Service, Sci. Technol. IJUNESST **9**(1), 349–362 (2016)
3. Liu, Z., Wang, X.: A PSO-Based Algorithm for Load Balancing in Virtual Machines of Cloud Computing Environment. ICS. Springer, Part I, LNCS 7331, pp. 142–147 (2012)
4. Xu, M., Tian, W., Buyya, R.: A survey on load balancing algorithms for virtual machines placement in cloud computing. Concurr. Computat Pract. Exper. **29**(12), e4123, 1–16 (2017)
5. Panda, N., Majhi, S. K.: How effective is the salp swarm algorithm in data classification. In: Computational Intelligence in Pattern Recognition, pp. 579–588. Springer (2020)
6. Kennedy, J., Eberhart, R.C.: Particle swarm optimization. In: Proceedings of the IEEE International Conference on Neural Networks pp. 1942–1948. Piscataway, NJ, USA (1995)
7. Bansal, J.C., Singh, P.K., Saraswat, M., Verma, A., Jadon, S.S., Abraham, A.: Inertia weight strategies in particle swarm optimization. Third World Congress on Nature and Biologically Inspired Computing. IEEE, pp. 640–647 (2011)
8. Alireza, A.L.F.I.: PSO with adaptive mutation and inertia weight and its application in parameter estimation of dynamic systems. Acta Automatica Sinica **37**(5), 541–549 (2011)
9. Masdari, M., Salehi, F., Jalali, M., Bidaki, M.: A survey of PSO-based scheduling algorithms in cloud computing. J. Netw. Syst. Manage. **25**(1), 122–158 (2016)

10. Huang, X., Li, C., Chen, H., An, D.: Task scheduling in cloud computing using particle swarm optimization with time varying inertia weight strategies. Cluster Computing. Springer (2019)
11. Zhou, Z., Chang, J., Hu, Z., Yu, J., Li, F.: A modified PSO algorithm for task scheduling optimization in cloud computing. Concurrency Computat Pract Experiment, pp. 1–11, Wiley (2018)
12. Agarwal, M., Srivastava, G.M.S.: A PSO algorithm based task scheduling in cloud computing. IJAMC **10**(4), 1–17 (2019)
13. Ebadifard, F., Babamir, S. M.: A PSO based task scheduling algorithm improved using a load balancing technique for the cloud computing environment. Concurrency Computat: Pract Exper, pp. 1–16. Wiley (2017)
14. Jacob, T.P., Pradeep, K.: A multi objective optimal task scheduling in cloud environment using cuckoo particle swarm optimization. Wireless Personal Communications. Springer (2019)
15. Saleh, H., Nashaati, H., Saber, W., Harb, H.M.: IPSO task scheduling algorithm for large scale data in cloud computing environment. IEEE Access **7**, 5412–5420 (2019)
16. Shi, Y., Eberhart, R.: A modified particle swarm optimizer, In: Evolutionary Computation Proceedings, 1998. IEEE World Congress on Computational Intelligence. IEEE International Conference, pp. 69–73 (1998)
17. Nickabadi, A., Ebadzadeh, M.M., Safabakhsh. R.: A novel particle swarm optimization algorithm with adaptive inertia weight. Appl. Soft Comput. **11**(4), 3658–3670 (2011)
18. Muthukumaraswamy, S.M., Rao, M.V.C.: On the performance of the particle swarm optimization algorithm with various inertia weight variants for computing optimal control of a class of hybrid systems. Discrete Dynamics in Nature and Society, Hindawi Publishing Corporation, vol. 2006, Article ID 79295, pp. 1–17 (2006)
19. Gao, Y.L., An, X.H., Liu, J.M.: A particle swarm optimization algorithm with logarithm decreasing inertia weight and chaos mutation. Int. Conf. Comput. Intell. Secur. **1**, 61–65 (2008)

Analysis of MATLAB-Based Segmentation and Thresholding of Satellite Image

Kamlesh Kumar Singh and Shantanu Trivedi

Abstract The shrinking of largest lakes is of worldwide problem. The Aral sea (endorheic lake) lying between Kazakhstan in north and Uzbekistan in south is continuously shrinking since 1960s after rivers that fed into it were diverted by soviet irrigation projects. Satellite image of Aral sea is not much clear to identify the region of water, land and iceland in the image. In this work, the Aral sea satellite image has been taken for the image segmentation. In image segmentation, many steps have been taken like color channel extraction, grayscale conversion, combined (RGB) grayscale conversion, thresholding then thresholding for final red, green and blue color extraction then region of interest segmentation and region of interest inverse of original image for clear human viewing of satellite image in order of water, land and iceland.

Keywords Image segmentation · Thresholding · Color channel extraction · Grayscale conversion

1 Introduction

Satellite color images carry a vast amount of information and undefined objects due to which analysis of image becomes complex for human viewing. To overcome this problem, image segmentation method is used [1]. Image segmentation plays an important and critical role in image analysis. It is defined by dividing an image into various parts based on some homogeneity criteria like color, gray, indexing, intensity, or texture, etc. In image segmentation method, color image segmentation carries more

K. K. Singh (✉)
Amity School of Engineering and Technology, Amity University Uttar Pradesh
Lucknow Campus, Lucknow, Uttar Pradesh, India
e-mail: kksingh@lko.amity.edu

S. Trivedi
IPE Global Center for Knowledge and Development, New Delhi, India
e-mail: shantanutrivedi3@gmail.com

S. Das and M. N. Mohanty (eds.), *Advances in Intelligent Computing and Communication*, Lecture Notes in Networks and Systems 202,
https://doi.org/10.1007/978-981-16-0695-3_19

information than gray image segmentation. Because gray picture conveys just the strength data, black at the feeblest strength and white at the most sturdiest strength [2]. But color image segmentation carries different objects or more information not only the strength yet in addition the chrominance of light. Human eyes can distinguish a great many shading, whenever while just a many gray scale are conceivable to be recognized in the meantime. Along these lines, we have first gone for color image segmentation where we find red, green and blue images of original image [3]. Then, second part is color channel grayscale conversion from original image where only single channel is extracted by not making others channel zero the image that will get produced will be a colorless images with respective red green and blue channels [4]. After that these RGB channels are merged so to get a combined RGB grayscale image from original image [5]. Third part will be the histogram analysis of satellite image showing the number and values of black and white areas present in a combined RGB grayscale image [6]. Fourth will be the thresholding of satellite image which produced only black and white colors of original image, black part shows the intensity of land and water while white part shows the only intensity of icelands presents in the image [7]. In the next part, satellite image of aral sea is internally processed and first converted into RGB grayscale image. After thresholding of this image doing final red color extraction, same process is done for green and blue color extraction. In the last part, region of interest segmentation for original satellite image and region of interest (inverse) for color image has been done which gives the enhanced and identified image in order of land water and iceland [6]. In the final inverse image, black parts show the iceland, gray part shows the land marks, and blue part shows the water area.

2 Color Channel Extraction

The figure demonstrates the RGB color extraction of aral sea satellite image where two of the respective channels out of three are put to zero for a single color extract generation [7] (Fig. 1).

Fig. 1 Color channel extraction of original image in red, green and blue parts (RGB)

3 Producing Single Grayscale Image Through Color Channel of Aral Sea Satellite Image

The figure demonstrates the red part, green part and blue part; grayscale images with a single channel in gray scale that is RGB with their respective grayscale color extract with no zeroing of the other channels while making a gray channel of a single color (Fig. 2).

4 Combined RGB Grayscale Image

Figure demonstrates the RGB grayscale image of original Aral sea satellite image with no color channel extraction [1] which is a mix of all three (RGB) gray channels for better enhancement and deep down study of Aral sea topography [8], its just a mixing of all the single-color channel grayscale extract together to produce a combined single grayscale RGB image (Fig. 3).

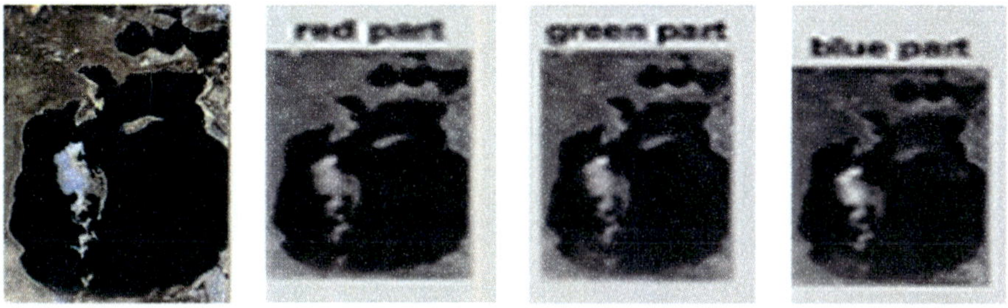

Fig. 2 Single grayscale conversion of original image in red, green and blue part images

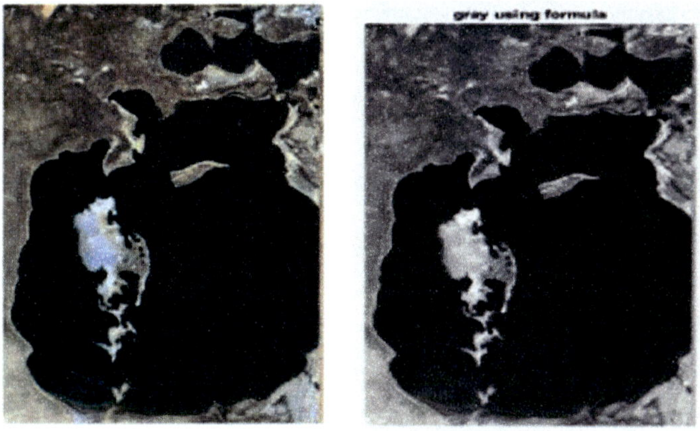

Fig. 3 Combined RGB grayscale image of original image

5 Histogram Analysis of Combined RGB Grayscale Image

The figure represents number of black and white parts present in the grayscale image produced through MATLAB which is very useful in pitching out the regions comprising of land, water and iceland [5]. Through this reproduction, we can clearly see the changes in water area depletion with time leaving shallow masses of land due to the reducing water level of the Aral sea (Fig. 4).

6 Thresholding of Satellite Image

It is most important powerful technique in segmentation [2]. Thresholding technique converts color image or grayscale image into binary image so here Aral sea satellite image is taken and is converted into its respective threshold image by doing so we can check different areas of interests with suitable markings and patterns different than previous images by adjusting the values there in the image pixels; we can highlight certain points and areas in the threshold image [9], and the region of interests which were not possible to study are now highlighted clearly such as the iceland areas in the land and water, white dotted parts of threshold image show the iceland present in the water and land areas while black part of threshold image shows the land and water areas (Fig. 5).

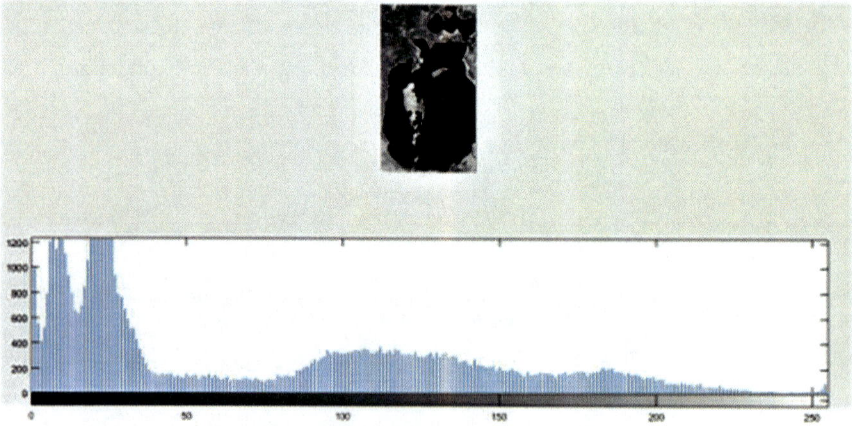

Fig. 4 Histogram of combined RGB grayscale image

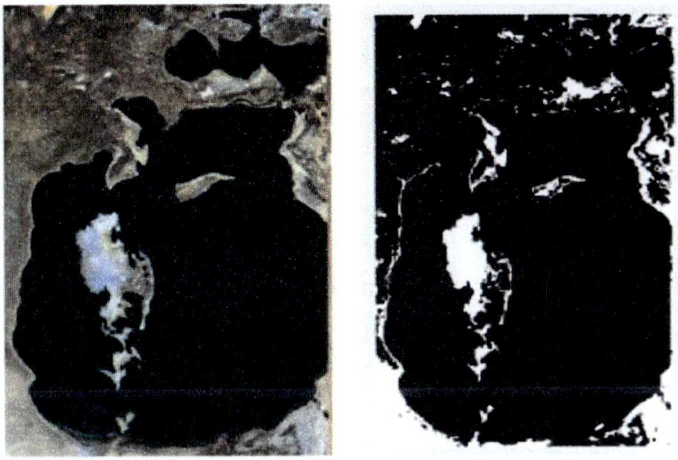

Fig. 5 Original image and thresholding of original image

7 Thresholding on Satellite Image with RGB Color Channel Extraction

The figure represents threshold imaging done upon the original satellite image with red (a), green (b) and blue (c) color channel extraction [9]. In this MATLAB-based process, original image has been taken as input which produced first grayscale image; then thresholding is done on the grayscale image. After this process, finally these three different threshold images red, green and blue produced due to which we can easily identify certain differences in the appearance of the imagery in the region of land, water and iceland [3]. These image analyses are based on the different colors present in the image after the thresholding [10]. Each color defines the different objects present in the satellite image (Fig. 6).

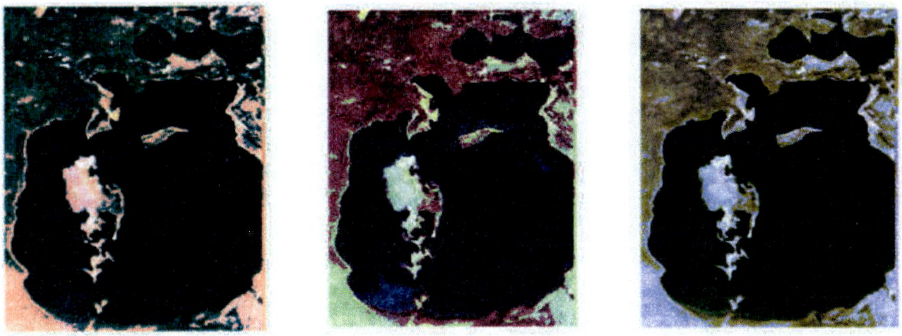

(a) (b) (c)

Fig. 6 Satellite image thresholding with **a** red color extraction, **b** green color extraction and **c** blue color extraction

(a) (b) (c)

Fig. 7 **a** Real satellite image, **b** region of interest segmented image and **c** region of interest inverse segmented image

8 Region of Interest Segmentation and Region of Interest Inverse for Color Image

In this MATLAB-based segmentation, original satellite image has been taken for the region of interest [11]. Region of interest segmentation gives the black and white images which identify the smallest areas of iceland, land and water present in image which was not properly analyzed in original image [12]. To enhance this analysis for every object individually, we have done the region of interest inverse for the color (original) image [4] (Fig. 7).

9 Analysis of Lake Satellite Image

The lake Aral sea has been declined to 10% of its original size and been splitting into four small lake. Image processing tools help in the detecting, extracting and monitoring the area of land water and iceland in the lake. With the help of image segmentation techniques, we can clarify how much area of the water in the lake has been shrinked and what is the rate of water shrinking in every 10 years. My paper talks about these aspects and the tool which is used for doing the above analysis is MATLAB.

10 Conclusion

In this paper, MATLAB-based work concluded all the information about the satellite image of Aral sea. Different segmentation techniques are applied on the satellite image for identify certain differences in the appearance of the imagery in the region of land, water and iceland areas. Each smallest part of the land, water and iceland are analyzed through the different colors of image. Every applied technique gives the unique clarification about the original satellite image.

References

1. Deepika, N.P., Vishnu, K.: Different techniques for satellite image segmentation. In: Online International Conference on Green Engineering and Technologies (IC-GET 2015)
2. Mukhopadhyay, A.: Interactive approach to multiobjective genetic fuzzy clustering for satellite image segmentation. In: IEEE Uttar Pradesh Section International Conference on Electrical, Computer and Electronics Engineering (UPCON) Indian Institute of Technology (Banaras Hindu University) Varanasi, India, 9–11 Dec 2016
3. Muangkote, N., Sunat, K., Chiewchanwattana, S.: Multilevel thresholding for satellite image segmentation with moth flame based optimization. In: 13th International Joint Conference on Computer Science and Software Engineering (JCSSE) (2016)
4. Xing, J., Sieber, R., Kalacska, M.: The challenges of image segmentation in big remotely sensed imagery data. Annals of GIS Taylor & Francis 21 Oct 2014
5. Ganesan, P., Sajiv, G.: User oriented color space for satellite image segmentation using fuzzy based techniques. In: International Conference on Innovations in information, Embedded and Communication Systems (ICIIECS) (2017)
6. Awad, M.M.: A new geometric model for clustering highresolution satellite images. International Journal of Remote Sensing Taylor & Francis 02 Apr 2012
7. Bora, D.J., Gupta, A.K.: AERASCIS: an efficient and robust approach for satellite color image segmentation. In: International Conference on Electrical Power and Energy Systems (ICEPES) Maulana Azad National Institute of Technology, Bhopal, India, 14–16 Dec 2016
8. Marpu, P.R., Neubert, M., Herold, H., Niemeyer, I.: Enhanced evaluation of image segmentation results. Journal of Spatial Science Taylor & Francis 09 Jul 2010
9. Pare, S., Bhandari, A.K., Kumar, A., Singh, G.K., Khare, S.: Satellite Image Segmentation based on Different Objective Functions using genetic algorithm: A Comparative Study. IEEE (2015)
10. Grozavu, N., Rogovschi, N., Cabanes, G., Troya-Galvis, A., Gançarski, P: VHR satellite image segmentation based on topological unsupervised learning. In: 14th IAPR International Conference on Machine Vision Applications (MVA) 18–22 May 2015. Miraikan, Tokyo, Japan
11. Protschky, V., Seifert, P., Feit, S.: Stop Line Detection Using Satellite-Image Segmentation. IEEE (2015)
12. Manjula, K.R., Dinesh Kumar, E.: Segmentation of Satellite Image by Enhanced Spatial Clustering Approach. IEEE (2015)

Design and Characteristics Mode Analysis of Compact DGS-Based UWB-MIMO Antenna

Vipan Kumar Gupta and Rikshit Baru

Abstract A compact ultrawide band (UWB) multiple input multiple output (MIMO) antenna is presented in this paper. Miniaturization is attained by using defected ground structure (DGS). Antenna is designed using two square-shaped patch and a vertical stub in the ground plane. MIMO antenna has overall dimensions of $48 \times 36 \times 1.57$ mm^3. Proposed structure excites several characteristics mode (CM) which contributes to wide band resonance. Characteristics mode analysis (CMA) theory is used to evaluate modal significance (MS) and explaining antenna physical behaviour. Antenna shows mutual coupling less than −15 dB from 3 to 11 GHz resonating band. Envelope correlation coefficient (ECC) less than 0.01 and diversity gain more than 9.94 show diversity performance of antenna. Simulated and measured results are well matched, and antenna performance makes it a suitable candidate for UWB-MIMO applications.

Keywords Characteristics mode · Defected ground · Diversity performance · Vertical stub · Mutual coupling

1 Introduction

UWB technology has captivated much consideration due to low power spectral density of −41.3 dBm/MHz, high data rate and wide bandwidth from 3.1 to 10.6 GHz [1]. It is widely used for short-range communication as it does not require carrier frequency [2]. However, multi-path fading caused by diffraction, reflection and channel interference over the wider spectrum degrades the performance. This problem can be regulated by MIMO technology. Intermingling of UWB-MIMO

V. K. Gupta (✉) · R. Baru
Department of Electronics and Communication Engineering,
Sri Sai College of Engineering and Technology, Badhani, Pathankot, India
e-mail: er.vipingupta14@gmail.com

R. Baru
e-mail: rikshit11@gmail.com

S. Das and M. N. Mohanty (eds.), *Advances in Intelligent Computing and Communication*, Lecture Notes in Networks and Systems 202,
https://doi.org/10.1007/978-981-16-0695-3_20

technology provides significant improvement in spectral efficiency, channel capacity and communication link reliability [3]. In MIMO technology, multiple antennas are placed in close proximity that leads to mutual coupling. Thus, MIMO antenna must have high isolation (>15 dB) to enhance diversity performance as well as compact size to integrate easily with the device [4].

Various UWB-MIMO antenna structures with different isolation techniques have been presented by researchers. Decoupling structures are added to reduce mutual coupling among antenna elements. Meander line parasitic structure [5], open-end slot in ground plane [6] and neutralization line with defected ground structure (DGS) at the top and bottom layers of antenna are used in [7]. However, these decoupling structures make antenna design complex and possess larger planar dimensions. Radiation pattern and polarization diversity is another method to obtain good isolation with compact size. Hexagonal slot with I-shaped arm slot at ground plane and L-shaped slits in feed line are introduced to excite two orthogonal E-field vectors [8]. Orthogonal fed with stepped ground stub [9], T-shaped parasitic strip between radiators [10] and cross-connected octagonal shape DGS [11] are used for polarization diversity and coupling enhancement. Antipodal slot antenna is presented in [12] to reduce cross-polarization and to improve diversity performance. Anti-parallel ring-shaped radiator with connected ground stub [13], unconnected cross-shaped ground stub [14], Yagi-like arc-shaped reflector [15] and fractal geometries [16, 17] are used to enhance isolation more than 15 dB. Simple antenna structures without additional isolation enhancement techniques are presented by researchers. Two Spema-shaped radiators are connected at the point of lowest surface current concentration [18]. Impedance matching and correlation between antenna elements are analysed for three different orientations [19]. Orthogonally placed semicircular radiator antenna with reduced ECC is designed in [20].

Study of literature reveals that limited physical insight has been provided for antenna geometry. Characteristic mode analysis is one of the ways to describe antenna physical behaviour. In this method, antenna total current is divided into number of orthogonal modes referred as characteristics mode for perfectly conducting bodies without any excitation [21]. A systematic design approach using CMA to develop an UWB antenna is presented in [22]. In [14] CMA theory is used to tune the UWB-MIMO antenna performance in terms of mutual coupling and bandwidth. In Zhao et al. [23], different characteristics modes are used for the two antenna elements which helps to enhance isolation. CMA theory is also used to design reconfigurable antenna structure [24], optimization and synthesis of antenna [25] and for the analysis of antenna behaviour [26].

In this paper, a compact antenna structure is proposed for UWB-MIMO applications. Miniaturization is attained by using defected ground structure. Characteristic mode analysis is focused to understand operating behaviour of antenna. Two antenna elements are placed anti-parallel to reduce mutual coupling. Structure is designed, simulated and analysed using CST Microwave Studio.

2 Characteristic Mode Analysis Theory

Theory of CM was introduced by Garbacz [21] and further refined by Harington and Mautz [27, 28]. They provide relation between characteristics current and incidence electric field on the conducting surface and computational methods to evaluate CMs. Characteristics modes depend upon geometric dimensions, shape and operating frequency of the PEC structure and independent of any feed or excitation. CMs are evaluated using impedance matrix, which is calculated from the surface current density. Generalized equations of characteristic modes are given as follows:

$$X(I_n) = \lambda_n R(I_n)$$

where I_n are the characteristics current and λ_n are the magnitude of eigen values of the CMs. R and X are the real symmetrical operators given by $Z = R + jX$, Z is the Hermitian symmetrical operator and follows reciprocity theorem. Value of λ_n varies from $-\infty$ to ∞ and directly related to the stored energy in the radiating body. $\lambda_n = 0$ represents resonating mode; negative value of λ_n represents stored electrical energy, and positive value gives stored mechanical energy. Modes with small eigen value are good radiator and with larger eigen values are poor radiator. In this way, magnitude of λ_n tells how much a characteristics mode contributes in radiation. Modal significance (MS) is the normalized form of eigen values which varies from 0 to 1. MS is given as $\text{MS}_n = |1/1 + \lambda_n|$. Resonance from a CM occurs if $\text{MS}_n = 1$ and 70.7% value of MS_n give upper and lower cut-off frequency of that mode.

3 Design Methodology of DGS-Based Antenna

Antenna is designed on Fr4 substrate with $\varepsilon_r = 4.3$ and loss tan = 0.2. Defected ground structure with rectangular monopole is used to achieve wide band resonance. Antenna design process for single element is shown in Fig. 1. Initially, partial ground microstrip patch antenna1 is taken which is resonating from 3.1 to 5.8 GHz as shown in Fig. 2. A vertical slot is etched in the ground plane just below the fee line in antenna2 which enhances the impedance matching of the antenna. In antenna3, another vertical slot is etched on the right side of partial ground. Etching of vertical slots introduces inductive effect and generates higher resonating modes that contributes both in impedance matching and wide band resonance. Finally, to obtain UWB a vertical slit is added on the left side of ground plane.

To design MIMO antenna, two identical antenna elements are placed in two orientations parallel and anti-parallel to each other. Surface current by port1 is shown in Fig. 3. It can be observed that significance power is coupled from element one to another when antennas are placed in parallel position. Minimum surface current on antenna2 is flowing when antenna1 is excited while anti-parallel

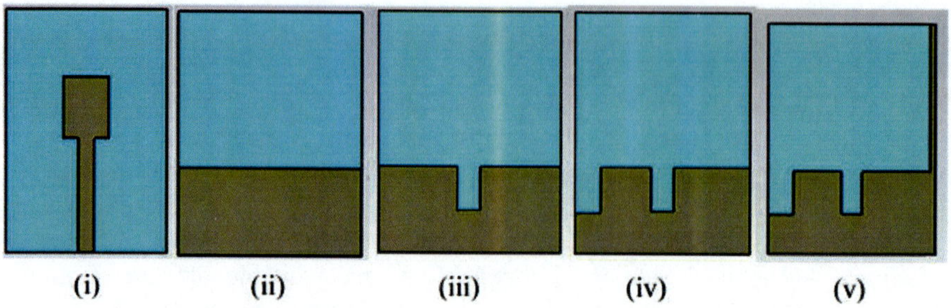

Fig. 1 Step wise evaluation of antenna **(i)** front view for all antennas **(ii)** back view antenna1 **(iii)** back view antenna2 **(iv)** back view antenna3 **(v)** back view antenna4

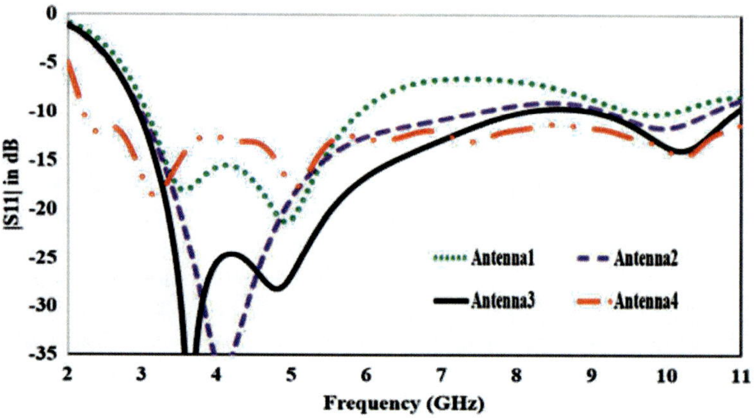

Fig. 2 Comparison of reflection coefficient for the designed structures

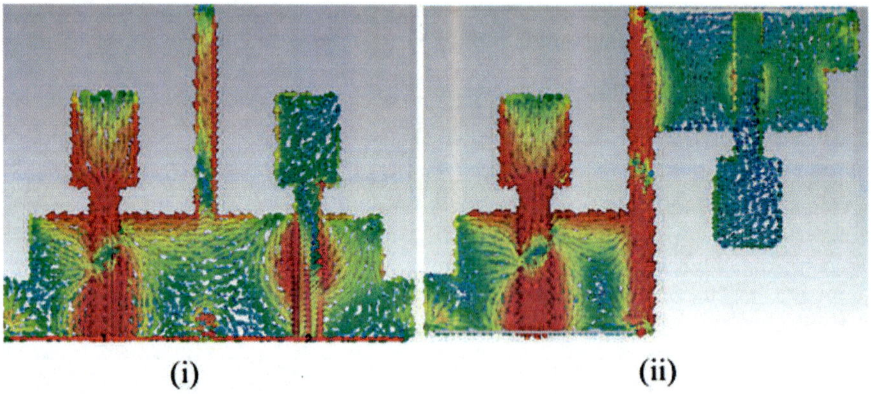

Fig. 3 Surface current distribution **(i)** parallel orientation **(ii)** anti-parallel orientation

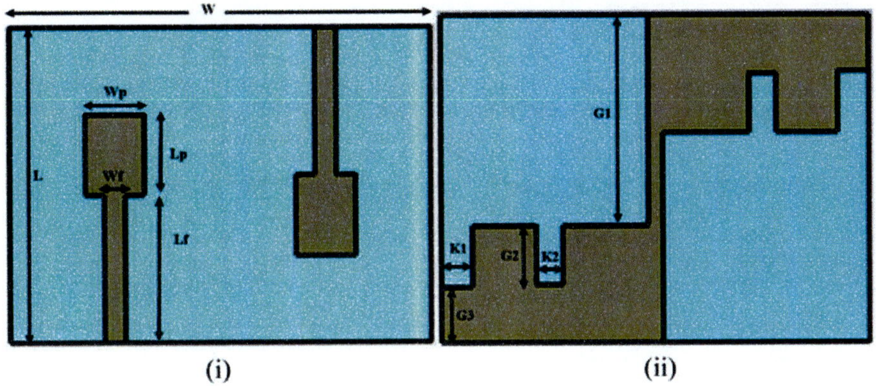

Fig. 4 Geometry of the proposed structure **(i)** front view **(ii)** back view

Table 1 Geometrical parameters of proposed structure

Parameter	Value (mm)	Parameter	Value (mm)
W	48	L	36
Wp	9.24	Lp	6.73
Wf	2.40	Lf	16.82
G1	23.07	G2	6.54
G3	6.12	K1, K2	3.5,3.0

orientation is used. Maximum isolation is achieved for the anti-parallel structure. Vertical slit has significant role in reducing mutual coupling by providing current path to the coupling current and offset the current from antenna1 to atenna2. Further common ground is achieved with this vertical slit. Geometry of anti-parallel UWB-MIMO structure is shown in Fig. 4. Table 1 lists geometrical dimensions of antenna.

4 Modal Significance

Characteristics modes are considered for better understanding of the design. MS analysis is carried out by integral equation solver in CST microwave studio. It signifies how the multiple CMs contribute to achieve wide band operation. Plots for modal significance and characteristics current distribution of CMs at their respective frequencies are shown in Figs. 5 and 6, respectively. Electrically small antennas can be characterized using only a few modes; therefore, significant six modes are presented here.

Mode 1 and Mode 2 dominate the input admittance of antenna at the lower-frequency ranges from 2.8 to 4.3 GHz. Mode 1 and 2 current is flowing in vertical upward direction, and null is occurring at the left edge of the ground plane.

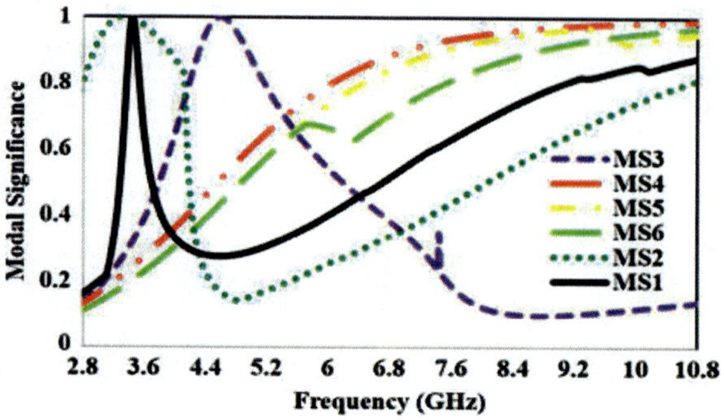

Fig. 5 Modal significance of antenna

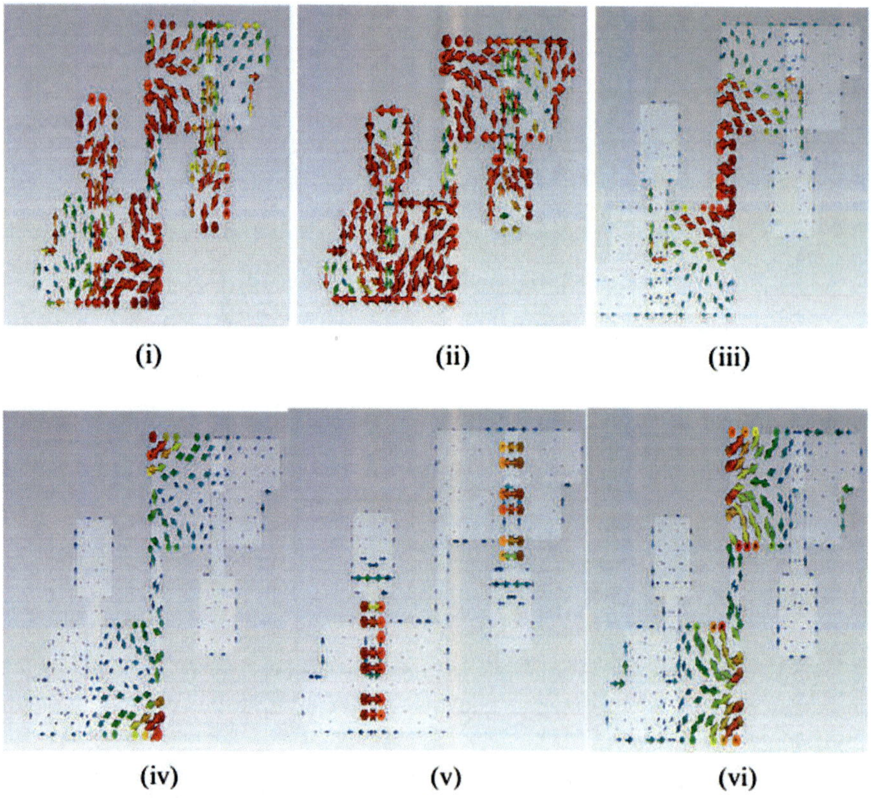

Fig. 6 Characteristics current distribution of CMs **(i)** mode 1 **(ii)** mode 2 **(iii)** mode 3 **(iv)** mode 4 **(v)** mode 5 and **(vi)** mode 6 at their respective frequencies

Mode 3 supports middle frequency band from 4.3 to 5.4 GHz. Maximum current density is concentrated at the common vertical ground slit. Mutual coupling is maximum at this band as the current is flowing in horizontal direction. Three upper modes have MS value close to each other and dominated the input admittance collectively from 5.4 to 10.6 GHz. In modes 4 and 6, current is concentrated at the lower left edge of the ground, and null is occurring at the centre of edge, providing better isolation between two antenna elements. Mode 5 maximum current is flowing in horizontal direction on the feed line. Maximum null area can be observed in this mode and the least dominating mode of antenna. Mode 2 has the minimum current null and the most dominating mode of antenna.

5 Results and Discussion

5.1 Return Loss and Isolation Between Ports

A prototype of the defected ground UWB-MIMO antenna is fabricated and measured using Anritsu MS2028C network analyser. Scattering parameters are measured to analyse impedance matching and level of mutual coupling in terms of reflection coefficients (S11 and S22) and transmission coefficients (S21 and S12), respectively. Simulated and measured S-parameters are shown in Fig. 7. Both antenna elements have impedance bandwidth more than 3.1 to 10.6 GHz which satisfies the requirement for ultrawide band spectrum specified by FCC. It can be observed that isolation is more than 15 dB throughout the entire spectrum. Slight

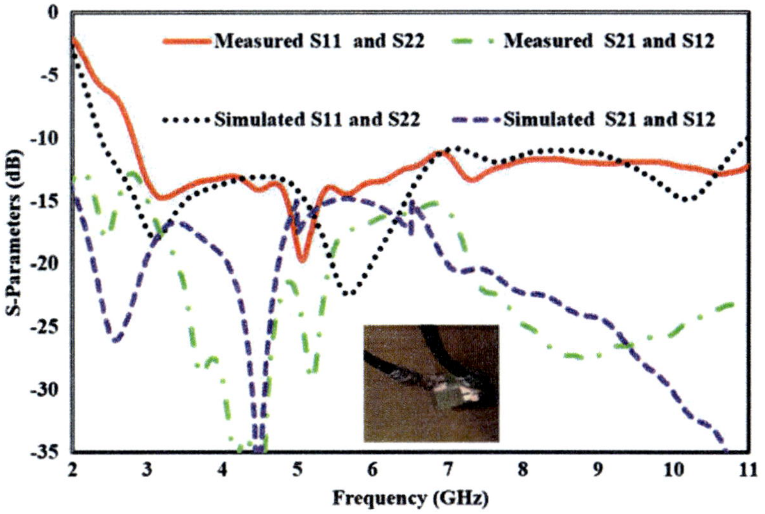

Fig. 7 Simulated and measured *S*-parameters

variation in simulated, and measured results are due to manufacturing tolerances and measurement set-up. Results obtained for impedance bandwidth (S11 < −10 dB) and mutual coupling (S21 < −15 dB) are acceptable in UWB-MIMO systems for practical applications.

5.2 Radiation Characteristics

The radiation characteristics of the proposed antenna have been measured with Port 1 excited. Simulated and measured radiation patterns at 3.5, 5.2 and 10 GHz are plotted in Fig. 8. Since the MIMO antenna consists of two identical elements located anti-parallel to each other, the patterns of Port 1 and 2 are almost similar with a 180° rotation. It can be observed that antenna has unidirectional radiation pattern in E-plane and dumb-bell-shaped pattern in H-plane at lower frequency. Due to anti-parallel position of antenna elements and unidirectional radiation pattern, MIMO structure can have orthogonal radiation characteristics that can reduce coupling effect between adjacent antenna elements. Few side lobes are observed in the H-plane at middle frequency and becomes quasi-omnidirectional type at higher frequency.

5.3 Diversity Performance

To support the MIMO capability of proposed antenna, two important diversity parameters such as envelope correlation coefficient (ECC) and diversity gain (DG) are analysed. ECC signifies the correlation of antenna radiation pattern with other antenna when operated in closed proximity. High value of ECC degrades

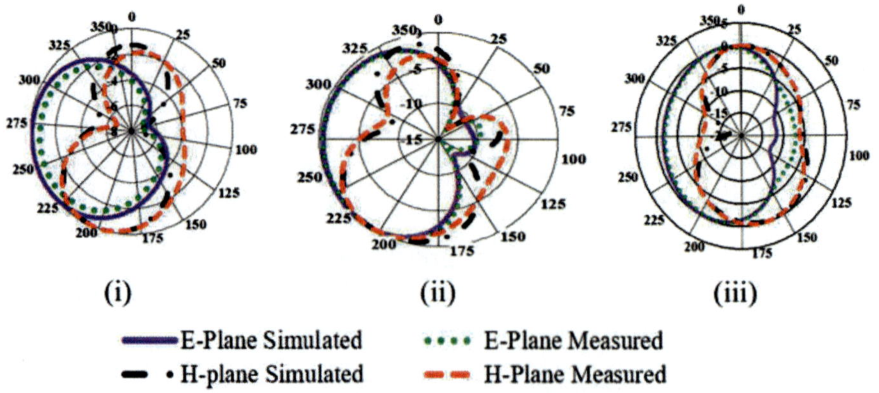

Fig. 8 Radiation pattern (i) 3.5 GHz (ii) 5.2 GHz (iii) 8.5 GHz

antenna MIMO performance. ECC value lower than 0.05 is acceptable for MIMO antennas. ECC can be calculated using S-parameters in uniform domain [16].

$$\rho_e = \frac{\left|S_{11}^* S_{12} + S_{21}^* S_{22}\right|^2}{\left(1 - \left(|S_{11}|^2 + |S_{21}|^2\right)\right)\left(1 - \left(|S_{22}|^2 + |S_{12}|^2\right)\right)}$$

Diversity gain shows accomplishment of diversity performance of MIMO antenna. For practical applications, DG should be higher than 9.9, and it can be evaluated using ECC.

$$\mathrm{DG} = 10\sqrt{(1 - |\mathrm{ECC}|)}$$

Simulated and measured ECC and DG plots are shown in Fig. 9. It can be observed that ECC results are well below 0.01, and DG is more than 9.94 throughout the UWB from 3.1 to 10.6 GHz.

In Table 2, proposed antenna is compared with existing UWM-MIMO designs described in literature. Various criteria such as dimension, design techniques and diversity performance are considered. It is clear that additional structure contributes to reduce mutual coupling but our structure has good diversity performance with simple geometry. Moreover, physical behaviour is analysed using characteristics mode theory.

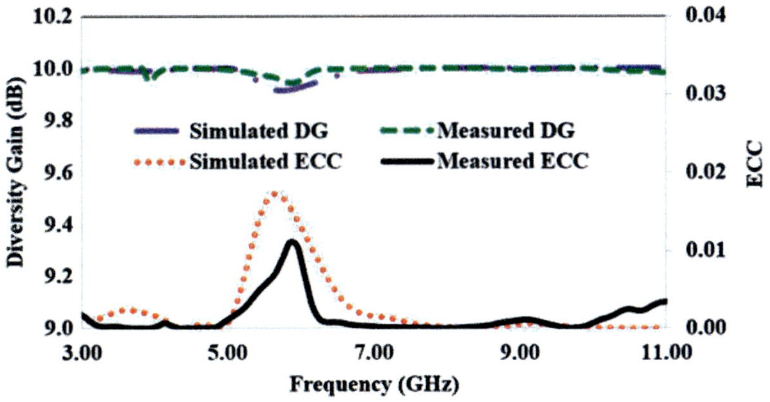

Fig. 9 Plots for ECC and DG

Table 2 Comparison of proposed structure with existing UWB-MIMO antennas

Refs.	Antenna arrangement	Design methodology	Dimension (mm^2)	Max. mutual coupling (dB)	ECC
[9]	Orthogonal	Connected ground and isolation using with T-shaped parasitic strip	39 × 39	−15	0.02
[13]	Anti-parallel	Connected ground using arc-shaped slits	25 × 45	−15	0.2
[15]	Parallel	Connected and used of Yagi-like reflector for isolation	50 × 80	−17	0.058
[17]	Orthogonal	Fractal octagonal and connected ground plane with L-shaped ground stub for isolation	40 × 25	−15	0.01
[18]	Opposite	Two connected Spema-shaped radiators	59 × 46	−13	0.02
[20]	Orthogonal	Truncated partial unconnected ground plane	26 × 55	−20	0.01
Proposed work	Anti-parallel	Connected defected ground and CMA theory	48 × 36	−15	0.01

6 Conclusion

In this article, a defected ground UWB-MIMO antenna is designed and analysed using characteristics mode theory. Two rectangular patch antennas are placed anti-parallel for better isolation. It is verified that several CMs are excited on the conducting surface supporting wide band resonance. Vertical stub has significant contribution in reducing mutual coupling in lower and upper frequency band. Mode 3 shows maximum mutual coupling near 5 GHz frequency. Less than −15 dB coupling is provided by the antenna over the entire UWB. Unidirectional radiation pattern with low side lobes is offered by the antenna. Antenna has less than 0.01 ECC and more than 9.94 diversity gain. Antenna features make proposed structure a potential candidate for the upcoming wireless technologies.

References

1. Federal Communications Commission: Revision of Part 15 of the Commission's Rules Regarding Ultra-Wideband Transmission System from 3.1 to 10.6 GHz, pp. 98–153. Federal Communications Commission, Washington, DC, ET Docket (2002)
2. Xu, H., Yang, L.: Ultra-wideband technology: yesterday, today, and tomorrow. IEEE Radio and Wireless Symposium, Orlando, FL, pp. 715–718 (2008)
3. Kaiser. T., Zheng. F., Dimitrov, E.: An overview of ultra-wide-band systems with MIMO. In: Proceedings of the IEEE, pp. 285–312 (2009)

4. Ben, I.M., Talbi, L., Nedil, M., Hettak, K.: MIMO-UWB channel characterization within an underground mine gallery. IEEE Trans. Antennas Propagat. 4866–4874 (2012)
5. Ren, J., Hu, W.: Compact printed MIMO antenna for UWB applications. IEEE Antennas Wirel. Propag. Lett. 1517–1520 (2014)
6. Lee, J.M., Kim, K.B., Ryu, H.,Woo, J.: A compact ultrawideband MIMO antenna with WLAN band-rejected operation for mobile devices. IEEE Antennas Wirel. Propag. Lett. 990–993 (2012)
7. Zhang, S., Pedersen, G.F.: Mutual coupling reduction for UWB MIMO antennas with a wideband neutralization line. IEEE Antennas Wirel. Propag. Lett. 166–169 (2015)
8. Saxena, S., Kanaujia, B.K., Dwari, S., Kumar, S., Tiwari, R.: A compact dual polarized MIMO antenna with distinct diversity performance for UWB applications. IEEE Antennas Wirel. Propag. Lett **16**, 3096–3099 (2017)
9. Kang, L., Li, H., Wang, X., Shi, X.: Compact offset microstrip-fed MIMO antenna for band-notched UWB applications. IEEE Antennas Wirel. Propag. Lett. 1754–1757 (2015)
10. Biswal, S.P., Das, S.: A low-profile dual port UWB-MIMO/diversity antenna with band rejection ability. Int. J. RF Microwave Comput.-Aided Eng. Wiley Periodicals, 1–11 (2017)
11. Arun, H., Alsath, M.N.: Octagonal DGS based dual polarized ring-shaped antenna for MIMO communications. Int. J. Electron. **6**(5), 756–769 (2019)
12. Lu, W., Cheung, S.W., Cheng, Y., Zhu, H.: Ultra-wideband antipodal slot antenna with improved radiation performance. Int. J. Electron. **9**, 1263–1279 (2012)
13. Mathur, R., Dwari, S.: Compact CPW-fed ultrawideband MIMO antenna using hexagonal ring monopole antenna elements. Int. J. Electron. Commun. (AEÜ)1–6 (2018)
14. Singh, H.V., Tripathi, S.: Compact UWB MIMO antenna with cross-shaped unconnected ground stub using characteristic mode analysis. Microwave Optical Technol. Lett. 1–8 (2019)
15. Jehangir, H., Sharawi, M.S.: A miniaturized UWB bi-Planar Yagi-like MIMO antenna system. IEEE Antennas Wirel. Propag. Lett. **16**, 2320–2323 (2016)
16. Gurjar, R., Upadhyay, D.K., Kanaujia, B.K., Sharma, K.: A novel compact self-similar fractal UWB MIMO antenna. Int. J. RF Microwave Comput.-Aided Eng. **3**, 216–232 (2018)
17. Tripathi, S., Mohan, A., Yadav, S.: A compact octagonal fractal UWB MIMO antenna with WLAN band rejection. MicrowOptTechnol Lett. **57**, 1919–1925 (2015)
18. Mondal, S., Mandal, K., Sarkar, P.P.: Design of MIMO antenna for ultra-wideband applications. IETE J. Res. **4**, 497–502 (2018)
19. Jusoh, M., Jamlos, M.F., Kamarudin, M.R., Erawan, A.: Correlation analysis on UWB MIMO tree-antenna orientations. Microwave Optical Technol. Lett. **3**, 688–692 (2013)
20. Toktas, A., Akdagli, A.: Compact multiple-input multiple-output antenna with low correlation for ultra-wide-band applications. IET Microwaves Antennas Propag. **9**, 822–829 (2015)
21. Garbacz, R.J.: Modal expansions for resonance scattering phenomena. Proc. IEEE , 856–864 (1965)
22. Wu, W., Zhang, Y.P.: Analysis of ultra-wideband printed planar quasi-monopole antennas using the theory of characteristic modes. IEEE Antennas Propag. Mag. **6**, 67–77 (2010)
23. Zhao, X., Yeo, S.P., Ong, L.C.: Planar UWB MIMO antenna with pattern diversity and isolation improvement for mobile platform based on the theory of characteristic modes. IEEE Trans. Antennas Propag. **1**, 420–425 (2017)
24. Wu, W., Wang, B.Z., Yang, X.S., Zhang.Y.: A pattern-reconfigurable planar fractal antenna and its characteristic-mode analysis. IEEE Antennas Propag. Mag. **3**, 68–75 (2007)
25. Liu, D., Garbacz, R.J., Pozar, D.M.: Antenna synthesis and optimization using generalized characteristic modes. IEEE Trans Antennas Propag. **6**, 862–868 (2010)
26. Cabedo-Fabres, M., Antonino-Daviu, E., Valero-Nogueira, A., Bataller, M.F.: The theory of characteristic modes revisited: a contribution to the design of antennas for modern applications. IEEE Antennas Propag. Mag. **5**, 52–68 (2007)
27. Harrington, R., Mautz, J.: Theory of characteristic modes for conducting bodies. IEEE Trans. Antennas Propag. **5**, 622–628 (1971)
28. Harrington, R., Mautz, J.: Computation of characteristic modes for conducting bodies. IEEE Trans. Antennas Propag. 629–639 (1971)

A Hybrid Model for Epileptic Seizure Classification Using Wavelet Packet Decomposition and SVM

Millee Panigrahi, Dayal Kumar Behera, and Krishna Chandra Patra

Abstract The synaptic disturbance in the prefrontal portion of the brain induces epileptic seizures. Electroencephalography is a noninvasive tool for diagnosing the different brain disorders including the seizures. Visual inspection of these signals for prognosis of a disease is very time consuming. Therefore, automated epileptic seizure detection methods are adopted by the medical practitioners for fast analysis and accurate detection of disease. An integrated method was proposed in this work, in which wavelet packet decomposition (WPD) method is implemented for time frequency transformation and then the features are extracted. After feature extraction, four different classification models are compared using the balanced train test split method, where 70% train dataset and 30% test dataset have been taken for validation of the model. The results reveal that WPD with SVM has an accuracy of 96% which outperforms other conventional models applied to the benchmark dataset of Bonn University's EEG signal for seizure prediction.

Keywords Electroencephalography · Epilepsy · Seizure prediction · Wavelet packet decomposition · SVM

1 Introduction

The physiological activities of the brain are recorded by electroencephalogram (EEG) [1]. EEG is a noninvasive method of recording information of all the electrical activities of the neurons inside the brain. It provides a lot of information regarding any dysfunction occurred inside the brain of the epileptic patients and patients with other brain disorders. World Health Organization estimates that about 50 million people are afflicted by epilepsy worldwide [2]. People with epilepsy

M. Panigrahi (✉) · K. C. Patra
Sambalpur University Institute of Information Technology, Odisha 768019, India
e-mail: millee.panigrahi82@gmail.com

D. K. Behera
Silicon Institute of Technology, Bhubaneswar, Odisha, India

S. Das and M. N. Mohanty (eds.), *Advances in Intelligent Computing and Communication*, Lecture Notes in Networks and Systems 202,
https://doi.org/10.1007/978-981-16-0695-3_21

normally experience intermittent seizures with no symptoms of weakness or the whole body collapse. Hence, sudden attack of seizures has a profound consequence on the patient and needs immediate attention with proper medication after diagnosis. Therefore, manual investigation of the disease may be time consuming and inaccurate, for which the patients may suffer.

EEG recordings contain non-Gaussian, nonlinear and non-stationary signals with a 10 microvolt to 100 microvolt amplitude and a frequency from 1 to 100 Hz [3]. These signals are measured by placing the electrodes on the patient's scalp externally, whose positions are determined by 10–20 electrode system decided by International Federation of Societies for Electroencephalography. It is a challenging task to extract the information related to the seizures from the original signal using existing signal processing techniques.

Previously, neurologists have put some effort in analyzing the normal and seizure activity through visual inspection which is a very tedious and prolonged task. There is no particular methodology to distinguish between the seizure occurrences except detecting the abnormal spikes in the signals which may be sometimes erroneous. Gradually, with the advent of brain computer interface, several automatic computational methods are proposed by researchers which helped the neurologists in the detection of seizure.

In past, many machine learning algorithms [4] are proposed by the researchers for the classification of EEG to predict seizures. The performance of the models is validated by using multiple datasets for the classification problem in different domains. Hence, wavelet packet decomposition (WPD) methods are very effective for extracting every minute transition in the EEG signals. The statistical features are extracted, and feature selection method is applied for dimension reduction. The classification of the features for the prediction of seizure and non-seizureclass involves algorithms like SVM, linear regression, decision tree, gradient boosting, K-nearest neighbor, etc. [5].

In this paper, a hybrid SVM-based classifier is employed with eight level WPD for time frequency analysis of binary class prediction problem for balanced train test data. Wavelet packet decomposition employs approximation as well as detailed coefficients for feature extraction as compared to discrete wavelet transform that employs only approximation coefficients for feature extraction. Hence, WPD provides better frequency resolution than any other decomposition methods.

High output by dimension reduction, based on WPD with feature extraction technique, is the key inspiration for the paper. The model would be tested with a dataset of 70% of the testing and 30% of the evaluations. This reduces the model's complexity and allows for a binary prevision class, minimizing the calculation time. The majority of the paper is structured as follows: Sect. 2 outlines the work that the researchers have previously performed. The benchmark dataset is addressed in Sect. 3. The proposed model is defined in Sect. 4. The study and discussion of findings are seen in Sect. 5, and work finishes in Sect. 6.

2 Related Work

Several automated methods of epilepsy diagnosis were proposed during the last decades to take account of the diverse EEG domains. There were two drawbacks to this domain-based approach. Firstly, these methods were unable to predict the non-stationary property of the EEG signal [6]. Secondly, the different artifacts and noise interference produce serious distortion of the EEG signals for which the prediction of the seizures becomes erroneous [7]. To address this issue, wavelet-based transformation techniques have been implemented by the researchers. An optimized wavelet coefficient in a continuous wavelet-based adaptive technique is employed for the improvement of the performance of the model [8]. In [9], the processing time is reduced by applying ideal channel selection method for multichannel EEG epochs taking the statistical features and adaptive method for decomposition. Frequency domain and the nonlinear methods [10] have also become an effective way for EEG extraction of features. Features dependent upon time frequency, such as wavelets, using DWT techniques allow quantitative study of EEG signals more powerful in terms of precision and efficacy in the extraction of features and EEG classification than other approaches [11]. Therefore, different wavelet-based approaches are recognized to denoise the EEG signals. In [12], MODWPT method is applied for EEG classification taking nonlinear entropy features. Recent studies show that EWT-based transformation technique is implemented for non-stationary signals like EEG using Hilbert transformation [13] and firefly Levenberg Marquardt (FLM) [14] to obtain time frequency analysis of the signals. The WPD employs division of sub-bands into packets for autoregressive models proposed by Zhang et al. [15]. Li et al. [16] classified ECG signals by using wavelet packet entropy on benchmark database of MIT BIH arrhythmia disease. This method provided a better performance than any other methods for identification of the disease. In [17], the IMFs of EEG signals provided by EMD and then function sets are defined by the LSSVM classifier in order to predict focal or non-focal seizure by different entropy-based characteristics. David et al. [18] implemented various entropy features on two different EEG datasets that outperformed the existing models in artifact removal and added a qualitative approach in performance.

3 Materials and Methods

The benchmark dataset for this work was considered by the University of Bonn [1]. There are five sets (A, B, C, D and E). These datasets consist of 100 separate channels, each 23.6 s long of 4096. These signals were divided into 1 s parts, each of which contained 178 marks. These datasets have been grouped into eight combinations of classes named as AE, BE, CE, DE, AB-E, AD-E, CD-E and ABCD-E where E is the seizure class and other groups are non-seizure class. Here, we modeled this as a combination of classes for feature extraction where even the minute seizure occurrence could not be missed and accurate prediction could be done easily. The EEG recording is done through 128 channel amplifier with a sampling frequency of 173 Hz.

4 Model Description

The model proposed was seen in Fig. 1. The dataset is compiled using the seizure class (E) with all other possible combination of data. Here we have taken eight possible combinations of data. After that eight level WPD of the dataset is done using Daubechies 4 (db4) mother wavelet function. The EEG data samples undergo dimension reduction keeping in view that there is no data loss by applying feature extraction technique. Here the statistical features are extracted from the coefficients of the wavelets and are appended for processing in the next step. Statistical features such as mean (*M*), standard deviation (σ), skewness (Skew), kurtosis (Kurt), average power (*AP*) and ratio of the mean (RM) are extracted considering all the wavelet coefficients. Formulas used to extract the features are represented in Eq. 1–6.

$$M(X) = \frac{1}{n}\sum_{i=1}^{n}|Y_i| \tag{1}$$

$$\sigma = \sqrt{\frac{1}{n}\sum_{i=1}^{n}(Y_i - M(X))^2} \tag{2}$$

$$\text{Skew} = \sqrt{\frac{1}{n}\sum_{i=1}^{n}\frac{(Y_i - M(X))^3}{\sigma^3}} \tag{3}$$

$$\text{Kurt} = \sqrt{\frac{1}{n}\sum_{i=1}^{n}\frac{(Y_i - M(X))^4}{\sigma^4}} \tag{4}$$

$$\text{AP} = \sqrt{\frac{1}{n}\sum_{i=1}^{n}Y_i^2} \tag{5}$$

$$\text{RM} = \frac{\frac{1}{n}\sum_{i=1}^{n}|Y_i|}{\frac{1}{n}\sum_{i=1}^{n}|Z_i|} \tag{6}$$

where '*n*' represents the periodic duration of the data series EEG signal. *Y*(*i*) contains the signal information at *i*th number of sample. $Y\{y_1, y_2, \ldots, y_i\}$ and $Z\{z_1, z_2, \ldots, z_i\}$ are two neighboring sub-bands (or IMFs).

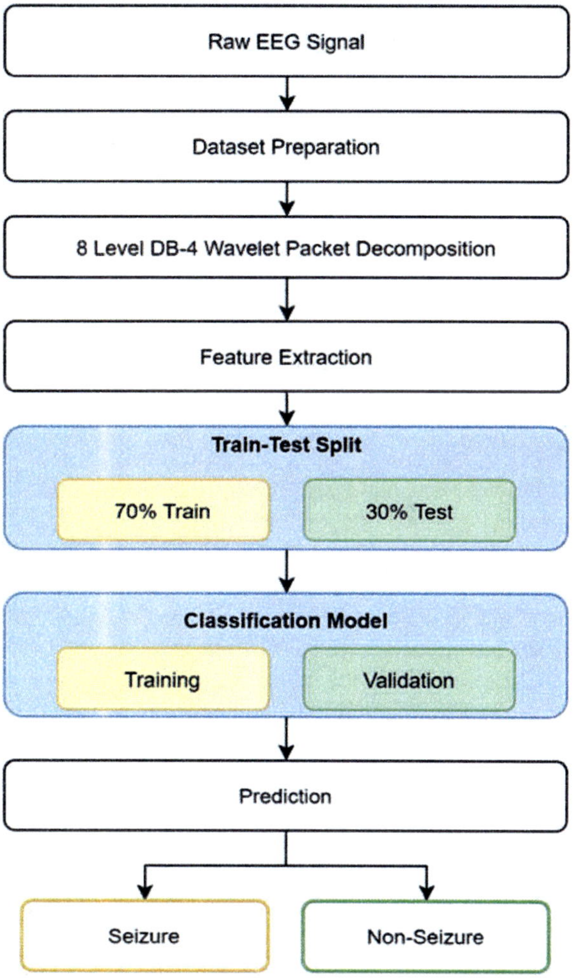

Fig. 1 Flowchart presentation of the proposed model

The feature samples are then divided into training and testing modules. The training data is taken to be 70%, and testing data is 30%. Then the data samples are balanced by synthetic minority oversampling (SMOTE) method [19]. This method effectively reduces the clock skew problem, and here this approach is used to increase the number of samples. The balanced feature set is then passed to the classifiers for seizure prediction. For seizure detection, the following classifications are applied, and classifying output is compared. In logistic regression (LR), a binary-dependent variable is represented using a logistic function. Decision tree (DT) is a managed learning process with a larger effect on the whole layout of the tree in the top nodes. In K-nearest neighbors (KNN), the nearest 'k' training samples are applied for the predictive analysis. Support vector machine (SVM) is the classifier which uses hyperplanes to separate the samples in a two-dimensional plane. The transformation of the samples is then implemented known as kernels. The classifications of the feature samples are then validated by implementing K-fold cross-validation method. The method reduces the global error by 4% compared to other evaluation techniques to separate the seizure and non-seizure class.

5 Result and Discussion

A comparative analysis with different classifiers is performed to quantify the evaluation of classification process. To obtain this, specific parameters are set for the classifiers. For KNN, neighbors are taken to be 5, weight is uniform, leaf size is 30, and p value is 2 with the distance metric to be “minkowski”. For decision tree classifier, criterion is “gini”, the minimum number of split samples is 2, and the minimum leaf sample is 1. Similarly for SVM, the “C” value is taken to be 1, kernel is radial basis function, and the degree is taken to be 3. The classifier performance is given in Table 1.

It has been observed that support vector machine outperforms other classifiers according to the performance parameters obtained in Table 1. The confusion matrix of various models is plotted in Fig. 2.

The results obtained are compared with all the classifiers taken into account for only statistical feature extraction. Here six statistical characteristics are taken into account for the extraction of features, and then the features are categorized by the above classifiers. Then the results obtained are compared with designed model where both statistical features and the WPD method are considered.

It is observed that when only statistical features are considered, the accuracy of LR, decision tree, KNN and SVM is found to be 0.552, 0.832, 0.912 and 0.943, respectively. But when both statistical features and WPD are considered, then the accuracy of LR, DT, KNN and SVM results is 0.541, 0.911, 0.937 and 0.967, respectively. The proposed model shows better accuracy than the conventional approach. The comparative performance analysis is exhibited in Fig. 3.

Table 1 Metrics of results with multiple classifiers

Classifier	F1Score	Accuracy	Precision	Recall
LR	0.295	0.541	0.207	0.510
DT	0.794	0.911	0.705	0.902
KNN	0.806	0.937	0.973	0.688
SVM	0.914	0.967	0.904	0.924

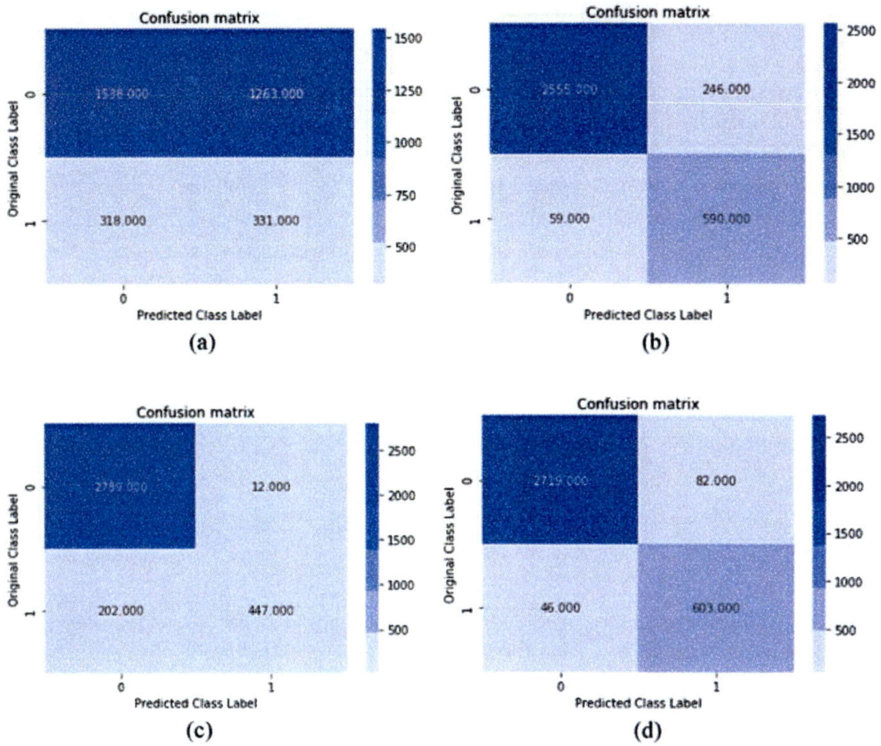

Fig. 2 Confusion matrix of **a** LR, **b** DT, **c** KNN and **d** SVM classifier

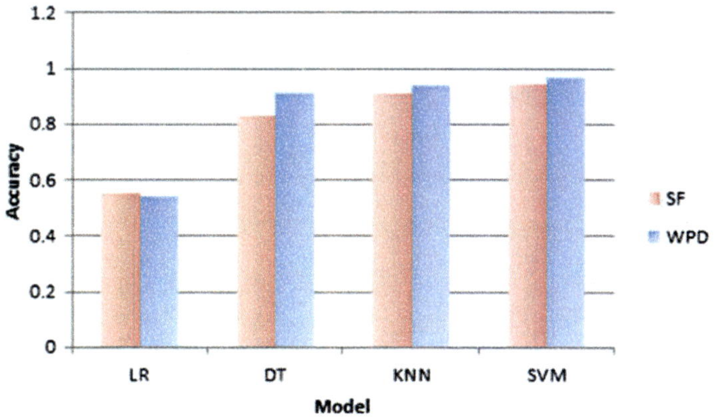

Fig. 3 Comparison of feature extraction methods for different classifiers

6 Conclusion

The diagnosis of epileptic seizures has been a problem for doctors to correctly predict the seizure and the non-seizure class. The main conflict is to differentiate between the normal EEG and the seizure occurrence. Hence, a binary classification approach has been modeled using a hybrid support vector machine with the WPD. This hybrid model outperforms other conventional models with high accuracy of about 96.7%. The WPD followed by the proper extraction of statistical features increases the performance of the model with improved accuracy. The predicted model has been implemented on the benchmark dataset available publicly and achieved greater classification accuracy. This allows the researchers to implement this proposed model to long-term EEG signal for prediction of seizure class in a more accurate way. In the future, deep learning algorithm may be implemented with multi-class prediction of seizures.

References

1. Andrzejak, R.G., Lehnertz, K., Rieke, C., Mormann, F., David, P., Elger, C.E.: Indications of nonlinear deterministic and finite dimensional structures in time series of brain electrical activity: dependence on recording region and brain state. Phys. Rev. E 64, 061907 (2001)
2. Fisher, R.S., Boas, W.V.E., Blume, W.: Response: definitions proposed by the international league against epilepsy (ILAE) and the international bureau for epilepsy (IBE). Epilepsia **46** (10), 1701–1702 (2005)
3. Karthick, P.A., Tanaka, H., Khoo, H.M., Gotman, J.: Prediction of secondary generalization from a focal onset seizure in intracerebral EEG. Clin. Neurophysiol. **129**, 1030–1040 (2018)
4. Tanveer, M.A., Salman, A.: Epileptic seizure classification using gradient tree boosting classifier. In: Proceedings of the 9th International Conference on Biomedical Engineering and Technology, pp. 71–77 (2019)
5. Shoeb, A., Guttag, J.: Application of machine learning to epileptic seizure detection. In: Proceedings of the 27th International Conference on Machine Learning (ICML-10), pp. 975–982 (2010)
6. Kevric, J., Subasi, A.: Comparison of signal decomposition methods in classification of EEG signals for motor-imagery BCI system. Biomed. Signal Process. **31**, 398–406 (2017)
7. Raghu, S., Sriraam, N.: Classification of focal and non-focal EEG signals using neighborhood component analysis and machine learning algorithms. Expert Syst. Appl. **113**, 18–32 (2018)
8. Bilal, K., Anam, H., Omar, F.: An automated system for epileptic seizure detection using EEG in book. advances in data and information sciences. In: Lecture Notes in Networks and Systems, Springer Nature Singapore Pte Ltd. pp. 147–149 (2020)
9. Panigrahi, M., Behera, D.K., Patra, K.C.: Classification of EEG recording in an epileptic patient using DWT and extreme gradient boosting classifier. Int. J. Adv. Sci. Technol. **29**(5), 4163–4180 (2020)
10. Abualsaud, K., Mahmuddin, M., Saleh, M., Mohamed, A.: Ensemble classifier for epileptic seizure detection for imperfect eeg data. The Scientific World Journal. (2015)
11. Bhattacharyya, A., Pachori, R.B.: A multivariate approach for patient-specific EEG seizure detection using empirical wavelet transform. IEEE Trans. Biomed. Eng. **64**(9), 2003–2015 (2017)

12. Hunyadi, B., Signoretto, M., van Paesschen, W., Suykens, J.A.K., van Huffel, S., de Vos, M.: Incorporating structural information from the multichannel EEG improves patient-specific seizure detection. Clin. Neurophysiol. **123**, 2352–2361 (2012)
13. Jacobs, J., Vogt, C., LeVan, P., Zelmann, R., Gotman, J., Kobayashi, K.: The identification of distinct high-frequency oscillations during spikes delineates the seizure onset zone better than high-frequency spectral power changes. Clin. Neurophysiol. **127**(1), 129–142 (2016)
14. Ahmadi, A., Tafakori, S., Shalchyan, V., Daliri, M.R.: Epileptic seizure classification using novel entropy features applied on maximal overlap discrete wavelet packet transform of EEG signals. In: 7th International Conference on Computer and Knowledge Engineering (ICCKE), pp. 390–395. IEEE (2017)
15. Zhang, Y., Liu, B., Ji, X., Huang, D.: Classification of EEG signals based on autoregressive model and wavelet packet decomposition. Neural Process. Lett. **45**, 365–378 (2017)
16. Li, T., Zhou, M.: ECG classification using wavelet packet entropy and random forests. Entropy **18**, 285 (2016)
17. Sharma, R., Pachori, R.B., Acharya, U.R.: Application of entropy measures on intrinsic mode functions for the automated identification of focal electroencephalogram signals. Entropy **17** (2), 669–691 (2015)
18. Cuesta–Frau, D., Miró–Martínez, P., Núñez, J.J., Oltra–Crespo, S., Picó, A.M.: Noisy EEG signals classification based on entropy metrics. Performance assessment using first and second generation statistics. Comput. Biol. Med. **87**, 141–151 (2017)
19. Fergus, P., Hignett, D., Hussain, A., Al-Jumeily, D., Abdel-Aziz, K.: Automatic epileptic seizure detection using scalp EEG and advanced artificial intelligence techniques. BioMed Res. Int. (2015)

Blockchain Technology and Its Implementation Challenges with IoT for Healthcare Industries

Ashish Kumar Mourya, Bhavya Alankar, and Harleen Kaur

Abstract Currently, the electronic health records are stored digitally, and the systems are centralized on a small scale. Internet of Things (IoT) is not related to one thing/device or one aspect alone, rather it enfolds and connects various associated innovative things such as blockchain, artificial intelligence (AI), big data, cloud storage, and hyper-computing. Besides, there are almost no untouched aspects of social fields like education, research, e-commerce, healthcare, defense, transportation, etc., which does not get affected by IoT and its associative supporting technology. The concept of blockchain in healthcare sounds so amazing, but not been successfully implemented by any healthcare sector, and since there is no proven use case, nobody wants to take risk. Nevertheless, to make blockchain technology successful, multiple fund providers, healthcare researchers, and health ministries will have to work together for the transformation of new collaboration, because once it would be done, it is so going to benefit the end users. This review has enlightened to all the research communities about the use of blockchain and IoTs devices in healthcare industries with major challenges.

Keywords Electronic health records (EHRs) · Crypto-currency · Smart contract · Transaction · Cryptography · Interoperability · Decentralized computing

1 Introduction

The demand of blockchain technology and the wide range of its applications provided a new edge to research in various systematic and scientific areas. Blockchain technology is still in its infancy stage, although it is considered as a revolutionary solution, dealing with recent technological issues like data ownership, decentralization, trust, identity, and data-driven decisions. However, the era of technology is

A. K. Mourya · B. Alankar (✉) · H. Kaur
Department of Computer Science and Engineering, School of Engineering Sciences and Technology, Jamia Hamdard, New Delhi, India
e-mail: bhavya.alankar@gmail.com

S. Das and M. N. Mohanty (eds.), *Advances in Intelligent Computing and Communication*, Lecture Notes in Networks and Systems 202,
https://doi.org/10.1007/978-981-16-0695-3_22

approaching to next level where innovative and efficient tools and techniques are being deployed to handle and process big data. The sudden increase of information that is being generated online by means of social media, Internet, and worldwide communications has increasingly rendered data-driven learning. A new study revealed that over 4 million queries are being received by Google every minute, e-mails' sent by users reaches the limit of 200 million messages, 72 h of videos are uploaded by YouTube users, 2 million chunks of content are shared over Facebook, 277,000 Tweets are generated every minute on Twitter, WhatsApp users share 347,222 photos, and Instagram users post 216,000 new photos every minute [1, 2]. The present age is the age of big data, where data is growing on a large scale than ever before. According to the Computer World, 70–80% of data is unstructured form in organizations [3]. The data, which derives from social media, form 80% of the data globally and report for 90% of big data. As stated by the International Data Corporations (IDC) annual digital universe study [4], the data are being produced too rapidly, and by the end of 2020, it would touch the range of 44 zeta bytes which would be ten times larger than it was in 2013 [5]. With the advancement in technology, there is a massive expansion of digital data in terms of both quantity and variety that are produced by machines as well as users. Although finding the optimal approach to store, process, and organize big data, the blockchain technology comes with effective and efficient ways to handle big data. The blockchain technology has dominated the digital market nowadays because of its huge number of applications [6]. However, blockchain technology is still novel, and in experimenting phase, not fully deployed in most of the research fields, it finds a place in the digital market industry as a ground-breaking solution, and tackles the concerns related to modern technology such as decentralization, reliance, distinctiveness, data ownership, and decisions that are data driven [7, 8].

The data on the cloud environment are usually safeguarded with the conventional access control mechanisms [9–11], but such methods have been broken several times in the past; hence, a system that handles the security and privacy of the sensitive data is highly anticipated. The access control mechanisms are usually applied by accomplishment of three subparts, identification of the users, authentication of whether they are legitimate, and authorization given to them based on the category they belong to. Along with these functionalities, it is the responsibility of the access control protocols to maintain the log of the operations or actions performed by the users and declare up to what extent the user can make the actions [12]. Therefore, the access control mechanism must be applied to the healthcare systems to protect the patient's data from being misused. By storing the patient's data over cloud, the system is providing the anytime accessing facility to the users. The user could be the patient's themselves or medical practitioners, insurance providers, or the pharmacists, and they might want to access the data from anywhere around the globe through the Internet [13–15].

2 Challenges in Implementation of Blockchain-Based Healthcare Data Management System

Currently, the electronic health records are stored digitally, and the systems are centralized on a small scale. The concept of blockchain in healthcare sounds so amazing, but not been successfully implemented by any healthcare sector, and since there is no proven use case, nobody wants to take risk. Nevertheless, to make blockchain technology successful, multiple fund providers, healthcare researchers, and health ministries will have to work together for the transformation of new collaboration, because once it would be done, it is so going to benefit the end users.

Health records privacy are respecting on the blockchain technology open token ledger because only transaction hashes and decoded data emerge and nothing connection can be prepared with the uses. Consequently, we could disagree that blockchain technology adds an addition layer of security and privacy to user information, but it does not make data records unhackable. Security and privacy of medical data are one of the most important concerns, because the data are extremely sensitive and the rate of hacking the data is increasing exponentially. If the blockchain architecture is implemented in the health care, it would use on-chain and off-chain which means that most of the data will not be on the blockchain; in fact, it would be stored on database or cloud, but blockchain uses the cryptography, so only the person having key can access the data. This means that blockchain have a layer of security over cloud and databases. Also, the other issue could be the loss of key. Since the patient will be having the key and if he somehow loses it, then it will cause a problem. Another issue could be with the patients that are not in the condition to control their data, they will have to share their key to someone else for accessing the data. There are also few patients who are not willing to share their personal data on a distributed network. Some patients who are having STDs or any other severe disease do not want everyone to know about it. So, they are against it. Also, some hospitals do not share the exact details of the medications to the insurance companies and the interoperability between the two organizations becomes difficult [16, 17].

3 Blockchain Technology

Blockchain, a new emerging technology, and cloud computing are the two related component of digital data processing business. However, blockchain can process, record, and authenticate. Machine leaning helps in pattern reorganization and decision making. Blockchain technology has a capability to switch over highly secure in other industries as well. Blockchain technology has the capability to make the data accessing and transfer easier and simpler among different nodes. It has maximum security and accountability of bandwidth network. One of the best scenarios of blockchain technology in medical domain is accessing clinical data of hospital and patients [18]. We have already known that in today scenario, millions of data stored in a decentralized architecture. Its processing in cloud is also a big challenge. In a recent white paper of World economic Forum that has been released in September 2015, the online business and stored product will be switched on blockchain technology. Blockchain was proposes for trading cryptocurrency (Bitcoin) on online. This technology was first described in a white paper report of Satoshi Nakamoto, who point out the challenges of cryptocurrencies and their solution in the form of blockchain technology proceeding to the completion of transaction. The traceability and transparency are the two main features provided by the blockchain to the distributed network. The Bitcoin blockchain was the first ever application based on blockchain, proposed by Nakamoto [19], which maintains the transactions of the crypto-currency Bitcoin. The blockchain network is based on peer-to-peer connections, where nodes are connected to one another and verifies the transactions before allowing them to make their way into the blockchain. The nodes that are responsible for the verification of the transactions are called miners. Whenever a transaction is requested, a new block is generated, and for the transaction to be completed, it needs to be verified by the miner nodes. The blocks are connected to one another as shown in Fig. 1.

The blockchain is decentralized network that does not require any type of central authority and thus making the transactions safe and the identity of users confidential. Table 1 gives a brief description of various application areas in healthcare domain where blockchain can be used beneficially, with respect to the properties provided by the blockchain technology.

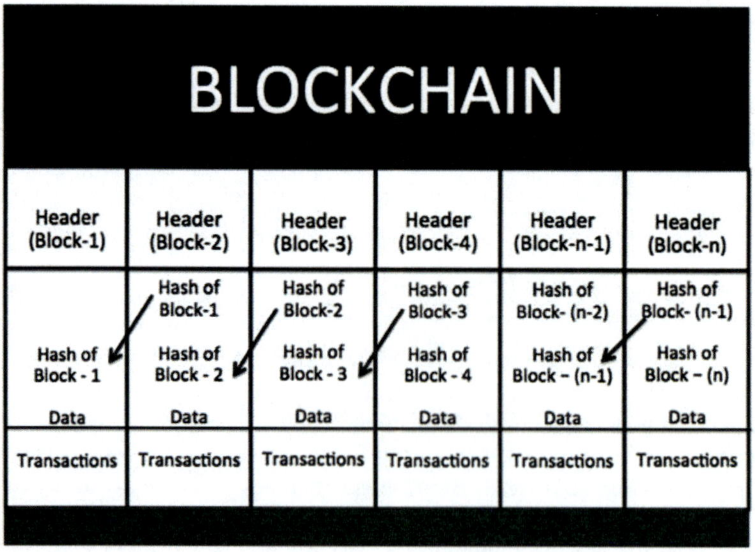

Fig. 1 Blocks in blockchain

Table 1 Blockchain use cases in various healthcare application areas

Blockchain property	EHRs management	Healthcare insurance management	Medical research
Decentralized network	Patient-centric management of the EHRs that give the power to the patients over the data [20]	Because of no intermediary present, the real-time health insurance claiming can be provided [21]	The data sharing is increased without losing the control over the data or resources [22]
Data security	The EHRs are secure as they are in encrypted form and can be only decrypted by rightful user [23]	The private information of the patient such as personal details, financial details, and medical details are kept secure [24]	The data of different types of users are shared among the institutes while preserving the security [25]
Data provenance	The EHRs are signed digitally by the source node and therefore proving its legitimacy [21]	The verification of the records for the processing of insurance claims is possible because of the distributed nature of blockchain [24]	The medical data are provided with the provenance for enhanced research [26]
Data availability	The data are stored in distributed manner; thus, no single party is holding the data, and hence, it is difficult to be attacked [12]	The data are accessible and available all the time from multiple storages [21]	The blockchain provides the real-time accessing of the data that improve the overall care provided, in case of emergency [12]
Immutability	The EHRs stored on the blockchain cannot be altered by anyone [13]	The detection of fraud is better in blockchain along with the audit facility [21]	The data generated in blockchain are time-stamped and can be tracked easily [25]

4 Blockchain Technology and Their Applicability in Different Domain

Before we discuss the inadequacies of this framework, we need to understand some basic terminology used in this technological framework. Nowadays, every end user is not using any smart device in its exclusivity. Each Internet-enabled device is connected through a chain of other such online active devices for attaining more optimized performance. Blockchain being one of the supporting associative technologies of the IoT came as an innovative distributed public ledger appended as a platform to facilitate virtual or cryptocurrency like Bitcoin in 2008 and provide a more secure peer-to-peer transfer electronic money though smart contracts or cryptocontract.

Don and Alex Tap Scott define it in his book blockchain revolutions as "The Blockchain is an incorruptible digital ledger of economic transactions that can be programed to record not just financial transactions but virtually everything of value." In simple terms, a blockchain is a zero-transaction charge decentralized platform to store and transfer information from one server to another in such a secured manner that it cannot be altered. In reference to blockchain-dependent IoT technical framework, a smart contract is a device to devise protocol on the Internet, stored on a blockchain platform waiting to be executed as when and where the pre-agreed terms of contract are met. Every node connected to this blockchain is traceable, and the transactions are verifiable in real time. Although, the unique feature of a blockchain-based smart contract is that every contract once executed at a particular point of time in the form of a block cannot be re-modified or altered retrospectively. That is why, chronologically, it is one of the safest measures of transferring secured data, which may be financial in nature or of any other kind. Each appended block on a blockchain will have the following characteristics for its unique identification and safety: Software Version: this contains the details of the version of the software used by the previous miner. Block Index: which contains a specific number of indexes in which block is stored. Last #: this contains the details about the previous block to reaffirm that it has already been validated. Root #: this contains all the details of the cryptic-sensitive data of the transaction between the sender and the receiver including miner's node. Time Seal: this contains specific block creation time of new appended block acting as a chronological tool of identification. New #: this contains the details about the next node where it is headed. This component confirms that how big or small a hash function should be to claim the message's next validity as a proof of work. Nonce: this is an additional security variable besides proof of work, where the miner must calculate a hash function smaller than its destination.

4.1 Major Security Concerns of Blockchain Platform Protocols for IoT

Despite the major categories of attacks namely on identity-theft through impersonation, crypto-analytic, reputation, manipulation, of an end user, some industry expert claim that a blockchain is still a best bargain for meeting the efficiency, privacy, confidentiality, integrity, availability, reliability, trust, safety, and security standards for IoT. Nevertheless, the true picture is far from clarity. A very few of such attacks are discussed as in the ensuing paragraphs. In reference to the attacks through identity-theft or impersonation, the hacker creates a number of fake anonymous identities, which he later attaches to any of the real user identity, using a malicious code, during an unprotected data access or data transfer portal. Then, the hacker uses this anonymous identity to replace the real one and executes it, as if he is the real end user. Thereby, he uses his tools to unlock the personal and sensitive information of the real user to his own advantage. In the cases of distributed denial of services (DDoS) attack, connectivity services or the access to a particular service like sign-in or login are strategically denied for a particular node or server or computer system on the grid, so that the hacker's objective is achieved by the delay during which DDoS was in execution. Broadly, the security concerns can be divided into three categories viz. technological, organizational, and environmental. Starting with the technological security, the first one is scalability. As more transaction load is being built up on exchanges set for Bitcoin currencies, the blockchain is getting to slow down in terms of storage for the creation of additional blocks as well as the processing.

4.2 Issues Rising from Blockchain and IoT Used in Various Sectors

With a never before boom in broadband Internet capacities, distributed computing, and decentralized data sharing, the blockchain engagement rules with IoT have also evolved to encircle multiple utilities in new fields like e-recruitment, intelligent traffic management, logistics, supply chain management, health care, defense, banking, etc. Let us discuss a few of the necessity like healthcare. The IoT when used with even a Blockchain platform, there is a certain possibility of unauthorized sharing of data viz. electronic medical record of a particular patient or category of patients. Now the individual data may not have extremely specific value unless and until the patient is a key celebrity or personality of some business or political association. However, if a database of all such category of patient falls into the hands of a pharmaceutical manufacturing entity, then they might use it to manipulate the drugs availability and its price. Besides, it will be extremely hard to put zoom in and seek a proof for such misuse of personal databases. A similar breach of trust and misuse of personal data and sensitive information can be traced to banking sector, where Know Your Customer (KYC) scheme at times becomes a mandatory clause to identify and classify a customer. But to know how many smart grids and how many nodes of IoT, such data or database is shared; nobody can give a clear response to that. Therefore, there is an urgent need to address such boiling issues, despite an efficient performance can be achieved with the judicious use of Blockchain dependent IoT, like in the fields of banking, transportation, aviation, healthcare, e-voting, e-recruitments, logistics, supply chain management, energy management, etc. But we should be warned that as we approach the 5G era of telecommunication, the previous address and value of the block can be identified and exposed from the block-header. This puts the data in a kind of vicious loop of chained exposure despite having encryption in place.

5 Conclusion

Technology is a multi-facet and sharp-edged tool for progress of man. But how that tool is used for serving a society better than yesterdays must be governed by a set of unanimous rules, protocols, and law. Just like technology, human society and law are always in a state of continuous improvement. Its only that whenever a new technology is implemented at a mass scale, the terms of its micro- and macro-implications should settle at an acceptable standard. So that, the rights, liabilities, and duties should be defined and implemented in such a manner that the end user of any such technology may not end being a victim of it. There is not a slightest doubt of any degree that every invention or innovation, be it AI, IoT, big data, blockchain, cloud Computing, etc., all are the creation of human ingenuity and intelligence. The man has tried his best to improvise his standard of living through his curiosity and perseverant spirit for the betterment in every field of the civilization. However, it is also most ironic to realize that in all his brave efforts to instill the humane attributes into his creations, on his quest for being the creator of such machines and technologies, he has somehow ignored to put the crucial element of human touch in them. Therefore, until the machines are not fully equipped and upgraded with more discretionary algorithms, about morality, privacy, security, and respect for the fellow humans, such a technology may prove to be a very costly affair for everyone.

References

1. Kemkarl, O.S., Dahikar, D.P.B.: Can electronic medical record systems transform health care potential health benefits, savings, and cost using latest advancements in ICT for better interactive healthcare learning. Int. J. Comput. Sci. Commun. Netw. **2**(3/6), 453–455 (2012)
2. Bordea, G., Jothi, N., Rashid, N.A., Husain, W.: Data mining in healthcare: a review. Proc. Comput. Sci. **72**, 306–313 (2015). https://doi.org/10.1016/j.procs.2015.12.145
3. Zhang, Y., Chen, M., Huang, D., Wu, D., Li, Y.: idoctor: Personalized and professionalized medical recommendations based on hybrid matrix factorization. Futur. Gener. Comput. Syst. (2016). https://doi.org/10.1016/j.future.2015.12.001
4. Fox News Health: 'Ransomware' Cyber-attack Cripples Hospitals Across England. Associated Press, New York (2017). May
5. Hashemi, S.T., Ebadati, O.M., Kaur, H.: Cost estimation and prediction in construction projects: a systematic review on machine learning techniques. SN Appl. Sci. **2**(10), 1–27 (2020)
6. Birjais, R., Mourya, A.K., Chauhan, R., Kaur, H.: Prediction and diagnosis of future diabetes risk: a machine learning approach. SN Appl. Sci. **1**(9), 1112 (2019)
7. Journal H (2018) Largest Healthcare Data Breaches of 2018. HIPAA J.
8. Baboota, R., Kaur, H.: Predictive analysis and modelling football results using machine learning approach for English Premier League. Int. J. Forecast. 35 (2), 741–755 (2019)
9. Chen, H., Bhargava, B., Zhongchuan, F.: Multilabels-based scalable access control for big data applications. IEEE Cloud Comput. **1**, 65–71 (2014)
10. Maw, H., Xiao, H., Christianson, B., Malcolm, J.: A survey of access control models in wireless sensor networks. J. Sens. Actuator Netw. **3**, 150–180 (2014)
11. Brucker, A.D., Hang, I., Lückemeyer, G., Ruparel, R.: Secure BPMN: modeling and enforcing access control requirements in business processes. In Proceedings of the 17th ACM Symposium on Access Control Models and Technologies (SACMAT'12), Newark, NJ, USA, 20–22 June 2012, pp. 123–126 (2012)

12. Kaur, H., Alam, M.A., Jameel, R., Mourya, A.K., Chang, V.: A proposed solution and future direction for blockchain-based heterogeneous medicare data in cloud environment. J. Med. Syst. **42**(8), 156 (2018)
13. Bodkhe, U., Tanwar, S., Parekh, K., Khanpara, P., Tyagi, S., Kumar, N., Alazab, M.: Blockchain for industry 4.0: a comprehensive review. IEEE Access **8**, 79764–79800 (2020)
14. Frizzo-Barker, J., Chow-White, P.A., Adams, P.R., Mentanko, J., Ha, D., Green, S.: Blockchain as a disruptive technology for business: a systematic review. Int. J. Inf. Manage. **51**, 102029 (2020)
15. Tandon, A., Dhir, A., Islam, N., Mäntymäki, M.: Blockchain in healthcare: a systematic literature review, synthesizing framework and future research agenda. Comput. Ind. **122**, 103290 (2020)
16. Taylor, P.J., Dargahi, T., Dehghantanha, A., Parizi, R.M., Choo, K.K.R.: A systematic literature review of blockchain cyber security. Digital Commun. Netw. **6**(2), 147–156 (2020)
17. Ghosh, A., Gupta, S., Dua, A., Kumar, N.: Security of Cryptocurrencies in blockchain technology: State-of-art, challenges, and future prospects. J. Netw. Comput. Appl. 102635 (2020)
18. Soltanisehat, L., Alizadeh, R., Hao, H., Choo, K.K.R.: Technical, temporal, and spatial research challenges and opportunities in blockchain-based healthcare: a systematic literature review. IEEE Trans. Eng. Manage. https://doi.org/10.1109/tem.2020.3013507 (2020)
19. Nakamoto, S.: Bitcoin: a peer-to-peer electronic cash system. Manubot (2019)
20. Ivan, D.: Moving toward a blockchain-based method for the secure storage of patient records. In: ONC/NIST Use of Blockchain for Healthcare and Research Workshop. Gaithersburg, MD, USA, ONC/ NIST (2016)
21. Yue, X., Wang, H., Jin, D., Li, M., Jiang, W.: Healthcare data gateways found healthcare intelligence on blockchain with novel privacy risk control. J. Med. Syst. **40**(10), 218 (2016)
22. Blough, D., Ahamad, M., Liu, L., Chopra, P.: MedVault: ensuring security and privacy for electronic medical records. In: NSF CyberTrust Principal Investigators Meeting. http://www.cs.yale.edu/cybertrust08/posters/posters/158medvaultposterCT08.pdf (2008). Accessed 20 Dec 2016
23. Attili, S., Ladwa, S.K., Sharma, U., Trenkle, A.F.: Blockchain: the chain of trust and its potential to transform healthcare—our point of view. In: ONC/NIST Use of Blockchain for Healthcare and Research Workshop. Gaithersburg, MD, USA, ONC/NIST (2016)
24. Culver, K.: Blockchain technologies: a whitepaper discussing how the claims process can be improved. In: ONC/NIST Use of Blockchain for Healthcare and Research Workshop. Gaithersburg, MD, USA, ONC/NIST (2016)
25. Vian, K., Voto, A., Haynes-Sanstead, K.: A blockchain profile for medicaid applicants and recipients. In: ONC/NIST Use of Blockchain for Healthcare and Research Workshop. Gaithersburg, MD, USA, ONC/NIST (2016)
26. Linn, L.A., Koo, M.B.: Blockchain for health data and its potential use in health IT and health care related research. In: ONC/NIST Use of Blockchain for Healthcare and Research Workshop. Gaithersburg, MD, USA, ONC/NIST (2016)

Smart Solution for Leaf Disease and Crop Health Detection

Alisha Dwari, Adyasha Tarasia, Adyasha Jena, Sukalpa Sarkar, Sourav Kumar Jena, and Sipra Sahoo

Abstract Agricultural sector is among the major sources of food for humans. About 60% of the country is involved in agriculture and other agricultural practices in a certain way. Being a major occupation of most Indians, significant developments in the field are strongly required to ensure higher yields and less crop damage. With the introduction of technology such as artificial intelligence and the Internet of Things, there have been major advances in the estimation and accuracy of tasks. Using AI and IoT, we can build advanced monitoring systems that can ensure crop health and prevent damage. This can also be implemented in the field of agriculture. The key steps that we addressed in this paper are image pre-processing, extraction of features, classification, and analysis of the results provided by the technique.

Keywords Artificial intelligence (AI) · Internet of things (IoT) · Support vector machine (SVM) · Red–green–blue (RGB) · Hue–saturation–intensity (HIS) · Multiplayer perception (MLP)

1 Introduction

Agriculture with its amalgamated divisions is the largest source of providing sustainable subsistence in India. Undoubtedly, India has been prominent in attaining grain self-sufficiency; however, it always needs to be taken care of the production of the crops which needs to be equally good, both quality and quantity wise. Acting as the vertebrae that make up the spine of the country, it is essential to maintain our

A. Dwari · A. Tarasia · A. Jena · S. Sarkar · S. K. Jena
Institute of Technical Education and Research, SOA (Deemed to be University), Bhubaneswar, India

S. Sahoo (✉)
Department of Computer Science and Engineering, Institute of Technical Education and Research, SOA (Deemed to be University), Bhubaneswar, India
e-mail: siprasahoo@soa.ac.in

S. Das and M. N. Mohanty (eds.), *Advances in Intelligent Computing and Communication*, Lecture Notes in Networks and Systems 202,
https://doi.org/10.1007/978-981-16-0695-3_23

crop health and keep a strict monitor on diseases that may or are affecting plants. Diagnosis of plant diseases is very essential to cure and control them and must be done at earlier stages for better results [1]. However, it is noteworthy to mention that along with the various types of diseases, there are various species of plants too, many of which might not have been discovered yet. We believe that to keep an active database of the same, teaching a machine and training algorithms to classify plants is necessary. In support of the same, we believe that leaf recognition is the way to go. Plant leaves are ideal specimens for image processing because they are two dimensional and one of their primary distinguishing factors, their shape [2].

2 Related Literatures

Studies shows that image processing technology is the first that is applied in monitoring crop after naked eye observation approach [3]. Gautam Kaushal and Rajni Bala [4] proposed image processing technique which is used to process the digital information from the images of the plant. Ch. Usha Kumari et al. [5] also proposed that image processing could be applied to identify the leaf spot. This approach involves image acquisition, segmentation of images, extraction, and classification of features. In his survey, Ghaiwat et al. [6] demonstrates that the K-nearest neighbor method is the simplest of all class prediction algorithms. Eftekhar Hossain et al. [7] suggested that the technique could be the K-nearest neighbor (KNN) classifier for plant leaf disease detection and classification. Dheeb Al Bashish et al. [8] suggested that the technique of clustering K-means can be used in the system of image processing. Basically, the approach of partitioning is used to evaluate and handle data observation as an entity based on the position and distance between different data points. Thus, color patch segmentation using K-mean is essential. A technique in which leaf disease is identified automatically was proposed by Jimita Baghel and Prashant Jain [9]. In this process, segmentation of defects is done. First, to detect defective leaf areas, K-means clustering is applied, then the color and characteristics of the leaves are clustered, and finally, the clustered blocks are combined into a particular number of regions. To solve low-level image segmentation, Tushar H. Jaware et al. [8] proposed a novel and improved K-means clustering technique. An unsupervised, programmed K-means clustering algorithm was suggested in [10] to perform the division assignment of the wheat leaf scab image regarding the laboratory color space. The use of this technique on photographs of wheat leaves resulted in the discovery of three essential diseases. Applying this technique on wheat leaf images led to discovery of three basic diseases. They find that the expanded precision in the division will help with taking after procedures (e.g., include extraction and distinguishing proof) for automatic disease detection and investigation of a wheat leaf [11]. R. N. Kadu et al. [12] developed a research using Otsu threshold and support vector machine (SVM). Otsu threshold was applied to separate the injured part and the healthy part. After feature extraction, Gabor filter was applied to segment image. Finally, that features were used as input to SVM classifier. The main advantage of SVM being high prediction accuracy and lesser complexity algorithm than neural networks. Pranjali B. Padol and Anjali A. Yadav [13] proposed a SVM classification technique to detect and classify leaf diseases of grape. First segmentation is done by K-means clustering followed by color and texture feature extraction. After that, the classification technique is applied for the detection of type of leaf disease. A system in which a disease called corynespora is distinguished from other diseases affecting the leaves of the rubber tree was proposed by Abdullah et al. (2007) [14]. It does not have segmentation involved. Primary component analysis is used here. It is applied to the RGB values of the leaves' low-resolution image. The first two main components are then supplied to a one hidden layer multilayer perceptron (MLP) neural network, the output of which indicates whether the sample is contaminated by the disease of interest or not.

3 Proposed Methodology

3.1 *Image Acquisition*

Image processing is a vital part. Images of leaves are taken in black/white background. The main part of crop disease detection is image processing. We capture the images of leaves in a natural background which then undergoes the image processing steps and infection is detected through classification.

3.2 *Image Pre-processing*

It includes:

1. K-means clustering for image segmentation.
2. Converting the segmented image into gray scale
3. Converting the gray scale image to binary
4. Soon after, the next stage—image segmentation is done.

3.3 *Image Segmentation*

It is the process of conversion of a digital image into simpler and smaller segments which are better known as super pixels. The super pixels are used to locate boundary lines, are easier to analyze, and hence, help in leaf disease detection.

Low Contrast: It is done to provide a softer or flat image. It removes highlights, shadows, and color difference in the image.

Contrast Enhancement: The image is processed, and the output is the enhanced image (Fig. 1).

Feature Extraction Conversion of Red-Green-Blue to Hue-Saturation-Intensity:

[Taking the acronym RGB as red–green–blue and HSI as hue–saturation–intensity].

Over here, we are taking the RGB image in the size, M-by-N-by-3. The three dimensions are accounting for three image planes (red, green, blue).

If all the three components are found equal, then the conversion is undefined. The pixel range of RGB is generally [0, 255].

The pixel range of HSI is generally [0, 1].

Conversion of pixel range can be done by calculating the components: hue, saturation, and intensity.

[Taking the acronym H as hue, I as intensity and S as saturation]

[Taking N = numerator, D = denominator, R = red, G = green, B = blue]

$$\text{For } H : N = 1/2\,[(R-G) + (R+B)]$$

$$D = ((R-G)*2 + ((R-B)*(G-B)))^{0.5}$$

$$H = a * \cos d * (N/D)$$

$$\text{For } S : S = 1 - (3R + G + B) * \min[R,\ G,\ B]$$

$$\text{For } I : I = (R + G + B) \times 3$$

On combining the above results into a single value, the HSI image is formed (Fig. 1).

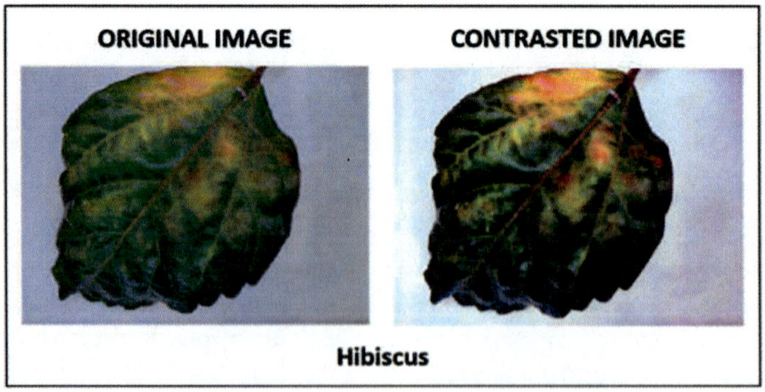

Fig. 1 Enhanced image figure

3.4 K-Means Clustering

K-means clustering is an unsupervised algorithm that is effective for quantization of vector data. It is effective when the input data or the data being fed into the model is not labelled. Used to derive the area of interest from background inputs, it calculates the centers (centroids to be precise) and loops until it finds the optimal centroid.

The major purpose of applying this algorithm is for image segmentation. It helps to segment the leaf image by dividing the leaf features into different '*K*' numbered clusters.

The image of the leaf is taken in dimensions of $X * Y$, where every data item present in the matrix is a combination of the primary colors: RGB (red, green, blue). The user specifies the '*K*' value. When the image data is being loaded, in the first iteration, the dataset is clustered with respect to the closest centroid. On further iterations, new centers are calculated by taking the average of each pixel associated with the clusters. It is done by calculating the Euclidean distance metric. Out of the clusters generated, the one that shows the infected region(s) is subjected to classification using the support vector machine (SVM) for further accuracy (Fig. 2).

3.5 Support Vector Machine (SVM)

SVM—a classification/regression algorithm that falls under the broad category of statistical learning techniques.

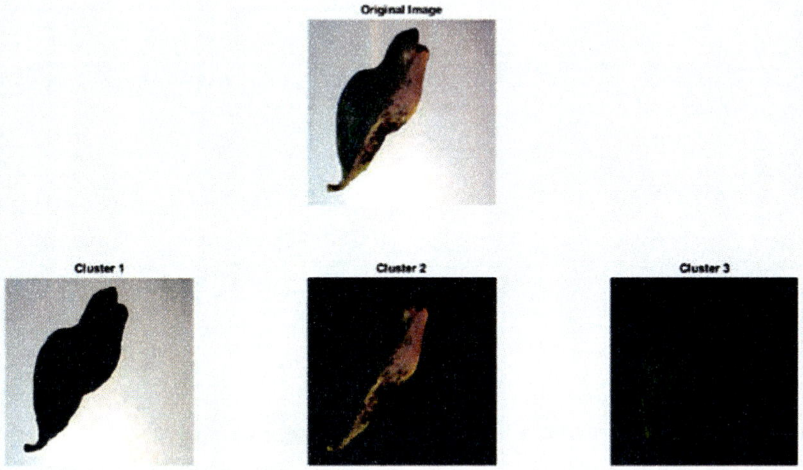

Fig. 2 Clustering

Defining its simple logic—assuming there is a dataset of dimension X with distinguishable data points belonging to two different categories. SVM tends to classify the data points by developing a plane in between the two set of points which is known as a "Hyperplane" (Fig. 3).

There are two steps to classification:

(a) To train the classification model for identification of the required weights: we divide our samples into training and testing datasets in the ratio 7:3, respectively. After thoroughly training our model using the training datasets, we test it against our testing samples.
(b) Checking the accuracy of the model using the test datasets.

Finally, we identify the infection index and draw conclusions from it (Fig. 4).

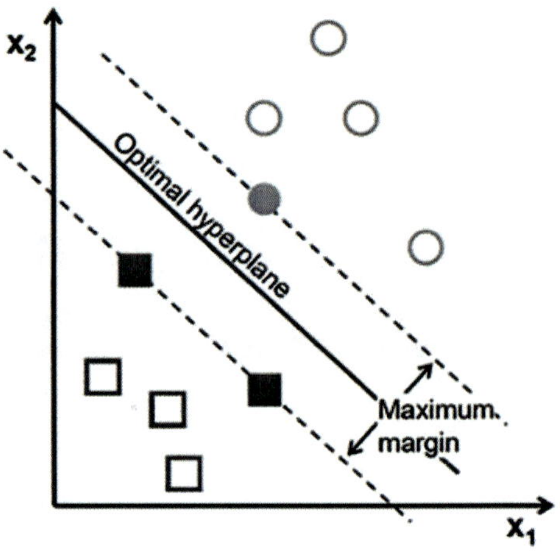

Fig. 3 Hyperplane (I)

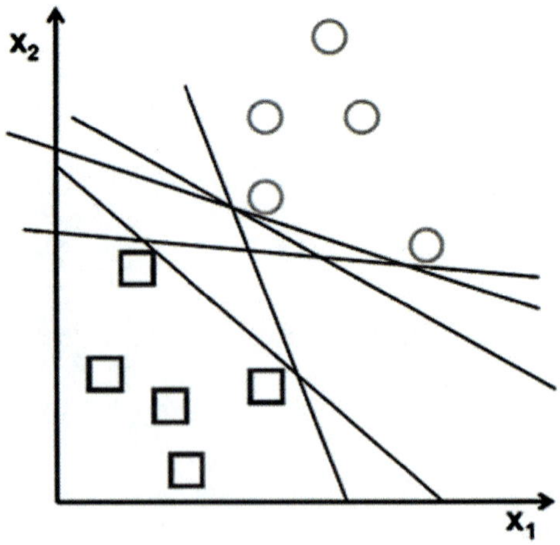

Fig. 4 Hyperplane (II)

4 Experimental Setup and Result

The crop fields are our primary region of focus. We plan to monitor the crop health and relay the images using a Raspberry Pi module. After processing, segmentation, and classification, the output is to be sent to the owner's smart device in a pre-determined format. For image acquisition, cameras are to be placed at every row which shall send the sample images periodically for processing. The array of cameras will capture leaf images and will feed the data through the RPI module to the image processing backend. In the backend, the contrast of the images is enhanced, and then, by applying K-means clustering, the segmented image is obtained. At the end, by applying SVM classifier, we can detect the infection. Feeding the image to a trained model (neural network) will also provide us the disease name.

5 Results and Discussion

We have taken the data samples of leaves with bacterial disease, fungal disease, and sunburns. To precisely specify the plant specimens, we have taken hibiscus, jasmine, lemon, and curry leaf plants. Each of Figs. 5, 6, 7, and 8 displays two images in which the first one is the original image while the second one represents the corresponding segmented image. When K-means clustering along with multiclass SVM is applied on the segmented images, the disease get classified with an accuracy varying for each specimen. The accuracy for hibiscus is 98.38%, for jasmine, it is 96.77%, for lemon, it is 95.16%, and for curry leaf, it is 96.77%. The upper mentioned accuracies can be clearly inferred from Table 1 and its corresponding histogram.

We also compared the proposed methodology with other techniques such as using minimum distance classifier as the classification algorithm along with

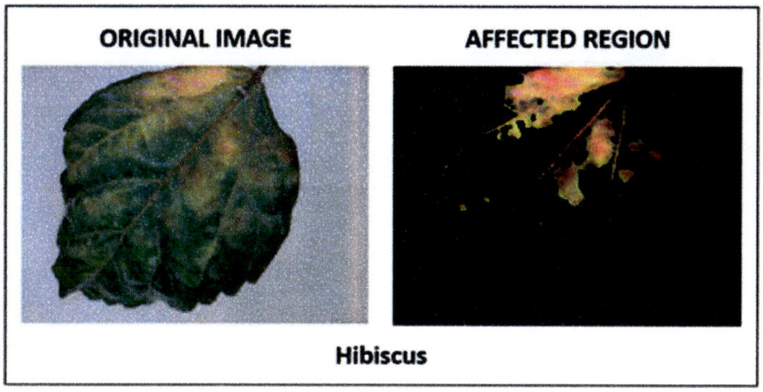

Fig. 5 Hibiscus with sunburn

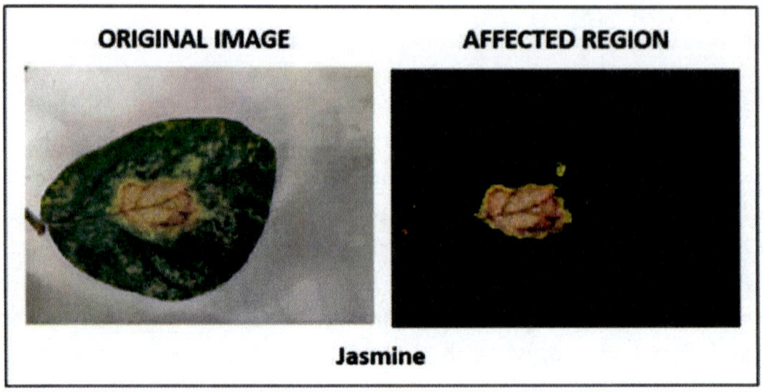

Fig. 6 Jasmine with fungal infection

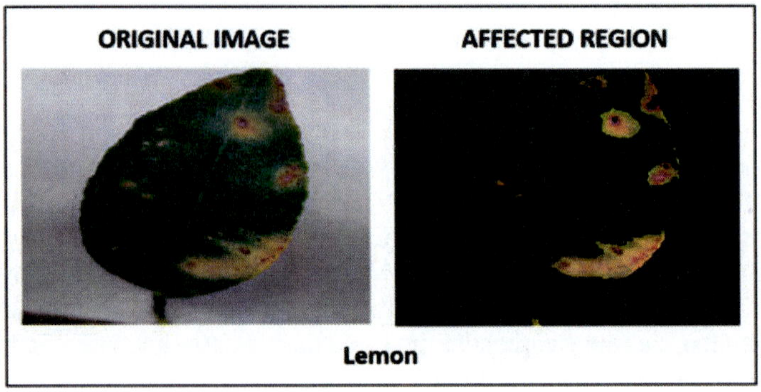

Fig. 7 Lemon with bacterial infection

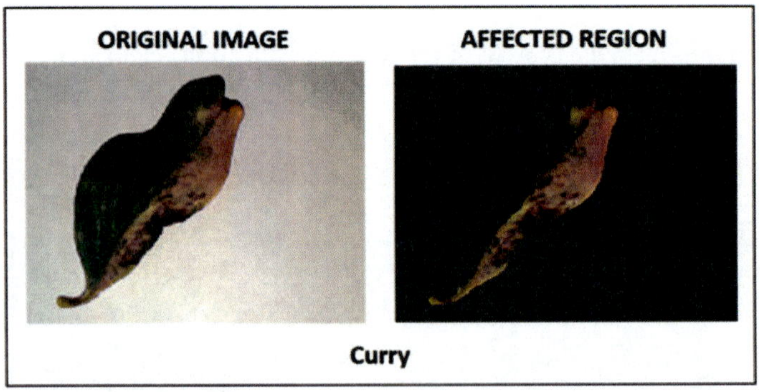

Fig. 8 Curry leaf with sunburn

Table 1 Accuracy and its affected data

Leaf specimen	Affected region (%)	Accuracy (%)
Hibiscus (*Hibiscus rosa-sinesis*)	15.0016	98.3871
Jasmine (*Jasminum*)	15.1601	96.7741
Lemon (*Citrus limon*)	15.0683	95.1613
Curry leaf (*Murraya koenigii*)	23.8172	96.7742

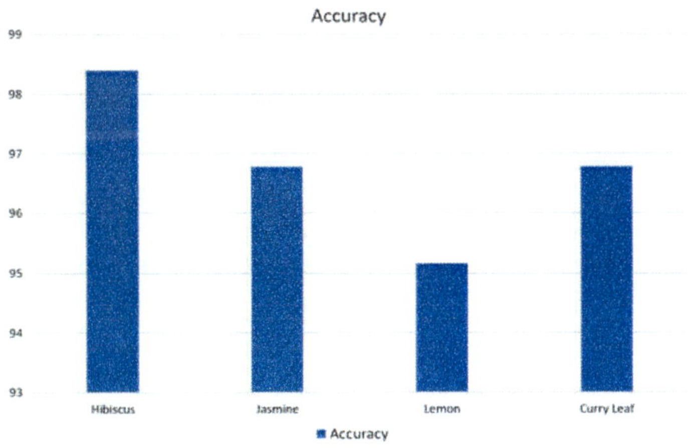

Fig. 9 Accuracy plot

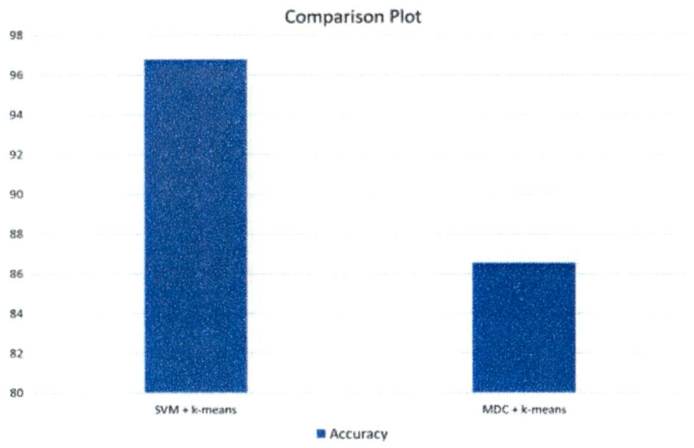

Fig. 10 Accuracy comparison plot

K-means. The overall accuracy obtained with the technique was 86.54 [15], whereas applying SVM with K-means provides result with an accuracy of 96.77% (Figs. 9 and 10).

6 Conclusion

From this study, it is observed SVM along with K-means clustering gives more accurate results. The other methods to perform the detection can be using minimum distance classifier (MDC) along with K-means clustering. The overall accuracy obtained using the latter is around 86.54%. But the accuracy obtained using the technique covered in this study is around 96.77%. We can effectively classify defected leaf samples and get better detection accuracy using this technique. Once the classification gets completed, we can reliably send the generated report to the concerned farmer(s) or the crop owner's smart device for them to take necessary steps to solve any problem if found. We believe that this analysis shall help the farmers to monitor their crop health accurately and reduce losses immensely.

References

1. Ghaiwat, S.N., Arora, P.: Detection and classification of plant leaf diseases using image processing techniques: a review. Int. J. Recent Adv. Eng. Technol. **2**(3), 1–7 (2014)
2. Kumar, M., Kamble, M., Pawar, S., Patil, P., Bonde, N.: Survey on techniques for plant leaf classification. Int. J. Mod. Eng. Res. **1**(2), 538–544 (2011)
3. Weizheng, S., Yachun, W., Zhanliang, C., Hongda, W.: Grading method of leaf spot disease based on image processing. In: Proceedings of the 2008 International Conference on Computer Science and Software Engineering, pp. 491–494. IEEE Computer Society (2008)
4. Kausha, G., Bala, R.: GLCM and KNN based algorithm for plant disease detection. Int. J. Adv. Res. Electr. Electr. Instrum. Eng. **6**(7), 5845–5852 (2017)
5. Kumari, C.U., Prasad, S.J., Mounika, G.: Leaf disease detection: feature extraction with K-means clustering and classification with ANN. In: 2019 3rd International Conference on Computing Methodologies and Communication (ICCMC), pp. 1095–1098. IEEE March (2019)
6. Ghaiwat, S.N., Arora, P.: Detection and classification of coffee diseases using image processing techniques. Review **2**(3) (2014)
7. Hossain, E., Hossain, M.F., Rahaman, M.A.: A color and texture based approach for the detection and classification of plant leaf disease using KNN classifier. In: 2019 International Conference on Electrical, Computer and Communication Engineering (ECCE), pp. 1–6. IEEE Feb (2019)
8. Grape Leaf Disease Detection Using K-means Clustering Algorithm. Int. Res. J. Eng. Technol. (IRJET) (2016)
9. Baghel, J., Jain, P.: K-means segmentation method for automatic leaf disease detection. Int. J. Eng. Res. Appl. **6**(3), 83–86 (2016)
10. Jaware, T.H., Badgujar, R.D., Patil, P.D.: Crop disease detection using image segmentation. In: National Conference on Advances in Communication and Computing, World Journal of Science and Technology, pp. 190–194, Dhule, Maharashtra, India (2012)

11. Niu, X.: Image segmentation algorithm for disease detection of wheat leaves. In: 2014 IEEE Proceedings of the 2014 International Conference on Advanced Mechatronic Systems, Japan, pp. 270–273 (2014)
12. Kadu, R.N., Kangane, S., Vikhe, S., Pandita, R., Inamke, V.: Leaf disease detection using Arm7 and image processing. Int. J. Eng. Res. Appl. **5**(2), 2248–9622 (2019)
13. Padol, P.B., Yadav, A.A.: SVM classifier based grape leaf disease detection. In: 2016 Conference on Advances in Signal Processing (CASP), pp. 175–179. IEEE (2016) June
14. Abdullah, N.E., Rahim, A.A., Hashim, H., Kamal, M.M.: Classification of rubber tree leaf diseases using multilayer perceptron neural network. In: 2007 5th Student Conference on Research and Development (2007)
15. Singh, V., Misra, A.K.: Detection of plant leaf diseases using image segmentation and soft computing techniques. Inf. Process Agric **4**(1), 41–49 (2017)

Implantable Insulin Delivery System Based on the Genetic Algorithm PI Controller (GA-PIC)

Akshaya Kumar Patra, Narayan Nahak, Bidyadhar Rout, and Anuja Nanda

Abstract The intention of this chapter is to formulate a GA-PIC for monitoring the blood glucose (BG) in TIDM patients. In this strategy, the conventional PIC is redesigned by use of the genetic algorithm (GA) concept to upgrade the control accomplishment. The endorsement of the upgraded control accomplishment of GA-PIC is set up by comparative outcome examination with other well-known control strategies. The overall results clearly uncover the better execution of the prescribed methodology to screen the TIDM elements inside a consistent territory concerning the robustness, exactness, and stability.

Keywords Diabetes · Carbohydrate · Insulin · GA-PIC

1 Introduction

Based on the WHO report, one among across the board ailments is diabetes mellitus and is come about because of the breaking down of the pancreas. This decreases insulin affectability influencing the *normo-glycaemic* scope of plasma glucose fixation (70–120 mg/dl) in a health person. Presently, various research ventures are embraced by a few specialists to dispose of this issue by concocting advanced clinical apparatus like mechanized MID. Till now, glucose level is manually

A. K. Patra (✉) · A. Nanda
Department of EEE, ITER, S'O'A University, Bhubaneswar 751030, India
e-mail: hiakp@yahoo.com

A. Nanda
e-mail: anuja.sus@gmail.com

N. Nahak
Department of EE, ITER, S'O'A University, Bhubaneswar 751030, India
e-mail: narayannahak@soa.ac.in

B. Rout
Department of EE, VSSUT, Burla 768019, India
e-mail: rout.bdr@gmail.com

S. Das and M. N. Mohanty (eds.), *Advances in Intelligent Computing and Communication*, Lecture Notes in Networks and Systems 202,
https://doi.org/10.1007/978-981-16-0695-3_24

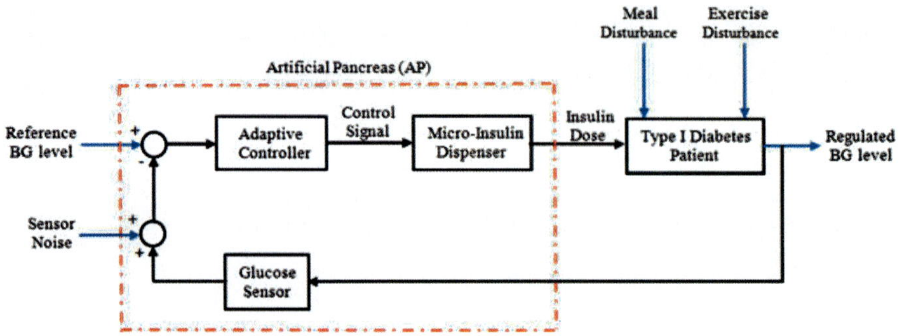

Fig. 1 The GA-PIC along with patient

regulated to embrace the open loop technique. The hypo-glycaemic or hyper-glycaemic disorders may advance because of the troubles in taking care of the inward framework variations and outside conflicts by realizing the control loop strategy. The advancement of embedded AP empowering the satisfactory insulin dose transfer proportional to the sensor estimation in the patient's body may give the way to consolidate the closed loop procedure. Figure 1 represents a closed loop system with an AP. The AP comprises of a controller, sensor, and MID. The sensor quantifies the plasma glucose level of the human being ceaselessly and imparts sign to the controller for creating the ideal control activities.

The patient's GM process states are pulled toward operating point based on the full state feedback approach through the controller to achieve the upgraded control action. The $u(t)$ produced by the controller likewise delineates the association of the structure parameters and disturbances. From that point, as indicated by the $u(t)$, the ideal dose of insulin is delivered into the VB by MID to accomplish the *normo-glycaemic* scope of plasma glucose level [1, 2]. To decide an ideal solution for the AP like structure, a suitable model of the composite plasma glucose administrative framework, various snags, and difficulties such as the impacts of nonlinear activity, time subordinate elements, the existence of numerous sources of Gaussian noise, model parameters, and lack of plasma glucose detecting are to be confronted. Also, the difficulties and constraints associated to control, explicitly for plasma glucose control design are significant glucose estimation delay, insulin assimilation delay in the wake of being infused, the irreversible activity of insulin, meal estimation, parameter deviation, irregular risk of excess BG level deviations and time-based control desires, and so forth [3]. Regardless of technical advancement and significant advancement on aforementioned issue, considerable enhancement is still essential in the control technique. Appraisal of the BG profiles subsequent the insulin dose regulation desires a control algorithm for plasma glucose guideline in AP and numerous researchers recommended the PID control as a reasonable solution [4]. However, because of the glucose detecting time delay, insulin activity and control variables, the ideal execution, guaranteeing high precision, dependability, and robustness could not be accomplished. The fuzzy [5, 6], LQG [7, 8],

$H\infty$ [9–11], SM [12–15], and MP [16–18] are very much familiar control algorithms to handle the BG control issues. The plasma glucose regulation in the patients inside the *normo-glycaemic* range instilling the above control techniques improved the exactness and heartiness somewhat contrasted with the PID. However, these control techniques are not completely insensitive to model uncertainties and disturbances in spite of the better response. Subsequently, ideal control variables setting for better execution and escaping slow action, the present work recommends an another new method executing the GA-PIC. The GA-PIC is an efficient approach for noise elimination without affecting the accuracy and robustness. Implementation of the suggested method to monitor the plasma glucose level in patient to certify a superior robust control in contrast with other existing well acknowledged strategies under both matched and mismatched uncertainties.

2 Problem Formulations

2.1 System Outlines

The AP with the patient compartmental model is displayed in Fig. 1, and it exposes the GM process. The GA-PIC is implemented and tested over the above model. The human being has kidney, CNS, liver, gastric emptying sub-system, peripheral cells, and heart/lungs as various useful parts and are appeared as six-compartments independently. The directions of blood stream are also shown in Fig. 1. The GA-PIC gets the deliberate blood vessel glucose esteem as contribution to give the ideal dose of insulin to be infused as a yield with the assistance of MID.

2.2 Clinical Support

A body with a drawn out more blood vessel glucose that surpasses 144 mg/dl is recognized to experience the ill effects of *hyper-glycaemia* malady. This kind of the clinical issue is likewise named as diabetes. The insulin lack alongside its resistance or insensitiveness autonomously or combinedly reason of *hyper-glycaemia* and are a portion of the significant causes as talked about in [12]. Diabetes may happen if the BG remains unutilized adequately under any typical life cycle. It is delegated type I and II according to the rate of insulin secretion in the body. In the type I diabetes, the insulin secretion by the pancreas is totally incomprehensible, however, in the event of the type II is at somewhat littler rate, which additionally gets inhibited in due course. Consequently, in the current circumstance, a prompt insulin infusion to re-establish the typical BG level is fundamental that additionally forestalls unified unfavorable effects on the ordinary activity of any body. Subject to various fooding, fasting, and exercise practices, the control of the glucose levels are not simple in the genuine life cycle. The outside sources like the sugar nourishments are processed down into glucose at various pieces of the stomach related framework (gut) and afterward filtered. The filtered glucose is transferred to the VB. Afterward, the liver stores this glucose in the wake of being moved from the VB as glycogen. At the point when the glucose level in the VB is below the 81 mg/dl, glycogen is re-infused as plasma glucose by the liver into VB. It is recognized as the interior source of glucose. These are the general marvels in the glucose breakdown process. The all out vitality required by GM organs regardless of insulin reliance gets from the amount of glucose use.

The secretion of insulin is because of the pancreas, and it regulates the plasma glucose level in the VB. A typical human with high-glucose level has two significant roles of insulin, initially, to permit the liver for the glucose retention alongside the putting away it as glycogen, explicitly under the meal ingestion situations. Therefore, the creation of abundance interior carbohydrate by the liver is halted. Also, insulin accelerates the plasma glucose assimilation in the muscles and satisfies the peripheral shell energy desires of the human body [12]. But the metabolic process becomes abnormal in diabetes patient due to partial or complete failure of the above-mentioned actions of insulin. In the diabetic patient, the cells do not consume glucose properly, and also internal glucose is created by the liver, as a result glucose level is raised unusually. Afterward, as the glucose level surpasses the RTG level of 162 mg/dl, the surplus amount of VB glucose is removed by the kidney.

2.3 *Numerical Structure*

Right now, with due respect of the activities of GI interactions, a few SIMULINK models are projected to regulate the concentration of glucose in the diabetic patients [12, 13]. Due to the improved structure and required estimation to the process of GM with diminished inaccuracy, the model projected by Gallardo [12] and Rmileh [13] are both the broadly recognized ones. The checking and authentication of the projected strategy utilizing Gallardo [12] model for glucose regulation is taken in the current investigation. Figure 1 depicts the TIDM compartmental model associated with the embedded AP. The controller processes the insulin infusion adds up to the VB at the 5 min interim utilizing an incorporated embedded insulin pump. Contingent upon the supporting tool and detecting innovation the sampling rate shifts when is being implemented in the real-time domain [12]. The meal intake in the gut and exercise in peripheral cells are taken as disturbance inputs to the process. The compartmental modeling technique in view of the conditions of the fast principle has been utilized to build up the GM process along with MID [12, 13]. A basic structure of the GI process in structure is synthesized with regard to insulin and glucose action as reflected in Fig. 1.

2.4 Investigation of Patient Activities

The patient activities and characteristics are investigated under various operational conditions like the noise, ingestion quantity of carbohydrates, and differing exercises, and so forth. The plasma glucose level and the insulin secretion rate in the TIDM structure with 60 gm meal intake at 600 min are outlined in Fig. 2a. The rate of glucose formation and utilization by the liver and rate of glucose consumption by the organs like gut, CNS, and cells are demonstrated in Fig. 2b. The control of glucose level in the VB takes place normally through the glucose consumption by the peripheral cells and the liver that solely depend upon the insulin. The insulin dependent organs consume the minimum quantity of glucose with the situation of deficiency of insulin. In this circumstance, plasma glucose increases unusually and exceeds 144 mg/dl. This type of irregular phenomena is known as the *hyper-glycaemia*. Some portion of the glucose is extracted from the patient's body through kidney under the situation when the glucose level ascents up to the RTG. The glucose extraction rate of kidney is directly proportional to the BG level in patient and demonstrated in Fig. 2c. Figure 2a–c illustrates the abnormal activities under several model parameters and disturbances. These irregular activities can be reduced by implementing the AP with the appropriate control methods.

3 Control Algorithms

The optimal GA-PIC control technique is demonstrated in this fragment. The patient activity with suggested control algorithm is studied. The regulator execution files such as settling time t_s(min), steady-state error e_{ss} (%), peak overshoot O_{Peak}, and peak undershoot U_{Peak} are furthermore surveyed and investigated with proper endorsement of the GA-PIC activity. The patient structure along with GA-PIC is exhibited in Fig. 1. The $e(t)$ is employed to formulate the $u(t)$. The TF of GA-PIC represents as [4]:

$$u(t) = K_p e(t) + K_i \int_0^{\infty} e(t)dt \tag{1}$$

$$T \cdot F = K_p + \frac{K_i}{s} \tag{2}$$

where K_p, and K_i are the parameter gains of the GA-PIC. Elementary stages of GA-PIC:

(1) Emerge K_p to reduce the e_{ss} (%).
(2) Emerge K_i to discard the e_{ss} (%).
(3) Adjust the control boundaries (K_p, and K_i) in such a manner to get the ideal yield.

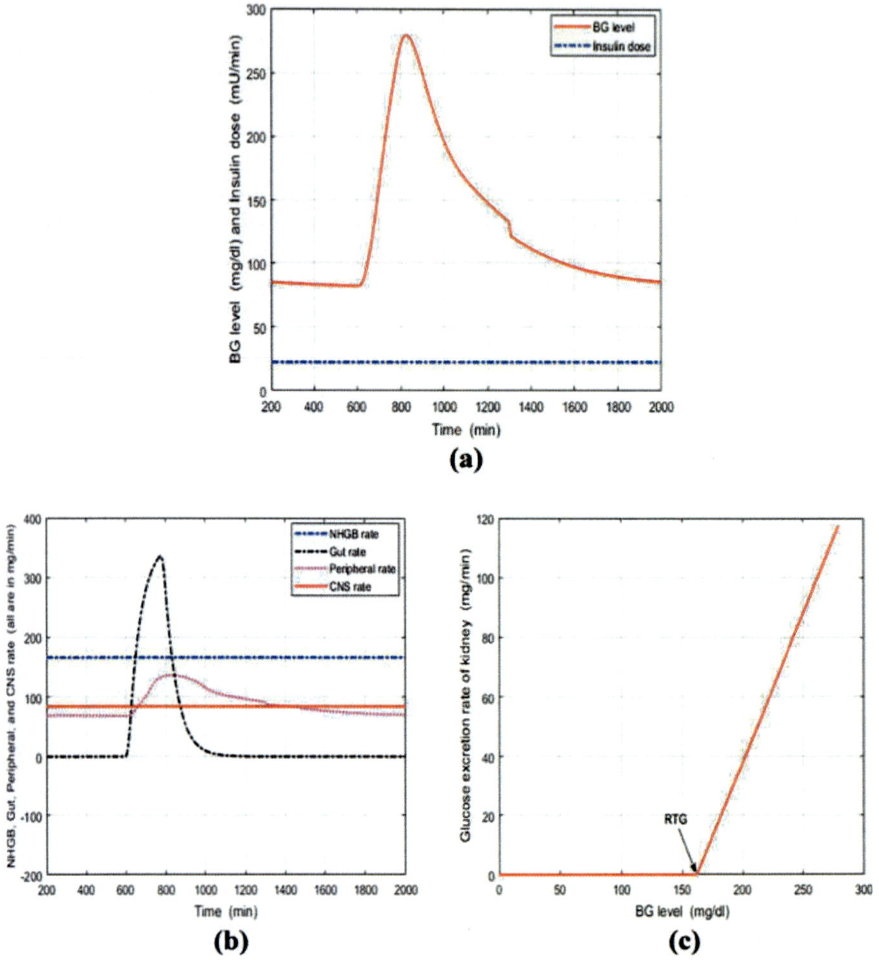

Fig. 2 **a** Plot of BG level verses insulin secretion rate; **b** plot of glucose secretion and intake rate by the organs; **c** plot of glucose removal rate through kidney verses glucose level

For monitoring of plasma glucose level in patient model, the GA-PIC factors are acquired by applying the GA improvement procedure. The accompanying boundaries are considered to discover the ideal increases of regulator as: iterations (30); population (45); factor of crossover (0.30); factor of mutation (0.09). The ideal estimations of the GA-PIC factors are adopted dependent on the GA strategy as spoken to in Table 1.

Table 1 The control gains

K_p	K_i
10	0.008

4 Result and Analysis

The outcomes of the patient model with GA-PIC is analyzed thoroughly in this stage. The GA-PIC is diverged from other notable control ways to deal by approving its redesigned activity.

4.1 *Investigation of Patient Activities with GA-PIC*

Right now, closed loop response of the patient with proposed GA-PIC is examined under different working circumstance variations of exercise, meal intake, and sensor noise. The rate of glucose formation and absorption by the GM organs in the closed loop patient model are represented in Fig. 3a–c. Compared to the open loop outcomes (Fig. 2a–c), the closed loop outcomes (Fig. 3a–c) evidence a more absorption of glucose by the GM organs like liver and peripheral cells. Because of this impact, the glucose level in VB reduces to 82.3 mg/dl. With the goal, the *hyper-glycaemia* issue is anticipated as appeared in Fig. 3a. The insulin self-governing organ (CNS) utilizes the carbohydrate at a fixed rate. Figure 3b delineates the fixed rate of carbohydrate consumption by CNS at 84 mg/min. As the glucose level is less than the RTG, the kidney does not remove any quantity of glucose from the VB. The extraction rate of glucose through the kidney stays fixed at zero level as shown in Fig. 3c. Therefore, all performances of patient structure are enhanced by utilizing the AP with proposed control methodology.

4.2 *Relative Outputs*

The GA-PIC is contrasted with different notable control procedures to approve its redesigned exhibitions. Figure 3 reflects the influence of carbohydrate intake in the plasma glucose level of the structure with GA-PIC. Table 2 gives a close to examination respect to steady-state error e_{ss} (%), t_s, O_{Peak}, and U_{Peak}. The impact of the $w(t)$ in the results of the patient executing above control techniques are likewise displayed in Table 2 as for the written works [4, 5, 11], and [14]. All comparative functioning circumstances are followed with a similar degree of noise level in all control procedures application for examination.

The glucose level in patient ingested to 60 gm carbohydrate under numerous control techniques is analyzed. The comparing consequences are exposed for the diverse control techniques along with GA-PIC concerning control subtleties, for

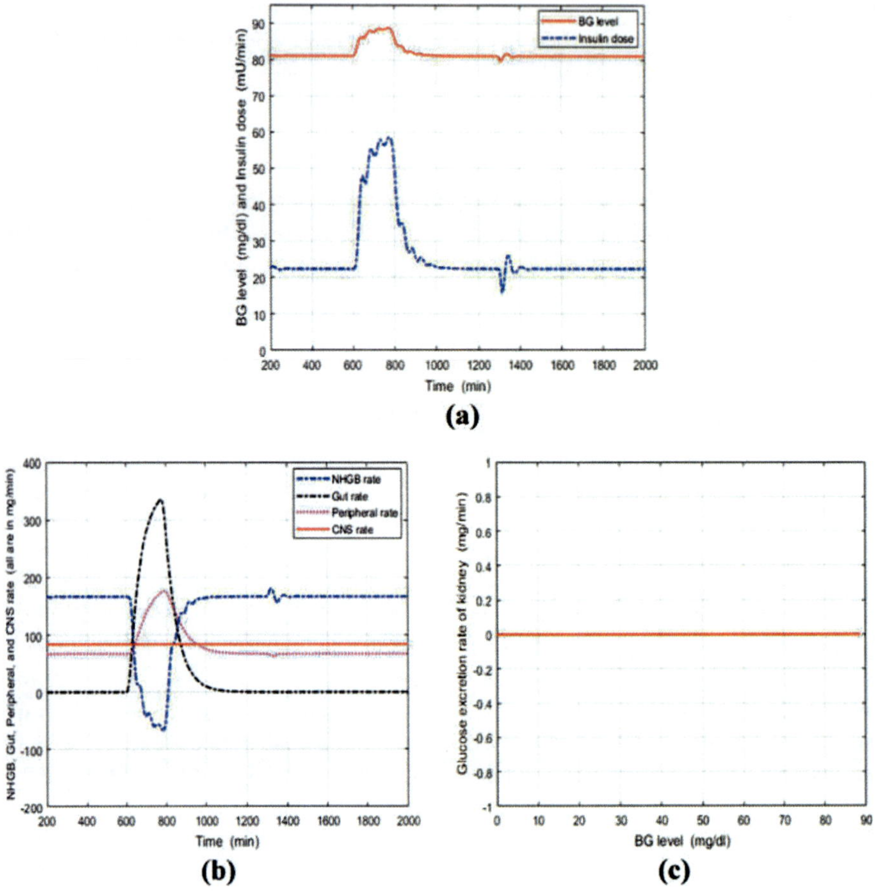

Fig. 3 Profiles of patient model with GA-PIC: **a** Plot of BG level verses insulin secretion rate; **b** plot of glucose secretion and intake rate by the organs; **c** plot of glucose removal rate through kidney verses glucose level

Table 2 Relative data associated to the BG level with regard to the different methods

Control techniques	PI [4]	Fuzzy [5]	$H\infty$ [11]	SM [15]	GA-PIC (Proposed)
Carbohydrate intake (gm)	60.0	60.0	60.0	60.0	60.0
Insulin secretion rate (mU/min)	60.5	60.4	60.2	60.2	60.1
t_s (min)	320	320	310	300	270
O_{Peak} (mg/dl)	6.6	6.4	6.3	6.1	4.4
U_{Peak} (mg/dl)	4.1	3.1	2.1	2.2	1.1

example, O_{Peak}, U_{Peak}, and t_s. These results suggest the superior control execution of the GA-PIC that outputs furthermore show the commotion concealment capacity with more heartiness of the GA-PIC. Generally, it is seen from above assessment that the revelations of the suggested GA-PIC inclinations are the higher exactness and dependability, more power, better capacity to smother commotion, and better capacity to screen model vulnerability underneath various anomalous circumstances and huge deviation of meal disturbances.

5 Conclusions

This investigation proposed a new approach (GA-PIC) for plasma glucose control in diabetes patient inside the steady range. For the construction of GA-PIC, a mathematical structure of the patient is reflected. In this approach, the GA is adopted to accomplish the improved action. The relative outputs obviously recreate that GA-PIC is land at favoured execution over the notable control methods with respect to steadiness, exactness, and robustness under various strange circumstances and unsettling influences. The improved activity of the GA-PIC to the extent better exactness, more robustness, improved stability, better capacity to dominate the noise, and vulnerability approve its continuous execution.

References

1. Chee, F.: Closed-loop glucose control in critically ill patients using continuous glucose monitoring system in real time. IEEE Trans. Inf. Technol. Biomed. **7**(4), 43–53 (2003)
2. Patra, A.K.: Kalman filtering linear quadratic regulator for artificial pancreas in type-I diabetes patient. Int. J. Model. Ident. Control **34**(1), 59–74 (2020)
3. Patra, A.K.: Design of artificial pancreas based on the SMGC and self-tuning PI control in type-I diabetic patient. Int. J. Biomed. Eng. Technol. **32**(1), 1–35 (2020)
4. Patra, A.K.: The fractional order PID controller design for BG control in type-I diabetes patient. In: Advances in Intelligent Computing and Communication, pp. 321–329. Springer, Singapore (2020)
5. Ibbini, M.: The PI-fuzzy logic controller for the regulation of blood glucose level in diabetic patients. J. Med. Eng. Technol. **30**(2), 83–92 (2006)
6. Patra, A.K.: Design of artificial pancreas based on fuzzy logic control in type-I diabetes patient. In: Innovation in Electrical Power Engineering, Communication, and Computing Technology, pp. 557–569. Springer, Singapore (2020)
7. Patra, A.K.: An automatic insulin infusion system based on LQG control technique. Int. J. Biomedical engineering and technology **17**(3), 252–275 (2015)
8. Patra, A.K.: Design of backstepping LQG controller for blood glucose regulation in type-I diabetes patient. Int. J. Autom. Control **14**(4), 445–468 (2020)
9. Chee, F.: Optimal H∞ insulin injection control for blood glucose regulation in diabetic patients. IEEE Trans. Biomed. Eng. **52**(10), 1625–1631 (2005)
10. Yasini, S.: Knowledge-based closed-loop control of blood glucose concentration in diabetic patients and comparison with H_∞ control technique. IETE J. Res. **58**, 328–336 (2012)

11. Patra, A.K.: Optimal H-infinity insulin injection control for blood glucose regulation in IDDM patient using physiological model. Int. J. Autom. control **8**(4), 309–322 (2014)
12. Gallardo, H.: High-order sliding-mode control for blood glucose: practical relative degree approach. Control Eng. Pract. **21**(5), 747–758 (2013)
13. Rmileh, A.: Wiener sliding-mode control for artificial pancreas: a new nonlinear approach to glucose regulation. Comput. Methods Programs Biomed. **107**, 327–340 (2012)
14. Patra, A.K.: Backstepping sliding mode Gaussian insulin injection control for blood glucose regulation in TIDM patient. J. Dyn. Sys. Meas. Control **140**(9), 091006-091006-15 (2018)
15. Patra, A.K.: Adaptive sliding mode Gaussian controller for artificial pancreas in TIDM patient. J. Process Control **58**, 23–27 (2017)
16. Patra, A.K.: Backstepping model predictive controller for blood glucose regulation in type-I diabetes patient. IETE J. Res. **66**(3), 326–340 (2020)
17. Patra, A.K.: Adaptive continuous-time model predictive controller for implantable insulin delivery system in type I diabetic patient. Optimal. Control Appl. Methods **38**(2), 184–204 (2017)
18. Patra, A.K.: An automatic insulin infusion system based on Kalman filtering model predictive control technique. J. Dyn. Syst. Meas. Contr. **143**(2), 021004-1–021004-11 (2020)

A Novel Random Vector Functional Link Neural Network for Identification of Nonlinear Dynamic Systems

Debashisa Samal, Hemanta Kumar Palo, Laxmipriya Samal, and Badri Narayan Sahu

Abstract This piece of work proposes a novel random vector functional link neural network (RVFLNN) to determine the nonlinear systems. It can benefit many vivid real-world application domains like design and analysis of control and power systems, broadband communications, industrial sectors, etc. For identification in system modelling, chaotic particle swarm optimization (CPSO) has been used for the optimization of the parameters. An extensive investigation has been formalized to validate the performance of the novel RVFLNN-CPSO model with conventional FLANN in terms of the convergence rate, error parameters, and computational complexity. The results reveal that the novel identification system modelling has outperformed the state-of-the-art system with lower error, better convergence, and faster response algorithm.

Keywords Nonlinear system · FLANN · RVFLNN · CPSO · Error performance · The convergence rate

1 Introduction

During the last few decades, the identification of nonlinear systems has been a major topic of discussion as it benefits the design and development of efficient control and power systems in industrial applications. An accurate analysis, synthesis, and identification of nonlinear systems remain an essential domain for the

D. Samal · H. K. Palo (✉) · L. Samal · B. N. Sahu
Siksha 'O' Anusandhan (Deemed to Be University), Bhubaneswar, Odisha, India
e-mail: hemantapalo@soa.ac.in

D. Samal
e-mail: debashisasamal@soa.ac.in

L. Samal
e-mail: laxmipriyasamal@soa.ac.in

B. N. Sahu
e-mail: badrinarayansahu@soa.ac.in

S. Das and M. N. Mohanty (eds.), *Advances in Intelligent Computing and Communication*, Lecture Notes in Networks and Systems 202,
https://doi.org/10.1007/978-981-16-0695-3_25

effective functioning of controlled structures governing the dedicated devices, components, sensors, actuators, and smart machines. The necessary nonlinear mathematical framework is mostly designed and developed based on linear operators, complex variable theory, linear algebra, and the theory of linear differential equations [1]. The adjustment of design parameters relies on stable adaptive axioms and theorems to provide guaranteed global stability of the nonlinear systems. The effectiveness of the overall relevant systems depends on the characteristics of the stable time-varying linear systems with known outcomes established in the past. However, for stability and accuracy, the nonlinear systems need to be evaluated system-by-system or component-by-component. Hence, it is not surprising that the analysis and synthesis of nonlinear systems determine the dynamic response, robustness, and stability that remains complex with adaptive mathematical concepts.

For suitable identification in system modelling, the artificial neural network (ANN) can learn complex patterns giving a superior performance on nonlinear systems [2]. Functional link artificial neural network (FLANN) which is one of the variants of ANN can be used for pattern classification. The network has a faster convergence rate with reduced computational complexity as compared to the popular multilayer perceptions (MLP). Other literature suggests the use of a functional basis perceptron system to control and identify nonlinear systems with improved outcomes [3]. With the selection of suitable parameters, the FLANN has shown to outperform the conventional MLP in the identification of plant [4, 5]. All these models require algorithms like least mean square (LMS) for learning the patterns. There are several variations of FLANN trained with these optimization techniques like the GA, PSO, and the CLPSO, Tabu Search, with considerable success in this field [6, 7]. Among these, the GA develops its range of operation with the optima which can be located more specifically. The major drawback of this algorithm is its slower response and convergence rate. The issues have been tackled using the GA PSO technique to study the social behaviour exhibited by schools of fish and flocks of birds [8]. The CLPSO is a variant of PSO introduced by the authors to mitigate the premature convergence issue concerned with the PSO technique [9]. The Chebyshev polynomial-based methodology of ANN shows a faster convergence rate than the MLP for functional approximation as observed in the literature [10]. The authors have suggested that the FLANN structures with exponential functional expansion are capable of suppressing the Gaussian noise with better performance in identifying nonlinear systems [11]. Similarly, the hybrid FLANN-AIS model a variant of the conventional FLANN has improved the recognition performance of nonlinear MIMO plants as claimed by the researchers [12].

In this paper, the authors attempt to apply another variation of FLANN named the RVFLNN for the identification of the nonlinear system. It further explores the capability of an efficient chaotic particle swarm optimization (CPSO) technique for adapting variables of RVFLNN. A simulation study illustrates RVFLNN-CPSO model performance in terms of better convergence rate, lower computational complexity, and errors over the existing FLANN model.

The article has been structured in the following way. Section 2 defines the system identification with the requisite block diagram. Section 3 explains the conventional FLANN and the proposed RVFLANN. The CPSO optimization algorithm has been briefed in Sect. 4. Section 5 provides the simulation results and comparative analysis of the proposed and the existing systems. Section 5 concludes the work.

2 The System Identification

System identification is a methodology used for the analysis and calculation of parameters of a dynamic system. It is a step by step process used to evaluate a mathematical model for an unknown system and inspecting its input–output relation [4]. This develops a compatible structure model to approximate the plant output corresponding to similar input, with a minimum mean square error (MSE). The basic structure of an identification system model is explained in Fig. 1.

2.1 *FLANN Based Model*

The FLANN consists of a single layer ANN structure having the ability to perform in a complex region by creating a nonlinear decision boundary. In this model, a set of linearly independent functions is created which acts as a functional link to be used over the entire process. The block diagram of FLANN is shown in Fig. 2 [5].

2.2 *The RVFLNN Based Model*

The RVFLNN is capable of modelling nonlinear processes quickly with better generalization ability. The basic RVFLNN structure is provided in Fig. 3.

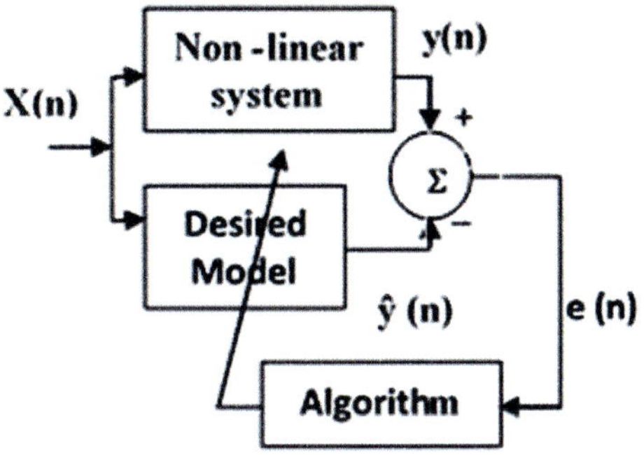

Fig. 1 Block diagram of system Identification

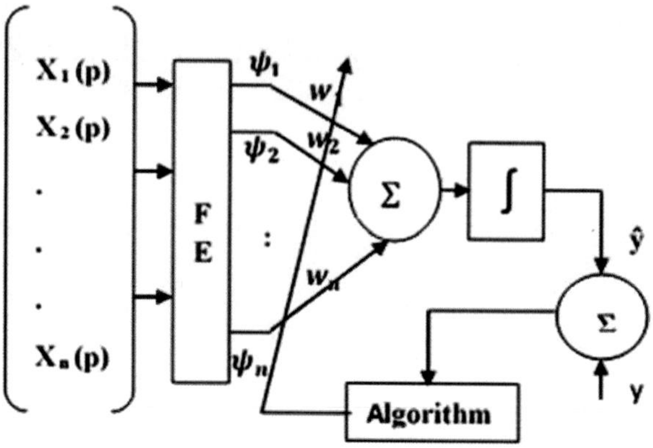

Fig. 2 Block diagram of FLANN

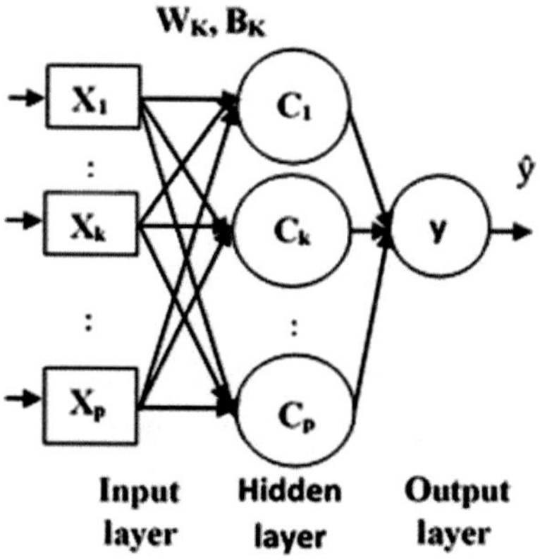

Fig. 3 Block diagram of RVFLNN structure

It has three layers such as the input, hidden, and output. The input layer is fed with a given training set of dimensions $d \times p$. It is connected to the hidden layers with the weight vectors W_k and bias vectors B_k as shown in Fig. 3. These weight and bias vectors are randomly selected based on the training samples at the beginning of the learning process. These vectors in an RVFLNN technique depend on a predefined probability distribution function. The input feature vector $X_{d,\ p}$, weight vector, and bias vector of RVFLNN are given below.

$$X_{d,p} = \begin{bmatrix} x_{1,1} & \cdots & x_{2,p} \\ \vdots & \ddots & \vdots \\ x_{d,1} & \cdots & x_{d,p} \end{bmatrix} \tag{1}$$

$$W_k = (w_1, w_2, \ldots w_k) \tag{2}$$

$$B_k = (b_1, b_2, \ldots b_k) \tag{3}$$

The hidden layer aims to incorporate an activation function for updating the network. In this work, the sigmoid activation function has been used and is given by

$$h_k = \frac{1}{1 + \exp(-w_k x_k + b_k)} \tag{4}$$

The hidden matrix is in the form as given by

$$H_{p,L} = \begin{bmatrix} h_{1,1} & \cdots & h_{L,1} \\ \vdots & \ddots & \vdots \\ h_{1,p} & \cdots & h_{L,p} \end{bmatrix} \tag{5}$$

The desired output is a P-dimensional vector and is represented as $Y_p = (y_1, \ldots y_j, y_p)$. The output of an RVFLNN can be described as

$$g(x) = \sum_{k=1}^{L} \beta_k h_k(x) \tag{6}$$

Equation (4) can be modified to provide output as

$$Y_p = H_{p,L}\beta_k \tag{7}$$

where $\beta_k = (\beta_1, \ldots \beta_k)$ denotes the output weights. The learning task can be accomplished by minimizing both the output weight norm and training errors.

$$\min \frac{1}{2} \left\| H_{P,L}\beta_K - Y_P \right\|^2 + \frac{\sigma}{2} \|\beta_K\|^2 \tag{8}$$

where σ is the smoothing parameter. The solution for β_K can be approximated as

$$\beta_K = \left(H_{P,L}^T H_{P,L} + \sigma I_L \right)^{-1} H_{P,L}^T Y_P \tag{9}$$

3 Chaotic Particle Swarm Optimization

PSO is another optimization algorithm that can provide information on the individual or particle population. In this case, each individual is governed to arrive at in a search space. In a multidimensional search space, the particles begin the learning process at a randomly chosen initial position. The transition of jth particle x_j relies on its velocity v_j as well as the location given by $p_j = (p_{j1}, p_{j2}, \ldots p_{jd})$. The neighbourhood equation is given by $lbest_j = (lbest_{j1}, lbest_{j2}, \ldots lbest_{jd})$.

The equation of PSO is given by

$$v_{jd} = w * v_{jd} + c_1\psi_1(p_{jd} - x_{jd}) + c_2\psi_2(lbest_{jd} - x_{jd}) \tag{10}$$

$$x_{jd} = x_{jd} + v_{jd} \tag{11}$$

where c_1 and c_2 are cognitive coefficients, whereas ψ_1 and ψ_2 are uniform random numbers.

To accomplish the desired search behaviour, the CPSO is introduced into the optimization algorithm. This work utilizes the tent map to create chaos parameters. The tent map can be described as

$$z_{p+1} = c * (1 - 2|z_p - 0.5|),\ 0<c<1 \tag{12}$$

The chaotic dynamics used for initialization for tent map ($c = 1$) are given by

$$z_p^{m+1} = c * \left(1 - 2\left|z_p^m - 0.5\right|\right),\ p = 1, 2, \ldots d \tag{13}$$

where z_p represents the chaotic parameters and m denotes the chaos iteration number. The chaos variable is mapped as given by

$$x_{mn} = x_{\min,n} + z_n^m\left(x_{\max,n} - x_{\min,n}\right) \tag{14}$$

By defining $x_m = (x_{m1}, x_{m2}, \ldots x_{md})$ where $m = 1, 2, \ldots N$, using the following steps.

1. Using the chaotic initialization method explained above, initialize the swarm ($r = 0$).
2. Evaluate the fitness f_j to update p_j and $lbest_j$ as and when required.
3. Update concurrently the velocity V_j of the particle and its position x_j.
4. A ranking of the swarm is carried out in an ascending order based on fitness. If the current iteration $r \le \frac{1}{2} r_{\max}$ and the current optimum $f(p_j)$ do not change with several iterations, proceed to step 5, else move to Step 6.
5. Incorporate the chaotic disturbance to choose the most suitable particle. In the swarm. Further, map these into the desired decision variables. The swarm is re-ranked by fitness to fetch the new most suitable particle.
6. When the desired stopping criterion is achieved, end the process, else choose $r = r + 1$ and return to step 2.

4 Simulation Results and Comparative Studies

Figure 4 shows the nonlinear input signal for the analysis in the proposed system identification task.

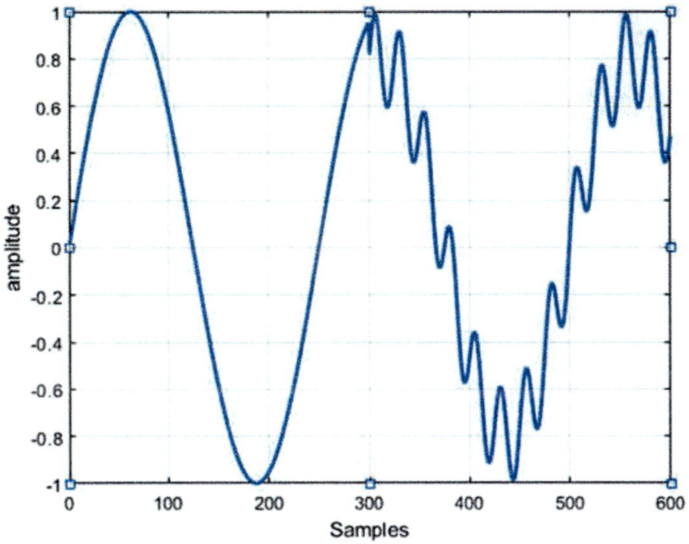

Fig. 4 Input signal

To demonstrate the nonlinear dynamic system identification, we have used a nonlinear model. This model is assumed to be of a second-order differential equation and is given by

$$y(k+1) = 0.3y(k) + 0.6y(k-1) + g[u(k)] \tag{15}$$

$$g[u(k)] = 0.6\sin[\pi u(k)] + 0.3\sin[3\pi u(k)] + 0.1\sin[5\pi u(k)] \tag{16}$$

The testing for the above model is carried out by taking input given as

$$u(k) = \begin{cases} \sin\left(\frac{2\pi k}{50}\right) & , K < 250 \\ 0.8\sin\left(\frac{2\pi k}{50}\right) & , K > 250 \end{cases} \tag{17}$$

Figure 5 shows the plant identification of nonlinear dynamics using the RVFLNN-CPSO algorithm. The desired, estimated, and the corresponding error is shown in this figure for a detailed understanding of the purpose.

To evaluate the performance of the model, three performance indices have been used. These are the (RMSE), mean absolute percentage error (MAPE), and mean absolute error (MAE). The equations of the performance indices are as given below.

$$\text{RMSE} = \sqrt{\frac{1}{K}\sum_{j=1}^{K}\left(y_j - \hat{y}_j\right)^2} \tag{18}$$

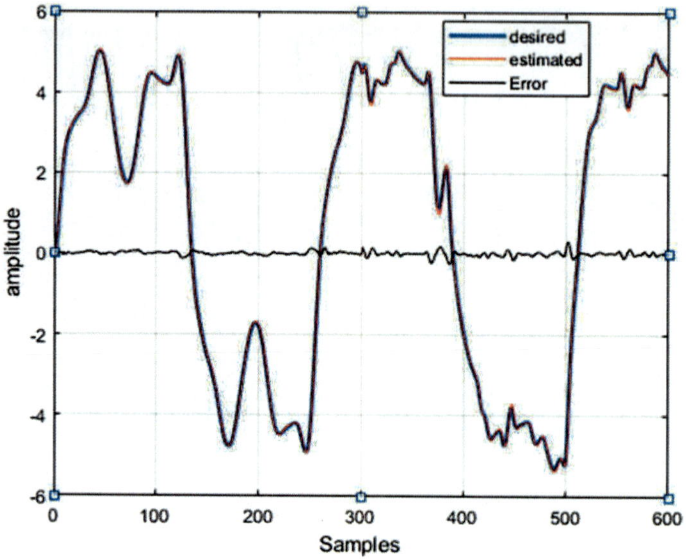

Fig. 5 Nonlinear dynamic plant identification using RVFLNN

$$\text{MAPE} = \frac{1}{K}\sum_{j=1}^{K}\left|\frac{y_j - \hat{y}_j}{y_j}\right|^2 \tag{19}$$

$$\text{MAE} = \sum_{j=1}^{K}\left(y_j - \hat{y}_j\right) \tag{20}$$

where $K, y_j, \hat{y}_j$ denotes number of samples, actual value, and predicted value, respectively. Figure 6 provides the RMSE error using the RVFLN-CPSO algorithm for the proposed non-linear dynamic plant identification using 100 epochs. The error is smoothened up as it progresses; thus, provides the desired convergence.

A comparison is made between the FLANN and RVFLNN-CPSO algorithms in terms of MSE in Fig. 7. As observed, the proposed FLANN-CPSO model provides lower MSE as compared to the conventional FLANN technique.

Table 1 compares different error performance indices between the FLANN and the proposed RVFLNN-CPSO algorithm in the identification of the nonlinear system. All the chosen performance indices such as the MAE, MAPE, and the RMSE are found to be lower with the proposed RVFLNN-CPSO than the conventional FLANN algorithm.

FLANN and the proposed RVFLNN-CPSO algorithm in the identification of the nonlinear system have been made in Fig. 8. As expected, the proposed model has a faster response in system identification than the existing FLANN system.

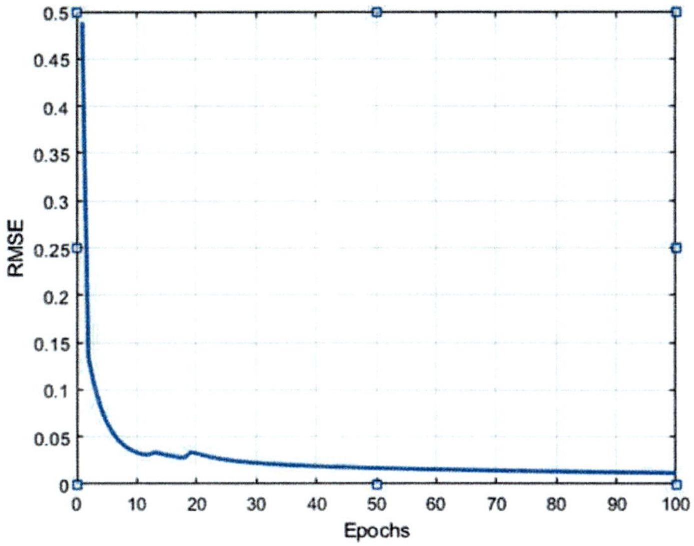

Fig. 6 Nonlinear dynamic plant identification using RVFLN with RMSE error

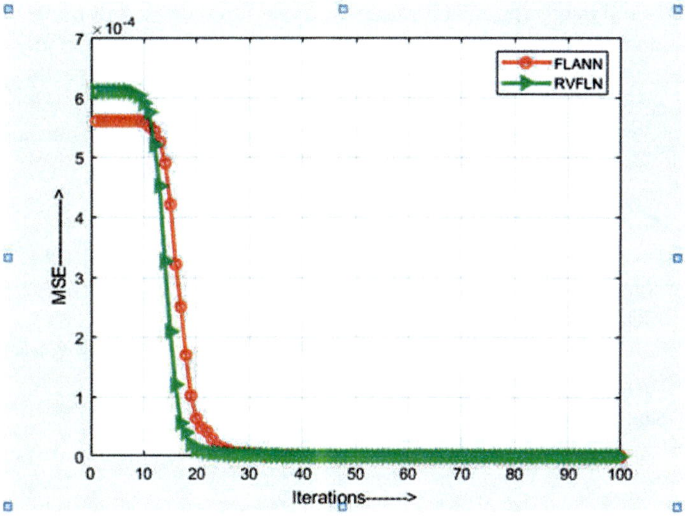

Fig. 7 Performance comparisons of MSE between FLANN and RVFLN

Table 1 Performance comparisons between FLANN and RVFLNN with the identified model

Chosen algorithms	FLANN			RVFLNN		
Performance indices	RMSE	MAE	MAPE	RMSE	MAE	MAPE
Magnitude	0.084	0.052	1.080	0.0028	0.0046	0.756

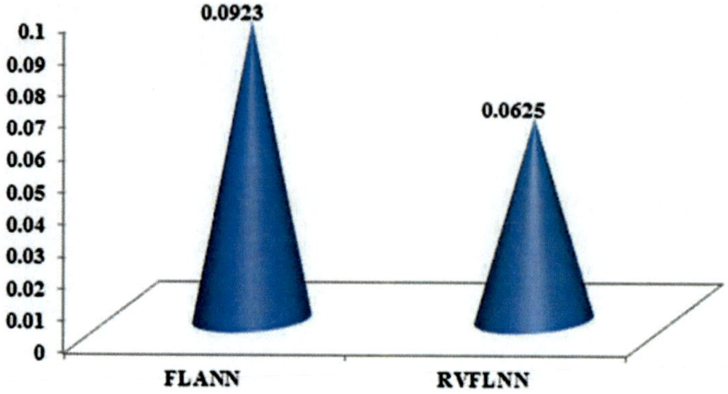

Fig. 8 Comparison of the computational time (in seconds) of FLANN and RVFLNN

5 Conclusions

In this work, a CPSO-based RVFLNN algorithm has been proposed for system identification. The proposed system has been compared to its performance based on different error performance indices and the computational complexity against the traditional FLANN algorithm. The proposed system has shown to outperform with lower error and computational time as revealed from our results. Other efficient optimization algorithms with compatible NN classifiers can provide better response and can be extended as future work.

References

1. Kumpati, S.N., Kannan, P.: Identification and control of dynamical systems using neural networks. IEEE Trans. Neural Netw. **1**(1), 4–27 (1990)
2. Pao, Y.H.: Adaptive Pattern Recognition and Neural Networks. Addison-Wesley Reading, MA (1989)
3. Sadegh, N.: An orthonormal neural network for function approximation. IEEE Tans. Neural Netw. **4**, 982–988 (1993)
4. Narendra, K.S., Parthasarathy, K.: Identification and control of dynamical systems using neural networks. IEEE Trans. Neural Netw. **1**, 4–26 (1990)
5. Patra, J.C., Pal, R.N., Chatterji, B.N., Panda, G.: Identification of nonlinear dynamic systems using functional link artificial neural network. IEEE Trans. Syst. Man Cybern. B **29**(2), 254–262 (1999)
6. Mishra, S.K., Panda, G., Majhi, R.: A comparative performance assessment of a set of multi objective algorithms for constrained portfolio assets selection, swarm and evolutionary computing. Elsevier **16**, 38–51 (2014)
7. Chu, S.-C., Tsai, P.-W.: Computational intelligence based on the behavior of cats. Int. J. Innov. Comput. Inf. Control **3**, 163–173 (2007)

8. Kennedy, J., Eberhart, R.C.: Particle swarm optimization. In: Proceedings of the IEEE International Conference on Neural Networks, pp. 1942–1948, Perth, Western Australia (1995)
9. Jiang, J.J., Qin, A.K., Suganthan, P.N., Baskar, S.: Comprehensive learning particle swarm optimizer for global optimization of multimodal functions. IEEE Trans. Evol. Comput. **10**, 281–295 (2006)
10. Lee, T.T., Jeng, J.T.: The Chebyshev polynomial based unified model neural networks for functional approximation. IEEE Trans. Syst. Man Cybern. B **28**, 925–935 (1998)
11. Mishra, S.K., Panda, G., Meher, S., Sahoo, A.K.1; Exponential functional link artificial neural Networks for denoising of image corrupted by Gaussian noise. In: IEEE International Conference on Advanced Computer Control (ICACC-2009), pp. 355–359. Singapore, 22–24 Jan 2009
12. Nanda, S.J., Panda, G., Majhi, B., Tah, P.: Improved identification of nonlinear MIMO plants using new hybrid FLANN-AIS model. In: IEEEInternational Conference on Advance Computing (IACC), pp. 141–146, (2009)

Object Detection using GM based Clustering Algorithms

Susmita Panda and Aman Kumar Agrawal

Abstract Object detection of a moving object under variable illumination condition is one of the inspirational tasks for the researchers. In order to do so, background subtraction plays a vital role. In this paper, we constructed the background model by computing the geometric mean (GM) of log-normal features of frames in temporal direction. This is followed by subtracting the background model with the input frame. Finally, for object detection, we have proposed three clustering algorithm, i.e. fuzzy C-mean (FCM), fast fuzzy C-mean (FFCM) and adaptive fuzzy C-mean (AFCM) algorithm. We have implemented the proposed algorithm on the standard PET data set. Partition coefficient (PC), partition entropy (PE) and Xie Beni index (XBI) are used to validate the performance. The simulation result of our proposed algorithm has been compared with the existing work, and by using different validity index, we concluded that GM-based AFCM is better than GM-based FFCM and GM-based FCM.

Keywords Log-normal feature · Background subtraction · Clustering · Object detection

1 Introduction

Background subtraction is also known as foreground detection which is mostly used in detection of moving objects in foreground by subtracting the background objects. Background subtraction is one of the central problems along with many applications in numerous fields which are used for research purpose for several years. Along this, it has many applications such as human detection, car detection, counting the running vehicle on the highway. Detection of objects from the background by subtraction has other application like object segmentation, action

S. Panda (✉) · A. K. Agrawal
Institute of Technical Education and Research, SOA Deemed to Be University, Bhubaneswar, Odisha, India
e-mail: susmitapanda@soa.ac.in

S. Das and M. N. Mohanty (eds.), *Advances in Intelligent Computing and Communication*, Lecture Notes in Networks and Systems 202,
https://doi.org/10.1007/978-981-16-0695-3_26

recognition, target tracking [1] and also estimates the speed, depth and height of the moving object [2]. However, with the help of this method we can also overcome problems like dynamic, fast changing environment along with essential parameters such as speed of detecting target, size of the detecting target relating to background, weather condition, illumination variation, background colour, clogging of objects, and the segmentation of various targets with respect to the background.

In video analysis, video segmentation plays a major role such that it will be first step that should be done before consequent operation like feature extraction, classification and description. Background subtraction is very first step that to be done for visual surveillance purpose [3]. The purpose of background subtraction is to obtain foreground objects from the background and done by background modelling in which frames are updated by using the background models to get updated frame to obtain results. However, background modelling is of two types that are recursive and non-recursive technique. In recursive method, it uses the input frame to update particular background model recursively, whereas in non-recursive, it uses frame by frame to distinguish the pixels as background or foreground. Though frame differencing is one of the easiest method used for modelling background were previous frame are being considered as background with respect to the current frame. And it has less computational complexity [4]. Frame differencing is applied in real-time detection of moving objects, and due to the advantage of the method, it has been used as temporal frame difference to detect the moving object in a more precise manner. Then, the resulting image is segmented into black and white image where each pixel belongs to background or foreground. After segmentation, image contains some noise component in it, and in order to remove them, it includes nonlinear filters like median filter, Wiener filter or blurring the image to reduce the image detailing leading to noise reduction.

In this research, we have focused on the problem of the background subtraction with a single static camera. Numerous data sets were used in testing of the algorithms which are subjected to detect the moving objects such as cars, humans and any moving object in the real-time scenario. In future, these algorithms may use for automation of vehicles like self-driven cars. In this paper, we determine that methods used for background subtraction problem are more specific, easy to understand alongside it is more precise and economical way. In the proposed method, background subtraction can tackle the problem with more accuracy even in varying background condition like illumination inconsistency by the use of continuous update and re-modelling of background frame pixel intensity using geometric mean as the data used were logarithmic in nature and the subtracted frame is clustered using fuzzy c-mean clustering, fast fuzzy c-mean and adaptive fuzzy c-mean algorithm and later passed through Wiener filter to remove the noise component if existed for more accurate results and the over view of the paper is illustrated in the form of a flow chart in Fig. 1.

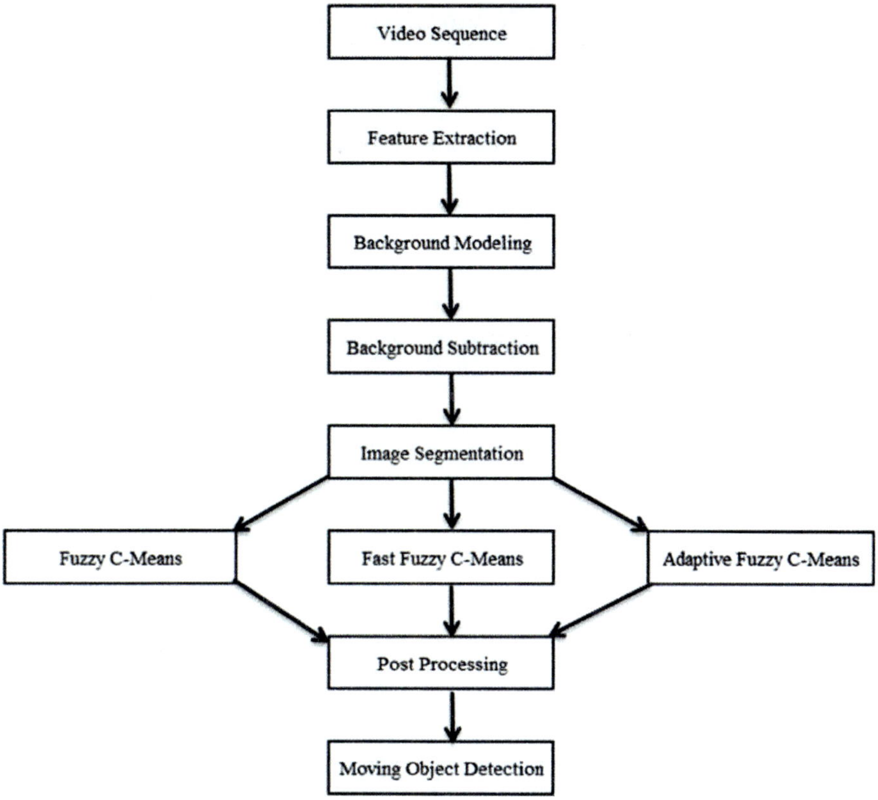

Fig. 1 Flow chart for the proposed methods

2 Related Work

In detection of moving objects, background subtraction is one of the crucial parts. In [5], authors have used recognized algorithm of mixture of Gaussian for modelling each pixel and update the model by using online approximation. However each pixel is distinguished as background or foreground with respect to Gaussian distribution model where pixels that do not match the model is considered as foreground else background. Additionally, it comprises two-pass linked components in order to relate the foreground regions. The model of MOG is also seen in [6] which are associated with the basic structure that has been portrayed in [5]. Along this, there is a new method that tackles the shadow detection along with background subtraction and is only utilized for the pixels that are categorized as foreground with minimum calculation. Moreover, [7] is the enhanced adaptive mixture model for background subtraction, whereas [8] is pixel-based background subtraction is repetitively restructured with the help of recursive equation of GMM. However, non-parametric adaptive density estimation method is stated and associated with the previous methods. In [9], author addressed the two problems that are formation of statistical model for the background and foreground scene that consist of moving objects. To form the background and foreground, it has used the non-parametric density estimation. In [10], author emphasizes on modelling the pixel intensities using time-based average of each pixel upon a small training sample, and modelling of pixel is done with log-normal probability density function which better accommodates the deviation among background pixels. In [11], authors have utilized the log-normal distribution of each pixel in a frame as a feature to model the background, and after subtraction, the resultant image is segmented using Otsu method. And the nonlinear filter is used to remove the noise component from the segmented result frame. In [12], authors have addressed the problem of background subtraction with fuzzy C-means clustering to obtain the segmentation threshold value for background subtraction in moving object detection and compared with Otsu method. In [13], authors have highlighted the most important medical imaging procedure in which segmentation is one of the most important parts in examination of MRI images. And FCM is usually used for the same as because of its fuzzy nature in which one pixel is found in many clusters leading to improved execution.

3 Proposed Algorithm

In this section, proposed methods were explained in details to have clear understanding. In order to get the foreground objects from a video sequence, firstly video was captured through a single static camera, and then converted to frames. Here, we have used the grey scale images for preforming the background subtraction because grey scale images improve the speed of processing the algorithms; however, most of the background subtraction make use of grey scale images for better results. Data that were used have varying weather condition that provides more information while using the log-normal distribution to model the background. And proposed methods were designed to overcome the environmental challenges like illumination, multiple moving objects and dynamic weather.

3.1 Pseudo Code

i. Take input of images and covert them to grey image if input is colour image
ii. Find the log normal of the input and store the result in another variable
iii. Calculate the geometric mean of above log-normal result
iv. Subtract the mean result with the current frame image
v. Segment the result using the fcm, ffcm, afcm algorithms
vi. If any noise exists, then pass through Wiener filter to remove them

3.2 Proposed GM-Based FCM Clustering Algorithm

Step-1: Video Sequence
Here, the video sequence is captured using a single static camera and converted into frames for further analysis.

Step-2: Feature Extraction (Log-Normal Feature)
All the input data samples were used to calculate the log-normal probability density function (pdf) for each pixel of the said samples individually as F_X shown in Eq. 1 [10],

$$F_X(\delta;\mu,\sigma) = \frac{1}{\delta\sigma\sqrt{2\pi}} e^{\frac{-(\ln\delta-\mu)^2}{2\rho^2}} \delta > 0 \tag{1}$$

Step-3: Background Modelling (Geometric Mean Averaging)
To model the background, we have used the geometric mean averaging which is defined as the n-th root of the product of n numbers. And along this geometric mean is more accurate while using the logarithmic data sets upon arithmetic mean.

Now, the average of data samples of n images are calculated as $G_{\substack{\prime\ \prime \\ x,y}}$ shown in Eq. 2,

$$G_{\substack{\prime\ \prime \\ x,y}} = \left(\prod_{i=1}^{n} x_i, y_i\right)^{\frac{1}{n}} n = \text{sample} \tag{2}$$

Here, x_i, y_i represents the individual pixel of the feature extracted image F_X.

Step-4: Background Subtraction
After estimation of background samples average $G_{\substack{\prime\ \prime \\ x,y}}$ in the above-said method and then the current pixel $C_{\substack{\prime\ \prime \\ x,y}}$ values were evaluated with $G_{\substack{\prime\ \prime \\ x,y}}$ and are stated in Eq. 3,

$$S_{ij} = \left| G_{\substack{\prime\ \prime \\ x,y}} - C_{\substack{\prime\ \prime \\ x,y}} \right| \tag{3}$$

Step-5: Image Segmentation (Fuzzy C-Means Clustering)
After the subtraction, segmentation is used to classify pixels as background and foreground. FCM [12, 14] is one of the well-known fuzzy clustering algorithms. It returns a membership matrix, which comprises the degree of membership of pixel relating to entire cluster. It tries to reduce the sum of square error.

Step-6: Wiener Filter
After segmentation of the subtracted frame, it is passed through a Wiener filter [15, 16] to reduce the noise component from the frame to obtain more precisely detecting objects. It mainly reduces the noise and minimizes the mean square error in the algorithm and smoothen the image.

3.3 Proposed GM-Based FFCM Clustering Algorithm

Repeat steps-1 to step-4 stated above.
Step-5: Image Segmentation (Fast Fuzzy C-Means Clustering)

After the subtraction, segmentation is used to classify pixels as background and foreground. FFCM [17] is the modified form of FCM clustering. However, in FCM, c membership grades were assigned to each pixel and updating membership matrix as $c \times n$ member which takes more time leading to reduction in the efficiency of the algorithm. In order to overcome the same, in MFCM, hard membership is introduced which is assigned to pixels for cluster centres updation at each iteration step.
Repeat step-6 as stated above.

3.4 Proposed GM-Based Adaptive FCM Clustering Algorithm

Repeat steps-1 to step-4 stated above.
Step-5: Image Segmentation (Fast Fuzzy C-Means Clustering)

After the subtraction, segmentation is used to classify pixels as background and foreground. AFCM clustering [15] is also modified form of fuzzy c-mean clustering. However, in FCM, the pixel classification is done by using the basic method without considering the spatial information as the information is considered as an important parameter for segmentation. In order to suppress the drawback, one method is to combine the neighbouring effect of pixel that to be classified, and there are numerous statistic estimators are there to do the same.
Repeat step-6 as stated above.

4 Performance Evaluation

In this section, we have discussed about the cluster validity [18, 19] parameters that are being used to validate the results.

i. Partition Coefficient (PC)
ii. Partition Entropy (PE)
iii. Xie Beni Index (XBI)

5 Results

In this section, we have discussed about the results of the proposed algorithms and are being tested upon numerous data sets that include variable speed moving objects and have varying lighting conditions within a single frame. And data sets sample size varies from 15 to 50 images for background modelling.

Results are tabulated below in Table 1, and frame number 15 shows that a car is moving in one side of road and has three pedestrian in the other side of the road, along with this, frames have multiple moving objects, and background is dynamic in nature along with uniform illumination and has good visibility. In the S. Diamantas et al. algorithm, detected objects are not precise; however, the designed algorithm detects the moving car and pedestrian accurately than the existing one. Frame number 19 shows that the moving cars along with two persons walking along the roadside, and these frames have multiple moving objects, and background is dynamic in nature along with non-uniform illumination. In the existing algorithm, detected objects are not precise; however in the designed algorithms detect the moving car and walking persons accurately than the existing one though the illumination is not uniform. Frame number 2748 shows that the moving cars in snowy weather and these frames have poor visibility, and the background is dynamic in nature along with non-uniform illumination. In the existing algorithm, detected objects are not precise; however, the designed algorithms detect the moving car accurately than the existing one though the illumination is not uniform. Frame number 1349 shows that a person is walking along with a stroller in snowy weather, and these frames have poor visibility, and the background is dynamic in nature along with non-uniform illumination. In the existing algorithm, detected objects are not precise; however, the designed algorithms detect the walking person with stroller accurately than the existing one though the illumination is not uniform (Fig. 2).

However, the detection of foreground is more accurate and precise in our proposed algorithm than other algorithms.

6 Conclusions

This work presents a geometric mean (GM)-based clustering method for background modelling and object detection by using log-normal features. Unlike other work, the three proposed methods, i.e. GM-based FCM, GM-based FFCM and GM-based AFCM, make use of GM of log-normal features in temporal direction. We have tested the proposed algorithms in four PET data sets having variable illumination condition. The method provides comparable results with selected works of the current state-of-the-art. Finally, we concluded that object detection using GM-based AFCM provides better result as compared to GM-based FFCM which is further better than GM-based FCM.

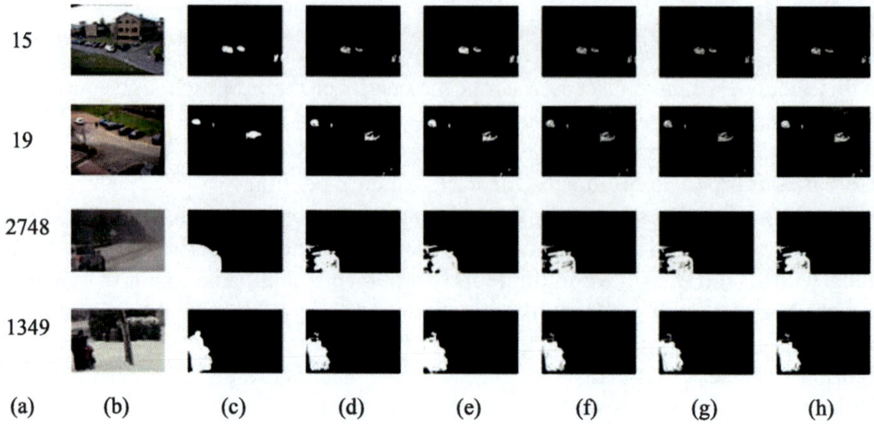

Fig. 2 Results obtained from the proposed algorithms; **a** Frame number; **b** Original frame; **c** Groud truth; **d** S. Diamantas et al. algorithm; **e** S. Panda et al. algorithm; **f** Proposed GM-based FCM; **g** Proposed GM-based FFCM; **h** Proposed GM-based AFCM

Table 1 Table for PC, PE and XBI on resultant image

Sl No.	Frame number	Method	PC	PE	XBI
1	15	S. Diamantas et al. algorithm	0.0168	0.1079	3.5200
		S. Panda et al. algorithm	0.0200	0.1260	3.0210
		Proposed GM-based FCM	0.8912	0.0412	2.1730
		Proposed GM-based FFCM	0.9029	0.0307	2.0010
		Proposed GM-based AFCM	0.9912	0.0257	1.5020
2	19	S. Diamantas et al. algorithm	0.0248	0.1221	2.1670
		S. Panda et al. algorithm	0.0253	0.1233	2.0010
		Proposed GM-based FCM	0.9050	0.0988	1.1771
		Proposed GM-based FFCM	0.9320	0.0696	0.9210
		Proposed GM-based AFCM	0.9750	0.0492	0.6520
3	2748	S. Diamantas et al. algorithm	0.0944	0.6244	4.4120
		S. Panda et al. algorithm	0.1127	0.5233	3.8900
		Proposed GM-based FCM	0.9590	0.1902	1.6256
		Proposed GM-based FFCM	0.9790	0.0821	1.5210
		Proposed GM-based AFCM	0.9980	0.0571	1.2560
4	1349	S. Diamantas et al. algorithm	0.1008	0.6567	5.0020
		S. Panda et al. algorithm	0.1185	0.5585	3.2120
		Proposed GM-based FCM	0.8845	0.4450	1.4465
		Proposed GM-based FFCM	0.9524	0.1986	1.0260
		Proposed GM-based AFCM	0.9945	0.0445	0.8920

References

1. Bardas, G., Astaras, S., Diamantas, S., Pnevmatikakis, A.: 3D tracking and classification system using a monocular camera. Wirel. Pers. Commun. **92**(1), 63–85 (2017). Jan 1
2. Diamantas, S., Astaras, S., Pnevmatikakis, A.: Depth estimation in still images and videos using a motionless monocular camera. In: 2016 IEEE International Conference on Imaging Systems and Techniques (IST), pp. 129–134. IEEE, 4 Oct 2016
3. Srinivasan, K., Porkumaran, K., Sainarayanan, G.: Improved background subtraction techniques for security in video applications. In: 2009 3rd International Conference on Anti-counterfeiting, Security, and Identification in Communication, pp. 114–117. IEEE, 20 Aug 2009
4. Ng, K.K., Delp, E.J.: Object tracking initialization using automatic moving object detection. In: Visual Information Processing and Communication, vol. 7543, p. 75430 M). International Society for Optics and Photonics, 18 Jan 2010
5. Stauffer, C., Grimson, W.E.: Adaptive background mixture models for real-time tracking. In: Proceedings of the 1999 IEEE Computer Society Conference on Computer Vision and Pattern Recognition (Cat. No PR00149), vol. 2, pp. 246–252. IEEE, 23 Jun 1999
6. KaewTraKulPong, P., Bowden, R.: An improved adaptive background mixture model for real-time tracking with shadow detection. In: Video-Based Surveillance Systems, pp. 135–144. Springer, Boston, MA (2002)
7. Zivkovic, Z.: Improved adaptive Gaussian mixture model for background subtraction. In: Proceedings of the 17th International Conference on Pattern Recognition, 2004. ICPR 2004, vol. 2, pp. 28–31. IEEE, 26 Aug 2004
8. Zivkovic, Z., Van Der Heijden, F.: Efficient adaptive density estimation per image pixel for the task of background subtraction. Pattern Recogn. Lett. **27**(7), 773–780 (2006). May 1
9. Elgammal, A., Duraiswami, R., Harwood, D., Davis, L.S.: Background and foreground modeling using nonparametric kernel density estimation for visual surveillance. Proc. IEEE **90**(7), 1151–1163 (2002). Nov 7
10. Diamantas, S., Alexis, K.: Modeling pixel intensities with log-normal distributions for background subtraction. In: 2017 IEEE International Conference on Imaging Systems and Techniques (IST), pp. 1–6. IEEE, 18 Oct 2017
11. Panda, S., Agrawal, A.K., Das, S.: Geometric mean based log-normal distribution for background subtraction. Int. J. Adv. Sci. Technol. **29**(5s), 1672–1677 (2020)
12. Soeleman, M.A., Hariadi, M., Purnomo, M.H.: Adaptive threshold for background subtraction in moving object detection using Fuzzy C-Means clustering. In: TENCON 2012 IEEE Region 10 Conference, pp. 1–5. IEEE, 19 Nov 2012
13. Choudhry, M.S., Kapoor, R.: Performance analysis of fuzzy C-means clustering methods for MRI image segmentation. Proc. Comput. Sci. **1**(89), 749–758 (2016). Jan
14. Duda, R.O., Hart, P.E., Stork, D.G.: Pattern Classification. Wiley, 9 Nov 2012
15. Ayech, M.W., El Kalti, K., El Ayeb, B.: Image segmentation based on adaptive Fuzzy-C-Means clustering. In: 2010 20th International Conference on Pattern Recognition, pp. 2306–2309. IEEE, 23 Aug 2010
16. Gonzalez, R.C., Woods, R.E.: Digital Image Processing, 2nd edn. (2002)
17. Biniaz, A., Abbasi, A., Shamsi, M.: Fast FCM algorithm for brain MR image segmentation. In: International Conference on Fuzzy Information and Engineering, vol. 1, pp. 1–8, 25 Oct 2012
18. Kumar, S., Kumar, P., Gupta, M., Nagawat, A.K.: Performance comparison of median and wiener filter in image de-noising. Int. J. Comput. Appl. **12**(4), 27–31 (2010). Dec
19. Wu, K.L., Yang, M.S.: A cluster validity index for fuzzy clustering. Pattern Recogn. Lett. **26** (9), 1275–1291 (2005). Jul 1

A Comparison of Oversampling and Transformation Techniques Used for Analysis of PAPR in a Real and Complex OFDM System for 5G Applications

Saumendra Kumar Mohanty, Asit Subudhi, and Santanu Kumar Sahoo

Abstract Peak-to-average power ratio (PAPR) brings a limit to the real and complex-valued massive or multi-user (MU) multiple-input-multiple-output (MIMO) orthogonal frequency division multiplexing (OFDM) system in terms of its computational complexity and usage of memory. This paper presents a PAPR evaluation of complex and real-valued massive MIMO systems. This evaluation scheme compares the discrete time with the continuous-time PAPR of real and complex OFDM systems. The result depicts two times or more sampled discrete PAPR best approximate the continuous-time PAPR. The analysis of this paper is not limited to a specific digital modulation scheme rather to all possible digital modulation schemes. Besides, knowledge of efficient hardware implementation is compared and realized using fast Fourier transform (FFT), discrete cosine transform (DCT), and discrete wavelet transform (DWT) techniques. The simulation results provides the tenacity and suitability of different modulation schemes with different transformation techniques and vice versa in the presence of high PAPR. These results also compares the robustness against multipath propagation, power consumption, and hardware implementation complexity.

Keywords FFT · DCT · DWT · Massive MIMO · OFDM · PAPR · Envelope power function · Real and complex OFDM

S. K. Mohanty (✉) · A. Subudhi · S. K. Sahoo
Department of Electronics and Communication Engineering, FET, ITER, Siksha O Anusandhan Deemed to Be University, Bhubaneswar, Odisha, India
e-mail: saumendramohanty@soa.ac.in

A. Subudhi
e-mail: asitsubudhi@soa.ac.in

S. K. Sahoo
e-mail: santanusahoo@soa.ac.in

S. Das and M. N. Mohanty (eds.), *Advances in Intelligent Computing and Communication*, Lecture Notes in Networks and Systems 202,
https://doi.org/10.1007/978-981-16-0695-3_27

1 Introduction

Today's wireless transmission that incorporates 5G massive MIMO requires a high spectral efficiency, robust against multipath propagation, power consumption, and implementation complexity. OFDM facilitates the transmission of high data rate over broadband radio channel subject to multipath fading [1–3]. OFDM does not need powerful channel equalization however sensitive to nonlinearity due to high PAPR [4]. The invention of MIMO facilitates multiple transmitter antennas to beamforming and multiple receiver antennas to diversity. While saying so, the limitations of PAPR and frequency offset may jeopardize the whole system. In particular, PAPR can increase the complexity of ADC and DAC and reduces the efficiency of radio frequency (RF) power amplifiers [5–7]. FFT in MIMO-OFDM can be used to simplify implementation. FFT helps to decompose any complex OFDM signal to a series of simple sinusoids. DFTs are used for real-time OFDM signal samples. FFTs are used for composite OFDM signals that the need for banks of oscillators and the modulators associated with it sub-carrier [8, 9]. The DCT is the most computationally efficient, and it hence is easy to implement in hardware that also has high compression performance. Also, it does not introduce discontinuities while imposing periodicity in the time signal. But DCT does not provide the ease of hardware implementation. However, DWT plays a major role in both time and frequency domain analysis simultaneously. This paper presents a comparative study of different modulation schemes with FFT, DCT, and DWT, which have been shown in Sect. 4. The PAPR of the sampled signal should not be larger than the continuous signal [10, 11]. The PAPR reduction of the sampled signal does not guarantee a similar PAPR reduction in continuous signal. In this paper, a comparison of PAPR is presented in Sect. 4.1 with different sampling rates. Simultaneously, comparison of PAPR has been made for different transformation technique with different modulation schemes that has been shown in Sects. 4.2 and 4.3. This paper introduces a practical technique for evaluating the continuous-time PAPR of OFDM signals using complex modulation. The proposed scheme exhibits that a minimum of two times over sampled discrete-time PAPR best approximates the continuous time PAPR even for complex OFDM signals. Secondly, it introduces a conventional OFDM system with limitations of their behavior with PAPR. The pioneering work of calculating PAPR of multicarrier BPSK-OFDM (real-value modulation) and multicarrier QPSK-OFDM, 16-QAM-OFDM (complex modulation) is achieved. Realization of hardware implementation is compared using fast Fourier transform (FFT), discrete courier transform (DCT), and discrete wavelet transform (DWT) techniques for high-speed wireless communication.

2 Peak-to-Average Power Ratio

One of the major drawbacks of multi-tone transmission is PAPR, which may go up to N times (where N is the number of sub-carriers). A high PAPR is very much undesirable in practice for which the ADC and DAC design becomes complex, and the efficiency of RF amplifier is reduced. When the analog signal is applied to highly nonlinear RF power amplifiers, it may produce in-band and out-band distortions. In-band distortion produces high BER, and out-band produces less spectral efficiency. To compute PAPR, the roots of the derivative of the envelope power function (EFP) are required. But, the derivative is a sum of sinusoids whose roots are difficult to find. Therefore, the EPF is converted to the sum of polynomials by using inverse cosine transformation. This approach consequently provides an accurate value of the peak of the envelope power.

2.1 *PAPR of a Multicarrier Signal: A General Prospective*

Let a vector $X = [X_0, X_1 \ldots X_{N-1}]^{\mathrm{T}}$ is the data block of length N. Any symbol X_k in the set X has a time duration of T and represents one of the sub-carriers set $\{f_n, n = 0, 1 \ldots\ldots N{-}1\}$. The N sub–carriers are orthogonal to each other, and thus, [12].

$$f_n = n\Delta f \text{ and } n\Delta f = \frac{1}{NT}$$

NT is the duration of the OFDM data block X. The complex data block for the OFDM signal is:

$$x(t) = \frac{1}{\sqrt{N}\sum_{n=0}^{N-1} X_n e^{j2\pi n\Delta ft}} \text{ for } \leq t \leq NT \tag{1}$$

The transmitted signal has the PAPR and is defined as:

$$\text{papr} = \frac{\max|x(t)|^2}{\frac{1}{NT}\int_0^{NT} |x(t)|^2 dt} \tag{2}$$

The PARP can be best reduced by reducing the $\max|x(t)|$ [13, 14]. Since discrete- time signals are dealt within most systems, many PAPR techniques are implemented to deal with amplitudes of various samples of $x(t)$. Due to symbol spaced output in the first equation, we find some of the peaks missing, which can be compensated by oversampling the equation by some factor to give the true PAPR value. But, to conserve bandwidth, we have considered two times sampling rate, which best approximates the over sampling rate, i.e., four times.

3 Mathematical Modelling of the Complex OFDM System

The OFDM signal with N sub-carriers can be expressed as [1]

$$x(t) = \left(\frac{1}{\sqrt{N}}\right)\sum_{n=0}^{N-1} S_n \exp(j2\pi nt) \tag{3}$$

$\{s_n\}$: are data symbols and
t: is normalized with respect to the OFDM symbol duration.
γ_c is the continuous-time PAPR and is given as

$$\gamma_c = \max|x(t)|^2 \tag{4}$$

γ_c provides the instant envelope greatest power of the baseband signal and denotes the highest PAPR. Tellambura's method [6] works only for the exceptional case of real-valued modulation. As a computationally feasible alternative, the discrete-time PAPR, γ_d, is frequently used instead of γ_c, which is defined as:

$$\gamma_d = \max_{0 \le k \le LN-1} |C_k|^2 \tag{5}$$

where

$$C_k = \left(\frac{1}{\sqrt{N}}\right)\sum_{n=0}^{N-1} S_n e^{j2\pi nk/LN} \tag{6}$$

with L being the over sampling rate.

Let $P_a(t) = |x(t)|^2$ is the instantaneous real-valued envelope power of the signal. Without loss of generality, no assumptions are made on the modulation scheme used to generate $\{S_n\}$. It can be easily shown that,

$$P_a(t) = 1 + \frac{2}{N}\sum_{k=1}^{N-1}\begin{bmatrix}\beta_k \cos(2\pi kt) + \\ \alpha_k \sin(2\pi kt)\end{bmatrix} \tag{7}$$

where β_k and α_k are defined as follows:

$$\beta_k = R\left\{\sum_{m=0}^{N-1-k} S_m S^*_{m+k}\right\},\ k = 1, 2, \ldots, N-1 \tag{8}$$

$$\alpha_k = I\left\{\sum_{m=0}^{N-1-k} S_m S^*_{m+k}\right\},\ k = 1, 2, \ldots, N-1 \tag{9}$$

with (·)∗denotes complex conjugation, and $R\{\cdot\}$ and $I\{\cdot\}$ represent the real and imaginary part of the bounded quantity. Clearly, a necessary condition for $P_a(t)$ to achieve its maximum at t^*, i.e., $\max_t P_a(t) = P_a(t^*)$ is

$$\frac{\partial P_a(t)}{\partial t} = 0\,\text{at}\,t = t* \tag{10}$$

Thus, a practical methodology to calculate $P_a(t*)$ is to first discover the roots of $\partial Pa(t)/\partial t$ followed by equating the values of $Pa(t)$ at only real roots. Following the approach of [2], we denote by $Tk(t) = \cos(k\cos - 1t)$ is the kth-order Chebyshev polynomial. For each k, $Tk(x)$ can be expressed as a kth-degree polynomial in terms of x, where $T_0(x) = 1$, $T_1(x) = x$ and $T_{k+1}(x) = 2xT_k(x) - T_{k-1}(x)$ for $k > 1$. Exploiting the equalities $Tk(\cos\theta) = \cos k\theta$ and $\sin(\theta) = \cos\left(\frac{\pi}{2} - \theta\right)$, we can rewrite (7) as

$$P_a(t) = 1 + \frac{2}{N}\sum_{k=1}^{N-1}\beta_k T_k(\cos(2\pi t)) + \frac{2}{N}\sum_{k=1}^{N-1}\alpha_k T_k\left(\cos\left(2\pi t - \frac{\pi}{2k}\right)\right) \tag{11}$$

Being different from the BPSK-OFDM systems considered in [15], the complex OFDM signal introduces the second term on the right-hand side of Eq. (11), which presents a major challenge in obtaining exact γc values.

4 Simulation Results and Discussion

This section displays and pronounces simulation results of three broad sections, namely sampling rate, transformation techniques, and modulation schemes of massive MIMO-OFDM. Here, in Sect. 1, a comparison of sampling for sampling rates (L) of 2, 4, and 8 are made and simultaneously, compared with continuous-time PAPR. In Sect. 2, different transformation techniques, namely FFT, DCT, and DWT, are compared pertaining to a fixed modulation scheme. Subsequently, in Sect. 3, different modulation schemes, namely BPSK, QPSK, and 16-QAM, are compared pertaining to a fixed transformation technique. This section uses MATLAB R2015a simulation software to verify the performance of the massive MIMO-OFDM. The simulation parameters are as follows in Table 1.

Table 1 Simulation parameters

Parameters	Value
Number of base station antennas	128
Number of users	8
Number of subscribers	1024
Number of multipath channels	6
Number of consecutive OFDM symbols	5
Channel characteristics	Raleigh
Frequency	6 GHz

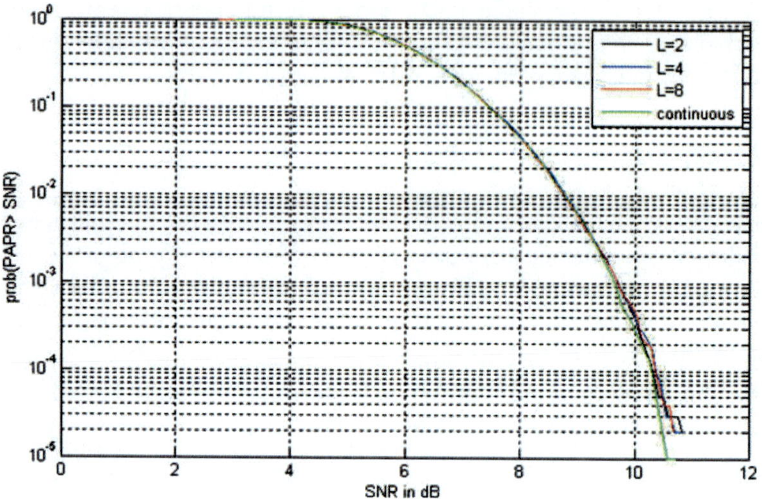

Fig. 1 PAPR probability curves of QPSK-OFDM for different sampling rates

4.1 Simulation of Sampling Rates for Discrete-Time and Continuous-Time PAPR

The results of discrete-time PAPR for different sampling rates are compared with continuous-time PAPR. As shown in Fig. 1, the PAPR probability curves of complex QPSK-OFDM (for different sampling rate) shows that for a oversampling greater than two times has the CDF of PAPR equal to CDF of PAPRs for sampling rate of 4 and 8 and also equal to the CDF of continuous-time signal. Thus, it is quite evident that a sampling rate of twice is sufficient to calculate the discrete-time PAPR, which advertently reduces the computational complexity and memory.

4.2 PAPR Simulation of Transformation Techniques for Real and Complex-Valued OFDM

This section brings different transformation techniques for different modulation schemes of MIMO-OFDM system onto one platform. Figures 2, 3, and 4 are the plots of PAPR probability curves of BPSK-OFDM, QPSK-OFDM, and 16-QAM-OFDM, respectively, considering FFT, DCT, and DWT. From Fig. 2, we see that for transmission of real-valued BPSK signal, FFT is more suitable compared to both DCT and DWT as the PAPR probability curve has less error rate. The BPSK-FFT probability error is less than 10^{-4} at SNR of 10 dB compared to error rate of BPSK-DCT and BPSK-DWT, which have error rates more than 10^{-4} at the same 10 dB. From Fig. 3, the probability error curve of QPSK-DWT is much less

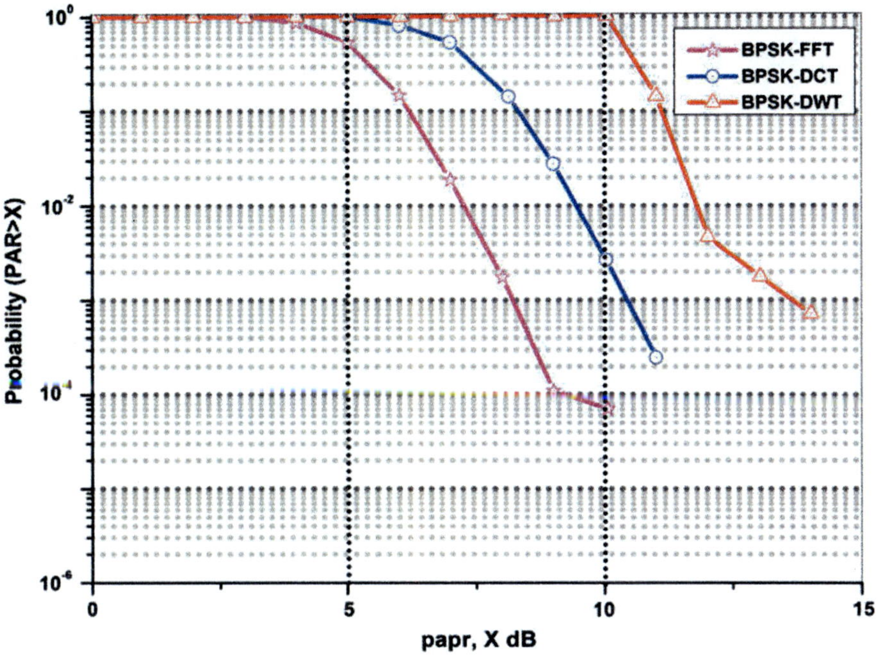

Fig. 2 PAPR probability curves of massive MIMO BPSK-OFDM with FFT, DCT, and DWT

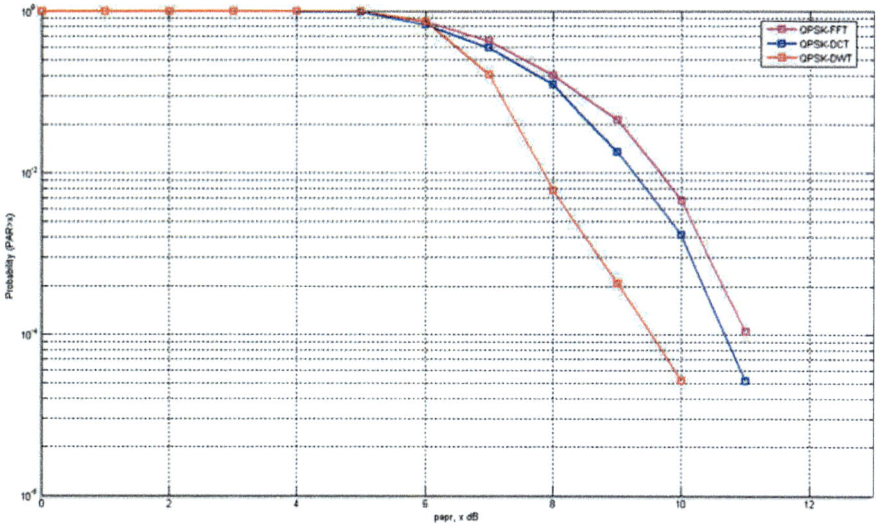

Fig. 3 PAPR probability curves of massive MIMO QPSK-OFDM with FFT, DCT, and DWT

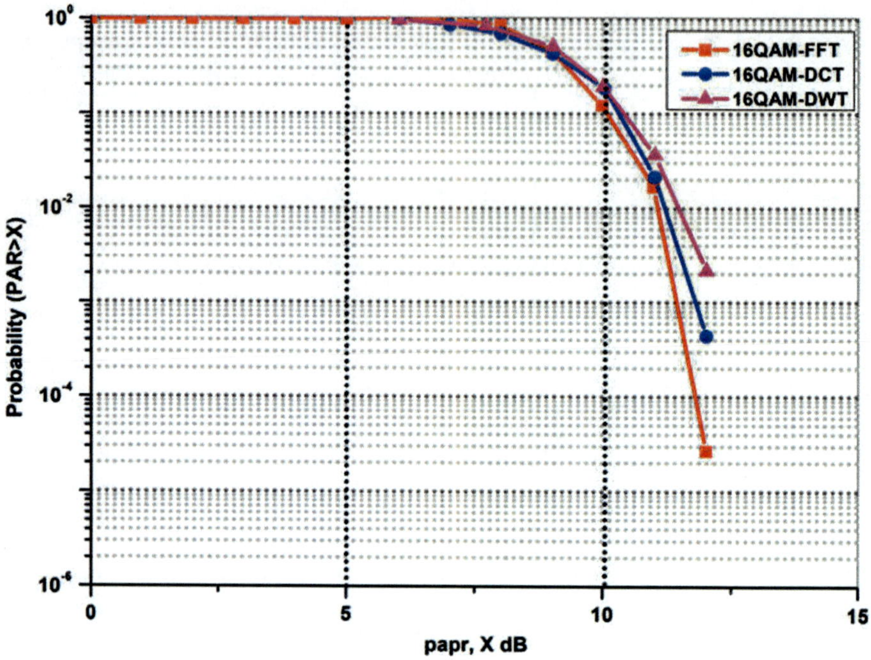

Fig. 4 PAPR probability curves of massive MIMO 16-QAM-OFDM with FFT, DCT, and DWT

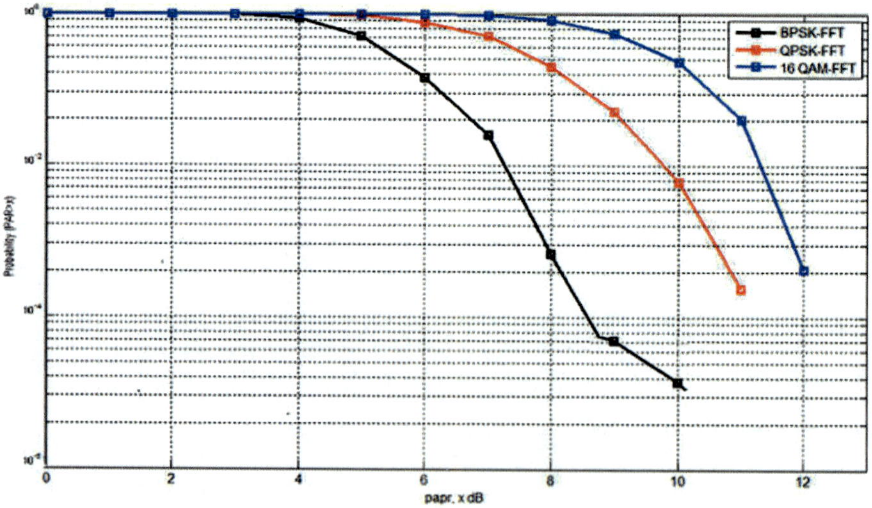

Fig. 5 PAPR probability curves of massive MIMO BPSK-OFDM, QPSK-OFDM, and 16-QAM-OFDM with FFT

than 10^{-4} at 10 dB. Comparing this with the QPSK-DCT and QPSK-FFT signal, the probability curves have errors more than 10^{-4}even at the same 10 dB SNR. Figure 4 depicts the plot of PAPR curves for 16-QAM with FFT, DCT, and DWT transformation techniques. 16-QAM-DWT has a probability error of nearly 10^{-5}compared to probability errors of 16-QAM-DCT and 16-QAM-FFT, which have errors more than 10^{-5} at same SNR, i.e., at 10 dB. We see that for transmission of complex-valued QPSK and 16-QAM massive MIMO-OFDM signal, DWT is more suitable compared to both DCT and FFT.

4.3 *PAPR Simulation of Real and Complex-Valued OFDM for Different Transformation Techniques*

This section brings different modulation schemes for different transformation techniques of MIMO-OFDM system onto one platform. Figures 5, 6, and 7 are the plots of PAPR probability curves of FFT, DCT, and DWT techniques, respectively, considering BPSK-OFDM, QPSK-OFDM, and 16-QAM-OFDM. Figure 5 shows PAPR curve of BPSK-OFDM, QPSK-OFDM, and 16-QAM-OFDM with FFT transformation. Here, we see that BPSK-FFT has the least error in the order of

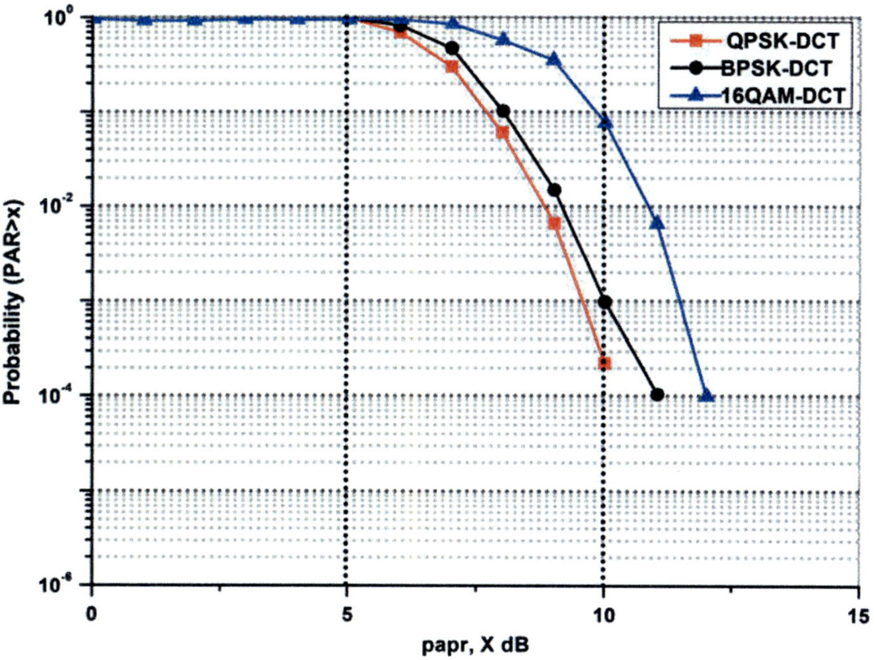

Fig. 6 PAPR probability curves of massive MIMO BPSK-OFDM, QPSK-OFDM, and 16-QAM-OFDM with DCT

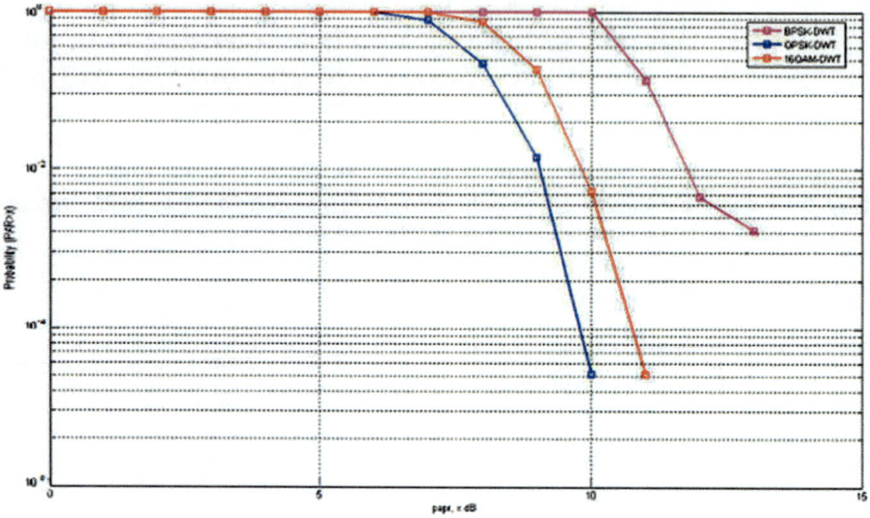

Fig. 7 PAPR probability curves of massive MIMO BPSK-OFDM, QPSK-OFDM and 16-QAM-OFDM with DWT

Table 2 Comparison of PAPR of different transformation techniques for a particular massive MIMO modulation scheme at SNR threshold of 10 dB

MIMO BPSK			MIMO-QPSK			MIMO-16QAM		
FFT	DCT	DWT	FFT	DCT	DWT	FFT	DCT	DWT
Least error	More error	Most error	Most error	More error	Least error	Most error	More error	Least error

Table 3 Comparison of PAPR of different massive MIMO modulation schemes for a particular transformation technique at SNR threshold of 10 dB

FFT			DCT			DWT		
BPSK	QPSK	16QAM	BPSK	QPSK	16QAM	BPSK	QPSK	16QAM
Least error	More error	Most error	More error	Least error	Most error	Most error	Least error	More error

nearly 10^{-5}compared to QPSK-OFDM and 16-QAM-OFDM, which has significant errors at 10 dB SNR.

From Fig. 6, the probability curves QPSK-DCT give probability error of nearly 10^{-4} at 10 dB compared to BPSK-DCT and 16-QAM-DCT, which have more errors compared to QPSK-DCT. Hence, QPSK-DCT outperforms the other two in terms of its BER performance and becomes the obvious choice. While comparing the different modulation schemes with DWT technique as shown in Fig. 7, QPSK-DWT outpaces the other two. QPSK-DWT has a probability error of nearly

10^{-5}compared with BPSK-DWT and 16-QAM-DWT, which have stiff errors compared to QPSK-DWT at the same 10 dB SNR. For the better understanding, two comparison tables have been shown as Tables 2 and 3 so that amicable distinguish between the different massive MIMO modulations schemes, and transformation techniques can be drawn. These tables help to find the adaptability and efficacy of MIMO schemes under a certain transformation technique and vice versa.

5 Conclusion

In this paper, various simulations based on sampling, transformation techniques, and different modulation schemes of massive MIMO-OFDM systems have been conducted. Out of seven simulation results, the first result compares various sampling rates of the complex MIMO-OFDM system. Based on the facts of the sampling rate, we see that two times or more sampled discrete-time PAPR approximates the PAPR of the continuous-time signal even for complex massive MIMO-OFDM systems such as QPSK-OFDM. This advertently reduces the computational complexity and memory of the complex MIMO-OFDM. The next three results are the comparison between FFT, DCT, and DWT for individual modulation scheme. Comparing the result of real-valued and complex-valued OFDM systems considering different transformation techniques, FFT is a suitable candidate for a real-valued massive MIMO-OFDM system, whereas DWT is more suitable for complex-valued systems. Different transformation techniques have their own suitability according to their applications. The last three results compare the suitable modulation scheme required for the MIMO-OFDM system in case of individual transformation technique. When different modulation schemes are compared considering an individual transformation technique, BPSK-FFT, 16-QAM DCT, and 16-QAM DWT are more suitable in the case of FFT, DCT, and DWT, respectively.

References

1. Wong, K.D., On Pun M., Poor, H.V.: The continuous-time peak-to-average power ratio of OFDM signals using complex modulation schemes. IEEE Trans. Commun. **56**(9), 1390–1393 (2008)
2. Han, S.H., Lee, J.H.: An overview of peak-to-average power ratio reduction techniques for multicarrier transmission. IEEE Trans. Wirel. Commun. **5**, 56–65 (2005)
3. Shin, C.E., Rim, K.S., Kim, Y.: A weighted OFDM signal scheme for peak to average power ratio reduction of OFDM signals. IEEE Trans. Veh. Technol. **62**(3), 1406–1409 (2013)
4. Bao, H., Fang, J.: An efficient Bayesian PAPR reduction method for OFDM-based massive MIMO systems. IEEE Trans. Wirel. Commun. **15**(6), 4183–4195 (2016)
5. Sakran, H., Nasr, O., Shokair, M.: An efficient adaptive technique with low complexity for reducing PAPR in OFDM-based cognitive radio. ISRN Signal Process.-Hindawi. Article ID 584941, 7p (2012)

6. Tellambura, C.: Computation of the continuous-time PAR of an OFDM signal with BPSK subcarriers. IEEE Commun. Lett. **5**(5), 185–187 (2001)
7. Jiang, T., Wu, Y.: An overview: peak-to-average power ratio reduction techniques for OFDM signals. IEEE Trans. Broadcast. **54**(2), 257–268 (2008)
8. Rajbanshi, R., Wyglinski, A.M., Minden, G. J.: Peak-to-average power ratio analysis for NC-OFDM transmissions. In: Proceedings of the IEEE 66th Vehicular Technology Conference, pp. 1351–1355, Oct (2007)
9. Slimane, S.B.: Reducing the peak-to-average power ratio of OFDM signals through precoding. IEEE Trans. Veh. Technol. **56**(2), 686–695 (2007)
10. Goff, S.Y., Khoo, L.B.K.C., Tsimenidis, C., Sharif, B.S.: A novel selected mapping technique for PAPR reduction in OFDM systems. IEEE Trans. Commun. **56**(11), 1775–1779 (2008)
11. Sharma, P.K., Basu, A.: Performance analysis of peak-to-average power ratio reduction techniques for wireless communication using OFDM signals. In: International Conference on Advances in Recent Technologies in Communication and Computing, pp. 89–95 (2010)
12. Pal, S., Sharmaet, N.: PAPR reduction in weighted OFDM system and it's BER analysis. Int. J. Digital Appl. Contemp. Res. **5**(6), 13–19 (2017)
13. Rahmatallah, Y., Mohan, S.: Peak-to-average power ratio reduction in OFDM schemes: a survey and taxonomy. IEEE Commun. Surv. Tutorials **15**(4), 1567–1592 (2013)
14. Sahu, S., Mohanty, S.K., Srivastav, R.: A comparative study of frequency offset estimations in real and complex OFDM systems using different algorithms. In: International Conference on Communication and Electronics System Design, pp. 87060–87061 (H-1), H1-H10 (2013)
15. Mohanty, M., Mishra, L., Mohanty, S.K.: Design of MIMO space-time code for high data rate wireless communication. Int. J. Comput. Sci. Eng. **3**(2), 693–696 (2011)

Spectrum Sensing Analysis Using Water-Fill Algorithm and Energy Detection Using GNU-Radio for Cognitive Radio Users

Shibanee Dash and Saumendra Kumar Mohanty

Abstract In this paper, the power spectrum density (PSD) has been considered based on energy detection problem by spectrum sensing for cognitive radios. It is shown that there is an improve in PSD performance of the receiving signal of interest from NI-USRP-2920 using the water-filling algorithm. Further the comparison has been shown using FFT/NFFT conventional periodogram and water-filling algorithm. The constellation pattern has been shown for identification of the multi signal channel (Presence of the primary signals in different bands). The simulation has been performed using real-time signal and QPSK demodulation.

Keywords Spectrum sensing · Cognitive radio · Water-fill algorithm · GNU-Radio · USRP

1 Introduction

Cognitive radio (CR) was first introduced in 1999 by Joseph Mitola and Gerald Q. Maguire to the radio community as a SDR extension, which was served by David A, who improved the overall act of the radio with relation to its interaction with the spectrum using a cognitive cycle [1]. Cognitive radio network is a next generation wireless network, where the efficiency of spectrum utilization can be improved through dynamic spectrum access [2]. This is introduced to work in a dynamic environment. Its designed transceiver can be used for best wireless channel in its vicinity, and this type of radio detects automatically the available channels in wireless spectrum, then for allowing more concurrent wireless

S. Dash · S. K. Mohanty (✉)
Institute of Technical Education and Research, S'O'A University,
Bhubaneswar, Odisha, India
e-mail: saumendramohanty@soa.ac.in

S. Dash
e-mail: shibaneedash.3@gmail.com

S. Das and M. N. Mohanty (eds.), *Advances in Intelligent Computing and Communication*, Lecture Notes in Networks and Systems 202,
https://doi.org/10.1007/978-981-16-0695-3_28

communications for one or more channels of a particular area, it accordingly changes its transmission and reception parameter [3, 4].

The real-time signal is being taken using universal software radio peripheral (USRP), which is a range of software-defined radios designed by Ettus research and its company, National Instruments [5–7]. GNU radio is used as simulation software which is a free and open-source software development toolkit which provides signal processing and many other blocks. These blocks may be used for radio related program design and modifications [8].

From the USRP, a real-time signal is taken, and to obtain band of interest, the signal has been passed through a low pass filter. For getting the corresponding PSD using FFT/NFFT technique of 1024 size length, the output is processed of the low pass filter. The Energy which was then obtained with a corresponding threshold and based on the results the prediction of Primary users can be obtained. Energy Detection is widely considered since there is no need of a priori knowledge of primary user and also due to its simple complexity [9–11].

In Fig. 1, the experimental setup has been used for performing the proposed research work based on energy detection in spectrum sensing.

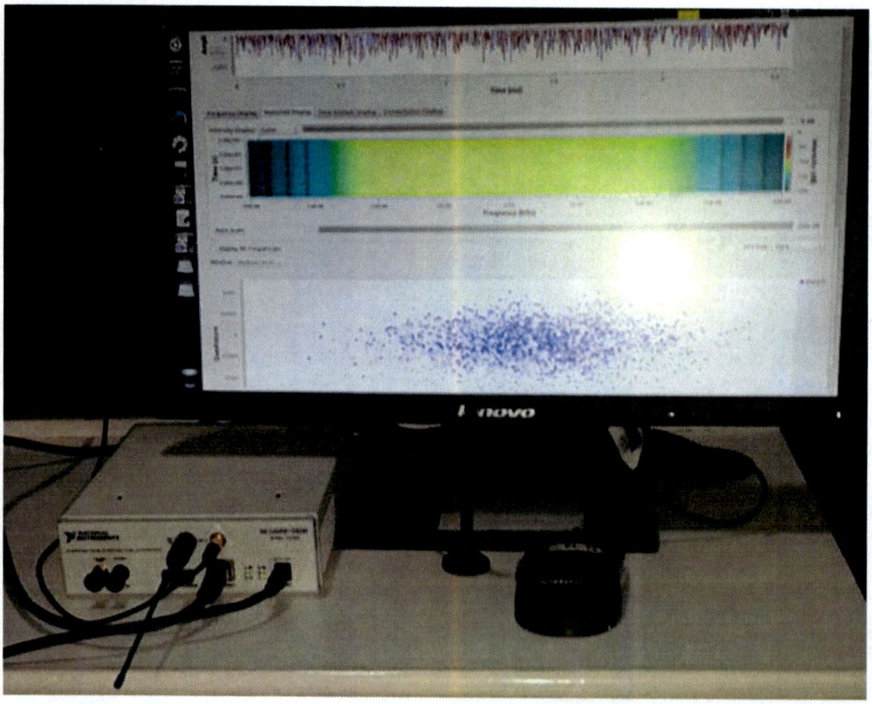

Fig. 1 Experimental setup using GNU-Radio and NI-USRP 2920e

2 Analysis of Energy Detection in Real-Time Environment

For efficient use of spectrum, the best available solutions are cognitive radio which works based on dynamic spectrum access model. The unused spectrum band can be accessed by the secondary or cognitive users which has been already allotted to the primary users or the licensed users in their absence [12]. The working of CR user's phenomenon deals with the non-interference with the primary users [13]. The spectrum sensing should be maximum accuracy as well as computationally fast and simple at the CR receiver in Eq. (1), and it is denoted by presenting hypothesis testing, where the presence of AWGN and alternate hypothesis and primary users' absence is being indicated by null hypothesis,

$$xi(t) = \begin{cases} wi(t), & H0, \\ hi(t)s(t) + wi(t), & H1 \end{cases} \tag{1}$$

where $H0$: absence of primary user;
$H1$: primary user is in operation

In this case, $xi(t)$ is the received complex baseband signal at ith CR, whereas $s(t)$, $hi(t)$ and $wi(t)$ are primary user signal, the complex channel gain of the sensing channel between the PU and the ith and additive white Gaussian noise (AWGN), respectively [14, 15].

As our basic model of spectrum sensing and analysis, here, we have implemented the energy detection technique. The threshold with the obtained energy is being compared in such a way that if the threshold is greater than the energy of the signal, then there is a absence of primary users and considering it as a false alarm case and vice versa indicates the presence of primary users [16].

The section below shows that the detection threshold λ is proportional to σ_w^2. In most practical situations, σ_w^2; (and λ consequently) needs to be estimated at the receiver.

Denoting the estimates by $\widehat{\sigma}_w^2$ and λ. Assume the error in estimating σ_w^2 is:

$$(1 - \varepsilon 1)\sigma_w^2 \le \widehat{\sigma}_w^2 \le (1 + \varepsilon 2)\sigma_w^2, \; 0 \le \varepsilon 1 < 1; \; \varepsilon 2 \ge 0 \tag{2}$$

$$Pd = P\left(Y > \frac{\lambda}{H1}\right) = Qm\left(\sqrt{2\gamma}, \sqrt{\lambda}\right) \tag{3}$$

$$Pf = P\left(Y > \frac{\lambda}{H0}\right) = \Gamma \frac{\left(m, \frac{\lambda}{2}\right)}{\Gamma(m)} \tag{4}$$

$$Pm = 1 - Pd \tag{5}$$

where P_m is probability of miss detection, and the SNR is denoted by Y. The sensing observation is shown as m = TW with respect to time bandwidth product. The generalized Marcum Q-function is denoted by Qm, Γ denotes the complete and

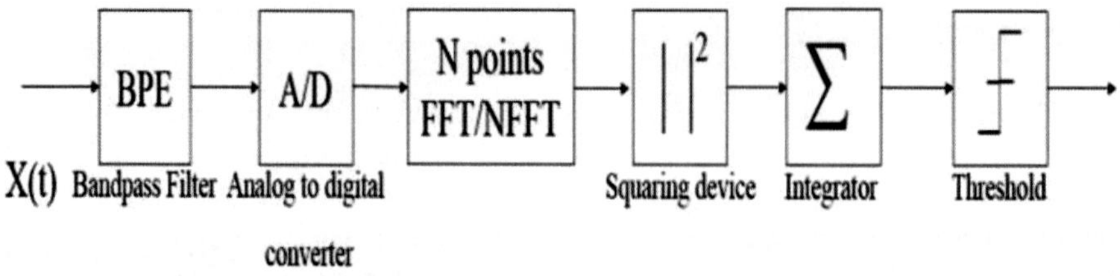

Fig. 2 Block diagram of energy detection

incomplete gamma functions. The audio output is continuous in nature, and thus, for PSD calculation converting it to discrete using FFT, the sampling frequency is changed to 400 MHz.

In both time domain and frequency domain, the energy detection process can be performed. A band pass filter is used for obtaining the signal of interest for a particular scanned band taken from USRP. The signal is the passed through integrator for calculation of signal power spectrum in time domain [17] (Fig. 2).

3 Implementation of Water Fill for Spectrum Sensing

In water-filling optimization technique, data rate is being maximized for a set of parallel sub-channels, when $R = \frac{b}{T}$, with 1/*T* as the symbol rate which is fixed. The maximization is required for $b = \sum_{n} b_n$ over ε_n and b_n. Maximized sum by the number of bits which is largest and can be transmitted using a parallel sub-channel set,

$$b = \frac{1}{2}\sum_{n=1}^{N} \log_2\left(1 + \frac{\varepsilon_n \cdot g_n}{\Gamma}\right) \tag{6}$$

When unit energy is being applied by the transmitter to the sub-channel, then shows the sub-channel SNR,

$$g_n = \left(\frac{|H_n|^2}{(\sigma_n^2)}\right) g_n \tag{7}$$

For a fixed function of channel, i.e., multitone, but for maximization of b, ε_n, is variable, subject to an energy constraint that,

$$\sum_{n=1}^{N} \varepsilon_n = N\bar{\varepsilon}_x \tag{8}$$

To maximize g_n, cost function considering the constraint shown in Eq. (8) with the help of Lagrange multipliers, we get,

$$\frac{1}{\ln(2)}\sum_n \ln\left(1+\frac{\varepsilon_n \cdot g_n}{\Gamma}\right)+\lambda\left(\sum_n \varepsilon_n - N\bar{\varepsilon}_x\right) \tag{9}$$

Now by differentiating with respect to ε_n, we get,

$$\frac{1}{2\ln(2)}\frac{1}{\frac{\Gamma}{g_n}+\varepsilon_n}=-\lambda \tag{10}$$

Now Eq. (7) is being maximized to Eq. (8), when,

$$\varepsilon_n+\frac{\Gamma}{g_n}=\text{constant} \tag{11}$$

The set of channels capacity or the maximum data rate is obtained when $\Gamma = 1(0\ \text{dB})$ which is shown using water-filling algorithm. Here, we construct the corresponding solution in a graphical manner. This is done by taking the inverted channel SNR curve. This is compared as filled with water or energy with respect to a constant line. Considering the case where $\Gamma \neq 1$ that is the water-fill optimization form carries the same number of bits on every sub-channel whose description is given as,

$$b_n = 0.5 * \log_2\left(1+\frac{\varepsilon_n \cdot g_n}{\Gamma}\right) \tag{12}$$

The optimum water-filling transmit energies, satisfies for the example of multitone,

$$\varepsilon_n = \Gamma \cdot \frac{\sigma_n^2}{|H_n|^2} = \text{constant} \tag{13}$$

The above equation, the constant is calculated for the case when $\sum_n \varepsilon_n = N\bar{\varepsilon}_x$ and $\bar{\varepsilon}_x \geq 0$. The water filling is applied from the analog curve of $\frac{\Gamma}{g_n}$. To a constant level, water will rise. The amount of energy in sub-channel is shown by different colors as the depth of water/energy at the corresponding point in the band. The convex function is being minimized, so for a unique optimum energy distribution considering their set of sub-channel data rates for each (Inter symbol interference) ISI with their multi-channel modulation.

4 Simulation and Results

A real-time signal has been taken from a USRP which is having capacity of 50 Hz–2.2 GHz, then it is passed into a low pass filter having a sample rate of 400 k and cutoff frequency at 115 k. when the signal is being passed to a QT GUI frequency sink having a bandwidth of 400 k to position the widget within the application.

It is a graphical sink to display multiple signals in frequency. The below FFT plot is power verses frequency plot having 1024 FFT size. It has been used to display in XY-plane(scatter) the input signal which identifies the complex nature for every symbol sampling instant (Fig. 3).

Now, when the signal is being passed to a QT GUI constellation sink. Further, the output in Fig. 4 also illustrates the possible symbols for a given modulation scheme (QPSK) may be selected in the plane. The constellation pattern graph is of quadrature phase shift keying is shown.

To get waterfall display, signal is passed to a QT GUI waterfall sink, and a waterfall output pattern is obtained which is a 3D plot with multiples curves for data.

These are the different spectra from the band which has been selected as shown in Fig. 5. Here, in this paper, we had used Waterfall plots for estimation of spectral density for successive intervals in time domain. The diagram represents the

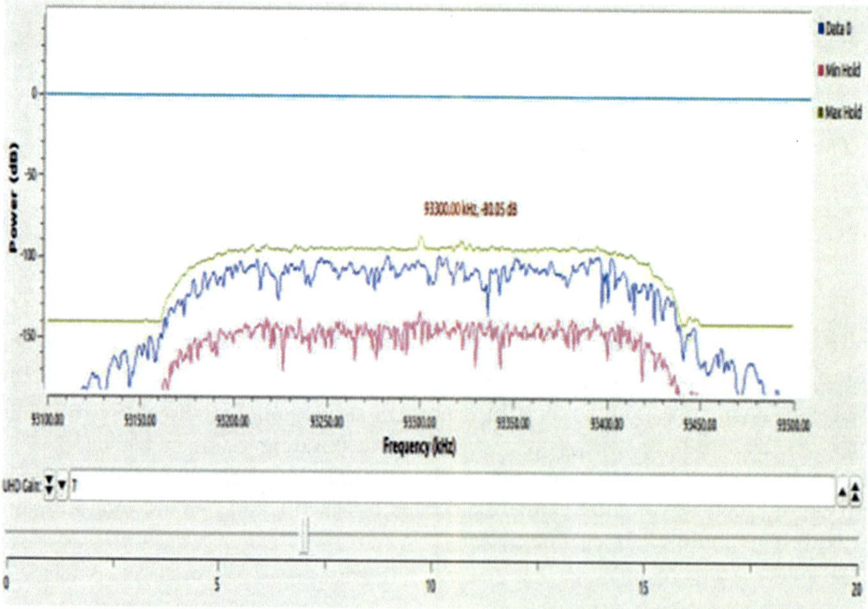

Fig. 3 FFT plot from USRP input

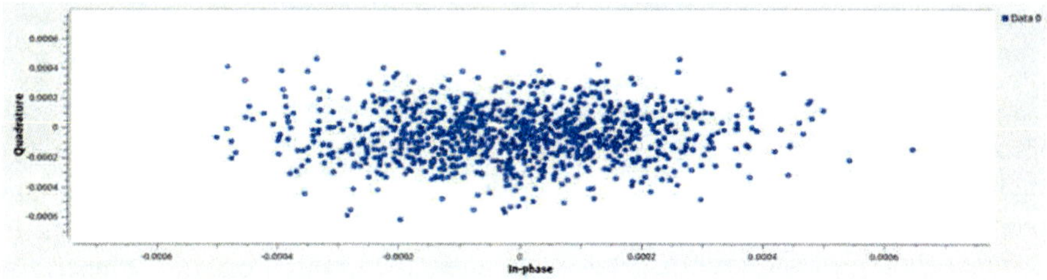

Fig. 4 Real-time signal's constellation plot FT plot

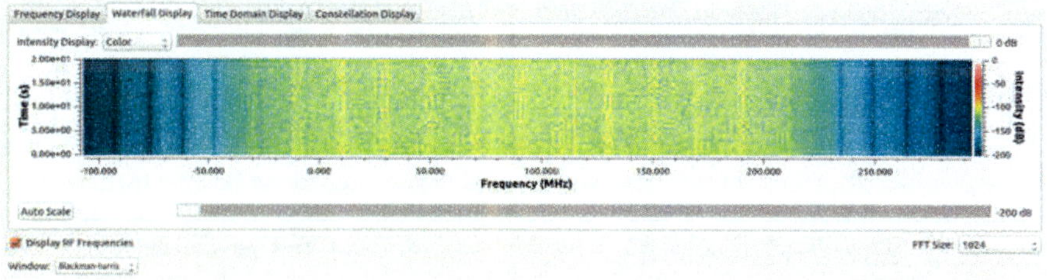

Fig. 5 Waterfall plot for the real-time signal

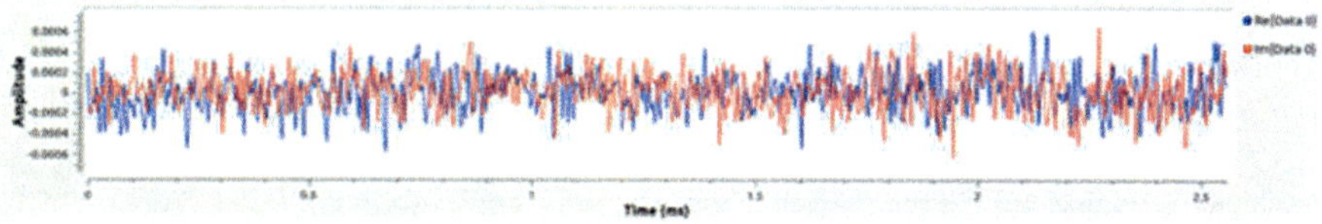

Fig. 6 Time verses amplitude plot of the real-time signal

waterfall display of the spectrum which is having a few different controls. This is a graph which is verses time, frequency, and intensity. Here, a Blackman–Harris window is being used.

To get multiple signals in time, the signal is being passed to a QT GUI time sink. It is a QT–based graphical sink that takes a set of complex streams and plot it in the time domain. The graph identifies the amplitude verses time plot.

Figure 6 signifies the real and imaginary nature of the signal whose energy is being found(continuous) in the time domain. The continuous distortion shows the presence of AWGN and the channel in Rayleigh fading channel and multipath propagation.

5 Conclusion

Energy detection using water-filling algorithm and constellation pattern detects the signal (sources) present with respect to their phases. Then, the real-time continuous amplitude variation in PSD has been presented that is shown in time domain, the complex signal. For getting better PSD estimation, the water-filling algorithm is used that shows the spectrum at successive intervals from the band in comparison with the conventional PSD calculation. Further, the work can be extended by calculation PSD and analyzing using Welch's method and calculating the corresponding BER for cognitive radio transmitters.

References

1. Mitola, J.: Software radios-survey, critical evaluation and future directions. In: IEEE National Telesystems Conference, pp. 13/15–13/23 (1992)
2. Akyildiz, I.F., Lee, W.-Y., Vuran, M.C., Mohanty, S.: Next generation/dynamic spectrum access/cognitive radio wireless networks: a survey. Int. J. Comput. Telecomm. Netw. 2127–2159 (2006)
3. Crohas, A.: Practical implementation of a cognitive radiosystem for dynamic spectrum access. Master of Science in Electrical Engineering Thesis. Notre Dame, Indiana (2008)
4. Ettus, M.: Universal software radio peripheral. http://www.ettus.com
5. Ma, J., Li, G.Y., Juang, B.H.: Signal processing in cognitive radio. Proc. IEEE **97**(5) (2009)
6. Yucek, T., Arslan, H.: A survey of spectrum sensing algorithms for cognitive radio applications. IEEE Comm. Surv. Tutorials **11**(1), 116–130 (2009)
7. Cabric, D., Mishra, S. M., Broderson, R.W.: Implementation issues in spectrum sensing for cognitive radios. IEEE Proc. Signals, Syst., Comput. 772–776 (2004)
8. Sahai, A., Hoven, N., Tandra, R.: Some fundamental limits on cognitive radio. In: Forty-second Allerton Conference on Communication, Control, and Computing, pp. 1662–1671. Monticello, IL, Oct 2004
9. Zhu, J., Xu, Z., Wang, F., Huang, B., Zhang, B.: Double threshold energy detection of cooperative spectrum sensing in cognitive radio. IEEE Proc. Crown Com. 1–5 (2008)
10. Farhang-Boroujeny, B.: Filter bank spectrum sensing for cognitive radios. IEEE Trans. Signal Process. **56**(5), 1801–1811 (2008)
11. Penna, F., Pastrone, C., Spirito, M.A., Garello, R.: Energy detection spectrum sensing with discontinuous primary user signal. In: 2009 IEEE International Conference on Communications, pp. 1–5
12. Gardner, W.A.: Signal interception: a unifying theoretical framework for feature detection. IEEE Trans. Comm. **38**, 897–906 (1988)
13. Shin, D.: A Dictionary of the GNU Radio Blocks (2005)
14. Sarijari, M.A. Rashid, R.A., Fisal, N., Lo, A.C.C., Yusof, S.K.S., Mahalin, N.H.: Dynamic Spectrum access using cognitive radio utilizing GNU radio and USRP. In: 26th Wireless World Research Forum (WWRF26), Doha, Qatar (2011)
15. Tanuja, S.D., Dina, S.: Spectrum sensing algorithm for cognitive radio networks for dynamic spectrum access for IEEE802.11 at standard. Int. Res. Rev. Wirel. Sens. Netw. (IJRRWSN) **2**,77–84 (2012)
16. Brodersen, R.W., Wolisz, A., Cabric, D., Mishra, S.M., Willkomm, D.: Corvus: a cognitive radio approach for usage of virtual unlicensed spectrum. (2004)
17. Abdulsattar, M.A., Hussein, Z. A.: Energy detection technique for spectrum sensing in cognitive radio: a survey. Int. J. Comput. Netw. Comm. (IJCNC) **4**(5) (2012)

An Empirical Study on Analysing DDoS Attacks in Cloud Environment

Iflah Aijaz, Sheikh Mohammad Idrees, and Parul Agarwal

Abstract A denial of service (DoS) is a computer crime where machines and other resources are made unavailable to its legitimate users by flooding the resources with unwanted requests when connected to the Internet. The DoS attack takes away the services that are being anticipated by legitimate users. Since most of the functions of the organization are automated and stored in servers, security must be taken care of at the highest level. Distributed denial of service (DDoS) represents a big challenge to the network. Since the network is the basic requirement in cloud computing or other latest technologies and DoS being distributive should be addressed properly. This paper gives the comparison of a healthy network with that of the DDoS network to outline the declination of the performance of a cloud network under this attack. In this specific scenario, the DDoS network shows how its output deteriorates. OPNET Modeler 14.5 is employed as a simulation tool.

Keywords Cloud computing · DoS · DDoS · EDoS · OPNET · Performance

1 Introduction

Cloud computing gives us a platform where different computers are connected in a network and serve their customers by operating together. It offers the business and individual users dynamically computing virtualized data and networking services via the Internet. Cloud computing was created in 2007 and provides potential users with several opportunities including communication tools and virtual devices, servers, and data storage. According to National Institute of Standards and Technology [1] "cloud computing is a model for enabling ubiquitous, convenient, on-demand network access to a shared pool of configurable computing resources (e.g. networks, servers, storage, applications and services) that can be rapidly provisioned and released with minimal management effort or service provider

I. Aijaz · S. M. Idrees (✉) · P. Agarwal
Department of Computer Science and Engineering, Jamia Hamdard, New Delhi, India
e-mail: sheikh.idrees99@gmail.com

S. Das and M. N. Mohanty (eds.), *Advances in Intelligent Computing and Communication*, Lecture Notes in Networks and Systems 202,
https://doi.org/10.1007/978-981-16-0695-3_29

interaction". Virtualization makes cloud computing possible, and all that a user needs is a web browser and Internet connectivity to use the services provided by the cloud providers. Cloud computing is a recent technology that intends to furnish access to resources dynamically as per the needs of its users. Cloud computing helps its users to make use of the resources that are widely distributed on the Internet to perform their desired work without having to install particular software. The advances in cloud computing are linked to the developing popularity of large data. Cloud computing delivers service online. Most customers use the cloud service via some untrustworthy network that renders the cloud more vulnerable to attack. Cloud computing is an important and dynamic system. It gets rid of the need for costly computer hardware, energy and applications. Massive growth has been reported in the amount of data produced by cloud computing. Big data is a complex and timely challenge that involves a large computer network to ensure the data is stored and interpreted effectively. The rapid economic advancement and technical obstacles in cloud computing have several beneficial dimensions. Cloud computing provides full ownership costs which help businesses to concentrate on the important strategies without caring about problems, including technology, accessibility and resource quality. It provides an extremely attractive system where scientists conduct experiments incorporating cloud computing and a wide range of software, network and storage. The vulnerability and number of attacks often grow with the rapid increase in cloud applications. Availability is one of the necessary factors in cloud computing to be maintained and should not be affected by the shortage of resources at any point in time. Such systems would achieve strong service demand levels, with reasonable turnaround time and low energy and maintenance costs. Availability suffers from various attacks among which one of the leading attacks is denial of service (DoS), where the attacker floods the target server with unwanted requests making it difficult to respond to legitimate requests. DoS attacker inundates the victim with traffic or transmitting some information that causes a system crash. It is driven to deplete the target framework or network resources to delay or interrupt the operation by accident, contributing to service failure. Attackers will initiate a variety of DDoS attacks from almost any part of the world, including resource-driven (i.e. network connectivity, RAM and CPU) and application-driven (i.e. web servers, database applications). Distributed denial of service (DDoS) attack uses several infected machines called zombies. The attack group attempts to create restrictions by using all the bandwidth accessible between the target and the Internet. Huge amounts of data are sent to a destination by a means of an amplification process or by some other form of mass traffic production, such as botnet requests. DDoS attacks are exponentially growing, and therefore proper mechanisms should be deployed. The DDoS attack may cause collateral damage [2] in the absence of an adequate defence mechanism, meaning that the co-located cloud services are also affected.

As stated earlier, the DoS attack troubles the availability of resources and prevents legitimate users from accessing the services. The attacker attempts to prevent authorized users from using resources. Since it involves the flooding of requests to a particular service or resource, the attacked resources fail to respond to its legitimate users. This threat can take various forms. It may involve wasting communication bandwidth, consuming system resources inefficiently or making use of SQL injection attack. An economic denial of sustainability (EDoS) attack, unlike traditional DDoS attack, primarily focuses not on the physical network and storage infrastructure, but the organization's financial resources [3, 4]. A massive economic denial of sustainability attack on Amazon EC2 in 2015 incurred regular $30,000 in financial losses [5]. In the DDoS attack, the flooding of requests originates from multiple sources. This attack is not possible to overcome by blocking a single attacker. It makes original traffic indistinguishable from compromised traffic. The critical aim of DDoS attacks is to stop the normal functionality of the system resources by overloading the target machines via sending bad requests such as sending the packet size above 65,525 bytes which is the default message size of IPv4 packet size. This paper studies the effects of DDoS on a network and evaluates its performances in the cloud concerning certain parameters.

2 Related Work

The authors in [6] discuss some security issues concerning service level agreements (SLAs). Here, different levels of security in SLA have been discussed and describe the procedure to make the consumer realize how the security policies are executed. Paper [7] addresses strategies for exploiting cloud computing to establish multiple DDoS attacks. The study discusses the high rate of DDoS attacks and all potential variations in a cloud network. For each form of DDoS attack, avoidance, identification and mitigation methods are defined. An algorithm introduced in [8] is simulated using the CloudSim simulator, and the results prove to be better in terms of performance and throughput. Iosup et al. analysed the performance of web applications and checked the performance of many task applications in the cloud [9]. Xiong et al. give the model that provides guidelines for computer services performance in cloud computing which would be useful in the design of computing paradigms [10]. Here, the probabilistic multivariate model is used for improving resource utilization for cloud providers on the proposed algorithm. The author in paper [11] clarifies different strategies for mitigation, and the efficiencies of the mitigation strategy are illustrated on a small network's topology. The author concluded that DoS attackers have many methods of executing it, and for all attacking systems there is no remedy. DDoS flooding attack problem is explored in [12] where the solution is provided by categorizing the attacks. Bonguet et al. introduced new types of XML and HTTP DoS attacks and justified possible detection and mitigation techniques [13]. The authors obtained and tested the current defence products according to the compulsory requirements and experimental design. A structure that incorporates avoidance and monitoring measures of DDoS attacks and the improved interactive turning checks to deter attacks from the group of attackers has been suggested in paper [14]. The authors in [7] summarize some security issues on cloud computing and how DDoS is a challenge to the accessibility of cloud computing and its resources. Besides exploring the high rate of DDoS attacks in a cloud environment and all potential variations, this paper also provides a survey for each form of DDoS attack and how to avoid, track and minimize it. Meng et al. indicated an anomaly detection method for discrete-time Markov chains. Normal user features are extracted based on user behaviour [15]. If the detected functions exceed a threshold value, judged to be a DDoS attacker, there is a threshold value in the model. The attacks on the DDoS in a cloud application system are distinct from the attacks on the conventional network infrastructure when initiating the attack strategy and the results of the attack [16]. The attackers leverage cloud computer's excellent capabilities to successfully conduct DDoS attacks [17].

3 Simulation

To accomplish the research goal, the analysis is split into two scenarios. The main aim of this research work is an evaluation of network performance when the network is put under a DDoS attack. OPNET Modeler 14.5 is used as a simulation tool, and the simulation is conducted for two hours. OPNET provides advanced components and protocols and works in different scenarios. OPNET libraries offer different pre-defined node models that can only be used when the necessary functions are imported and allowed [18, 19]. This research covers two scenarios with the same network topology. Scenario 1 depicts a standard and efficient network which is used to distinguish with the second scenario under DDoS attack concerning certain parameters. Both scenarios have the same configuration, but the second scenario consists of attackers in place of legitimate users. In the first scenario, the application definition of the server is configured in three rows. The first is the database (high load), the second is the HTTP server (high browsing), and the third is FTP (high load). The profiles are configured as database_user, HTTP_user and FTP_user. These profiles are consigned to an application supported profiles of the database, HTTP, FTP, respectively. Two routers are connected by IP_cloud. Main_office is set to 200 workstations. The second scenario is duplicated from the first scenario where the IP_32 cloud is subjected to DDoS attacks from three attacking machines by assigning source and destination addresses to an attack, using the IP ping traffic connection to the application. Two rows of packet size, 65,527 packets, near-maximum IPv4 permissible packet size, infinite count and a second timeout interval are configured in the IP ping attribute specification. Figure 1 shows the designed network for scenario 1. In scenario 2, normal_user gets replaced by three attackers.

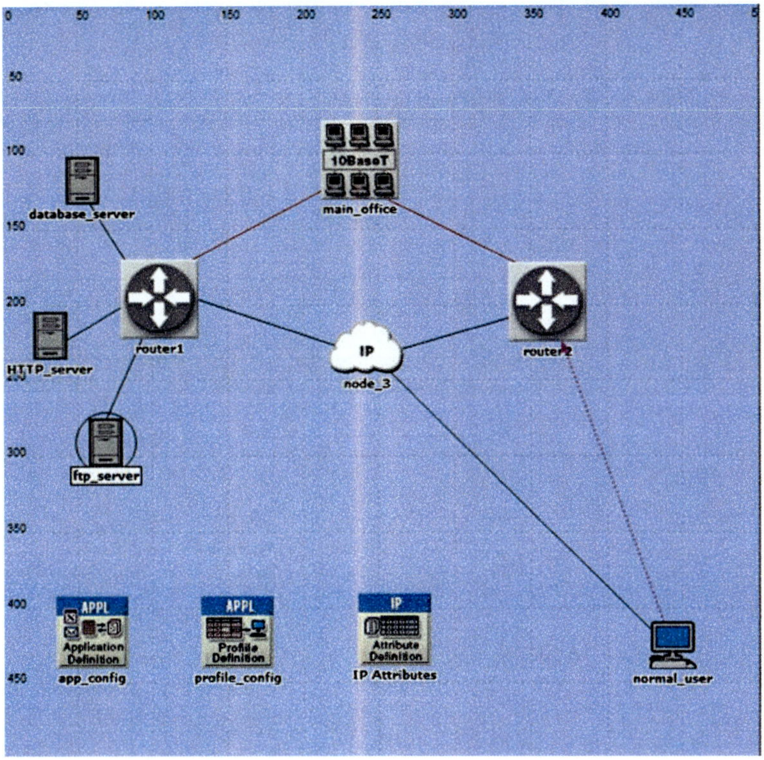

Fig. 1 Simulation scenario for designed network

4 Results and Discussion

This segment gives the simulated outcomes of the network under DDoS attack compared with the normal network. The results of scenario 1 are given by blue colour, and the results of the attacked scenario are displayed in red colour in the below-mentioned graphs. The efficiency of the network can be measured in terms of many factors that comprise transit time and response time. Transit time is the time needed to transfer a message from one device to another. Network performance relies on many factors, which include the number of users, the type of transmission medium, linked hardware capabilities and the software efficiency [20]. Two networking metrics—performance and delay—are often evaluated. The optimal network needs more efficiency and less time. When we attempt to send more data to the network, we can increase the performance, but, due to network traffic interference, we increase the delay. Figure 2 shows the database query response time also called elapsed time. Query response time is defined as the amount of time a server takes to answer to a request. This is the time that occurs between the client for the registration of the database and the request from the server. As mentioned

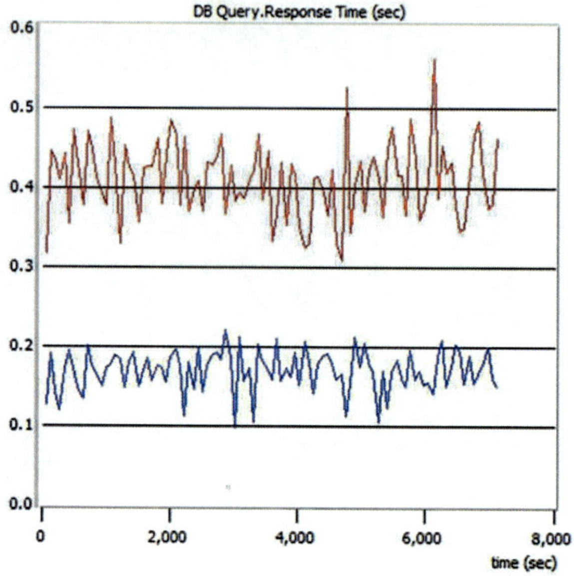

Fig. 2 DB response time

above, the response time for an ideal network is lower, while the response time for the DDoS network is higher, because the buffer is waiting to deal with several unexpected requests. As such, each request takes more time to be processed. The impact of a DDoS attack can be seen in Fig. 3 while comparing the two scenarios based on FTP upload response time. FTP upload response time is the time between submitting a message to the FTP server from the client application and getting back the response packet. While normal network gives desirable results, the network under attack gives poor performance because the FTP server is flooded with multiple requests from the attacker. Additional factors like capability and traffic can affect the response time. Any change in response time is a serious condition since it indicates that the network exceeds its capacities. Figure 4 depicts the performance of average traffic sent in bytes per second. The results of both scenarios are comparable. Nonetheless, the amount of bytes sent in the DDoS network is much smaller than that of the normal network. If the traffic received parameter is evaluated in Fig. 5, which supports the disruption of cloud performance in the DDoS network, the effects are similar.

Throughput in a network is defined as the maximum rate at which requests can be processed. In networking, this metric is defined as counting the actual data packets that are sent to a specific location or distributed over a specified time [20]. It can be analysed from Fig. 6 that throughput in a healthy cloud network is better as compared to a network where the cloud is under attack and throughput of cloud in this scenario is negligible. Network utilization, usually specified in percentage, is the amount of traffic on the network compared to the peak amount that the network

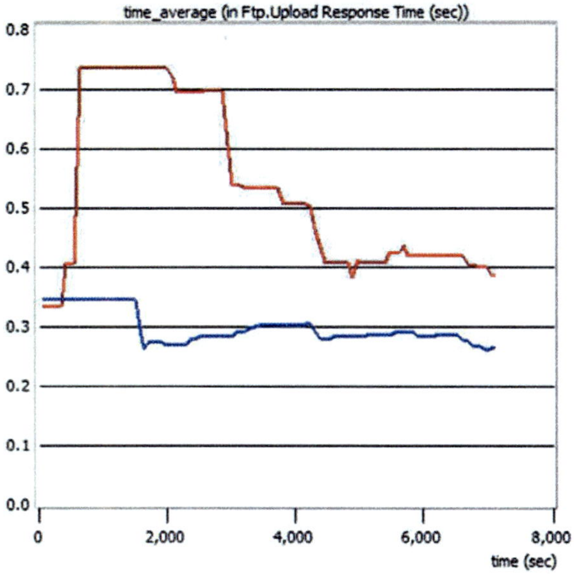

Fig. 3 FTP upload response time

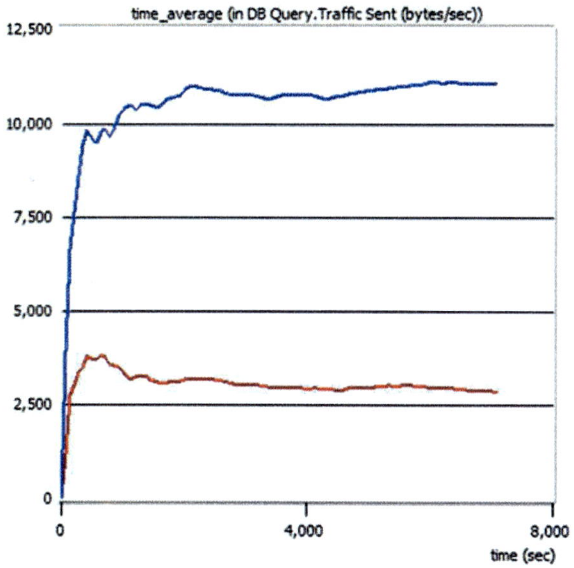

Fig. 4 Traffic sent

can support. We may recognize whether the network is active, regular or idle through tracking network usage. It is clear in Fig. 7 the resources are not utilized in the DDoS network, whereas in the normal network resources are utilized to 60% or 70%. Below is the performance analysis of both scenarios in Table 1.

Fig. 5 Traffic received

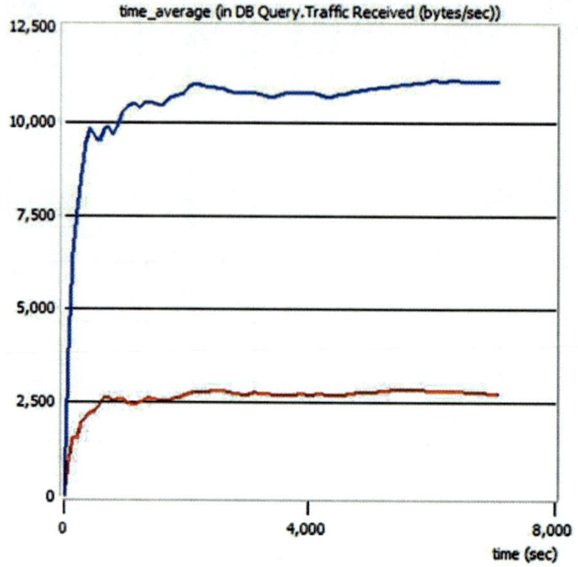

Fig. 6 Throughput

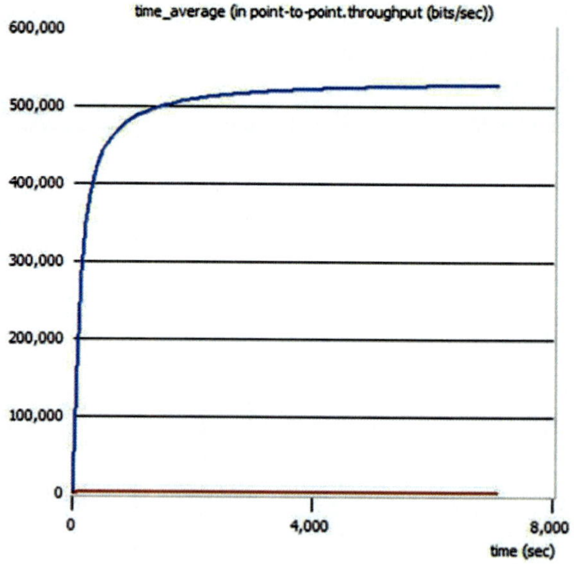

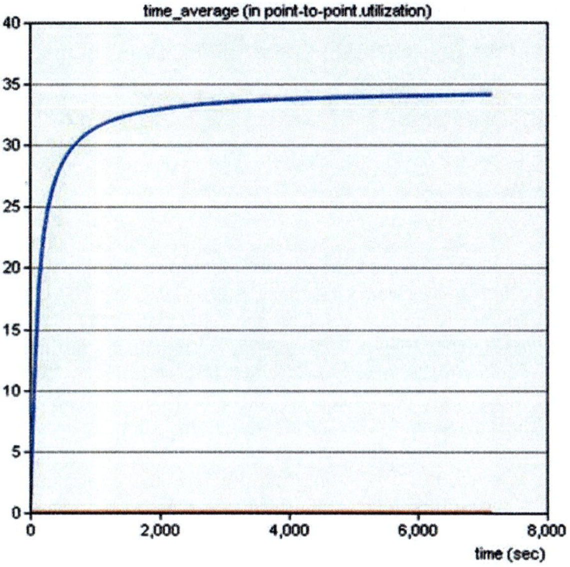

Fig. 7 Utilization

Table 1 Comparison of healthy and DDoS network

Parameters	Healthy network	DDoS network
DB response time	0.2 s	0.58 s
FTP upload response time	0.35 s	0.73 s
Traffic sent	11,000 bytes/s	2555 bytes/s
Traffic received	11,000 bytes/s	2500 bytes/s
Throughput	50,000 packets/s	0 packets/s
Utilization	70%	0%

5 Conclusion

The graphs of the simulated network in the above section clearly show how DDoS affects the normal functioning of service. Cloud performance measured in terms of throughput and utilization declines significantly in the network under DDoS attack. It can be seen the effect of DDoS becomes more severe when the sources of the attack increase, and also the attackers are hard to detect. It can also be concluded that business organizations should deploy proper mechanisms against DDoS attacks. Security software, hardware sellers and security experts should look at and collaborate in preventing first-hand methods of DDoS attacks that could affect global network infrastructure, government agencies, corporations and individuals [4, 21]. Network managers should actively monitor their networks to reduce attack risks and secure networked computers for use in other attacks. We focus on incorporating, evaluating and minimizing counteractions in both DoS and DDoS attacks in our future work.

References

1. Mell, P., Grance, T.: The NIST definition of cloud computing (2011)
2. Somani, G., Gaur, M.S., Sanghi, D., Conti, M.: DDoS attacks in cloud computing: collateral damage to non-targets. Comput. Netw. **109**, 157–171 (2016)
3. Al-Haidari, F., Sqalli, M., Salah, K.: Evaluation of the impact of EDoS attacks against cloud computing services. Arab. J. Sci. Eng. **40**(3), 773–785 (2015)
4. Mourya, A. K., Idrees, S. M.: Cloud computing-based approach for accessing electronic health record for healthcare sector. In: Microservices in Big Data Analytics, pp. 179–188, Springer, Singapore (2020)
5. Munson, L. (2015). Greatfire. org faces daily $30,000 bill from ddos attack
6. Kandukuri, B.R., Paturi, V.R, Rakshit, A.: Cloud security issues. In: 2009 IEEE International Conference on Services Computing (2009). https://doi.org/10.1109/scc.2009.84
7. Agrawal, N., Tapaswi, S.: Defense mechanisms against DDoS attacks in a cloud computing environment: state-of-the-art and research challenges. IEEE Commun. Surv. Tutorials **21**(4), 3769–3795 (2019)
8. Lakra, A.V., Yadav, D.K.: Multi-objective tasks scheduling algorithm for cloud computing throughput optimization. Proc. Comput. Sci. **48**, 107–113 (2015)
9. Iosup, A., Ostermann, S., Yigitbasi, M.N., Prodan, R., Fahringer, T., Epema, D.: Performance analysis of cloud computing services for many-tasks scientific computing. IEEE Trans. Parallel Distrib. Syst. **22**(6), 931–945 (2011)
10. Xiong, K., Perros, H.: Service performance and analysis in cloud computing. In: 2009 Congress on Services-I, pp. 693–700, July. IEEE (2009)
11. Cheung, S.: Denial of service against the domain name system: threats and countermeasures. In: SRI International. CSL Technical Report SRI-CSL-05-02 27 July 2005 (2005)
12. Zargar, S.T., Joshi, J., Tipper, D.: A survey of defense mechanisms against distributed denial of service (DDoS) flooding attacks. IEEE Commun. Surv. Tutorials **15**(4), 2046–2069 (2013)
13. Bonguet, A., Bellaiche, M.: A survey of denial-of-service and distributed denial of service attacks and defenses in cloud computing. Future Internet **9**(3), 43 (2017)
14. Saravanan, A., Bama, S.S., Kadry, S., Ramasamy, L.K.: A new framework to alleviate DDoS vulnerabilities in cloud computing. Int. J. Electr. Comput. Eng. **2088–8708**, 9 (2019)
15. Meng, B., Andi, W., Jian, X., Fucai, Z.: DDOS Attack Detection System Based on Analysis of Users' Behaviors for Application Layer. In: 2017 IEEE International Conference on Computational Science and Engineering (CSE) and IEEE International Conference on Embedded and Ubiquitous Computing (EUC), vol. 1, pp. 596–599. IEEE, July (2017)
16. Cong Luong, N., Hoang, D.T., Wang, P., Niyato, D., Han, Z.: Applications of economic and pricing models for wireless network security: a survey (2017). arXiv preprint arXiv:1707.07846
17. Yan, Q., Yu, F.R., Gong, Q., Li, J.: Software-defined networking (SDN) and distributed denial of service (DDoS) attacks in cloud computing environments: a survey, some research issues, and challenges. IEEE Commun. Surv. Tutorials **18**(1), 602–622 (2015)
18. Lu, Z., Yang, H.: Unlocking the Power of OPNET Modeler. Cambridge University Press, Cambridge (2012)

19. Iflah, A., Idrees, S. M.: Performance evaluation of multi-protocol label switching-traffic engineering schemes COGCOM EAI (2020). https://doi.org/10.4108/eai.8-10-2020.166550
20. Coombs, C.A., Forouzan, B.A., Fegan, S.C.: Data Communications and Networking. McGraw-Hill, Boston (2001)
21. Dong, S., Abbas, K., Jain, R.: A survey on distributed denial of service (DDoS) attacks in SDN and cloud computing environments. IEEE Access **7**, 80813–80828 (2019)

COVID-19 Pandemic Analysis and Prediction Using Machine Learning Approaches in India

Abhilash Pati, Manoranjan Parhi, and Binod Kumar Pattanayak

Abstract An extremely infectious disease, coronavirus disease 2019 (COVID-19) nowadays, is continuously threatening the whole world by its omnidirectional spreading. The aggressiveness and impact of COVID-19 disease worldwide have made the World Health Organization (WHO) to declare it as global pandemic. Machine learning (ML), an application of AI, can be implemented efficiently to track COVID-19, predict the increase of the pandemic, and design approaches to limit its spread. In this paper, three ML regression techniques, namely linear regression, polynomial regression, and support vector regression, are used to propose a model. Some experiments are performed on the online time series data from the dashboard of Johns Hopkins University sourced from Github repository using Python language to predict the affected people in the next 60 days. This article will help out in creating awareness among people toward the prevention of spread of infection due to COVID-19.

Keywords Coronavirus disease (COVID-19) · Machine learning (ML) · Artificial intelligence (AI) · Linear regression (LR) · Polynomial regression (PR) · Support vector regression (SVR)

1 Introduction

The novel coronavirus (COVID-19), reported first in Wuhan of China on December 31, 2019, continues to spread randomly throughout world thereby affecting people severely [1]. The increasing frequency of the causative virus (SARS-Cov-2) is

A. Pati (✉) · M. Parhi · B. K. Pattanayak
Department of Computer Science and Engineering, ITER, Siksha 'O' Anusandhan (Deemed to Be University), Bhubaneswar, Odisha, India
e-mail: er.abhilash.pati@gmail.com

M. Parhi
e-mail: manoranjanparhi@soa.ac.in

B. K. Pattanayak
e-mail: binodpattanayak@soa.ac.in

S. Das and M. N. Mohanty (eds.), *Advances in Intelligent Computing and Communication*, Lecture Notes in Networks and Systems 202,
https://doi.org/10.1007/978-981-16-0695-3_30

growing swiftly by affecting all the countries on earth approximately with USA, India, and Brazil, are being among the largely affected countries [2]. It has been transformed to a global pandemic in all the areas incorporating health industries, financial sectors, manufacturing units, businesses, etc. The coronavirus outbreak has been proclaimed as a pandemic on March 11, 2020, by the World Health Organization (WHO) because of its random and swift spread [3]. In India, the first case was recorded from Kerala, a student who returned from Wuhan, on January 30, 2020 [4]. The total collective number of confirmed cases reported is more that 60 lakhs (i.e., 6 Millions) across India as on September 27, 2020. Initially, three cases were reported on COVID-19 by February 3, 2020, in India. After that, there were no cases approximately for a month. From the starting of March 2020, it has been noticed to have increased in a random manner [5]. As reported by various researches, it is observed that COVID-19 circulates in some feral animals as well. COVID-19 has the ability to spread to humans from feral animals. The symptoms due the virus generally cause respiratory complications in humans, cold, and fever, etc. [6]. But, symptoms are not limited, varying randomly, recorded from some researches. There are no specific medicines yet developed, but the infection can be avoided by maintaining the social distances.

The spread of coronavirus for any country can be categorized into four stages, i.e.,

Stage 1: The cases reported for the public who traveled to or from infected countries.
Stage 2: The cases recorded from the family members, friends, and relatives who were in contact with the infected people of Stage 1.
Stage 3: The cases reported from those who neither belong to Stage 1 (no travel history) nor Stage 2 (not related to Stage 1 people) as well as not having any connections with them, i.e., the undetectable cases. This stage imposed the countries to lock down.
Stage 4: This is the stage where no nation can practically control the virus spread, and it may lead to clusters of infections.

In this paper, three ML regression techniques, namely linear regression, polynomial regression, and support vector regression, are used to propose a model. Some experiments are performed by considering evaluative measures, namely mean absolute error (MAE), mean-squared error (MSE), and root mean-squared error (RMSE), on the online time series dataset sourced from Github repository using Python language to predict the number of affected people in the next 60 days.

This study may consider the following research questions (RQs):

RQ1. Is it possible to forecast 100% of the COVID-19 spread in India using ML regression models?
RQ2. To what extent the regression techniques used can predict the COVID-19 spreading?

The rest of the paper has been organized as follows: Sect. 2 considers the spectrum of research work conducted in this field. Some information regarding methodologies and datasets used has been collected in this paper in Sect. 3. Section 4 gives an understanding of the proposed work. The performance evaluations in the field of analysis and a discussion on results have been covered in Sect. 5. Section 6 concludes the paper with a summary of findings and future scope.

2 Related Work

Tuli et al. [7] applied an enhanced mathematical approach to analyze and predict the possible risks of COVID-19 in countries globally and concluded that to obtain a better fit in developing a prediction framework; it must include iterative weight to fit generalized inverse Weibull distribution. Salgotra et al. [8] developed prediction model based on genetic programming (GP) that focuses on three mostly affected states of India, namely Maharashtra, Delhi, and Gujarat, claiming that this model uses simple linkage methods, and it is extremely consistent for time series prediction of coronavirus cases in India. Malki et al. [9] proposed various regression models to figure out the association between various factors and spreading rate of coronavirus with a conclusion that the important features of coronavirus mortality rate prediction are temperature and humidity. Burdick et al. [10] developed a model based on ML algorithms to evaluate the performance for invasive mechanical ventilation prediction of coronavirus infected people within 24 h of a first counter and concluded that this model helps the clinical procedures efficiently triage infected people and allotment of resources. Sun et al. [11] used gradient boosting algorithm for estimating the risk of coronavirus infectivity in nursing homes (NHs) having at least one confirmed case and observed that the NHs mostly predict the infection rate of the country in spite of other predictors including population mass, Medicare center's history, etc. Hasan [12] proposed a hybrid approach incorporating ensemble empirical mode decomposition (EEMD) to generate sub-signals and build original data de-noised, and artificial neural network (ANN) architecture incorporated in training the de-noised data and concluded that this model is helpful for coronavirus pandemic prediction. Wadhwa et al. [13] proposed a model based on regression techniques for predicting the lockdown extension in order to prevent coronavirus cases in India and concluded based on assumptions that coronavirus confirmed cases will be limited based on recovery rate and death rate within 3 months if lockdown is properly followed. Wang et al. [14] used logistic paradigm for fitting the limit of pandemic movement and then supplied the limit value into a time series ML prediction technique, FbProphet model, to obtain the pandemic curvature and forecast the pandemic movement and concluded that the cumulative infected cases may be up to 14.12 million, the peak value, in late October 2020. Yadav et al. [15] proposed a novel vector regression technique to analyze five different tasks of COVID-19 and claimed to have achieved better efficiency and classification accuracy comparatively. Ghosh et al. [16] focused on the affected people in each state, which are mostly affected, and make three growth models, namely exponential, logistic, and susceptible infectious susceptible (SIS) model with daily infection rate (DIR) for affected people prediction for those states in the next 30 days and reported accordingly. Das et al. [17] designed a convolutional neural network (CNN) model by considering the dataset of x-ray images of chest and detected the presence of coronavirus on human body. Mohapatra et al. [18] used support vector machine (SVM) classifier with different kernels considering accuracy, precision, and recall as evaluative measures on a randomly generated dataset, thereby concluding that SVM linear provides better results as compared to others.

3 Methodologies and Datasets

3.1 *Dataset Used*

There are various real-time datasets available for COVID-19 worldwide. In this paper, the dataset used has been derived from the official repository of Johns Hopkins University. We have sorted out time series summarized tables in .csv format, which are available in Github repository, having four tables for confirmed cases, number of deaths, number of recovered cases, and latest cases of coronavirus [19]. The six attributes, namely country/region, province/state, last update, confirmed, recovered cases, and death, are used for prediction purpose.

3.2 *Platform Used: Python*

Python, a user-friendly as well as interpreted programming language, can be used in multiple programming formats. Python is the most popular language worldwide as it is enhancing its various open-source libraries on a regular basis. In this proposed approach, Python has been used for the experimental purpose as it provides open-source libraries.

3.3 *Regression Techniques Used*

Regression techniques are considered based on ML for a computational process to forecast the COVID-19 transmission.

3.3.1 Linear Regression (LR)

It is the simplest, popular, and well-understood technique in ML and statistics. It is also known as linear model; input variables (x) and the single output variable (y) are linearly related, especially intended from linearly combined input variables (x). This technique is attractive as its presentation is very simple. The different LR learning models are simple linear regression, ordinary least squares, gradient descent, regularization, etc. The equation for LR with a single predictor is

$$Y = \theta_0 + \theta_1 x \tag{1}$$

Here, target is Y, predictor is x, bias is θ_0, and weight is θ_1.

3.3.2 Polynomial Regression (PR)

It is a special type of LR in which polynomial equation can be fitted on the data where target variable and the independent variables are nonlinearly related as well as modeled as an nth degree polynomial, denoted $E(y|x)$. PRs are mostly included for nonlinear occurrences like tissue growth rates, pandemic disease progresses, etc. The equation for PR with a polynomial equation of nth degree is

$$Y = \theta_0 + \theta_1 x + \theta_2 x^2 + \theta_2 x^3 + \cdots + \theta_n x^n \tag{2}$$

Here, target is Y, predictor is x, bias is θ_0, and weights are $\theta_1, \theta_2 \ldots \theta_n$.

3.3.3 Support Vector Regression (SVR)

Support vector machines (SVMs) are used for classification, whereas SVRs are used for regression problems. SVR provides acceptable flexibility with errors in any model and can be used in defining the appropriate line in fitting the data. It is a powerful technique in regression analysis.

4 Proposed Model

The working procedure of this proposed approach for COVID-19 pandemic prediction is elaborated stepwise as follows and detailed diagram in Fig. 1:

STEP 1: The first step in the proposed model is the collection of data. We considered the time series real-time dataset from the Github repository;
STEP 2: This step involves data preprocessing, analysis, and cleaning of data;

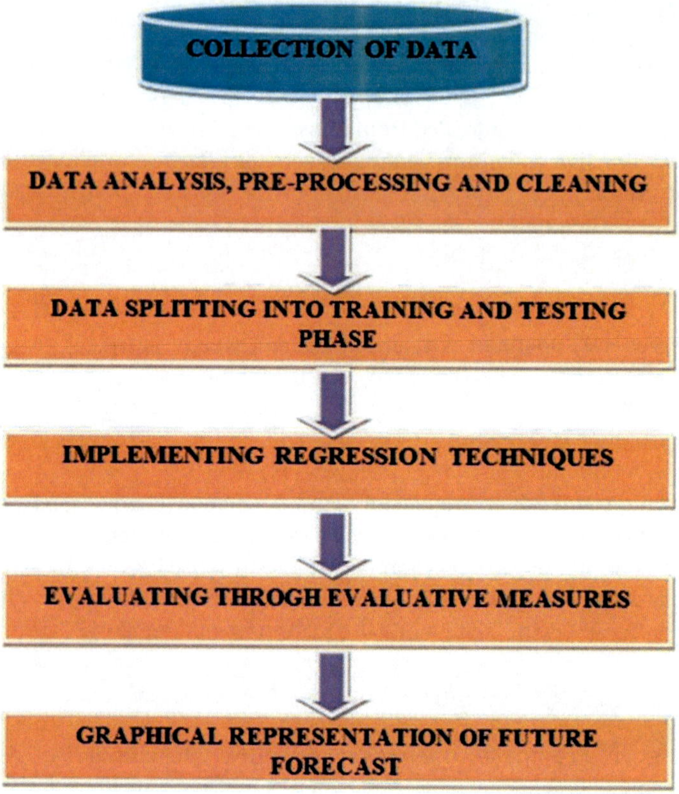

Fig. 1 Block diagram of proposed model

STEP 3: In this step, the dataset has been split into training dataset and testing dataset with a ratio of 4:1;
STEP 4: This step is intended to implement different regression algorithms;
STEP 5: In this step, the graphical representation of analysis and prediction are performed using different evaluative measures achieved by the models;
STEP 6: This step is used to conclude the best fit regression algorithm based on the results achieved.

5 Results and Discussion

The most significant objective of performance evaluation in prediction of COVID-19 is to forecast based on the real-time dataset used. The different evaluative measures on which predictions can be carried out in this study are briefly discussed in Table 1.

Table 1 Evaluative measures with meanings and formula

Evaluative measure	Meaning	Formula
Mean absolute error (MAE)	Average of all absolute errors	$MAE = \frac{1}{n}\sum_{i=0}^{n} \lvert x_i - x \rvert$
Mean-squared error (MSE)	Average of a set of errors	$MSE = \frac{1}{n}\sum_{i=0}^{n} (y_i - \hat{y}_1)^2$
Root mean-squared error (RMSE)	Standard deviation of the prediction errors	$RMSE = \sqrt{\frac{1}{n}\sum_{i=0}^{n} \left(y_i - \hat{y}_1\right)^2}$

Absolute error (AE), the amount of error in the estimations, is the difference between the estimated value and actual value. The MAE is the average of all AEs. The MSE is used in closing a regression line to a set of points by considering the distances from the points to that line and then squaring them to remove any negative signs. In simple terms, MSE is the average of a set of errors. The RMSE is the standard deviation of the prediction errors, i.e., residuals and is used to measure spread of these residuals. The RMSE is generally used in climatology, forecasting, and regression analysis, etc. Once, the data are thoroughly preprocessed, the regression algorithms are applied for regression analysis. In this study, the data were divided into two parts, i.e., training data and testing data in the ratio of 3:1 next. Then, the different evaluative measures are calculated, and accordingly the proposed method is implemented. Later, comparisons are performed among the three regression techniques based on the above-said evaluative measures as depicted in Table 2.

The future forecasting of three regression techniques like LR, PR, and SVR in the next 60 days is shown in Figs. 2, 3 and 4, respectively. From the results, the PR technique generates a minimum RMSE amount over the other two techniques like LR and SVR and shows actuations in trending the confirmed cases in predicting the spread of COVID-19 in India, and at the same time, it fits to the proposed model. However, if the transmission continues as the movement of the PR technique, then it would be the reason behind a huge number of deceases.

6 Conclusions

In this paper, regression analysis is conducted on COVID-19 transmission in India. The purpose of the proposed approach is to forecast the COVID-19 transmission in the next 60 days using regression models. It uses three regression techniques, namely LR, PR, and SVR, for the pandemic regression analysis and forecasting based on the evaluative measures like MAE, MSE, and RMSE. Then, the results are compared based on experiments performed on the online time series dataset sourced from Github repository using Python language to predict the number of possibly affected people in the next 60 days. In consequences, the PR technique generates a

Table 2 Comparative study of different regression techniques

Regression techniques	MAE	MSE	RMSE
Linear regression (LR)	2712745.23	8994845807371.16	2999140.84
Polynomial regression (PR)	132387.45	21303596745.27	145957.52
Support vector regression (SVR)	487521.31	260388554076.10	510282.82

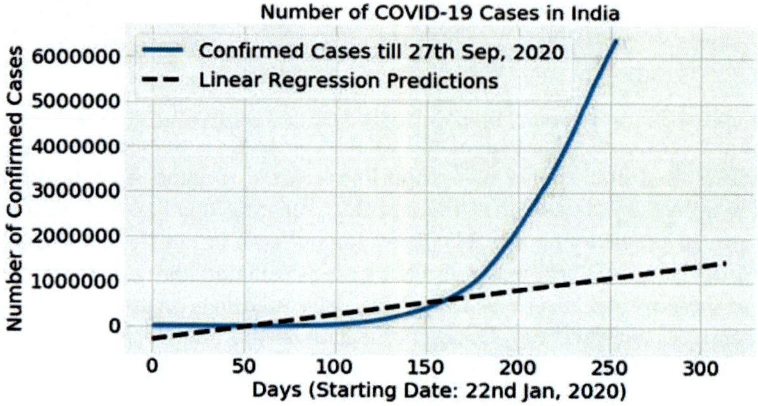

Fig. 2 Linear regression predictions of COVID-19 cases in India

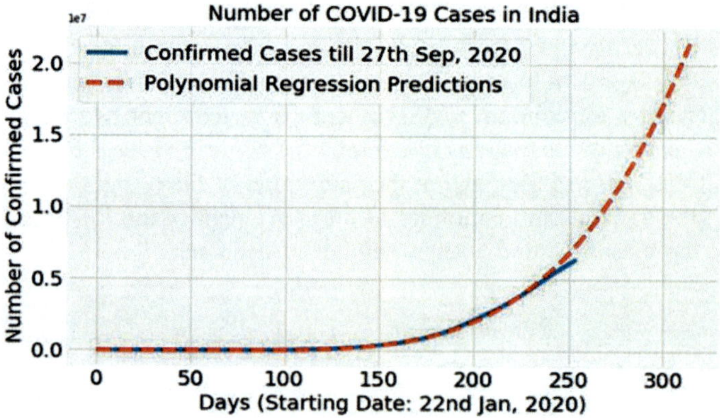

Fig. 3 Polynomial regression predictions of COVID-19 cases in India

minimum RMSE amount over the other two techniques like LR and SVR and shows actuations in trending the confirmed cases in predicting the spread of COVID-19 in India. However, if the transmission continues as the movement of the PR technique, then it would be the reason behind a huge number of deceases. The increased cases of COVID-19 can be limited by shrinking the number of susceptible

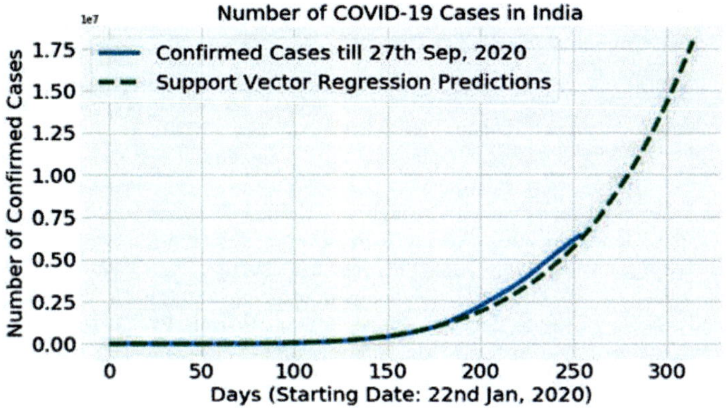

Fig. 4 Support vector regression predictions of COVID-19 cases in India

people from affected individual, as perceived in China. As there are neither any effective medicines made nor the vaccine, hence, social distancing and public awareness are the only weapons for coronavirus.

Furthermore, this article will be useful in designing approaches to limit this COVID-19 pandemic and help out in creating awareness among the people toward the prevention in the number of infected cases due to COVID-19.

References

1. Wang, C., Horby, P.W., Hayden, F.G., Gao, G.F.: A novel coronavirus outbreak of global health concern. Lancet **395**, 470–473 (2020)
2. Coronavirus—worldometer. Retrieved from https://www.worldometers.info/coronavirus/
3. Li, G., DeClercq, E.: Therapeutic options for the 2019 novel coronavirus (2019-ncov)
4. Kerala reports first confirmed coronavirus case in India—India News. Retrieved from https://www.indiatoday.in/india/story/kerala-reports-first-confirmed-novel-coronavirus-case-in-india-1641593-2020-01-30. 28 March 28 2020
5. Mele, M., Magazzino, C.: Pollution, economic growth, and COVID-19 deaths in India: a machine learning evidence. Eniviron. Sci. Pollut. Res. 1–9 (2020)
6. Zhu, W., Xie, K., Lu, H., Xu, L., Zhou, S., Fang, S.: Initial clinical features of suspected Coronavirus Disease 2019 in two emergency departments outside of Hubei, China. J. Med. Virol. (2020)
7. Tuli, S., Tuli, S., Tuli, R., Gill, S.S.: Predicting the growth and trend of COVID-19 pandemic using machine learning and cloud computing. Internet Things **11** (2020)
8. Salgotra, R., Gandomi, M., Gandomi, A.H.: Time series analysis and forecast of the COVID-19 pandemic in India using genetic programming. Chaos, Solitons Fractals **138** (2020)
9. Malki, Z., Atlam, E., Hassanien, A. E., Dagnew, G., Elhosseini, M. A., Gad, I., Association between weather data and COVID-19 pandemic predicting mortality rate: Machine learning approaches. Chaos, Solitons and Fractals. Vol. 138 (2020)
10. Burdick, H., Lam, C., Mataraso, S., Siefkas, A., Braden, G., Dellinger, R.P., McCoy, A., Vincent, J., Green-Saxena, A., Barnes, G., Hoffman, J., Calvert, J., Pellegrini, E., Das, R.:

Prediction of respiratory decompensation in COVID-19 patients using machine learning: the READY trial. Comput. Biol. Med. **124** (2020)

11. Sun, C.L.F., Zuccaralli, E., Zerhouni, E.G.A., Lee, J., Muller J., Scott, K.M., Lujan, A.M., Levi, R.: Predicting COVID-19 infection risk and related risk drivers in nursing homes: a machine learning approach. J. Am. Med. Directors Assoc. (2020)
12. Hasan, N.: A Methodological approach for predicting COVID-19 epidemic using EEMD-ANN hybrid model. Internet Things **11** (2020)
13. Wadhwa, P., Aishwarya, Tripathy, A., Singh, P., Diwakar, M., Kumar, N.: Predicting the time period of extension of lockdown doe to increase in rate of COVID-19 cases in India using machine learning. In: Materials Today: Proceedings (2020)
14. Wang, P., Zheng, X., Li, J., Zhu, B.: Prediction of epidemic trends in COVID-19 with logistic model and machine learning technics. Chaos, Solitons Fractals **139** (2020)
15. Yadav, M., Perumal, M., Srinivas, M.: Analysis on novel coronavirus (COVID-19) using machine learning methods. Chaos, Solitons Fractals **139** (2020)
16. Ghosh, P., Ghosh, R., Chakraborty, B.: COVID-19 in India: state-wise analysis and prediction. JMIR Public Health Surveillance **6**(3) (2020)
17. Das, A., Mohanty, M. N.: Covid-19 detection from x-ray images using convolutional neural network. Int. J. Adv. Sci. Technol. **29**(8) (2020)
18. Mohapatra, S.K., Mohanty, M.D., Mohanty, M.N.: Corona virus infection probability classification using support vector machine. Int. J. Adv. Sci. Technol. **29**(8) (2020)
19. Novel Coronavirus (COVID-19) Cases, provided by JHU CSSE. Retrieved from https://github.com/CSSEGISandData/COVID-19

Bandwidth Enhancement of a Parasitic Path Loaded Antenna Using PSO Algorithm

Vipan Kumar Gupta, Simranjit Kour, and Dinesh Kumar

Abstract In this paper, particle swarm optimization algorithm is used to widen the impedance bandwidth of parasitic resonator-based dual-band antenna. Fractional bandwidth of 43.6% is achieved which covers C-band from 4.6 GHz to 6.9 GHz. Particle swarm optimization (PSO) algorithm is implemented for the goal function on the |S11| and radiation efficiency of a dual-band antenna. Frequency range for optimization is considered from 4 to 8 GHz; 57 iterations are performed with 7 swarms to optimize the antenna dimensions. Good impedance matching and radiation efficiency is maintained after optimization for the achieved bandwidth.

Keywords C-band · Parasitic resonator · PSO · Wide bandwidth

1 Introduction

Microstrip patch antenna (MPA) structures are getting more attraction in the modern wireless communication system. Increasing demand of patch antenna is due to their lightweight, low profile, low cost, and high gain features [1]. Conventional rectangular and circular patch antenna suffers from the issue of narrow impedance (−10 dB) bandwidth due to its inherent property of high Q-factor resonance and multiple harmonic radiation [2]. Typical bandwidth of conventional microstrip antenna structures is less than 5% [3, 4]. Various modifications have been

V. K. Gupta (✉) · S. Kour
Department of Electronics and Communication Engineering, Sri Sai College of Engineering and Technology, Badhani, Pathankot, India
e-mail: er.vipingupta14@gmail.com

S. Kour
e-mail: simramkour76@gmail.com

D. Kumar
Department of Computer Science and Engineering, Sri Sai College of Engineering and Technology, Badhani, Pathankot, India
e-mail: dineshgarg82@gmail.com

S. Das and M. N. Mohanty (eds.), *Advances in Intelligent Computing and Communication*, Lecture Notes in Networks and Systems 202,
https://doi.org/10.1007/978-981-16-0695-3_31

performed by researchers in the ground plane, feed topology, and radiator to enhance the bandwidth of patch antenna.

Differential fed [5–7] and proximity coupling fed [8] mechanisms are widely used in literature. In [5], thick dielectric substrate along with differential fed technology is investigated. However, limited improvement of 9% (12.36–13.57 GHz) in bandwidth is obtained. In [6], vertical slots and shorting pins are added with differential fed radiator to achieve the fractional bandwidth of 13% (1.8–2.14 GHz). Further in [7], differential fed triangular radiator with multiple shorting pins is used to achieve the wide -10 dB bandwidth of 50.46% (2.5–4 GHz). Capacitive coupling through multiple patches and proximity-fed technology is used in [8] to enhance the bandwidth 2.5 times wider from the conventional patch. In these structures, limited bandwidth is enhanced at the cost of complex and large geometry. Defected ground approach with slots in ground plane is investigated in [9–13]. Ring-shaped ground [9, 10], stepped-slot with orthogonal fed [11], modified H-shaped slot [12], and array of 12 rectangular slot [13] are some recently used bandwidth enhancement methods using defected ground technology. Planar dimensions of DGs structure are relatively larger that makes them unsuitable to integrate with compact wireless devices. parasitic patch is added vertically above the main radiator to broaden the impedance bandwidth up to 95% [14] and 67% [15]. However, these structures have the drawback of large thickness and high cost as they required double layers of substrate materials and they could not be considered as planar antennas. Two simultaneous resonance modes (TM_{10} and TM_{20}) are excited by etching V-shaped slot in the patch to obtain 32.2% fractional bandwidth [16], and two additive resonance modes (TM_{01} and TM_{02}) are used in [17], still limited bandwidth of 18% is achieved. Zero order [18] and λ/4 parasitic resonators [19] are the alternative approaches for widening the impedance bandwidth that also maintains the size of antenna compact.

In this paper, wideband antenna is designed and optimized for C-band applications using CST microwave solver. Optimization is an effective way to tune the antenna for desired performance as compared to other methods studied in literature. Antenna has covered the wide impedance bandwidth of 2.3 GHz. Proposed structure has simple geometry, low volume, and good radiation performance. Antenna performance is characterized in terms of return loss, radiation efficiency, VSWR, and directivity.

2 Antenna Design and Bandwidth Enhancement Using PSO Algorithm

Proposed antenna is designed on FR-4 substrate with relative permittivity of 4.4, loss tan of 0.025 and thickness of 1.57 mm. Antenna has planar dimensions of 40 mm × 30 mm. Geometry of the designed antenna is shown in Fig. 1. Initially, a parasitic patch loaded antenna is designed that gives resonance at dual-band frequencies at 5.0 and 7.5 GHz. Partial ground plane and inverted L-shaped radiator

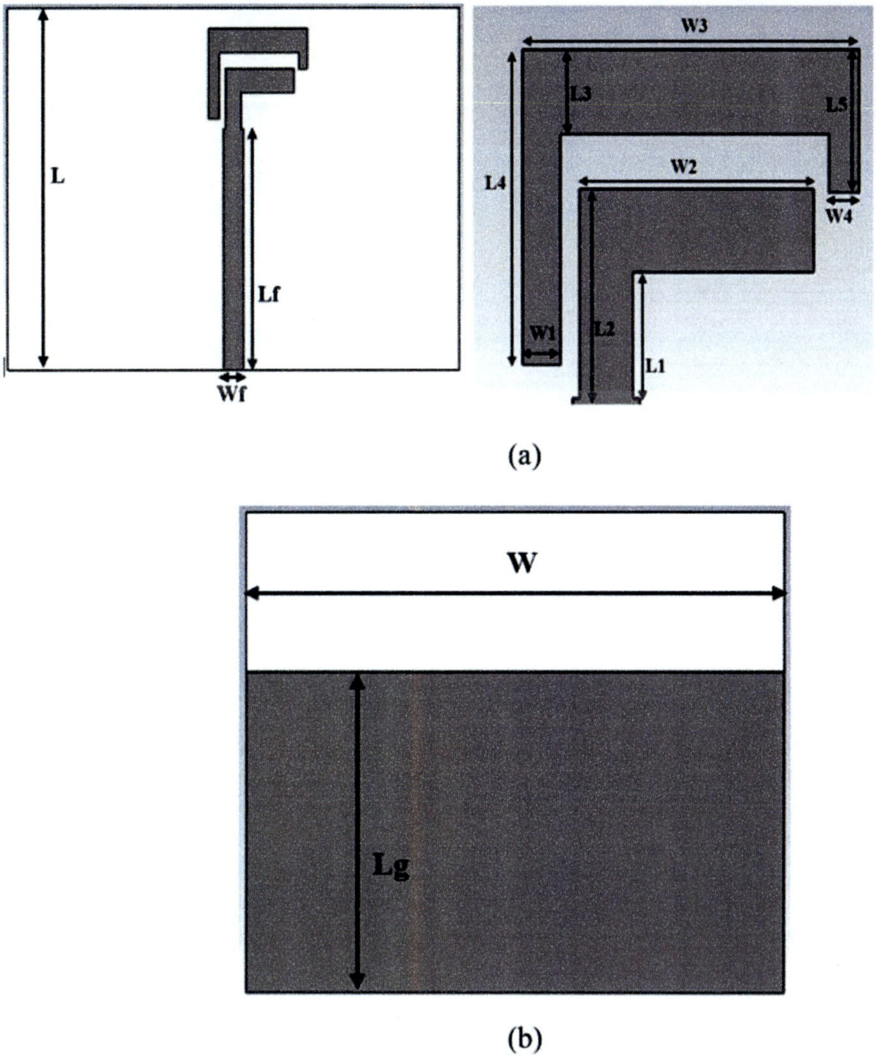

Fig. 1 Geometry of the proposed antenna, **a** front view, **b** back view

contribute to the generation of lower band at 5.0 GHz. Capacitive coupling between the radiator and parasitic elements excite the higher resonance mode and tuned the antenna to resonate at higher band of 7.5 GHz. Surface current distribution at the two operating frequencies is shown in Fig. 2. At lower band, maximum current is concentrated only in the main radiator element whereas at upper band maximum current is coupled at the parasitic resonator. Further, merging the dual resonance frequency and enhancing the impedance bandwidth optimization are performed on antenna dimensions.

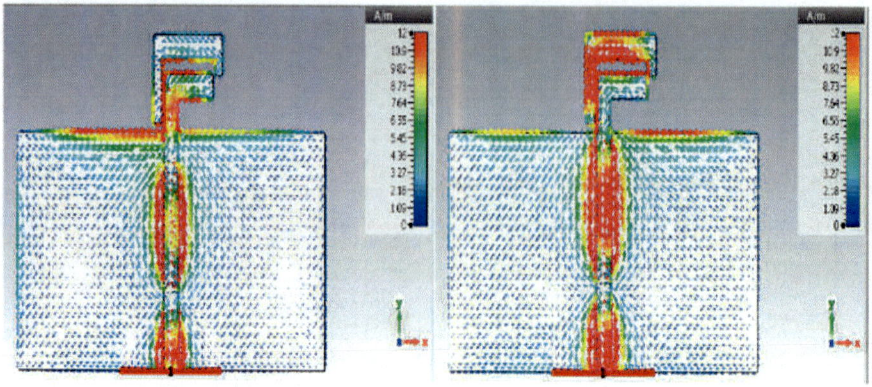

(a) (b)

Fig. 2 Surface current distribution (**a**) at 5.0 GHz (**b**) at 7.5 GHz

Particle swarm optimization (PSO) algorithm is used to optimize the structure. Electromagnetic solver CST has integrated optimization tool that allows designers to set the goal function value for antenna parameters. The particle swarm optimization (PSO) is a parallel computation and optimization technique. During the PSO process, each potential solution is represented as a particle and optimization is initiated by randomly allocating particle positions and velocities in the solution space. Each particle is assigned a randomized velocity and is iteratively moved through the problem space. It is attracted toward the location of the best fitness achieved so far by the particle itself and by the location of the best fitness achieved so far across the whole population.

The PSO algorithm includes some tuning parameters that greatly influence the algorithm performance [20–22]. Assume there are M particles in a swarm and N geometrical parameters to be optimized. Position of these particles (*X*) and their velocities (*V*) are stored in two *M***N* size matrices. After the completion of each iteration, velocity of each particle is updated according to the distance from its current position to the P_{best} and G_{best} location. Velocities of these particles after each iteration are calculated according to the given equation

$$V_t = wV_{t-1} + C_1 n_1 (P_{t-1} - X_{t-1})\,\Delta t + C_2 n_2 (G_{t-1} - X_{t-1})\,\Delta t \qquad (1)$$

X_t = Particle position, it is updated by
$X_t = X_{t-1} + V_t \Delta t$
Δt = time interval between two consecutive iterations
$C_1 = C_2 = 2$

n_1, n_2 uniformly distributed random variable within ranges from 0 to 1 to inject the unpredictability in the swarm behavior. P_{best} gives the best fitness value of individual particle up to the current iteration. G_{best} gives the best fitness value attained by any particle. Figure 3 represents the flow diagram of optimization process.

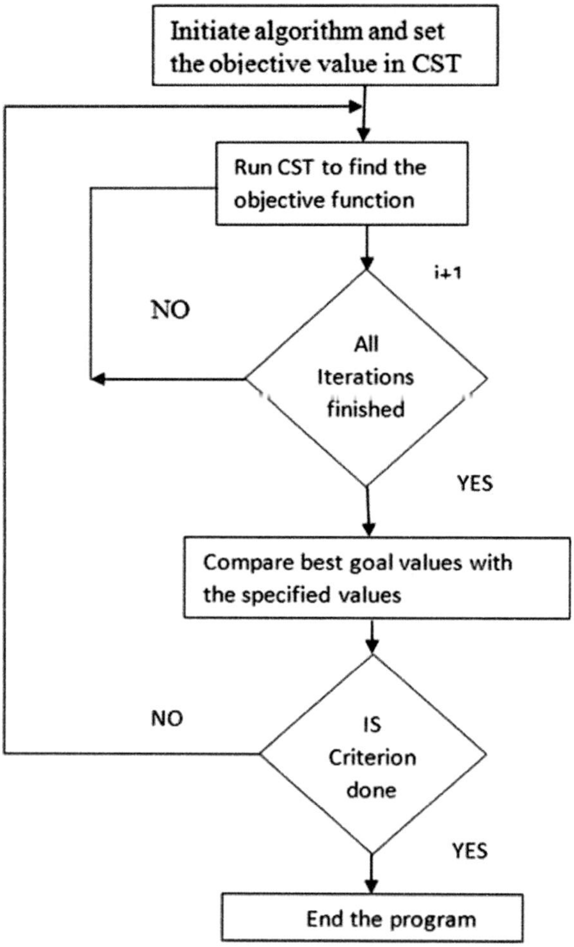

Fig. 3 Flowchart of PSO optimization process

Some restrictions must be followed while optimization of antenna.

- Feed position of antenna should be less than half of the patch length so that feed cannot go away from the patch.
- Length and width should not exceed the dimensions of ground plane; it will generate an invalid structure of antenna

Antenna is optimized for return losses (S parameters) and radiation efficiency. Goal values for these parameters are defined according to the simulated result of purposed antenna. Objective functions for optimization can be defined as

$$F_1 = \frac{1}{N}\sum_{f_1}^{f_2} |\max(0, S_{11}(f) - S_{obj}|^2;\ S_{obj} < -30\ \text{dB} \tag{2}$$

$$F_2 = \frac{1}{N}\sum_{f_1}^{f_2} |\max(0, \eta_{rad}(f) - \eta_{rad}|^2;\ \eta_{rad} > 0.9\ (\text{linear}) \tag{3}$$

A fitness equation is designed to achieve the best performance of antenna. Fitness equation for PSO is defined as:

$$F = w_1 F_1 + w_2 F_2 \tag{4}$$

The fitness equation (F) is the sum of the objective function of the input impedance, return loss, and radiation efficiency. N is the number of sampling points. The resonant band is fixed from 4 to 8 GHz. Value of weight factors w_1 and w_2 is taken 0.5 A and reasonably a large number of iterations are also necessary for the particles to get converged. In this rectangular patch antenna design, a 7-particle swarm and a maximum of 57 iterations are utilized to fully test the performance of the optimizer using different fitness functions. Table 1 listed the value of input parameters (geometrical dimensions of antenna) before and after optimization.

3 Results and Analysis

Return loss plot for the proposed antenna is represented in Fig. 4. It can be clearly observed that before applying PSO algorithm antenna has dual resonance with operating bandwidth of 1.2 GHz (4.6–5.8 GHz) and 0.6 GHz (7.1–7.7 GHz) for the lower and upper resonance, respectively. After optimizing the antenna for wide impedance bandwidth, antenna has wideband resonance from 4.6 to 6.9 GHz (2.3 GHz). Optimization was performed to tune the antenna for C-band application. Optimization contributes to shift the upper resonance band leftward and add the two modes and modified the structure from a dual-band antenna to wideband antenna. Radiation efficiency is another important factor to analyze the antenna performance. All the antenna parameters depend upon antenna dimension, and there can be possibility of reducing antenna impedance matching while optimization. Therefore, radiation efficiency is also optimized with the return loss. Figure 5 shows the plot for radiation efficiency for both the cases. 10% improvement in efficiency is achieved after optimization. More than 70% efficiency is maintained over the resonating band. Maximum value of 84% is achieved at 5.5 GHz frequency.

Table 1 Antenna geometrical parameters

Parameter	W	L	Wf	Lf	Lg	W1	W2	W3	W4	L1	L2	L3	L4	L5
Value before optimization (mm)	40	30	2.4	20	20	1.0	6.2	8.9	0.8	3.0	5.0	2.0	5.8	3.4
Value after optimization (mm)	40	30	2.4	20	20	0.852	5.85	8.59	0.86	3.24	4.2534	2.298	5.8253	3.398

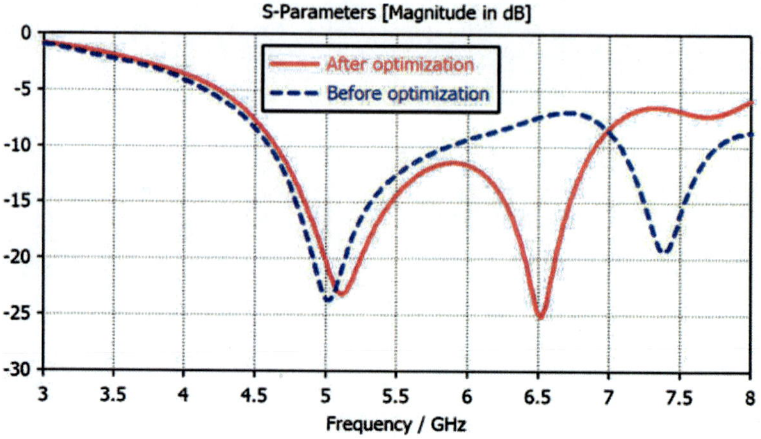

Fig. 4 Return loss plot of the proposed antenna

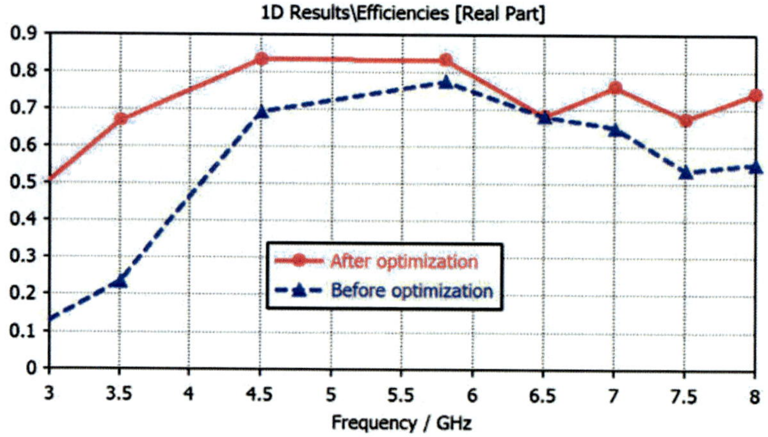

Fig. 5 Plot for radiation efficiency for the proposed antenna

To justify the impedance matching characteristics, plot for voltage standing wave ration is represented in Fig. 6. Antenna shows less than 2 value of VSWR for the operating range. 3D directivity plot at 5.0 and 6.5 GHz is shown in Fig. 7. At 5.0 GHz, radiations are concentrated toward negative X-axis and depict broadside characteristics while at 6.5 GHz vertical dipole like radiation can be observed. Directivity is the function of radiation efficiency; therefore, directivity is higher at the lower band as compared to the higher band.

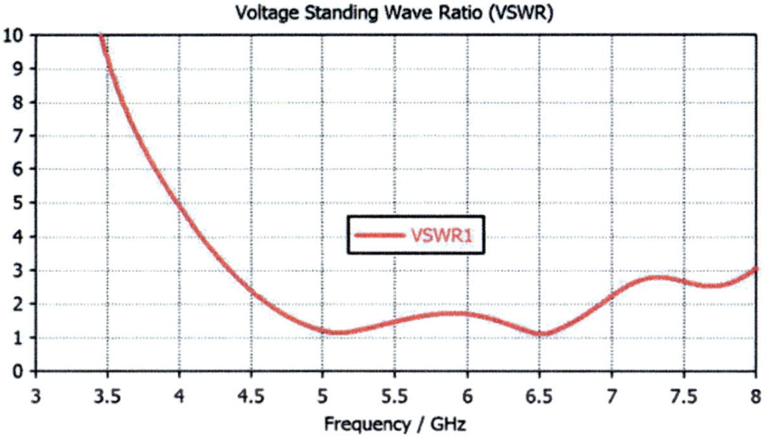

Fig. 6 Plot for VSWR for the optimized antenna

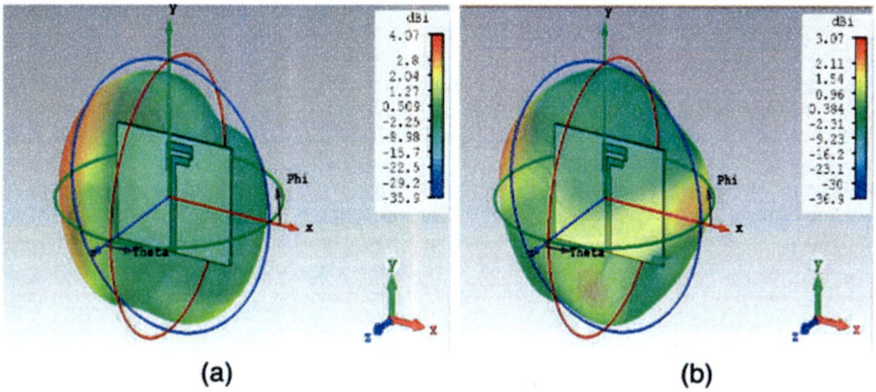

Fig. 7 Plot for radiation efficiency for the proposed antenna

4 Conclusion

A compact planar antenna is designed and optimized for widening the impedance bandwidth. Antenna covered wide impedance bandwidth from 4.6 to 6.9 GHz and has compact dimensions of 30 mm × 40 mm. High radiation efficiency of 84% and directivity of 4.07 dBi are achieved at 5.0 GHz. Voltage standing wave ratio is well below 2 for the operating frequency range. Proposed antenna is a good candidate for C-band communication applications.

References

1. Pozar, D.M.: Microstrip antennas. Proc. IEEE **80**(1), 79–91, Jan (1992)
2. Lee, K.F., Luk, K.M.: Microstrip Patch Antennas. Imperial College Press, London, UK (2011)
3. Mandal, K., Sarkar, P.P.: High gain wide-band U-shaped patch antennas with modified ground planes. IEEE Trans. Antennas Propag. **61**, 2279–2282 (2013)
4. Lin, C.-C., Kuo, L.-C., Chuang, H.-R.: A horizontally polarized omnidirectional printed antenna for WLAN applications. IEEE Trans. Antennas Propag. **54**(11), 3551–3556, Nov (2006)
5. Jin, H.Y., Chin, K.-S., Che, W.Q., Chang, C.-C., Li, H.-J., Xue, Q.: Differential-fed patch antenna arrays with low cross polarization and wide bandwidths. IEEE Antennas Wirel. Propag. Lett. **13**, 1069–1072 (2014)
6. Liu, N.W., Zhu, L., Choi, W.-W.: A differential-fed microstrip patch antenna with bandwidth enhancement under operation of TM 10 and TM 30 modes. IEEE Trans. Antennas Propag. **65** (4), 1607–1614 (2017)
7. Wang, J., Liu, Q., Zhu, L.: Bandwidth enhancement of a differential-fed equilateral triangular patch antenna via loading of shorting posts. IEEE Trans. Antennas Propag. **65**(1), 36–43 (2016)
8. Yasin, T., Baktur, R.: Bandwidth enhancement of meshed patch antennas through proximity coupling. IEEE Antennas Wirel. Propag. Lett. **16**, 2501–2504 (2017)
9. Gupta, A., Kansal, A., Chawla, P.: Design of a patch antenna with square ring-shaped-coupled ground for on-/off body communication. Int. J. Electron. **106**(12), 1814–1828 (2019)
10. Kumar, V.: Rectangular DR-based dual-band CP-MIMO antenna with inverted Z-shaped slot. Int. J. Electr. 1–15 (2020)
11. Lian, R., Wang, Z., Yin, Y., Wu, J., Song, X.: Design of a low-profile dual-polarized stepped slot antenna array for base station. IEEE Antennas Wirel. Propag. Lett. 362–365 (2016)
12. Wang, C., Chen, Y., Yang, S.: Bandwidth enhancement of a dual-polarized slot antenna using characteristic modes. IEEE Antennas Wirel. Propag. Lett. **17**(6), 988–992 (2018)
13. Al-Ahmadi, A., Khraisat, Y.S.H.: Bandwidth enhancement of microstrip patch antenna. Appl. Phys. Res. **11**(1), 35–40 (2019)
14. Ge, L., Luk, K.M.: A wideband magneto-electric dipole antenna. IEEE Trans. Antennas Propag. **60**(11), 4987–4991, Jan (2012)
15. Cui, Y., Wu, L., Li, R.: Bandwidth enhancement of a broadband dual-polarized antenna for 2G/3G/4G and IMT base stations. IEEE Trans. Antennas Propag. **66**(12), 7368–7373 (2018)
16. Wong, H., So, K., Gao, X.: Bandwidth enhancement of a monopolar patch antenna with V-shaped slot for car-to-car and WLAN communications. IEEE Trans. Vehicular Techn. **3**, 1130–1136, Mar (2016)
17. Liu, J., Xue, Q., Wong, H., Lai, H.W., Long, Y.: Design and analysis of a low-profile and broadband microstrip monopolar patch antenna. IEEE Trans. Antennas Propag. 11–18, Jan (2013)
18. Peng, L., Mao, J.-Y., Li, X.-F., Jiang, X., Ruan, C.-L.: Bandwidth enhancement of microstrip antenna loaded by parasitic zeroth-order resonators. Microwave Opt. Technol. Lett. **59**(5), 1096–1100 (2017)
19. Zhang, J.-D., Zhu, L., Wu, Q.-S., Liu, N.-W., Wu, W.: A compact microstrip-fed patch antenna with enhanced bandwidth and harmonic suppression. IEEE Trans. Antennas Propag. **64**(12), 5030–5037 (2016)
20. Jin, N., Rahmat-Samii, Y.: Parallel particle swarm optimization and finite-difference time-domain (PSO/FDTD) algorithm for multiband and wide-band patch antenna designs. IEEE Trans. Antennas Propag. 53(11), 3459–3468 (2005)

21. Khan, T., De, A., Uddin, M.: Prediction of slot-size and inserted air-gap for improving the performance of rectangular microstrip antennas using artificial neural networks. IEEE Antennas Wirel. Propag. Lett. **12**, 1367–1371 (2013)
22. Li, Y.-L., et al.: An improved PSO algorithm and its application to UWB antenna design. IEEE Antennas Wirel. Propag. Lett. **12**, 1236–1239 (2013)

Brain Image Classification Using the Hybrid CNN Architecture

Pranati Satapathy, Sateesh Kumar Pradhan, Sarbeswara Hota, and Rashmi Ranjan Mahakud

Abstract Accurate diagnosis of brain diseases is a challenging task for the physicians. The manual detection of brain disorders is infeasible and tedious task. Magnetic resonance imaging (MRI) of brain images is one of the advanced neuroimaging techniques used in recent days. The research reveals that the machine learning techniques have significant contributions for the classification of brain MRI images. Extreme learning machine (ELM) has been used in image classification. In this paper, an attempt has been made to hybridize the convolutional neural network (CNN) model. The proposed CNN-ELM model integrates the CNN model for feature engineering and the ELM method for classification of brain MRI images. For simulation study, brain tumor MRI image dataset is considered and the results suggest that the proposed CNN-ELM model outperformed the CNN model in classifying the brain images in the dataset.

Keywords Magnetic resonance image · Convolution · Classifier · Max pooling

1 Introduction

The analysis and classification of brain MR images using machine learning techniques have become pervasive in recent years [1]. In the field of medical imaging and diagnostics, improvement in the classification accuracy is the prime objective

P. Satapathy · S. K. Pradhan
Utkal University, Bhubaneswar, Odisha, India
e-mail: satapathy.pranati@gmail.com

S. K. Pradhan
e-mail: sateesh.cs@utkaluniversity.ac.in

S. Hota (✉) · R. R. Mahakud
Siksha O Anusandhan Deemed to Be University, Bhubaneswar, Odisha
e-mail: sarbeswarahota@soa.ac.in

R. R. Mahakud
e-mail: rashmiranjanmahakud@soa.ac.in

S. Das and M. N. Mohanty (eds.), *Advances in Intelligent Computing and Communication*, Lecture Notes in Networks and Systems 202,
https://doi.org/10.1007/978-981-16-0695-3_32

of the researchers as the study prevails. MRI is one of the advanced imaging technique used in identifying the different brain disorders [2]. The manual study and analysis of the different brain MR images are time consuming and tedious. This may also lead to erroneous conclusions and treatments [3].

The literature study reveals that the uses of deep learning are useful and outperformed the other statistical models for the classification of brain MR images [4]. The authors in [5] used different pretrained CNN models for the classification of brain MRI images. From the results, it is found that the ResNet-50 produced higher classification accuracy. The authors in [6] utilized transfer learning with CNN architecture for the classification of Alzheimer's disease. They obtained the classification accuracy of 95.73% during testing set. In [7], Resnet-50 model of CNN architecture is considered as the base model for identification of brain tumors in brain images. The authors in [8] proposed the CNN-ELM model combining the CNN architecture and ELM technique for ECG image classification. The simulation study suggested that the authors achieved the classification accuracy of 97.5%. In [9], a hybrid architecture involving CNN and ELM was developed for the age and gender classification. The role of CNN was for feature extraction, and the role of ELM was for classification. The purpose of this paper is to develop the CNN-ELM architecture for the classification of brain MR images. The CNN model is used for feature engineering; i.e., it determines the optimal features using various layers. Then ELM is used for the classification purpose. Different performance measures are used to validate the performance of the proposed CNN-ELM architecture. The performance of the proposed architecture is compared with CNN for the classification of Glioma dataset. The organization of the paper is as follows. The methods used in this work are described in Sect. 2. The simulation study is shown in Sect. 3, and Sect. 4 concludes this work.

2 Methodologies

In this section, the ELM technique and the CNN model used in this work are described.

2.1 Extreme Learning Machine

ELM is a suitable learning technique proposed as proposed in [10] for single hidden layer feedforward neural network. ELM is applied in sequential, batch and incremental learning not only due to its efficient learning speed, but also of its fast convergence ability [11, 12]. ELM assigns the weights between the input layer and hidden layer randomly. The Moore–Penrose generalized inverse is used to determine the hidden layer weights analytically. The basic steps of ELM are as follows:

i. Consider a dataset $M = \{(X_i, y_i)| \in R^d, y_i \in R, i = 1, 2, \ldots, n\}$.
ii. Assign the weight w_i and bias $b_i(i = 1, 2, \ldots, N)$ between the input and hidden layer randomly.
iii. Compute the output of the hidden layer as H
iv. Find out the output weight β, where$\beta = H^{\dagger}T$, where $H^{\dagger}$ is the MP generalized inverse of H and $T = (y_1, y_2, y_3, \ldots, y_n)^T$.

2.2 Convolutional Neural Network (CNN)

Deep learning (DL) helps automated learning of high-level abstractions through hierarchical structures from the input images. CNN has been recognized as the most efficient model among the DL methods for the analysis of various types of medical images. The structure of a basic CNN model is shown in Fig. 1. CNN is a composition of three different network layers. The three main layers of CNN model are listed as follows:

- Convolutional layer
- Pooling layer
- Fully connected layer.

Apart from the above important layers, the input layer is the initial layer that contains the images to be processed. The convolutional layer and pooling layer perform the feature extraction process. The fully connected layer performs the classification task. The functions of each layer are described in the following sections.

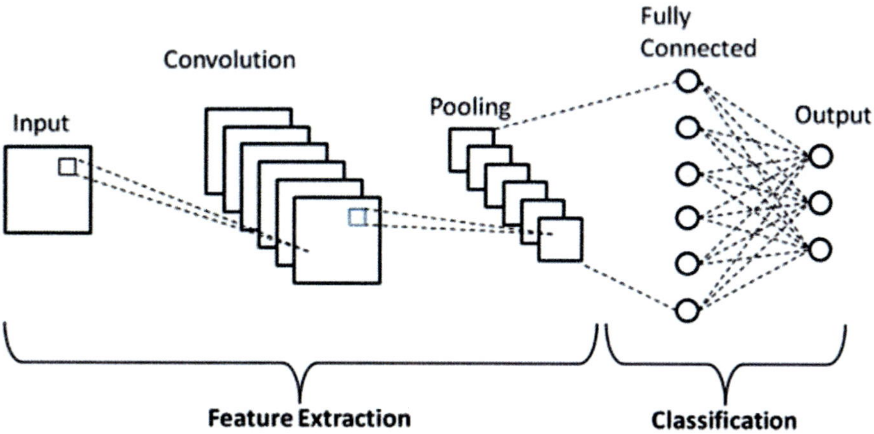

Fig. 1 Architecture of convolutional neural model

Convolutional Layer Convolutional (Conv) layer is the first layer in the basic CNN model that extracts the representative features from the input images. It also helps in learning the pixel relationship in the input images (local connectivity). This layer uses different kernels that convolve with the input image to generate the feature maps. The use of kernels in the Convo layer will reduce the number of parameters (weight sharing) of the network in this layer; each neuron is linked with the specific region of the image, and same weights are assigned with each neuron.

The Convo layer takes an input image of size $IW_1 \times IH_1 \times ID_1$ and produces an output of size $OW_1 \times OH_1 \times OD_1$ using the following hyperparameters. Here, IW_1, IH_1, ID_1 represents the width, height, and depth of the input image and OW_1, OH_1, OD_1 represents the width, height and depth of the output image.

- Number of filters (F_n)
- Size of filter (F_s)
- Amount of zero padding (P)
- Stride (S_d)

The output image is produced using Eqs. (1) to (3).

$$OW_1 = \frac{IW_1 - F_s + 2P}{S_d} + 1 \tag{1}$$

$$OH_1 = \frac{IH_1 - F_s + 2P}{S_d} + 1 \tag{2}$$

$$OD_1 = F_n \tag{3}$$

The rectified linear unit (ReLU) layer is positioned after the Convo layer in the basic CNN architecture, but this ReLU layer is not a separate layer. It maps the negative values to zero and retains the positive values. In simple words, it maintains the nonlinearity in the network.

Pooling Layer The pooling layer is positioned in between the adjacent convolutional layers. The primary purpose of the pooling layer is to decrease the size the feature map and to reduce the number of network parameters. This process is also known as down-sampling. Thus, the complexity of the network structure is decreased. The pooling layer does not involve any training parameters. The common pooling techniques are max pooling and average pooling. The various literature suggests the better performance and effectiveness of max pooling in CNN architectures.

Fully Connected Layer The fully connected layer (FC Layer) is positioned after the convolutional and pooling layers in the CNN architecture. It works in the same manner as the ANN model. The feature maps obtained after the convolutional and pooling layer are flattened so that it acts as inputs to the FC layer. The FC layer is basically used for classification task. The soft max function is applied after the final FC layer, and it assigns a probability score for each class. The FC layer can be substituted with a similar structure that performs the classification task with reduced complexity.

2.3 Proposed CNN-ELM Model

In the CNN model, backpropagation algorithm is used for the training purpose in the FC layer. Due to the computational complexity of the optimal weight determination, the ELM model is proposed after the convolutional and pooling layer of the CNN model. The purpose of the ELM model is to improve the classification performance. The proposed model is shown in Fig. 2.

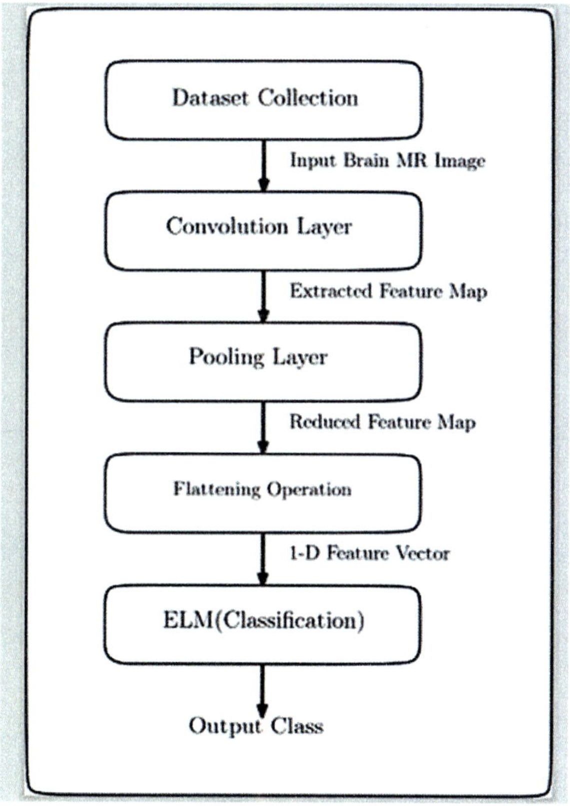

Fig. 2 Block diagram of proposed CNN-ELM model

3 Experimental Study

In this section, the dataset collection, training, and testing of the proposed model are discussed.

3.1 *Dataset Collection*

The brain tumor dataset used in this work is collected from kaggle. There are 253 images collected for this work. The MR images consist of both tumorous and non-tumorous.

3.2 *Training and Testing the Model*

During training phase, the proposed model is trained with 75% of total images. Figures 3 and 4 show the training accuracy and the training error convergence curve, respectively. In this work, the following sequence of layers are applied in the CNN architecture.

- Convo layer1
- Max pool
- Convo layer2.

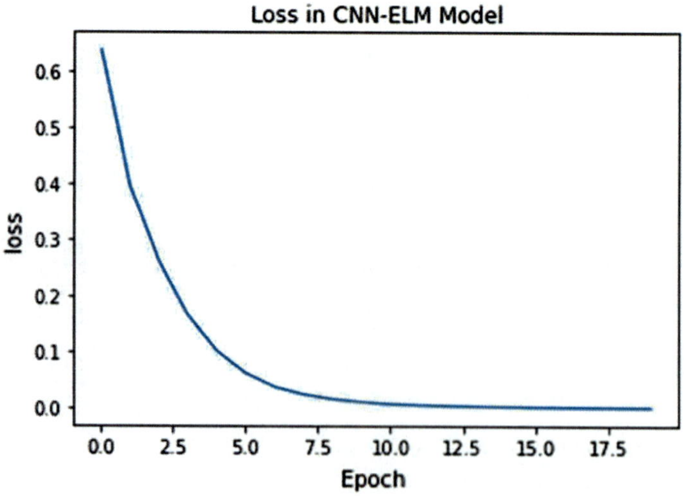

Fig. 3 Loss during training

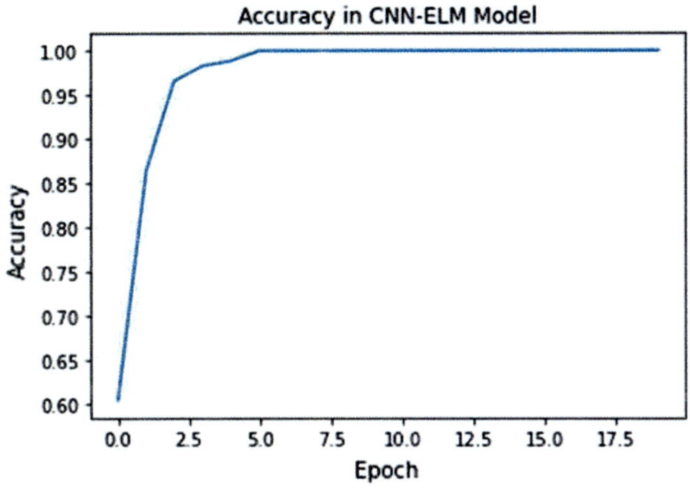

Fig. 4 Training accuracy

For the input image size of [36 × 36 × 1], the output of Convo layer1 is of image size [34 × 34 × 1] using a kernel size of [3 × 3]. Then, using max pool layer, the image size becomes [17 × 17 × 1]. Now, using the Convo layer 2, the image size becomes [15 × 15 × 1]. The flattening operation then generates 225 input features that act as input to the ELM model. So, the number of neurons in the input layer is 225 and in the hidden layer is 500.

After the training phase, the proposed CNN-ELM model is validated with the testing data, i.e., 25% of the images. During testing, classification accuracy is considered as the performance measure. The confusion matrix is shown in Table 1. The accuracy values along with the sensitivity and specificity are shown in Table 2.

From Table 2, it is concluded that the CNN-ELM architecture attended higher classification accuracy for the brain MR image classification of brain tumor dataset. So, the proposed model outperformed the CNN model.

Table 1 Confusion matrices for both the models

CNN model		CNN-ELM model	
TP: 40	FP: 4	TP: 41	FP: 3
FN: 2	TN: 18	FN: 1	TN: 19

Table 2 Performance measures

Model	Accuracy	Sensitivity	Specificity
CNN	90.625	95.238	81.81
CNN-ELM	93.75	97.619	86.34

4 Conclusion

This work proposes the CNN-ELM architecture to classify the brain images with reference to brain tumor dataset. The dataset consists of 253 images. The use of convolutional and pooling layers in CNN architecture is to extract and reduce the number of features of the brain MR image. The ELM component is used to classify the images. The simulation results suggest that the proposed architecture obtained higher classification accuracy as compared to the basic CNN model.

References

1. Kim, Y.K., Na, K.S.: Application of machine learning classification for structural brain MRI in mood disorders: critical review from a clinical perspective. Prog. Neuropsychopharmacol. Biol. Psychiatry **80**, 71–80 (2018)
2. Grover, V.P., Tognarelli, J.M., Crossey, M.M., Cox, I.J., Taylor-Robinson, S.D., McPhail, M. J.: Magnetic resonance imaging: principles and techniques: lessons for clinicians. J. Clin. Exp. Hepatology **5**(3), 246–255 (2015)
3. Mohan, G., Subashini, M.M.: MRI based medical image analysis: survey on brain tumor grade classification. Biomed. Signal Process. Control **39**, 139–161 (2018)
4. Bernal, J., Kushibar, K., Asfaw, D.S., Valverde, S., Oliver, A., Martı, R., Lladó, X.: Deep convolutional neural networks for brain image analysis on magnetic resonance imaging: a review. Artif. Intell. Med. **95**, 64–81 (2019)
5. Talo, M., Yildirim, O., Baloglu, U.B., Aydin, G., Acharya, U.R.: Convolutional neural networks for multi-class brain disease detection using MRI images. Computerized Med. Imaging Graph. **78** (2019)
6. Jain, R., Jain, N., Aggarwal, A., Hemanth, D.J.: Convolutional neural network based Alzheimer's disease classification from magnetic resonance brain images. Cogn. Syst. Res. **57**, 147–159 (2019)
7. Cinar, A., Yıldırım, M.: Detection of tumors on brain MRI images using the hybrid convolutional neural network architecture. Med. Hypotheses 109684 (2020)
8. Zhou, S., Tan, B.: Electrocardiogram soft computing using hybrid deep learning CNN-ELM. Appl. Soft Comput. **86**, 105778 (2020)
9. Duan, M., Li, K., Yang, C., Li, K.: A hybrid deep learning CNN–ELM for age and gender classification. Neurocomputing **275**, 448–461 (2018)
10. Huang, G.B., Zhu, Q.Y., Siew, C.K.: Extreme learning machine: theory and applications. Neurocomputing **70**(1–3), 489–501 (2006)
11. Huang, G.B., Zhou, H., Ding, X., Zhang, R.: Extreme learning machine for regression and multiclass classification. IEEE Trans. Syst., Man, Cybern. Part B (Cybernetics) **42**(2), 513–529 (2011)
12. Huang, G.B., Wang, D.H., Lan, Y.: Extreme learning machines: a survey. Int. J. Mach. Learn. Cybernet. **2**(2), 107–122 (2011)

Classification of Soil and Prediction of Total Nitrogen Content Present in Soil by Using Hyperspectral Imaging

Manoj Kumar Behera, Kanti Mahanti Sai Kishore, and S. Chakravarty

Abstract In this paper, the focus is to design a machine learning model that will classify the type of soil and predict the value of total nitrogen (TN) content of soil for an input soil image based on hyperspectral imaging (HSI) technology. To design the model, a total of 176 soil image samples are collected, where each HSI image consists of 870–1735 nm wavelength. The samples belong to three different types of soil, i.e., alluvial soil, red soil and coastal saline soil. For selection of effective wavelengths in spectrum, successive projections algorithm (SPA) has deployed. Then, from the gray-scale images of soil, different texture features are extracted at the effective wavelengths. Finally, the classification followed by prediction model based on the extracted features is performed using support vector machines (SVM) and partial least squares regression (PLSR), respectively. For achieving better result, first SVM is used to classify the type of soil for an input image. Then, to predict the value of TN content, a PLSR model is applied depending on the type of soil. By using the combination of extracted features and effective wavelengths, an average classification accuracy of 91.2% is achieved. The results indicate that HSI technology can be used for classification and prediction of soil nutrient contents, if the features considered correctly at effective wavelengths from full spectrum of the soil image.

Keywords Total nitrogen (TN) · Hyperspectral imaging (HSI) · Successive projections algorithm (SPA) · Support vector machines (SVM) · Partial least squares regression (PLSR)

M. K. Behera · K. M. S. Kishore · S. Chakravarty (✉)
Centurion University of Technology and Management, Bhubaneswar, Odisha, India
e-mail: sujata.chakravarty@cutm.ac.in

M. K. Behera
e-mail: manojbehera@cutm.ac.in

K. M. S. Kishore
e-mail: skishore46@gmail.com

S. Das and M. N. Mohanty (eds.), *Advances in Intelligent Computing and Communication*, Lecture Notes in Networks and Systems 202,
https://doi.org/10.1007/978-981-16-0695-3_33

1 Introduction

The issues regarding environmental impacts by using high amounts of nitrogen fertilizers are rising rapidly in recent years. It is quiet important to have all the information about the total nitrogen (TN) of the cultivating land. The conventional chemical analysis methods to find TN are pretty much out dated and complicated, time consuming and generate low quality results in real time. A swift and productive ways has to be developed which can be a major step toward precision farming.

From early twenties, near-infrared (NIR) spectroscopy has been used as an efficient tool for the analysis of soil properties. Various studies on the soil TN have been described by using this method [1–3]. However, many factors such as moisture content, organics, soil texture still make the prediction process complicated for the models in soil near-infrared (NIR) spectroscopy analysis [4, 5]. The spectral predictions may differ from sample to sample due to differences in characteristics. Hence, soil has to be classified into groups, and then, a local spectral model has been assigned for each soil type which can improve the model accuracy on predictions [6]. While on the other side, soil classification is one of the basic pillars for soil science, which gives the base for exploration and technological management of the soil system. Many researchers had adopted NIR spectroscopy to classify different soil types [7, 8]. The correct classification will improve the accuracy greatly [9].

Initially, hyperspectral imaging (HSI) was developed from remote sensing technology, but nowadays, HSI is used in other fields like agriculture, medical science and many more areas [10–12]. The measurements done through HSI set up. This also has been useful for classification of fruit ripeness, plant disease intensity and fish freshness [13–15]. As the soil samples will be similar in spectral features, the image texture features will play a significant role in recognition of soil sample, and also, by composition of both the spectral and image texture data, we can increase the classification accuracy.

Hyperspectral imaging produces a huge amount of data due to which the data may have some noise and redundancies which may create a major challenge for the analysis of hyperspectral images. Suitable wavelength selection, which carries most of the crucial information with a less noise and redundancy from the whole spectrum, can reduce the incoming data and computational task, which will be helpful in creating a simple and effective model. The successive projections algorithm (SPA) is a widely used tool for selection of wavelength in a multi-variable classification and calibration. This algorithm can select sets of spectral wavelengths with minimum differences in the wavelengths. The gray level co-occurrence matrix (GLCM) is a widely used method in computer vision systems for the analysis of texture features. GLCM and SPA are used for extraction of texture features and selection of suitable wavelengths, respectively.

2 Methodology

2.1 Soil Samples Collection

The study area is located near the village of Beguru (Shivamog, Karnataka, India). The Shivamog region mainly consists of hills and mountains and dominated by red soil. The geography of this region also contains a river network plain and coastal plain, and hence, the major soil types are alluvial soil and coastal saline soil.

A total of 192 soil samples were collected from the area out of which 87 alluvial soil, 60 red soil and 45 coastal saline soil. To decrease the impact of soil moisture on the sample, they were placed on a plate and left to dry at a temperature of 85 °C for 56 h. At 42nd hour and 56th hour, four samples were chosen randomly for weighing. The change in weight was negligible, then the sample was filtered, and only fine particles were taken out through sieving in which particles having diameter less than 1 mm were collected and again sent for drying for another 46 h at 62 °C for decreasing the effect of moisture during storage. A little amount of each sample was sent to laboratory for chemical analysis, and the remaining sample was utilized for the HSI measurement.

2.2 Hyperspectral Imaging

A near-infrared HSI system was used for the capturing of hyperspectral images of the soil samples, with the wavelength between 870–1735 nm and 256 bands. This system was consisting of an image spectrograph, a CCD camera, two 150 W quartz tungsten halogen lamps and a conveyor belt which was driven by a stepper motor for movement of samples. All the samples were placed on petri dishes, and these dishes were placed on the conveyor for procurement of the image, both the spectral and image information can be obtained from these hyperspectral images. The conveyor belt speed was set at 25 mm/s, time before camera lens was 2.7 m/s, and the height between sample and lens of camera was 28.8 cm. The acquisition process is shown in Fig. 1.

The corrected image (I) was calculated from the raw hyperspectral image (RI) with white (W) and dark (D) as the reference images, respectively. The equation for the same can be written as

$$I = \frac{\mathrm{RI} - D}{W - D} * 100 \tag{1}$$

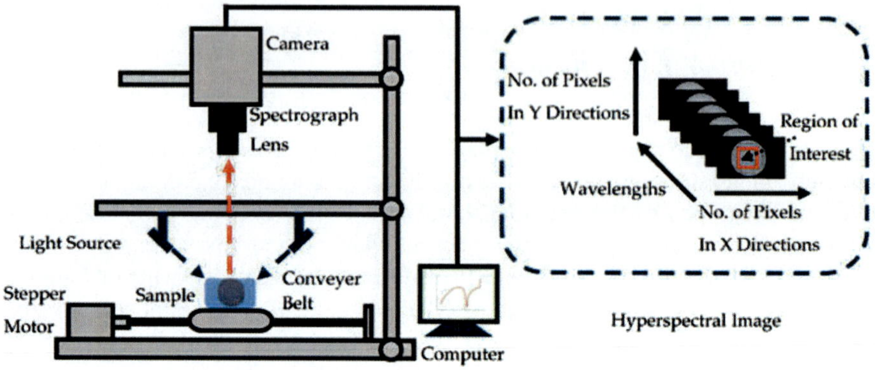

Fig. 1 Image acquisition using HSI

2.3 *Preprocessing and Selection of Effective Wavelength*

The region of interest (ROI) of soil sample was the region that covered petri dish without any edge. The average of the reflectance values of all the pixels of ROI was generated to obtain one mean spectrum. The wavelengths of 970–1735 nm (205 bands) were used further for model development, this same process was reiterated for the rest of the remaining ROI images, and a complete spectrum matrix of 175 samples X 205 bands was generated. The SPA method was used to select the suitable wavelengths.

2.4 *Texture Value Extraction*

A GLCM was created by taking the directions of 0°, 30°, 60° and 90°, and farness of one pixel was applied, after that, the texture variables namely energy, entropy, contrast and homogeneity were calculated for each direction on the basis of GLCM. The mean of the four directions was calculated, and then, an average was taken out of these mean values were obtained for one gray-scale image of ROI. As the gray-scale images with continuous wavelength bands were already available with the hyperspectral image of the soil samples, a sum of 205 images with minimal differences obtained for one sample of soil. Hence, a huge amount of data with redundancy in it will be generated if we use gray-scale image for the extraction of texture features which was may not be suitable for modeling. Therefore, only those gray-scale images were used for texture feature extraction with suitable wavelengths.

2.5 *Defining the Classification and Regression Model*

After the preprocessing of the acquired hyperspectral image of the soil samples, the images are ready for the further use. Now, these soil samples were split randomly into training and testing set with the ratio of nearly 70:30 for developing a classification model. The training set was consisting of 50 samples of alluvial soil, 34 red soil samples and 32 coastal saline soil sample, whereas the testing set was consisting of 26 alluvial soil samples, 18 red soil samples and 16 coastal saline soil samples. Then, a SPA was deployed, and in SPA, the reference output was set to be the category value, i.e., 1, 2, 3 for different soil type where the category value 1 was assigned for alluvial soil, 2 was assigned for red soil and 3 for coastal saline soil. After choosing the suitable wavelength, by using GLCM, the texture features were extracted. A widely used classification model the support vector machines (SVM) method was used on the basis of suitable wavelength and texture features. The kernel function was selected as the radial basis function kernel (RBF), which in general is the most effective among all.

After classification, local models are being used for prediction of TN content depending on the type of soil. The partial least square regression (PLSR) was used for defining the prediction model to make the predictions more accurate for the calculation of soil TN content using suitable. The different steps involved in the proposed model have shown in Fig. 2.

2.6 *Performance Evaluation*

The evaluation of the model was done by using performance measuring parameters like root mean squared error prediction (RMSEP), the residual predictive deviation (RPD) and the coefficient of determination (R2) for the testing set. In general, a large value for R2 and RPD where as a small value for RMSEP indicates a decent performance.

3 Results and Discussion

The three soil samples RGB images are provided in Fig. 3, and on the first look, we can observe that the surface of the coastal saline soil sample is very rough as compared to the samples of alluvial soil and red soil.

A principal component analysis (PCA) was applied on spectral data. The scores generated by the PCA were given as the inputs for Fisher's linear discriminant analysis (LDA). Figure 4 reveals the score plot of Fisher's LDA which distinguishes in between the soil samples. By seeing the figure, we can draw the conclusions that grouping done for the samples of alluvial soil and coastal saline soil is

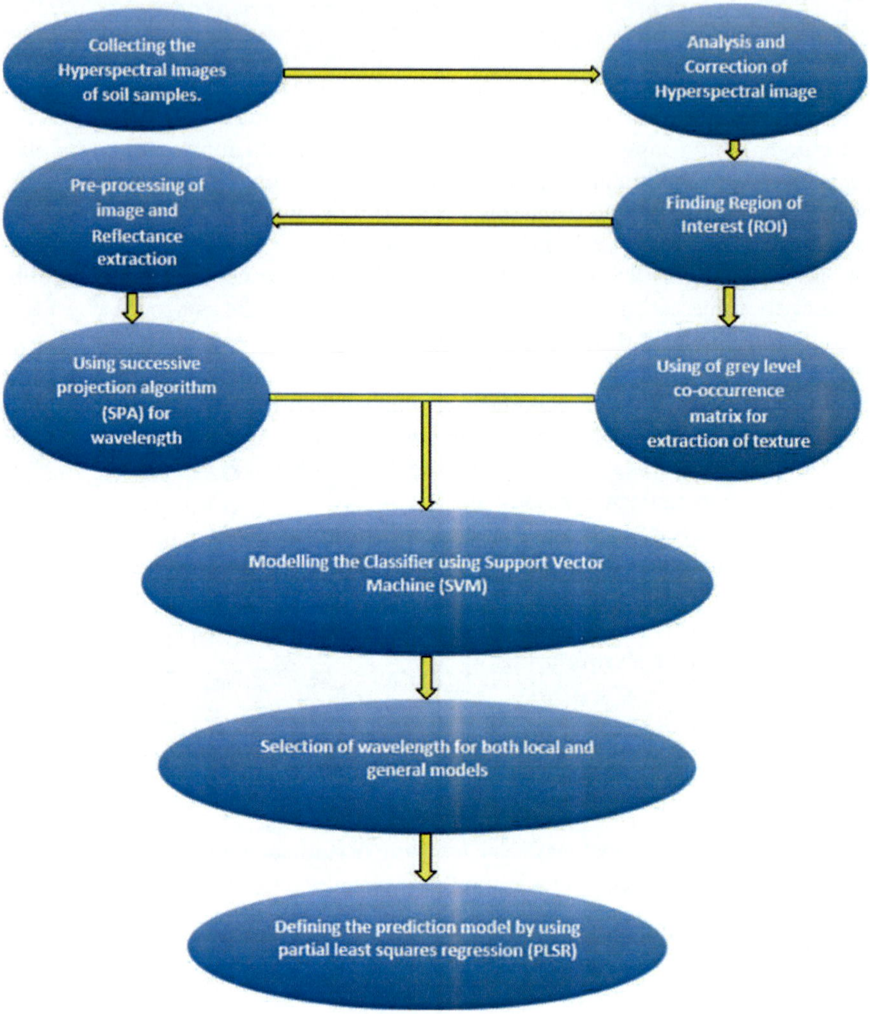

Fig. 2 Flow diagram of the proposed model

very prominent but the red soil samples were mixed with both the soil samples of alluvial as well as coastal saline.

For identifying the effective variables, SPA was performed on the entire spectrum. Let us consider the minimum value of the RMSECV series will be the Rmin. By comparing all the values of RMSECV, seven variables were selected out by relevant F-test criteria with $\alpha = 0.25$ whose values were not greater than Rmin. Figure 5 gives us a brief idea about the selected variables over the raw spectra. Some of the selected variables near the wavelength of 1395–1400 nm were overtone due to assimilation of water characteristic, while the remaining variables near

Fig. 3 RGB images of different soil types (alluvial, red, coastal saline, respectively)

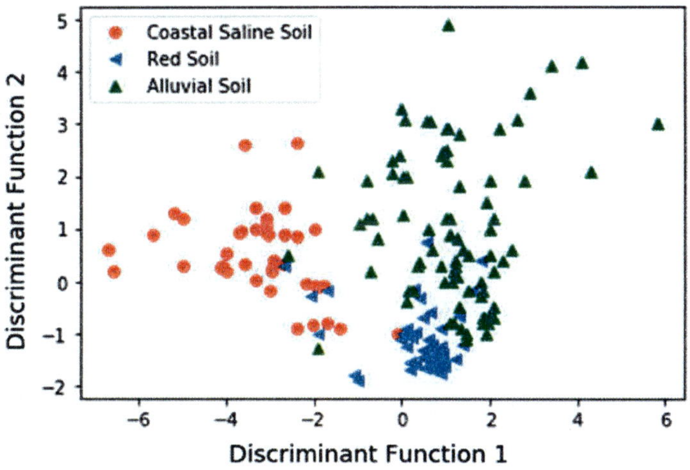

Fig. 4 Grouping of 192 soil samples based on Fisher's LDA

930–1045 nm were overtone of aromatics of C–H bond and amine N–H bond in organics

Earlier the ROI was the middle of the sample which was chosen as a rectangular area with 45 × 45 pixels. Texture features such as energy, contrast, homogeneity and entropy using GLCM were extracted at seven suitable wavelengths, producing a sum of 28 texture features (4 texture features × 7 wavelengths) obtained from the ROI of the soil sample.

Finally, by using SVM, the proposed model archives an overall accuracy of 91.2%. The same can derived from the confusion matrix listed in Table 1 for a test set of 61 soil samples which includes all three types of soils.

After classification, the predicted figures of soil total nitrogen (TN) content for the test set in each local model are listed in Table 2.

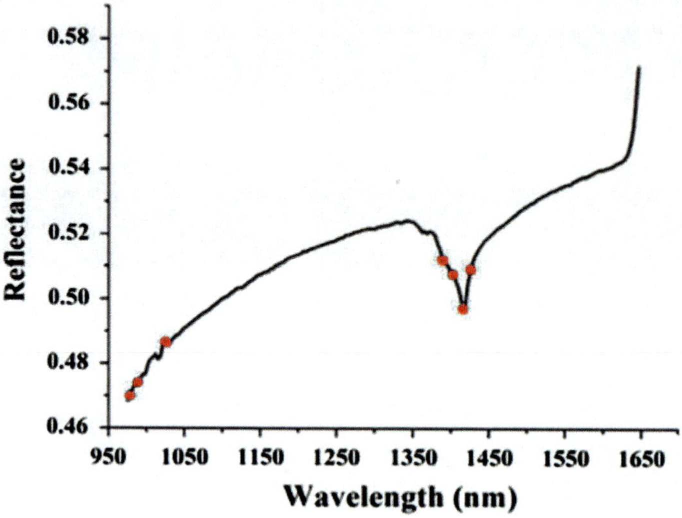

Fig. 5 Seven selected variables corresponding to raw spectra

Table 1 Confusion matrix

Actual versus predicted	Alluvial soil	Red soil	Coastal saline soil	Accuracy (%)
Alluvial soil	27	1	0	96.38
Red soil	3	16	0	84.24
Coastal saline soil	0	1	13	92.85
Overall accuracy of the classifier				**91.2**

Table 2 Predicted figures of total nitrogen (TN) in the local models

Soil type	Number of samples	Range (%)	Mean (%)	Standard deviation
Alluvial soil	28	0.124–0.255	0.176	0.043
Red soil	19	0.102–0.215	0.175	0.028
Coastal saline Soil	14	0.055–0.178	0.136	0.029

4 Conclusion and Future Work

A HSI system having a spectral range of 970–1735 nm has been used for classification and prediction of soil total nitrogen (TN) in different types of soil. The technique of SPA was used for selection of suitable wavelength from the full spectrum, and from the gray-scale images, texture features and contrast were extracted using suitable wavelengths. For classification SVM and for prediction PLSR were used to the proposed model. With test data size 61 samples the overall accuracy of the model is 91.2%. In future, the accuracy of the classifier can be raised by using other machine learning techniques like random forest and deep learning.

References

1. He, Y., Huang, M., García, A., Hernández, A., Song, H.: Prediction of soil macronutrients content using near-infrared spectroscopy. Comput. Electron. Agric. **58**, 144–153 (2007)
2. Yang, H., Kuang, B., Mouazen, A.M.: Quantitative analysis of soil nitrogen and carbon at a farm scale using visible and near infrared spectroscopy coupled with wavelength reduction. Eur. J. Soil Sci. **63**, 410–420 (2012)
3. Vohland, M., Emmerling, C.: Determination of total soil organic C and hot water-extractable C from VIS-NIR soil reflectance with partial least squares regression and spectral feature selection techniques. Eur. J. Soil Sci. **62**, 598–606 (2011)
4. Jia, S., Li, H., Wang, Y., Tong, R., Li, Q.: Recursive variable selection to update near-infrared spectroscopy model for the determination of soil nitrogen and organic carbon. Geoderma **179**, 211–219 (2016)
5. Lacerda, M., Demattê, J., Sato, M., Fongaro, C., Gallo, B., Souza, A.: Tropical texture determination by proximal sensing using a regional spectral library and its relationship with soil classification. Remote Sens. **8**, 701–721 (2016)
6. Zhang, C., Guo, C., Liu, F., Kong, W., He, Y., Lou, B.: Hyperspectral imaging analysis for ripeness evaluation of strawberry with support vector machine. J. Food Eng. **179**, 11–18 (2016)
7. Neittaanmaki-Perttu, N., Gronroos, M., Tani, T., Polonen, I., Ranki, A., Saksela, O., Snellman, E.: Detecting field cancerization using a hyperspectral imaging system. Lasers Surg. Med. **45**, 410–417 (2013)
8. Dai, Q., Cheng, J.-H., Sun, D.-W., Zhu, Z., Pu, H.: Prediction of total volatile basic nitrogen contents using wavelet features from visible/near-infrared hyperspectral images of prawn (*Metapenaeus ensis*). Food Chem. **197**, 257–265 (2016)
9. Li, J., Tian, X., Huang, W., Zhang, B., Fan, S.: Application of long-wave near infrared hyperspectral imaging for measurement of soluble solid content (SSC) in pear. Food Anal. Methods **9**, 3087–3098 (2016)
10. Mollazade, K.: Non-destructive identifying level of browning development in button mushroom (agaricus bisporus) using hyperspectral imaging associated with chemometrics. Food Anal. Methods **10**, 2734–2754 (2017)
11. Haralick, R.M., Shanmugam, K., Dinstein, I.H.: Textural Features for Image Classification. IEEE Trans. Syst. **3**, 610–621 (1973)
12. Jia, S., et al.: Hyperspectral imaging analysis for the classification of soil types and the determination of soil total nitrogen. Sensors 17(10), 2252 (2017)
13. Mendoza, F., Aguilera, J.M.: Application of image analysis for classification of ripening bananas. J. Food Sci. **69**, E471–E477 (2004)
14. Li, S.X., Zhang, Y.J., Zeng, Q.Y., Li, L.F., Guo, Z.Y., Liu, Z.M., Xiong, H.L., Liu, S.H.: Potential of cancer screening with serum surface-enhanced Raman spectroscopy and a support vector machine. Laser Phys. Lett. **11**, 065603 (2014)
15. Langeron, Y., Doussot, M., Hewson, D.J., Duchêne, J.: Classifying NIR spectra of textile products with kernel methods. Eng. Appl. Artif. Intell. **20**, 415–427 (2007)

Adaptive Beamforming Using LMS Algorithm for Planar Arrays and Subarrays

Aditya Peddireddy, B. S. Sachin, H. R. Rohit, and B. S. Premananda

Abstract The importance of adaptive algorithms to recognize the existence of an interference is increasing since the adaptive antennas adjust the pattern automatically. One such adaptive algorithm is the least mean square (LMS), where the SNR of the required signal is optimized in the specified direction. The LMS adaptive algorithm is achieved by directing the beam toward the desired direction and generation of complex weights. This paper shows the need for the LMS algorithm to suppress the interference and the implementation of the algorithm on uniform planar arrays of size 8 × 8 with interelement spacing $\lambda/2$ and at subarray level. In subarray level, the 8 × 8 planar array is divided into 2 × 2 subarrays.

Keywords Adaptive · LMS · Planar · Subarrays · Beamforming

1 Introduction

Smart antennas are used in both commercial and military applications, and modern wireless communication depends on the smart antennas that are adaptive to provide maximum clarity and range. The set of weights on the antenna array can be updated to optimize the radiation pattern by combining the incident signals with the direction of arrival of the signal. Arrays of antennas direct the radiated power toward the desired angle. The technique that uses an array of antennas for maximum signal reception in the desired direction and rejects the signals of the same frequency is called as adaptive beamforming. LMS beamforming algorithm is a prominent method in array signal processing because of its low complexity [1].

LMS algorithm is stochastically execution of the steepest descent algorithm and is a non-blind algorithm as it uses a training or reference signal to update its weight vector. The transmitter emits the training signal. Beamforming at the receiver uses

A. Peddireddy (✉) · B. S. Sachin · H. R. Rohit · B. S. Premananda
Department of Electronics and Telecommunication, R.V. College of Engineering, Bengaluru, India
e-mail: aditya.peddireddy@gmail.com

S. Das and M. N. Mohanty (eds.), *Advances in Intelligent Computing and Communication*, Lecture Notes in Networks and Systems 202,
https://doi.org/10.1007/978-981-16-0695-3_34

this signal to compute the new weights [2]. The minimum mean square error is obtained in the negative direction of the gradient vector by successively updating the weight vector.

The objective of this paper is to implement the LMS algorithm for uniform planar arrays at element level and at the subarray level. Uniform linear arrays that are currently in use are traditional antennas where the elements are placed along a line. Individual radiators when positioned across a plane become a planar array. The advantage of the planar array over linear arrays is that additional variables are used to control the array pattern. In an antenna array, a set of elements is excited by a single source. Such an elementary group is called subarray [3]. In the subarray, for every newly selected direction of the beam, the jammer signals and the phase shifter settings of the antenna are combined for the new beam to adapt in the direction. This can be accomplished in a short interval of time so that the radar can be operated in real time. The final subarray output is obtained by weighting the vector and summing it [4]. The advantages of subarrays are that it requires less source, the number of amplifiers and phase shifters are reduced and if the interelement spacing is kept small, there is a possibility of achieving 100% aperture efficiency.

In this paper, the beam pattern for LMS algorithm for a uniform linear array (ULA) is implemented in MATLAB. It is then implemented for a uniform planar array (UPA) and at subarray level for the designed UPA.

The paper is organized as follows: The literature review on LMS discussed in Sect. 2. The implementation of LMS for linear, planar and subarray level are discussed in Sect. 3. The implementation result is illustrated in Sect. 4 and conclusion derived in Sect. 5.

2 Literature Review of LMS Algorithm

Adaptive algorithms control the antenna radiation pattern in adaptive beamforming. According to the change in the electromagnetic environment, the algorithm optimizes the radiation pattern in the required direction for maximum gain [5]. LMS algorithm is an adaptive beamforming method that was introduced in the year 1959 by Windrow and Hoff, which uses the gradient-based method of steepest descent technique. Unlike other adaptive algorithms, LMS is simple because there is no matrix inversion and calculation of correlation functions [6].

In the LMS algorithm, the weight vector estimation is based on the minimum squared error (MSE). The error with respect to the reference signal $d(t)$ is minimized by using the minimum mean squared error algorithm (MMSE).

The actual output $y(n)$ is given by (1).

$$y(n) = w^H(n)x(n) \tag{1}$$

where

$w^H(n)$ is the conjugate transpose of weight matrix.
$x(n)$ is the input vector.

The error can then be written as given in (2).

$$e(n) = d(n) - y(n) \tag{2}$$

where

$d(n)$ is the required output from the antenna array.

In Fig. 1, for optimal gain in the direction of the target, the corresponding weights are scaled and are linearly added with the outputs of individual sensors [7]. The linear combination of the input vector and weight vector is the output of uniform linear antenna $y(n)$ at any time 'n' is given by (1) [8], where

$$w^H(n) = [w_1, w_2, \ldots, w_N]^H \tag{3}$$

$$w(n+1) = w(n) + \mu x(n)e^*(n) \tag{4}$$

where

$e^*(n)$ is the complex conjugate of the error signal $e(n)$.

The convergence of the LMS algorithm and the step size parameter μ are directly proportional to each other. Depending on the step size, there is the underdamped and overdamped case. The overdamped case is when the step size is too small and thus the convergence will be slower [9].

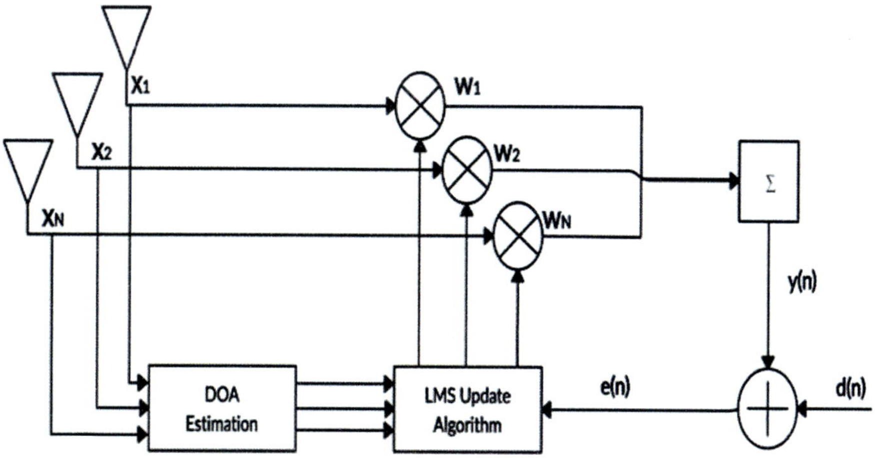

Fig. 1 LMS adaptive array system [7]

If the convergence is slow when compared to the change in the angle of arrival, the adaptive algorithm may not detect the signal of interest. In the underdamped case, the step size is too high that the algorithm overshoots the optimum weights [7].

If convergence is too fast, the desired solution will not be accurately tracked because the weights will vary around the optimum weights. Thus, it is crucial to select a step size within a range to convergence [7]. The step-size value μ is greater than 0 and lesser than $1/\lambda_{\max}$. $\lambda_{\max}$ is the largest eigenvalue of the autocorrelation matrix $x(n)$.

3 Implementation of LMS Algorithm

LMS algorithm is used for the extraction of radar signal when a jamming interference source is present. The flow of the LMS algorithm is given in Fig. 2. The input signal is convoluted with the weights, and the convoluted result is subtracted with the desired signal.

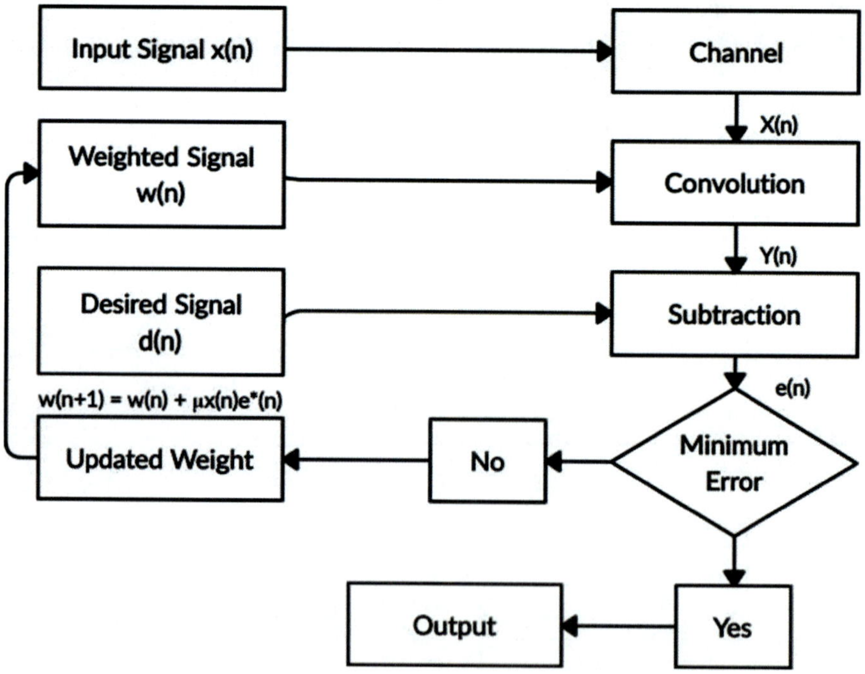

Fig. 2 Flowchart of the LMS algorithm [7]

If an error is present, the weights are updated and then convoluted with the input signal. If an error is not present, then the obtained signal is considered as the final output.

The LMS algorithm can be summarized using (1), (2) and (4). Equation (1) calculates the output of beamformer by multiplying received signal with algorithm weights and (2) calculates the error between the out and the desired signal. Equation (4) is used for updation of the filter weights, and these weights are calculated by changing the previous weight and the converging factor. The time signals are discrete, and discrete-time (DT) signal model is given by (5).

$$x(n) = \sqrt{M}v(\phi_s)s(n) + w(n) \tag{5}$$

Array response vector $v(\phi)$ is obtained using (6) and M is the total number of array elements.

$$v(\phi) = \frac{1}{\sqrt{M}}\left[1e^{-j2\pi F_c \tau_{2(\phi)}} \ldots e^{-j2\pi F_c \tau_{M(\phi)}}\right] \tag{6}$$

The normalized spatial frequency is listed in (7) [3].

$$u \triangleq \frac{U}{U_s} = \frac{d \sin\phi}{\lambda} \tag{7}$$

where U_s is the spatial sampling frequency given by $U_s = 1/d$

U is the frequency of the spatially propagating signal given by $U = \sin\phi/\lambda$.

The modeled UPA is of size 8 × 8 with interelement spacing $\lambda/2$ and is shown in Fig. 3. This modeled planar array is divided into subarrays of each 2 × 2 size. Hence, 16 subarrays are obtained as shown in Fig. 4. The intersubarray distance is taken as thrice the interelement distance. The received signal and the interference signal are modeled using MATLAB. The value of converging factor (μ) = 0.00001. The weights are approximated for 100 cycles. Operating frequency is 2 GHz. The beamformer is implemented using MATLAB.

4 Results and Discussions

Spatial signals are those signals which propagate through space and originate from a source which travels through a propagation medium. A simple spatial matched filter cannot be used if there is an interference signal. In Fig. 5, the signal consists of both noise and interference. It is not possible to observe the signal of interest. The total elements in the array are 20, and the desired signal is present for a single time sample. The array SNR is 50 dB and at $\varphi = 0°$, the target is at 100th time sample which is an impulse.

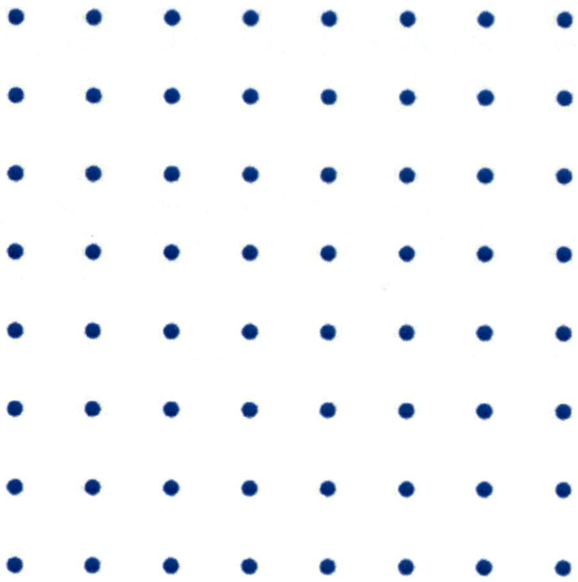

Fig. 3 Modeled 8 × 8 planar array

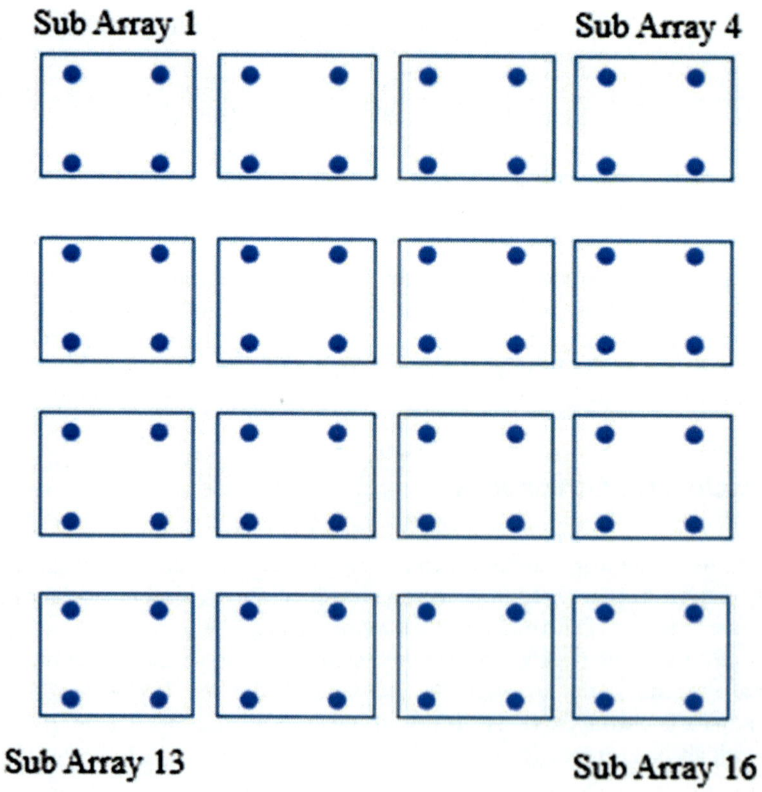

Fig. 4 16 subarrays of each 2 × 2 size

The jammer signal is generated for a total of 200 samples with unit power additive sensor thermal noise. At an angle $\varphi_i = 30°$, a jammer is present and it transmits high power, uncorrelated waveform (white noise) of 50 dB strength. The output of the LMS algorithm is shown in Fig. 6 where a peak is obtained at 100th sample indicating the presence of a jammer signal. The power is around 45 dB.

Figure 7 shows the output of ULA beamformer with steering angle 0°. The beam pattern has one main lobe, and the first side lobe has a height of −11 dB for 100 cycles of approximation. The interference is at 30° with interference-to-noise ratio 40 dB. The effect of this interference has been mitigated.

A taper is used to minimize the side lobe levels and results in the suppression of jammer signals as shown in Fig. 8. There is need for the beamformer to adapt to the response automatically in practical situations and mainly to reduce the noise present in the received signal.

Figure 9 is the beam pattern in 3D of the modeled UPA. The steering angle is 0° which means the main lobe is formed at 0° since the beamformer is not steered and then the side lobes are formed. For ordinary planar array, the main lobe width

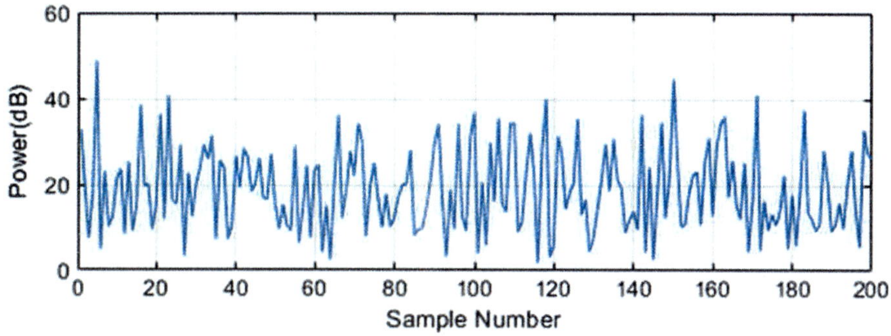

Fig. 5 Signal with interference

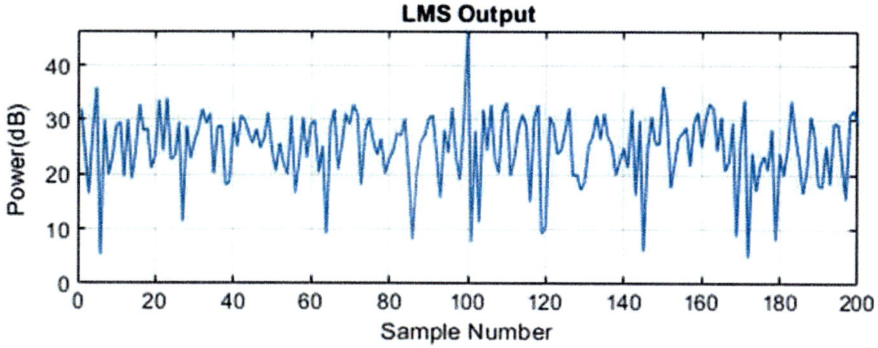

Fig. 6 LMS algorithm output

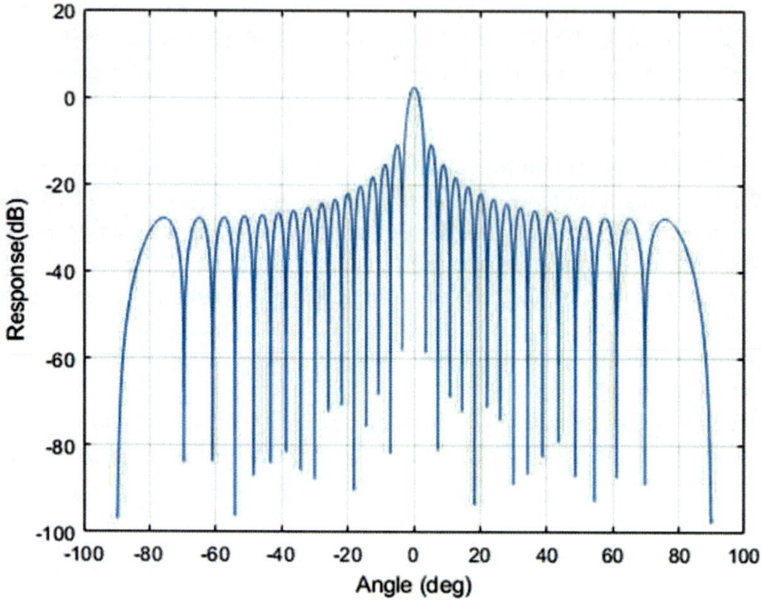

Fig. 7 LMS ULA beam pattern

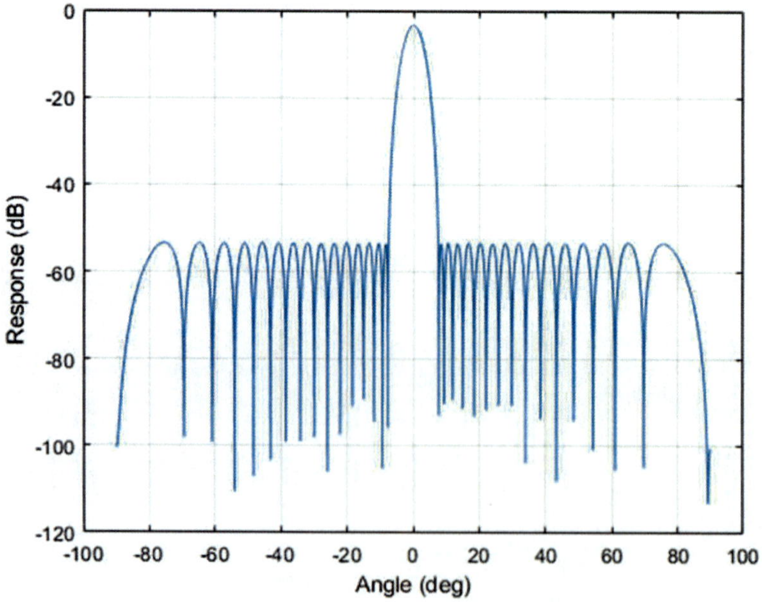

Fig. 8 LMS ULA beam pattern with taper

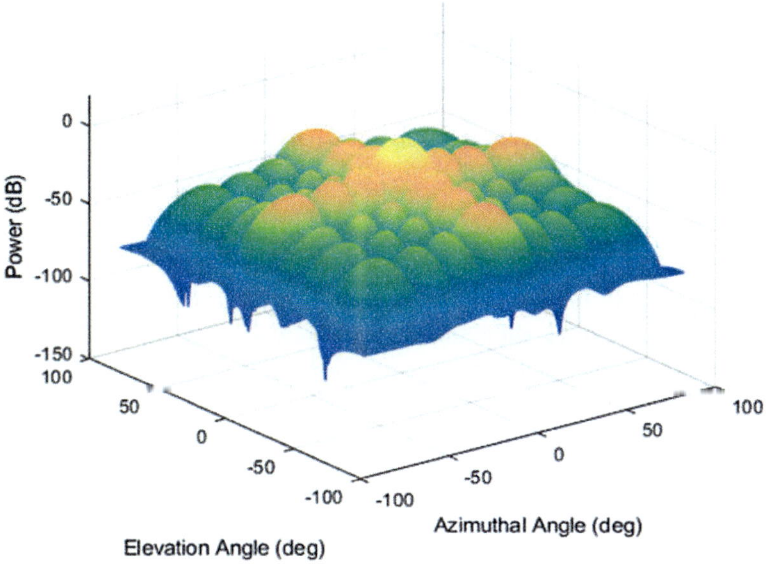

Fig. 9 LMS UPA beam pattern

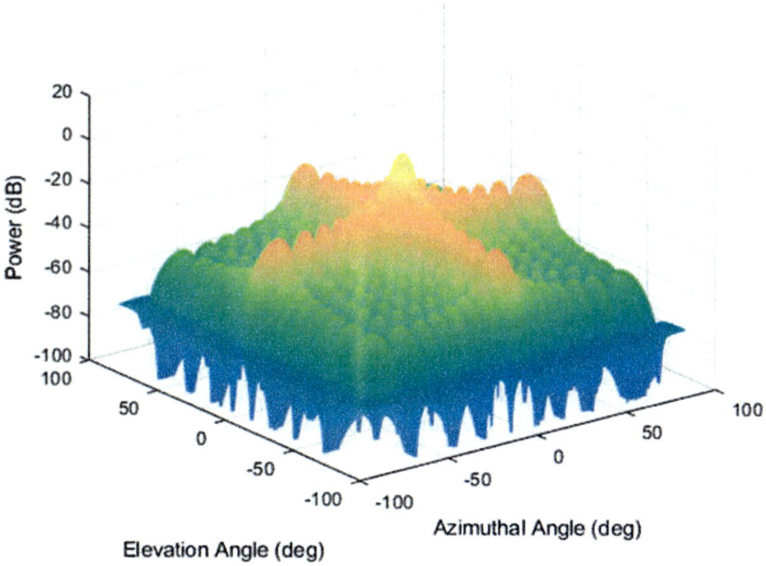

Fig. 10 2 × 2 subarray LMS beam pattern

decreases and the number of side lobes increases with the increase in the number of antenna elements. The effect of interference is suppressed. Figure 10 is the beam pattern in 3D of the modelled subarray of size 2 × 2 each.

The output of individual subarray is combined to get the complete array response. A center lobe called the main lobe is formed at 0° indicating that the beamformer is not steered. The main lobe is followed by the side lobes, and the effect of interference is suppressed.

5 Conclusion

UPAs have high directivity and narrow main lobe. Adaptive arrays which are based on active receiving arrays with subarrays are proven to be an efficient countermeasure against multiple jammers. Subarray-level designing can reduce the manufacturing cost and the complexity of the system. LMS algorithm is used for the extraction of radar signal when a jamming interference source is present. The LMS algorithm executed in this paper is the basic algorithm which can suppress the effect of interference. LMS algorithm is preferred over other adaptive algorithms because it is simple and less complex. It is stable, but the convergence speed is slow compared to other adaptive algorithms like sample matrix inversion and recursive least square.

References

1. Veerendra, Bakhar, M.: A novel LMS beamformer for adaptive antenna array. In: 7th International Conference on Advances in Computing and Communications, pp. 94–100 (2017)
2. Thakur, R., Mourya, K.S.: Performance comparison of blind and non-blind adaptive beamforming algorithms in mobile communication. IOSR J. Electron. Commun. Eng. **12**(6), 51–57 (2017)
3. Rohit, H.R., Sachin, B.S., Aditya, P., Bhishm, T., Premananda, B.S.: Performance evaluation of various beamforming techniques for phased array antennas. In: 11th IEEE International Conference on Computing, Communication and Networking Technologies. IEEE (2020)
4. Wirth, W.D.: Radar Techniques Using Array Antennas, 2nd edn. Institution of Engineering and Technology (IET) (2013)
5. Ismail, M.M., Bashar, B.S., Eliasand Pyliasvskyi, Q.: Study and analysis of an adaptive beamforming for smart antenna using LMS algorithm. Telecommun. Radio Eng. **79**, 399–411 (2020)
6. Jose, A.: Simulation of an adaptive digital beamformer using MATLAB. Int. J. Adv. Res. Basic Eng. Sci. Technol. **2**(12), 10–18 (2016)
7. Smitha, B., Dwivedi, V.: An LMS adaptive antenna array. Int. J. Adv. Res. Eng. Technol. **4**(6), 166–174 (2013)
8. Azabaze, J., Mwangi, E., Konditi, D.B.: Performance analysis of LMS adaptive algorithm for Adaptive Beamforming. Int. J. Appl. Eng. Res. **12**(22), 12735–12745 (2017)
9. Leong, W.Y.: Angle-of-arrival estimation: beamformer-based smart antennas. In: 3rd IEEE Conference on Industrial Electronics and Applications, pp. 1593–1598. IEEE (2008)

Performance Analysis of Cloud Computing Encryption Algorithms

Mhamad Bakro, Sukant K. Bisoy, Ashok K. Patel, and M. Adib Naal

Abstract Cloud computing has become an essential need in our daily life, especially in our present days after the pandemic of our era COVID-19 and the need to conduct all business and meetings from our home and remotely. The emergence of the Internet and the concepts of parallel distributed/networked computing have changed the face of the world. Also, the current situation indicates that the future of information technology will be closely related to the use of the cloud. The need has become more urgent to access resources, applications, massive storage capabilities and data processing from anywhere, flexibly and securely. To send data securely from one place to another, the data has to send in an encrypted form. In this paper, we provide a sufficient theoretical background of cloud computing and analyze the performance of the encryption algorithms (ECC and RSA) of cloud computing. Our simulation result shows that ECC performs better than RSA in total time.

Keywords Cloud computing (CC) · Cloud service provider (CSP) · Steganography · Hashing · Cryptography · RSA · ECC · Blockchain

---

M. Bakro (✉) · S. K. Bisoy · A. K. Patel
Department of Computer Science and Engineering, C. V. Raman Global University, Bhubaneswar 752054, India
e-mail: mhwb14794@gmail.com

S. K. Bisoy
e-mail: sukantabisoyi@yahoo.com

A. K. Patel
e-mail: ashokbecse2006@yahoo.com

M. A. Naal
Department of Computer Science and Engineering, Kalinga Institute of Industrial Technology, Bhubaneswar 751024, India
e-mail: blackstaradib92@gmail.com

S. Das and M. N. Mohanty (eds.), *Advances in Intelligent Computing and Communication*, Lecture Notes in Networks and Systems 202,
https://doi.org/10.1007/978-981-16-0695-3_35

1 Introduction

The huge technological developments of the various sectors have produced rapid inflation in the volume of data. It was confirmed by the Cisco Global Cloud Index, that has expected more than 59% of current internet users are to use cloud storage in 2020 [1] . The proportion is expected to be even higher, especially after COVID-19. With the importance of easy, costly and flexible access to resources on demand anytime, anywhere, cloud computing emerged as an unlimited global concept. In the use of cloud computing, security of user data storage are needed more important concern. There are many concepts suggested for cloud computing, but the concept the National Institute of Standards and Technology (NIST) has introduced is the most agreed commonly: "Cloud computing is a paradigm for offering omnipresent, easy, on-demand network access to a shared pool of configurable computing resources (e.g., networks, computers, storage, software and services) that can be easily provided and released with minimum management effort" [2].

1.1 Characteristics of Cloud Computing

CC has different key features based on this definition [2].

- On-demand resources: The cost of asscessing the resources depending on the length and volume of processing, as they are using "pay-as-your-using."
- Broad network access: This offers consumers flexibility as they can connect and exchange cloud resources with anyone irrespective of where they are based or the platforms they use to connect cloud services.
- Resource pooling: Provider's resources are gathered using a multi-tenant model to support multiple clients.
- Rapid elasticity: Resources are supplied to the client flexibly and immediately published according to his request and desire.
- Measured service: Those metrics create the cost of the "pay-as-you-go" form for cloud services. Thus, transparency can be provided.

1.2 The Architecture of Cloud Computing

The CC architecture has two parts: front and back end. They are both connected to each other over the Internet. The front part is what the customer uses and sees, e.g., browsers. And the back end is the cloud itself and consists of multiple resources and services such as unlimited data storage and virtual devices. Also, it can be categorized into four layers: equipment (datacenter), infrastructure, platform and application. Every one of them may be called a supplier for the above layer and a consumer for the layer below [3].

In the cloud, there are five major participants shown in Fig. 1 [4]:

- CSP: It is responsible for everything needed to deliver a cloud service.
- Cloud consumer: A CC user is either a customer for service or owner of it.
- Cloud broker: It is a middleware layer between CSP and cloud users.
- Cloud Checker (Auditor): It checks the Cloud Services Group to provide an evaluation of the privacy and security, and makes sure that the corresponding service level agreements (SLAs) are complete.
- Cloud Carrier: It serves as a means of providing and transporting cloud service and resources between cloud clients and CSP (via the Internet).

1.3 Classification of Cloud Computing

The taxonomy is done on two sides [3]:

A. **Deployment model**:
   Cloud computing provides different deployment models for the delivery of cloud resources. Such deployment models reflect varying cloud environments and scenarios of use:

   1. Public Cloud:
      The provider hosts cloud infrastructure in its own institutions, the client does not have any hegemony over where the infrastructure exists, so is

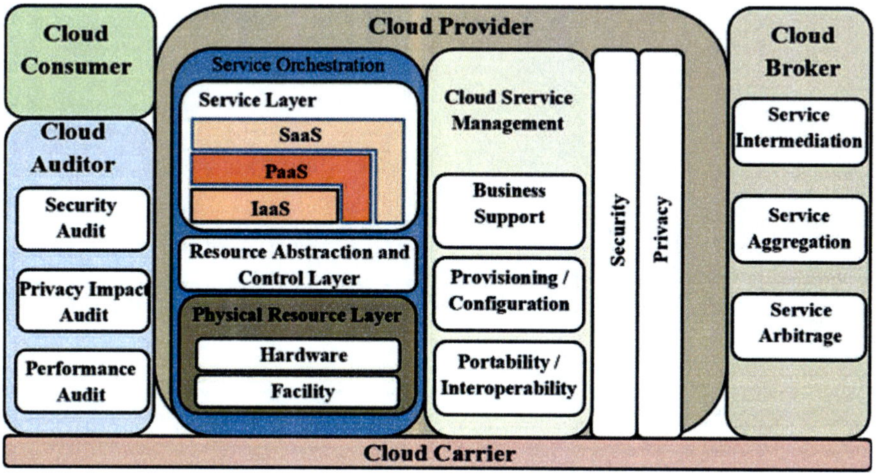

Fig. 1 More clarification of the architecture of cloud computing [4]

much less safe than other cloud models. Due to that, a reliable cloud service provider like AWS, Microsoft Azure, Google Cloud, Rackspace, etc., has to be used.

2. Private Cloud:
   The cloud infrastructure in a private cloud model is provided for only one (one customer) organization as it is not shared by several organizations, and therefore it is safer and easier to maintain, and provides more control over deployment and use without impacting the management process, but it is more costly than public clouds.
3. Hybrid Cloud:
   It is the most common design, where the infrastructure is a combination with 2 clouds or more (private, community or public), linked together by some standard or private technologies, which provides good benefits for organizations/users to use for multiple purposes according to the standard technology [5].
4. Community Cloud:
   The infrastructure is here for special utilization by a set of organizations/consumers of the local community from institutions that have common concerns so that they can share the same infrastructure (security, compliance, identical legal authority, etc.). For example, all government agencies in Washington can engage in the collecting and processing of data from people residing in the same state in America [6].

B. **Service model**:
   According to NIST, service models can be categorized into three basic layers based on different requirements of customers. SPI is an abbreviation for the service models identified as SaaS, PaaS and IaaS as shown in Fig. 2.

   1. Software-as-a-Service (SaaS):
      It is used online to rent services like software, which removes the necessity to purchase, install and run applications on an end-user system. SaaS is very similar to application service providers (ASPs) so that multiple copies of one application run. The end-user does not monitor or control the fundamental cloud resources.

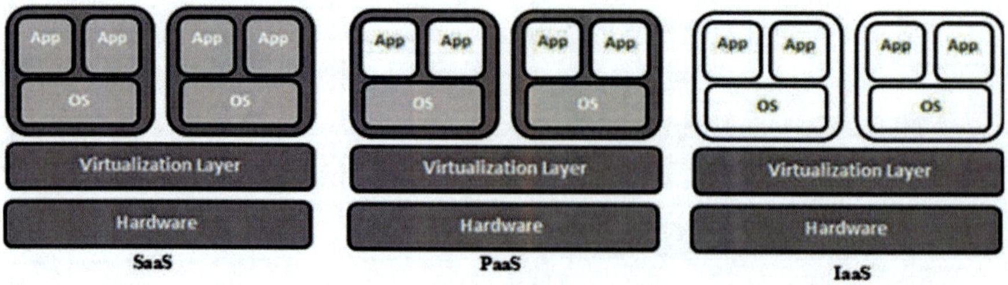

Fig. 2 Abstract architectures for service models [7]

2. Platform-as-a-Service (PaaS):
   PaaS is the means used to create applications via the Internet so that a geographically dispersed team works together to develop software, and this not just raises efficiency but also economizes the costs and increases benefits for customers due to great flexibility. In this model, the customer only has control over the applications.
3. Infrastructure-as-a-Service (IaaS):
   It represents the back of the cloud, is used to develop and deploy the SaaS and PaaS applications, and provides its users (such as system developers and system administrators) with infrastructure resources of low levels (such as DB data centers, processing), where random apps can be tested and published and may involve programs and operating systems.

1.4 Features and Disadvantages of Cloud Computing

1. Features of Cloud Computing:
   Notwithstanding potential privacy and security threats, cloud infrastructure has advantages that IT departments in the public sector and government continue to take advantages of it; in brief, these are [8]: Location and device independence, reduced cost, increased storage, highly automated and changing resources in the cloud are possible and easily implemented more quickly than traditional computing methods, efficiency improvement, fewer problems and ease of implementation, the stability of service and compatible with document formatting and most operating systems.
2. Disadvantages of Cloud Computing:
   Although several advantages of utilizing the cloud and certain limitations exist which typically differ according to the service provider's policies, the issue of data security and its privacy in the cloud and not being exposed to hacker attacks remain one of the greatest fears that make the clients anxious [9], as we said cloud security includes maintaining confidentiality, integrity and availability of data stored in the cloud [10]:

   - Confidentiality is the protection of customer data and the ability to hide it against unauthorized parties through using cryptography and steganography algorithms.
   - Integrity is to keep the data stored in the CPS completely through using hashing algorithms, to maintain its integrity against any unauthorized modification or access.
   - Availability is the user's ability to successfully and easily access his information and resources upon request and use it through using blockchain technology.

Therefore, it was necessary to face the different types of attacks and end the customers' concerns related to their data being stored in the cloud, whether this breach was for the purpose of deletion, leakage or otherwise. As we know, the cloud consists of three layers, which are SaaS, PaaS and IaaS, each layer has its own security flaws in addition to the challenges that affect them all [10], and we noted that achieving security for data requires providing several things (steganography, hashing, cryptography, etc.), in addition to the fact that the most important factor in securing data protection is the nature of the relationship between the cloud service provider and the customer, this is known as service level agreement (SLA) and this is the most important clause between the two parties. In addition to the previous traditional solutions, blockchain technology has been added. In our paper, we will focus on cryptography algorithms (ECC and RSA) to achieve the confidentiality.

2 Cryptography

It is a mix between mathematics and computer science, in which data is obfuscated and hidden during transfer to be confidential between the two parties (sender/receiver) and cannot be read for unauthorized people who do not have key decryption. The cryptography algorithms have two types as shown in Fig. 3.

- Symmetric Key Encryption:
  Symmetric Key Encryption also known as secret key algorithms is fast algorithms, in which the sender performs the encryption process with the same key that the receiver uses for decryption. The AES algorithm is the most stable, secure, fast, overcome the brute force attack, and most used, now that it has been adopted as the main standard [12].
- Asymmetric Key Encryption:
  Asymmetric Key Encryption also known as public-key encryption algorithms, where two interconnected keys are used, one of which is a public key for the encryption process, and the other is a private key for the decryption process, and there is a relationship mathematical so that the private key is generated from the public key. Asymmetric encryption was suffering from some problems in administration and distribution until their Diffie–Hellman came and solved these

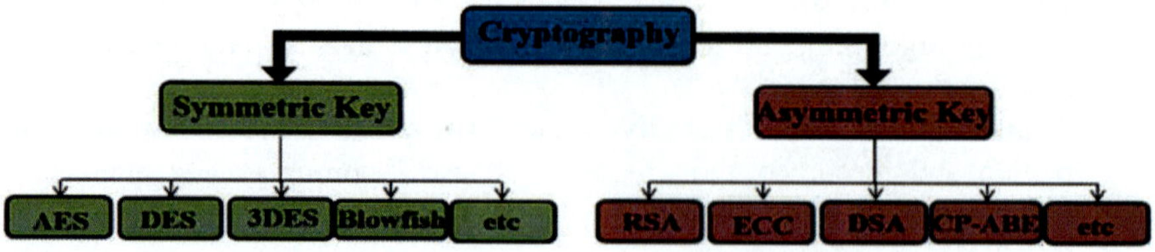

Fig. 3 Classification of cryptography algorithms [11]

problems and ensured the integrity and confidentiality of the message. We will go over some important points about the two most common asymmetric algorithms that are RSA and ECC.

A. Rivest–Shamir–Adleman (RSA):
RSA is the first asymmetric encryption system, named after its inventors. RSA consists of one round in which the plain text is encrypted and decrypted. It consists of three steps, known as the process of generating the public key and the process of encryption and the process of decryption. The private key is generated according to mathematical relationships from the public key. The key is variable in length, and the encrypting block is variable in length as well. The encryption process in RSA is more effective than the decryption process, so researchers have proposed a model to improve decryption time so that the Chinese remainder theorem (CRT) is adopted [13]. RSA has some weak points for being subject to brute force attacks, mathematical attacks and encrypted text attacks [11], so the digital signature with RSA came to increase its security [12]. Having better data security means greater key sizes which lead to an increased load on the cloud system in addition to consuming more memory, and the memory issue is very important for small devices shared in the cloud, so RSA becomes the second option [13].

B. Elliptic Curve Cryptography (ECC):
It is announced by Miller and Koblitz, and it is based on the theory of elliptic curve (EC) over finite fields [12]. ECC security is due to its dependence on the discrete logarithm problem (DLP), in addition to that the public key generated from complex algebraic and geometric equations [11]. The public key is used in the process of encryption and signature generation, and the private key is used in the process of decryption and verification of signatures, as it is used as a supportive algorithm for other algorithms such as ECC–Diffie–Hellman and ECC-DSA [11]. ECC is more suitable for low-memory mobile devices such as smartphones and others, where it requires less and relatively smaller parameters than RSA during encryption and decryption. ECC was designed to improve performance, to reduce battery resource consumption and thus to reduce the power consumed in computing [11], while maintaining at the same level of safety [13]. ECC is also used within ad hoc wireless networks to generate short keys, which leads to fast encryption and less energy consumption, as the security efficiency of the ECC with a length of 160-bit key equals the same as that provided by RSA along with the key of 1024 bits [12], and as a result of the characteristics distinctive that ECC has it, the most important of which is that it is able to offer the same level of security that RSA does, but by using keys with shorter lengths. Thus, ECC is a faster, more efficient and safer model for cloud applications [13].

3 Implementation and Result Analysis

The experiments are done on Android 9.0 on Intel (R) Core (TM) i5-8250U CPU@1.60 GHz 1.80 GHz with 8 GB DDR2 RAM under MS Windows platform. It should be noted that our recorded results are average for 100 experiments and a benchmark to maintain the stability of our results, which may change with changing the speed of the frequency of the processor's work. Noting that after taking the results several times, the values may sometimes change depending on the state of the device and the processor, but the data in theory and according to the values of previous studies is correct.

We will present a comparative analysis of RSA and ECC where the encryption and decryption intervals are found on two samples of input data from 64 bits as shown in Table 2, and 256 bits as shown in Table 3 with random keys according to NIST recommendation as shown in Table 1, and based on experiments we find that the decryption process in RSA takes much more time than the encryption process time, but the ECC is slow in encryption and it is faster in decryption, as shown in Figs. 4, 5, 6, 7, 8 and 9. Thus, ECC is better than RSA in terms of operational efficiency and security with lower parameters.

Table 1 NIST recommended security bit level [13]

Security level L_k, where L_k presents the length of a symmetric key k.	80	112	128	192	256
ECC key length (bits)	160	224	256	384	512
RSA key length (bits)	1024	2048	3072	7680	15,360

Table 2 64 bits—encryption, decryption and total time (in seconds)

Size of input data 64 bits						
Security bit level	Enc time in secs		Dec time in secs		Total time in secs	
	RSA	ECC	RSA	ECC	RSA	ECC
80	0.3248	2.1151	5.7785	5.11518	6.1033	7.23028
112	1.0056	11.0157	18.2957	7.6381	19.3013	18.6538
128	1.3948	14.6942	67.866	9.0143	69.2608	23.7085
192	7.5114	37.1584	882.5009	26.0152	890.0123	63.1736
256	24.6279	51.1694	5831.5827	38.4382	5856.2106	89.6076

Table 3 256 bits—encryption, decryption and total time (in seconds)

Size of input data 256 bits						
Security bit level	Enc time in secs		Dec time in secs		Total time in secs	
	RSA	ECC	RSA	ECC	RSA	ECC
80	0.4855	3.1098	6.4975	6.1437	6.9830	9.2535
112	1.1401	23.8851	39.0435	9.9547	40.1836	33.8398
128	1.8734	41.8559	141.7839	28.5694	143.6573	70.4253
192	8.9996	63.3564	972.5675	36.5612	981.5671	99.9176
256	26.1674	86.3601	6805.1167	41.2014	6831.2841	127.5615

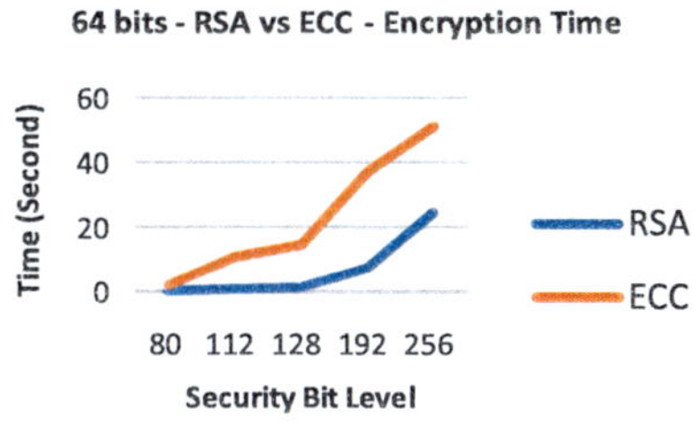

Fig. 4 64-bit Enc time (in secs)

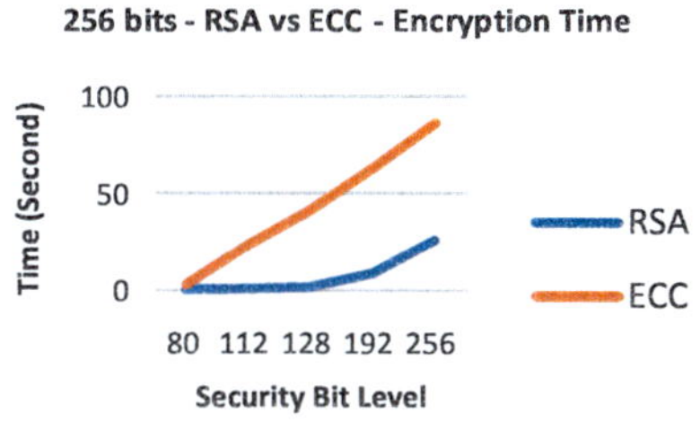

Fig. 5 256-bit Enc time (in secs)

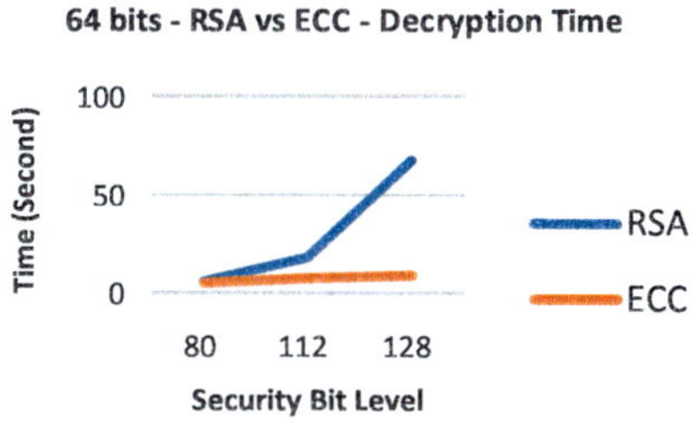

Fig. 6 64-bit Dec time (in secs)

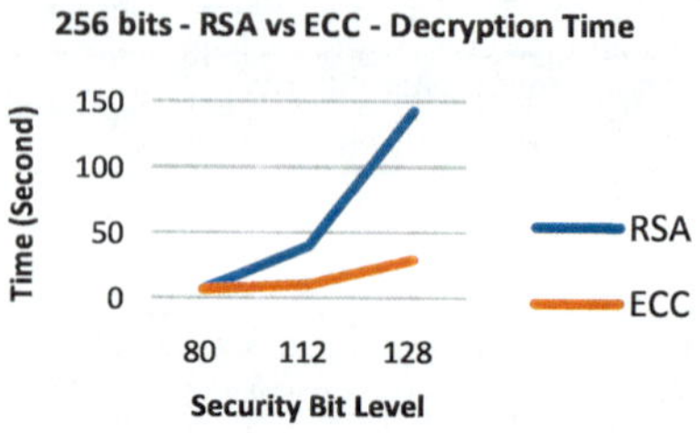

Fig. 7 256-bit Dec time (in secs)

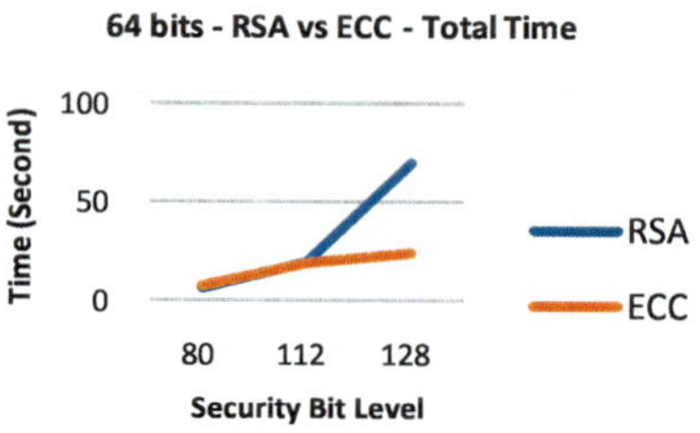

Fig. 8 64-bit total time (in secs)

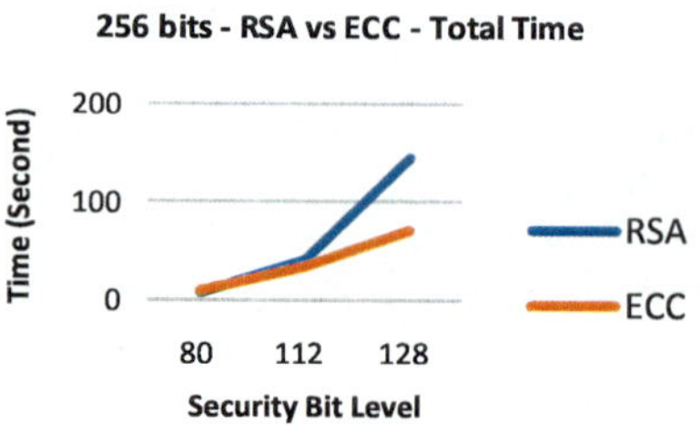

Fig. 9 256-bit total time (in secs)

4 Conclusion

In this paper, we discussed the structure of the cloud service and deployment models, and the many advantages of using the cloud, such as ease of use and payment accordingly. But despite the recent rapid and increasing growth, the flaws, especially related to security aspects of data stored (like its confidentiality, integrity and availability), have made this growth slow at times and put the user in a state of hesitation in front of using the cloud or not, so we have suggested about cryptography as a solution to these security challenges.

In this paper also, we analyzed the performance of ECC and RSA using JAVA. Depending on implementation, we find that the decryption process in RSA takes much more time than the encryption process time, but the ECC is slow in encryption, but it is faster in decryption. Thus, ECC is better than RSA in terms of operational efficiency and security with lower parameters.

References

1. Barnett, T.: Cisco Global Cloud Index 2015–2020. Cisco Knowledge Network (CKN) Session, Nove (2016)
2. Mell, P., Grance, T.: The NIST definition of cloud computing (2011)
3. Darwish, M.A., et al.: Privacy and security of cloud computing: a comprehensive review of techniques and challenges. Int. J. Eng. Technol. **7**(4.29) (2018)
4. Zhang, Q., Cheng, L., Boutaba, R.: Cloud computing: state-of-the-art and research challenges. J. Internet Serv. Appl. **1**(1), 7–18 (2010)
5. Saleh, M.E., Aly, A.A., Omara, F.A.: Data security using cryptography and steganography techniques. Int. J. Adv. Comput. Sci. Appl. (IJACSA) **7**(6), 390–397 (2016)
6. Mishra, B., Jena, D.: Securing files in the cloud. In: 2016 IEEE International Conference on Cloud Computing in Emerging Markets (CCEM). IEEE (2016)
7. Zhang, T.: Detection and mitigation of security threats in cloud computing. Princeton University, Diss (2017)
8. Avram, M.G.: Advantages and challenges of adopting cloud computing from an enterprise perspective. Procedia Technol. **12**, 529–534 (2014)
9. Pavithra, S., Ramya, S., Prathibha, S.: A survey on cloud security issues and blockchain. In: 2019 3rd International Conference on Computing and Communications Technologies (ICCCT). IEEE (2019)
10. Basu, S., et al.: Cloud computing security challenges and solutions—A survey. In: 2018 IEEE 8th Annual Computing and Communication Workshop and Conference (CCWC). IEEE (2018)
11. Yassein, M.B., et al.: Comprehensive study of symmetric key and asymmetric key encryption algorithms. In: 2017 International Conference on Engineering and Technology (ICET). IEEE (2017)
12. Semwal, P., Sharma, M.K.: Comparative study of different cryptographic algorithms for data security in cloud computing. In: 2017 3rd International Conference on Advances in Computing, Communication and Automation(ICACCA) (Fall). IEEE (2017)
13. Mahto, D., Yadav, D.K.: RSA and ECC: a comparative analysis. Int. J. Appl. Eng. Res. **12**(19), 9053–9061 (2017)

Comparative Analysis Among Five Stochastic Search-Based Optimization Algorithms

Shreya Pandey, Biswakalpita Routray, Barsha Bijayini Samal, Niranjan Nayak, and Pravat Rout

Abstract Evolution of biological events is represented with help of five stochastic search-based optimization algorithms which is: particle swarm optimization, differential evolution, teaching, and learning optimization technique, shuffled frog leaping optimization technique, and gravitational search optimization. They help us to overcome large-scale optimization problems and convert it to near optimum solutions. The explanation in the paper explains the framing and outcomes of five such techniques along with their respective pseudocode to facilitate smooth implementation. This paper also contains a comparison of a benchmark among the entire algorithm with respect to the quality of results, processing time, and convergence speed. The comparative result of EAs is discussed in this paper along with guidelines to interpret the best operators for each algorithm.

Keywords Evolutionary algorithms · Differential evolution · Teaching and learning-based algorithm · Shuffled frog leaping algorithm · Gravitational search algorithm · Optimization

S. Pandey (✉) · B. Routray · B. B. Samal · N. Nayak · P. Rout
Department of EEE, ITER, Siksha 'O' Anusandhan University, Bhubaneswar, Odisha, India
e-mail: im.shreyapandey@gmail.com

B. Routray
e-mail: imbiswakalpita@gmail.com

B. B. Samal
e-mail: barshasamal19@gmail.com

N. Nayak
e-mail: niranjannayak@soa.ac.in

P. Rout
e-mail: pkrout_india@yahoo.com

S. Das and M. N. Mohanty (eds.), *Advances in Intelligent Computing and Communication*, Lecture Notes in Networks and Systems 202,
https://doi.org/10.1007/978-981-16-0695-3_36

1 Introduction

Evolutionary algorithms (EAs) are stochastic searched-based optimization algorithms that have given us the way to overcome large-scale optimization solutions by using a minimum optimization technique. The problem related to the optimization of large-scale manufacturing problem leads to another solution. Generally, two types of programming are there: linear-programming (LP) techniques and dynamic programming (DP) techniques. To solve the above problems, researchers have suggested the EA for seeking the nearest optimum solutions. It is a hypothetical search algorithm that resembles the symbol of the biological evolution of body. Conduct of these bodies is directed by studying, adaption, and evolution. To decrease the processing time and advanced quality, mainly to escape the trap in local optima. Recently, five developing techniques are present that inspire the many natural processes, they are: particle swarm optimization (PSO) [1], gravitational search algorithm (GSA) [2], shuffled frog leaping algorithm (SFLA) [3], teaching and learning-based algorithm (TLBA) [4], and differential evolution (DE) [5].

In this paper, five EAs based algorithms are discussed to enable its execution. The performance of five algorithms is further compared and presented for different benchmark multi-variable functions that are certain constraint problems.

2 Five Stochastic Search-Based Optimization Algorithms

2.1 Particle Swarm Optimization (PSO)

It is an optimizational technique that was established by James Kennedy and Russell Eberhart. It is stimulated by the conduct of the group of wandering birds fluttering in an unidentified location. In this optimization, an individual 'bird' in the group is known as a 'particle.' The presented article is similar to the population member as a chromosome in the genetic algorithm [6]. However, the development process in PSO does not replicate rather they develop their social behavior and change together. Each bird is responsible for finding their destination by itself (local search) or with the help of its other mates (global search). Whereas, 'swarm' is represented as the unsystematic collection, i.e., population of individuals that incline to be a group together while every individual is changing in different directions [7].

The PSO code contains number of particles N, a position at a point in a space of D-dimension is represented by ith a particle that takes three values in its account: (a) current position (X_i), (b) best position (P_{best}), and (c) flying velocity (V_i). PSO code is based on how a swarm of bird search their food. Every element in PSO code represent a bird among the group of birds while searching their food. The position and velocity of each element are represented by a bird. The necessary requirements are extracted through its own best positional data (P_{best}) as well as the globally best positional data (G_{best}) to enhance its process of search and space. This algorithm is a continuous procedure. After performing certain loops, the code starts to give best results by converging. The computation is as follows.

$$\begin{cases} vel_i = \omega v_i^k + c_1 r_1 \left(P_{best} - x_i^k\right) + c_2 r_2 \left(G_{best} - x_i^k\right) \\ x_i^{k+1} = x_i + v_i^{k+1} \end{cases} \quad (1)$$

Here, x_i^k and v_i^k are the location and speed of ith element for kth iterative step. P_{best} is the ith elements individual position, and in the entire flock, G_{best} is the universal best position.

Further, the inertial factor if weight is ω that influence how better does the search techniques perform. c_1 is cognitive and social factor c_2 value of which commands the progress of search, r_1 and r_2 are uniformly distributed random numerical value in [0,1], which solves the random nature of algorithm.

2.2 *Differential Evolution (dE)*

DE algorithm requires the user to create NP encrypts the D-dimensional parameter vector that candidate solutions as $X_{i,G} = \left\{X_{i,G}^1, \ldots, X_{i,G}^D\right\}$, $i = 1, 2, \ldots,$ NP to the universal optimum solution.

(a) Mutation operation

Through this operation, we create a mutant vector $V_{i,G}$ based on each target vector $X_{i,G}$. For example:

$$V_{i,G} = X_{\text{best},G} + F * \left(X_{r_1^i,G} - X_{r_2^i,G}\right) + F * \left(X_{r_3^i,G} - X_{r_5^i,G}\right)$$

where r_1^i, r_2^i, r_3^i, r_5^i random integers which are mutually exclusive produced in range of 1 to NP.

(b) Crossover operation

To produce a trial vector, crossover of mutation is applied to every group of target vector and equivalent vector. The crossover rate (CR) is represented as the user-defined constant. The CR is used to control some of the portions of parameter values that are derived from the mutant vector else is derived from the target vector. However, another operator is present, i.e., exponential crossover operator which is used for trial vector parameters that are derived from the mutant vector. The extra parameters of the trial vector are then further derived from the equivalent target vector.

(c) Selection operation

The re-initialization of parameters is required for the recently produced trial vector and if it exceeds the identical upper and lower limit, thereafter, our selection process is done. The main functional value of the target vector is further contrasted along with equivalent trial vector in the recent population. If the given value is lesser than or equivalent to the target vector. Then, the trial vector is substituted by target vector and the population of the next-generation is taken into account.

2.3 Teaching and Learning-Based Algorithm (TLBA)

It is a method based on the teaching process and its influence on the learners in the class [4]. The outputs in this process are considered according to the results and grades. Implementation of TLBO is given by the following steps:

(a) Initialization of parameters like no. of generation (G_n), population size (P_n), no. of design variables (D_n), and limits of design variable (UL, LL)
(b) Initialization of population is carried out through population matrix which requires the generation of random population and no. of design variables. In TLBA, no. of learners are indicated by population size and subjects are indicated through design variables.
(c) In the teacher phase, the calculation of mean of the population MD $= [m_1, m_2, \ldots, m_D]$ is required. The best solution among them will perform as a teacher for that process. A new mean will also be created when a teacher will try to shift the mean from M_D toward X_{teacher}. The difference between them will be given by

$$\text{Difference}(D) = r(M\_\text{new}_D + T_F M_D)$$

(d) In the learner's phase, they should voluntarily increase their interaction skills to increase expand their knowledge.
(e) Terminate if the maximum number of generations is attained else repeat the process from step 3.

In this technique of optimization, the solution comes through selecting two results and comparing it with one another. However, the rules that are set for the process of selection are as follows:

1. In case one is possible and another is not at that time, the most possible is chosen as a solution.
2. In case both are possible at that time, the improved objective function is chosen.
3. In case both are impossible, at that time the minimum constraint violation is chosen.

The above rules are executed at the termination of the learner phase and teacher phase. Alternatively, if improved function value at termination is achieved then the afore-mentioned rules are applied for the selection of a new solution [8].

2.4 Shuffled Frog Leaping Algorithm (SFLA)

It is a scheme which copies the evolution of several frogs while in search of a landing place having a large number of foods. This optimization technique develops an algorithm that contains elements of global search and local search. It contains possible solutions for a population that is defined by the sets of simulated frogs that are divided into subsets known as memeplexes that include frogs of different kinds. All the memeplexes perform a search locally and a specific frog inside every memeplex has thoughts that are prejudiced and developed by the thoughts of another frog in the set through a memetic evolution process [9]. The memetic evolution is dependent upon the theory that survival of the fittest and most efficient. The creation of the fitness function for each frog is related to the position vector, $X_i = [X_{i1}, X_{i2}, \ldots, X_{iN}]$ where N = number of variables required for design.

In each memeplex, the frogs' position of having the best fitnesses is denoted by X_b and the worst fitnesses is denoted by X_w, respectively. And the location of the frog with globally best fitness is denoted by X_g. To get frogs' position with worst fitness, we need to approach a process similar to the process of PSO

$$x_W^{\text{new}} = x_w^{\text{old}} + D_i\left(D_{i,\min} \le D_i \le D_{i\max}\right) \tag{2}$$

$$D_i = \text{rand}(X_b - X_w) \tag{3}$$

where

D_i represents the change in the frog location, rand represents the random number $D_{i\min}$ and $D_{i\max}$ represents the step size minimum and maximum for the position of the frog.

This process will replace the worst frog and will lead to better solution Eqs. (2) and (3) will be repeated but only in case when the X_b will be replaced by X_g.

2.5 Gravitational Search Algorithm (GSA)

This optimization technique is created based on gravity law. In this technique, the mediums are treated as objects. All the objects appeal to one another by the law of gravity that is the gravitational force, and this force is responsible for a universal movement of all the objects to another object having a heavier medium. So, the medium is combined with the use of communication through gravitational force.

The heavy masses move slowly as compared to lighter masses and heavy masses are considered to be a good solution. In GSA, each mass (medium) has four specifications which are as follows [2]:

1. Position (solution of the problem)
2. Mass of inertia
3. Gravitational mass (active)
4. Gravitational mass (passive).

Its inertial and gravitational medium is resolute by using the fitness function means each medium signifies a solution and the algorithm is executed by the proper alteration of the inertial and gravitational mediums. By interval of time, it is assumed that the masses (lighter masses) are captivated by the heaviest masses. This mass is the best solution in the search space. By considering a system of a particle with N masses and position is ith are as follows:

$$X_i = \left(x_i^1, \ldots, x_i^d, \ldots, x_i^n\right) \quad \text{for } i = 1, 2, 3, \ldots, N$$

x_i^d represents the position of the ith mass in the dth dimension

At a specific time ‘t’, we define the force acting on mass ‘I’ from mass ‘j’

$$F_{ij}^d(t) = G(t)\frac{M_{pi}(t) \times M_{aj}(t)}{R_{ij}(t) + \varepsilon}\left(x_j^d(t) - x_i^d(t)\right) \tag{4}$$

$$R_{ij}(t) = \left\|x_i(t), x_j(t)\right\| \tag{5}$$

M_{aj} represented as active gravitational mass
M_{pi} represented as gravitational mass which is passive in nature with respect to i
$G(t)$ represented as constant of gravitation at the time t
R_{ij} represented as Euclidian between two masses i and j
ε represented as a small constant.

3 Benchmark Functions

All the evolutionary algorithms are presented above are properly coded by using the MATLAB scripting language. However, five recognized nonlinear benchmark functions are considered to check the proficiencies of the EAs: Rastrgin function, Ackley function, Sphere function Griewank function, and Rosenbrock function as given in Table 1. The optimized function is nonlinear as well as non-separable which takes X_i as the number of variables. The details of the afore-mentioned functions are given below.

Table 1 Benchmark function

Function	Formula	Sketch
Rastrigin	$f(x) = \sum \left(x_i^2 - 10\cos(2\prod x_i\right) + 10$ Range value from −5.12 to 5.12	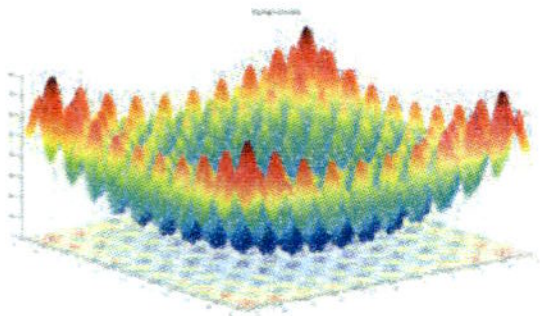
Ackley	$f(x,y) = -20e^{[-0.2\sqrt{0.5}(x^2+y^2)]_{-e}[0.5(\cos 2\prod x + \cos 2\prod y)]_{+e^{20}}}$ Range value from −5.12 to 5.12	
Sphere	$f(x) = \sum_{i=1}^{n} x_i^2$ Range value from −5.12 to 5.12	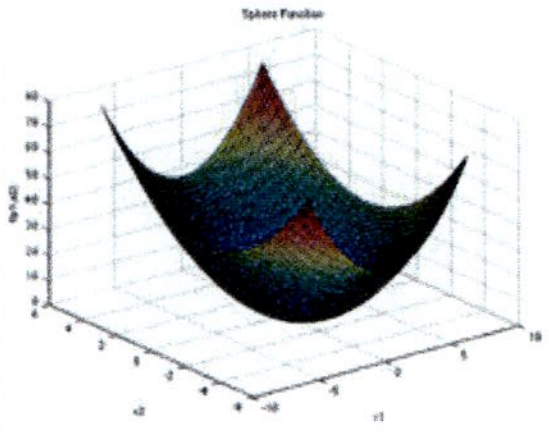

(continued)

Table 1 (continued)

Function	Formula	Sketch
Griewank	$f(x) = \frac{1}{4000}\sum_{i=1}^{n} x_i^2 - \prod_{i=1}^{n}\cos\left(\frac{x_i}{\sqrt{i}}\right) + 1$ Range value from −600 to 600	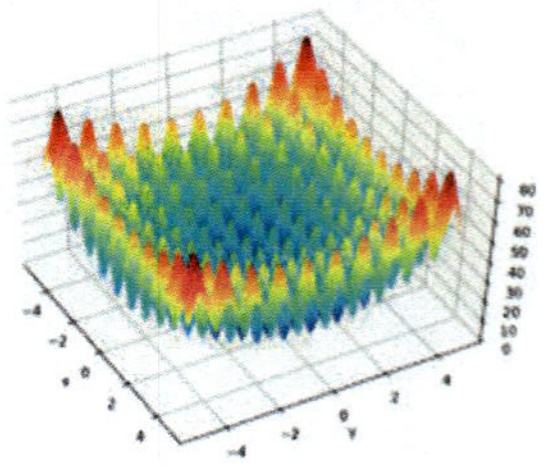
Rosenbrock	$f(x) = \sum\left[100\left(x_{i=1} * x_i^2\right)^2 + (x_i - 1)^2\right]$ Range value from −30 to 30	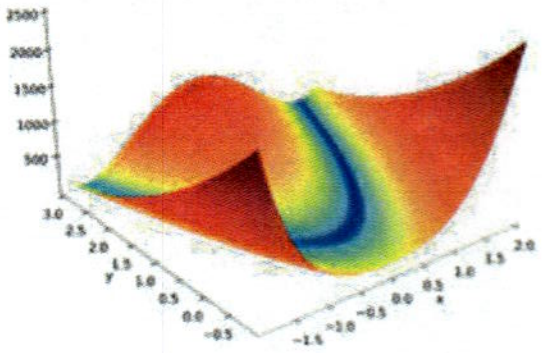

4 Parameter Setting

All the evolutionary algorithm that has been discussed in the paper has a specific parameter that determines its execution in regards with time processing and quality of solution. For calculation of the best value, mean value, standard deviation, and time is taken we have taken 2 and 5 variables. As all the functions considered have given a global minimum value at 0, therefore, any value less than 10 has been considered 0. Each time the population size change, as a result, the quality of solution and speed get changed. The parameter settings for the five EAs based algorithms are as follows:

1. *Particle swarm optimization algorithm*
   Size of the population is taken as 20 and maximum weight factor is taken as 0.9 and the minimum weight factor is set as 0.4.
2. *Differential evolution algorithm*
   The population size is taken as 20. The first scaling factor is set as 0.5 and the second scaling factor is set as 0.3. No. of generations is taken as 1000.
3. *Teaching and learning-based optimization algorithm(TLBA)*
   Size of the population is taken as 20 and the no. of generations is taken as 1000.
4. *Shuffled frog leaping optimization algorithm*
   Here, the population size is taken as 200, the number of memeplexes is set as 10, the frogs per memeplex is taken as 20, and the no. of generations is set as 1000.
5. *Gravitational search algorithm*
   Here, the population size is set as 50 and the no. of generations is taken as 1000.

5 Result Analysis and Discussion

Tables 2 and 3 give the result for the five benchmark functions. The results are analyzed by using the five optimization techniques. PSO technique gives the best result as compared to other optimization techniques that failed to attain the desired value. Further, the optimization techniques are compared based on several variables, i.e., for 2 number of variables and 5 number of variables as depicted in Figs. 1 and 2.

Table 2 For 2 number of variables

Fn. No.		PSO	DE	TLBO	GSA	SFLA
1	Best	0	0	0	7.95967	6.58015
	Mean	0	0	0	0	0
	Std Dev	0	0	0	0	0
	Elapsed time	2.9583	36.136	0.6423	15.0806	5.5320
2	Best	0	8.88178	4.4408	2.3168	0
	Mean		0.0580	5.1514	2.3168	5.1514
	Std Dev		0.4548	1.4580	2.3168	5.1514
	Elapsed time	2.889406	34.9363	0.4855	15.398	7.2360
3	Best	0	5.5516	1.88832	1.26221	6.08255
	Mean	0.0086	0.0786	9.1456	1.26221	6.08255
	Std Dev	6.0859	0.9126	3.5255	1.26221	6.08255
	Elapsed time	2.635850	35.566	0.4178	14.928	4.9724
4	Best	0	1.29003	3.964	0.158691	1.64671
	Mean	238.924	5.9928	3.9643	3.9643	3.9643
	Std Dev		58.3789	2.5408	3.9643	3.9643
	Elapsed time	2.8994	34.565	0.3014	14.900682	5.2166
5	Best	0	0	0	0.02	4.03843
	Mean	0	0.016	0	0	0
	Std Dev	0	0.0456	0	0	0
	Elapsed time	2.8759	34.861	0.35908	15.3385	7.066

Table 3 For 5 number of variables

Fn. No		PSO	DE	TLBO	GSA	SFLA
1	Best	0	0	0	0	0
	Mean	0.241	0.1052	0	0	0
	Std Dev	1.5798	0.8496	0	0	0
	Elapsed time	2.414	35.413	0.3378	14.387	5.135
2	Best	0	0	0	0	0
	Mean	6.111	0	2.207	2.207	2.207
	Std Dev	4.5400	0	6.041	6.0418	6.018
	Elapsed time	2.611	37.4154	0.439	15.085533	6.029561
3	Best	0	0	0	0	0
	Mean	6.532	4.2902	0	0	0
	Std Dev	3.892	6.0036	0	0	0
	Elapsed time	2.875	38.308	0.8404	15.0926	6.524
4	Best	0	0	0	0	0
	Mean	6.9389	0	0	0	0
	Std Dev	8.3304	0	0	0	0
	Elapsed time	2.985	36.927	0.4204	6.470	14.98
5	Best	0	0	0	0	0
	Mean	1.3249	0.001	0	0	0
	Std Dev	2.4119	0.011	0	0	0
	Elapsed time		36.64	0.784	6.594	14.80

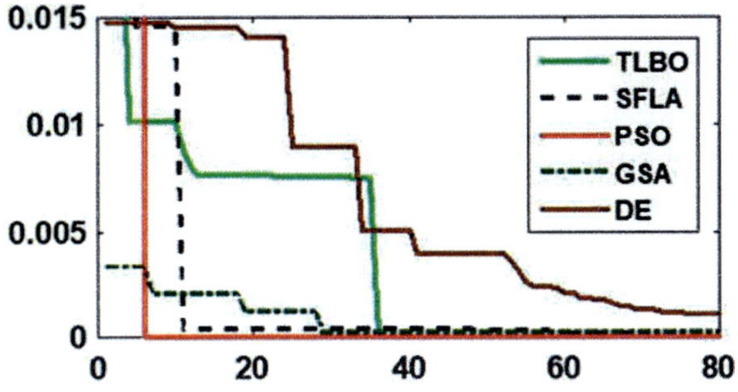

Fig. 1 Comparison of optimization techniques for 2 variables

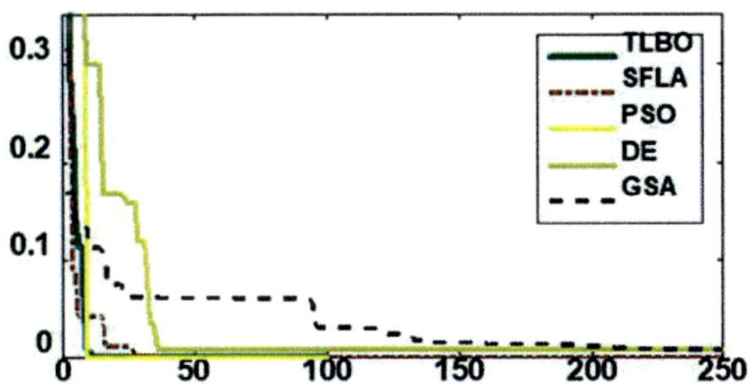

Fig. 2 Comparison of optimization techniques for 5 variables

6 Conclusion

In this paper, five EAs based stochastic search methods are presented. This includes PSO TLBO, GSA, SFLA, and DE. A brief discussion about each optimization technique or algorithm is represented to facilitate their execution. Five benchmark function is tested by using these afore-mentioned algorithms. A comparative result based on several variables is also presented. PSO is very effective for the high-dimensional problem as compared to another algorithm.

References

1. Elbeltagi, E., Hegazy, T., Grierson, D.: Comparison among five evolutionary-based optimization algorithms. Adv. Eng. Inf. **19**(1), 43–53 (2005)
2. Rashedi, E., Nezamabadi-Pour, H., Saryazdi, S.: GSA: a gravitational search algorithm. Inf. Sci. **179**(13), 2232–2248 (2009)
3. Zhen, Z., Wang, D., Liu, Y.: Improved shuffled frog leaping algorithm for continuous optimization problem. In: 2009 IEEE Congress on Evolutionary Computation, pp. 2992–2995. IEEE (2009)
4. Rao, R.V., Savsani, V.J., Vakharia, D.P.: Teaching–learning-based optimization: a novel method for constrained mechanical design optimization problems. Comput. Aided Des. **43**(3), 303–315 (2011)

5. Qin, A.K., Huang, V.L., Suganthan, P.N.: Differential evolution algorithm with strategy adaptation for global numerical optimization. IEEE Trans. Evol. Comput. **13**(2), 398–417 (2008)
6. Krishnanand, K.R., Nayak, S.K., Panigrahi, B.K., Rout, P.K.: Comparative study of five bio-inspired evolutionary optimization techniques. In: 2009 World Congress on Nature and Biologically Inspired Computing (NaBIC), pp. 1231–1236. IEEE (2009)
7. Trelea, I.C.: The particle swarm optimization algorithm: convergence analysis and parameter selection. Inf. Process. Lett. **85**(6), 317–325 (2003)
8. Rao, R.V., Savsani, V.J., Vakharia, D.P.: Teaching–learning-based optimization: an optimization method for continuous non-linear large scale problems. Inf. Sci. **183**(1), 1–15 (2012)
9. Eusuff, M.M., Lansey, K.E.: Optimization of water distribution network design using the shuffled frog leaping algorithm. J. Water Resour. Plann. Manag. **129**(3), 210–225 (2003)

Comparative Study of Machine Learning Algorithms for Predicting Lung Cancer

Imlee Rout, Monarch Saha, Soumen Nayak, Nirjharini Mohanty, and Vishal Baral

Abstract Lung cancer is a significant health concern globally. This leads to a burning need to find a technique that predicts lung cancer with the highest accuracy. Classification is a method to predict the target class from the data used as input. A large population of data is used in the classification that works as the training set. This training set can be used for training the machine learning algorithms. Clinical data analytics pose a mammoth task of predicting lung cancer with the highest precision and accuracy. But since there are so many deaths caused due to lung cancer, finding an algorithm that can predict if a person has lung cancer or not with high accuracy and precision becomes the need of the hour. This work's focal point is to examine the lung cancer dataset to find a classification technique that predicts if a person has lung cancer or not with the highest accuracy possible. The work conducts a comparative study among two classification techniques using various factors to determine which one of the classification techniques can predict lung cancer with the highest possible accuracy. In this paper, random forest and extra tree classifier are compared and conclude it using all the measuring factors.

Keywords Lung cancer · Machine learning algorithms · Classification

I. Rout · M. Saha · N. Mohanty · V. Baral
Department of Computer Science and Information Technology,
ITER, S'O'A (deemed to be University), Bhubaneswar, India
e-mail: imlyrout7@gmail.com

M. Saha
e-mail: monarchsaha98@gmail.com

N. Mohanty
e-mail: nirjharini99@gmail.com

V. Baral
e-mail: vishal.baral19@gmail.com

S. Nayak (✉)
Department of Computer Science and Engineering,
ITER, S'O'A (deemed to be University), Bhubaneswar, India
e-mail: soumennayak@soa.ac.in

S. Das and M. N. Mohanty (eds.), *Advances in Intelligent Computing and Communication*, Lecture Notes in Networks and Systems 202,
https://doi.org/10.1007/978-981-16-0695-3_37

1 Introduction

There are many medical cases every year. On average of 10 million people die due to cancer annually worldwide. Lung cancer does not stand far from being lethal and deadly. There were reporting of two million cases of lung cancer from 2018 to 2019. This shows how fatal lung cancer is. The increase in cancer cases and their reports leads to a log of the database in medical records in a large amount, which screams for a need to analyze the data for medical research work and other activities related to the data accumulated over time. Data analysis has also been performed on provided medical reports of subjects that reach heart disease hospitalization. Then, various algorithms are operated to visually represent the data using graphs and tables [1].

To examine the data, it needs to be refined, and only the essential information should be used to improve the medical facility and research standard. The application of machine learning and deep learning to help maintain and analyze such a large amount of data is necessary for bioinformatics to retrieve essential data and predict the desired output. In the past, as we can see, a lot of implementation as such has been noticed while predicting diseases using the algorithms and deep learning methods like cancer, [2] kidney diseases [3], chronic diseases [4], brain tumor segmentation [5] among other studies. The implementations even helped people study medical reports and data conditions to understand patients' requirements and assistance when medical experts are not available [6]. It has also been anticipated for machine learning to be a biomedicine component with companies affirming to exert actions farther in this domain to assist and reduce human engagement [7] by various contributors and researchers in this sensitive yet mattering part regardless of the sphere of medicine.

This paper's central idea is to predict whether a person is affected by lung cancer or not by using some of the medical data or a dataset, with precise outcomes while conducting the comparative study, to know which classifier would be better in terms of prediction on the dataset used. The two classifiers compared are namely extra trees and random forest classification. Both are almost the same as models; both use many separate decision trees to predict the outcomes while having their differences, we will be discussing in further sections.

2 Related Work

Machine learning implementation has always been studied and has been one of the major focus areas of researchers for a long time and is even widely accepted in the engineering and biomedical spheres.

The research work conducted by Shailaja et al. [8] describes how machine learning techniques have proven to be a support system for the healthcare sector because of its ability to recognize patterns and predict diseases efficiently. The paper adequately concluded that machine learning is suitable for predicting and assisting in the prognosis and diagnosis of numerous deadly diseases using various large datasets.

Data mining is the process of gathering information from vast sets of data and then organizing this information into an understandable structure. There are considerable datasets in health care needed to collect information to understand any pattern created and predict future cases following the same way. The research paper by Kesavaraj et al. [9] throws light on how various classification techniques can be used to conduct data mining, i.e., gather structured information from different massive medical datasets.

The research paper by Pradeep et al. [10] details how machine learning classification techniques like support vector machine (SVM), Naive Bayes, and C4.5 can be used in analyzing and predicting the survivability rate of a person who has lung cancer using the electronic health records (EHRs) as a basis to train and test the classification techniques. This research work is performed to help the doctors to increase the cancer survivability rate.

Lynch et al. [11] describe various classification techniques on surveillance, epidemiology, and end results (SEER) database that contain large datasets to find an algorithm that predicts the survival rate of cancer patients using various factors. In this work, the gradient boosting machines (GBMs) have proven to be the most accurate classification technique while predicting cancer's survival rate among patients.

Discharging patients when they are ready to get released is a crucial process in the healthcare system. This also a scary and difficult phase as it is challenging to know when a patient is ready to be discharged. Failing in distinguishing if a patient is ready to get released or not can lead to various complexities like chances of readmission if discharged without sufficient treatment and financial problems if kept admitted for a longer time than required. The research work carried on by Gramaje et al. [10] makes efforts to acknowledge and eradicate this issue by using various classification algorithms to analyze the training and testing data and predict whether the patient is ready to be discharged or not. In this work, decision trees and probabilistic classifiers can anticipate the lowest error rate from the dataset.

In the research work performed by Jayaraj et al. [12], it is described how random forest can be used on computer topography (CT) images to predict lung cancer among patients and classify patients with or without cancer CT scan images. The research concluded that random forest is a highly accurate classification algorithm while classifying CT scan images as 'Normal' or 'Abnormal,' where the 'Abnormal' class of images depicts lung cancer cases.

3 Background

Refining data and analyzing it using some statistical, logical, or any kind of technique to evaluate the data and use these observations to predict/decision making the outcomes of the future input is data analysis. Many data analysis techniques exist to examine the data, interpret it, and conclude based on the needs. Data analysis has many techniques like data mining, machine learning, and statistics to explore the data [13].

Classification comes under the branch of supervised learning of machine learning. It is implemented to differentiate the data elements into respective discrete classes, which show the most similar feature. The data element specifies the class best used when the output has finite and discrete values [1]. Classification involves the prediction of the outcome based on the data that is provided to the classifier. The source data is split into training and a testing dataset. Then it is decided which classification technique should be applied to get better output with higher precision. The classification algorithms can be used to both structured and unstructured data.

In this paper, two distinct classification algorithms are being used to predict the given input's output. The two classifications are as follows:

1. **Random Forest**: It is a combination of many separate decision trees that operate as an ensemble, known to be estimators (Fig. 1) [14].
2. **Extra Tree Classifier**: It is an ensemble learning strategy that integrates the outputs of several de-correlated decision trees in the 'forest' to return the necessary results. The concept of the extra tree classifier is similar to the random forest classifier. Still, the only difference is in the decision trees' construction part in the forest (Fig. 2).

3.1 Comparison Between Extra Trees and Random Forest Classifier

Similarities:
The two classifications have a lot of similar properties like the fact that both of the ensembles are composed of many decision trees, where the final result is considered based on the prediction of each individual trees.

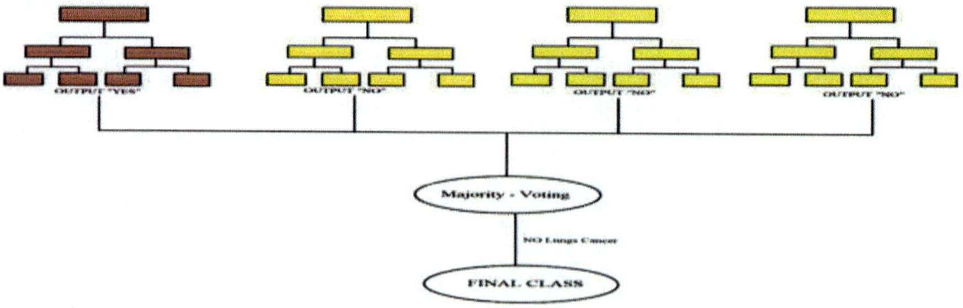

Fig. 1 Random forest classification structure

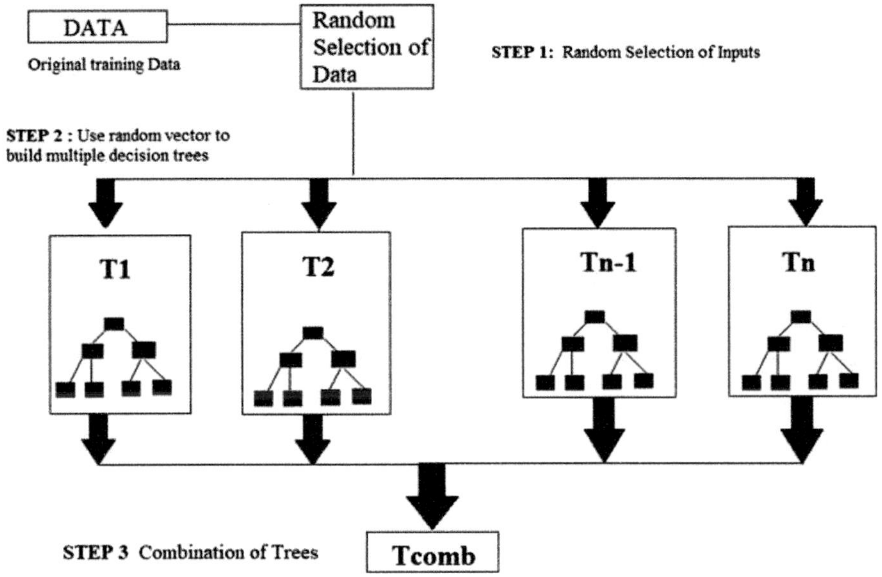

Fig. 2 Extra tree classification structure

Differences:
One of the main differences between both the algorithms is that extra trees work on the original sample, whereas random forest uses the bootstrap replicas, i.e., the subsamples in the input data with replacement. In extra trees implementation, there is an optional parameter to use the bootstrap replicas, but it uses the entire training dataset which might increase the variance due to the diversified which is created by bootstrapping.

4 Result and Discussion

The machine learning classification algorithms were implemented on the lung cancer dataset that was available for the public. The dataset contains 309 entries with 16 numbers of attributes. Each tuple representing a specific person with only their gender in the first column, followed by the activities the respective person carries out (like smoking, consumption of alcohol, etc.), symptoms that are diagnosed or noticed (like yellowish fingers, coughing, wheezing, etc.), is among many others, where the model is made based on a binary classification model. The columns from the third to the fifteenth are filled up with two values, which represent:

1. no (the person does not have the respective attribute)
2. yes (the person has the respective attribute)

While the first and second columns are filled up with values as follows:

1st: Gender: **M** or **F**
2nd: Age: In **numbers**

The target class named LUNG_CANCER indicates: If the person has a lung cancer or not: **YES** or **NO.**

The used model is split into 85% as train and 15% as testing. As displayed in Table 1, the corresponding values show the result of the classifications.

Precision (P) is the measurement of the Avg Precision of the predictions, i.e., the fraction of true positive by total positive case.

$$P = P_T/P_T + P_F \tag{1}$$

where P_T is true positive while P_F denotes false positive.

Recall (R) is used to measure the true positive rate (F_n is false negative)

$$R = P_T/P_T + F_n \tag{2}$$

Average Precision (AP) is the area under the precision–recall curve and the value lies between the range of 0 and 1 because the values of precision and recall have a range of [0,1].

$$\mathrm{AP} = \int_0^1 p(r)\mathrm{d}r \tag{3}$$

When represented graphically, Fig. 3 shows the difference between the two classifications.

As deduced from the graph above, the attributes precision and recall give a better value when it comes to the extra tree classifier which are 0.933 and 0.969 against 0.924 and 0.961 of random forest. But the average precision of random forest performs better than that of the extra trees with a value of 0.967 over 0.965 as shown in Table 2.

Accuracy is one standard of measurement for evaluating the performance of a classification algorithm. In Layman's terms, we can say that, the accuracy is the value up to which our model got the prediction right. If we put it into a formula then, it can be said that:

Table 1 Average precision, accuracy and precision values of the classification

Classifier	Avg. precision	Precision	Recall
Random forest	0.967	0.924	0.961
Extra tree	0.965	0.933	0.969

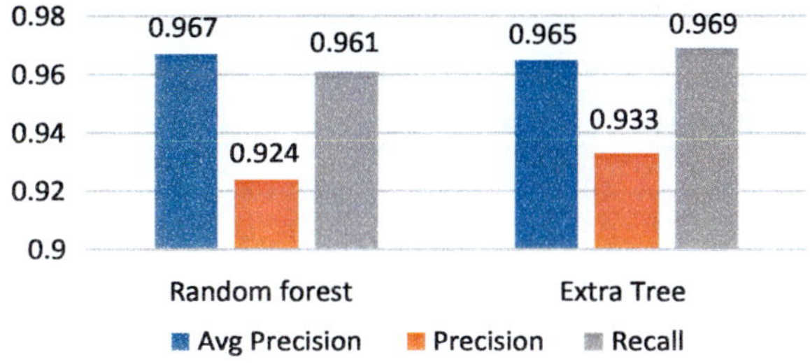

Fig. 3 Average precision, precision, and recall of the classifiers

Table 2 Accuracy of each classification

Classifiers	Accuracy
Random forest	0.897
Extra tree	0.912

$$\text{Accuracy} = \frac{\text{Number of correct predictions}}{\text{Total number of predictions}} \tag{4}$$

For binary classification, accuracy can also be calculated in terms of positives and negatives as follows:

$$\text{Accuracy} = \frac{(P_T + N_T)}{(P_T + N_T + P_F + N_F)} \tag{5}$$

Where P_T = True positive, N_T = True negatives, P_F = False positive, and N_F = False negative

According to Fig. 4, the accuracy of predicting the correct output for some given data is higher in extra tree which is up to 0.015 more than that of random forest classifier. In terms of accuracy, extra tree clearly performs better.

The other attributes of the classifications are F_1, log loss, ROC AUC, normalized Gini coefficient, and model build time (in sec) as given in Table 3 are also been mentioned for the comparative analysis of the classifiers.

The time taken for the classification techniques to build a model using the data is called the classification algorithm's build time. In this case, the random forest classifier takes a longer build time(3 s) than that of the extra trees classifier (1 s).

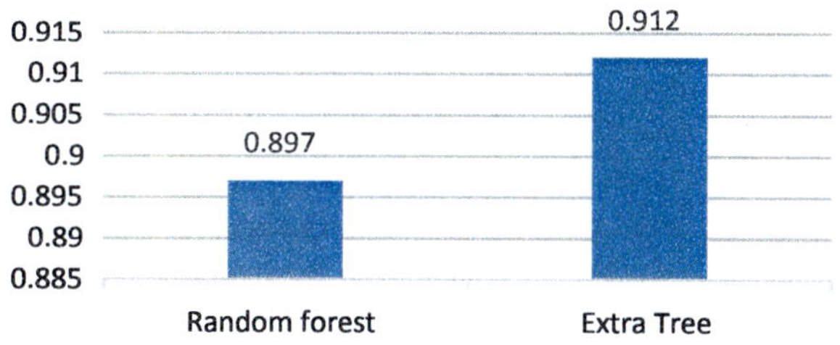

Fig. 4 Accuracy values of the two selected models

Table 3 Other performance factors of the classifications

Classifiers	F1	Log loss	Gini coefficient	ROC AUC	Build time
Random forest	0.942	0.721	0.387	0.861	3
Extra tree	0.951	0.822	0.489	0.861	1

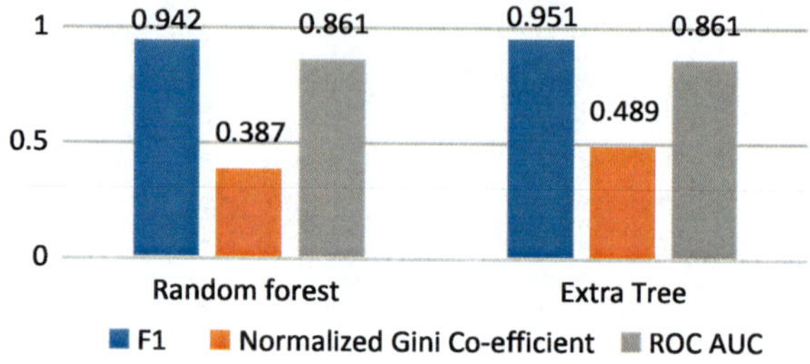

Fig. 5 Representing some of the other factors of the classification

Comparing other factors for both the classifiers like F1, log loss, normalized Gini coefficient, and ROC AUC, we find that extra tree has a higher value for F1 than that of the random forest classifier. It is also found that both classifiers have equal values for ROC AUC. From the table, it is evident that random forest has lesser values for log loss and normalized Gini coefficient as represented in Fig. 5.

After analyzing all the graphs and tables, it can be deduced that among the two machine learning classification techniques, the extra trees classifier has shown higher accuracy than that of the random forest classifier on predicting whether a person has lung cancer or not using the lung cancer dataset to train and test the model. It is evident from above that the extra trees classifier's accuracy value is 0.912 (91.2%), which is greater than the random forest classifier's accuracy value, which is 0.897 (89.7%).

5 Conclusion

At the end of the comparison study between the two classifiers, it can be concluded that the accuracy of two classification algorithms is the random forest classifier (0.912) and extra trees classifier (0.897) for using the lung cancer dataset. As training and testing, data are created for classification model that can predict if a person has lung cancer or not, with highest possible accuracy, it is found that extra trees classifier having an accuracy percentage of 91.2% is the classification model that surpasses the random forest classifier having an accuracy percentage of 89.7%. There is always room for improvement in predicting target classes, mainly in the healthcare sector, as a classification model with the highest possible accuracy is required as much as desired. This requirement can be served in various ways, one of which can be multiple models to create a hybrid model that predicts the required result with higher accuracy.

References

1. Ranganatha, S., Anusha, C., Vinay, S. K., Pooja Raj, H. J.: Medical data mining and analysis for heart disease dataset using classification techniques. National Conference on Challenges in Research & Technology in the Coming Decades (CRT 2013) (2013)
2. Ravì, D., et al.: Deep Learning for Health Informatics. IEEE J Biomed Health Inform **21**(1), 4–21 (2017)
3. Jena, L., Patra, B., Nayak, S., Mishra, S., Tripathy, S.: Risk Prediction of Kidney Disease Using Machine Learning Strategies. In: Intelligent and Cloud Computing. Smart Innovation, Systems and Technologies, vol. 153. Springer, Singapore (2021)
4. Jena, L., Nayak, S., Swain, R.: Chronic Disease Risk (CDR) Prediction in Biomedical Data Using Machine Learning Approach. In: Advances in Intelligent Computing and Communication. Lecture Notes in Networks and Systems, vol. 109. Springer, Singapore (2020)
5. Lefkovits, L., Lefkovits, S., Vaida, M. F., Emerich, S., Măluţan, R.: Comparison of Classifiers for Brain Tumor Segmentation. IFMBE Proceedings, 195–200 (2017)
6. Singh, C., Cheggoju, N., Satpute, V. R.: Implementing Classification algorithms in Medical Report Analysis for Helping Patient During Unavailability of Medical Expertise. 2018 9th International Conference on Computing, Communication and Networking Technologies (ICCCNT) (2018)
7. Bhardwaj, R., Nambiar, A. R., Dutta, D.: A Study of Machine Learning in Healthcare. 2017 IEEE 41st Annual Computer Software and Applications Conference (COMPSAC) (2017)
8. Shailaja, K., Seetharamulu, B., Jabbar, M. A.: Machine Learning in Healthcare: A Review. 2018 Second International Conference on Electronics, Communication and Aerospace Technology (ICECA) (2018)
9. Kesavaraj, G., Sukumaran, S.: A study on classification techniques in data mining. 2013 Fourth International Conference on Computing, Communications and Networking Technologies (ICCCNT) (2013)
10. Pradeep K R, Naveen N C: Lung Cancer Survivability Prediction based on Performance Using Classification Techniques of Support Vector Machines, C4.5 and Naive Bayes Algorithms for Healthcare Analytics. Procedia Comput. Sci. **132**, 412–420 (2018)
11. Lynch, C.M., Abdollahi, B., Fuqua, J.D., de Carlo, A.R., Bartholomai, J.A., Balgemann, R. N., van Berkel, V.H., Frieboes, H.B.: Prediction of lung cancer patient survival via supervised machine learning classification techniques. Int. J. Med. Informatics **108**, 1–8 (2017)
12. Jayaraj, D., Sathiamoorthy, S.: Random Forest based Classification Model for Lung Cancer Prediction on Computer Tomography Images. 2019 International Conference on Smart Systems and Inventive Technology (ICSSIT), (2019)
13. Kalyankar,G. D., Poojara, S. R., Dharwadkar, N. V.:"Predictive analysis of diabetic patient data using machine learning and Hadoop," 2017 International Conference on I-SMAC (IoT in Social, Mobile, Analytics and Cloud) (I-SMAC), Palladam, pp. 619–624 (2017)
14. Breiman, L.: Random forests. Mach. Learn. **45**(1), 5–32 (2001)

An Interactive Machine Learning Approach for Brain Tumor MRI Segmentation

Sachikanta Dash, Rajendra Kumar Das, Subhankar Guha, Surendra Nath Bhagat, and G. K. Behera

Abstract These days, image processing in health care is the most challenging and developing field. Tumor detection of MR image is suppose to be the most significant task in this field. This paper depicts the proposed methodology to distinguish the tumor edges of the brain from patient's magnetic resonance images (MRI) of brain. At initial stage, this strategy incorporates some noise expulsion capacities improving features that give better qualities of medical images for solid diagnosis utilizing anisotropic filtering technique for removal of noise from the image. The consequence of second phase is exposed to image segmentation utilizing support vector machine (SVM) classification technique. And finally, morphological operation is carried out for tumor segmentation. During the experimental examination, we utilized images with or without tumor. We have used MATLAB programming for the segmentation of brain tumor. The experimental outcome shows that the used classifier obtained a accuracy of 96.04%. Finally, the segment part of the image is outlined for identification.

Keywords MRI · Tumor · SVM · Anisotropic filter · Segmentation

S. Dash (✉)
Department of CSE, DRIEMS Autonomous Engineering College, Tangi, Cuttack 754 022, Odisha, India
e-mail: sachikanta_dash@rediffmail.com; dr.sachikanta@driems.ac.in

R. K. Das
Department of ENTC, DRIEMS Autonomous Engineering College, Tangi, Cuttack 754 022, Odisha, India
e-mail: drrkd67@gmail.com

S. Guha · S. N. Bhagat · G. K. Behera
Department of CSE, GIET University, Gunupur, Odisha, India
e-mail: subhankar.guha78@gmail.com

S. N. Bhagat
e-mail: suren.bhagat@gmail.com

G. K. Behera
e-mail: gkbehera.cse@gmail.com

S. Das and M. N. Mohanty (eds.), *Advances in Intelligent Computing and Communication*, Lecture Notes in Networks and Systems 202,
https://doi.org/10.1007/978-981-16-0695-3_38

1 Introduction

The human brain is one of the biggest and generally mind boggling organs in human body. It comprised of more than hundred billion nerves that impart in trillions of associations called neural connections. It might be influenced by tumor. An uncommon development of tissue in the brain is known as brain tumor. In the existing system like 1) man-made reasoning system and 2) AI System, several issues are there in image processing. Existing system utilizes algorithms like bounding calculations which is utilized to order images to various two groups where one group will contain brain tumor and the other group is the one which do not have tumor that is destructive. Additionally these current frameworks made use of steps like element extraction that is critical to acquire and recover the fundamental data from the information that is unique that is finished by utilizing different methods that can be utilized when the size of the image is not in determined range or then again the image that is huge. The following stage that is utilized in the current framework is SVM algorithm with anisotropic filter. The principle disadvantage of this current framework is that these do not utilize preprocessing step, due to which this will result into efficiency toward the noise. These noises are present in preparing dataset. When there is enormous dataset, this current framework does not function admirably. Another disservice of this current framework is that the real shape, size, and sort of the tumor cannot be predicted.

2 Literature Review

Akram et al. [1] have used an automated approach with thresholding and morphological operation for segmentation of the MR images and got an accuracy of 97%. In 2012, puspa et al. [2] have used various features and applied SVM for the image segmentation. They got high accuracy in the classification with 96%. They have used very large dataset for the segmentation. Also, Dahab et al. [3] have used probalistic NN for classification with accuracy of 79%. Ciresan et al. [4] have addressed a automatic segmentation process that eliminate errors with a large margin ang give accuracy of 83%. Bengio et al. [5] have proposed backpropagation and gradient-based optimization technique for the detection of tumor and got successful result. In 2013, Bjoern et al. [6] have proposed a survey report on brain tumor segmentation and Roy et al. have proposed a multiscale CNN that produce good results including feature extraction. In 2016, Vijay et al. [7] in their study projected an automatic EDPSO technique for tumor segmentation that take advantage of existing PSO technique in optimization. In 2017, Soltaninejad et al. [8] have applied a automated technique on clinical dataset and BRATS dataset for the segmentation and get the result with 89.48% and 88.09% of accuracy. Convolutional neural net cannot physically accept images of different resolutions. However, if the network is fully-convolutional, then it should be able to accept images of any dimension. In 2018, Reshmi et al. [10] in their survey paper proposed different segmentation and detection methods of tumor cell from MRI. Their objective was to automate the system for segmentation and detection. In 2019, Alqazzaz et al. [11] in their study used a fully automated method for segmentation. They have integrated Segnet model which was separately trained to produce features after post-processing. Also, they have utilized the decision tree for the classification of brain tumor MR images. For their study, they have used BRATs 2017 and achieved a score of 85, 81, and 79% for different full tumor, core part, and enhancing tumor part, respectively. Shahriar et al. [12] have included enhancement approach initially for the automated segmentation process. Various filtration techniques have been used for better segmentation. Finally, the information from the expert was considered to enhance the accuracy label of segmentation process.

3 Proposed Method

In our proposed method, we have used six different steps. Those includes (1) image acquisition, (2) image filtering with anisotropic filter, (3) classification with SVM classifier, (4) bounding boxing, (5) erode detection, and (6) final segmentation of brain tumor with outline. Following is the block diagram of our proposed work in Fig. 1.

3.1 Dataset

We have used the kaggle dataset for our experiment for the segmentation of brain tumor MRI. The dataset contains 98 no. of images with no tumor and 155 images with tumor. There are varying dimensional images used for the experiment. Using MATLAB program, we have classify the image dimension with 224 × 224.

3.2 Anisotropic Filter

Anisotropic filtering (AF) is a strategy for improving the image nature of surfaces on surfaces of computer illustrations that are at diagonal review edges regarding the camera where the projection of the surface (not the polygon or other crude on which it is delivered) has all the earmarks of being non-symmetrical (subsequently the inception of: "a" for not, "iso" for same, and "tropic" from tropism, identifying with heading; anisotropic filtering does not filter the equivalent toward each path).

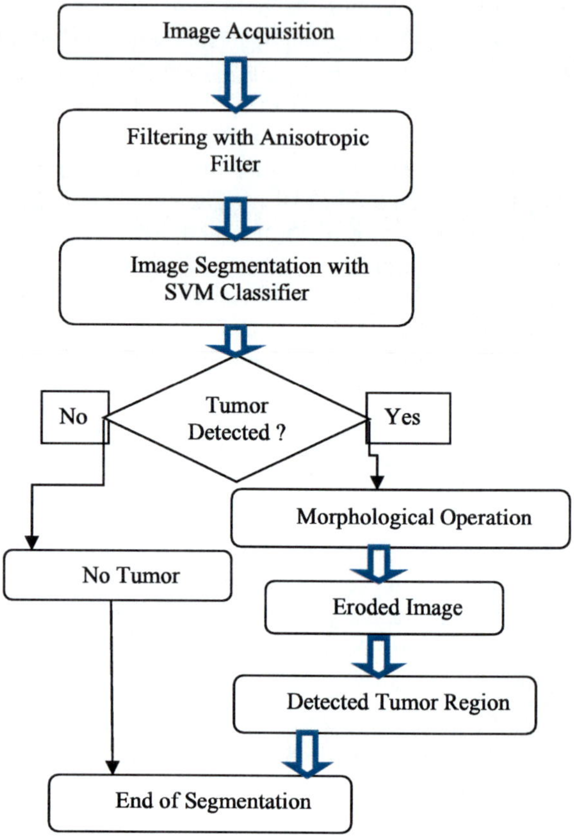

Fig. 1 Proposed flowchart for brain tumor MRI segmentation

Formally, let $\Omega \subset R^2$ represent a plane subset and $I(.,t) : \Omega \rightarrow R$ represents the grayscale images family, then the anisotropic filter diffusion can be defined as follows

$$\frac{\partial I}{\partial t} = \text{div}(c(x,y,t)\nabla I = \nabla c \cdot \nabla I + c(x,\ y,\ t)\Delta I \tag{1}$$

where ∇ represents the gradient, Δ represents the Laplacian, the divergence operator represented by Div (…), and c(*x*, *y*, *t*) represented as diffusion coefficient. c(*x*, *y*, *t*) is typically selected as a function of the image gradient that controls the rate of diffusion so as to preserve edges in the MRI. The functions for the diffusion are explained in "Eqs. (2) and (3)" as follows:

$$c(\| \nabla I \|) = e^{-(\|\nabla I\|/R)^2} \tag{2}$$

And also

$$c(\| \nabla I \|) = \frac{1}{1 + (\| \nabla I \| / R)^2} \tag{3}$$

where R is a constant that controls the edge sensitivity and generally R is chosen as a function of noise in the preprocessed image.

3.3 Segmentation Through SVM Classifier

Recently, with new hypotheses and new techniques showing up continually, increasingly more adjusted SVM techniques have accomplished great impact in image segmentation. The SVM classifier is trained after the denoising operation by the use of anisotropic filter.

This paper proposes an interactive segmentation process by the use of SVM classifier. This is the method for autosegmentation of the brain tumor MR image where SVM classifier is used. Here after randomly selecting the training and testing dataset the classification is applied. Linear kernel function is used to train an SVM classifier. Then MATLAB function "svmclassify" is used to classify the test set. However, it automatically avoids the overfit data. Here, the main objective of the SVM is to discover the hyperplane that boosts the base separation between any data centers. Finally, the performance is calculated through the accuracy and precision.

3.4 Morphological Operations

The changed over image contains flaws since paired image delivered by thresholding is misshaped by noise and surface. Morphological activities recognize articles and limits in an image. It is a legitimate activity dependent on pixel neighborhood correlation. They are performed on double images. The pixel which has high intensity is assembled and portioned as one cluster which go about as tumor recognized region.

Morphological operation is a mechanism to remove image features valuable in the presentation and legation of region shapes, for example, limits, skeletons, and arched structures. For morphological activity, there is requirement of organizing component. The organizing component utilized practically is commonly very smaller than that of the image frequently a 3 × 3 grid. The morphological method is applied on to the segmented image. The two significant activities of morphology are: (a) dilation: that works on object extension, gap filling lastly adding all the disjoint articles and (b) erosion: it shrivels the item. Frontal area pixel foundation is dissolved away in the twofold image by disintegration activity. Morphological method is applied on to the input image after transforming it into binary image after segmentation. It requires a binary masked window to segment the tumor area from the filtered image. Typically, higher intensity contrasting and other encompassing tissues are characterized through an irregular brain tumor MR image. By the application of the tumor cover on enlarged brain tumor MR image, the last image is acquired with distinguished tumor. Figure 5 shows the resultant images of morphological activity with recognized tumor in the MR image.

4 Result Analysis and Performance Evaluation

In this paper, our aim was to detect and segment brain tumor in a MR Image. To achieve this, we have used anisotropic filter for the noise removal then we classify the image using SVM and then we have tested an image whether tumor is found or not. If we found a tumor in the image, then different morphological operations are carried out for the identification of the tumor in the MR Image.

4.1 *Performance Evaluation of Anisotropic Filter*

We have applied the following formula for the evaluation of performance of anisotropic filter. A comparison has been made between different filter techniques on original image, and the respective PSNR and MSE values are evaluated as per the following.

$$\mathrm{MSE} = \frac{1}{mn}\sum_{i=0}^{m-1}\sum_{i=0}^{n-1} \| p(i,j) - q(i,j) \|^2 \quad (4)$$

$$\mathrm{PSNR} = 20\log 10\left(\frac{\mathrm{MAX}_f}{\sqrt{\mathrm{MSE}}}\right) \quad (5)$$

where

p represents the matrix data of input image

q represents the data matrix of the degraded image

m denotes row number of the intensity values of the images and row index is represented by i.

n denotes the intensity column number values of the images and column index is represented by j.

MAX$_f$ represents the availability of maximum signal in the input original image which is "known to be good" image.

The after calculation of SNR and MSE values using different filter and comparing with anisotropic filter gives good result [ref].

4.2 Evaluation of Confusion Matrix

Finally, after the removal of noise from the input image, the SVM classifier is used which segments the brain tumor image properly and after that the morphological procedures are carried out. At the end of this process, the accuracy is measured which defines the successful performance of the classification. The measurement of accuracy is carried out through the following (Table 1).

After the SVM classification and morphological operation, the accuracy was measured. The accuracy can be calculated by the above confusion matrix as follows.

$$\text{Accuracy} = \frac{\text{Total number of Correctly identified Image samples}}{\text{Total number of Image samples}} \times 100 \quad (6)$$

Here, we have taken total of 253 images from the kaggle dataset [13]. Our dataset contains 155 images with tumor and 98 images without tumor. After segmentation process, our proposed method has correctly classified 243 images out of 253 images.

Hence

$$\text{Accuracy} = \frac{\text{TP} + \text{TN}}{\text{TP} + \text{TN} + \text{FP} + \text{FN}} \times 100 = \frac{148 + 95}{253} \times 100 = 96.04$$

$$\text{Precision} = \frac{\text{TP}}{\text{TP} + \text{FP}} = \frac{148}{148 + 3} = 0.9801$$

The following figure depicts the output of segmented MR images with or without tumor.

Figure 2 depicts the detected normal brain image and Fig. 3 depicts the segmented brain image with tumor.

Table 1 Confusion matrix representation

		Prediction value	
		Negative	Positive
Real value	Negative	True negative (TN)	False positive (FP)
	Positive	False negative (FN)	True positive (TP)

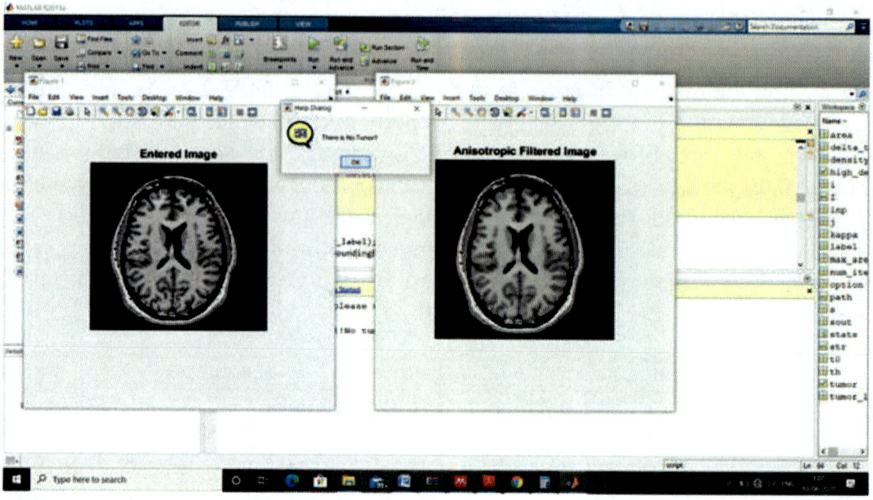

Fig. 2 Normal brain image without tumor

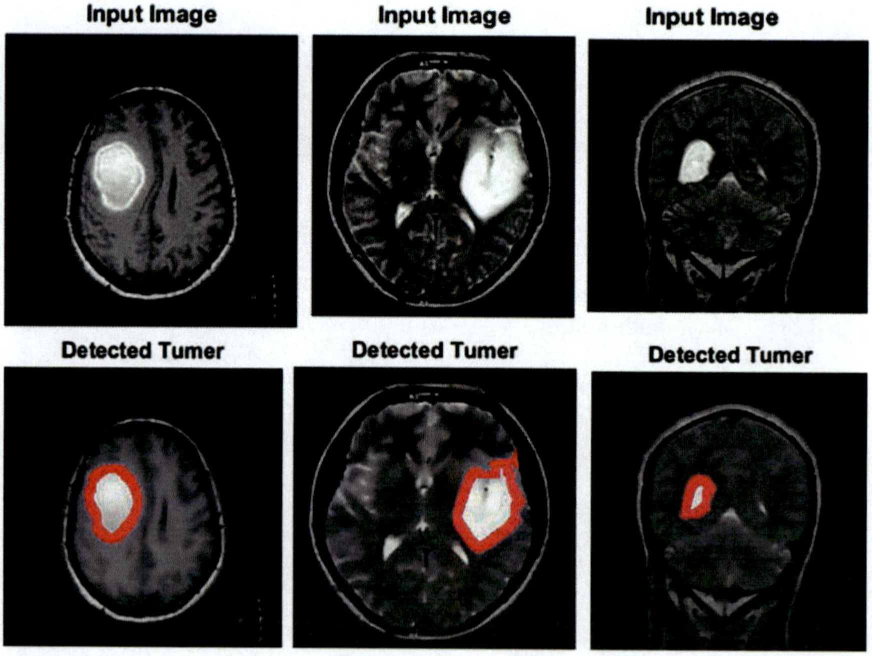

Fig. 3 Segmented brain image with detected tumor

5 Conclusion

In this proposed work, we perform an automatic detection technique for the presence or absence of the brain tumor in MRI. We follow the steps like image collection, normalization, classification using SVM classifier, and the anisotropic filter image classification using anisotropic filter. The method for segmentation was suggested in this paper and is also used for removal of noise from the image. The results of the filter are compared with other available techniques to justify the method. The result we get from the above proposed process gives accuracy of 96.04% and precision of 0.9801. Here, the MSE was calculated and tried to minimize the error rate as well as increase the classification accuracy. Thus, the proposed method performs better as compared to other available techniques discussed in the literature review. The scope of this work is to apply this method in complex images to detect and segment for finding out the type of tumor cell in the image.

References

1. Akram, M.U., Akram', M.U., Usman, A.: Computer aided system for brain tumor detection and segmentation Bio Medical Imaging View project International Journal of Future Computer and Communication-[IJFCC]. ieeexplore.ieee.org. (2011). https://doi.org/10.1109/ICCNIT.2011.6020885
2. Pushpa Rathi, V.P.G.: Brain Tumor MRI Image Classification with Feature Selection and Extraction using Linear Discriminant Analysis. Int. J. Inf. Sci. Tech. **2**, 131–146 (2012). https://doi.org/10.5121/ijist.2012.2413
3. Dahab, D., Ghoniemy, S., Selim, G.: Automated brain tumor detection and identification using image processing and probabilistic neural network techniques. Int. J. image Process. Vis. Commun. **1**, 1–8 (2012)
4. Cireşan, D.C., Giusti, A., Gambardella, L.M., Schmidhuber, J.: Deep neural networks segment neuronal membranes in electron microscopy images. In: Advances in Neural Information Processing Systems, pp. 2843–2851 (2012)
5. Bengio, Y.: Practical recommendations for gradient-based training of deep architectures. Lect. Notes Comput. Sci. 7700 LECTU, 437–478 (2012). https://doi.org/10.1007/978-3-642-35289-8-26
6. Bjoern Menze Stefan Bauer, A.J.: The Multimodal Brain Tumor Image Segmentation Benchmark (BRATS). ieeexplore.ieee.org. (2013)
7. Vijay, V., Rebecca, Sr.: Automated Brain Tumor Segmentation and Detection in MRI using Enhanced Darwinian Particle Swarm Optimization(EDPSO). Procedia - Procedia Comput. Sci. 92, 475–480 (2016). https://doi.org/10.1016/j.procs.2016.07.370
8. Soltaninejad, M., Yang, G., Lambrou, T., Allinson, N., Jones, T.L., Barrick, T.R., Howe, F. A., Ye, X.: Automated brain tumour detection and segmentation using superpixel-based extremely randomized trees in FLAIR MRI. Int. J. Comput. Assist. Radiol. Surg. **12**, 183–203 (2017). https://doi.org/10.1007/s11548-016-1483-3
9. Shelhamer, E., Long, J., Darrell, T.: Fully Convolutional Networks for Semantic Segmentation. IEEE Trans. Pattern Anal. Mach. Intell. **39**, 640–651 (2017). https://doi.org/10.1109/TPAMI.2016.2572683

10. Reshmi, S., Salagar, R.D., Veni, S.S.: A survey on brain tumor detection and segmentation from mri images. Int. J. Comput. Sci. Eng. 6, 776–778 (2018). https://doi.org/10.26438/ijcse/v6i10.776778
11. Alqazzaz, S., Sun, X., Yang, X., Nokes, L.: Automated brain tumor segmentation on multi-modal MR image using SegNet. Comput. Vis. Media. **5**, 209–219 (2019). https://doi.org/10.1007/s41095-019-0139-y
12. Shahriar Sazzad, T.M., Tanzibul Ahmmed, K.M., Hoque, M.U., Rahman, M.: Development of Automated Brain Tumor Identification Using MRI Images. In: 2nd International Conference on Electrical, Computer and Communication Engineering, ECCE 2019 (2019). https://doi.org/10.1109/ECACE.2019.8679240
13. Navoneel Chakrabarty, kaggle dataset, https://www.kaggle.com/navoneel/brain-mri-images-for-brain-tumor-detection

Performance Analysis of Fractional Order Filter Using Fractional Order Elements

Tapaswini Sahu, Madhab Chandra Tripathy, Kumar Biswal, and Sanjeeb Kumar Kar

Abstract This article focuses on performance analysis of band pass filter in fractional order using generalized impedance converter and fractional order capacitor. The band pass filter is analyzed by considering resistor, a fractional order inductor with order lies between $0 < \alpha \leq 2$ and a fractional order capacitor with order lies between $0 < \beta \leq 1$.The realization of inductor with fractional order is achieved through generalized impedance converter circuit where fractional capacitor is employed as a element in it. Finally, investigation on performance analysis has been done in fractional band pass filter and differentiated with respect to conventional order band pass filter through MATLAB environment.

Keywords Fractional order band pass filter · Fractional order capacitor · Fractional order inductor · Generalized impedance converter

T. Sahu (✉)
BPUT, Rourkela, Odisha, India
e-mail: tapaswini123sahu@gmail.com

M. C. Tripathy
Department of Instrumentation and Electronics Engineering, CET, BPUT, Rourkela, Odisha, India
e-mail: mctripathy@cet.edu.in

K. Biswal
School of Electronics Engineering, Kalinga Institute of Industrial Technology (DU), Bhubaneswar, Odisha, India
e-mail: biswalfet@kiit.ac.in

S. K. Kar
ITER, S'O'A (deemed to be University), Bhubaneswar, Odisha, India
e-mail: sanjeebkar@soa.ac.ins

S. Das and M. N. Mohanty (eds.), *Advances in Intelligent Computing and Communication*, Lecture Notes in Networks and Systems 202,
https://doi.org/10.1007/978-981-16-0695-3_39

1 Introduction

In recent years, fractional calculus is an important research topic which deals with derivatives and integrals of randomly real order [1]. Introducing these idea in various electronics circuits which is emerged the field of fractional order filter (FOF) design. The order of the filters are determined by the key components i.e. resistors and capacitors, it may be passive or active. The powers of 's' in transfer function (TF) of conventional analog type filters are integer. Fractional filter order is characterized by slope of attenuation which is similar to filters having integer order (IO). The main distinction between an integer order filter (IOF) and FOF is the transfer function, which contains extra parameters specified by gradient of attenuation of that given TF-. Therefore, the slope of the attenuation of the TF of a FOF is given by: $20(n+\alpha)$ (dB/decade), where the unsigned integer is 'n' (i.e. 1 < n < 10) [2, 3], and α is a real number (i.e.$0 < \alpha < 1$). Generally two types of techniques are widely used for designing fractional order filters i.e. 1st method is used for a particular FOC with impedance $Z_c(S) = \frac{1}{C_F S^\beta}$ [4] or inductor with impedance $Z_L(s) = L_F s^\alpha$ [5], which are then substituted with appropriate passive components in the layout of a traditional filter. These fractal passive elements are usually indicated as FOEs [6]. Filter circuits are usually categorized as 1st, 2nd, … nth order networks where 'n' is an integer number. In [7], the general TF of a FOF whose characteristic equation depends on two FC with different orders α and β is expressed by:

$$\mathrm{TF}(S) = \frac{dS^\gamma}{S^\beta + aS^\alpha + b} \tag{1}$$

where a, b, d are constants and α, β and γ are fractional orders. In this article, the performance analysis of fractional order band pass filter (FOBF) is investigated and compared with integer order BPF through simulation. The rest part of this work is arranged i.e. In Sect. 2, a general idea on FI, FC and GIC are given. In Sect. 3, analysis of FOBF using GIC is presented. In Sect. 4 Simulation results are discussed and finally, this article concludes with remarks and recommendations in Sect. 5.

2 Brief Idea of FC, FI and GIC

2.1 *Fractional Capacitor (FC)*

The relationship between voltage and current of FC is given by:

$$i(t) = C_F \frac{d^\beta v(t)}{dt^\beta} \tag{2}$$

where $0 < \beta < 1$. When $\beta = 1$, then it is called Integer Order Capacitor. A fractional order capacitor (FOC) is a passive element which shows constant phase response in between 0° and −90° with change in frequency [8]. The FC impedance is given by

$$Z(S) = \frac{1}{C_F S^\beta} = \left|\frac{1}{C_F S^\beta}\right| < \left(-\frac{\pi}{2}\beta\right) \tag{3}$$

In the same way, the behavior of amplitude characteristics of an standard fractional order capacitor is a straight line with gradient −20β dB/decade, whereas, for all frequencies the phase angle is constant.

2.2 *Fractional Inductor (FI)*

FI is a passive element which provides steady phase angle in the range of 0° and +90° with change in frequency. The FI impedance is given by

$$Z(S) = L_F S^\alpha = L_F \omega^\alpha < \left(\frac{\pi}{2}\alpha\right) \tag{4}$$

Here L_F = inductance of fractional order α. The behavior of amplitude characteristics of an ideal FI is a approximate linear with gradient 20α dB/decade, whereas, for all frequencies the phase angle is constant.

2.3 *Generalized Impedance Converter (GIC)*

The input impedance of generalized impedance converter is an active two port network, which is a function of load impedance and complex frequency variable. Generally, GICs are extremely not sensitive to component deviation [9]. The GICs are designed to simulate inductors for synthesis of active filter. From the various configurations, one that has gained prominence is the generalized impedance converter as shown in Fig. 3, which can be used to simulate inductance. The GIC enables impedances to be achieved with values that may be difficult to attain with passive components. For example, a GIC can convert a capacitor to inductor and vice versa, scale impedance and represent a negative resistance. These circuits are assumed to be in sinusoidal steady-state and can be analyzed using frequency-domain method. Figure 1 illustrates a GIC, which is designed by employing five impedances along with two operational amplifiers. The input impedance with respect to ground of GIC is given by $\frac{Z_1 Z_3 Z_5}{Z_2 Z_4}$.

The realization of fractional inductor with different orders (α) through GIC employing FCs is obtained. Initially, all the fractional inductors are characterized in

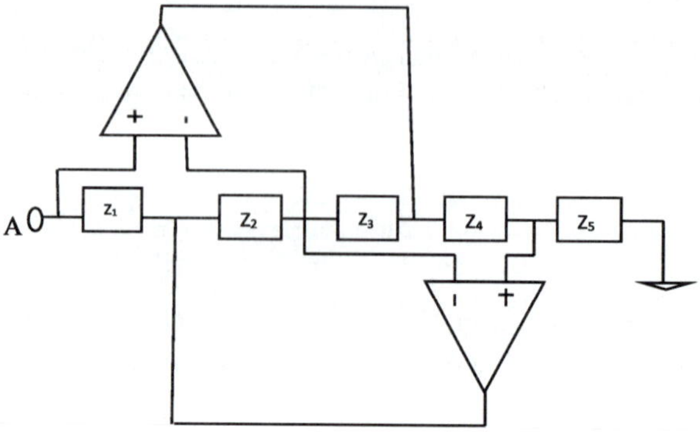

Fig. 1 GIC employed for realization of FI

frequency domain analysis. The order (α) of fractional inductor as shown in Fig. 3 can be changed from 0 to 2 by employing various combinations of elements as given below:

i. To set the order of fractional inductor less than one i.e. $\alpha < 1$, the combinations are taken in GIC as:$Z_1(S) = R_1$,$Z_2(S) = R_2$, $Z_3(S) = R_3$, $Z_5(S) = R_5$ and $Z_4(S) = \frac{1}{C_F S^{\beta}}$, and β lies between 0 and 1. The impedance value of GIC is shown as $Z_{L_F}(S) = L_F S^{\alpha}$ where, $L_F = \frac{R_1 R_3 R_5 C_F}{R_2}$, and $\alpha = \beta$.
ii. To set the order of fractional inductor greater than one i.e. $\alpha > 1$, the combinations are taken in GIC as: $Z_1(S) = R_1$, $Z_2(S) = \frac{1}{sC}$, $Z_3(S) = R_3$, `$Z_5(S) = R_5$ and $Z_4(S) = \frac{1}{C_F S^{\beta}}$, where the resistance elements are R_1, R_3, R_5 and the value of capacitance of conventional capacitor is C. The equivalent fractional inductance magnitude is L_F where $L_F = R_1 R_3 R_5 C C_F$ and the relationship between α, β is $\alpha = (1 + \beta)$.

GIC can be used to simulate inductors. While considering Z_2 or Z_4 as a capacitor and rest of the impedances are resistors, then only the impedance of an inductor will be the Z_{in}. The reason of using GIC as a simulator for inductors because it is practically impracticable to assemble an inductor in IC, while resistors, capacitors and operational amplifiers are comparatively easy to incorporate [10].

3 Study of Fractional Band Pass Filter Circuit Using GIC

As in Fig. 2, the filter uses one fractional order inductor, L_F of order α $(0 < \alpha \le 2)$ and one FOC of order β $(0 < \beta \le 1)$. In frequency domain, the relationship between voltage and current can be expressed in fractional orders as follows.

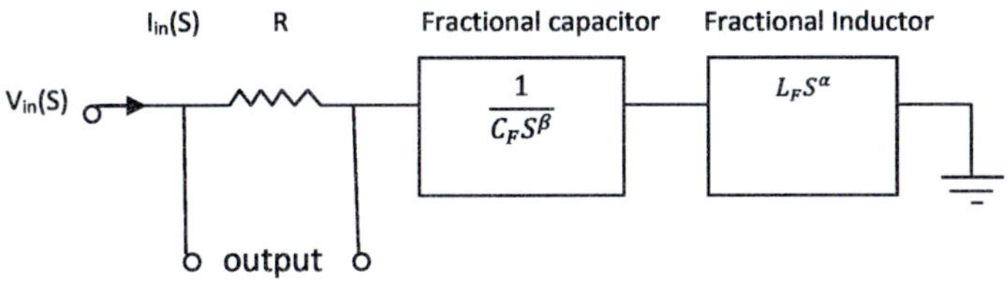

Fig. 2 Band pass filter circuit with fractional orders

$$V_{in}(S) = \left(R + L_F S^\alpha + \frac{1}{C_F S^\beta}\right) I_{in}(S) \tag{5}$$

where, the impedances of FOI and FOC are $L_F S^\alpha$ and $\frac{1}{C_F S^\beta}$. From the Eq. (5), the admittance *Y(S)* is given as follows.

$$\frac{I_{in}(S)}{V_{in}(S)} = Y(S) = \frac{\frac{S^\beta}{L_F}}{S^{\alpha+\beta} + aS^\beta + b} \tag{6}$$

Here, $a = \frac{R}{L_F}$ and $b = \frac{1}{L_F C_F}$

By putting $S^\alpha = w^\alpha e^{\frac{j\pi\alpha}{2}}$ and $S^\beta = w^\beta e^{\frac{j\pi\beta}{2}}$, in eq. (6), it becomes

$$|Y(j\omega)| = \frac{\omega^\beta}{L_F\sqrt{\left(\omega^{2(\alpha+\beta)} + \frac{R^2}{L_F^2}\omega^{2\beta} + \frac{p}{L_F^2}\omega^{\alpha+2\beta} + \frac{q}{L_F^2}\omega^\beta + \frac{r}{L_F^2}\omega^{\alpha+\beta} + \frac{1}{L_F^2 C_F^2}\right)}} \tag{7}$$

where, $p = 2RL_F cos\left(\frac{\pi\alpha}{2}\right)$ $q = \frac{2R}{C_F}\cos\left(\frac{\pi\alpha}{2}\right)$, $r = \frac{2L_F}{C_F}\cos\left(\frac{\pi(\alpha+\beta)}{2}\right)$.

From the Eq. (6) it is observed that the admittance is complex in nature for the band filter with fractional order. Again, considering imaginary part of *Y(s)* is zero, the real part admittance of FOF is obtained from (Eq. 6) at frequency $\omega_{real} = \left(\frac{\sin\left(\frac{\pi\beta}{2}\right)}{L_F C_F \sin\left(\frac{\pi\alpha}{2}\right)}\right)^{\frac{1}{\alpha+\beta}}$ [11] and can be obtained as given below.

$$Y_{real} = \frac{1}{R + L_F\omega^\alpha \cos\left(\frac{\pi\alpha}{2}\right) + \frac{\cos\left(\frac{\pi\beta}{2}\right)}{C_F\omega^\beta}} \tag{8}$$

4 Simulation Results and Discussions

The real admittance of the tuned filter (Eq. 8) is simulated in MATLAB/ SIMULINK environment keeping resistance (R), inductance (L_F) and capacitance (C_F) same for both FOBF and integer order BPF as shown in Fig. 2. It is observed that a sharper tuning characteristic has been obtained in fractional tuned filter though the component value remains same. Subsequently, it is also noticeable that tuned filter parameters are related with α and β, again these fractional orders are supplemented an additional design independence. So, a tuned filter having narrow band can possible through the help of fractional order capacitor and inductor. It is distinguished that the admittance at tuned frequency of 2nd order tuned filter is less as compared to its fractional order counterpart. This is because of cancellation of capacitive and inductive reactance at tuned frequency. For a fractional order (FO) tuned filter, a negative reactance is created in the denominator of Eq. (8), as the FOI is assumed to be α, which is greater than one, i.e., $1 < \alpha \leq 2$. It is noticeable from Fig. 4 that, the real admittance of 2nd order band pass filter is the reciprocal of the resistance ($\frac{1}{R}$), which is frequency independent. However, it is frequency dependent for a FO tuned filter. For this present work for simulation, we have chosen R = 100 Ω, $L_F = 0.0903 \left[\frac{mH}{s^{0.31}}\right]$ and $C_F = 2.11 \times 10^{-6} \frac{F}{s^{0.313}}$].

As per above Fig. 3, it can be observed that the tuned frequency decreases with the increase in α for a particular value of β. By keeping α constant if β is reduced, the tuned frequency of the FOBF increases. As a result, the needed tuned frequency of FOBF can be handled through varying the FOs (i.e. α, β), while in the case of IOBF; the tuned frequency response can be changed only through altering the traditional capacitors and inductors values employed. So, the benefits of FOBF provides supplementary adaptability in assortment of the tuned frequency as differentiated to conventional band pass filter.

Figure 4 demonstrates the deviation of real admittance with change in frequency. It is observed that the value of real admittance is constant for integer order band pass filter (i.e. $\alpha = 1$ and $\beta = 1$), when $\alpha = 1.2$ and $\beta = 0.8$ (fractional

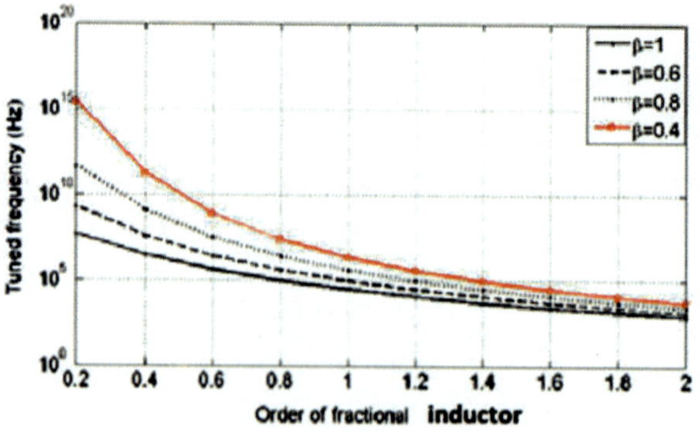

Fig. 3 Tuned frequency responses of FO and IO band pass filter for $0 < \alpha \leq 2$ with $\beta \leq 1$

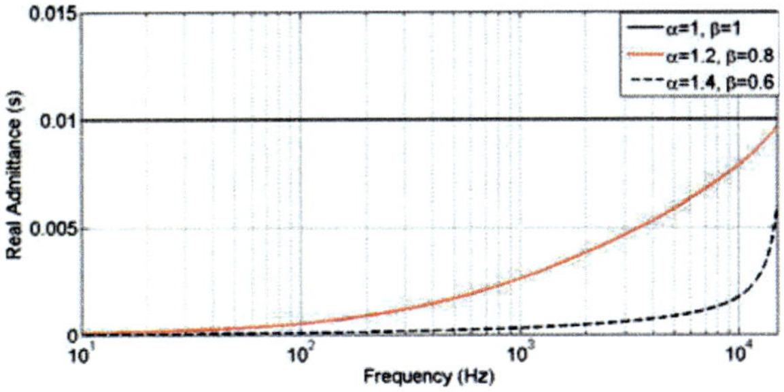

Fig. 4 Responses of real admittance of FOBF increases with frequency for $\alpha \geq 1, \beta \leq 1, \alpha + \beta = 2$

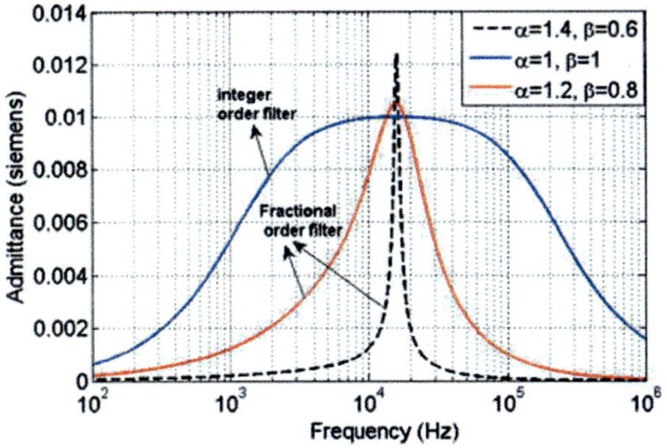

Fig. 5 Responses of Admittance of FOBF with respect to frequency for $1 \leq \alpha < 2, 0 < \beta \leq 1$

order), the real admittance increases as frequency increases, when $\alpha = 1.4$ and $\beta = 0.6$ (fractional integer) real admittance increases as frequency increases.

As shown in Fig. 5, when $\alpha = \beta = 1$, admittance is constant for a range of frequencies. When $\alpha = 1.2$ and $\beta = 0.8$, admittance increases and when $\alpha = 1.4$ and $\beta = 0.6$ admittance increases sharply towards resonance frequency. Further, admittance will enhance considerably the filtering characteristics and decrease loss in power. Also, Tuning characteristics can be changed through changing the order of fractional inductor (α) or fractional capacitor (β) which supplements more degree of independence in design. Since it is possible to alter the frequency characteristics of filter without changing the component value, filter characteristics is independent of the time constant of the circuits. Fractional order filters can provide additional flexibility such as gain and frequency adjustment flexibility, wider stability range, meet exact design requirements, characterized by infinite memory, exact order (up to decimal point) and cut off rate of any value (dB/decade). Again without varying the element R and C, BPF can be employed for very high value frequency or very low value frequency depending on fractional order 'α' which is easy to achieve a sharp cut- off frequency, high speed of response, more precise control over the stop band characteristics, more optimization techniques.

5 Conclusions

Through this paper, we introduced generalized impedance converter to simulate band pass filter using fractal elements (FOI and FOC). The design of various filters using basic LC filters is impractical to realize at minimum frequencies, inductors size becomes bulking with more number of turns. So the use of GIC for realizing various filters are explained which eliminates the use of such bulky inductor. The basic idea of fractal devices has also been described and analyzed. This type of filters are used at low frequencies where it is required to have maximally flat response. It has been found that by using two parameters of fractional order i.e. α and β, which provides supplementary degrees of pattern independence are obtainable for superior flexibility design of band pass filter.

References

1. Elwakil, A.S.: Fractional-order circuits and systems: an emerging interdisciplinary research area. IEEE Circuits Syst. Mag. **10**(4), 40–50 (2010)
2. Freeborn, T.J.: Comparison of (1 + α) Fractional Order Transfer Functions to Approximate Lowpass Butterworth Magnitude Responses. Circuits, Syst., Signal Process. **35**, 1983–2002 (2016)
3. Biswal, K., Tripathy, M.C., Kar, S. K.: "Performance Analysis of Fractional Order Low-pass Filter", Lecture Notes in Networks and Systems (ICAC) Springer, Vol. 109, pp. 224–231 (2020)
4. Carlson, G., Halijak, C.: Approximation of Fractional Capacitors (1/s) (1/n) by a Regular Newton Process. IEEE Trans. Circuit Theory **11**, 210–213 (1964)
5. Verma, R., Pandey, N., Pandey, R.: Electronically Tunable Fractional Order Filter. Arab. J. Sci. Eng. **42**, 3409–3422 (2017)
6. Tripathy, M.C., Mondal, D., Biswas, K., Sen, S.: Experimental Studies on Realization of Fractional Inductors and Fractional-Order Bandpass Filters. Int. J. Circuit Theory Appl. **43**, 1183–1196 (2015)
7. Soltan, A., Radwan, G. A., Soliman, M. A.: "CCII based KHN fractional order filter," IEEE 56th International Midwest Symposium on Circuits and Systems (MWSCAS) (2013)
8. Biswal, K., Tripathy, M.C., Kar, S.K.: "Performance Analysis of Fractional Order High-pass Filter", Lecture Notes in Electrical Engineering (IEPCCT) Springer, Vol. 630, pp. 511–519, (2020)
9. Franco, S.: Design with operational amplifier and analog integrated circuits. Tata MCgraw-Hill, New Delhi (2002)
10. Krishna, M. Sivrama, Das, S., Biswas, K., Goswami, B.: "B. Fabrication of a Fractional Order Capacitor with Desired Specifications: A Study on Process Identification and Characterization," IEEE Transactions on Electron Devices, vol 58, pp. 4067–4073 (2011)
11. Radwan, A.G., Salama, K.N.: Passive and Active Elements Using Fractional $L_\alpha C_\beta$ Circuit. IEEE Trans. Circuits Syst. I Regul. Pap. **58**(10), 2388–2397 (2011)

Execution of Adaptive Transverse Filter for Power Quality Improvement

Buddhadeva Sahoo, Sangram Keshari Routray, and Pravat Kumar Rout

Abstract This paper suggests an improved reference current extraction algorithm for a three-phase shunt active harmonic filter (SAHF) application. For improving the power quality (PQ), an adaptive transverse filter (ATF)-based control algorithm is executed. The proposed ATF design and the optimized parameter are selected through the particle swarm optimization (PSO) algorithm. The developed model is simulated through MATLAB software, and the test results justify the importance of the proposed controller requirement by reducing the harmonics during nonlinear load application. In addition to that, the DE-based traditional controller output is compared with the proposed PSO-based ATF results, to show the effectiveness of the controller for PQ improvement.

Keywords Power quality (PQ) · Particle swarm optimization (PSO) · Differential evolution (DE) · Shunt active filter (SAF) · Adaptive transverse filter (ATF)

1 Introduction

The excess use of nonlinear/sensitive load application increases the riskiness of PQ in power system applications [1, 2]. To circumvent the PQ problems, one of the important solutions, an active power filter (APF) is gaining interest. For eliminating voltage and current nonlinearity, providing reactive power support, DC-link voltage regulation, reducing flickers, and improving voltage stability, APF-based dynamic compensating device is chosen for three-phase applications [3, 4]. In topology point of view, APFs are classified into three types such as series, shunt, and hybrid active

B. Sahoo (✉)
Department of Electrical Engineering, Siksha 'O' Anusandhan University, Bhubaneswar, Odisha, India
e-mail: buddhadeva14@gmail.com

S. K. Routray · P. K. Rout
Department of Electrical and Electronics Engineering, Siksha 'O' Anusandhan University, Bhubaneswar, Odisha, India

S. Das and M. N. Mohanty (eds.), *Advances in Intelligent Computing and Communication*, Lecture Notes in Networks and Systems 202,
https://doi.org/10.1007/978-981-16-0695-3_40

filter. For economic, reactive power, and harmonic compensation, SAHF becomes the first choice of many power engineers [5, 6]. However, to improve the SAHF performance, high-accuracy controller and fast harmonic detection algorithm are required for real-time applications. Therefore, it is necessary to focus on a better filter and improved controller design regarding handling the PQ issues.

In SAHF operation, the most difficult and important part of the controller is to design the detection circuit, by which the system eliminates the harmonics and provides reactive power support effectively [7]. The percentage of harmonic mitigation and amount of reactive power support is completely depending on the filter design, inductor and capacitor sizes, and the capacity of energy storage devices [8, 9]. To solve the above problems, many detection algorithms have been discussed in the literature. From the presented techniques, discrete Fourier transform (DFT)-based control algorithm is selected due to easier operation and design [10]. However, the DFT-based controller lags its performance during non-stationary signal and requiring additional PLL. To resolve the above problems, the instantaneous reactive power theory (IRPT)-based controller is widely selected due to the direct sensing of $\alpha\beta$ voltage and current component for computing appropriate active and reactive power requirements [11, 12]. However, the controller performance was badly affected by nonlinear source voltage applications. In [13], the synchronous reference frame (SRF)-based control technique is used through the power and current controller for detecting *dq* voltage and current component. SRF-based control approach is less sensitive to nonlinear source voltage applications. However, the requirement of additional transformation (two) and synchronizing tool decreases the controller performance by increasing the complexity. In [14, 15], a peak detection algorithm (PDA) is suggested, in which the used filter is forced to track the reference signal produced from the designed controller. The proposed technique shows its superiority by eliminating the harmonic current directly and avoiding additional transformation requirements. However, PDA lags its performance due to the synchronizing tool requirement and also creating low-frequency DC-link voltage fluctuation problems. Furthermore, all the above-mentioned methods are generally open-loop systems, and the performance is affected due to parameter and operating condition variations.

To solve the detection problem, adaptive detection methods are proposed in [12, 16, 17]. In [12, 16], to avail better active filter performance, an adaptive method based on noise interference elimination approach is suggested. The above method is suitable only when the nonlinear load contains both harmonic and fundamental components. However, the controller performance is decreased during the presence of inter-harmonics. To deal with the inter-harmonic component, in [17], the adaptive algorithm is modified by using two multi-stage adaptive algorithms. However, the modified adaptive algorithm-based controller approach performance is decreased due to the complex structure and difficulty in implementation. Therefore, to avoid the design complexity and facilitate better operation, there is a necessity to improve the adaptive control strategy for real-time applications.

The major contribution of the paper is discussed below.

- A novel closed-loop detection algorithm based on the ATF theory is proposed for the three-phase SAHF application for regulating the harmonic current and improve the power quality.
- ATF control theory is operated by a self-adaptation technique and able to change the weights according to the system input conditions.
- To achieve optimum performance, the ATF design, and parameter selection, PSO-based evolutionary technique is used.
- The proposed algorithm is suitable for eliminating nonlinearity, inter-harmonics, and noise in the nonlinear load current application.
- It facilitates advanced detection techniques during the change in load and frequency variations.

2 Three-Phase SAHF Modeling and Control

The complete test modeling of the proposed SAHF is illustrated in Fig. 1. In the complete test system, a three-phase six switch-based voltage source inverter is connected to the source parallelly and two nonlinear loads are connected to the source in series. The nonlinear load consists of two diode-based inductive devices with a phase shift of 30 degrees. Initially, the delta-delta transformer is connected and after 10 cycles, the delta–wye transformer is connected to increase the nonlinear load from 6-pulse to 12-pulse.

In this proposed approach, by using the PLL, the SAHF generates a reference sinusoidal source current which is in-phase and the RMS value of source current is the same as the load current. The current error between the nonlinear load current

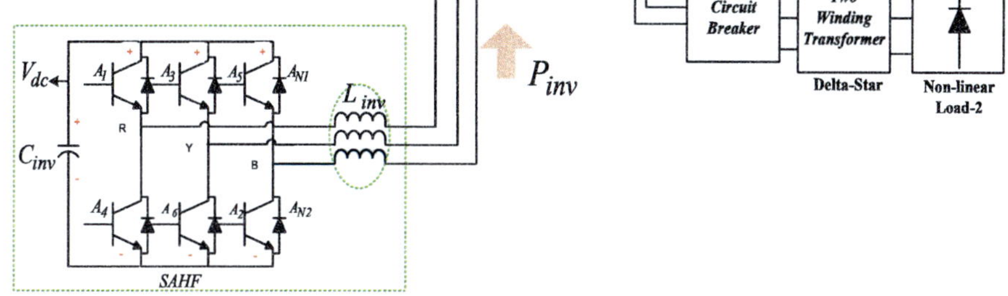

Fig. 1 SAHF-based test system model

and reference current generated by the VSI through the proposed ATF-based control strategy. The SAHF aims to inject this current error at the PCC in order to match the source current as possible with the reference current by reducing the harmonic current.

3 ATF Control Design

In this proposed manuscript, a better reference current estimation method based on closed-loop adaptive linear combiner (ALC) with three-phase inputs named as ATF is proposed. ATF-based control technique can monitor the operating conditions of the system continuously by self-evaluating and correcting from the starting to ending operation [10, 18].

(a) **ATF working principle**: The complete working principle of ATF is illustrated in Fig. 2. To equalize the filter output (O_k) with the desired vector (D_k), the weight vectors (W_k) of the ATF are regulated. The above objective is accomplished by comparing the output signal with the desired signal, to generate the error signal (E_k) and regulating the weight vectors to reduce the error. The regulation of weight vector is based on an adaptive technique named as least mean square (LMS) technique. The related equations of filter output, error, and weights are presented as follows [19].

$$O_k = I_k^T W_k = \sum_{n=0}^{L} W_{nk} I_{k-L} \tag{1}$$

$$E_k = D_k - O_k \tag{2}$$

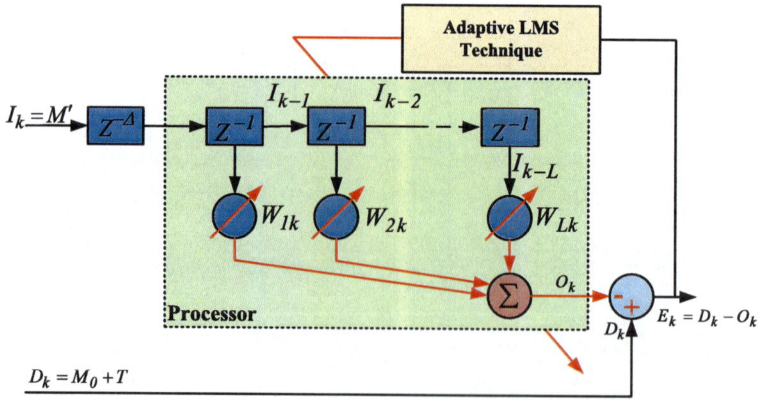

Fig. 2 Complete working principle of ATF

$$W_{k+1} = W_k + 2\mu E_k I_k \quad (3)$$

where I_k is the input vector, O_k is the filter actual output vector, W_{k+1} is the next weighing vector, and μ is the adaptive constant.

(b) **Proposed filter operation**: The detailed control operation regarding ATF operation is discussed in this section. The desired vector ($D_k = M_O + T$) is considered as the obtained nonlinear load current (I_L) comprised of fundamental, nonlinearity, and inter-harmonic components. The actual fundamental noise signals denoted as M_O and the eliminating current represents the desired signal (T). The input vectors are the pure sine wave with equal load current having rated frequency (60 Hz), and different amplitude and phase angle, which is termed as noise (M). The adaptive technique will regulate the filter weights by which the error between the actual and required signal become lesser. After the completion of the adaptation process, the actual output (O_k) is similar to the fundamental current component. Therefore, the error signal is computed by differentiating the distortion current and fundamental current and obtains the nonlinear (I_n) current signal. The amplitude of I_n is regulated based on the DC component I_{dc} generated from the DC voltage controller to generate the fundamental active and reactive current component. The obtained active and reactive current components are converted to the RYB component (I_{RYB}) through *dq/RYB* transformation. The obtained I_{dc}, I_n, and SAHF generated filter current (I_{SAHF}) is used to generate the error current (I_e), and the related equation is presented as follows.

$$I_e = I_{RYB} - I_{SAHF} \quad (4)$$

The complete control diagram of the proposed test system is illustrated in Fig. 3. Here *RYB/dq* and *dq/RYB* transformation is used to convert DC-AC or AC-DC conversion. After generating the appropriate error signal, it is used to generate the pulses for the proposed SAHF through the PWM technique.

(c) **ATF design and PSO optimization**: During the ATF design, there is a relation between the steady-state error and adaptation speed of filter to study the filter performance. The steady-state error of the proposed system is computed through the 'mis-regulation (M)' parameter. In 'M', the mean of the steady-state square error is determined. In the meantime, the adaptation speed of the filter is computed through the filter settling time 'T_s'. Therefore, by using the above two factors, filter performance is significantly improved. A detailed explanation about the parameter computation is presented in [14, 20]. To improve the stability and performance, three important factors are necessary to evaluate such as (1) delay, (2) order of the filter (L), and adaptation parameter. To compute these factors, PSO-based optimization technique is used to compute

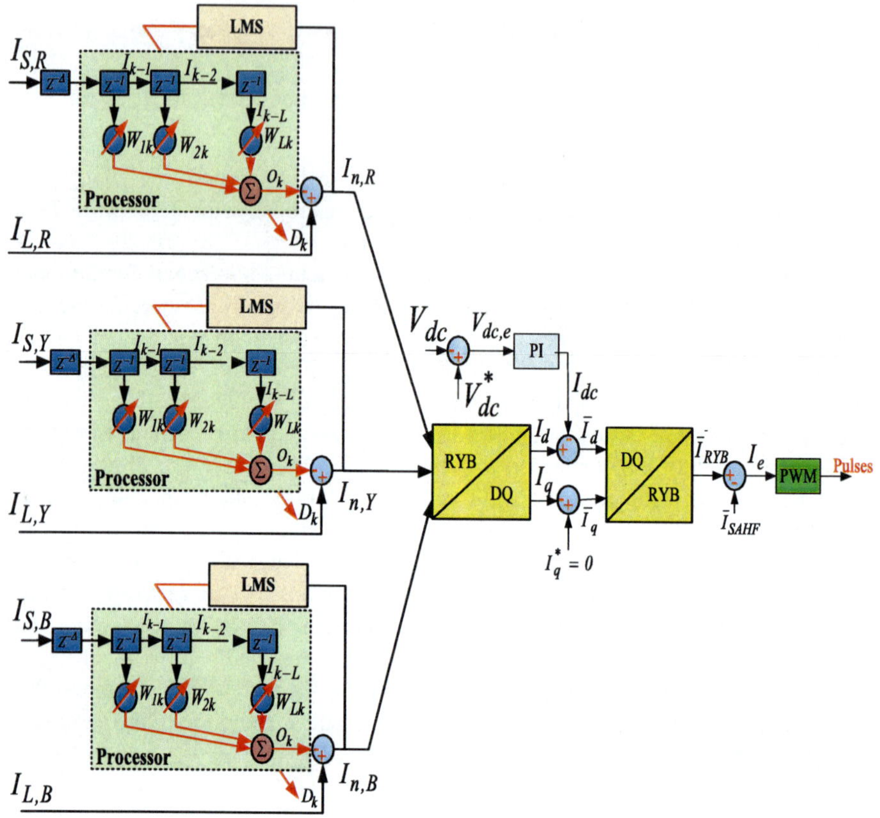

Fig. 3 Overall test system control diagram

the reference current for achieving the better performance of the filter. The detailed PSO working model is illustrated in Fig. 4 [20].

4 Result Analysis

To justify the superiority of ATF-based SAHF performance, different test conditions are performed by using the MATLAB platform. The different test conditions are presented as follows.

- Initially, the superiority of both optimization techniques is evaluated, and the optimized parameters of ATF are computed.
- The performance of the proposed PSO-ATF results is obtained and compared with traditional filter results.

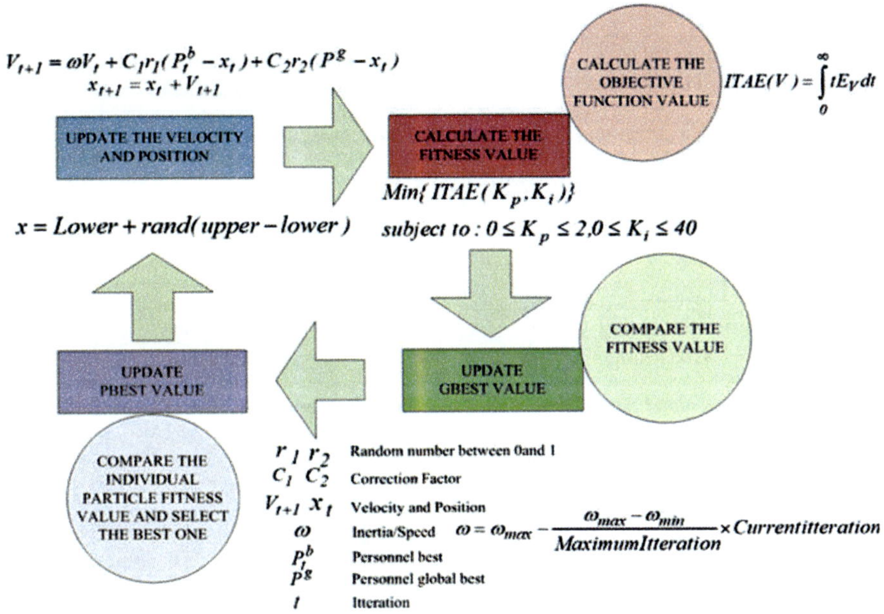

Fig. 4 Complete working principle of PSO

- A harmonic study is done for the above testing conditions, and the results are compared with each other to find a better solution for power quality improvement.

Study-1: Effectiveness of PSO Over DE

By considering the generated load current, the optimized parameter filter values are computed. To show the effectiveness of the chosen PSO algorithm over DE, different objective functions (OF) such as ITAE, IAE, and ITSE are selected. The superiority of the selected optimization technique is evaluated by the best minimum value and time consumption. The convergence speed results of the DE- and PSO-based optimization technique are illustrated in Fig. 5a–b. The detailed parameter value of the system is presented in Table 1, and it is found that ITAE-based OF shows excellent performance in both of the optimization cases. However, by minutely visualizing the table values, it is found that PSO-based ITAE method, the system achieves lesser minimum value at lesser time durations. After computing the best OT method, this is used to find the optimized parameter value and the corresponding result is presented in Table 2.

Study-2: Proposed and Traditional System Result

In Fig. 6a, the traditional filter-based test system result is illustrated. From the illustrated figure, it is visualized that the obtained source current contains more nonlinearity as per the IEEE-1541 standards. To circumvent, the above problem,

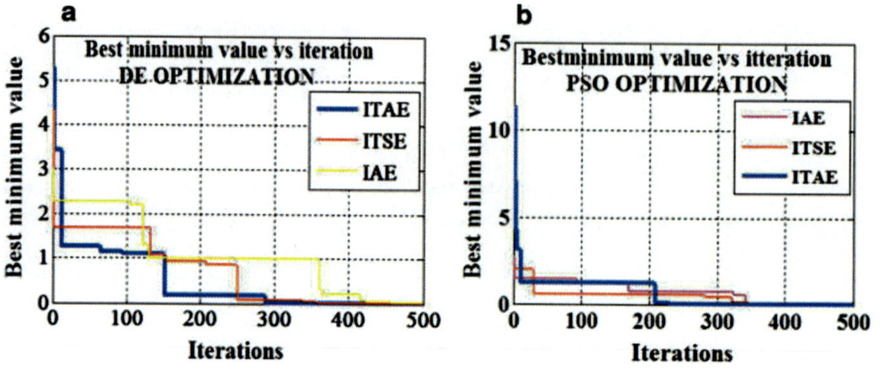

Fig. 5 Convergence speed results by using different optimization technique (OT) **a** DE and **b** PSO

Table 1 Comparative study

OT	OF	Time consumption	Optimum gbest
DE	**ITAE**	**0.13467**	**0.00063456**
	ITSE	0.19531	0.00485674
	IAE	0.15987	0.02546093
PSO	**ITAE**	**0.14321**	**0.00232315**
	ITSE	0.146731	0.00543562
	IAE	0.197434	0.02954289

Table 2 Optimized parameter value

Parameter	Value
Delay	22 samples
Order of the filter	29 samples
Adaptation parameter	0.00061
Sampling frequency	3745 sample sec^{-1}

the proposed PSO-ATF-based SAHF results are illustrated in Fig. 6b–k. Figure 6b–c shows the nonlinear load voltage and current results. As shown in Fig. 6b, at $t = 0$ s, only the delta–delta transformer-connected diode rectifier is connected. Therefore, the current rating of the load is achieved to 2000A. At $t = 5/60$ s, the load current is changed to 2000A to 3000A and at $t = 10/60$ s, the delta–wye transformer-based diode rectifier is connected to the circuit, which increases the load current from 3000A to 6000A. Figure 6d–e shows the source current and voltage results. In Fig. 6d–e, the linearity of the obtained responses is improved significantly. As per the load current demand, the source current correctly flows from the source end with reduced harmonics and improved power quality. It is visualized in Fig. 6f–g that by using the proposed approach the used SAHF produces appropriate current and voltages by which the nonlinearity of the system is eliminated. Figure 6h–i shows that the DC-link voltage and current of the SAHF are correctly regulated at its rated value. The nonlinear-load-1 and 2 DC-link voltages are illustrated in Fig. 6j–k. From the above results, the superiority of the proposed approach is correctly evaluated and suggested to use for real-time application. For better analysis, the harmonic study of the obtained outcomes is evaluated and the comparative study of the proposed and traditional approach is discussed in the next study.

Study-3: Harmonic Study
The test system harmonic analysis results are illustrated in Fig. 7a–h. In Fig. 7a–c, the load current THD results are illustrated and the values are 22.42, 20.56, and 21.87% at $t = 0$ s, 0.085 s, and 0.17 s respectively. From the above THD results, it can be evaluated that the nonlinear load current percentages contained excess harmonics and affect the test system performance. By using the DE-based traditional filter, the source current harmonic current percentage is reduced to 7.13, 10.63, and 5.82% for a similar time interval. The related conventional approach-based source current results are illustrated in Fig. 7d–f. Undoubtedly, the harmonic percentage is reduced to a lower value, but it is unacceptable for real-time applications according to IEEE-1541 standards. By using the PSO-ATF-based SAHF, the source current harmonic percentage is significantly reduced to a lesser value and lies well within the prescribed limit. Figure 7g–k shows the reduced harmonic source current results, and the computed values are 1.66, 1.42, and 1.15% for a similar time interval. The comparative results of the obtained variable are presented in Table 3. The percentage of improvement is also computed and shows the importance of the proposed controller for a real-time application point of view.

5 Conclusion

In this proposed approach, the improved reference current extraction algorithm is successfully developed and implemented for harmonic filtering operation. The proposed ATF-based controller improves the PQ significantly, and it is verified from the above result analysis. In addition to that, the PSO-based optimization technique accurately computes the optimized value and provides a faster settling time to get a stable response. The developed model is simulated through MATLAB software, and the test results justify the importance of the proposed controller requirement by reducing the harmonics during nonlinear load application. In addition to that, the DE-based traditional controller output is compared with the proposed PSO-based ATF results, to show the effectiveness of the controller for PQ improvement. The obtained and analyzed results serve as a basis for designing an appropriate system and controller for real-time applications.

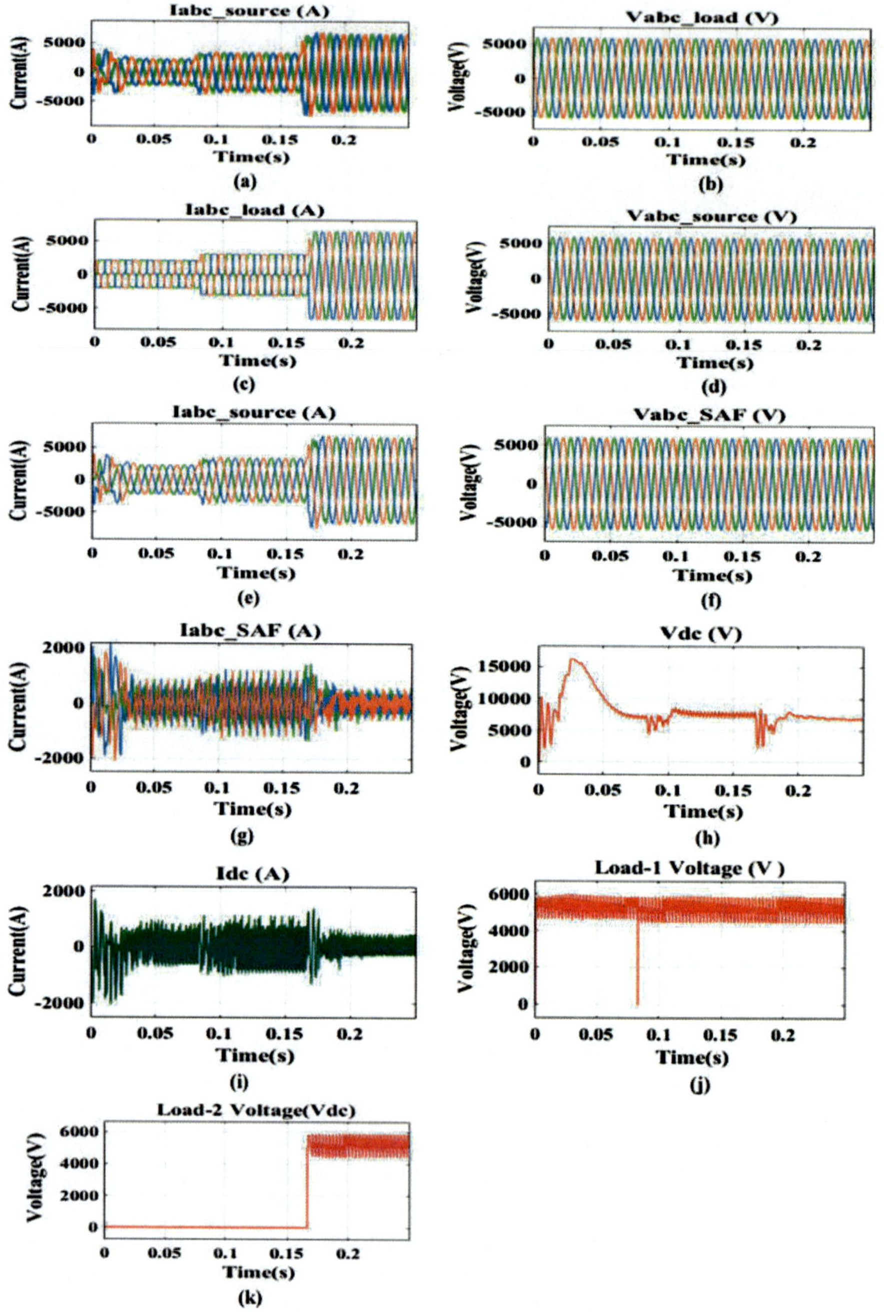

Fig. 6 **a** Conventional source current, **b** load voltage, **c** load current, **d** proposed source current, **e** proposed source voltage, **f** SAHF voltage, **g** SAHF current, **h** DC-link voltage, **i** DC-link current, **j** DC-load voltage-1, **k** DC-load voltage-2

Study-3: Harmonic study

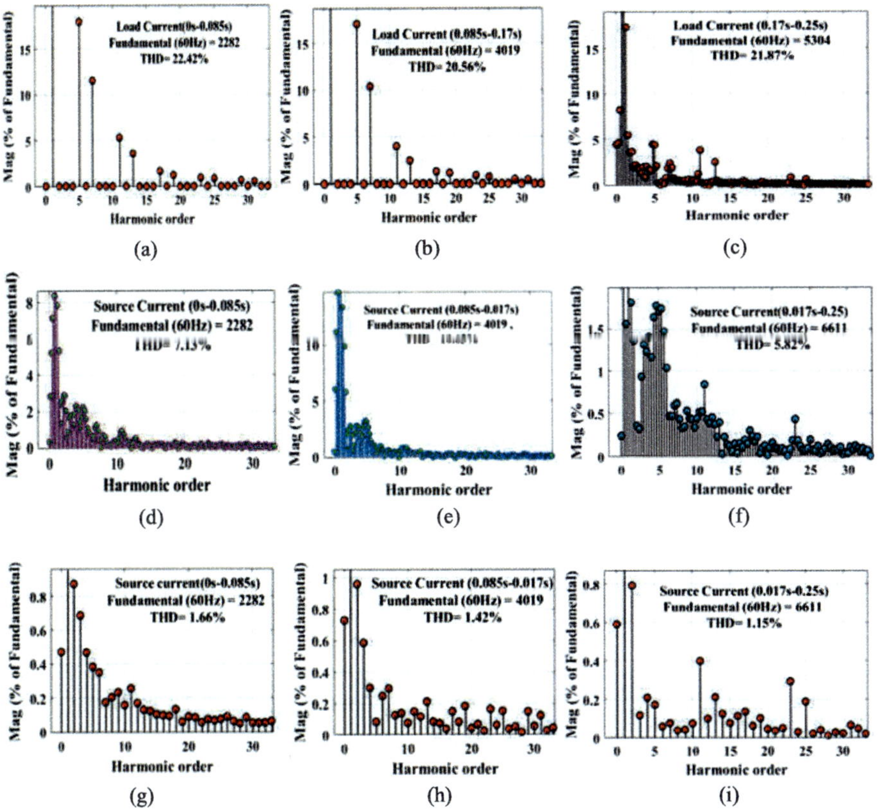

Fig. 7 **a** Load current (0–0.085 s), **b** load current (0.085–0.17 s), **c** load current (0.17–0.25 s), conventional: **d** source current (0–0.085 s), **e** source current (0.085–0.17 s), **f** source current (0.17–0.25 s), proposed: **g** source current (0–0.085 s), **h** source current (0.085–0.17 s), **i** source current (0.17–0.25 s)

Table 3 Comparative study

Parameters	THD (%)		
	0–0.085 s	0.085–0.17 s	0.17–0.25 s
Load current	22.42	20.56	21.87
DE + Traditional APF	7.13	10.63	5.82
Percentage of improvement	**68.20**	**48.30**	**73.39**
PSO-ATF + SAHF	1.66	1.42	1.15
Percentage of improvement	**92.60**	**93.09**	**94.74**

Appendix

Linear voltage (V_S): 4160 × $(2/3)^{1/2}$ V, 60 Hz, Source inductance = $1e^{-5}$, Resistance, inductance (Line): 8 Ω, 7 mH, SHAF integrating inductor: 2 mH, Load resistance and inductance: 26.66 Ω, 10 mH, SHAF dc voltage and capacitor: 6000 V, 2500 μF, Switches:6.

References

1. Sahoo, B., Keshari, S., Rout, P.K.: A new topology with the repetitive controller of a reduced switch seven-level cascaded inverter for a solar PV-battery based microgrid. Eng. Sci. Technol., Int. J. **21**(4), 639–653 (2018)
2. Biricik, S., Redif, S., Özerdem, Ö.C., Khadem, S.K., Basu, M.: Real-time control of shunt active power filter under distorted grid voltage and unbalanced load condition using self-tuning filter. IET Power Electron. **7**(7), 1895–1905 (2014)
3. Sahoo, B., Keshari Routray, S., Rout, PK.: Integration of wind power generation through an enhanced instantaneous power theory. IET Energy Systems Integration (2020)
4. Akagi, H.: New trends in active filters for power conditioning. Ind. Appl. **32**(6), 1312–1322 (1996)
5. Sahoo, B., Keshari Routray, S., Rout, PK.: Application of mathematical morphology for power quality improvement in microgrid. Int. Trans. Electr. Energy Syst. **30**(5), e12329 (2020)
6. Akagi, H.: New trends in active filters for improving power quality. In: Proceedings of International Conference on Power Electronics, Drives and Energy Systems for Industrial Growth, vol. 1, pp. 417–425. IEEE (1996)
7. Sahoo, B., Keshari Routray, S., Rout, PK.: Repetitive control and cascaded multilevel inverter with integrated hybrid active filter capability for wind energy conversion system. Eng. Sci. Technol., Int. J. (2019)
8. Sahoo, B., Keshari Routray, S., Rout, PK.: Artificial Neural Network-Based PI-Controlled Reduced Switch Cascaded Multilevel Inverter Operation in Wind Energy Conversion System with Solid-State Transformer. Iran. J. Sci. Technol., Trans. Electr. Eng., 1–21 (2019)
9. Sahoo, B., Keshari Routray, S., Rout, PK.: A novel sensorless current shaping control approach for SVPWM inverter with voltage disturbance rejection in a dc grid–based wind power generation system. Wind Energy
10. Yan, J.,, et al.: A novel phase performance evaluation method for particle swarm optimization algorithms using velocity-based state estimation. Appl. Soft Comput. **57**, 517–525 (2017)
11. Sahoo, B., Keshari Routray, S., Rout, PK.: A Robust Control Approach for the Integration of DC-grid based Wind Energy Conversion System. IET Energy Systems Integration (2020)
12. Nguyen, B.H.: Evolutionary computation for feature selection in classification (2018)
13. Sahoo, B., Keshari Routray, S., Rout, PK.: A novel control strategy based on hybrid instantaneous theory decoupled approach for PQ improvement in PV systems with energy storage devices and cascaded multi-level inverter. Sādhanā **45**(1), 1–13 (2020)
14. Badal, F.R., Das, P., Sarker, S.K., Das, S.K.: A survey on control issues in renewable energy integration and microgrid. Prot. Control. Mod. Power Syst. **4**(1), 8 (2019)
15. Sahoo, B., Keshari Routray, S., Rout, PK.: A novel centralized energy management approach for power quality improvement. International Transactions on Electrical Energy Systems, e12582 (2020)

16. Mishra, M.: Power quality disturbance detection and classification using signal processing and soft computing techniques: A comprehensive review. Int. Trans. Electr. Energy Syst. **29** (8), e12008 (2019)
17. Sahoo, B., Routray, S.K., Rout, P.K.: Robust control approach for stability and power quality improvement in electric car. Int. Trans. Electr. Energy Syst. (2020). https://doi.org/10.1002/2050-7038.12628
18. Sahoo, B., Keshari Routray, S., Rout, PK.: Robust Control and Inverter Approach for Power Quality Improvement. Green Technology for Smart City and Society, (2020)
19. Tan, K.T., et al.: Control and operation of a dc grid-based wind power generation system in a microgrid. IEEE Trans. Energy Convers. **31**(2), 496–505 (2015)
20. Kato, T., Tamura, K., Matsuyama, T.: Adaptive storage battery management based on the energy on demand protocol. 2012 IEEE Third International Conference on Smart Grid Communications (SmartGridComm). IEEE (2012)

Simultaneous Optimal Placement and Sizing of D-STATCOMs Using a Modified Sine Cosine Algorithm

Subrat Kumar Dash and Sivkumar Mishra

Abstract The optimal allocation of D-STATCOM using the modified sine cosine algorithm is presented in this paper. Oppositional-based learning is implemented to modify the performance of the sine cosine algorithm. In this work, both capacities, as well as placements of the D-STATCOMs, are carried out simultaneously to minimize the real power loss and to improve the voltage profile. To validate the effectiveness of the proposed approach, two standard test systems, namely 33-bus and 69-bus radial distribution networks, are considered. The results obtained are also compared with recent approaches and are found to be encouraging.

Keywords Electrical distribution networks · DSTATCOM · Optimal allocation of DSTATCOM · Modified sine cosine algorithm

1 Introduction

A typical electrical distribution system (EDS) often features radial topology, increased losses, poor voltage profile, and reduced voltage security margin. Lack of reactive power support at the distribution level following a rise in reactive power demand is the major cause behind such performance of the EDS. To address the above issues, different strategies have been proposed by practicing engineers and researchers which include inserting voltage regulators, connecting compensating devices such as shunt capacitors, distributed generators, and applying network reconfiguration, etc. Nowadays, the insertion of dynamic voltage regulators (DVRs), distribution static compensators (D-STATCOMs), and unified power quality conditioners (UPQCs) which are FACTS devices for distribution systems

S. K. Dash
Government College of Engineering Kalahandi, Bhawanipatna, Odisha, India
e-mail: dashsubratkumar@gmail.com

S. Mishra (✉)
Centre for Advanced Post Graduate Studies, BPUT, Rourkela, Odisha, India
e-mail: sivmishra@gmail.com

S. Das and M. N. Mohanty (eds.), *Advances in Intelligent Computing and Communication*, Lecture Notes in Networks and Systems 202,
https://doi.org/10.1007/978-981-16-0695-3_41

and considered as viable alternatives to improve the performance of the EDS. Being a shunt connected device, D-STATCOM is relatively easier to connect and also has a local control feature which makes it a strong contender among other D-FACTS devices to be considered for integration into the existing EDS. D-STATCOM is expected to play a crucial role not only for enhancing the voltage profile, improving voltage security margin, and curbing losses of EDS but also alleviating the effect of voltage sag, nonlinear loads, and voltage fluctuations by providing continuous reactive (both inductive and capacitive) power support [1]. Therefore, it is essential to find the optimal size as well as an optimal location for D-STATCOM placement to harness maximum benefit. In the last decade, a lot of progress has been reported for determining the optimal allocation of D-STATCOMs (OAD). Most of these techniques rely on the sensitivity approach to pre-locate the placement for D-STATCOMs, followed by determining the optimal D-STATCOM capacity using different metaheuristic techniques. Popular sensitivity approaches suggested by researchers are loss sensitivity factor (LSF) [1], index vector method (IVM) [2], voltage stability index (VSI) [3], new voltage stability index (NVSI) [4], and power stability index (PSI) [5]. However, sensitivity-based approaches are not reliable and foolproof [6]. Therefore, several metaheuristic approaches are proposed [7–11] for obtaining OAD.

In this paper, a modified sine cosine algorithm (MoSCA) is proposed for obtaining the OAD. The remaining of the paper is organized as follows. In Sect. 2, the modeling of D-STATCOM is presented. The problem formulation is stated in Sect. 3, which encompasses the objective function formulation and related constraints. Section 4 describes the sine cosine algorithm and the necessary modifications incorporated for OAD. The detailed results and their analysis are presented in Sect. 5. Finally,the conclusions of the paper are presented in Sect. 6.

2 Distribution Static Compensator (D-STATCOM)

D-STATCOM is a shunt connected device that consists of a voltage source converter (VSC), DC capacitor bus, coupling transformer, and tuned filter and may or may not contain energy storing devices. VSC allows the D-STATCOM to act as a generalized impedance converter by importing or exporting reactive power at the point of common coupling (PCC). A D-STATCOM with an energy storage device can also inject active power at the PCC but in the present study, D-STATCOM is used to regulate reactive power only.

A two-bus equivalent of the distribution network is shown in Fig. 1. Applying KVL to the network in Fig. 1 yields

$$V_{t+1} < \theta_{t+1} = V_t < \theta_t - (R_m + jX_m) I_m < \delta_m \tag{1}$$

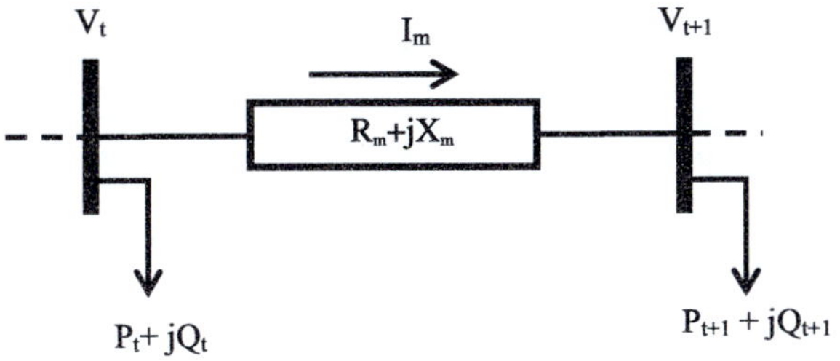

Fig. 1 Two node equivalent of a distribution system

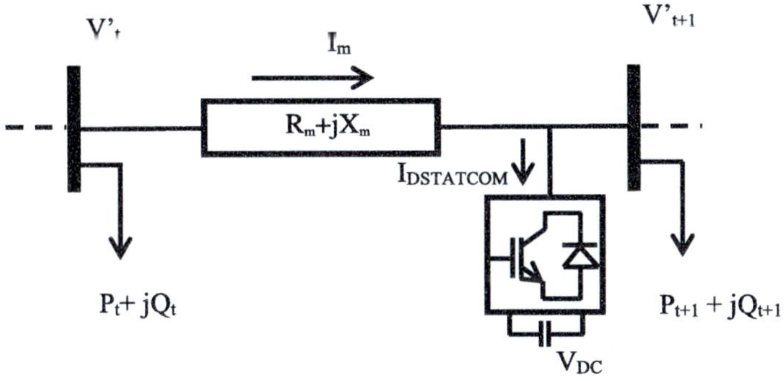

Fig. 2 Distribution system with D-STATCOM placed at $(t + 1)$th node

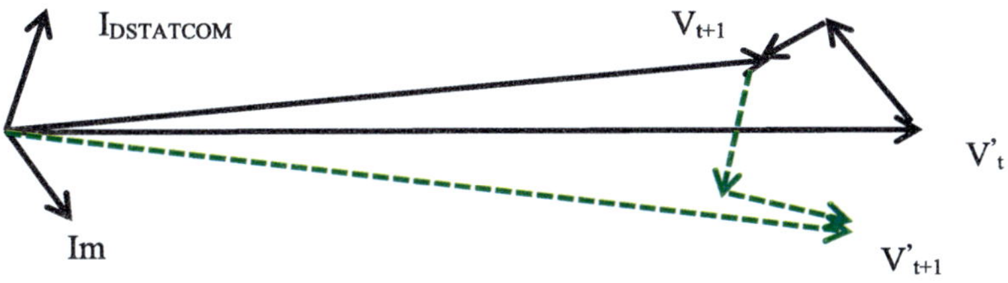

Fig. 3 Phasor diagram of Eq. (2)

The compensated network, i.e., the network when D-STATCOM is connected to the bus $t + 1$, is shown in Fig. 2. Figure 3 shows the phasor diagram of the compensated network. From the phasor diagram, one can write:

$$V_{t+1} < \theta_{t+1} = V_t < \theta_t - (R_m + jX_m)(I_m < \delta_m + I_{Dstatcom} + \psi) \tag{2}$$

Since D-STATCOM is modeled as a reactive power source, the phase angle of the D-STATCOM current will be in quadrature to the bus voltage of the compensated node. Hence,

$$\psi = \frac{\pi}{2} + \theta'_{t+1} \tag{3}$$

Therefore, power injected by the DSTATCOM can be expressed as:

$$-jQ_{DSTATCOM} = V'_{t+1} < \theta'_{t+1} I^{*}_{DSTATCOM} < \left(\frac{\pi}{2} + \theta'_{t+1}\right) \tag{4}$$

3 Problem Formulation

Here, the modified sine cosine algorithm is used to determine the optimal locations as well optimal capacities of the multiple D-STATCOM for active power loss minimization of the EDS while satisfying all the operational constraints of the system as formulated below.

3.1 Objective Function (OF)

Distribution utilities are more often concerned about curbing the real power loss other than attaining to voltage profile or stability of the EDS as it directly links with economic benefits. Therefore, in the present work, OAD is determined for minimizing the real power loss of the EDS. The real power loss can be expressed as:

$$ploss = \sum_{m=1}^{Nb} |I_m|^2 * R_m \tag{5}$$

Hence, the objective function can be mathematically written as:

$$J = \min(ploss) \tag{6}$$

3.2 Constraints

The following operational constraints are formulated while optimizing Eq. (6).

Bus Voltage Constraint
The bus voltage of the network is allowed to vary within ±5% of the nominal voltage.

$$V_{\min} \leq V_i \leq V_{\max} \tag{7}$$

Branch Flow Constraint

The inclusion of D-STATCOMs can modify the branch flows of the network. So, to restrict the branch flow when D-STATCOMs are inserted, within the prescribed limits, the following constraint is defined.

$$|I_m| \leq |I_m^{Max}| \tag{8}$$

D-STATCOM Position Constraint

The nodes for D-STATCOM placement are generated without any repetition

$$2 \leq DSTAT_{loc} \leq N_n \tag{9}$$

D-STATCOM Capacity Constraints

The sizes of D-STATCOMs must be within its capacity limit as defined in Eq. 10

$$\sum_{i=1}^{N_{DSTATCOM}} Q_{DSTATCOM,i} \leq \sum_{t=1}^{N_n} Q_{LOAD,t} \tag{10}$$

4 Optimal Allocation of DSTATCOM

The OAD problem is solved using a MoSCA in this paper.

4.1 Sine Cosine Algorithm (CSA)

The sine cosine algorithm (SCA) is a popular metaheuristic optimization technique first proposed by Mirjalili in 2016 [12]. Different applications, shortcomings, and modifications are suggested to improve the performance of the SCA in [13]. Similar to other metaheuristic approaches, it also begins with randomly generating an initial population that spread uniformly across the search space. Two simplified equations using the basic sine and cosine trigonometric functions are framed which guides the initial population toward the target solution. Mathematically, these equations are expressed as:

$$U_{i,k+1} = U_{i,k} + \Omega \sin(\lambda) |\alpha U_{best} - U_{i,k}| \tag{11}$$

$$U_{i,k+1} = U_{i,k} + \Omega \cos(\lambda) |\alpha U_{best} - U_{i,k}| \tag{12}$$

A random number *rand* is generated between (0, 1) which selects any one of the updating equations between (11) or (12) as follows.

$$U_{i,k+1} = \begin{Bmatrix} U_{i,k} + \Omega \sin(\lambda)\left|\alpha U_{best} - U_{i,k}\right|, & rand < 0.5 \\ U_{i,k} + \Omega \cos(\lambda)\left|\alpha U_{best} - U_{i,k}\right|, & rand \geq 0.5 \end{Bmatrix} \quad (13)$$

The factor Ω plays a crucial role in the exploration and exploitation of the search space. Initially, Ω assumes a higher value which allows searching the space outside of the region between the present solution and the target solution, whereas a relatively lower value of Ω facilitates to exploit of the region between the present solution and the target solution. To strike the right balance between exploration and exploitation, Ω is varied linearly as per the following equation.

$$\Omega = 2 - 2\left(\frac{k}{T}\right) \quad (14)$$

The term λ decides the movement of the current solution either toward the best solution or away from the best solution. If the value assigned to α is more than one $(\alpha > 1)$, then exploration becomes dominant or else if α is less than one $(\alpha < 1)$ exploitation sets in. The factor α also helps in getting rid of premature convergence.

4.2 Modified Sine Cosine Algorithm (MoSCA)

Though SCA has been successfully implemented for several real optimization problems, it is learned to exhibit a slow convergence rate, getting trapped in local optima and poor exploration of the search space [14, 15]. Therefore, authors here proposed to inculcate oppositional learning mechanism one at the initial population generation phase and other during the updation phase into the conventional SCA. The possibility of exploiting the search space closer to the optimal solution increases if one uses the opposite population rather than a mere random population [16]. Hence, this modified SCA (MoSCA) implements opposite numbers at the time of initial population generation as well as during the intermediate stage to generate a new population which helps in avoiding local optima and speeding up the convergence characteristic and maintaining a wide diversity.

- *Opposition-based population generation*

In the absence of any prior knowledge about the probable solution, it is better to start with a population and its opposite population which will have a better probability of initializing a more deserving solution, and then, from the pool, best *Np* candidates can be selected. In this regard, let us first understand the concept of the opposite point. An opposite point $U_{op} = \left(U^1_{op}, U^2_{op}, U^3_{op}, \ldots U^n_{op}\right)$ in n-dimensional space of a point $U = (U^1, U^2, U^3, \ldots U^n) \in R^n$ and $U^j \in [a^j, b^j]$ can be expressed as:

$$U^j_{op} = a^j + b^j - u^j \tag{15}$$

- *Opposition-based generation jumping*

Based on a jumping rate (J_r), the current population can be forced to jump to a completely new set of individuals using the opposition learning mechanism. Then, the best *Np* candidates from the pool of the opposite population and the current population are selected for the next stage. But unlike the opposition-based population generation, here, the dynamic range of the candidates is used for the generation of the opposite points as expressed in Eq. (15).

$$U^j_{op,i} = \max\left(U^j_i\right) + \min\left(U^j_i\right) - U^j_i \tag{16}$$

4.3 Implementation of the Proposed Work

The detailed algorithm for determining the optimal location and size of the D-STATCOM for active power loss minimization using the modified sine cosine algorithm is outlined as below.

Algorithm

1. The system line ane load data are fed.
2. Initialize the parameters of the MoSCA (N_p= 30, T = 100, J_r= 0.6)
3. Generate the initial population.

$$U_i = \left[loc_{i,1} \cdots loc_{i,\text{Ndstat}}, Q^{\text{Dstat}}_{1,i} \cdots Q^{\text{Dstat}}_{i,\text{Ndstat}}\right] \quad i = 1, 2, 3, \ldots N_p \tag{17}$$

4. Calculate the opposite population OU_i using Eq. (15)
5. Compute the fitness of each individual in the population U and OU using Eq. (6)
6. Evaluate the best fittest individual from the set $\{U \cup OU\}$
7. Select the best individual, U_{best}
8. Initialize the iteration counter k = 0.
9. ***while*** $k < T$

10. ***If*** rand < J_r
11. Generate opposite population OU_k using Eq. (15)
12. Evaluate the fitness of OU_k using Eq. (6)
13. Select best N_p from the set $\{U_k \cup OU_k\}$
14. Update the best individual, U_{best}
15. ***else***
16. ***for*** i = 1: N_p
17. Update each individual, $U_{i,\ k+1}$ using Eq. (13)
18. Compute the fitness of each updated individual using Eq. (6)
19. Update the best individual, U_{best}
20. ***end for***
21. ***end if***
22. Update Ω using Eq. (14)
23. Increment iteration counter, $k = k + 1$
24. ***end while***
25. Output the best individual, U_{best}

5 Results

The proposed method for OAD is validated on two standard distribution systems, i.e., a 33-bus and 69-bus. The effectiveness of the suggested approach is verified by considering three different scenarios. In the first case, only a single D-STATCOM is allocated. Two and three D-SATCOMs are considered in case-2 and case-3, respectively. For each case, the best result for ten independent runs of the algorithm is reported. The proposed algorithm is implemented on Intel(R) Core (TM) i3-6006U CPU @ 2.00 GHz, 4 GB RAM in the MATLAB environment.

5.1 33-Bus Test System

The line and load data for the 33-bus test system is taken from [17]. This system has a total connected load of 3715 + j 2300 kVA. The performance of the system without D-STATCOM and with D-STATCOM(s) ais reported in Table 1. It can be noticed from Table 1 that without D-STATCOM allocation, the performance of the system is very poor as it has a real power loss of 202 kW and voltage magnitude is as less as 0.9038 p.u. The real power loss of the system in the presence of D-STATCOM(s) has improved significantly. The real power loss of the system is 143.5, 137.8, and 135.2 kW for one D-STATCOM, two DS-TATCOMs, and three DS-TATCOMs, respectively. The results for OAD for different methods are also compared in Table 1 which suggests that the proposed method is better than other approaches. Figure 4 shows the convergence characteristic curve of MoSCA for

Table 1 Effect of different no. of D-STATCOM allocation on 33-bus system

Methods	$N_{dstatcom}$	Location	Size, MVAr	Ploss, kW	Vmin (P.U.)	VSI
Base case	0	–	–	202.0	0.9038	0.6951
Proposed	1	30	1.3060	143.5	0.9255	0.7338
GA [8]	1	12	1114.2	173.9	–	–
IA [8]	1	12	0.9624	171.8	–	–
DE [10]	1	30	1.2527	143.5	0.9256	–
Proposed	2	30	0.6645	137.8	0.9393	0.7783
		10	1.0561			
Proposed	3	30	0.7713	135.2	0.9420	0.7874
		4	0.9933			
		11	0.4251			

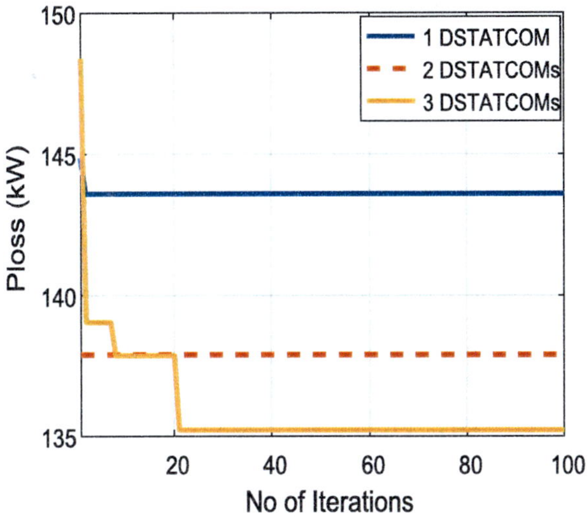

Fig. 4 Convergence graph of MoSCA for different no of D-STATCOM allocation in 33-bus system

OAD in the 33-bus system. It can be concluded that three D-STATCOM placements provide better system performance than one or two D-STATCOM allocations. The voltage profile and branch power loss for the base case and different no. of D-STATCOM allocations are shown in Figs. 5 and 6, respectively. These figures also confirm that three D-STATCOMs must be placed for obtaining the best performance in a 33-bus system.

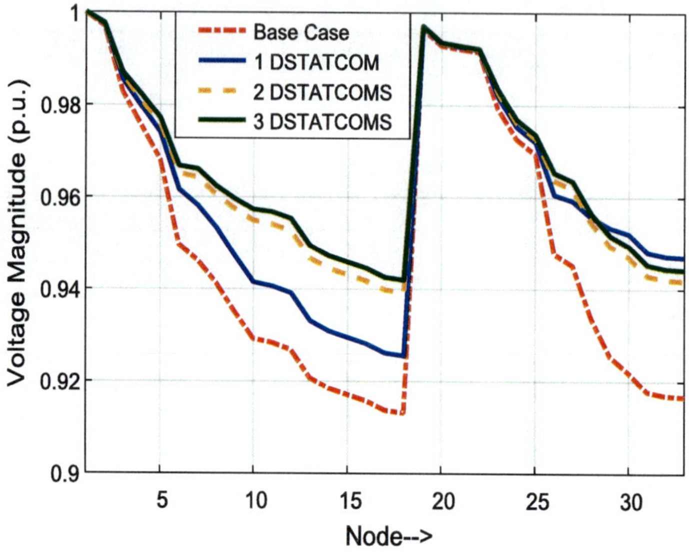

Fig. 5 Voltage profile of 33-bus system with and without D-STATCOM(s)

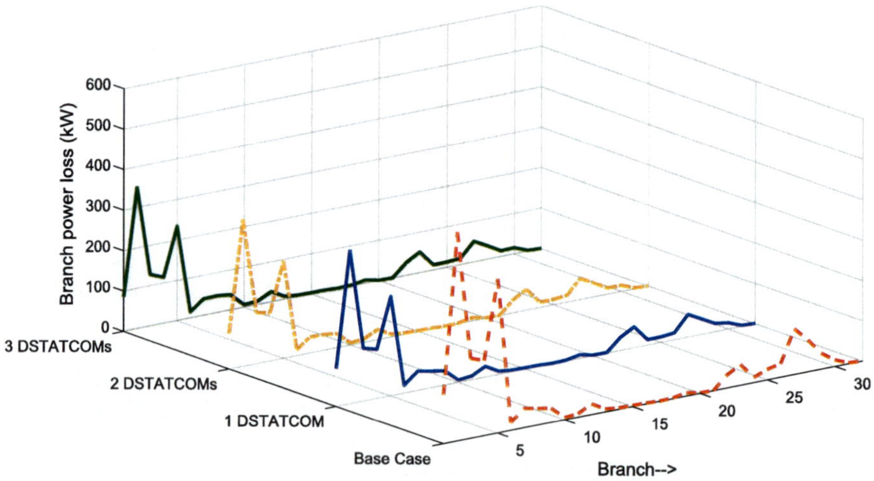

Fig. 6 Branch power loss of 33-bus system with and without D-STATCOM(s)

5.2 *69-Bus Test System*

The line and load data for 69-bus test system is taken from [17]. This system has a total connected load of 3791.9 + j 2694.1 kVA. The performance of the system without D-STATCOM and with D-STATCOM(s) is reported in Table 2. It can be

Table 2 Effect of different no. of D-STATCOM allocation on 69-bus system

Methods	$N_{dstatcom}$	Location	Size, MVAr	Ploss, kW	Vmin (P.U.)	VSI
Basecase	0	–	–	224.9	0.9092	0.6833
Proposed	1	61	1.4459	152.0	0.9310	0.7514
PSO [2]	1	61	0.9011	167.9	–	–
GA [8]	1	61	1.9183	165.4	0.9392	–
IA [8]	1	61	1.7044	157.5	–	–
DE [10]	1	61	1.3121	152.0	0.9338	–
GSA [11]	1	61	0.966	152.0	–	–
Proposed	2	5	1.6202	145.7	0.9360	0.7677
		61	1.8462			
Proposed	3	4	1.4480	149.1	0.9392	0.7780
		63	0.9004			
		55	1.5896			

noticed from Table 2 that without D-STATCOM allocation, the performance of the system is very poor as it has a real power loss of 224.9 kW and voltage magnitude is as less as 0.9092 p.u. The real power loss of the system in the presence of D-STATCOM(s) has improved significantly. The real power loss of the system is 152.0 kW, 145.7 kW, and 149.1 kW for one D-STATCOM, two D-STATCOMs, and three D-STATCOMs, respectively. From Table 2, it can also be noticed that the minimum voltage magnitude and VSI are marginally better for three D-STATCOM placement than two D-STATCOM placement. However, in terms of real power loss minimization, allocation of two D-STACOMs may be preferred over three D-STATCOM. Therefore, for 69-bus test system, the optimal no of D-STATCOMs is two. The results for OAD for different methods are also compared in Table 2 which suggests that the proposed method is better than other approaches. Figure 7

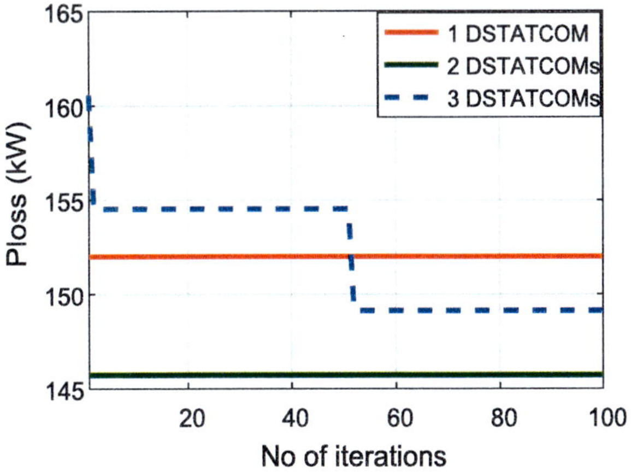

Fig. 7 Convergence graph of MoSCA for different no. of D-STATCOM allocation in 69-bus system

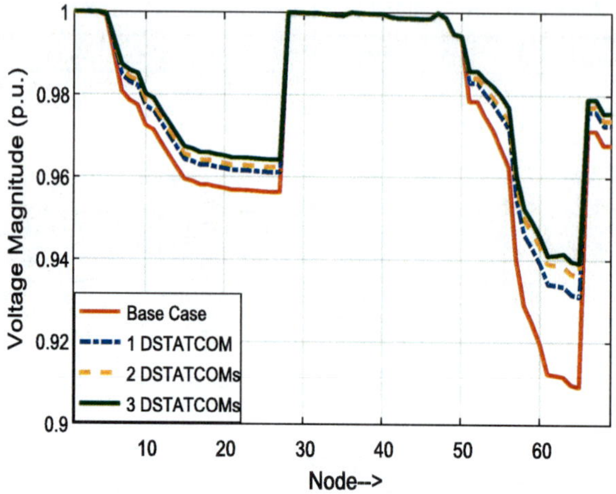

Fig. 8 Voltage profile of 69-bus system with and without D-STATCOM(s)

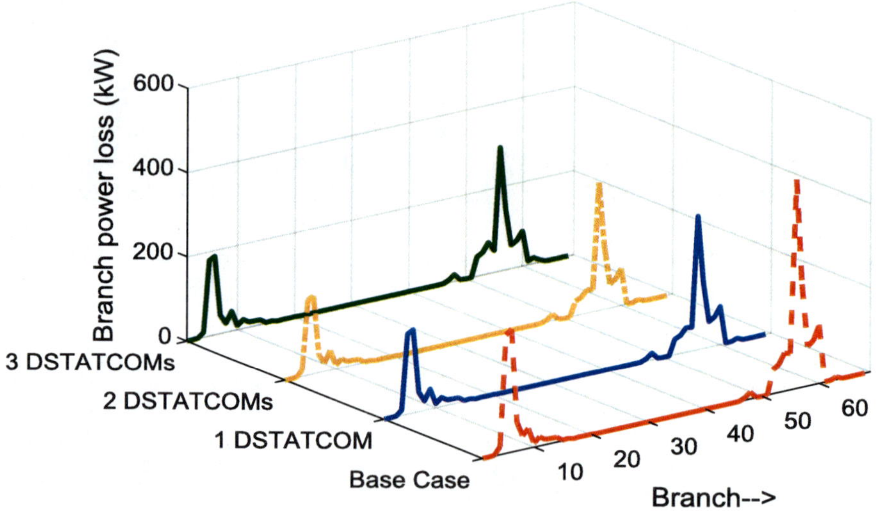

Fig. 9 Branch power loss of 69-bus system with and without D-STATCOM(s)

shows the convergence characteristic curve of MoSCA for OAD in 69-bus system. The voltage profile and branch power loss for the base case and different no. of D-STATCOM allocations are shown in Figs. 8 and 9, respectively. Thes figures also confirm that two no. of D-STATCOMs must be placed for obtaining the best performance of 69-bus system.

6 Conclusion

In this paper, MSCA has been applied to determine the optimal allocations (placement and sizing both) of D-STATCOM for minimizing the real power loss of the EDS while satisfying the system operational constraint. The proposed method was validated on two standard test systems, namely 33-bus and 69-bus systems. OAD was carried out for three different scenarios, i.e., single D-STATCOM, two D-STATCOMs. and three D-STATCOMs allocation. The results revealed that the proposed method is effective in solving OAD. It was also found that for 33-bus system three DS-TATCOMs of optimal size must be placed at the strategic locations to improve the performance of the EDS, whereas for 69-bus system, allocation of two D-STACOMs performs better as compared to three D-STATCOMs in minimizing the real power loss.

References

1. Devi, S., Geethanjali, M.: Electrical Power and Energy Systems Optimal location and sizing determination of Distributed Generation and DSTATCOM using Particle Swarm Optimization algorithm. Int. J. Electrical Power and Energy System **62**, 562–570 (2014)
2. Gupta, A.R., Kumar, A.: Energy saving using D-STATCOM placement in radial distribution system under reconfigured network. Energy Procedia. **90**, 124–136 (2015)
3. Yuvaraj, T., Ravi, K., Devabalaji, K.R.: DSTATCOM allocation in distribution networks considering load variations using bat algorithm. Ain Shams Engineering J. **8**(3), 391–403 (2017)
4. Gupta, A.R., Kumar, A.: Impact of various load models on D-STATCOM allocation in DNO operated distribution network. In 6th International Conference on Smart Computing and Communications, ICSCC, pp. 862–870 (2018)
5. Amin, A., Yu, J.: Developed Analytical Technique for Optimal Placement and Sizing of DG and DSTATCOM in Radial Distribution Systems. Twentieth. Int. Middle East Power System. Conference, 1087–1091(2018)
6. Kanwar, N., Gupta, N., Niazi, K.R., Swarnkar, A.: Improved Cat Swarm Optimization for Simultaneous Allocation of DSTATCOM and DGs in Distribution Systems. J. of Renew. Energy (2015)
7. Taher, S.A., Afsari, S.A.: Electrical Power and Energy Systems Optimal location and sizing of DSTATCOM in distribution systems by immune algorithm. Int. J. Electrical. Power and Energy Sys. **60**, 34–44 (2014)
8. Yuvaraj, T., Devabalaji, K.R., Ravi, K.: Optimal placement and sizing of DSTATCOM using Harmony Search algorithm. Energy Procidia. **79**, 759–765 (2015)
9. Sanam, J.: Optimization of planning cost of radial distribution networks at different loads with the optimal placement of distribution STATCOM using differential evolution algorithm. Soft. Comput. **24**, 13269–13284 (2020)
10. Arya, A.K., Loss, D.A.G.: Analysis of Distribution System with D-STATCOM by Gravitational Search Algorithm (GSA). J. Ins. of Eng. (India) Series B. **100**(1) (2019)
11. Srijani, R., Jordehi, A.R.: Optimal placement and sizing of distribution static compensator (D-STATCOM) in electric distribution network: A Review. Renew. Sust. Energy Review. **77**, 688–694 (2017)

12. Mirjalili, S.: SCA: a sine cosine algorithm for solving optimization problems Knowledge-Based System. **96**, 120-133 (2016)
13. Mirjalili, S.M., Mirjalili, S.Z., Saremi, S., Mirjalili, S.: Sine Cosine Algorithm: Theory, Literature Review, and Application in Designing Bend Photonic Crystal Waveguides. Nature-Inspired Optimizers. 201–217(2019)
14. Gupta, S., Deep, K.: A hybrid self-adaptive sine cosine algorithm with opposition based learning. Expert System with Application. **119**, 210–230 (2019)
15. Raut, U., Mishra, S.: An Improved Sine Cosine algorithm for simultaneous network reconfiguration and DG allocation in power distribution systems. Appl. Soft Comput. **92**, 1–25 (2020)
16. Rahnamayan, S., Tizhoosh, H.R., Salama, M.M.A.: Opposition versus randomness in soft computing techniques. App. Soft Computing J **8**(2), 906–918 (2008)
17. Mishra, S., Das, D., Paul, S.: A comprehensive review on power distribution network reconfiguration. Energy Systems. **8**(2), 227–284 (2017)

Big Data Analytics for Smart Grids, the Cyberphysical System in Energy—A Bibliographic Review

Sivkumar Mishra and Namita Dehury

Abstract This paper presents a detailed systematic bibliographic review of big data analytics in smart grids, which are fast emerging as the cyberphysical systems in energy. This review includes papers published in several reputed journals only and presents a concise bibliographical survey concerned to various aspects of big data analytics for smart grids using graphical illustrations.

Keywords Big data analytics · Smart grid · Smart meters · Phasor measurement units

1 Introduction

The adoption of modern intelligent control and monitoring technologies and growing trend of consumers becoming prosumers has led to the transformation of conventional power system to smart grid and finally emerging as a cyber physical system (CPS) in energy [1–10] as shown in Fig. 1. This will lead to gathering of huge amount heterogenous data at varying rates leaving a great potential for big data analytics (BDA), which is one of the prominent enabling technologies to make the electric power affordable, sustainable, reliable and qualitative [10–30].

Extensive research concerned to BDA for SG is carried out, and many useful review papers are also reported in literature. A chronological review of these review papers with their specific contributions is listed in Table 1, which point out that the research concerning BDA in SG has picked up pace in the last decade. In the context of all these developments, the objective of this paper is to present an extensive bibliographic review by revisiting various concerned aspects.

S. Mishra (✉) · N. Dehury
Centre for Advanced Post Graduate Studies, BPUT, Rourkela, Odisha, India
e-mail: sivmishra@gmail.com

N. Dehury
e-mail: namitadehuri94@gmail.com

S. Das and M. N. Mohanty (eds.), *Advances in Intelligent Computing and Communication*, Lecture Notes in Networks and Systems 202,
https://doi.org/10.1007/978-981-16-0695-3_42

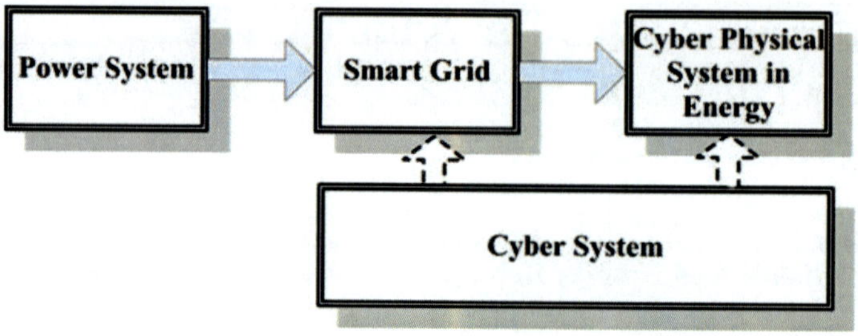

Fig. 1 Emergence of smart grids to cyberphysical systems in energy

Keeping in view, the developments as explained and analyzed, the rest of the paper is organized as follows. In Sect. 2, the characteristics of energy big data and its sources in various components of smart grid are discussed. The various big data analytic methods suitable for smart gird data are presented in Sect. 3. The various common applications of the BDA in SG are mentioned in Sect. 4. In Sect. 5, the future research directions are stipulated. The review concludes in Sect. 6 with some observations.

2 Source of Big Data in Smart Grid

The sources of big data in SG can be broadly classified as structured, semi-structured, and unstructured with the 5 V's and 3E's characteristics as depicted in Fig. 2 [13, 16, 17, 22–52].

3 Big Data Methods in Smart Grid

The big data available from a variety of sources stated in the last section undergo various methods so as to result some useful intelligence, and these methods with compatible frameworks and broad classification with the specific methods in each category are presented in Fig. 3 [13–20, 27, 28, 31, 45, 53–55].

4 Energy Big Data Applications

Typical BDA applications in SG are found in energy management with phasor measurement units (PMUs), management of active power distribution networks and its automation including micro-PMUs, distributed generation and associated storage management, advance metering infrastructure (AMI) including smart meters with

Table 1 A chronological review of various review papers published on BDA for SG

Author(s), Year	Contributions
Simmhan et al. 2013 [1]	Reviews the scalable software platforms for SGs with BDA
Diamantoulakis et al. 2015 [31]	Highlights various issues and challenges with future research directions
Schuelke-Leech et al., 2015 [12]	Reviews the issues and opportunities of the use of big data for electric utilities
Hu and Vasilakos 2016 [13]	Reviews various energy big data characteristics and methods to analyze
Wu et al., 2016 [14]	Reviews applications of BDA for green technologies in SG
Zhu et al., 2016 [15]	Reviews various standard based software frameworks to address various BDA-SG issues
Zhou et al., 2016 [16]	Reviews various energy big data characteristics with a focus to the present ongoing research directions
Jiang et al., 2016 [17]	Reviews various energy big data characteristics, applications and future directions with a focus for security concerns
Daki et al., 2017 [18]	Reviews various BDA methods for successful applications in SG
Ma et al., 2017 [19]	Review is focused on various BDA approaches for SG applications
Munshi and Muhammad,2017 [20]	Presents the importance BDA software frameworks for various SG applications
Ji et al., 2017 [21]	Reviews BDA for resilience of SGs
Tu et al., 2017 [22]	Review of big data applications in smart grid
Chen et al., 2018 [23]	Reviews BDA issues and challenges concerned to SG
Zhang et al., 2018 [25]	Reviews sources of SG big data its characteristics and applications
Ghofrani et al., 2018 [26]	A overall survey of BDA in SG applications
Bhattarai et al. 2019 [27]	A detailed overall review of all aspects of BDA in SG applications identifying the research gaps including methods, applications and future research directions
Ghorbanian et al. 2019 [28]	A thorough review on applications, standards, issues and challenges concerning BDA in SG
Hossain et al., 2019 [29]	This paper focuses on security issues relating BDA-based SG operations and the possible methods to with it
Kezunovic et al., 2020 [30]	Comparatively recent paper with a comprehensive review on energy big data characteristics, trends in BDA, future trends, etc.
Sundararajan et al., 2020 [31]	Presents a review focusing on BD standards and capacity maturity models meant specifically for SGs
Kumari and Tanwar, 2020 [33]	Thoroughly reviews the secure data analytics for energy big data

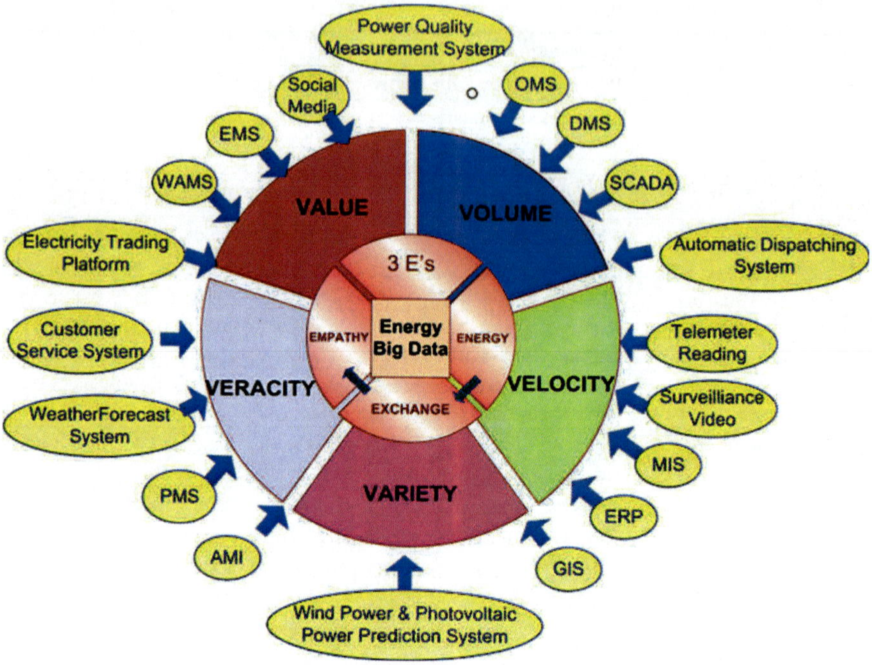

Fig. 2 Characteristics and sources of energy big data

advance infrastructure for communication and system to manage big data [32–68]. Some of these applications are identified and shown in Fig. 4. All these applications utilize the intelligence out of the big data and used in specific applications such as management of outages using social data, diagnostic and prognostics-based power and distribution transformer health monitoring, PMU data-based transient stability evaluation and preventive measures, advise to consumers, better tariff information, model rate managing demands, monitoring power thefts, etc.

5 Future Research Directions

A quite diversified research direction can be observed in case of BDA in SG applications as shown in Fig. 5. With enormous amount of data pouring in from wide number of sources and being utilized for several SG applications, there is a growing concern regarding data privacy and security [69, 70]. So as shown in

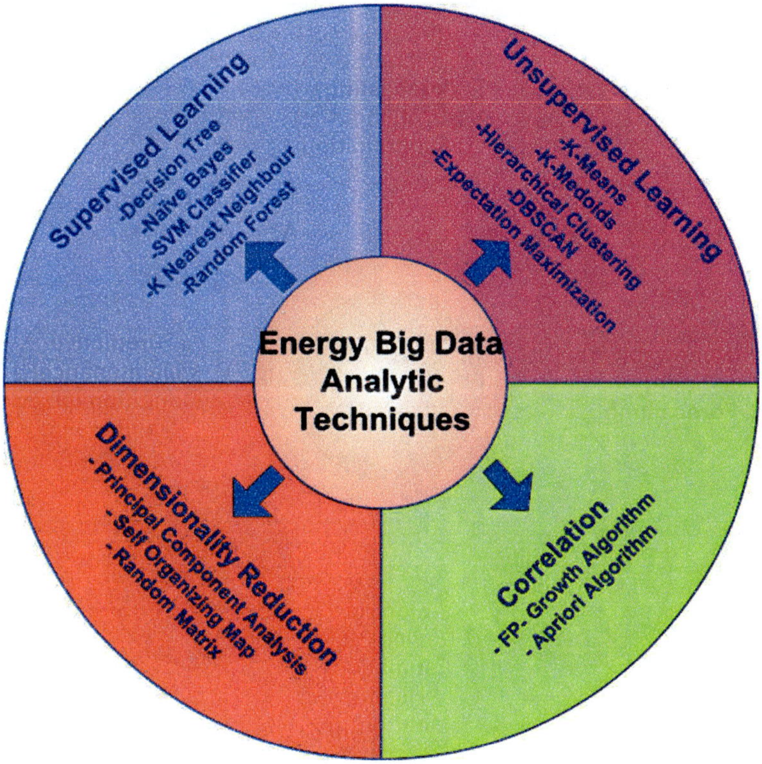

Fig. 3 Energy big data analytic techniques

Fig. 5, there are distinct research directions concerned to utilize the wide range of heterogenous data with new compatible architecture and platforms with a clear emphasis on real-time monitoring and control. More efficient data analytic methods are being proposed which ensure interoperability among various components of SG and equipped with powerful visualization. Finally, to the make the future smart grid a secure, reliable cyberphysical system, a lot of research are being carried out for developing robust standards and regulatory frameworks.

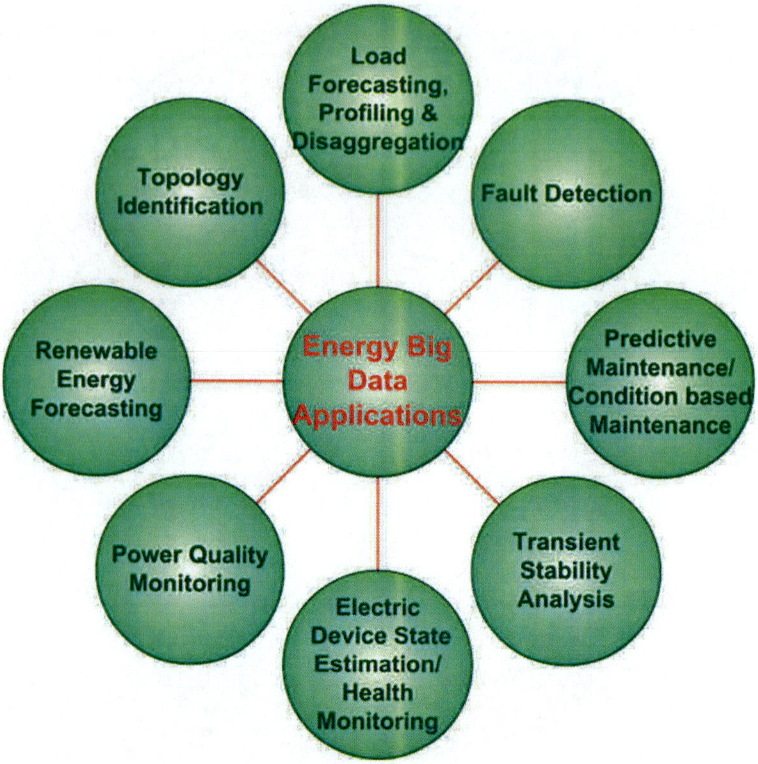

Fig. 4 Energy big data applications

6 Conclusion

In this paper, a review of most of the important review papers on SG with BDA is presented first, which clearly outlines state-of-the-art research on this field. The authors then have presented a concise bibliographical review on the various related aspects with the help of figure-based illustrations clearly pointing out the future research directions in BDA for SG applications.

Fig. 5 Big data for smart grid—future research directions

References

1. Simmhan, Y., Aman, S., Kumbhare, S., Liu, R., Stevens, S., Zhou, Q., Prasanna, V.: Cloud based software platform for big data analytics in smart grids. Comput. Sci. Eng. **15**(4), 38–47 (2013)
2. Tsai, C.-W., Lai, C.-F., Chao, H.-C., Vasilakos, A. V.: Big data analytics: a survey. J. Big Data **2**, 21 (2015)
3. Yu, X., Xue, Y.: Smart Grids: A Cyber-Physical Systems Perspective. Proceeding of the IEEE **104**(5), 1058–1070 (2016)
4. Xue, Y., Yu, X.: Beyond Smart Grid—A Cyber–Physical–Social System in Energy Future. Proceeding of the IEEE **105**(12), 2290–2292 (2017)
5. Rekaa, S.S., Dragicevic, T.: Future effectual role of energy delivery: A comprehensive review of Internet of Things and smart grid. Renew. Sustain. Energy Rev. **91**, 90–108 (2018)

6. Shi, L., Dai, Q., Ni, Y.: Cyber–physical interactions in power systems: A review of models, methods, and applications. Electr. Power Syst. Res. **163**, 396–412 (2018)
7. Cheng, L., Yu, T.: Smart dispatching for energy internet with complex cyber-physical-social systems: A parallel dispatch perspective. Int. J. Energy Res. **43**(8), 3080–3133 (2019)
8. Dileep, G.: A survey on smart grid technologies and applications. Renew. Eng. **146**, 2589–2625 (2020)
9. Gavriluta, C. Boudinet, C., Kupzog, F., Gomez-Exposito, A., Caire, R.: Cyber-physical framework for emulating distributed control systems in smart grids, International Journal of Electrical Power and Energy System, **114** (2020)
10. Das, L., Munikoti, S., Natarajan, B., Srinivasan, B.: Measuring smart grid resilience: Methods, challenges and opportunities. Renew. Sustain. Energy Rev. **130**, 59–69 (2020)
11. Yin, L., Gao, Q., Zhao, L., Zhang, B., Wang, T., Li, S., Liu, H.: A review of machine learning for new generation smart dispatch in power systems. Engineering Applications of Artificial Intelligence **88** (2020)
12. Schuelke-Leech, B., Barry, B., Muratori, M., Yurkovich, B.J.: Big Data issues and opportunities for electric utilities. Renew. Sustain. Energy Rev. **52**, 937–947 (2015)
13. Hu, J., Vasilakos, A.V.: Energy Big Data Analytics and Security: Challenges and Opportunities. IEEE Transaction on Smart Grid **7**(5), 2423–2436 (2016)
14. Wu, J., Guo, S., Li, J., Zeng, D.: Big Data Meet Green Challenges: Big Data Toward Green Applications. IEEE Syst. J. **10**(3), 888–900 (2016)
15. Zhu, J., Zhuang, E., Fu, J., Baranowski, J., Ford, A.: and. J. Shen A Framework-Based Approach to Utility Big Data Analytics IEEE Transaction on Power Systems **31**(3), 2455–2462 (2016)
16. Zhou, K., Fu, C., Yang, S.: Big data driven smart energy management: From big data to big insights. Renew. Sustain. Energy Rev. **56**, 215–225 (2016)
17. Jiang, H., Wang, K., Wang, Y., Ghao, M., Zhang, Y.: Energy big data: A survey. IEEE Access **4**, 3844–3861 (2016)
18. Daki, H., Hannani, A. E., Aqqal, A., Haidine, A., Dahbi, A.: Big Data management in smart grid: concepts, requirements and implementation. J. Big Data **4**, 13 (2017)
19. Ma, Z., Xie, J., Li, H., Sun, Q., Si, Z., Zhang, J., Guo, J.: The Role of Data Analysis in the Development of Intelligent Energy Networks. IEEE Network **31**(5), 88–95 (2017)
20. Munshi, A.A., Mohamed, Y.A.-R.I.: Big data framework for analytics in smart grids. Electric Power Sys. Res. **151**, 369–380 (2017)
21. Ji, C., Wei, Y., Poor, H. V.: Resilience of Energy Infrastructure and Services: Modeling, Data Analytics, and Metrics Proceedings of IEEE **105**(7):1354–1366 (2017)
22. Tu, C., Hea, X., Shuaia, Z., Jiang, F.: Big data issues in smart grid—A review. Renew. Sustain. Energy Rev. **79**, 1099–1107 (2017)
23. Chen, K., He, Z., Wang, S.X., Hu, J., Li, L., He, J.: Learning-based Data Analytics: Moving Towards Transparent Power Grids. CSEE J. Power and Energy Sys. **4**(1), 67–82 (2018)
24. Hong, T.: Big Data Analytics: Making the Smart Grid Smarter. IEEE Energy and P. Magazine **16**(3), 67–73 (2018)
25. Zhang, Y., Huang, T., Bompard, E. F.: Big data analytics in smart grids: a review. Energy Informatics **1**(8) (2018)
26. Ghofrani, M., Steeble, A., Barrett, C., Daneshnia, I.: Survey of Big Data Role in Smart Grids: Definitions, Applications, Challenges, and Solutions. Open Electr. & Electron. Eng. J. **12**, 86–97 (2018)
27. Bhattarai, B.P., Paudyal, S., Luo, Y., Mohanpurkar, M., Cheung, K., Tonkoski, R., Hovsapian, R., Myers, K.S., Zhang, R., Zhao, P., Manic, M., Zhang, S., Zhang, X.: Big data analytics in smart grids: state-of-the-art, challenges, opportunities, and future directions. IET Smart Grid **2**(2), 141–154 (2019)
28. Ghorbanian, M., Dolatabadi, S.H., Siano, P.: Big Data Issues in Smart Grids: A Survey. IEEE Syst. J. **13**(4), 4158–4168 (2019)

29. Hossain, E., Khan, I., Un-Noor, F., Sikander, S. S., Sunny, M.S.H.: Application of Big Data and Machine Learning in Smart Grid, and Associated Security Concerns: A Review IEEE Access **7**, 13960–13988 (2019)
30. Kezunovic, M, Pinsonb, P, Obradovicc, Z, Grijalvad, S, Honge, T, Bessa, R (2020) Big data analytics for future electricity grids. Electric Power System Research, **189** (2020)
31. Sundararajan, A., Hernandez, A.S., Sarwat, A.I.: Adapting big data standards, maturity models to smart grid distributed generation: critical review. IET Smart Grid **3**(4), 508–519 (2020)
32. Diamantoulakis, P.D., Kapinas, V.M., Karagiannidis, G.K.: Big data analytics for dynamic energy management in smart grids. Big Data Research **2**(3), 94–101 (2015)
33. Kumari, A., Tanwar, S.: Secure data analytics for smart grid systems in a sustainable smart city: Challenges, solutions, and future directions. Sustainable Computing: Informatics and Systems, 28 (2020)
34. Ma, J., Zhang, P., Fu, H., Bo, B., Dong, Z.: Application of Phasor Measurement Unit on Locating Disturbance Source for Low-Frequency Oscillation. IEEE Transactions on Smart Grid **1**(3), 340–346 (2010)
35. Catterson, V.M., McArthur, S.D.J.: Data Analytics for Transmission and Distribution. Smart Grid Handbook, 1–19. (2016) https://doi.org/10.1002/9781118755471.sgd085
36. Zhou, D., Guo, J., Zhang, Y., Chai, J., Lei, H., Liu, Y., Huang, C., Gui, X., Liu, Y.: Distributed Data Analytics Platform for Wide-Area Synchrophasor Measurement Systems. IEEE Transactions on Smart Grid **7**(5), 2397–2405 (2016)
37. Chu, L., Qiu, R., He, X., Ling, Z., Liu, Y.: Massive Streaming PMU Data Modelling and Analytics in Smart Grid State Evaluation based on Multiple High-Dimensional Covariance Test. IEEE Trans. on Big Data **4**(1), 55–64 (2018)
38. Shahsavari, A., Farajollahi, M., Stewart, E., Cortez, E., Mohsenian-Rad, H.: Situational Awareness in Distribution Grid Using Micro-PMU Data: A Machine Learning Approach. IEEE Transactions on Smart Grid **10**(6), 6167–6177 (2019)
39. Kerdphol, T., Matsukawa, Y., Watanabe, M., Mitani, Y.: Application of PMUs to monitor large-scale PV penetration infeed on frequency of 60 Hz Japan power system: A case study from Kyushu Island Electric Power System Research **185** (2020)
40. Meegahapola, L.G., Bu, S., Wadduwage, D.P., Chung, C.Y., Yu, X.: Review on Oscillatory Stability in Power Grids with Renewable Energy Sources: Monitoring, Analysis, and Control using Synchrophasor Technology. IEEE Transactions on Industrial Electronics, 1 (2020). https://doi.org/10.1109/tie.2020.2965455
41. Saric, A.T., Transtrum, M.K., Stankovic, A.M.: Data-Driven Classification, Reduction, Parameter Identification and State Extension in Hybrid Power Systems. IEEE Transactions on Power Systems, 1 (2020). https://doi.org/10.1109/tpwrs.2020.3027249
42. Liu, Y., Wu, L., Li, J.: D-PMU based applications for emerging active distribution systems. Electric Power Systems Research **179** (2020)
43. Siano, P.: Demand response and smart grids—A survey. Renew. Sustain. Energy Rev. **30**, 461–474 (2014)
44. Wolf, T.: Meter Data Collection, Management, and Analysis. Smart Grid Handbook, 1–26 (2016) https://doi.org/10.1002/9781118755471.sgd003
45. Liu, X., Golab, L., Golab, W., Ilyas, I.F., Jin, S.: Smart Meter Data Analytics: Systems, Algorithms, and Benchmarking ACM Transactions on Database Systems. **42**(1) (2016)
46. Alahakoon, D., Yu, X.: Smart Electricity Meter Data Intelligence for Future Energy Systems: A Survey. IEEE Transaction on Industrial Informatics **12**(1), 425–436 (2016)
47. Sun, Q., Li, H., Ma, Z., Wang, C., Campillo, J., Zhang, Q., Walin, F., Guo, J.: A Comprehensive Review of Smart Energy Meters in Intelligent Energy Networks. IEEE Internet of Things Journal **3**(4), 464–479 (2016)
48. Wen, L., Zhou, K., Yang, S., Li, L.: Compression of smart meter big data: A survey. Renew. Sustain. Energy Rev. **91**, 59–69 (2018)

49. Lin, Y.-H., Hu, Y-C.: Residential Consumer-Centric Demand-Side Management Based on Energy Disaggregation-Piloting Constrained Swarm Intelligence: Towards Edge Computing Sensors **18**(5), 1365 (2018)
50. Khan, Z.A., Jayaweera, D., Alvarez-Alvarado, M.S.: A novel approach for load profiling in smart power grids using smart meter data. Electr. Power Syst. Res. **165**, 191–198 (2018)
51. Wang, Y., Chen, Q., Hong, T., Kang, C.: Review of Smart Meter Data Analytics: Applications, Methodologies, and Challenges. IEEE Transaction on Smart Grid **10**(3), 3125–3148 (2019)
52. Wilcox, T., Jin, N., Flach, P., Thumim, J.: A Big Data platform for smart meter data analytics. Comput. Ind. **105**, 250–259 (2019)
53. Yen, S.W., Morris, S., Ezra, M.A.G., Huat, T.J.: Effect of smart meter data collection frequency in an early detection of shorter-duration voltage anomalies in smart grids. Electrical Power and Energy Systems **109**, 1–8 (2019)
54. Wu, W., Peng, M.: 'A data mining approach combining-means clustering with bagging neural network for short-term wind power forecasting'. IEEE Internet Things J. **4**(4), 979–986 (2017)
55. Tureczek, A., Neilsen, P.S., Madsen, H.: Electricity Consumption Clustering Using Smart Meter Data. Energies **11**, 859 (2018)
56. Munshi, A.A., Mohamed, R.I.: Data Lake Lambda Architecture for Smart Grids Big Data Analytics IEEE Access **6**, 40463–40471 (2018)
57. Wang, C., Zhu, Y., Shi, V., Chang, P., Vijayakumar, B., Liu, Y., Mao, J., Wang, Y.: A Dependable Time Series Analytic Framework for Cyber-Physical Systems of IoT-based Smart Grid. ACM Transactions on Cyber-Physical Systems **3**(1) (2018)
58. Kuralkar, S., Mulay, P., Chaudhari, A.: Smart Energy Meter: Applications, Bibliometric Reviews and Future Research Directions. Science & Technology Libraries **39**(2),165–188 (2020)
59. De Souza, M.A., Pereira, J.L.R., Alves, G., de Oliveira, B.C., Melo, I.D., Garcia, P.A.N.: Detection and identification of energy theft in advanced metering infrastructures. Electric Power Systems Research, 182 (2020)
60. Black, J., Hofmann, A., Hong, T., Roberts, J., Wang, P.: Weather data for energy analytics: From Modeling Outages and Reliability Indices to Simulating Distributed Photovoltaic Fleets. IEEE Energy and Power Magazine **16**(3), 45–53 (2018)
61. Ludwig, N., Feuerriegel, S., Neumann, D.: Putting Big Data analytics to work: Feature selection for forecasting electricity prices using the LASSO and random forests. Journal of Decision Systems **24**(1), 19–36 (2015)
62. Jokar, P., Arianpoo, N., Leung, V.C.M.: Electricity Theft Detection in AMI Using Customers' Consumption Patterns. IEEE Transactions on Smart Grid **7**(1), 216–226 (2016)
63. Sarwat, A.I., Amini, M., Domijan, A., et al.: Weather-based interruption prediction in the smart grid utilizing chronological data. J. Mod. Power Syst. Clean Energy **4**(2), 308–315 (2016)
64. Singh, S., Yassine, A.: Mining energy consumption behavior patterns for households in smart grid. IEEE Transactions on Emerging Topics in Computing, pp. 1 (2018)
65. Cao, Y., Song, H., Kaiwartya, O., Zhou, B., Zhuang, Y., Cao, Y., Zhang, X.: Mobile Edge Computing for Big-Data-Enabled Electric Vehicle Charging. IEEE Commun. Mag. **56**(3), 150–156 (2018)
66. Luo, X., Zhu, X., Lim, E.G.: A hybrid model for short term real-time electricity price forecasting in smart grid. Big Data Analytics **3**, 8 (2018)
67. Wang, K., Xu, C., Zhang, Y., Guo, S., Zomaya, A.Y.: Robust Big Data Analytics for Electricity Price Forecasting in the Smart Grid. IEEE Transaction on Big Data **5**(1), 34–35 (2019)
68. Anthony Jnr, B., Petersen, S.A., Ahlers, D., Krogstie, J.: API deployment for big data management towards sustainable energy prosumption in smart cities-a layered architecture perspective. Int. J. Sustain. Energy **39**(3), 263–289 (2020)

69. Wu, J., Ota, K., Dong, M., Li, J., Wang, H.: Big Data Analysis-based Security Situational Awareness for Smart Grid. IEEE Transaction on Big Data **4**(3), 408–417 (2018)
70. Chin, W.-L., Li, W., Chen, H.-H.: Energy Big Data Security Threats in IoT-Based Smart Grid Communications. IEEE Communications Magazine **55**(10), 70–75. (2017). https://doi.org/10.1109/mcom.2017.1700154

Present and Future Challenges: Evaluating Security of 6LoWPAN Networks in the Internet of Things

Harleen Kaur, Mozammil Hassan, and Bhavya Alankar

Abstract Every now and then a new technology is introduced to the world that becomes a prominent feature of our lives. Internet of things (IoT) has become one of the biggest trends, and it is shaping the coming future of technology with the global rise of smart technologies and "smart" device ownership. The Internet is already being connected to the daily objects like TV, cars, geysers, watches, light bulbs, AC, and refrigerators and more to come as they are in development stages. The IoT is a latest prototype for devices as well as Internet networks. The extensive use of devices and other objects can be alluded to as things. The proposed work is predominantly deals with an concept of IoT, their design, and background. This paper discussed the types of attacks on IoT systems, various challenges and their security issues in IOT layers. We have demonstrated the performance and analysis of 6low-power wireless personal area networks (6LoWPANs) protocol with Cooja Network Simulator on Contiki OS.

Keywords Layers · Internet of things (IoT) · Threats · 6Low-power wireless personal area networks (6LoWPANs) protocol · Cooja Network Simulator

1 Introduction

Internet of things (IoT) enable lots of devices to interconnect with all other through Internet. This ensures that the devices are keenly intelligent and dispatch the data to the central hub, which then monitors and takes actions in accordance with the work

H. Kaur · M. Hassan · B. Alankar (✉)
Department of Computer Science and Engineering, School of Engineering Sciences and Technology, Jamia Hamdard, New Delhi, India
e-mail: bhavya.alankar@gmail.com

H. Kaur
e-mail: harleen.unu@gmail.com

M. Hassan
e-mail: mozammilh11@gmail.com

S. Das and M. N. Mohanty (eds.), *Advances in Intelligent Computing and Communication*, Lecture Notes in Networks and Systems 202,
https://doi.org/10.1007/978-981-16-0695-3_43

done. IoT need to be used in several fields which include health care, transport system, entertainment, power framework, and smart buildings. The fast growth of Internet linked devices, which range from basic captors to extremely composite cloud servers, forms the Internet of things [1]. In technical prospective, the architecture relies on DCTs, mainly radio frequency identification (RFID tagged items), to provide the IoT with IT facilities that facilitate the secure and reliable exchange of "things" [2]. The evolving trends in IoT systems will change the interaction and operation of companies, regimes, and customer in a data-controlled economy [3]. In the following 15 years, the industrial Internet of things (IoT) will exceed 60 trillion by 2020 and IoT's market size by 2015 was approximately 900 M in annual revenue from IoT vendors as reported by Forbes [4]. It is one of the most important technological revolutions and is anticipated that its use will increment over the next few years [1, 5]. In IoT, various standards are integrated and different sensing, connectivity, storage, computation and other technology is supported [3, 6–8]. Today the world depends plenarily on the information provided by images or texts in the cyber world. The main involution of a individual in the amassment of information is limpidly designated. However, the problem with humans involvement is that they have short time and also not so much accuracy, which results in inopportune and inconsistently erratic data. This shows a system that can captures and transmits data automatically to web without interaction between people and the machine [9]. With a sizably voluminous number of contrivances linked to the web and astronomically immense data related with it, security concerns already exist there. By security, we understand that the level of IoT infrastructure and applications resistance or aegis. There are lots of contrivances are facile intrusion selected. They are predicated on very circumscribed external resources and often unattended. If there will be some problem occurs in network layers, gaining control and malevolently utilizing a contrivance is very facile for a hacker and also assailing contrivances through the node which are near. We all know that it is easy to attack those devices which can be run online. These contrivances with no aegis against viruses and malware may be utilized as "worms" transmitting malicious programs so that other contrivances will be infected [1]. Another issue of interest is the confidentiality of personal details taken by IoT machines, as the automated flow of data must be fully understood and controlled by the generic end user [10, 11].

Physical attack, side channel attack, cryptanalysis attack, software attack, and the network attack are the types of attacks which takes place in IoT system. In physical attacks, there is interfering with the hardware parts. In cryptanalysis attacks, the ciphertext is the main focus of such attacks and endeavors to break the encryption. The target of this type of attacks is the cipher text, and the encryption is breached. Software attack uses their own communication interface to exploit implementation vulnerabilities, and network attacks are basically divided into active and passive [3]. This paper is organized into several sections: Sect. 2 has background of the paper, Sect. 3 includes Security Layers in IoT And Their Issues, Sect. 4 has the security challenges, Sect. 5 has performance and analysis of 6LoWPAN protocols using Contiki–OS with Cooja Network simulator and at the last section is conclusion of the work.

In order to know about IOT safety concerns, we must study about IoT network components (native networks and ecumenical networks) and their interoperability. IoT ecosystem components are coordinator, controller, IoT devices, sensor bridges, IoT services. In the Fig. 1, we have shown the operational models of IOT ecosystem [12].

1. IoT device: Here the IoT contrivances are like sensors, OS, the actuators, services, communication interface, lastly applications. The main task of a perspicacious thing is to AC cumulate related information utilizing the sensors then take important measures utilizing actuators.
2. Coordinator: It functions like a contrivance controller. There will be operating single device or may be more than that under a only one coordinator. It is utilized to monitor health and perspicacious thing. It is also used to report their several operations and occurrence to the service provider. For example in a smart home, when the kineticism sensor camera discovers any human actions, it sends an order lock/unlock to the keenly intellective door lock.
3. Sensor bridge: Sensor bridge is also known to be IOT gateway. It acts as a bridge in the middle of the limited IoT web and IoT cloud services.
4. IoT services: IoT process automatic, tool management, conclusion, etc., are done under IoT services.
5. Controllers: The controllers are to be used in order to control the IOT devices. Controllers like smart mobile, tablets. For example, any person can use his smart mobile in order to give instructions to devices.

2 Security Layers in IoT and Their Issues

(1) Security Layers in IOT

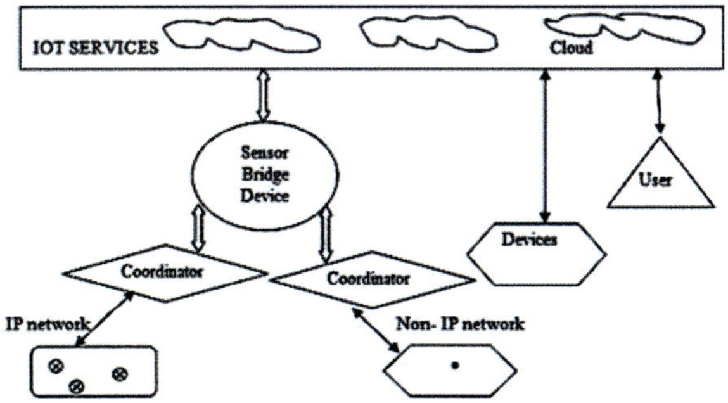

Fig. 1 IoT device interoperability

Fig. 2 Security layers of IOT

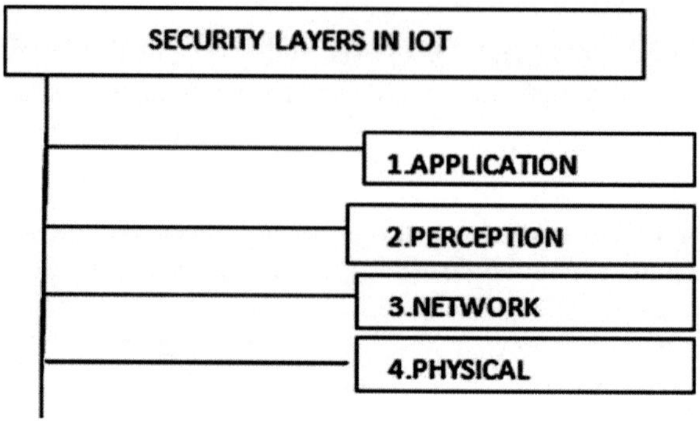

The most paramount layers as shown in the Fig. 2 for achieving an IoT objective are as follows [1].

(a) Application Layer: It includes different applications and accommodations provided by IoT. Keenly intellective cities, astute home surveillance, conveyance facilities, and health care are applications.
(b) Perception Layer: It comprises of several sensory technologies such as temperatures, sensors, pressure sensors, and RFIDs, which sanction contrivances to feel other elements. It is composed of different sensor technologies.
(c) Network Layer: The layer has software for network communication. This layer's major task is to transmit data between different devices.
(d) Physical Layer: It includes certain rudimentary hardware, for example, physical components, perspicacious appliances also the important one which is power stations which form the backbone for networking perspicacious objects.
(2) Major Security Issues of the Key Layers
(i) Security Issues in Application layers: Due to security difficulties in this layer, applications can be easily stopped and compromised. The applications were therefore not able to perform the services, and they were scheduled for or were not able to perform authenticated services in the wrong method. Malicious attacks on that layers may trigger the application failure by errors in the program code. This is a highly hazardous problem based on the number of appliances classified as application [1].
(ii) Major Security Issues in the Perception Layer: The security threats comes at the nodes in this layer. As the nodes consist of sensors, the hackers want to replace their own device software with the primary targets. The majority of threats in this layer have external, mainly in relation to sensors and other data collection tools [1].
(iii) Security Issues in the Network Layer: Because of its great number of data, the network layer is highly susceptible to attack, which causes a lot of "network congestion." The main security challenges in this laying are with regard to the integrity and authentication of data carried over the network. An assailment by

hackers, maleficent nodes compromising network contrivances are the very dangerous problem [1].

(iv) Security Issues in the Physical Layer: This layer also has many security problems. New technology is needed to protect energy sources and mechanisms of physical security. Devices from weather and individual perspectives must be secured against physical attacks. In the case of a urban grid or power interruption, it must also be power-efficient and able to rely on battery power. For a sufficient period of time, batteries must hold charges and recharge quickly to keep the device going [1].

3 Security Challenges

The widespread distribution of an IoT nodes and personnel information or data are collected and passed through IoT devices have posed a major challenge to security. Today there are lots of challenges IoT are facing in the field of security [3, 12]. Some of the IoT security challenges are followings:

(a) Authentication—Authentication in the IoT domain enables various IoT appliances which are used in different contexts to be integrated. It involves in the two ways. Authentication of routing peers participating in data transfer and in the data route source. The challenge of IoT device authentication is effective key deployment and also the effective key management. Any key creation and key interchange in encryption must not lead to a significant above of IoT nodes. Additionally, in the lack of an assured Certificate Ascendancy (CA), different methods are required to validate encryption keys and ensuring safety of key transfers.

(b) Authorization and Access Control—The authorization shall include the setting of access rights of various resources, while the access control operation should ensure access rights are available only to authorized resources. Every IoT node can only carry mechanisms for verification of access which may differ in the same node from other connected objects. Therefore, there will be a challenges in a heterogeneous IoT network for deploying and managing a number of authorizations and access control operations customized to various node potentiality.

(c) Privacy—The use of autonomous IoT objects sensing private data (e.g., in health) represents a new threat to the privacy of individuals. In contrast to conventional scenarios where users need to do certain things in order to protect their seclusion, nodes of IoT collect personal data without them being even aware. Centric user privacy, content-driven confidentiality, or context-driven confidentiality are available. However, IoT networks include separate nodes that collect data and need object-oriented confidentiality models. In addition, the major part of the privacy regulations' mandate informs users of the handling and administration of their private data. In heterogeneous IoT networks, identifying nodes which have access to privacy information collected passively are a major challenge.

(d) Secure architecture—It is not trivial to build an architecture that overcomes the above-mentioned IoT security challenges. Not just security problems, but challenges that are brought about by IoT device deployment through the software-defined networks (SDN) and cloud infrastructures, should be addressed by any IoT architecture. The majority of problems in SDN and cloud environment inherent in the underpinnings of IoT sensors would be unavoidable. Furthermore, many unique safety issues would be arising from complexity involving safe links between object-oriented IoT networks with data-based cloud infrastructure.

4 Performance and Analysis of 6LoWPAN Protocols Using Contiki-OS with Cooja Network Simulator

Contiki is one of the most widely used IoT operating programs for various sensor and RFID types. It has a BSD license with the programming language code C. Contiki programming takes place using the Cooja Network Simulator. They are used for programming and monitoring of remote IoT devices. After opening the Contiki os application, we will first see the login screen. The default password will be "user." After typing password, press the enter key.

There is the process of loading Cooja simulator in the Contiki os. There will take plenty of time to load. So you to wait for some moment to load.

4.1 Creating a New Network in the Cooja Simulator

From the file menu option of Cooja, choose "New simulation" option and proceed.

Now as we can see in Fig. 3, there is the creating of a new simulation in the Cooja [13, 14]. The simulation process starts from here. For creating new simulation, you have to choose new. If you already done the simulation and saved somewhere, then you can opt for open simulation. After creating new simulation, there is requirement of filling some basic information about simulation. So we have the setup of some basic properties of simulation. As you see in the dialog box, there will be setting of basic network parameters. We have to fill the simulation name, then the radio medium, after that mote startup delay and the random speed. Now chose create option.

Now after completing the layout and working environment is done, importing of the RFID tags, sensors nodes or may be different wireless devices, that will be required to get connected, make communication in IoT. They are known as "motes" in wireless network and in IoT. Large numbers of motes are available and that could

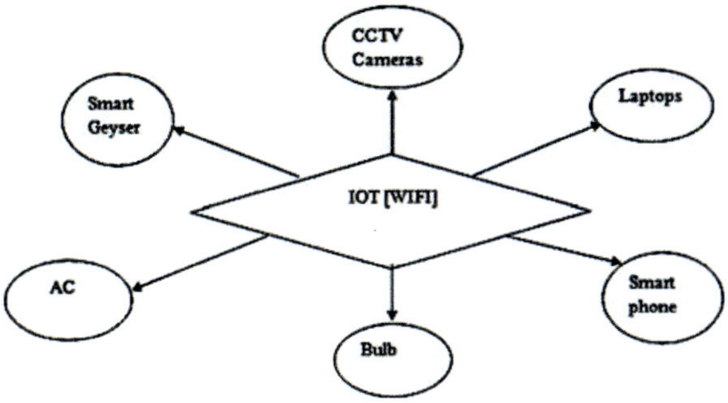

Fig. 3 IoT security provision

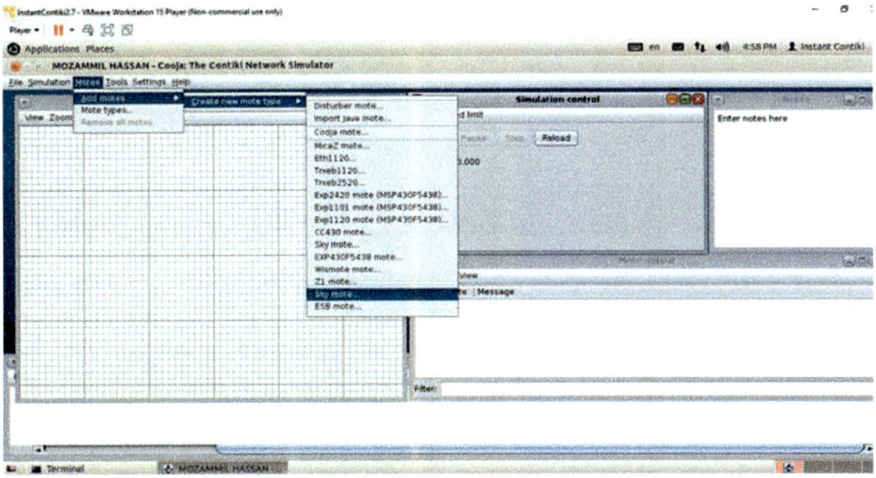

Fig. 4 Invoking of wireless and RFID motes in Cooja simulator

be programmed if necessary. We can also connect physical motes using ports on the computer system in order to do real-time interfacing as shown in Fig. 4.

In Fig. 5, there is importing of source code C for the recompilation process. Now, in order for real-time interfaces to be done, physical motes can even be connected through the system's port. Every motes consist of base properties and programming api's. It is given in the source code C. They are located at the back end of the Cooja. And all of this sources could be customized or change, and then they are recompiled in order to get the desired results or new results from these motes. Motes in Cooja. Now after the process of compilation is completed, the number of virtual motes which would be imported into simulation area in order to see and analyze the transmission of radio signals.

We need to compile C source code for desired behavior of wireless motes in Cooja simulator. Figures 6 and 7 show wireless motes displayed in Cooja with

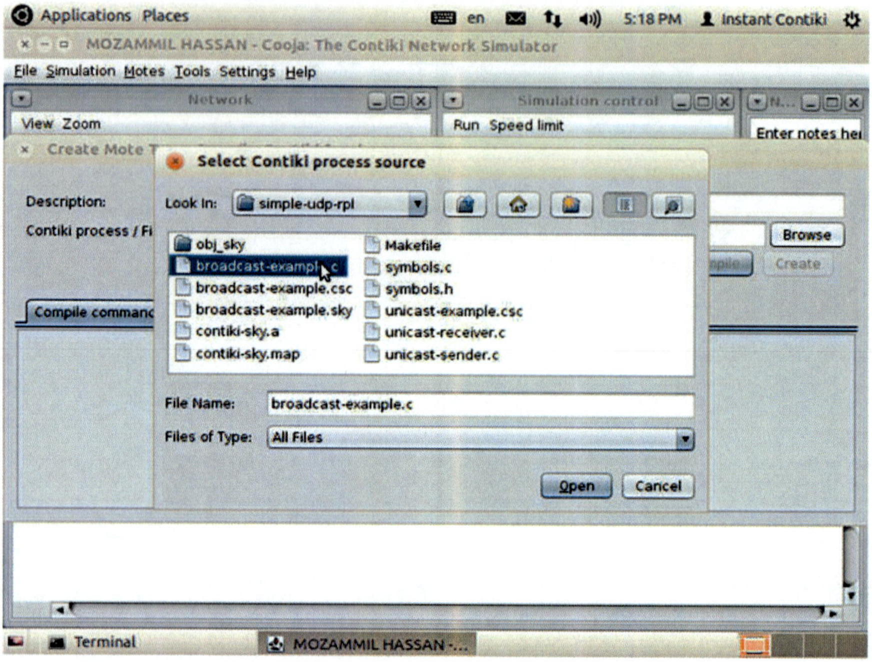

Fig. 5 Importing of C source code for recompilation in Cooja simulator

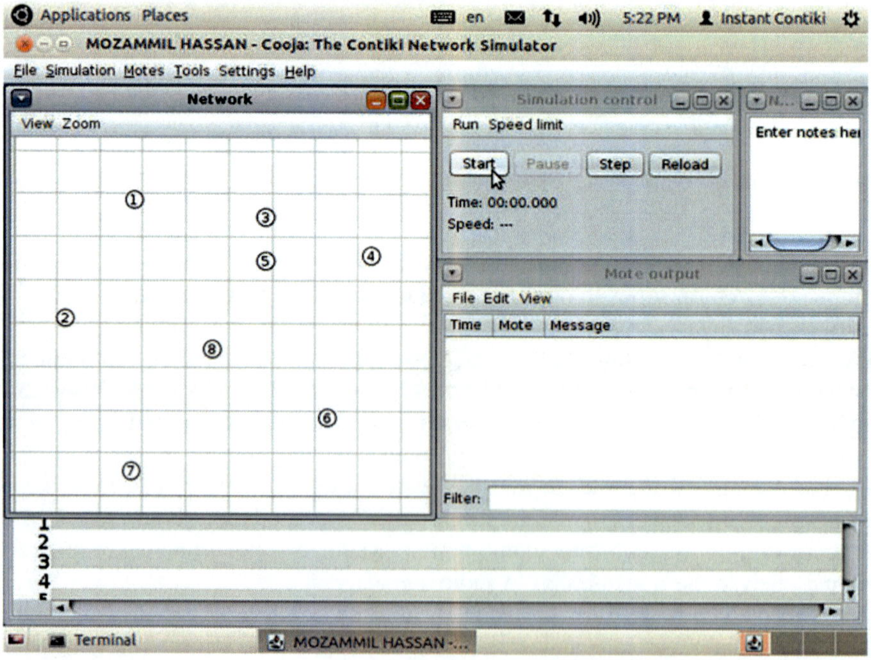

Fig. 6 Wireless motes displayed in Cooja with positions

positions. Now, after the process of compilation is completed, the number 3 of virtual motes which would be imported into simulation area in order to see and analyze the transmission of radio signals.

In dynamic key exchange scenario between motes, a simulation of the IoT network generates and authenticates the dynamic security key for communication. In the field of IoT security, the protocols and algorithms must be designed and implemented to prevent infringement by enforcing total privacy and security in communications. Because the IoT is used for military use, highly secure key exchange algorithms and dynamic security key cryptography are required. The network log files, including sources and destinations motes, the time, total simulation activities, will then be analyzed; once the simulation is completed, we can analyze using machine learning the log data in mote output window.

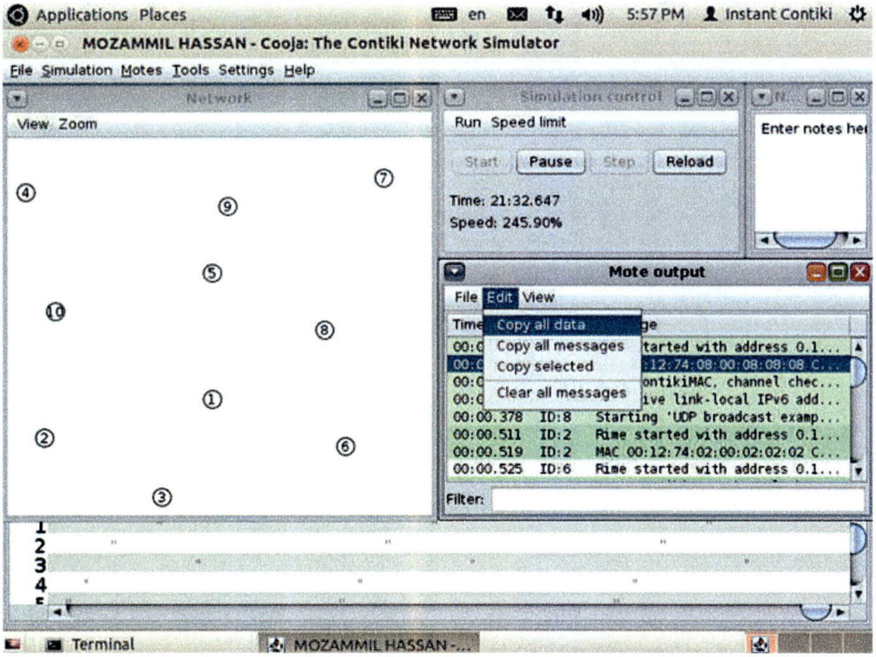

Fig. 7 Running of the simulation and behavior analytics of motes

5 Conclusion

For cyber criminals, IoT is the hot target. This paper generally discussed the IoT including its designed. IoT is very popular and approaching. There are lots of work to be done on IoT in order to make it secure for everyone which is very important. We have understood that that IoT would be easily and broadly used when it is safe. New security protocols and approaches are being chalked out by various researchers. There are lots of research work is going on in this field. We have performed a simulation of IoT network in the Internet of things environment and for desired behavior of wireless motes in Cooja simulator.

References

1. Kumar, S.A., Vealey, T., Srivastava, H.: Security in Internet of Things: Challenges, Solutions and Future Directions. In:49th Hawaii International Conference on System Sciences (HICSS), Koloa, HI, pp. 5772–5781 (2016) https://doi.org/10.1109/hicss.2016.714
2. Weber, R.H. Internet of Things—New security and privacy challenges. Elsevier, pp. 23–30 (2010)
3. Sicari, S., Rizzardi, A., Grieco, L.A., Coen-Porisini, A.: Security, privacy and trust in internet of things: The road ahead. Comput. Netw. **76**, 146–164 (2015)
4. Ammar, M., Russello, G., Crispo, B.: Internet of things: A survey on the security of iot frameworks. J. Inf. Secur. Appl. **38**, 8–27 (2018)
5. Kaur, H., Alam, M. A., Jameel, R., Mourya, A. K., & Chang, V.: (2018). A proposed solution and future direction for blockchain-based heterogeneous medicare data in cloud environment. J. Med. Syst. **42**(8)
6. Alem Colakovic, M.H.: Internet of Things (IoT): A review of enabling technologies, challenges, and open research issues. Elsevier, vol. 144, pp. 17–39 (2018)
7. Suchitra, V.C.: Internet of Things and Security Issues. Int. J. Comput. Sci. Mob. Comput. **5**(1), 133–139 (2016)
8. Xu, L.D., He, W., Li, S.: Internet of things in industries: A survey. IEEE Trans. Industr. Inf. **10**(4), 2233–2243 (2014)
9. Bhabad, M.A., Bagade, S.T.: Internet of Things: Architecture, Security Issues and Countermeasures. Int. J. Comput. Appl. **125**(14), 1–4 (2015)
10. Frustaci, M., Pace, P., Aloi, G., Fortino, G.: Evaluating Critical Security Issues of the IoT World: Present and Future Challenges. IEEE Internet of Things J. **5**(4), 2483–2495 (2018). https://doi.org/10.1109/JIOT.2017.2767291
11. Alankar, B., Sharma, G,, Kaur, H., Valverde, R., Chang, V.: Experimental Setup for Investigating the Efficient Load Balancing Algorithms on Virtual Cloud. Sensors. **20**(24), 7342 (2020)
12. Hossain, M.M, Fotouhi, M., Hasan, R.: Towards an Analysis of Security Issues, Challenges, and Open Problems in the Internet of Things. 2015 IEEE World Congress on Services, New York, NY, pp 21–28 (2015) https://doi.org/10.1109/SERVICES.2015.12
13. Dunkels, A.: The ContikiMAC Radio Duty Cycling Protocol. SICS, Tech. Rep. (2011) (Online). Available: http://soda.swedish-ict.se/5128/1/contikimac-report.pdf
14. Vincent, A., Francis, F., Ayyappadas, P.S.: Security Aspects in 6lowpan Networks: A Study. OSR J. Electron. Commun. Eng. (IOSR-JECE), 2278–2834

Design and Analysis of Grasshopper Evolutionary Technique Tuned Fractional-Order Proportional Integral Derivative for Magnetic Levitation System

Sunita S. Biswal, Dipak R. Swain, and Pravat Kumar Rout

Abstract In this paper, an optimized PID controller based on fractional calculus is applied to a magnetic levitation (MAGLEV) system to improve its controller. The objective of the controller is to provide precisely levitate the magnetic ball in a required electromagnetic field in MAGLEV. In addition to that, a recently population-based metaheuristic search strategy inspired by the swarming behaviors of grasshoppers known as grasshopper evolutionary technique (GET) is used to search the optimal gain parameters of the FOPID controller to enhance the performance in terms of reducing the system error with faster damping and controllability. The objective function is formulated considering the intention to reduce the rise time, settling time, and overshoot of the system response under disturbance conditions. The superiority of this optimized controller is tested by comparing the results with some modern heuristic optimization techniques such as particle swarm optimization (PSO) and genetic algorithm (GA). To justify the performance of the proposed controller for the MAGLEV model, it has been evaluated against 1 and 2DOF PID and FOPID controllers. The simulated comparative result reveals the fact that the proposed approach GET-FOPID controller predominates other controllers in terms of overshoot, settling time, and rise time. The designed GET-FOPID controller applied to the MAGLEV system exhibits a superior closed-loop response and ensures good control performance concerning reference input with improved system stability.

Keywords Magnetic levitation (MAGLEV) · PID controller · Fractional-order PID controller (FOPID) · Grasshopper evolutionary technique (GET)

S. S. Biswal (✉) · P. K. Rout
Department of Electrical and Electronics Engineering, S'O'A (deemed to be) University, Bhubaneswar, Odisha, India
e-mail: sunitasbiswal@gmail.com

D. R. Swain
Department of Electrical Engineering, CET, Bhubaneswar, Odisha, India

S. Das and M. N. Mohanty (eds.), *Advances in Intelligent Computing and Communication*, Lecture Notes in Networks and Systems 202,
https://doi.org/10.1007/978-981-16-0695-3_44

1 Introduction

The MAGLEV system plays an important role in the control system engineering in which a device is levitated in the electromagnetic field [1]. This system is based on electromagnetic attraction or repulsion to provide contact with less support. Looking at its dynamics and operation, it also has strong nonlinear, sensitive, and low noise features. Due to its nonlinear and unstable behavior, it is a challenging task to control and bring it to stable operation with the existence of any disturbance. Apart from that due to its extensive application in industrial sectors and many other engineering devices, it is necessary to control through a robust controller [2].

In the real world, MAGLEV has numerous applications such as space crafts, maglev transportation, power generating units, product display, and biomedical application. That why this issue is an open challenge for the researcher to design an efficient controller that must be capable of handling the complexity of the system dynamics [3]. This motivates us to consider this problem and investigate further to design for a robust controller.

The most commonly used 1DOF PID controller has been consistently employed for an extended period in control applications because of its simple configuration, trouble-free operation, and ability to deliver adequate performance [4, 5].

Even though the controller is delivering adequate performance along with good robustness criteria in many cases, till the associated major limitations like narrow region of robust performance, the requirement of re-tuning the gain parameters with the change of system conditions, and the existence of steady-state error need to be rectified. This issue can be handled either by modifying the conventional controller or through a better control approach. Looking at the first option, it is possible to achieve an additional better result from the PID controller by altering its initial configuration. With this objective, the current study tries to extend the capability of the PID controller further.

Recently, various authors suggest the modification by modifying the existing configuration from 1DOF to 2DOF [6, 7], and PID to 1DOF fractional-order PID [8], etc. In this work, the conventional PID control configuration has been modified and the MAGLEV system has been designed for 1DOF and 2DOF PID and 1DOF FOPID controller [9]. A fractional calculus-based controller has a similar configuration to that of a classical PID controller. Apart from that its derivative and integral orders are represented through fractional numerals and that help to formulate a better controller. The initiative of the fractional calculus was very old in mathematics and also establishes its implementation in the area of control engineering. As the FOPID controller [10] is the extended version of the PID controller, therefore it is characterized by three gains of PID and two extra parameters.

Still, fractional-order PID controller (FOPID) has limited real-time applications due to the complexity in establishing in real time with the help of proper digital and computational devices [11]. Apart from that, although FOPID controllers can able to improve the performance of the closed-loop system, still tuning these controllers is another limitation. As a result of which these controllers are complicated to establish in the industrial application until now. For this reason, many modern-day researchers have been focusing to develop new strategies for tuning FOPID controllers. From the review of the literature, it is found that the performance of FOPID controllers is completely determined by the best tuning method to set an optimal control parameter. To set a tuning method for these controllers, in this work a new algorithm known as grasshopper evolutionary technique (GET) has been applied. The GET [12] is a recently proposed efficient nature-inspired, population-based, single objective, stochastic, metaheuristic algorithm. In this optimization technique, the behavior of grasshopper is used as the optimizer to determine the finest gain parameters of the FOPID controller [4]. As, this method is uninfluenced by the irregular nature and/or level of a dilemma, where generally other optimization methods illustrate more earlier convergence [13], whereas this searches the finest decision skilfully with quicker convergence. Because of the above good points of this optimization method, a GET-based 1DOF FOPID controller is proposed for a MAGLEV system in this work.

To validate the proposed method's efficacy and robustness [14], various comparative outputs of the discussed controller are shown with PSO-FOPID and GA-FOPID controller in the MAGLEV system. Validating the proposed controller's efficacy and robustness, the results of the proposed controller are compared and shown with 1DOF PID, 2DOF PID, and 1DOF FOPID controller keeping the system conditions the same. This paper presents a new and novel GET tuning methodology for parameter tuning of a 1DOF FOPID controller. The remaining part of this study is structured below. Section 2 briefs the dynamic model of MAGLEV. Section 3 explains the design configuration of the 1DOF FOPID controller. Section 4 is keen to the explanation of the GET algorithm. The simulation results are presented and discussed in Sect. 5. At last, the concluding Sect. 6 presents the findings and effectiveness of the proposed controller.

2 Dynamic Model of MAGLEV

MAGLEV is based on the theory where an object is hanging with no support in the air. Only magnetic forces are applied in this system which also eliminates the effects of gravitational force and some other forces. In this work, the model number 33–210 [5] is selected as a maglev plant which was developed by Feedback Instruments Ltd. Basically, this system is well known for its energy efficiency, high-speed, low noise, high performance, and good driving span. The MAGLEV structure is shown in Fig. 1 and the system model parameter with their values is represented in Table 1.

As shown in Fig. 1a when current i flow through the electromagnetic actuator the system gets magnetized and due to this the steel ball is attracted in the upward direction. The steel ball moves in the downward direction due to the earth gravitational force. This up and down action of the ball is observed by IR sensors.

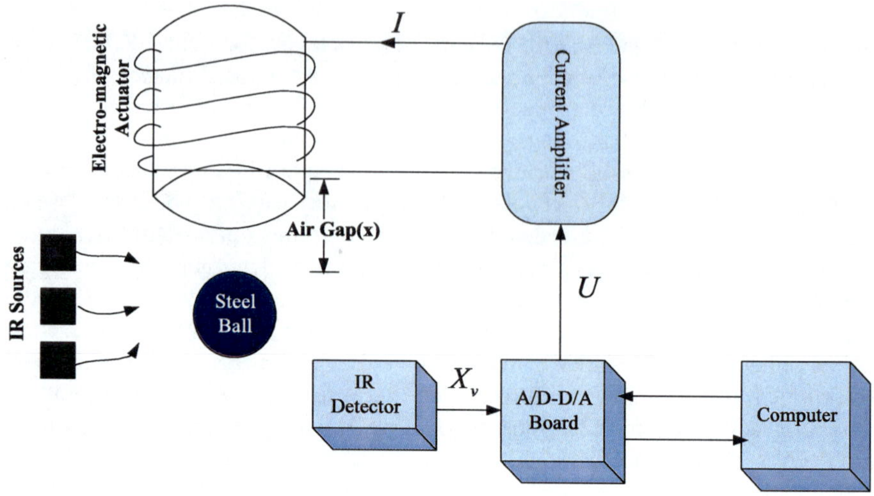

Fig. 1 MAGLEV block diagram representation

Table 1 Parameter values of the MAGLEV model

Description of parameter	Symbol	Values of parameters
Mass of the steel ball	m	0.02 kg
Gravitational force	g	9.81 m/s^2
At equilibrium current value	I_0	0.8A
At equilibrium position value	X_0	0.009 m
Control voltage to the coil current gain	a_1	1.05 V/A
Sensor gain and offset	a_2, η	1.43.48 V/m, −2.8 V
Input control voltage level	U	±5 V
Sensor output voltage level	X_v	±1.25 to −3.75 V

The nonlinear model of the maglev system in terms of the electromagnetic coil current I and position of the ball X is denoted as:

$$m \times \ddot{X} = m \times g - a \times \frac{I^2}{X^2} \tag{1}$$

where m denotes the mass of the steel ball, g denotes the gravitational force, and a denotes a constant which relies on the coil parameters. This nonlinear model can be adjusted to:

$$\Delta\ddot{X} = -\left\{\frac{\partial f(I,X)}{\partial I}|I_0,\ X_0\Delta I + \frac{\partial f(I,X)}{\partial X}|I_0,\ X_0\Delta X\right\} \quad (2)$$

where ΔX *and* ΔI signify the small variation from the current I_0 and position X_0 equilibrium points. After simplifying and applying Laplace transform on Eq. (2), the transfer function can be represented as follows:

$$\frac{\Delta X}{\Delta I} = -\frac{A_I}{s^2 - a_X} \quad (3)$$

where $A_I = \frac{2g}{I_0}$ and $a_X = \frac{2g}{X_0}$. As X and I are proportional to X_v and U, the Eq. (3) can be rewritten as $\frac{\Delta X_v}{\Delta U}$ and the modified transfer function can be represented as:

$$G_q(s) = \frac{\Delta X_v}{\Delta U} = -a_1 a_2 \frac{A_I}{s^2 - a_X} = \frac{-3581.85}{s^2 - 2180} \quad (4)$$

where X_v denotes sensor output voltage and U denotes the current amplifier input. Equation (4) has one pole on the right half of the s-plane that constitutes an unstable system. Therefore, an appropriate maglev controller design is required to get a stable and linear system. The current development in the research field shows that the stability of maglev has been improved extensively by employing several new control strategies.

3 Configuration of 1-DOF FOPID Controller

The classical PID controllers are well proved in a wide range of real-time applications because of their fine characteristics like structural simplicity, small overshoot, and quick response. This controller may be regarded as an extraordinary condition of the additional standard fractional-order PID controller (FOPID). The FOPID controllers are more accurate, robust, and flexible than the classical controller. These controllers are effectively implemented for the designing of a competent Maglev system because of its simplicity, reliability, and adequate performance. The basic block diagram of the FOPID controller is represented in Fig. 2

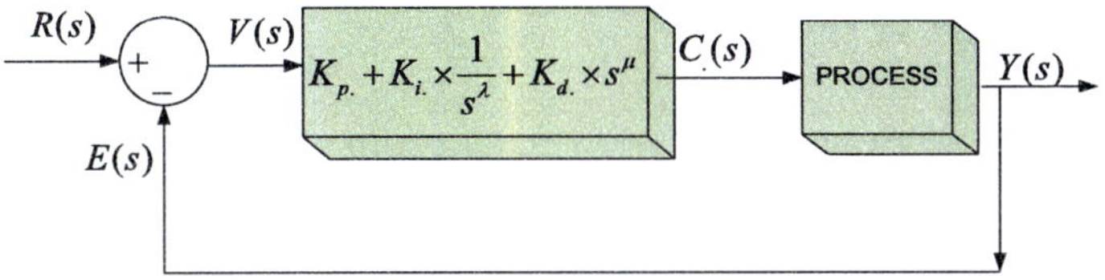

Fig. 2 1DOF FOPID controller block diagram representation

with three gains, i.e., proportional, integral, and derivative ($K_{p.}$, $K_{i.}$, and, $K_{d.}$), respectively.

As shown in the fig, including PID parameters there are two extra parameters to be adjusted (K_p, K_i, λ, K_d, and μ). The two extra parameters λ and μ are real numbers, and they representing the order of an integrator and order of a differentiator. The transfer function of fractional PID has the form:

$$C(s) = K_p + K_i \times \frac{1}{s^{\lambda}} + K_d \times s^{\mu} \quad (5)$$

where the values of λ and μ are varied between 0 and 2. For the classical PID controller, the values of λ and μ are 1. For the PI and PD controllers, the values of $\lambda = 1$, $\mu = 0$ and $\lambda = 0$, $\mu = 1$, respectively. Therefore, all these discussed controllers are particular instances of the FOPID controller. However, the conventional 1DOF PID controller when employed to an extremely unstable system establishes inadequate zeros in the closed-loop transfer function that results in high overshoot, and as an effect robustness of the system becomes less. To overcome this problem, the 2DOF PID controller has been configured to get suitable zeros in the closed-loop transfer function. The extra two parameters q_1 and q_2 help out to add zeros in the appropriate position that promotes to find improved performance and greater robustness. The involvement of this type of PID controller yields several applications in research fields. Few of them are accurate trajectory tracking, to know the behavior of a robust system, etc. Maglev with the 2DOF PID controller is represented in Fig. 3.

The three gains with two additional parameters ($q_1 = K_i$ and q_2) are taken in Fig. 3. But the main difficulties of this controller are to minimize the error. To reduce errors caused by 1 and 2DOF PID and FOPID controllers, a new evolutionary technique has been discussed in the next section which gives optimum parameter values and improves the effectiveness of these controllers applied in the MAGLEV system.

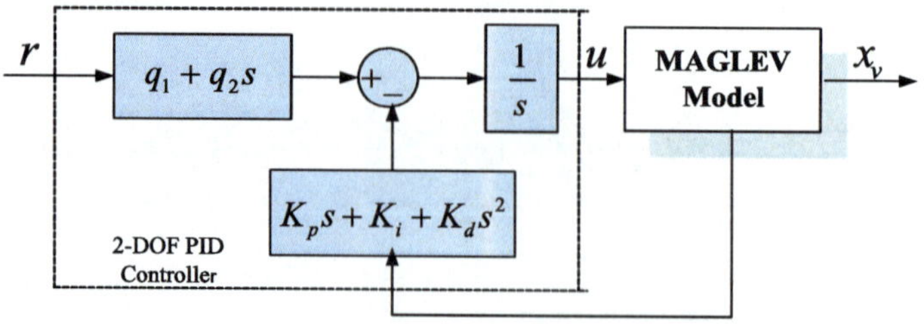

Fig. 3 MAGLEV applying 2DOF PID controller

4 Grasshopper Evolutionary Technique (GET)

Nowadays, it is essential to the extent of the performance of PID by introducing improved modern configuration. GET is a "nature-inspired" swarm algorithm which is based on the behavior of grasshopper. The life process of grasshoppers consists of three steps: egg, nymph, and adulthood. In the nymph stage, they slowly move with a minor increase (known as exploitation) and have plants get within their pathway. During the adulthood stage, they migrate with long-range and abrupt movements (known as exploration). Based on these behaviors, mathematical model is formulated by considering the grasshopper position and is written as follows:

$$X_n = S_n + G_n + A_n \tag{6}$$

where X_n denotes the position of the *nth* grasshopper, S_n denotes the social interaction, G_n denotes the gravitational force on the *nth* grasshopper, and A_n denotes the wind speed. The S_n in Eq. (6) is defined as:

$$S_n = \sum_{j=1,n\neq j}^{M} s(d_{nj})\hat{d}_{nj}, \quad d_{nj} = |x_n - x_j| \tag{7}$$

where d_{nj} denotes the distance connecting the *n*th and *j*th Grasshoppers. However, Eq. (6) cannot able to resolve the optimization difficulty, so, after reformulating the equation is:

$$X_n = c\left(\sum_{j=1,n\neq j}^{M} c\frac{a-b}{2} s(|x_j - x_n|)\frac{x_j - x_n}{d_{nj}}\right) + \hat{T}_d \tag{8}$$

where a and b denote the upper and lower bounds of the search space, respectively; T_d denotes the best solution value originate. However, Eq. (6) G_n is not measured, and the A_n is measured to be $\hat{T}_d$. Here c is a shrinking coefficient to balance exploration and exploitation. The shrinking parameter c decides the exploration or global search (when c becomes larger) and exploitation or local search (when c becomes smaller) and is given as follows:

$$c = c_{mx} - t\left[\frac{c_{mx} - c_{mn}}{t_{mx}}\right] \tag{9}$$

where c_{mx} denotes the maximum value of *C*, i.e., 1 and c_{mn} denotes minimum value of *C*, i.e., 0.00001; t denotes to the present iteration, and t_{mx} denotes the maximum number of iterations which have been used for this study. The optimization method starts with a set of random solutions. Grasshopper positions are updated by Eq. (8). In each iteration, the best target is updated and the position is updated iteratively till it reaches to stop criterion. This recently proposed algorithm has been applied to

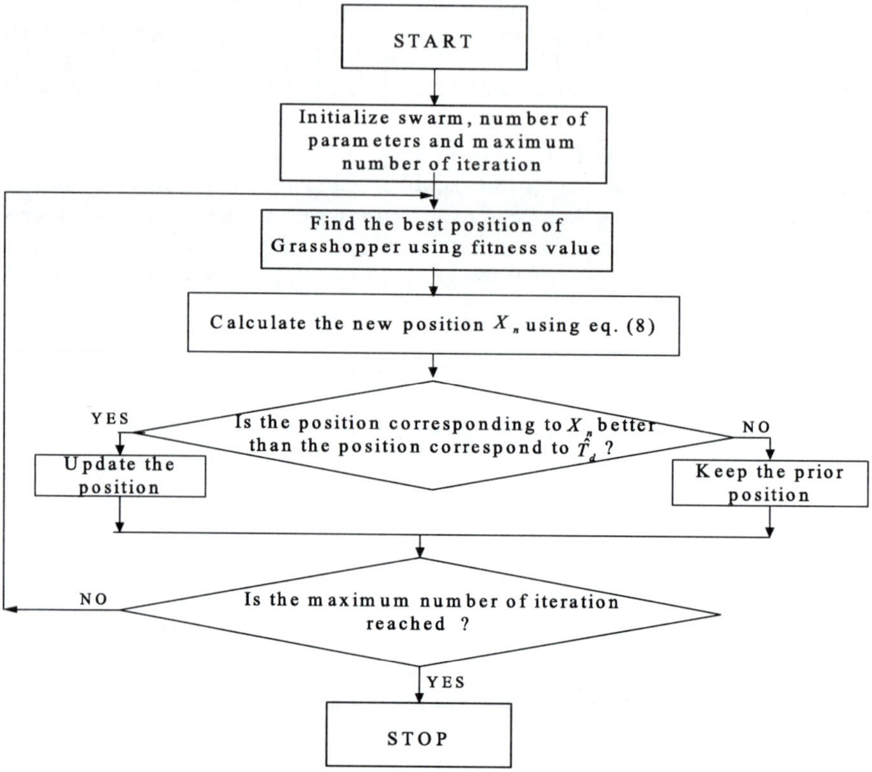

Fig. 4 Flowchart of the GET algorithm

various areas for its simple implementation and can efficiently solve all the global optimization problems. The flowchart for GET is represented in Fig. 4.

GET offers superior outcomes in terms of convergence speed and accuracy over various well-known nature-inspired evolutionary techniques such as GA and PSO [4]. This is poles apart from GA and PSO. In each step, GA changes its initial position, and use mutation, recombination and finds the survival of fittest value [5], whereas PSO changes its position according to the current position and uses personal best, global best. In GET, the position of every grasshopper is employed to characterize the location of search agents. Therefore, it can utilize all the search assistants in the population to search for the target. So, all agents involved in the optimization procedure, and the search competence become superior. Therefore in the search process of GET, each grasshopper upgrades the position by its existing position, global best, and also by all other grasshoppers' positions.

5 Analysis with Simulation Outcomes

The proposed GET-tuned FOPID controller is presented in this section to formulate a better transient response of the MAGLEV system. It has a perfect response and static characteristics than 1 DOF PID and 2 DOF PID. The discussed techniques refer to the precise optimal design that improves the dynamic response of the maglev model. The finest outcomes are given in the allied tables. The significant observations and analysis of this study are presented in the following subsections.

5.1 Analysis Based Upon Transient Response

In the first step, the conventional PID controller parameters $K_p = -2.3973$, $K_i = -2.7749$, $and\, K_d = -0.4451$ were approximated by using the Ziegler–Nichols (ZN) method [10], and controller gains are negative and the cause can be validated by the ZN method.

Applying MATLAB simulation to the transfer function of MAGLEV with respective controllers given in the earlier section, the best possible values of the 1DOF, 2DOF PID, 1DOF FOPID, and GET-FOPID controller parameters obtained are presented in Table 2.

It has been very clearly found out from the above-stated table and also from Fig. 5 that GET-FOPID gives a better response than 1DOF PID, 2DOF PID, and 1DOF FOPID.

To analyze and measure the increased error in the system, a performance index known as the integral of absolute value of error (IAE) has been used and given as follows:

$$IAE = \int_0^\infty |e(t)|\, dt \tag{10}$$

The performance of GET has been compared with the genetic algorithm (GA) and particle swarm optimization (PSO). From Table 3, it can be clearly observed that GET provides finer results than the other two methods.

5.2 Analysis Based upon Robustness

The robustness of the maglev system can be proved only when it can able to withstand system disturbances and ecological changes in real-time applications. In this study, robustness of maglev using all the described controllers has been measured through gain and phase margin, cross over frequency, and sensitivity.

Table 2 Comparison of performance indices between PID and FOPID controllers

Type of controller	Value of K_p	Value of K_i	Value of K_d	Value of λ	Value of μ	Max. overshoot (%)	Peak-time (S)	Rise-time (S)	Settling time (S)
1-DOF PID	−2.3973	−2.7749	−0.4451	1	1	11.95	1.266	0.713	5.581
2-DOF PID	−0.15	−2.5	−2.75	1	1	8	1.288	0.709	5.2
1-DOF FOPID	−4	−3	−0.098	0.98	0.99	5	1.362	0. 612	4
GET-FOPID	−3.5	−4.91	−0.083	0.875	0.997	0	0	0.54	3.434

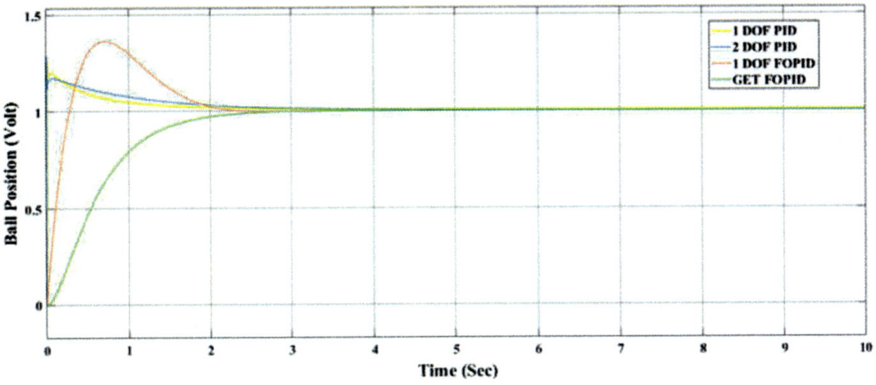

Fig. 5 Transient response of MAGLEV using different controllers

Table 3 Comparison of GET with GA and PSO through FOPID controller

Different optimization methods	Value of K_p	Value of K_i	Value of K_d	Value of λ	Value of μ	IAE
GA	−2.95	−5.49	−0.051	0.836	0.716	3.7515
PSO	−3.75	−5.17	−0.075	0.892	0.695	3.5823
GET	−3.5	−4.91	−0.083	0.875	0.997	2.3510

Table 4 Comparison of robustness through four-different controllers

Type of controller	Gain_margin	Phase_margin	The phase angle slope around the gain crossover frequency	Sensitivity
1-DOF PID	−23.73	35.7749	3.9123×10^{-6}	1
2-DOF PID	−21.15	38.5	3.5153×10^{-6}	1.21
1-DOF FOPID	−15.83	52.7	2.483×10^{-8}	1
GET-FOPID	−15.1	55.1	2.083×10^{-8}	1.1

Table 4 compares the robustness of all four controllers. It should be pointed out that the transfer function used for 1DOF and 2DOF controller remains the same. The table given below depicts that the value of the gain margin is negative in all four cases, whereas the phase margin is positive and has superior value in comparison to the gain margin. Therefore, the maglev system is said to be closed-loop stable for all four controller structures. Here, sensitivity lies in between 1 and 2 which ensures a robust and error-free MAGLEV system. To prove the robustness of the maglev with four different controllers, a noise signal of magnitude 10% of the average of the input signal has been employed as system disturbance. Figure 6 represents the transient response of the MAGLEV with pulse disturbance. The aforementioned

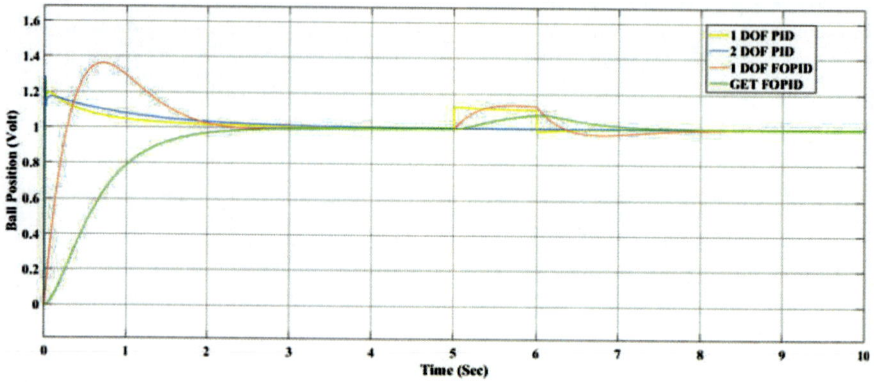

Fig. 6 Transient response of MAGLEV with pulse disturbance

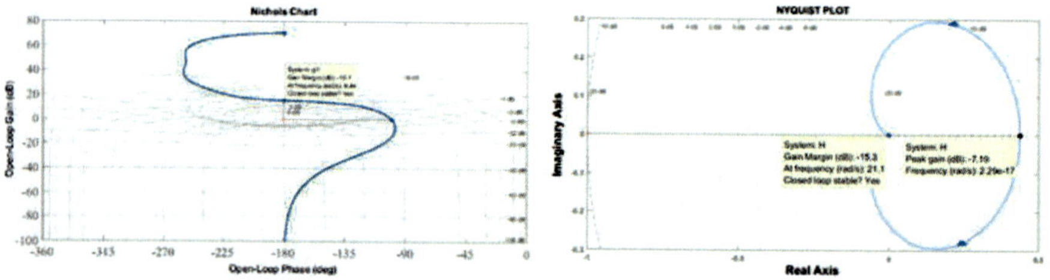

(a) (b)

Fig. 7 **a** Nichols plot and **b** Nyquist plot of the maglev system with GET-FOPID

results and analysis show that the MAGLEV system with the GET-FOPID controller has proved to be more robust than the other mentioned controllers with the same system.

It can be clearly seen from Table 4 that all the gain margins are negative in nature. Even though the gain margins are not positive, still an unstable maglev system along with negative gain associated with an optimized controller results in a stable response. That is obvious from both plots presented in Fig. 7a and b.

According to the stability analysis results of these plots, it can be observed that all the closed-loop poles are lying in the left half of the *s*-plane. Also, peak gain, gain margin, and phase margin are obtained from Nichols plot analysis are given in Fig. 7a and in Table 4. Therefore, it can be concluded that the system is stable and has a very good frequency response. Plots of Fig. 7 provide the required information about the stability of the MAGLEV model. Therefore, it can be clearly said that the GET-FOPID-MAGLEV system is considered to be a stable system.

6 Conclusion

In this paper, the performance of 1DOF, 2DOF PID, 1DOF FOPID, and the proposed GET-FOPID controller have been compared. Using the recently evolved GET Algorithm 1 DOF FOPID controller parameters is optimized for the MAGLEV. Based on the transient response of different MATLAB simulation results, the final analysis concludes that when the controller configuration shifts from 1 to 2 DOF, the performance of the MAGLEV gets enhanced. Analysis based upon robustness shows that the 1 and 2DOF PID and 1-DOF FOPID and the proposed GET-FOPID controllers designed for the MAGLEV are robust and fairly perform under different conditions. It is found that the performance of the GET algorithm is based on the objective function and also on the range of the parameter. In the coming years, the Maglev system can be designed through simulation and real time by using more robust and adaptive controllers. The proposed GET-FOPID may be a better option for this MAGLEV system controller in comparison to other similar suggested PID modified approaches.

References

1. Prasad, N., Jain, S., Gupta, S.: Electrical components of Maglev systems: emerging trends. Urban Rail Transit, 1–13 (2019)
2. Sun, Y., Xu, J., Qiang, H., Wang, W., Lin, G.: Hopf bifurcation analysis of maglev vehicle–guideway interaction vibration system and stability control based on fuzzy adaptive theory. Comput. Ind. **108**, 197–209 (2019)
3. Sun, Y., Xu, J., Qiang, H., Chen, C., Lin, G.: Adaptive sliding mode control of maglev system based on RBF neural network minimum parameter learning method. Measurement **141**, 217–226 (2019)
4. Wai, R.J., Lee, J.D., Chuang, K.L.: Real-time PID control strategy for maglev transportation system via particle swarm optimization. IEEE Trans. Industr. Electron. **58**(2), 629–646 (2010)
5. Safaei, F., Suratgar, A.A., Afshar, A., Mirsalim, M.: Characteristics optimization of the maglev train hybrid suspension system using genetic algorithm. IEEE Trans. Energy Convers. **30**(3), 1163–1170 (2015)
6. Ghosh, A., Krishnan, T.R., Tejaswy, P., Mandal, A., Pradhan, J.K., Ranasingh, S.: Design and implementation of a 2-DOF PID compensation for magnetic levitation systems. ISA Trans. **53**(4), 1216–1222 (2014)
7. Sain, D.: Real-Time implementation and performance analysis of robust 2-DOF PID controller for Maglev system using pole search technique. J. Ind. Info. Int. **15**, 183–190 (2019)
8. Dulău, M., Gligor, A., Dulău, T.M.: Fractional order controllers versus integer order controllers. Procedia Engineering **181**, 538–545 (2017)
9. Rojas-Moreno, A., Cuevas-Condor, C.: Fractional order PID control of a MAGLEV system. In 2017 Electronic Congress (E-CON UNI) (pp. 1–4). IEEE, November (2017)
10. Podlubny, I., Tavazoei, M. S., Vinagre, J. B., Xue, D., Chen, Y., Haeri, M.: A Special Issue in ISA Transactions" Fractional Order Signals, Systems, and Controls: Theory and Application. ISA transactions **82**(1) (2018)

11. Valério, D., Da Costa, J.S.: Tuning of fractional PID controllers with Ziegler–Nichols-type rules. Sig. Process. **86**(10), 2771–2784 (2006)
12. Saremi, S., Mirjalili, S., Lewis, A.: Grasshopper optimisation algorithm: theory and application. Adv. Eng. Softw. **105**, 30–47 (2017)
13. Mafarja, M., Aljarah, I., Heidari, A.A., Hammouri, A.I., Faris, H., Ala'M, A.Z., Mirjalili, S.: Evolutionary population dynamics and grasshopper optimization approaches for feature selection problems. Knowl.-Based Syst. **145**, 25–45 (2018)
14. Biswal, S.S., Swain, D.R., Rout, P.K.: Control of parallel AC voltage source converter high-voltage DC system Using an adaptive PI evolutionary-based controller. Arabian J. Sci. Eng. 1–15 (2020)

Automatic Classification of Life-Threatening Cardiac Arrhythmias Using Empirical Mode Decomposition

Monalisa Mohanty, Santanu Sahoo, and Asit Subudhi

Abstract Ventricular fibrillation (VF) and ventricular tachycardia (VT) are the main causes of sudden cardiac deaths. It can be prevented by the appropriate application of automated external defibrillators (AEDs) by recognizing the severity of cardiac arrhythmias using complex algorithms. In this paper, we propose a novel algorithm based on empirical mode decomposition (EMD) to decompose the electrocardiogram (ECG) signal. Initially, we have applied the EMD technique to decompose VF and VT signals into five sub-bands, respectively. A total of 15 time–frequency features were extracted from each sub-band, and the classification via regression (CVR) and multilayer perceptron (MLP) classifier were used for the classification of the VT/VF conditions. This feature set has been validated with the projected data in the open database gotten from PhysioNet database. The anticipated algorithm results in an accuracy of 90.86% in MLP classifier and 95.61% in CVR classifier. The results demonstrate that the time–frequency data representations that are provided to the classifiers deliver greater performance values.

Keywords Ventricular fibrillation · Ventricular tachycardia · Electrocardiogram · Empirical mode decomposition · Feature extraction · Classification

1 Introduction

Thousands of deaths occur annually due to sudden cardiac arrest [1]. Rapid ventricular arrhythmias are the main cause of death. Ventricular fibrillation (VF) and ventricular tachycardia (VT) are the main two types of ventricular arrhythmias.

M. Mohanty · S. Sahoo · A. Subudhi (✉)
ITER, Siksha 'O' Anusandhan (Deemed to be University), Bhubaneswar, India
e-mail: asitsubudhi@soa.ac.in

M. Mohanty
e-mail: monalisamohanty@soa.ac.in

S. Sahoo
e-mail: santanusahoo@soa.ac.in

S. Das and M. N. Mohanty (eds.), *Advances in Intelligent Computing and Communication*, Lecture Notes in Networks and Systems 202,
https://doi.org/10.1007/978-981-16-0695-3_45

The ventricular fibrillation is the initial rhythm which occurs in almost all patients who suffer from sudden cardiac arrest in out-of-hospital patients [2]. Automated external defibrillator (AED) is the most effective method for delivering electrical shocks that improve the life-saving opportunity in case of the shockable/non-shockable heart rhythms [3]. Application of defibrillator at an early stage along with chest compression (CC) is important for patients suffering from cardiac arrest for the recovery of spontaneous circulation [4]. But the artifact components due to the CCs in the electrocardiogram (ECG) signal cause inaccuracy and adversely affect the decision of the AED [5]. Therefore, various methods have been proposed based on machine learning (ML) techniques for accurate detection of ventricular arrhythmias [6, 7].

However, the effectiveness of these techniques depends on the identification of the most informative features accompanied by a feature selection (FS) algorithm [8]. Wavelet transform (WT) method has been effectively adapted for studying non-stationary signals like ECG analysis for automatic detection of atrial fibrillation on Holter recordings [9]. The wavelet transforms along with fuzzy neural network are proposed to detect shockable rhythms based on features derived from a frequency band [10]. An effective algorithm based on the threshold crossing sample count (TCSC) has been reported to classify VT into the non-VF group, but it fails as it does not emphasize the shape of the ECG signal [11]. The empirical mode decomposition (EMD) technique is used effectively to remove undesirable frequency in order to improve the detection rate of the ECG signal analysis [12]. Anas et al. [13] suggested a novel method based on EMD that significantly improved the detection of the life-threatening cardiac arrhythmias. Li et al. [14] presented a VT/VF classification method using 14 time–frequency-based features extracted from 5s ECG signal. The feature was selected based on genetic algorithms and classified with the support vector machine (SVM) classifier. It achieved an accuracy of 98.1%. An effective method is reported for recognition of life-threatening ventricular arrhythmias based on the total of 13 features consisting of temporal (morphological), spectral, and complexity features derived from the ECG signal and classified with the SVM classifier. In another study, they have considered 11 ECG parameters such as TCI, standard exponential, complexity measurement, VF filter (VFleak), spectral algorithm, phase space reconstruction (PSR), and Hilbert transform (HT) for the recognition of life-threatening arrhythmias using the SVM classifier [14]. The obtained results showed that the VFleak, HT, and PSR features improve the recognition pace. In our previous work, we have used 13 features based on temporal, spectral, and statistical characteristics only to classify VT and VF arrhythmias using C4.5 algorithms [15].

Owing to the non-stationary nature of VF signals and complex morphology of the ECG signal, it is a challenging task to distinguish the VT and VF arrhythmias effectively using time–frequency domain features. Hence, we present an efficient automated method 15 time–frequency-based features for recognition of VF and VT episodes. The empirical mode decomposition technique has been used to reduce the noisy signal and to segregate the VT, VF, and normal signals, and the important features have been extracted and classified with CVR and MLP classifiers.

2 Methods

2.1 ECG Signals

We have used the CUDB and the VFDB arrhythmia database from PhysioNet repository for evaluation of our algorithm. The CUDB consists of 35 files each of approximately 8 min duration long. The VFDB contains 22 files each of approximately 30 min duration long. A 5s window length has been chosen for both CUDB and VFDB databases in order to extract features in the temporal, spectral, and statistical domain. The algorithm proposed in this study consists of three major steps, as shown in Fig. 1. The first step is to decompose the ECG using empirical mode decomposition (EMD) method, the second step is to extract features, and the last step is to classify for VF/VT detection.

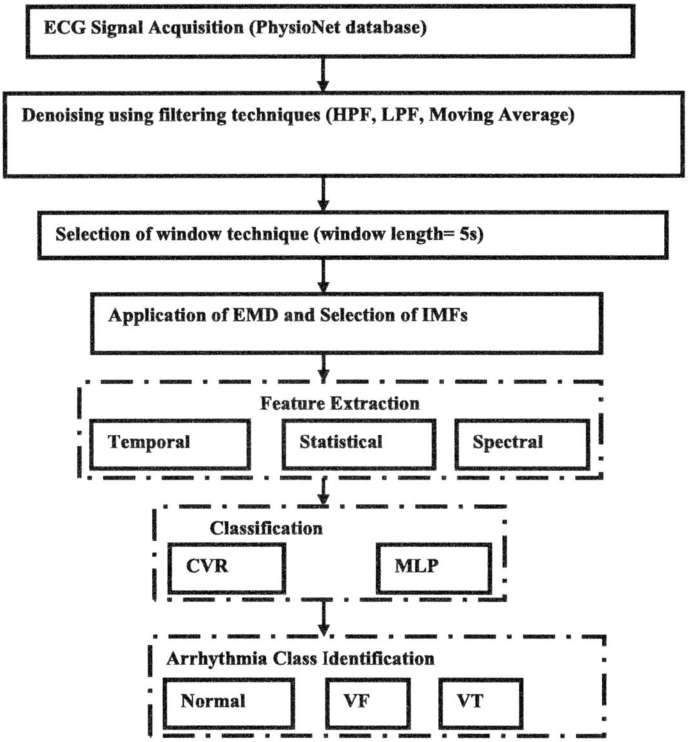

Fig. 1 Flowchart of EMD based ventricular arrhythmia recognition process

2.2 Feature Extraction and Selection

Fifteen different features (as listed in Table 1) were extracted in temporal, spectral, and time domain [16]. These features are LKG, FSMN, TCSC, AB, VAL, MEA, CROSS, CO, KUR, FB, C3, C1, HURST, TH_ENT, and LOG_ENT. Then the performance analysis of the extracted features was done using ranker search algorithm for correlation attribute evaluation method [16].

2.3 Empirical Mode Decomposition (EMD)

The EMD technique is an essential step for reduction of any specified data into a set of intrinsic mode functions (IMFs) [12]. The EMD decomposes a signal x of duration l as a sum of N oscillatory components known as intrinsic mode functions (IMFs). The process of finding an IMF is called sifting, and the process is given as below:

1. Identify the positions of all the extrema of the variable $x(t)$.
2. Interpolate between all the minima/maxima to obtain the lower/upper envelope to find the minima, i.e., $xmin\ (t)/xmax\ (t)$.
3. Calculate the local mean value as follows:

$$m = \frac{xmin(t) + xmax(t)}{2} \tag{1}$$

Table 1 Feature selection using ranker search algorithm

Features	Rank	Weight	ACC	SE	SP	PPV
LKG	1	0.529	79.51	67.47	84.68	66.48
FSMN	2	0.37	74.19	60.51	81.03	59.64
TCSC	3	0.367	71.23	56.33	78.81	56.20
AB	4	0.339	65.43	47.39	74.18	44.66
VAL	5	0.327	69.10	51.45	76.19	52.66
MEA	6	0.37	69.10	51.45	76.19	52.66
CROSS	7	0.327	69.10	51.45	76.19	52.66
CO	8	0.307	66.26	49.09	74.87	47.42
KUR	9	0.304	75.98	62.55	82.33	62.48
FB	10	0.298	73.21	57.98	79.26	64.36
C3	11	0.28	73.54	58.54	79.84	58.14
C1	12	0.261	71.21	52.11	77.48	52.66
HURST	13	0.24	66.70	47.42	74.01	45.14
TH_ENT	14	0.226	70.76	54.02	77.21	62.86
LOG_ENT	15	0.211	69.19	52.11	76.05	54.28

4. The difference between the data *x*(*t*) and mean *m*(*t*) is the known as the first component h_1:

$$h_1 = x(t) - m \tag{2}$$

5. In the consequent sifting procedure, $h1$ is considered as a proto-IMF. In the subsequent step, $h1$ is calculated as follows: $h_{11} = h_1 - m_{11}$
6. After k number of shifting, $h1$ becomes an IMF and is given as follows:

$$h_{1k} = h_{1(k-1)} - m_{1k} \tag{3}$$

7. Then, the first IMF component of the data will be given as follows:

$$C_1 = h_{1k} \tag{4}$$

The stoppage criteria of the sifting process occur when a sum of the difference, SD, is smaller than a previously set value.

$$\mathrm{SD}_k = \sum_{t=0}^{T} \frac{|h_{k-1}(t) - h_k(t)|^2}{h_{k-1}^2(t)} \tag{5}$$

2.4 *Classifiers*

Multilayer perceptron (MLP): The perceptron calculates a sole output from numerous real-valued inputs by creating a linear combination according to its input weights and putting the output through some nonlinear activation function [17]. It can be given mathematically as follows:

$$y = \varphi\left(\sum_{i=1}^{n} v_i a_i + b\right) = \varphi\left(V^T A + b\right) \tag{6}$$

where V represents the vector of weights,

A represents the vector of inputs,

b represents the bias, and

φ represents the activation function.

Classification via regression (CVR): It is used to perform classification using regression methods. Here the class is binarized, and one regression model is built for each class value [18].

3 Experimental Results

The experiments for the detection of ventricular arrhythmias were done using the ECG signals obtained from CUDB and VFDB standard databases as described in the method section. In the preprocessing stage, the signals were de-noised for eliminating motion artifacts and power line noise as shown in Fig. 2. A window length of 5 s was chosen for feature extraction. The following 15 important features were considered for evaluation: LKG, FSMN, TCSC, AB, VAL, MEA, CROSS, CO, KUR, FB, C3, C1, HURST, TH_ENT, and LOG_ENT. The ranking approach was used for each attribute in order to increase the classification accuracy which was performed by a correlation attribute evaluation method. The ranker search algorithm has been used for individual evaluations of the extracted features. The method was evaluated for detection of VT/VF conditions in terms of sensitivity (SE), specificity (SP), accuracy (ACC), and positive predictivity (PPV). Table 1 shows the rank of features sorted by ranker search method with their respective weight related to the arrhythmia conditions. The results indicated that the LKG has a maximum weight and LOG_ENT has a minimum weight. Table 2 shows the

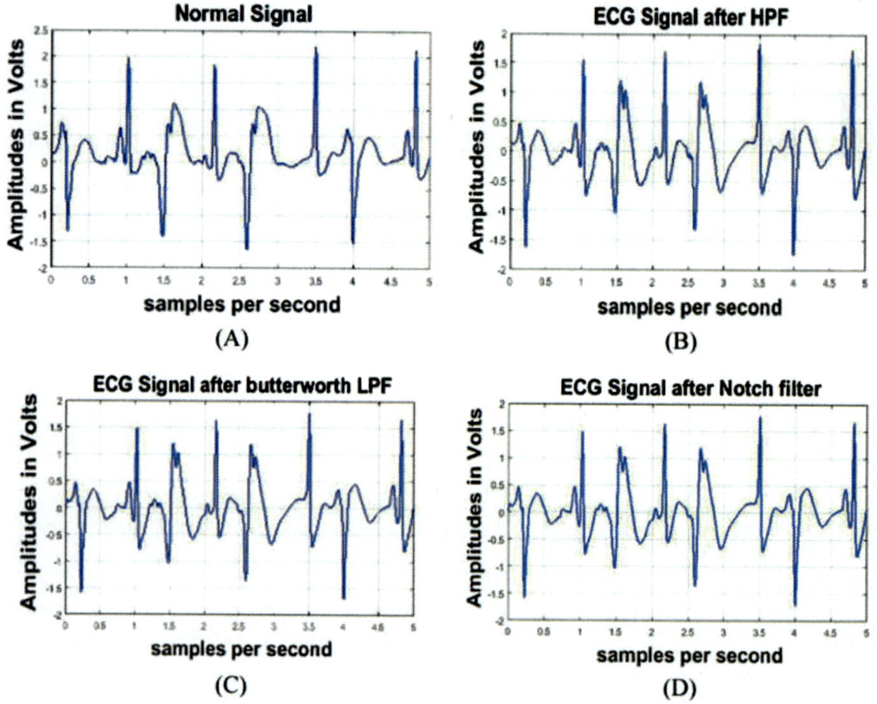

Fig. 2 Preprocessing output **a** Original signal, **b** High-pass filter signal, **c** Low-pass filter output, **d** Notch filter output

Table 2 Performance analysis of CVR classifier using 15 features for a window length of 5 s

No. of features	ACC	SE	SP	PPV
1	77.96	64.62	83.59	64.34
1, 2	83.05	72.79	87.44	72.74
1, 2, 3	86.12	77.87	89.82	78.28
1, 2, 3, 4	87.13	79.45	90.64	80.33
1, 2, 3, 4, 5	87.77	80.50	91.11	80.98
1, 2, 3, 4, 5, 6	87.77	80.50	91.11	80.98
1, 2, 3, 4, 5, 6, 7	87.77	80.50	91.11	80.98
1, 2, 3, 4, 5, 6, 7, 8	87.77	80.50	91.11	80.98
1, 2, 3, 4, 5, 6, 7, 8, 9	87.60	80.04	91.05	80.49
1, 2, 3, 4, 5, 6, 7, 8, 9, 10	90.30	84.54	93.02	85.22
1, 2, 3, 4, 5, 6, 7, 8, 9, 10, 11	90.14	84.13	92.89	84.37
1, 2, 3, 4, 5, 6, 7, 8, 9, 10, 11, 12	90.00	83.67	92.78	83.68
1, 2, 3, 4, 5, 6, 7, 8, 9, 10, 11, 12, 13	90.30	84.19	93.02	84.20
1, 2, 3, 4, 5, 6, 7, 8, 9, 10, 11, 12, 13, 14	95.21	92.08	96.46	92.28
All 15 features	95.61	92.81	96.76	92.92

performance evaluation of CVR algorithm of all 15 features for a window length of 5 s, and the best accuracy rate found was 95.61% with Se = 92.81%, Sp = 96.76%, and PPV = 92.92%, respectively. The performance evaluation of the MLP algorithm was also done for comparative analysis, and the best accuracy rate found was 90.86% with Se = 82.83%, Sp = 92.02%, and PPV8 = 2.74%, respectively. Figure 3 was plotted for accuracy. Figure 4 indicates the comparative analysis of evaluated parameters for both the classifiers.

4 Discussion and Conclusion

A novel method based on empirical mode decomposition for the identification of life-threatening ventricular arrhythmias has been presented in this paper. The efficiency of the proposed method has been tested using PhysioNet databases for a wide range of ECG records. The decomposition was done for extracting informative features from a windowing signal of 5 s. This tested algorithm confirmed superiority due to the validated performance of the CVR classifier using extracted 15 features as the input for detection of VF/VT rhythms. As a result, the ranked feature allows an efficient search for informative features in order to increase the final classification accuracy. The VF/VT conditions were detected with good accuracy in the CVR classifier with an accuracy of 95.61%. It has been shown that the proposed algorithm based on EMD technique and 15 different extracted feature set can detect

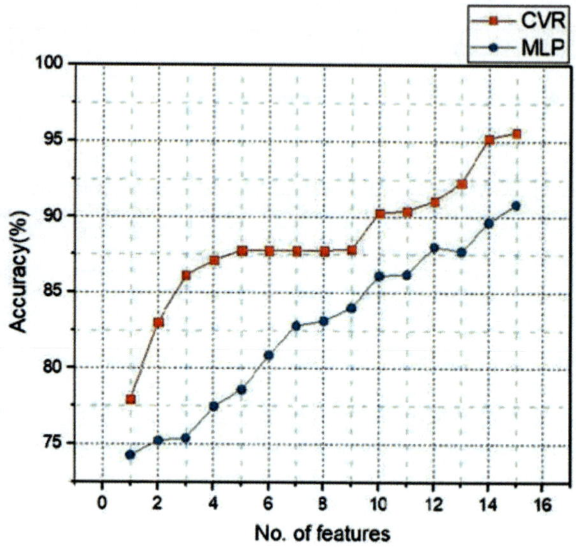

Fig. 3 Results of accuracy for different feature combinations using classifiers

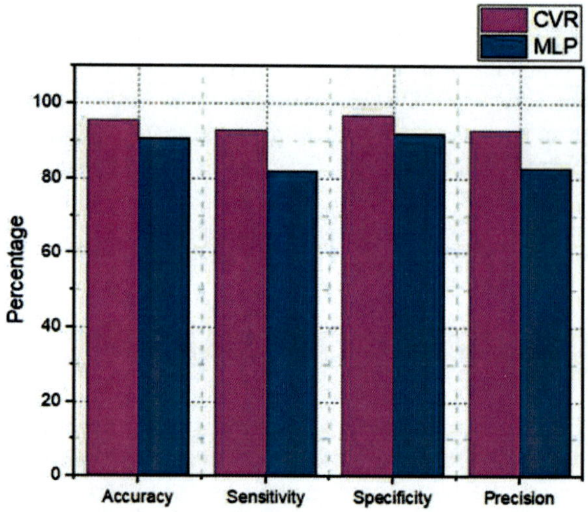

Fig. 4 Comparative analysis of evaluated parameters of classifiers

the VT and VF signals accurately, and the obtained accuracy level was higher than that of other techniques. The algorithm may be used for automated detection of ventricular arrhythmias and may be suitable for practical application of defibrillators.

References

1. Israel, C.W.: Mechanisms of sudden cardiac death. Indian Heart J. **66**(Suppl 1), S10–S17 (2014)
2. Waalewijn, R.A., de Vos, R., Koster, R.W.: Out-of-hospital cardiac arrests in Amsterdam and its surrounding areas: results from the Amsterdam resuscitation study (ARREST) in Utstein style. Resuscitation **38**, 157–167 (1998)
3. White, R.D.: Technologic advances and program initiatives in public access defibrillation using automated external defibrillators. Curr. Opin. Crit.Al Care **7**, 145–151 (2001)
4. Ristagno, G., et al.: Amplitude spectrum area to guide defibrillation a validation 1617 patients with ventricular fibrillation. Circulation **131**(5), 478–487 (2015)
5. van Alem, A.P., Sanou, B.T., Koster, R.W.: Interruption of cardiopulmonary resuscitation with the use of the automated external defibrillator in out-of-hospital cardiac arrest. Ann. Emerg. Med. **42**, 449–457 (2003)
6. Mohanty, M., Biswal, P., Sabut, S.: Ventricular tachycardia and fibrillation detection using DWT and decision tree classifier. J. Mech. Med. Biol. **19**, 1950008 (2019)
7. Alonso-Atienza, F., Morgado, E., Fernandez-Martınez, L., GarcıaAlberola, A., Rojo-Alvarez, J.L.: Detection of life-threatening arrhythmias using feature selection and support vector machines. IEEE Trans. Biomed. Eng. **61**, 832–840 (2014)
8. Mohanty, M., Subudhi, A.K., Biswal, P.K., Sabut, S.: An efficient classifier-based approach for early arrhythmia detection with feature reduction using ranker search algorithm. Lect. Notes Netw. Syst. **109**, 313–320 (2020)
9. Duverney, D., Gaspoz, J.M., Pichot, V., et al.: High accuracy of automatic detection of atrial fibrillation using wavelet transform of heart rate intervals. Pacing Clin. Electrophysiol. **25**, 457–462 (2002)
10. Shyu, L.-Y., Wu, Y.-H., Hu, W.: Using wavelet transform and fuzzy neural network for VPC detection from the Holter ECG. IEEE Trans. Biomed. Eng. **51**, 1269–1273 (2004)
11. Arafat, M.A., Chowdhury, A.W., Hasan, M.K.: A simple time domain algorithm for the detection of ventricular fibrillation in the electrocardiogram. SIViP **5**, 1–10 (2009)
12. Sahoo, S., Mohanty, M., Behera, S., Sabut, S.K.: ECG beat classification using empirical mode decomposition and mixture of features. J. Med. Eng. Technol. **41**, 652–661 (2016)
13. Anas, E., Lee, S., Hasan, M.: Sequential algorithm for life-threatening cardiac pathologies detection based on mean signal strength and emd functions. BioMedical Eng. OnLine **9**, 43–64 (2010)
14. Li, Q., Rajagopalan, C., Clifford, G.D.: Ventricular fibrillation and tachycardia classification using a machine learning approach. IEEE Trans. Biomed. Eng. **61**, 1607–1613 (2014)
15. Alonso-Atienza, F., Morgado, E., Fernández-Martínez, L.: Combination of ECG parameters with support vector machines for the detection of life-threatening arrhythmias. Proceeding Comput. Cardiol. **39**, 385–388 (2012)
16. Mohanty, M., Sahoo, S., Biswal, P., Sabut, S.: Efficient classification of ventricular arrhythmias using feature selection and C4.5 classifier. Biomed. Signal Process. Control. **44**, 200–208 (2018)
17. Raji, C.G., Chandra, S.S.V.: Long-Term Forecasting the Survival in Liver Transplantation Using Multilayer Perceptron Networks. IEEE Trans. Syst., Man, Cybern.: Syst. **47**, 2318 (2017)
18. Hou, C., Nie, F., Yi, D., Wu, Y.: Efficient Image Classification via Multiple Rank Regression. IEEE Trans. Image Process. **22**, 340 (2013)

Execution and Fumes Discharge Investigation of a Diesel Motor Utilizing Karanja Oil with Counterfeit Neural Organization

Pragyan P. Patnaik, S. Beura, C. K. Sethi, and S. K. Acharya

Abstract The examination shows impact of ANN on execution and outflow attributes of a mono-chamber diesel motor with variable pressure proportion by utilizing Karanja oil. Four motor working boundaries mix, level of burden, pressure proportion and force were utilized as the info boundary for this displaying work. The presentation and outflow yield reactions for biodiesel are brake explicit energy utilization (BSEC), brake warm effectiveness (BTE), carbon monoxide (CO), hydrocarbon discharge (HC), oxides of nitrogen (NO_x), fumes temperature (text), and smoke opacity. Utilizing the trial information, an ANN model is created for the motor, set up on standard back spread calculation. The relationship coefficient (R) esteem for execution and outflow yields is 0.98647, 0.97966, 0.9728, 0.9334, 0.9875, 0.99693, and 0.96386; separately, ANN results shows a sensible arrangement between the ANN and the trial discoveries for a few motor presentation and qualities of discharge.

Keywords Karanja oil · Artificial neural network · Engine execution · Emission

1 Introduction

The artificial neural network (ANN) methods provide a rapid high performing many calculations using computers. The practice of ANN is grown at a remarkable rate along the high-performance computers. Currently, ANN has been applied to car designing, in anticipating of motor attributes, for various working conditions. This strategy is additionally used to gauge inside burning motor execution.

Alternatively, an exact model for the maneuver of a CI engines with biodiesel is tough to improve due to the compound processes tangled. Motor execution and fumes discharges can be exhibited by utilizing ANNs as another option. This

P. P. Patnaik · S. Beura · C. K. Sethi · S. K. Acharya (✉)
Department of Mechanical Engineering, SOA Deemed to Be University, Bhubaneswar, Odisha, India
e-mail: sarojacharya@soa.ac.in

S. Das and M. N. Mohanty (eds.), *Advances in Intelligent Computing and Communication*, Lecture Notes in Networks and Systems 202,
https://doi.org/10.1007/978-981-16-0695-3_46

displaying technique is applied to assessment of the picked yield factors when adequate trial information is given. They have been used in shifted entries in control frameworks, mechanical technology, design acknowledgment, estimating, medication, power frameworks, assembling, and streamlining, signal handling social, and mental sciences [1–5].

Yusuf et al. performed an experiment in a CI motor which was fueled with mix of compacted oil gas and diesel. This method was utilized to figure power, brake express fuel use, and motor deliveries, etc. It is viewed a good relationship among envisioned and the preliminary outputs with the association coefficient loosening up some place in the scope of 0.92 and 0.99 [6]. They utilized ANN to compute the introduction and arrivals of a diesel motor, considering of motor speed, imbuement weight and gag position quite far and motor force, power, express fuel use close by floods quite far, and perceived a fair relationship among the primer and test respects. The mean relative error was found within 10% [7]. They utilized ANN for the appearing of an oil motor to discover BSFC, BTE, and fume discharges [8]. They utilized ANN to show execution cutoff points and spreads of biodiesel motor utilizing leftover cooking oil as a substitute fuel. They found that the fall away from the faith respects was incredibly near solidarity for an enormous piece of the cutoff points and expected that ANNs can be utilized for the displaying of a biodiesel motor [9]. This mastery was similarly used for the presentation of valve timing in a SI engine [10]. Ismail et al. [11] foretold nine distinctive motor yield reaction to be specific CO, CO_2, NO_x, unburned hydrocarbon (UHC), greatest weight (P_{max}), location of P_{max}, heat discharge rate (HRR_{max}), location of HRR, combined warmth discharge rate (CHRR), and four appropriate motor activity boundaries that are motor speed, yield force, mass stream pace of fills and mixes as information. Tasdemir et al. [12] utilized ANN demonstrating for a gas motor to figure the motor force, force, and explicit fuel utilization and hydrocarbon outflow. ANN can be utilized to manage the issue with incomplete and erroneous information. Ghobadian et al. [13] saw that biodiesel offers extensive decrease in sulfur oxides (SO_x) discharges and impressive decrease in CO, hydrocarbon residue, and particulate issue; however, there is slight increment in NO_x. By the utilization of ANN it was found that, the relationship coefficient (R) values are 0.9487, 0.999, 0.929, and 0.999, for the motor force, explicit fuel utilization, carbon monoxide, and hydrocarbon emanation respectively. Shivkumar et al. [14] saw that BTE for the serious just as hindered infusion timing was reduced than that for the typical infusion timing (27° BTDC). Smoke, unburned hydrocarbon emanation was diminished for cutting-edge infusion timing, while NO_x discharge was expanded. By great control of infusion timing and pressure, proportion outflow can be very much arranged and execution impediments can be improved. ANN displaying was applied to conjecture the exhibition and outflow trademark. The mean mistake esteem was inside 5 and 8% for execution and outflow separately which is inside worthy breaking point.

The utilization of ANN for displaying of both the presentation boundaries and emanation highlights and their plan is a push to broadly comprehend the demonstrating capacity of ANN for these sorts of issues. The utilization of ANN is to help in better forecast with the accessible trial information.

2 Experimental Investigation

Information announced in this article is taken from the trials, led in laboratory. The tests were directed to look at the exhibitions and emanations of a solitary chamber, four-stroke diesel motor with various slick Karanja oil, and diesel fuel mixes.

2.1 Oil Preparation

Karanja oil after extraction from seeds using mechanical expeller is filtered by a filter; then, it is blended with diesel in volume/volume basis. Ten, 20% of oil is blended with 90 and 80% of diesel, respectively, and known as K10, K20. The mixture of Karanja oil and diesel is stirred by a mechanical stirrer fitted with electric motor. The prepared oil blends are then tested in the engine at varying compression ratio for analysis.

2.2 Experimental Setup

Figure 1 shows the authentic diagram of the test setup. The experimental setup consists of a variable weight extent 4 stroke diesel engine which have a single chamber and a dynamometer having capacity of 3.5 kW power at 1500 rpm associated with whirlpool current dynamometer is a Kirlosker mech. Programming "EngineSoftLV" is working for online execution and start evaluation. Piezo-sensor and wrench point sensor which evaluates the consuming weight and the contrasting wrench point independently are mounted into the engine head. The yield shaft of the whirlpool current dynamometer is fixed to a strain check-type load cell for evaluating intelligent weight to the motor. Type K-Chromel (nickel–chromium alloy)/alumel (nickel–aluminum alloy) thermocouples is utilized to gauge gas temperature at the motor fumes, calorimeter fume, water narrows of calorimeter and

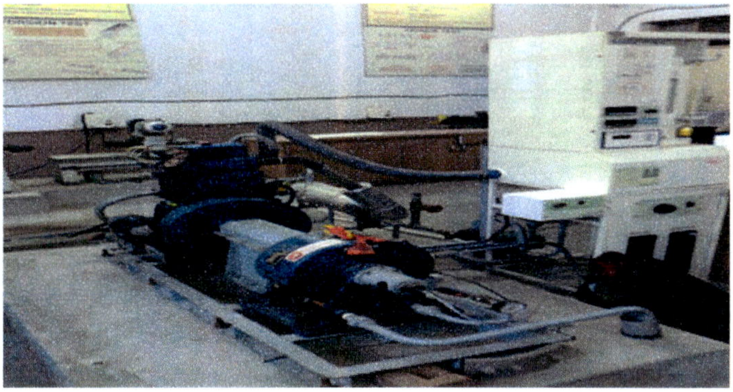

Fig. 1 Actual engine setup

water wellspring of calorimeter, motor cooling water source and climatic temperature. The fuel stream is assessed by the usage of 50 ml burette and stopwatch with level sensors. The fumes transmissions are evaluated by using multi-gas analyzer (AVL-444 model), which is good for taking care of CO, HC, CO_2, and NO_x focuses in the fume. Smoke indefinite quality is surveyed with smoke meter (AVL-437 model) (figuring range 0–100% with objective 0.1%). A modernized information procurement framework is utilized to collect, store, and take a gander at the information during the appraisal by utilizing different sensors.

3 Construction of ANN Model

3.1 Oil Preparation

In back proliferation calculation, scaling of data sources and yields significantly influences execution of ANN. As logistic sigmoidal exchange work was utilized in this investigation, one of the attributes of this capacity is that solitary an incentive somewhere in the range of 0 and 1 can be delivered. Info and yield information collection are standardized before preparing and testing measure. In ANN, standardization can be performed somewhere in the range of 0 and 1 or −1 and 1.

3.2 ANN Modeling for Performance and Emission for CI Engine Using Diesel and Neat Karanja Oil

Consistent state exploratory information was utilized for ANN displaying. For execution and discharge, there are four data sources and seven yields in the trial tests. The four sources of info factors are mix, level of burden, force and pressure proportion. The seven yields of motors are BSEC, BTE, exhaust temperature, CO, HC, NO_x, and smoke. Thus, the information layer comprises of four neurons and yield layer comprises of seven neurons. Out of 126 information, 70% (88) were utilized in the preparation set and staying 30% (38) were utilized for testing. To ensure that every information gives a comparable responsibility in ANN, the duties to the model were preprocessed and scaled into a normal numeric reach [−1, 1]. The associating loads were subjectively prepared for the chief use at the beginning of the planning cycle. In the planning cycle, the learning extent (α) and getting ready speed (β) coefficient were used as 0.25 and 0.3, independently. Association with two hid layers was used with the inception work in the covered layer as tan sigmoid and direct because of yield layer. Standard back expansion figuring has been used for setting up the association. The yield of the association was differentiated, and the ideal yield at each presentation and the mix-up was enrolled. This screw up was at that point back caused to the association and used for changing the heaps so much that the slip-up reduces with accentuation. The amount of ages shows up at the set parameter. From the outset, association was set up by picking subjectively some number of neurons in the covered layer. At a given point number the neurons were either extended or reduced, so that MSE showed up in any event. The number of neurons for which the MSE is least was picked as the ideal number of neurons in the covered layer. Net contribution of the organization can be discovered; summation work is given by condition,

$$\mathrm{NETi} = \sum_{j=1}^{n} wijxj + wbi$$

Loads ($w1$, $w2$, … wn) are the qualities which decide the impact of information set or another preparing component in past layer on the handling component.

The equation of the logistic sigmoidal exchange work is as per the following:

$$f(NETi) = \frac{1}{1 + e^{-N}}$$

ANN design with single hidden layer and seven outputs is as given in Fig. 2;

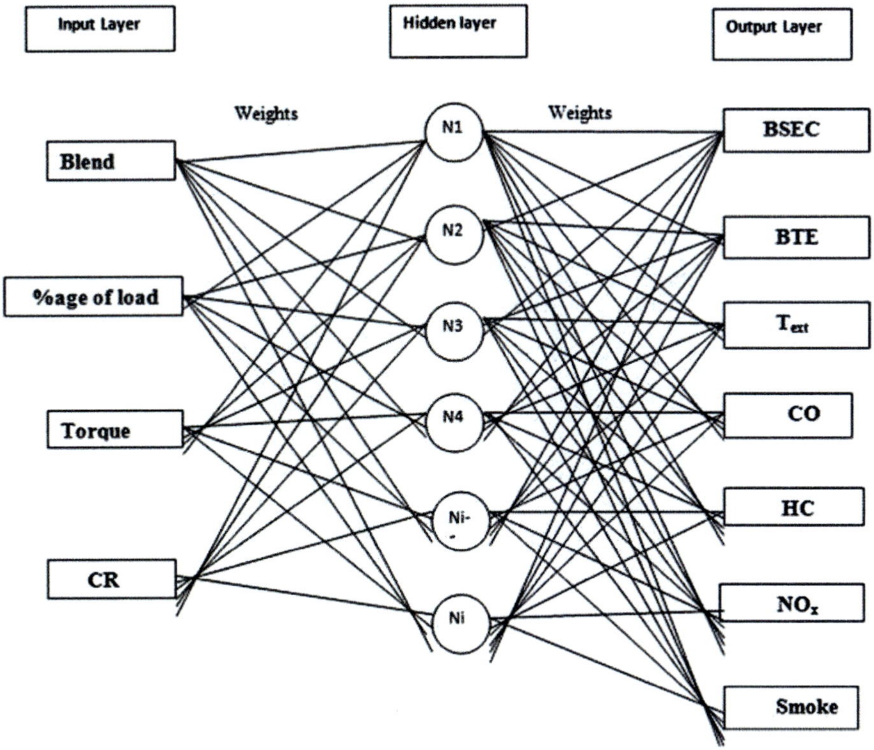

Fig. 2 ANN design

4 Results and Discussion

An ANN model was established to estimate BSEC, BTE, CO, HC, NO, exhaust temperature smoke opacity for neat Karanja oil. The variations of test esteems with those of ANN value are given in Figs. 3, 4, 5, 6, 7, 8, and 9.

ANNs estimated values of BSEC, BTE, CO, HC, NO_x, text, smoke opacity for neat Karanja were assessed among assessed and exploratory information. The ANN model has four inputs variables such as blend, percentage of load, torque and compression ratio. The seven outputs of engines are BSEC, BTE, exhaust temperature, CO, HC, NO_x, and smoke opacity. Some set of training data and some set of testing data were taken. This test was appeared to decide the limit of the made ANN model in estimating the motor yield reaction inside the decided scope of informational collection utilized during its encouraging. The developed ANN is found to be successful in predicting all engine output response accurately. The performance and emission output responses for neat Karanja oil are BSEC, BTE, CO, HC, NO_x, text, and smoke opacity. The correlation coefficient (R) values for performance and emission outputs are 0.98647, 0.97966, 0.9728, 0.9334 0.9875, 0.99693, and 0.96386, respectively.

When mathematical and numerical methods fail, at that time accurate, reliable, and fast predictions of results are given by ANN. For reducing the time and cost, ANN is very effective.

Fig. 3 Correlation for brake explicit energy utilization

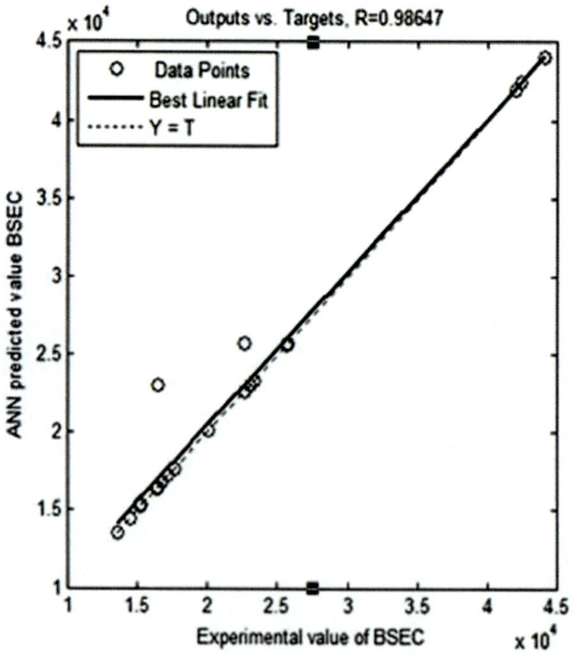

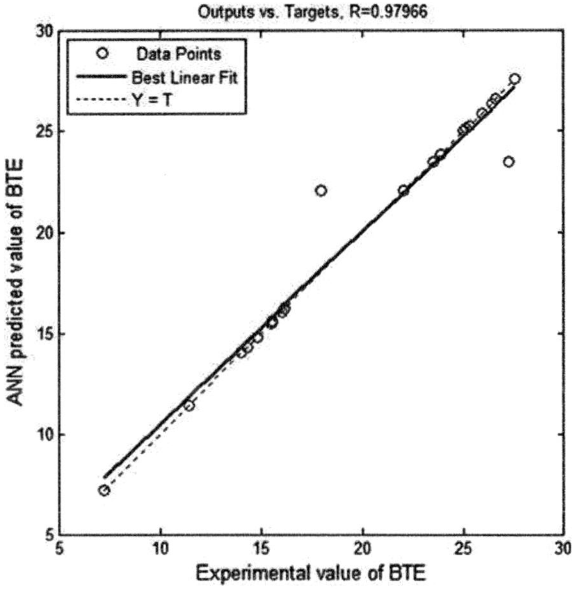

Fig. 4 Comparison for brake thermal efficiency

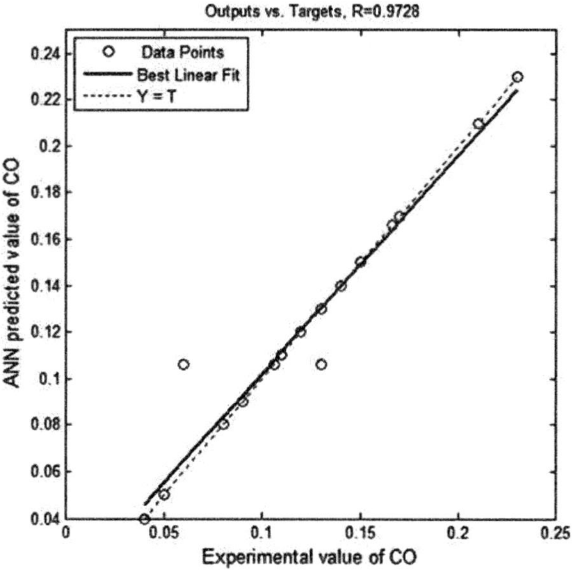

Fig. 5 Comparison for carbon monoxide

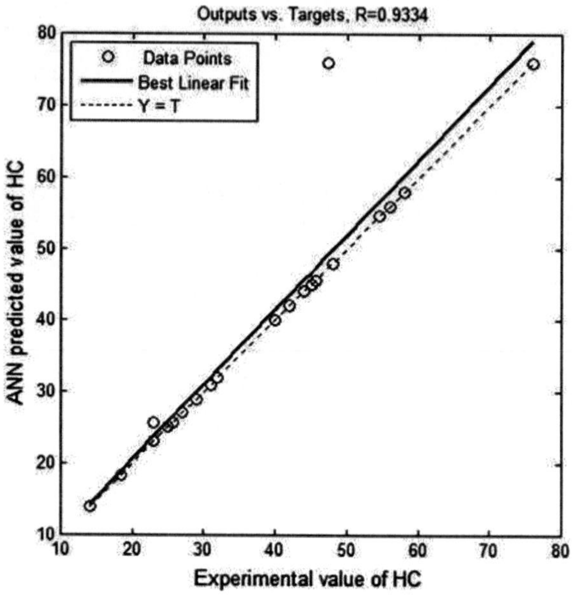

Fig. 6 Comparison for hydrocarbon

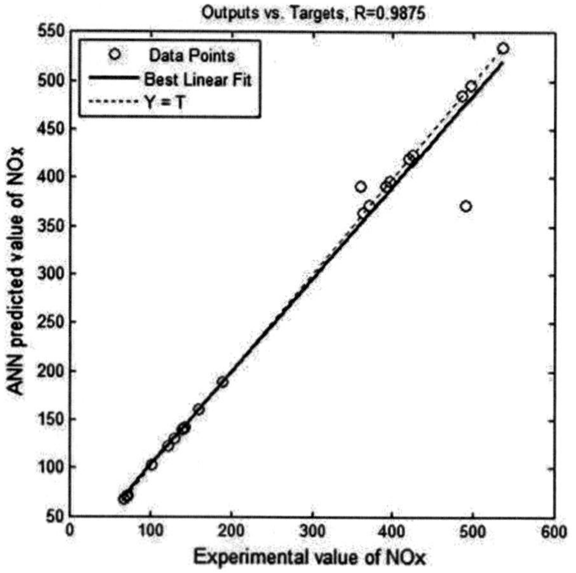

Fig. 7 Comparison for nitrogen oxides

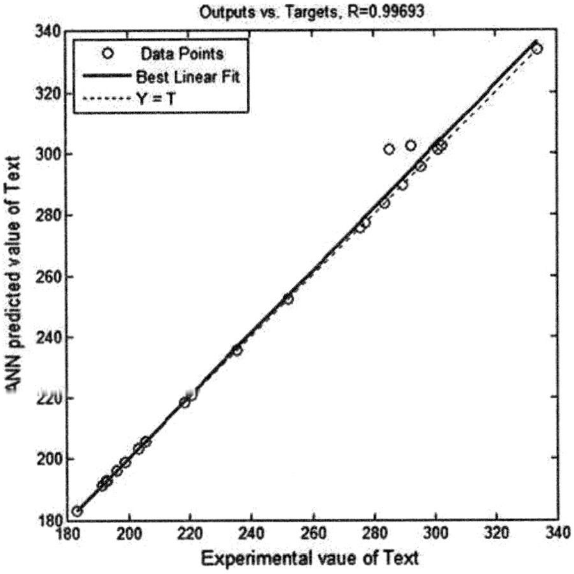

Fig. 8 Comparison for exhaust temperature

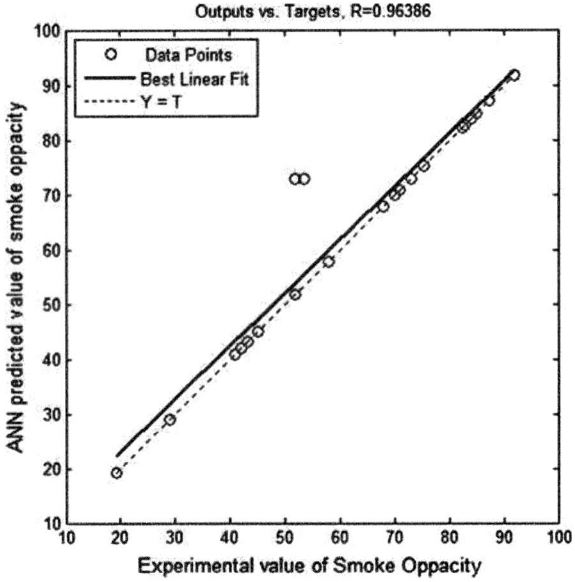

Fig. 9 Comparison for smoke opacity

5 Conclusions

As setup with least number of information boundaries, ANN was used to demonstrate the intricate communication between motor control factors and motor yield reaction. After the execution of ANN model it was found that R technique can also be used with ANN for the better improvement of evaluation process. The outcome demonstrated that back spread neural organization and mix of tan sigmoid exchange work. The planned ANN had the option to honestly estimate all motor yield reactions with serious extent of exactness. The capacity of ANN as a cheerful displaying apparatus with financially savvy and low reproduction run time was shown here. An ideal ANN model is needed for least number of trial cases, applied in the motor testing stage. Subsequent to building up this, motor execution and emanation conduct at various information/working conditions can be anticipated by utilizing ANN model. When input yield relationship is required, then the created approach is pertinent for other ignition framework. Car industry, primarily in diminishing the time and cost identified with the motor and advancement cycle, gets profited utilizing this model.

References

1. Prieto, M.M., Montanes, E., Menendez, O.: Power plant condenser performance forecasting using a nonfully connected ANN. Energy **26**, 65–79 (2001)
2. Bechter, H., Browne, M.W., Bansal, P.K.: Neural networks—a new approach to model vapor compression heat pumps. Int. J. Energy Res. **25**, 591–599 (2001)
3. Lucas, A., Duran, A., Caramona, M.: Modelling diesel particulate emission with neural networks. Fuel **81**, 1963–1970 (2002)
4. Esantes Jose, M.D., Lopez Jose, J., Garcia Jose, M.: Application of neural networks for prediction and optimization of exhaust emissions in a H.D. diesel engine. SAE 2002 1144, 1993–2001
5. Alonso Jose, M., Fernando, A.: Combining ANN with genetic algorithm to reduce diesel engine emission. IEEE Trans. Evolut. Comput. **11**, 46–55 (2007)
6. Yusuf Talal, F., Buttsworth, D.R., Saleh Khalid, H., Yousif, B.F.: CNG-diesel engine performance and exhaust emission analysis with the aid of artificial neural network. Appl. Energy **87**, 1661–1669 (2010)
7. Arcaklioglu, E., Celikten, I.: A diesel engines performance and exhaust emissions using ANN. Appl. Energy. **80**, 11–22 (2005)
8. Cenk, S., Metin, E.H.: Performance and exhaust emissions from a gasoline engine using ANN. Appl. Therm. Eng. **27**, 46–54 (2007)
9. Mustafa, C., Ahmet, E., Erol, A.: Performance and exhaust emissions of a biodiesel Engine. Appl. Energy **83**, 594–604 (2006)
10. Golci, M., Sekmen, Y., Erduranli, P., Salman, M.S.: Artificial neural network based modeling of a variable valve timing in a spark ignition engine. Appl. Energy **81**, 187–197 (2005)
11. Ismail, H., Ng, H.K., Queck, C.W., Gan, S.: Artificial neural networks modeling of engine-out responses for a light duty diesel engine fuelled with biodiesel blends. Appl. Energy **92**, 769–777 (2012)
12. TasdemirS, S., CinivizM, A.: Artificial neural network and fuzzy expert system comparison for prediction of performance and emission parameter on a gasoline engine. Expert Syst Appl **38**, 13912–13923 (2011)

13. Ghobadian, B., Rahim, H., Nikbakth, A.M., Najafi, G., Yusaf, T.F.: Diesel engine performance and exhaust emission analysis using waste cooking biodiesel fuel with an artificial neural network. Renew. Energy **34**, 976–982 (2009)
14. Shivakumar, Pai, P.S., Rao, B.R.S.: Artificial Neural Network based prediction of performance and emission characteristics of a variable compression ratio CI engine using WCO as a biodiesel at different injection timing. Appl. energy **88**, 2344–2354 (2012)

An Empirical Analysis of PCA-SVM Model for Cancer Microarray Data Classification

Swati Sucharita, Barnali Sahu, and Tripti Swarnkar

Abstract Cancer is nowadays becoming the main reason for increasing the death toll worldwide. The death percentage is increasing due to the unavailability of a proper method to identify cancer at an early stage. Machine learning plays a vital role in deploying an automated model which will help in better diagnosis. The conventional cancer diagnosis is fully based on biopsy data which misses the important aspects of cancer such as proliferation rate, certain behavior of tissues. Hence to indemnify the cancer subtypes, genetic data may play an important role. The microarray data contain the genetic information of a patient with a huge amount of genetic information as compared to the number of samples. To overcome this limitation, the dimensionality reduction technique needs to be applied before applying any kind of ML algorithm for classification. In this paper, we have used the principal component analysis (PCA) as a dimensionality reduction technique. After that supervised machine learning algorithm, SVM is used as the classifier due to its efficiency in high-dimensional datasets and kernel trick. The above-said methodology has been applied to four cancer datasets such as breast cancer, prostate cancer, colon cancer, cervical cancer. Finally, the performance of PCA-SVM and SVM without emphasizing the usage of gene selection technique before the classification of any microarray data with different kernels has been compared in terms of accuracy, precision, and specificity.

Keywords Principal component analysis · Support vector machines · Microarray · Dimensionality reduction · Cancer classification

---

S. Sucharita · B. Sahu (✉)
Department of Computer Science and Engineering, Siksha'O' Anusandhan (Deemed to Be University), Bhubaneswar, India
e-mail: sahu.barnali08@gmail.com

S. Sucharita
e-mail: swatisucharita08@gmail.com

T. Swarnkar
Department of Computer Application, Siksha'O' Anusandhan (Deemed to Be University), Bhubaneswar, India
e-mail: triptiswarnakar@soa.ac.in

S. Das and M. N. Mohanty (eds.), *Advances in Intelligent Computing and Communication*, Lecture Notes in Networks and Systems 202,
https://doi.org/10.1007/978-981-16-0695-3_47

1 Introduction

Cancer remains a challenge for the oncology department. Approximately 19 million cases were registered in the NCI and WHO statistical study in 2018, out of which about 9.6 million people died due to a lack of proper diagnostic methods [1]. Therefore, cancer classification is emerging as a major research area in the medical field. The conventional model of diagnosis is based on a collection of biopsy data that has minimal knowledge of cancer tissues, thereby restricting proper treatment at a different level. So, for accurate classification, the genetic information regarding the tissue of a cancer patient is required and the microarray data contains the same information. A large amount of cancer gene expression is included in the DNA microarray data with limited sample information defined as high dimensionality [2]. Various models have been developed by considering biopsy data. But, due to the limitation of the biopsy dataset, the accuracy of the model is very hard to accept. Thus, the microarray dataset is being considered as the input to the machine learning model to increase the accuracy of classification. In microarray data classification, the high dimensionality of the feature space degrades the generalization performance of the classifier and increases its computational complexity [3]. So, dimensionality reduction techniques are used before applying a classification model. There are several dimensionality reduction techniques available in the literature such as principal component analysis (PCA) [3], linear discriminant analysis (LDA) [3], independent component analysis (ICA) [4], etc. These techniques are independent of the classification algorithm with a low computational cost.

In the current study, PCA has been taken as the dimensionality reduction technique. SVM has been used as a classifier because of its key advantages, such as high performance in high-dimensional data where the number of dimensions is high relative to the number of samples. The kernel trick and the hyperparameter adds more precision level to it.

In the following manner, the rest of the paper is organized. Section 2 deals with the associated cancer classification work. The dataset used and the description of the PCA-SVM model with different kernels for classification are included in Sect. 3. Section 4 includes experimental outcomes and debate. Subsequently, the Conclusion is included in the last section of the paper.

2 Literature Survey

In Ref. [5], the author has used the genetic algorithm for human cancer classification. Seven cancer data have been taken as the input parameter, and the GA technique has been implemented to classify the data. In Ref. [6], the author has proposed a hybrid method based on the SVM, RFE, and Based Bayes Error Filter (BBEF) for feature selection, and then, the SVM is being used to classify the data. The author in Ref. [7] has used GA with KNN and SVM classification techniques for the cancer classification. A multilayer perception or MLP-based NN has been implemented to classify the cancer dataset in Ref. [8]. In Ref. [9], the author has developed a mutual-based gene selection model with different classifiers such as NN and RF to tackle the microarray data. The author in Ref. [10] has worked on the microarray for cancer classification and shown that the SVM has high accuracy as compared to others. Astuti et al [11] have used KNN as the classifier with ICA as the dimensionality reduction technique to compare the result that the hybrid approach obtains high accuracy as compared to the KNN when applied to cancer microarray dataset. The author in Ref. [12] has developed a hybrid model based on F-score and SVM and compared the results with the ANN to show the high accuracy. In Ref. [13], the author has used PCA and MBP method for reducing the dimensionality and then implemented LR as the classification technique to show the accuracy as 96%.

3 Materials and Methods

3.1 Cancer Dataset

Four types of cancer microarray dataset from the UCI Machine Learning public repository [14] have been taken. The breast cancer gene expression data contains 3476 gene details with 35 samples. Colon cancer gene expression data is having 2254 gene details with 66 sample details. Cervical cancer gene expression data consists of 1792 gene expression details with 38 samples, whereas the prostate cancer dataset consists of 2798 gene expression details with 57 sample sizes.

3.2 Support Vector Machine

Support vector machine or SVM [7] is being widely used as a supervised machine learning algorithm for classification and regression purposes. The objective of the SVM is to bring an optimum hyperplane so that the two classes can be separated. This optimal hyperplane separates the two classes and also maximizes the margin between the same two classes [15]. Margin can be defined as the distance between the hyperplane and the SVs. Let $\{x1, x2, \ldots, xn\}$ are the support vectors with class label as $y \in \{1, -1\}$. The hyperplane equation can be defined as

$$F(x) = wxi + b \tag{1}$$

where the w is a weighted vector. The decision boundary can be defined by solving the optimization problem $\frac{1}{2}\|w\|^2$. The Lagrange of the given optimization problem can be defined as follows:

$$L = \frac{1}{2}\|w\|^2 - \sum_{i}^{\infty} i(yi(W^T xi + b) - 1)) \quad (2)$$

Furthermore, the SVM was being developed for solving the linear data, but by applying the kernel trick the data can be projected into a higher dimension so that the data can be separated linearly. Various kernel functions are available for SVM such as RBF, linear, polynomial, and sigmoid. In this research paper, all types of kernel for comparing the result on different available cancer datasets have been taken. There are two available hyperparameters known as cost function c and gamma parameter which can be fine-tuned to achieve high accuracy. Both c and gamma need to be set before the training to provide a curvature weight of the decision boundary. C is the cost of misclassification, and the gamma is there to handle the nonlinear classification. Lowering c will get less accuracy, whereas lowering gamma will obtain high accuracy.

3.3 Proposed Model

Four types of cancer data such as breast cancer, prostate cancer, colon cancer, cervical cancer datasets are taken for study. PCA has been used to reduce the dimension as the microarray data suffers from a high dimensionality problem. The PCA has been applied for feature reduction. Scenario PC or principal component contains a different number of features obtained from PCA. For the current analysis, four scenarios were taken and labeled as PC1, PC2, PC3, and PC4 for 40, 30, 20, and 10% of large eigenvalues, respectively. All datasets are partitioned into training and test set with 0.2 distribution. After that, SVM is trained with the training data with reduced features and validated using test data, and evaluated using some significant parameters such as precision, recall, F1-score, and accuracy. Figure 1 shows the schematic representation of the PCA-SVM model.

3.4 Algorithm (PCA-SVM)

Step 1: Load microarray dataset (breast cancer, colon cancer, cervical cancer, and prostate cancer).
Step 2: Apply PCA for dimension reduction and generate four sets of samples with different dimensions as follows.

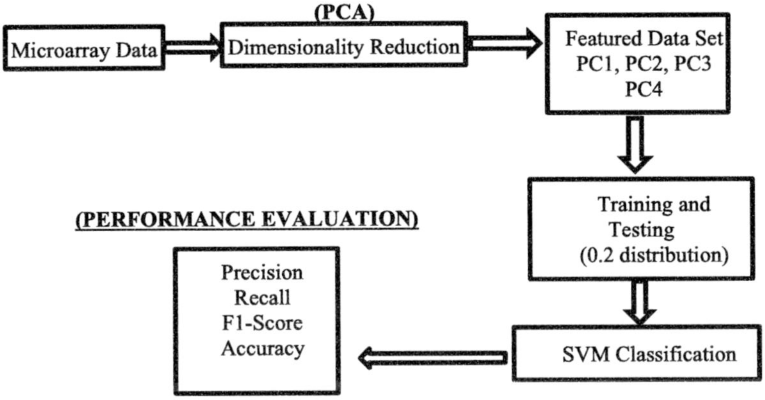

Fig. 1 Schematic representation of PCA-SVM classification model

a. Calculate the covariance matrix for the given microarray dataset.
b. Calculate eigenvalue and eigenvector for the obtained covariance matrix
c. Keep K features depending upon the k highest eigenvalue.

Step 3: Generate featured dataset from PCA for training and testing split with 0.2 distribution factor.
Step 4: Specify hyperparameter for SVM such as cost, gamma, and the kernel trick for better performance.
Step 5: Train SVM with training data.
Step 6: Test SVM with test data.
Step 6: Measure the performance in terms of Precision, Recall, F1 score, and Accuracy.
Step 7: Step 2–6 is repeated for each kind of cancer microarray dataset.

4 Experimental Result and Analysis

The current work has been implemented using Anaconda and Python 3.0.3 environment. After importing all the required libraries, the dataset has been loaded using panda. For PC1 1390 gene expression for breast cancer, 901 gene expression for colon cancer, 717 gene expression for cervical cancer, and 1119 genes for prostate cancer have been taken. For PC2 1042,676, 358 and 839 numbers of genes have been considered for breast cancer, colon cancer, cervical cancer, and prostate cancer, respectively. Similarly, PC3, PC4 (695,451, 358, 560), and (348,225,179,280) numbers of the gene have been considered for breast cancer,

Table 1 Accuracy of SVM with different kernel

	Kernel	Precision	Recall	F1 score	Accuracy
Breast cancer	RBF	0.86	0.87	0.85	0.88
	Linear	0.85	0.85	0.83	0.87
	Polynomial	0.85	0.83	0.832	0.87
	Sigmoid	0.84	0.84	0.84	0.85
Cervical cancer	RBF	0.85	0.83	0.88	0.87
	Linear	0.85	0.84	0.845	0.86
	Polynomial	0.85	0.84	0.84	0.86
	Sigmoid	0.84	0.84	0.82	0.81
Colon cancer	RBF	0.85	0.83	0.83	0.84
	Linear	0.85	0.85	0.84	0.84
	Polynomial	0.83	0.82	0.82	0.83
	Sigmoid	0.83	0.82	0.82	0.83
Prostate cancer	RBF	0.84	0.84	0.83	0.84
	Linear	0.84	0.84	0.83	0.84
	Polynomial	0.83	0.82	0.827	0.83
	Sigmoid	0.83	0.81	0.817	0.79

colon cancer, cervical cancer, and prostate cancer, respectively. After that, the SVM is applied to the featured matrix with its hyperparameter cost (C) and gamma. The cost parameter balances the misclassification of the training data concerning the decision boundary, whereas the gamma will define how far the influence of the training data reaches. The high value of C will result in low accuracy due to the high variance and low bias. So, the cost parameter needs to be taken as low with a high value of gamma. Table 1 shows the results of SVM without emphasizing dimensionality reduction and Tables 2 and 3 show the experimental results of the PCA-SVM model in four different scenarios such as PC1, PC2, PC3, and PC4. The performance of SVM with the PC4 scenario is shown in Fig. 2, where SVM gives better results in comparison to other scenarios for all datasets.

As per the results shown in Table 1, the SVM without emphasizing dimensionality reduction gives the best accuracy as 88% accuracy with RBF kernel for the breast cancer dataset. It is evident from Table 3 and Fig. 2 that the PCA-SVM model with RBF kernel gives the best result, i.e., 99.8% accuracy with 10% of total features. It can be seen from Tables 2 and 3 that the model shows an average accuracy of 96% for all types of kernels and different feature sets for cervical cancer. For colon cancer and prostate cancer datasets, the model offers various kernels and feature sets with an average accuracy of 95%. The result shows that the accuracy level is increasing with reduced dimensionality.

Table 2 Accuracy of PCA-SVM for PC1

		Kernel	Precision	Recall	F1 score	Accuracy
PCA-SVM Result (PC1)—40%	Breast cancer	RBF	0.98	0.94	0.95	0.987
		Linear	0.95	0.95	0.94	0.962
		Polynomial	0.94	0.94	0.93	0.95
		Sigmoid	0.91	0.89	0.92	0.92
	Cervical cancer	RBF	0.95	0.94	0.94	0.952
		Linear	0.94	0.93	0.934	0.95
		Polynomial	0.94	0.93	0.934	0.95
		Sigmoid	0.9	0.9	0.92	0.91
	Colon cancer	RBF	0.95	0.94	0.94	0.952
		Linear	0.94	0.92	0.92	0.93
		Polynomial	0.94	0.93	0.92	0.93
		Sigmoid	0.92	0.93	0.92	0.9
	Prostate cancer	RBF	0.94	0.93	0.935	0.93
		Linear	0.95	0.93	0.935	0.942
		Polynomial	0.94	0.93	0.934	0.95
		Sigmoid	0.92	0.9	0.925	0.917

Table 3 Accuracy of PCA-SVM for PC2, PC3, and PC4

			Precision	Recall	F1 score	Accuracy
PCA-SVM result (PC2)—30%	Breast cancer	RBF	0.98	0.98	0.983	0.989
		Linear	0.97	0.98	0.96	0.97
		Polynomial	0.97	0.96	0.98	0.97
		Sigmoid	0.95	0.95	0.95	0.94
	Cervical cancer	RBF	0.97	0.96	0.964	0.96
		Linear	0.96	0.93	0.944	0.96
		Polynomial	0.94	0.93	0.934	0.95
		Sigmoid	0.94	0.93	0.934	0.95
	Colon cancer	RBF	0.95	0.94	0.94	0.952
		Linear	0.94	0.94	0.944	0.95
		Polynomial	0.94	0.95	0.95	0.941
		Sigmoid	0.93	0.93	0.94	0.95
	Prostate cancer	RBF	0.96	0.934	0.941	0.936
		Linear	0.945	0.93	0.935	0.942
		Polynomial	0.94	0.93	0.934	0.95
		Sigmoid	0.94	0.92	0.92	0.93

(continued)

Table 3 (continued)

			Precision	Recall	F1 score	Accuracy
PCA-SVM for (PC3)—20%	Breast cancer	RBF	0.99	0.98	0.983	0.992
		Linear	0.98	0.98	0.975	0.97
		Polynomial	0.97	0.98	0.98	0.972
		Sigmoid	0.95	0.95	0.95	0.94
	Cervical cancer	RBF	0.97	0.96	0.97	0.966
		Linear	0.97	0.94	0.95	0.96
		Polynomial	0.95	0.94	0.94	0.944
		Sigmoid	0.94	0.93	0.934	0.95
	Colon cancer	RBF	0.95	0.94	0.94	0.952
		Linear	0.94	0.95	0.95	0.945
		Polynomial	0.94	0.95	0.95	0.945
		Sigmoid	0.93	0.94	0.93	0.938
	Prostate cancer	RBF	0.96	0.94	0.94	0.94
		Linear	0.95	0.94	0.94	0.946
		Polynomial	0.94	0.94	0.93	0.95
		Sigmoid	0.94	0.93	0.93	0.935
SVM-PCA for (PC4)—10%	Breast cancer	RBF	0.99	0.98	0.99	0.998
		Linear	0.97	0.97	0.96	0.97
		Polynomial	0.97	0.96	0.98	0.97
		Sigmoid	0.95	0.95	0.95	0.947
	Cervical cancer	RBF	0.971	0.95	0.96	0.96
		Linear	0.96	0.94	0.94	0.96
		Polynomial	0.95	0.93	0.934	0.954
		Sigmoid	0.95	0.94	0.94	0.95
	Colon cancer	RBF	0.95	0.94	0.94	0.952
		Linear	0.95	0.94	0.95	0.956
		Polynomial	0.94	0.95	0.95	0.941
		Sigmoid	0.93	0.93	0.94	0.95
	Prostate cancer	RBF	0.96	0.94	0.93	0.94
		Linear	0.95	0.93	0.94	0.946
		Polynomial	0.95	0.93	0.94	0.953
		Sigmoid	0.94	0.94	0.93	0.94

5 Conclusion

We have focused in this study on the significance of the PCA-SVM classification model for the classification of cancer microarray data. The PCA-SVM model is applied to breast cancer, colon cancer, prostate cancer, and cervical cancer microarray datasets. PCA-SVM and SVM were then compared with different

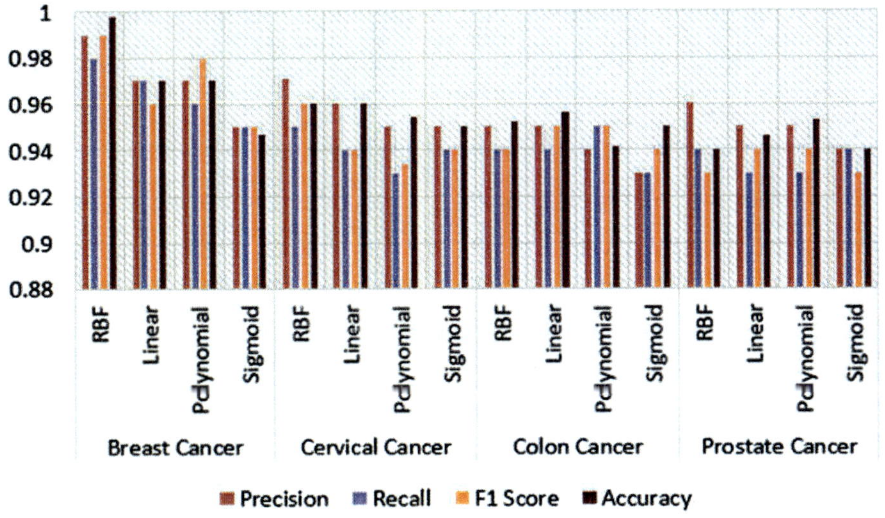

Fig. 2 Precision, recall, F1 score, and accuracy of SVM-PCA for PC4

kernels, and the results show that the PCA-SVM model with RBF kernel is showing better accuracy for breast cancer data set with 10% of a total feature. In terms of precision, PCA-SVM performs better than conventional SVM classifiers for all cancer datasets taken in the current study.

References

1. Alshamlan, H., Badr, G., Alohali, Y.: mRMR-ABC: a hybrid gene selection algorithm for cancer classification using microarray gene expression profiling. Biomed. Res. Int. (2015)
2. Furey, T.S., Cristianini, N., Duffy, N., Bednarski, D.W., Schummer, M., Haussler, D.: Support vector machine classification and validation of cancer tissue samples using microarray expression data. Bioinformatics **16**(10), 906–914 (2000)
3. Sharma, A., Paliwal, K.K.: Cancer classification by gradient LDA technique using microarray gene expression data. Data Knowl. Eng. **66**(2), 338–347 (2008)
4. Adiwijaya, W.U., Lisnawati, E., Aditsania, A., Kusumo, D.S.: Dimensionality reduction using principal component analysis for cancer detection based on microarray data classification. J. Comput. Sci. **14**(11), 1521–1530 (2018)
5. Kashyap, D., Somani, A., Shekhar, J., Bhan, A., Dutta, M.K., Burget, R., Riha, K.: Cervical cancer detection and classification using Independent Level sets and multi SVMs. In: 2016 39th International Conference on Telecommunications and Signal Processing (TSP), pp. 523–528. IEEE
6. Majumder, S.K., Ghosh, N., Gupta, P.K.: Support vector machine for optical diagnosis of cancer. J. Biomed. Optics **10**(2), 024034 (2005)
7. Sánchez, A., D., V.: Advanced support vector machines and kernel methods. Neurocomputing **55**(1–2), 5–20 (2003)

8. Liu, X., Krishnan, A., Mondry, A.: An entropy-based gene selection method for cancer classification using microarray data. BMC Bioinf. **6**(1), 1–14 (2005)
9. Salem, H., Attiya, G., El-Fishawy, N.: Classification of human cancer diseases by gene expression profiles. Appl. Soft Comput. **50**, 124–134 (2017)
10. Bennet, J.; Ganaprakasam, C.; Kumar, N. A hybrid approach for gene selection and classification using support vector machine. Int. Arab J. Inf. Technol. (2015)
11. Yeh, J.-Y., Wu, T.-S., Wu, M.-C., Chang, D.-M.: Applying data mining techniques for cancer classification from gene expression data. In: Proceedings of the 2007 International Conference on Convergence Information Technology (ICCIT 2007), pp. 703–708, Gyeongju, South Korea, 21–23 November 2007
12. Bouazza, S.H., Hamdi, N., Zeroual, A., Auhmani, K.: Gene-expression-based cancer classification through feature selection with KNN and SVM classifiers. In: Proceedings of the 2015 Intelligent Systems and Computer Vision (ISCV), pp. 1–6, Fez, Morocco, 25–26 March 2015
13. Paul, A., Sil, J., Mukhopadhyay, C.D.: Gene selection for designing optimal fuzzy rule base classifier by estimating missing value. Appl. Soft Comput. **55**, 276–288 (2017)
14. https://archive.ics.uci.edu/ml/index.php
15. Chu, F., Wang, L.: Applications of support vector machines to cancer classification with microarray data. Int. J. Neural Syst. **15**(06), 475–484 (2005)

A Stable Path-Based Cross-Layer Approach Utilizing Neighbours RSS Knowledge in MANET

Gyanendra Kumar Pallai, Meenakshi Sankaran, and Amiya Kumar Rath

Abstract Quality of service guarantee (QoS) is much more challenging in mobile ad hoc networks (MANETs), mostly because of node mobility and resource limitations. MANET protocols essentially find workable routes that are more susceptible to repeated failures and restoration of connectivity. Our study proposes an algorithm for providing a strong network with a stable route formulated on AODV (SP-AODV) taking into account cross-layer communication through physical, MAC and network layers. We measure three levels of mean RSS threshold values and dynamically adjusting the transmission power of all nodes at the physical layer. The study uses extensive NS2 simulations to analyse the effects of SP-AODV and AODV routing schemes under mobility. The findings of simulation indicate the effectiveness of the suggested SP-AODV protocol by proving its supremacy over the performance measures examined such as packet delivery ratio, delay, packet loss, packet drop, overhead and MAC collision compared to AODV.

Keywords QoS · AODV · Cross-layer · Physical layer · Network layer · MAC layer · Received signal power · RSS · Free-space · Two-ray ground

G. K. Pallai (✉)
Department of Computer Science & Engineering, SOA Deemed to be University, Bhubaneswar, India
e-mail: gyan.pallai@gmail.com

M. Sankaran
VIT Business School, VIT Bhopal University, Bhopal, India
e-mail: meenakshi.sankaran@vitbhopal.ac.in

A. K. Rath
Department of Computer Science & Engineering, VSSUT, Burla, India
e-mail: amiyakumarrath@gmail.com

A. K. Rath
National Assessment and Accreditation Council, Bengaluru, India

S. Das and M. N. Mohanty (eds.), *Advances in Intelligent Computing and Communication*, Lecture Notes in Networks and Systems 202,
https://doi.org/10.1007/978-981-16-0695-3_48

1 Introduction

Inherent features of MANETs [1], such as dynamic topology, higher mobility and decentralized control, have major QoS performance limitations. In addition to the channel, bandwidth limitations and insufficient battery power ultimately affect the operation of ad hoc wireless networks. Repeated route failures and process of restoration cause severe interference and massive overhead that ultimately degrades the performance of the protocol. MANET transmission power plays a larger role in influencing the operation of ad hoc networks. A node's transmission power literally measures the average distance covered by the data packet to meet the target. A comprehensive coordination among all three layers, i.e. physical, MAC and network layers [2] is needed while fine-tuning the transmission power of a node. It is therefore essential that protocols incorporate QoS metrics for effective data communication in finding a stable link and route maintenance. We modelled our work on ad hoc on-demand distance vector(AODV) [3–5] SP-AODV for establishing more stable path during route discovery. Node's transmission power is tuned based on the three distinct mean RSS threshold values at the physical layer. Remaining work is systematized as follows: Sect. 2 addresses the works associated on cross-layer approach. Section 3 provides detailed explanation on proposed SP-AODV and its algorithm. Simulation parameters and comprehensive analysis of SP-AODV and AODV have been discussed in Sect. 4. We summarized with the future scope of the proposed work in Sect. 5.

2 Related Study

Ramachandran and Shanmugavel [6] implemented a cross-layer scheme by measuring RSS for path loss estimation and energy minimization. A study [7] by Zoulikha and Amal improved the performance of AODV by establishing strong link estimating RSS parameter and considering cross-layer approach in shadowing propagation model. A cross-layer method [8] over AODV to determine reliable path by regulating the power based on RSS is studied by Sarfaraz et al. An approach [9] that used energy level of nodes and link stability of the path between source to destination is designed for improving energy usage and reliability. Jamali demonstrated a multipath multicast QoS protocol [10] that provided stable and reliable routes and minimizes the overhead with increased network life span. A QoS-based path selection approach considering queue length parameters including lower bandwidth and power usage is proposed by Chen et al. [11] for high speed network. A study has been conducted to discover stable path [12] on improving the QoS metrics like throughput, packet delivery and latency.

3 Stable Path AODV (SP-AODV)

3.1 Determination of Stable Path Over Cross-Layer Architecture

The conventional routing protocol finds working routes utilizing the shortest path available. These routes are not reliable and are more susceptible to loss of connectivity. Our proposed model SP-AODV considers cross-layer strategy and transmission power adjustment is done dynamically according to the three distinct mean RSS threshold values. Free-space [13] and two-ray ground [13] propagation models have been used to compute the RSS of a node at the physical layer and stored in the routing table. Minimum RSS represents a higher communication distance, whereas maximum RSS represents the lower communication region. The node receives the RREQ packet and determines whether to rebroadcast or discard the packet. If the RSS value (P_r) lies between MinPr and MaxPr, it is considered to be a stable node. Only reliable nodes are permitted to retransmit the packet that produces the RREP packet to the source.

3.2 SP-AODV Algorithm

Definitions: N—node, N_t—transmitter node, N_r—receiver node, $d(N)$—node degree, P_r—node RSS, P_t—transmission power, L-system loss (*1*), $d(n)$—neighbours with minimum threshold power, $d(m)$—neighbours with maximum threshold power.

1. Each node sends "HELLO" packets to assess neighbor knowledge.
2. Estimate distance d between transmitter and receiver

$$d(x, y) = \sqrt{(x_2 - x_1)^2 + (y_2 - y_1)^2}$$

3. Cross-over distance dc ($H_t = H_r = 1.5m$, $\lambda = 0.3282$)

$$dc = \frac{4 \pi H_t H_r}{\lambda}$$

4. If d < dc Then P_r (d) using free-space model ($G_t = G_r = 1$, $L = 1$)

$$P_r(d) = \frac{P_t G_t G_r \lambda^2}{(4 \pi d)^2 L}$$

   Else P_r (d) using two-ray ground model

$$P_r(d) = \frac{P_t G_t G_r H_t^2 H_r^2}{d^4 L}$$

   End if
5. Three threshold values AvgPr, MinPr, MaxPr are computed based on P_r
6. Average received signal power AvgPr is calculated as:

```
Initialize : sum-avg = sum-min = sum-max = 0.
For i = 1 to d(N)
    sum-avg = sum-avg + Pr(Ni)
End For
AvgPr = sum-avg / d(N)
```

7. Average MinPr and MaxPr calculated as:

```
For i = 1 to d(N)
    If Pr(Ni) < AvgPr Then
        sum-min = sum-min + Pr(Ni)
    Else if Pr(Ni) > AvgPr Then
        sum-max = sum-max + Pr(Ni)
    End If
End For
```

8. MinPr = sum-min / d(n), MaxPr = sum-max / d(m)
9. Node's transmission power is tuned based on distinct threshold values
10. N_r decides whether to rebroadcast or discard RREQ packet based on P_r

```
If ( Pr > MinPr ) && ( Pr < MaxPr ) Then
        Process the RREQ packet.
Else
        Discard RREQ
End If
```

4 Performance Analysis

4.1 Simulation Parameters

We examine and evaluate the performance of both protocols at different speeds using NS-2 [14]. Testing is carried out for 100 nodes with topology area of 1000 × 1000 m, using a random way point mobility model [15]. We range the speed between 10–30 m/s and simulations were performed for 900 s with zero pauses. Maximum of 10 connection pairs are randomly chosen as network traffic loads. The CBR traffic sending rate is 4 packets/sec with 512 bytes. We have taken six performance measures to assess both SP-AODV and AODV schemes in terms of packet delivery ratio (PDR), end-to-end delay (ENE), packet loss, packet drop, routing overhead (ROH) and MAC collision.

4.2 Performance Analysis

The effect of speed adjustments on PDR is seen in Fig. 1. Findings of the simulation reveal that the SP-AODV PDR performance surpasses the standard AODV. The lower speed network is more likely to be stagnant and degree of neighbours per node is higher. Exchange of packets suffers from unwanted interference with the participating nodes and the real bandwidth of the network is impaired. In contrast, speed between 25 and 30 m/s has a predictably advantageous effect on the performance. Overall, the simulation outcome demonstrates that the average PDR performance is 99% for SP-AODV and 97% for AODV.

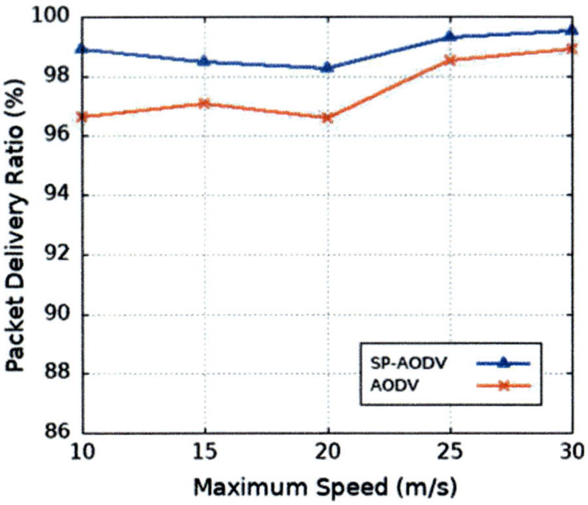

Fig. 1 Packet delivery ratio versus speed

Figure 2 displays ENE performance opposed to variant speed. Delay performance gets affected due to several reasons like delay in route discovery, queuing packet time, propagation and transmission latency. Network with lower speed experiences higher packet delays due to severe interferences of neighbouring nodes. Whereas, greater speed in node mobility (i.e. 25 and 30 m/s) lowers the number of hop counts for the destination. SP-AODV reveals its dominance by reducing the delay performance by 20% compared to AODV.

Figure 3 displays the ROH performance of both the protocols. Lower node mobility (10–20 m/s) has a negative effect on both the protocols. Overhearing between the nodes grow significantly which results increased congestion and performance deterioration. However, greater speed implies considerably lower

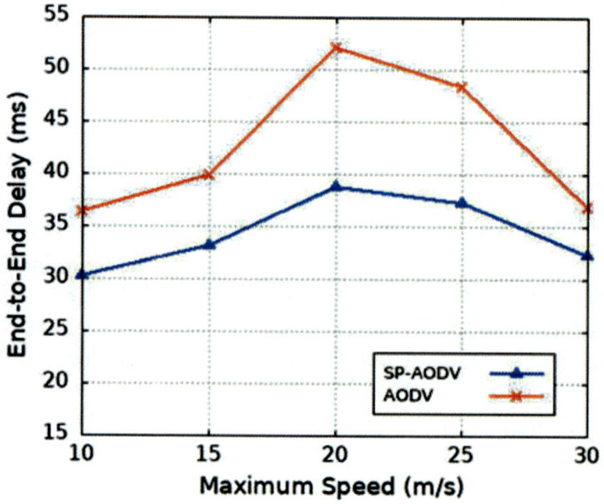

Fig. 2 End-to-end delay versus speed

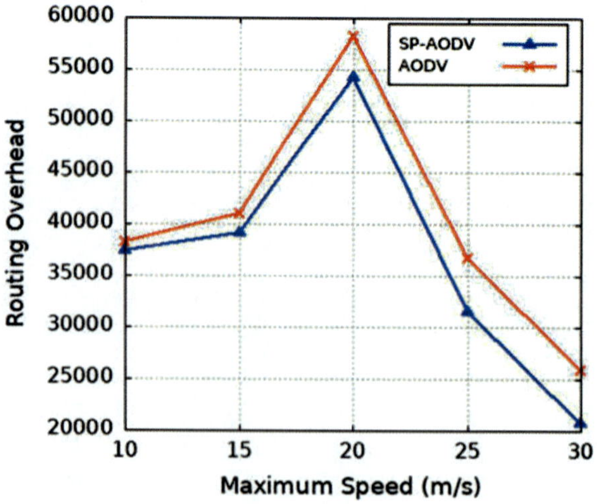

Fig. 3 Routing overhead versus speed

overhead. The results of the simulation indicate that SP-AODV minimizes overhead by approximately more than 22% relative to AODV.

Impact of speed on packet loss and packet drop performance is seen graphically in Figs. 4 and 5. Higher packets are lost and dropped as network suffers from higher congestion and interference at low mobility (10–20 m/s). Whereas, a substantial improvement in performance is seen in terms of packet loss and drop metrics at a speed of 25 m/s. The proposed SP-AODV method decreases the amount of packet loss and drop by 47 and 22% relative to AODV.

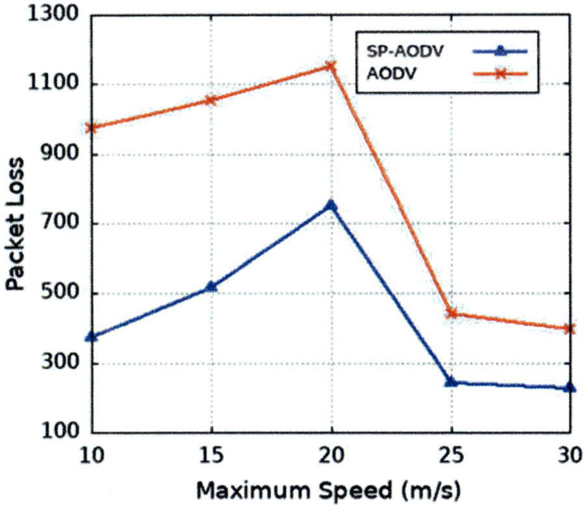

Fig. 4 Packet loss versus speed

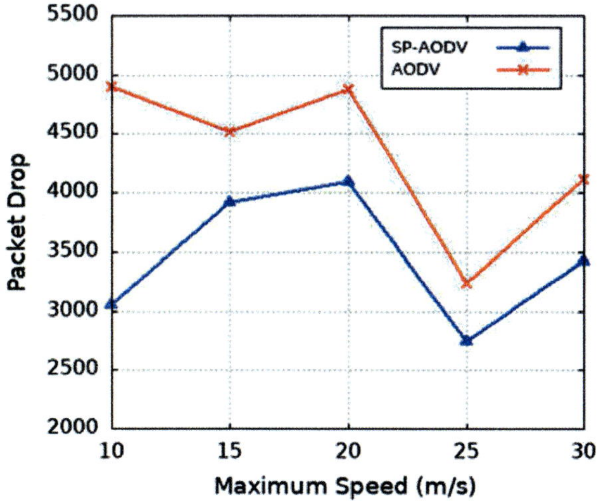

Fig. 5 Packet drop versus speed

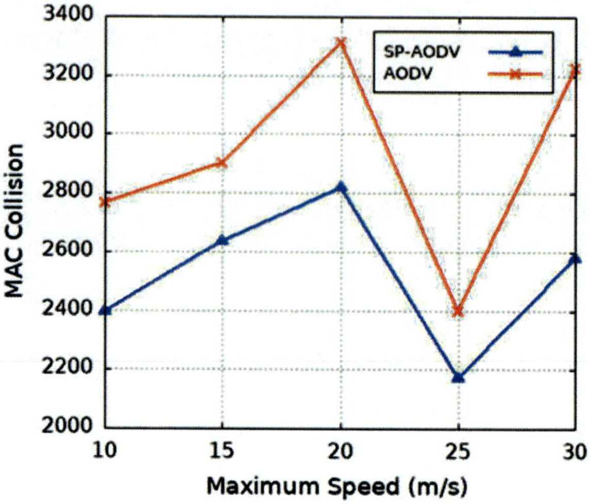

Fig. 6 MAC collision versus speed

The influence of speed on collisions is described in Fig. 6. Collision rate of both protocols increases between speed of 10–20 m/s due to intensified channel contention and interference. The effective bandwidth for individual nodes is obstructed, contributing to higher network collisions. On the other hand, both protocols show the optimum performance on MAC collisions at speed of 25 m/s. Compared to AODV, our proposed protocol SP-AODV minimizes rate of packet collision by more than 24%.

5 Conclusion

Speed differentials have a substantial impact on the effectiveness of routing schemes. Major objective of this study is to build an algorithm to establish a stable path that is less vulnerable to connectivity loss and re-establishment. Extensive simulations using NS-2 are made and analysed to examine both routing schemes at different speeds. Simulation findings specifically indicate that the suggested SP-AODV has proven its dominance in lowering ENE by 20%, MAC collision by 24%, ROH by 22%, packet loss by 47% and packet drop by 22% relative to AODV. Additional essential routing criteria, such as bandwidth, path loss and minimal energy usage would be included in the further expansion of this study.

References

1. Toh, C.K.: Ad hoc mobile wireless networks: protocols and systems. Pearson Education (2001)
2. Kawadia, V., Kumar, P.R.: Principles and protocols for power control in wireless ad hoc networks. IEEE J. Sel. Areas Commun. **23**(1), 76–88 (2005)
3. Perkins, C.E., Royer, E.M.: Ad-hoc on-demand distance vector routing. In: Proceedings WMCSA'99. Second IEEE workshop on mobile computing systems and applications, pp 90–100. IEEE (1999)
4. Perkins, C.E., Royer, E.M., Das, S.R., Marina, M.K.: Performance comparison of two on-demand routing protocols for ad hoc networks. IEEE Pers. Commun. **8**(1), 16–28 (2001)
5. Pallai, G.K., Meenakshi, S., Rath, A.K., Majhi, B.: Comparative study of on-demand and table-driven routing protocols in MANET. In: Computational intelligence in data mining, vol 2, pp. 369–379. Springer, New Delhi
6. Ramachandran, B., Shanmugavel, S.: Received signal strength-based cross-layer designs for mobile ad hoc networks. IETE Techn Rev **25**(4), 192–200 (2008)
7. Zoulıkha, M., Amal, B.: Cross-layer approach among physical, MAC and routing layer in a shadowing environment. Adhoc Sensor Wirel Netw **21**(1) (2004)
8. Ahmed, A.S., Kumaran, T.S., Syed, S.S.A., Subburam, S.: Cross-layer design approach for power control in mobile ad hoc networks. Egypt Inf J **16**(1), 1–7 (2015)
9. Benatia, S.E., Smail, O., Boudjelal, M., Cousin, B.: ESMRsc: energy aware and stable multipath routing protocol for ad hoc networks in smart city. In: International conference in artificial intelligence in renewable energetic systems, pp. 31–42. Springer, Cham (2018)
10. Jamali, M.A.J.: A multipath QoS multicast routing protocol based on link stability and route reliability in mobile ad-hoc networks. J Ambient Intell Human Comput **10**(1), 107–123 (2019)
11. Chen, Z., Zhou, W., Wu, S., Cheng, L.: An adaptive on-demand multipath routing protocol with QoS support for high-speed MANET. IEEE Access **8**, 44760–44773 (2020)
12. Moila, R.L., Velempini, M.: Enhancing AODV protocol using QoS constraints to improve routing in Cognitive Radio Ad Hoc Networks. In: 2020 international conference on artificial intelligence, big data, computing and data communication systems (icABCD), pp. 1–5. IEEE (2020)
13. Stepanov, I., Herrscher, D., Rothermel, K.: On the impact of radio propagation models on MANET simulation results. In: Proceedings of 7th international conference on mobile and wireless communications networks (MWCN 2005), Marrakech, Morocco (2005)
14. Fall, K., Varadhan, K.: The ns manual (formerly ns notes and documentation). VINT Project **47**, 19–231 (2005)
15. Camp, T., Boleng, J., Davies, V.: A survey of mobility models for ad hoc network research. Wireless Commun. Mobile Comput. **2**(5), 483–502 (2002)

Multi-model Face Recognition Pipeline and Anti-spoofing for a Generic Attendance System

Niladri Bihari Mohanty, Debabrata Singh, and Ashok Kumar Hota

Abstract Face recognition has many use cases, and the attendance system is one of the most promising areas of them. But choosing the right face detection and recognition model for the right purpose is always a billion-dollar decision before rolling out a real-time computer vision-based project. Although it is a very common understanding that face recognition is a closed problem, in reality it still has a lot of areas for improvements in the context of implementations. Development of a general-purpose face recognition-based attendance system is completely dependent on the availability of generalized face detection and recognition algorithms which can keep balance between speed, accuracy and anti-spoofing at the higher side of the benchmark. In some of the use cases, speed affects the usability and in some other use cases accuracy affects the usability. So in this paper, we have tried to find face detection and recognition models which can address all types of use cases and in the absence of such a model we tried to create alternative architecture to achieve the best out of the existing model by right positioning them in a pipeline.

Keywords Deep learning · Farce detection · Face recognition · Attendance system · CNN · Anti-spoofing · COVID-19

1 Introduction

For the development of general-purpose face recognition-based attendance systems which can address the heterogeneous use cases, we first need to identify the functionality requirements for benchmarking. No machine learning-based face

N. B. Mohanty
National Informatics Centre, Bhubaneswar, Odisha, India

D. Singh (✉)
Department of CSIT, ITER, Siksha O Anusandhan University, Bhubaneswar, Odisha, India
e-mail: debabratasingh@soa.ac.in

A. K. Hota
National Informatics Centre, Raipur, Chhattisgarh, India

S. Das and M. N. Mohanty (eds.), *Advances in Intelligent Computing and Communication*, Lecture Notes in Networks and Systems 202,
https://doi.org/10.1007/978-981-16-0695-3_49

detection and recognition model can be called as best due to various benchmarking keeping usability in focus [1]. Usability is the parameter which cannot be compromised and that varies from use case to use case. We will study the critical functional requirements of the various use cases and create a common benchmark matrix [2].

We will evaluate SSD MobileNet V1, tiny face detector, multitask cascaded convolutional neural networks, ResNet-34, emotion detector model, etc., for different types of use cases [3]. Without compromising with usability, it has been boldly attempted to use multiple models in a single pipeline for extracting the best from each participating model. Attendance system being in the center of the research and keeping the generic nature of the problem, it has been planned to not to touch too much detail of the use cases but to benchmark only the necessary components [4]. Face recognition systems can easily be fooled by using fake photographs or videos by simply placing the picture of the face of an authentic person in front of the camera. So, liveliness detection is also a matter of concern which will be addressed in this paper. Here also we have made a balance between the real-time aliveness detections with speed and accuracy.

In addition to face detection and recognition-based approach, we have also explored the possibility of using geo-fencing to solve the problem with machine intelligence. Image processing algorithms and programs are normally considered to be very CPU intensive, and running them in the edge is always a challenging job. In this paper, we have used a hybrid mechanism of using multiple models for face recognition based on different algorithms along with a stream processing pipeline using the message broker mechanism to achieve speed, accuracy, anti-finagle and ultimately the usability. We have also discussed various heterogeneous use cases which can use our architecture for a production ready environment [5].

2 Different Use Cases and Their Unique Requirements

2.1 Fixed Location Employee Attendance

A typical use case of attendance system is employees marking their presence in the office using biometrics devices which have many bottle necks like long queues and a single device are used by many people leading to an overall slowing down the process. Facial recognition is a promising tool for this use case. Employees can use their smartphones to lodge their attendance using cameras, GPS available in the mobile and face recognition-based technologies [6]. This is one of the simplest use cases requiring the below-mentioned features to be coupled with the solution.

i. Office campus should be geo-fenced
ii. Basic liveliness detection
iii. Small distance recognition

iv. Low lighting condition supported
v. GPS enabled mobile device and accurate GPS location tracking
vi. Employee ID and mobile number mapped.

2.2 COVID-19 Quarantine People Attendance

In the context of COVID-19, people under suspicion and confirmed cases are advised to stay in home isolation but keeping track of them is always a challenging task [7]. Many attempts have been made to develop mobile app to ensure they are in their home location using geo-fencing but it can only ensure their device is in location not the person. So, we proposed this as a use case where face recognition coupled with geo-fencing can ensure that the quarantine people are still staying in the home isolation. This can drastically reduce the human effort to monitor home isolation. This use case although looks similar to other face recognition based attendance system, but it has some exceptions like randomized attendance taking, liveness detection, edge computing etc [8]. Where in the previous use case employees are required to give attendance twice a day here, quarantine people need to give attendance on a random period of interval through notification to avoid predictability. We need below-mentioned unique features required for this special use case in addition to regular.

i. Strong liveliness detection
ii. Speed
iii. Highly scalable solution
iv. High level of accuracy
v. Strong geo-fencing limited circuit.

2.3 People Released on Conditional Bail and Parole

There are use cases where people are released on parole from jail for social grounds or may be released on conditional bail but advised to give attendance to the nearest police station on a daily basis [9]. In this use case, the face recognition-based geo-tagged attendance can be proved to be useful only if certain features as mentioned below can be achieved.

i. Strong liveliness detection
ii. Offline face recognition
iii. Synchronization module
iv. High level of accuracy
v. Loose geo-fencing to a large area.

2.4 Counting Heads and Monitoring Emotion in Conference/Seminar/Institution

In many use cases like in a conference, counting heads in any talk is a typical attendance requirement which does not involve individual face recognition [10]. In such scenarios, identification of small faces or partly covered faces from short as well as long distance is very important. Similar use cases can be counting students in a midday meal scheme.

i. High speed
ii. Small face detection
iii. Emotion or expression detection
iv. Low lighting condition.

2.5 Anganwadi Worker Attendances

Anganwadi is a type of rural child care center in India. In the rural part of India, child care is always community driven as most of the members of the family are daily wager so it is supported by the government for child nourishment and basic education. Extensive government machinery is involved in monitoring this Anganwadi Kendra and accountability of the financial support provided to them. In this use case, the attendance of the Anganwadi worker is not the only requirement but the head count of the children is also very vital. Here taking the face recognition-based attendance, children are not possible because of their changing nature [11, 12]. But head count will be helpful in correlating facts and finance. This use case requires features like below in addition to other basic features which other systems require.

A. Small face detection and head count up to 300 at one go
B. Low lighting condition
C. Basic-level liveliness detection
D. Strong geo-fencing.

3 Picking the Right Model for the Pipeline Building

3.1 Face Detection Models

For face detection, Single Shot Multibox Detector (SSD) is based on MobileNet V1 [13]. The neural net will compute the locations of each face in an image and will return the bounding boxes together with its probability for each face. This face detector is aiming toward obtaining high accuracy in detecting face bounding boxes instead of low inference time. This has a larger footprint, so it is more suitable for server-side implementation [14].

For implementation on the edge device, we need a smaller footprint for that tiny face detector seems to be the right choice as depicted in Fig. 1. This is a real-time face detector, which is much faster, smaller and less resource consuming compared to the SSD MobileNet V1 face detector; in return, it performs slightly less well on detecting small faces [15]. The disadvantages of this model if deployed on the edge will come while doing head count of large audience but will work perfect while taking individual attendance and small number of head counts. The size of the quantized model is only 190 KB (tiny_face_detector_model).

3.2 Face Recognition

After face detection, we can furthermore predict the facial landmarks for each detected face in the pipeline both at the client end for first-hand validation and in the server end for second-level validation for security [16]. We have implemented an ultra-lightweight and fast, yet accurate 68-point face landmark detector named face_landmark_68_tiny_model having a size of 80 KB. Although there is another higher accuracy version of the same model face_landmark_68_model, due to larger footprint this is difficult to use on the edge. For back-end secondary processing, we will use ResNet-34 which is a feature vector with 128 values having a footprint of around 6.28 MB as depicted in Fig. 2. By design, it has been decided to use the higher accuracy models that will be used in the server side with batch processing [17].

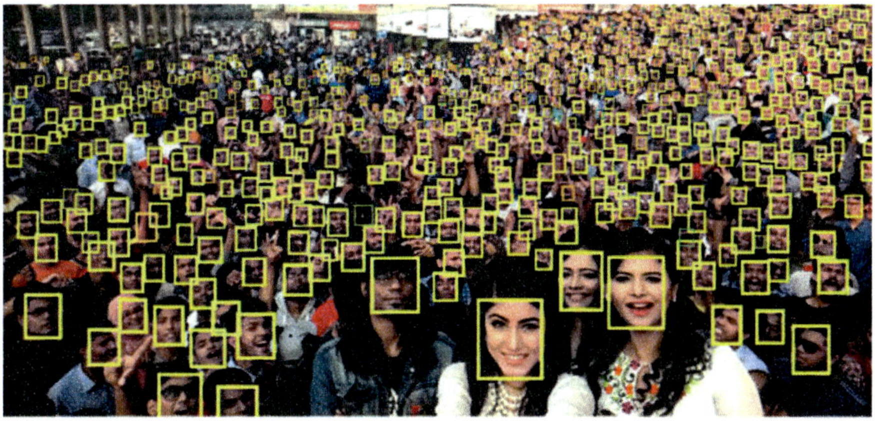

Fig. 1 Total 732 faces detected using tiny face detector model and 123 faces missed

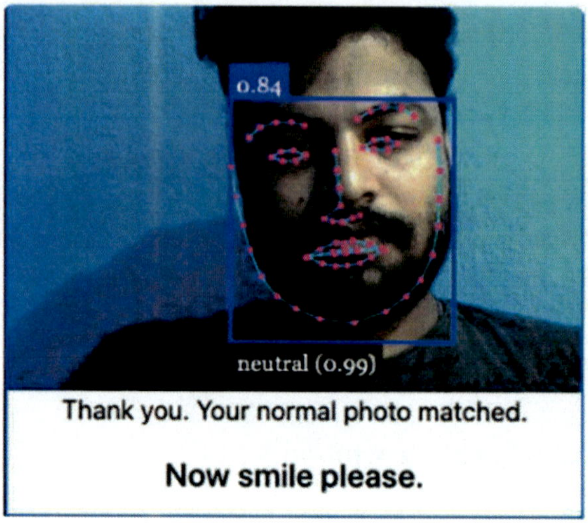

Fig. 2 68-point face landmark detection along with emotion detection

3.3 Liveliness Detection

Face recognition alone cannot help out in taking the correct attendance system as it can be easily be spoofed by using image or video in front of the camera. For liveliness check, we are asking the attendee to smile so that the expression recognizer model can detect the liveliness on the first hand on the edge device and deep analysis of the liveliness will be done on the server side using OpenCv and liveliness Net model [18]. Liveliness Net model is a CNN-based deep learning model to detect fake faces from the image in higher accuracy at the server side [19].

3.4 Geo-fencing

A geo-fence is a virtual perimeter for a real-world geographic area. A geo-fence could be dynamically generated as in a radius around a point location, or a geo-fence can be a predefined set of boundaries. In the attendance system, matching of the geolocation will be done in both client side and server side for enhanced security.

4 Establishing the Avenger Pipeline

From the above studies, we can clearly find that different use cases need different requirements which is very difficult to be addressed by any single model or approach. To build a generic deployment architecture which can address all types of use cases, we need to achieve high quality of speed, accuracy and integrity. So, we have chosen to explore the possibility of computing on the edge. The CPU-intensive ML-based computation if can be achieved on the client end, then user experience will be fast and rich but for executing machine learning models need a runtime. On the browser end, the only runtime available by default is the JavaScript engine. So, we have taken the help of TensorFlow Lite to build our model in JavaScript. It is planned and executed to do all tasks related to ML at both the end client and server. The computation at the client end will be focused on first-hand validation without deep analysis and only through lightweight approaches. The captured images will be evaluated on the client side for immediate observation but will be sent to the server side for detailed observation. On the server side, received images were pushed into Kafka cluster with zookeeper cluster manager, and to improve the scalability, we have deployed many worker processes on the server side which processes the images in batches fetching them from the Kafka cluster as depicted in Fig. 3.

Liveliness detection models need computation on the higher side, so we decided to keep this component on the server end which will evaluate the image taken directly from a real face or from another image. The evaluation will be based on various aspects like lighting condition, 3D and many others. But for the rapid evaluation in the client side, we have used the expression or emotion detection lightweight models to track the change in expression to detect liveliness. Although the client side liveliness detection can easily be spoofed using video call in front of the camera, it is still useful for the first-hand validation as depicted in Fig. 4.

To achieve high scalability in the server end, we deployed multiple worker processes and Kafka cluster. A load balancer manages the traffic between the

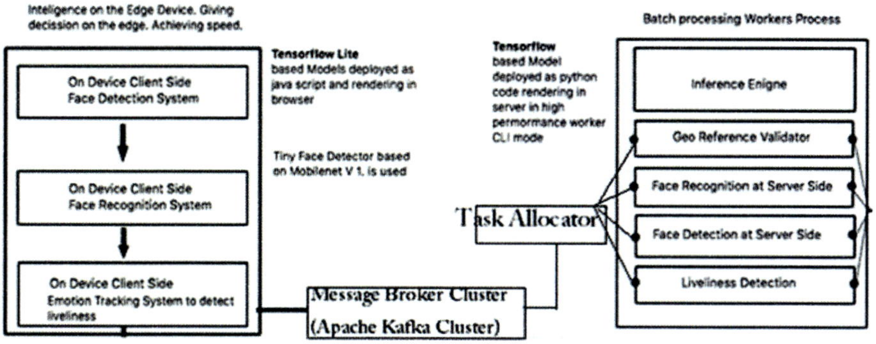

Fig. 3 Batch processing image fetching through Kafka cluster

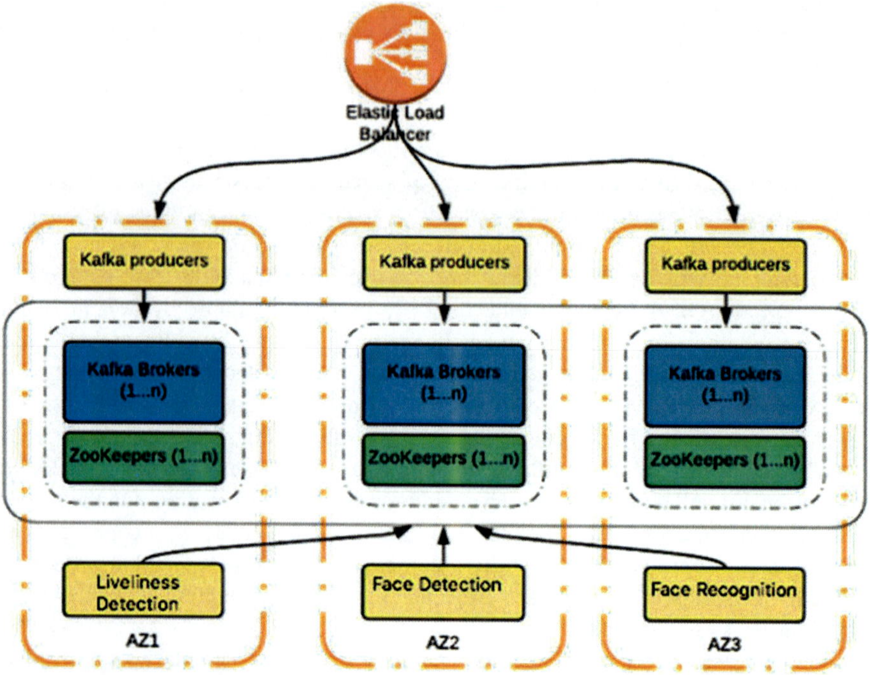

Fig. 4 Client-side liveliness detection

multiple Kafka producers interacting with the cluster. All the ML-based worker processes run on different group Ids to isolate processes and ensure no image is processed twice by the same worker process.

4.1 Client-Side Configurable Mobile Application

The mobile application has been developed using progressive web application technology using service workers running in the background to enhance offline as well as caching mechanisms. Here, we have used TensorFlow Lite-based models for face landmark detection and matching with preregistered images available in the server over the Asynchronous JavaScript & Xml (AJAX).

4.2 Server-Side Web Module

For attendance, the users need to be registered in the web application by giving their photographs in multiple angles for better accuracy along with their geolocation

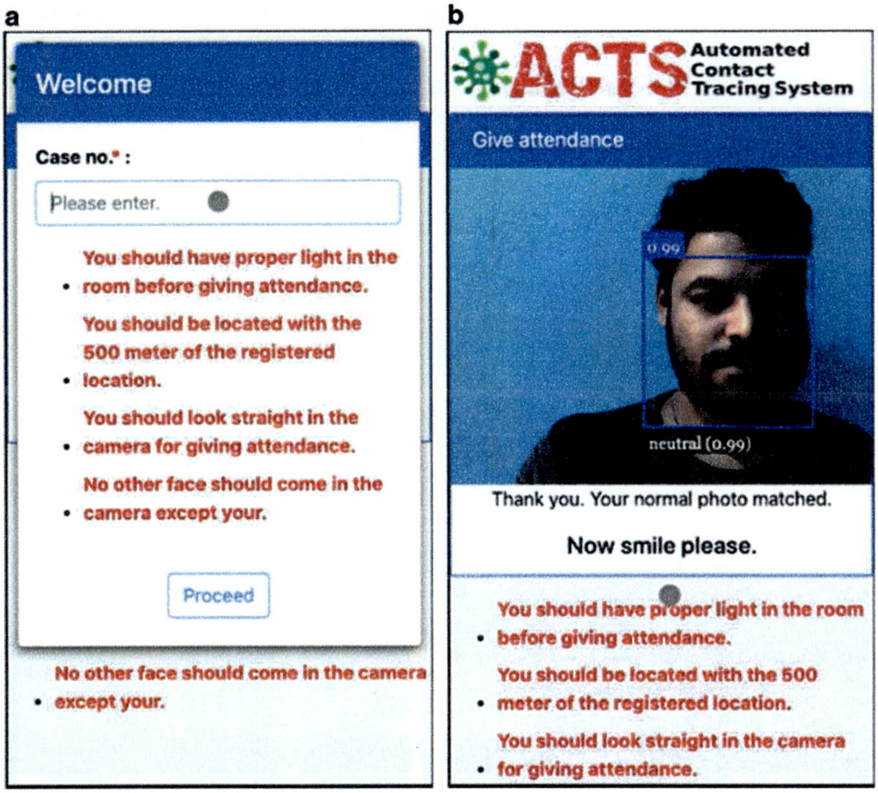

Fig. 5 **a** Face landmark detection and matching method and **b** face landmark detection and matching method and giving attendance

where to register their attendance. System will generate the geo-fencing automatically as depicted in Fig. 5a and b. API-based integration mechanism is depicted in Fig. 6a, where we can add attendance and find the location of the attendee.

But on many occasions, the system may need to interact with any legacy system instead of asking their users to register again so we have provided an open API-based integration mechanism for extensibility and interoperability as depicted in Fig. 6b.

5 Conclusion and Future Scope

We have used multiple models executed in a pipeline at both client side and server side to achieve speed, accuracy and integrity without compromising the usability. This hybrid architecture is supported by a highly scalable Kafka cluster, we have

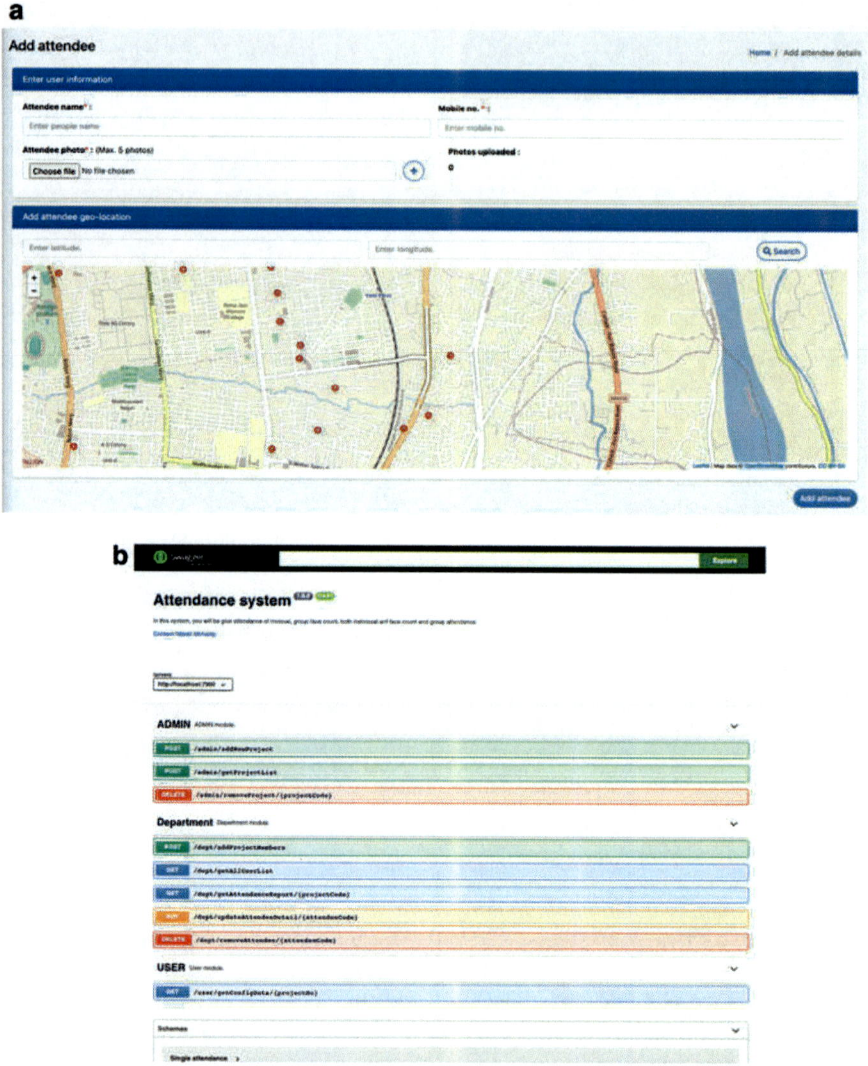

Fig. 6 **a** API-based integration mechanism and **b** attendance system implementation

achieved batch processing on various models, and then combining all the results in the inference engine we have achieved higher accuracy.

Instead of focusing on speed to get the final result, we have used lighter models for delivering ad hoc results on the edge device whereas batch processing the input with a more robust and heavier CNN network for higher accuracy at the server end. We also developed a configurable product based completely on open-source technologies and ML models like node JS, Angular, TensorFlow Lite, Python, TensorFlow, Kera, etc.

References

1. Taskiran, M., Kahraman, N., Erdem, C.E.: Face recognition: past, present and future (a review). Dig. Signal Process. 102809 (2020)
2. Son, N.T., Anh, B.N., et al.: Implementing CCTV-based attendance taking support system using deep face recognition: a case study at FPT polytechnic college. Symmetry **12** (2):307 (2020)
3. Kumar, A., Kaur, A., Kumar, M.: Face detection techniques: a review. Artif. Intell. Rev. **52** (2), 927–948 (2019)
4. Tian, L.-L., Teng, W.-L., Bian, H.-H.: Research on preprocessing algorithm of two-camera face recognition attendance image based on artificial intelligence. In: International conference on multimedia technology and enhanced learning, pp. 178–191. Springer, Cham (2020)
5. Sutar, R., Rokadia, S., Shah, A.: A survey on face recognition technologies and techniques. Int. J. Technol. Comput. (IJTC) **2**(7). Techlive Solutions (2016)
6. Castillo, R.V.C., Catanaoan, M.-J.T., Obliopas, J.F.E., Linsangan, N.B.: Class attendance generation through multiple facial detection and recognition using artificial neural network. In: Proceedings of the 2018 International Conference on Machine Learning and Machine Intelligence, pp. 38–42 (2018)
7. Botes, W.M., Thaldar, D.W.: COVID-19 and quarantine orders: a practical approach. SAMJ: S. Afr. Med. J. **110**(6), 1–4 (2020)
8. Chew, C.B., Mahinderjit-Singh, M., Wei, K.C., Sheng, T.W., Husin, M.H., Malim, N.H.A.H.: Sensors-enabled smart attendance systems using NFC and RFID technologies. Int. J. New Comput. Archit. Appl **5**, 19–29 (2015)
9. Burgess, E.W.: Is prediction feasible in social work? An inquiry based upon a sociological study of parole records. Soc. Forces **7**(4), 533–545 (1929)
10. Friedman, H.S., Miller-Herringer, T.: Nonverbal display of emotion in public and in private: self-monitoring, personality, and expressive cues. J. Person. Social Psychol. **61**(5), 766 (1991)
11. Heseltine, T., Pears, N., Austin, J.: Three-dimensional face recognition using combinations of surface feature map subspace components. Image Vision Comput. **26**(3), 382–396 (2008)
12. Ouarda, W., Trichili, H., Alimi, A.M., Solaiman, B.: Face recognition based on geometric features using Support Vector Machines. In: 2014 6th International Conference of Soft Computing and Pattern Recognition (SoCPaR), pp. 89–95. IEEE (2014)
13. Arya, S., Pratap, N., Bhatia, K.: Future of face recognition: a review. Procedia Comput. Sci. **58**(2), 578–585 (2015)
14. Fdhila, R., Ouarda, W., Alimi, M.A., Abraham, A.: A new scheme for face recognition system using a new 2-level parallelized hierarchical multi objective particle swarm optimization algorithm. J. Inf. Assur. Secur. **11**(6), 385–394 (2016)
15. Sharma, M., Thakur, P., Kumar, N.: Automated Attendance System Using Face Recognition (2018)
16. Kakadiaris, I.A., Toderici, G., Theoharis, T., Passalis, G.: Automated method for human face modeling and relighting with application to face recognition. U.S. Patent 8,090,160, issued Jan 3, 2012
17. Mery, D., Mackenney, I., Villalobos, E.: Student attendance system in crowded classrooms using a smartphone camera. In: 2019 IEEE Winter Conference on Applications of Computer Vision (WACV), pp. 857–866. IEEE (2019)
18. Leu, J.-S., Lin, W.-H., Yu, M.-C., Tzeng, H.J.: Recognition assisted dynamic surveillance system based on OSGi and OpenCV. In: 13th International Conference on Advanced Communication Technology (ICACT2011), pp. 83–87. IEEE (2011)
19. Sahu, B., Panigrahi, A., Pani, S., Swagatika, S., Singh, D., Kumar, S.: A crow particle swarm optimization algorithm with deep neural network (CPSO-DNN) for high dimensional data analysis. In: 2020 International Conference on Communication and Signal Processing (ICCSP), pp. 0357–0362. IEEE (2020)

ABC Versus PSO: A Comparative Study and Analysis on Optimization Aptitude

Sushree Sasmita Dash, Subrat Kumar Nayak, and Debahuti Mishra

Abstract Many real-world optimization problems can be solved very efficiently by evolutionary algorithms. There exist many solutions in this field to solve multi-dimensional optimization problems, which is an important issue in today's world. Swarm-based algorithms are very popular to solve these types of problems because of its effectiveness and quality of solutions. All swarm-based algorithms have different advantages and disadvantages. This paper has discussed two such mostly used algorithms, i.e., particle swarm optimization (PSO) and artificial bee colony (ABC). A fair comparison has been done between these two algorithms by implementing those to minimize five different standard benchmark functions. Results have been recorded and the performance of both the algorithms is analyzed with corresponding to the quality of solutions. It has been observed from the simulation results that ABC outperforms PSO in most of the cases corresponding to solution quality but it requires more time to converge than PSO.

Keywords Artificial bee colony · Particle swarm optimization · Exploration · Exploitation · Benchmark function

1 Introduction

Over the past few decades, the development in the field of computational intelligence (CI) has grown very fast. CI is based on a nature-inspired computational approach that can determine the solutions to many complex problems [15]. The

S. S. Dash · S. K. Nayak · D. Mishra (✉)
Department of Computer Science and Engineering, Institute of Technical Education and Research, Siksha 'O' Anusandhan (Deemed to Be) University, Bhubaneswar, Odisha, India
e-mail: debahutimishra@soa.ac.in

S. S. Dash
e-mail: sasmitadash83@gmail.com

S. K. Nayak
e-mail: subratnayak@soa.ac.in

S. Das and M. N. Mohanty (eds.), *Advances in Intelligent Computing and Communication*, Lecture Notes in Networks and Systems 202,
https://doi.org/10.1007/978-981-16-0695-3_50

artificial neural network, fuzzy logic, evolutionary computation, and swarm intelligence comes under CI. Swarm intelligence (SI) models are the cooperative nature of social swarms that includes, foraging of honey bees [8], flocking of birds [4], and the colonies of ants [3]. The agents present in a swarm interact locally among themselves and with their environment. It is inspired by the biological system [11]. Their interactions with each other lead to the introduction of intelligent behavior accepted globally, which is previously not known to any individual [17]. In 1995, Kennedy and Eberhart [4] introduced particle swarm optimization (PSO), which simulates bird flocking social behavior. In 2005, Karaboga [8] proposed an algorithm that is influenced by the foraging nature of honey bees known as the artificial bee colony (ABC) algorithm. Dorigo et al. [3] proposed a new nature-inspired optimization technique, ant colony optimization (ACO) to solve hard computational problems. PSO and ABC are two famous swarm-based optimization algorithms. Because of the presence of many optimization algorithms, researchers often asked some questions about these such as:

1. In SI-based algorithms, how do the agents learn to achieve the optimal point?
2. How does the learning of different agents affect exploration and exploitation?
3. Though different algorithms have different learning mechanisms and different policies to update the population, what are the similarities and the differences exist between different SI-based algorithms?

All these above-said areas are discussed in this paper with the two most popular algorithms of SI namely PSO and ABC.

Rest of the manuscript is organized in the following manner. Section 2 explains ABC and PSO, Sect. 3 discusses benchmark functions, in Sect. 4, and 5 experimental setup and results are presented respectively followed by the conclusion in Sect. 6.

2 Optimization Algorithms

2.1 ABC

ABC is modeled to simulate the foraging behavior of honey bees. The swarms of real honey bees work in a smart way collectively for searching the food. Honey bees have some special qualities like they can store information in memory and share them with others and they also can take decisions depending upon that [2]. Due to this intelligent behavior, researchers are attracted to give attention to the foraging activity of the honeybees.

In ABC, the community of bees is mainly categorized into three different groups: employed bees, onlooker bees, and scouts. In ABC, the required solutions are represented as the food sources of honey bees. The number of employed bees and onlooker bees is the same as the number of food sources. A bee searches for a specific food source when needed. From it, the bee gathers the total profitability facts which consist of the nectar amount, the effort needed to extract it, the direction and distance of the food origin from the hive as one unit. After getting the information, the employed bee returns to their hive, unloads the honey, and shares the information with the unemployed bees with a certain probability. Onlooker bees compiled the information got by observing the dance of employed bees and select the origin of food for herself. Scout bees looking for a new food source randomly around the hive, when an existing food source got exhausted. On average, in the colony of honey bees, 50% of bees are employed and 50% are onlooker bees and only 5–10% of the total is scouts [4].

ABC algorithm comprises of four phases [6]

1. Initialization phase
2. Employed bees phase
3. Onlooker bees phase
4. Scout bees phase.

Initialization Phase. In this phase, the algorithm generates uniformly dispersed points as the swarm of SN solutions, x_i $(i = 1, 2, 3, \ldots, SN)$ and different parameters are being set. Here, x_i is a vector having d dimension. Here, d represents the number of variables in the problem. Each solution can be represented as follows.

$$x_i = l_i + \text{rand}(0, 1) * (u_i - l_i) \tag{1}$$

Here, l_i and u_i are boundary parameters for x_i as the lowest and highest positions.

Employed Bees Phase. The employed bees inspect for a new food origin v_i around the neighborhood of the food source x_i to get more amount of nectar. This can be obtained by the following formula.

$$v_i = x_i + \phi_i(x_i - x_j) \tag{2}$$

Here, x_j is an arbitrarily selected neighbor solution and $\varnothing_i$ is a randomly selected number between [−1, 1]. After getting it, the employed bee exploits the food source. The bee renews her position when a better solution has been found by comparing the fitness value with the old one.

Onlooker Bees Phase. The onlooker bees get the fitness details and the position information from the employed bees. Afterward, they analyze this knowledge and find a food origin with a probability P_i. It can be estimated using the following equation.

$$P_i = \frac{fit_i}{\sum_{n=1}^{SN} fit_n} \tag{3}$$

Here, fit_i is the fitness value of the ith solution.

Like the employed bee, the onlooker bee also modifies her position by comparing the previously memorized fitness value and the value found from the newly searched position.

Scout Bees Phase. This phase starts when the position of the food origin memorized by the employed or onlooker bee is not updated for a predefined number of periods. In this case, that solution is discarded and the bee-related to that solution turns into a scout bee [9]. Then, the scout bee replaces the abundant solution by arbitrarily choosing a new one within the exploration space. In ABC, the limit value is the control parameter that specifies the predefined number of cycles for abandonment.

Phases from 2 to 4 are iterated until the completion condition has arrived. From the above 4 phases, it is clearly visible that the ABC algorithm has three constraints [7]. First is the population size, second is the limit value for the abandonment of a food source which helps in exploration, and third is the maximum number of cycles. The ABC algorithm is summarized in Algorithm 1.

2.2 PSO

PSO is a computational technique used for optimization [10]. It follows the behavioral habit of bird flocking or fish schooling. It combines the self-experiences of the birds with social experiences of them to move forward for searching food. In PSO, the collection of particles is known as swarm. Birds are represented as particles. Birds move toward a promising area to get the food. Here, food represents the global optimum value. The location of a particle served as a solution within the search space with d dimension. The ith particle is positioned as $x_i = (x_{i1}, x_{i2}, \ldots, x_{id})$. In this algorithm, each *i*th particle has to keep track of two positions.

1. Personal best: This is the coordinates of the best solution based on particles' fitness that has achieved so far.

$$p_i = (p_{i1}, p_{i2}, \ldots, p_{id})$$

2. Global best: This is the best value achieved so far by all particles in the population.

The PSO algorithm starts by initializing the population within the search area. Then, the fitness values for each particle are calculated and both the personal and global best values are updated. The third step of PSO is to determine the velocity of the particle. With that velocity, it moves from one position to the other. The fourth step is updating the position for each particle. *i*th particle's velocity is defined as $(v_i = v_{i1}, v_{i2}, \ldots, v_{id})$. The formula for updating the velocity and position for *i*th particle in the basic PSO is given below.

$$v_{id}^{t+1} = v_{id}^{t} + c_1 \cdot \text{rand}(0,1) \cdot (p_{id}^{t} - x_{id}^{t}) + c_2 \cdot \text{rand}(0,1) \cdot (g_{d}^{t} - x_{id}^{t}) \tag{4}$$

$$x_{id}^{t+1} = x_{id}^{t} + v_{id}^{t+1} \tag{5}$$

Here, t represents the time interval, $c1$ and $c2$ are acceleration coefficient that affects the amount of exploration and exploitation while searching. According to the formula, the velocity v_{0069} of the ith element is mainly composed of three parts. The first part is the momentum. It consists of the past velocity to move the element in the direction it has traveled so far. The second part is the cognitive component, i.e., $c_1 \cdot \text{rand}(0,1) \cdot (p_{id}^{t} - x_{id}^{t})$. It has the nature of returning to the best position held so far by ith particle. The third part is the social component, i.e., $c_2 \cdot \text{rand}(0,1) \cdot (g_{d}^{t} - x_{id}^{t})$. It has a nature to move toward the best position found in the neighborhood. After finding out the velocity, the velocity is clamped to a limiting value [13]. The second to fourth steps are iterated until the completion condition has arrived. For each member of the swarm, personal and global best positions are updated in each iteration which leads to convergence by bringing all particles close to the particle with the best solution. The exploitation occurs if the differences, i.e., $(p_{id}^{t} - x_{id}^{t})$ and $(g_{d}^{t} - x_{id}^{t})$ are both small, and exploration happens if either or both of them are large [12]. PSO has already proved its efficiency in optimizing different high dimensional problems by searching the entire problem space. The PSO algorithm is depicted in Algorithm 2. PSO is empirically studied later on by Shi and Eberhart 3 years later from the introduction of PSO [14]. They have written the following formula for velocity computation.

$$v_{id}^{t+1} = w \cdot v_{id}^{t} + c_1 \cdot \text{rand}(0,1) \cdot (p_{id}^{t} - x_{id}^{t}) + c_2 \cdot \text{rand}(0,1) \cdot (g_{d}^{t} - x_{id}^{t}) \tag{6}$$

Inertia weight (w) is the new parameter introduced in their paper. It helps in regulating the influence of inertia generated by the previous velocity [16]. At first, a high value of w helps PSO for searching globally. Then, a decreasing value of w in every iteration helps PSO for giving attention toward local search. Inertia weight helps to increase the speed of convergence and it also helps to get a better-optimized solution.

3 Benchmark Function

To compare different optimization algorithms fairly in terms of reliability, efficiency, and validation, mainly some accepted standard benchmark functions are needed. The performance of the two algorithms cannot be compared fairly if the test problems are specialized enough without any diversifications [5]. So, for the evaluation of an algorithm, the test should be added with a mixture of different problems like unimodality, multimodality, separability, non-separability, and multi-dimensionality. A function is called separable if it can be written as the sum of functions from one variable [16]. The complicacy increases when the function is also multimodal. A function having extra local optima rather than one is known as a multimodal function. The escaping ability from the local minimum of an algorithm can be tested using these types of functions. An algorithm with a poor exploration process is not able to search the whole landscape of a function effectively and normally got stuck to the local minimum [5]. The dimension of the search area is one more important characteristic which adds another level of difficulty to the function for finding the global optimum [1]. Table 1 shows five benchmark functions that are used in this paper to compare the ability of both algorithms.

Table 1 Benchmark function

Sl. no.	Function	Formula	Optimum value	Range	Dimension	Properties
1	Sphere	$f(x) = \sum_{i=1}^{d} x_i^2$	0	[−5.12, 5.12]	10 20 30	Unimodal Separable
2	Rosenbrock	$f(x) = \sum_{i=1}^{d-1} [100(x_{i+1} - x_i^2)^2 + (x_i - 1)^2]$	0	[−5,10]	10 20 30	Unimodal Inseparable
3	Rastrigin	$f(x) = 10d + \sum_{i=1}^{d} [x_i^2 - 10\cos(2\pi x_i)]$	0	[− 5.12, 5.12]	10 20 30	Multimodal Separable
4	Ackley	$f(x) = -20\exp\left(-0.2\sqrt{\frac{1}{d}\sum_{i=1}^{d} x_i^2}\right) - \exp\left(\frac{1}{d}\sum_{i=1}^{d}\cos(2\pi x_i)\right) + 20 + \exp(1)$	0	[−32.768, 32.768]	10 20 30	Multimodal Inseparable
5	Griewank	$f(x) = \sum_{i=1}^{d} \frac{x_i^2}{4000} - \prod_{i=1}^{d} \cos\left(\frac{x_i}{\sqrt{i}}\right) + 1$	0	[−600,600]	10 20 30	Multimodal Inseparable

Algorithm 1 The ABC algorithm

1. Initialization: Maximum Population (SN), dimension (d), maximum iteration $(It_{\max})$, and lower limit (l), upper limit (u)
2. Repeat $i = 1$ to SN
3. Initialize x_i as
$x_i = l_i + rand(0,1) * (u_i - l_i)$
4. Evaluate the objective value of x_i
5. While $It \leq It_{\max}$
6. For $i = 1$ to SN
7. A candidate solution v_i is generated for x_i by equation
$v_i = x_i + \phi_i(x_i - x_j)$
8. Evaluate $f(v_i)$
9. If $f(v_i) < f(x_i)$
10. Replace x_i by v_i
11. $Count(i) = 0$
12. Else
13. $Count(i) = Count(i) + 1$
14. Calculate P_i for each x_i by
$P_i = \frac{fit_i}{\sum_{n=1}^{SN} fit_n}$
15. $t = 1, i = 1$
16. While $t \leq SN$
17. If $Rand_{num} \leq P_i$
18. A candidate solution v_i is generated for x_i by equation
$v_i = x_i + \phi_i(x_i - x_j)$
19. Evaluate $f(v_i), t = t + 1$
20. If $f(v_i) < f(x_i)$
21. Replace x_i by v_i
22. $Count(i) = 0$
23. Else
24. $Count(i) = Count(i) + 1$
25. $i = i + 1$
26. If $i = SN + 1$
27. $i = 1$
28. Remember the best solution found till now
29. Find out the solution $x_{\max}$ with maximum $Count$ value
30. If $Count(\max) > \text{limit}$
31. Randomly generate a new solution and replace it with x_i by
$x_i = l_i + rand(0,1) * (u_i - l_i)$
32. $Count(\max) = 0$
33. $It = It + 1$
34. Remember the best solution found till now

Algorithm 2 The PSO algorithm

1. Initialization: Max Population (N), dimension (d), max iteration $(It_{\max})$, and lowermost limit (l), uppermost limit (u) for both the velocity and position
2. For $i = 1$ to N
3. Initialize particle's positions as x_i, where $x_l \leq x_i \leq x_u$
4. Initialize *pbest* to its initial position i.e., $pbest = x_i$
5. Initialize *gbest* to the minimal value of the swarm: $gbest = \min(f(x_i))$
6. Initialize particle's velocity as v_i, where $v_l \leq v_t \leq v_u$
7. While $It \leq It_{\max}$
8. For $i = 1\ to\ N$
9. Update velocity by $v_{id}^{t+1} = w.v_{id}^{t} + c1.rand(0,1).(p_{id}^{t} - x_{id}^{t}) + c2.rand(0,1).(g_{d}^{t} - x_{id}^{t})$
10. Update position by $x_{id}^{t+1} = x_{id}^{t} + v_{id}^{t+1}$
11. If $f(x_i) < f(pbest)$
12. $pbest = x_i$
13. If $f(x_i) < f(pbest)$
14. $gbest = x_i$
15. $It = It + 1$
16. Output: *gbest* that found the global optimum solution

4 Experimental Setup

Experimental settings are very much important and the experimental outcomes can be greatly influenced by it. To get a fair comparison between these two algorithms, the setting up values for control parameters used in both of these are essential. The size of the population and the number of maximum iterations are the common control parameter in both the algorithm. In this experiment, for different dimensions of 10, 20, and 30, the total iterations taken are 1000, 2000, and 3000, respectively, and the population size is 100. The minimization of different objective functions has been accomplished using the ABC algorithm and PSO algorithm by simulating it in MATLAB with a system having CORE i5 CPU, 4 GB RAM. The algorithms have run for twenty times to get the standard deviation and mean value. The control parameter values are set as 1, 2.5, and 2 for inertia weight, cognitive acceleration coefficient, and social acceleration coefficient respectively in PSO. The swarm size used in ABC is 100. The number of food source taken is 50% of the swarm size. The employed bee discards the solution if the solution is not updated at all after a limit value that is population_size × dimension.

5 Result and Discussion

In our experiment, the performance of optimization techniques that is ABC and PSO are evaluated on five different benchmark functions. The mean and standard deviation (SD) of 20 trial runs have been found by running the algorithms with different parameter sets having dimensions of 10, 20, and 30 are recorded, and the results are presented in Tables 2, 3, and 4.

Table 2 Results obtained by ABC and PSO for dimension = 10

Function/algorithm	Standard		Mean	
	ABC	PSO	ABC	PSO
Sphere	2.63E−17	5.78E−16	8.58E−17	1.31E−16
Rastrigin	0.00E+00	3.00E+00	0.00E+00	8.11E+00
Ackley	1.30E−15	8.08E−09	7.46E−15	4.15E−09
Griewank	2.70E−03	7.89E−02	1.10E−03	1.34E−01
Rosenbrock	2.80E−02	2.29E+00	1.99E−02	5.70E+00

Table 3 Results obtained by ABC and PSO for dimension = 20

Function/algorithm	Standard		Mean	
	ABC	PSO	ABC	PSO
Sphere	4.24E−17	7.95E−06	2.79E−16	5.07E−06
Rastrigin	0.00E+00	6.66E+00	0.00E+00	2.59E+01
Ackley	3.51E−15	3.07E−01	1.92E−14	7.79E−02
Griewank	2.20E−03	5.15E−02	4.93E−04	4.82E−02
Rosenbrock	2.35E−02	2.36E+01	1.85E−02	2.67E+01

Table 4 Results obtained by ABC and PSO for dimension = 30

Function/algorithm	Standard		Mean	
	ABC	PSO	ABC	PSO
Sphere	2.80E−17	1.20E−03	5.05E−16	1.30E−03
Rastrigin	0.00E+00	1.52E+01	0.00E+00	4.10E+01
Ackley	3.75E−15	7.31E−01	3.36E−14	1.24E+00
Griewank	5.58E−17	2.68E−01	4.44E−17	6.58E−01
Rosenbrock	1.19E−02	4.29E+01	9.70E−03	6.98E+01

Tables 2–4 show the results respectively for 10, 20, and 30 dimensions. According to the reported result in all the table, ABC performed better than PSO for both unimodal and multimodal benchmark functions. From this, it can be concluded that the ABC algorithm can search out for the global minimum by overcoming any local minimum. It means the exploration is better at ABC than PSO. This happens because of the probabilistic modeling used in the PSO algorithm for changing a part of the solution while looking for a better solution. Small randomness is added in the cognitive and social component at the time of finding velocity in Eq. 3. It provides the exploration effect in finding an improved solution. But in the case of the ABC algorithm, scout bees are responsible for the exploration process by removing an unproductive population and randomly initialize an entirely new population to search for better solutions which is better for global optimization. Artificial onlookers and employed bees efficiently handle the exploitation process in ABC which leads to local optimization. Due to this property, the ABC algorithm properly converges to the globally optimized value in a better way than PSO does. From the result, it also can be viewed that PSO has faster convergence. Because the exploitation rate is more than the exploration in PSO. But, ABC has a balance exploration and exploitation to achieve the global optimum. Due to the slower convergence of ABC than PSO, ABC is not recommended for the applications which need to be solved quickly.

Figures 1, 2, 3, 4, 5, 6, 7, 8, 9, 10, 11, 12, 13, 14, and 15 show the simulations result of different objective functions by testing once for minimization. The simulations represented in the below figures are executed for dimension = 10, 20, and 30 for all the functions.

Fig. 1 Simulation result of Ackley for dimension = 10

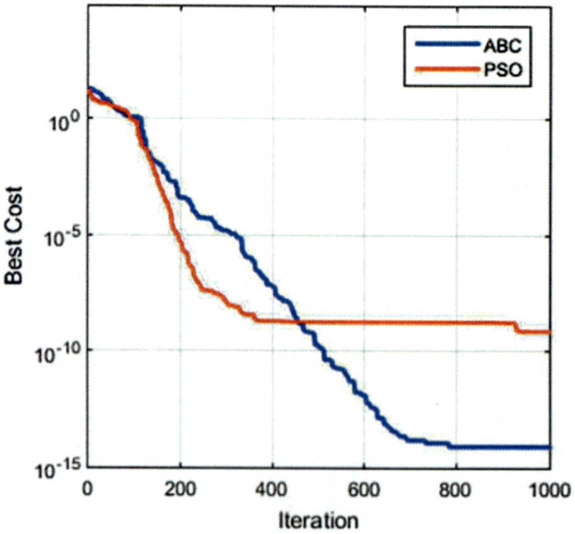

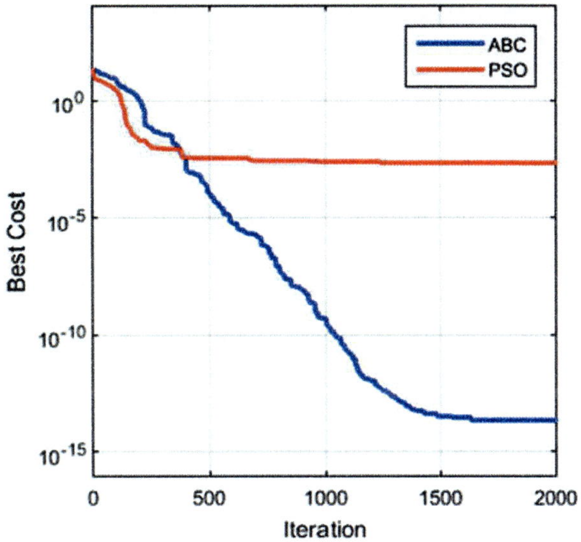

Fig. 2 Simulation result of Ackley for dimension = 20

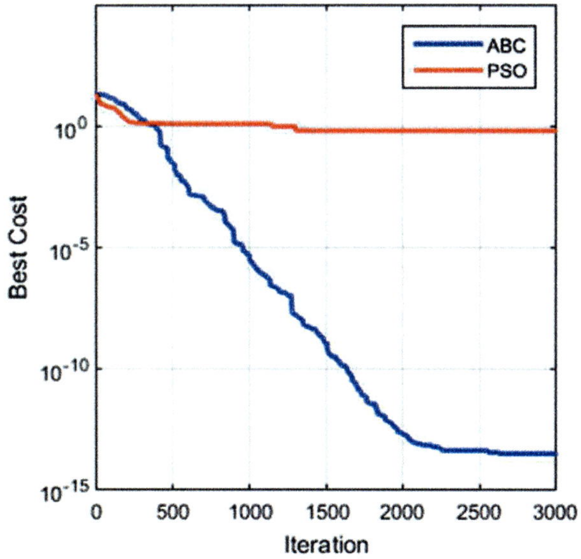

Fig. 3 Simulation result of Ackley for dimension = 30

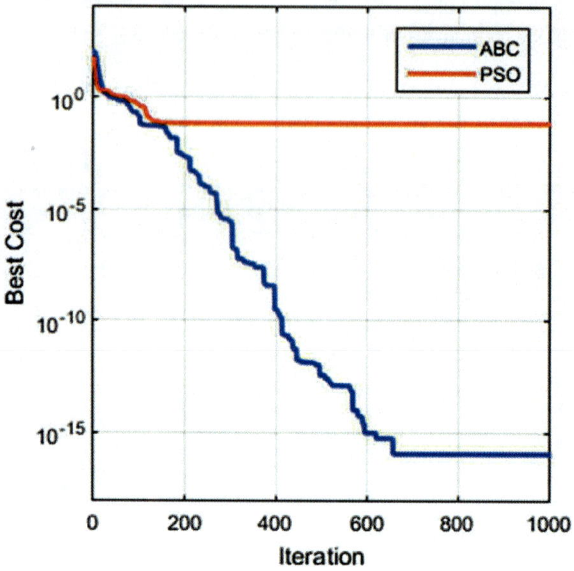

Fig. 4 Simulation result of Griewank for dimension = 10

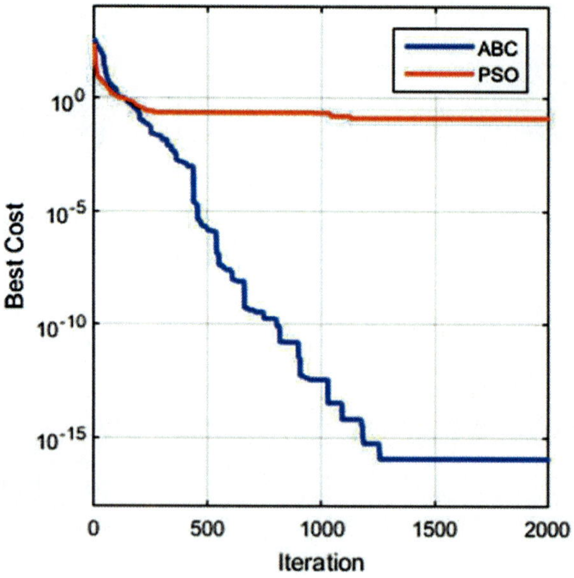

Fig. 5 Simulation result of Griewank for dimension = 20

Fig. 6 Simulation result of Griewank for dimension = 30

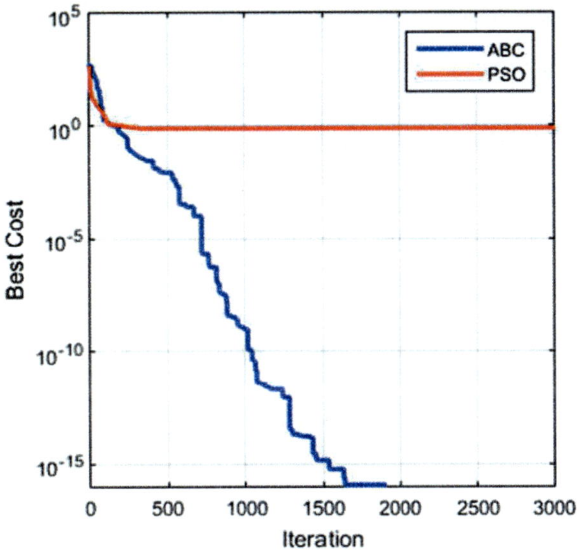

Fig. 7 Simulation result of Rastrigin for dimension = 10

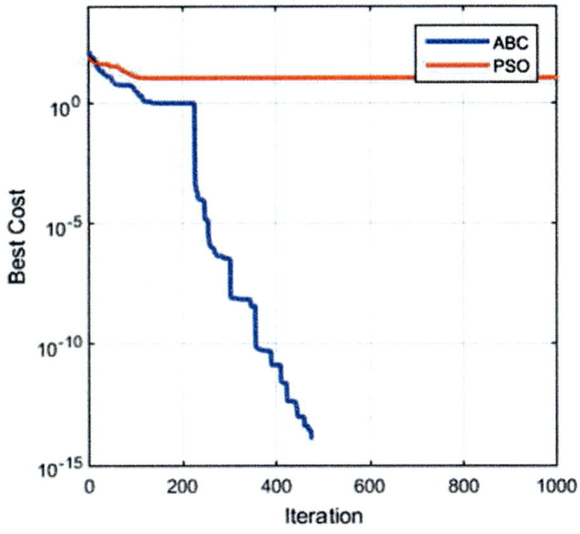

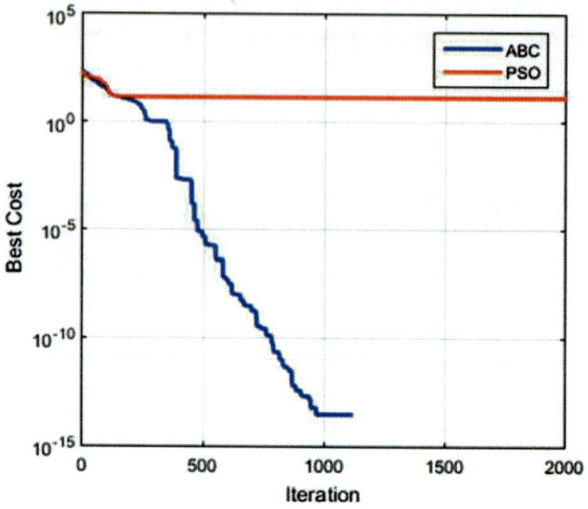

Fig. 8 Simulation result of Rastrigin for dimension = 20

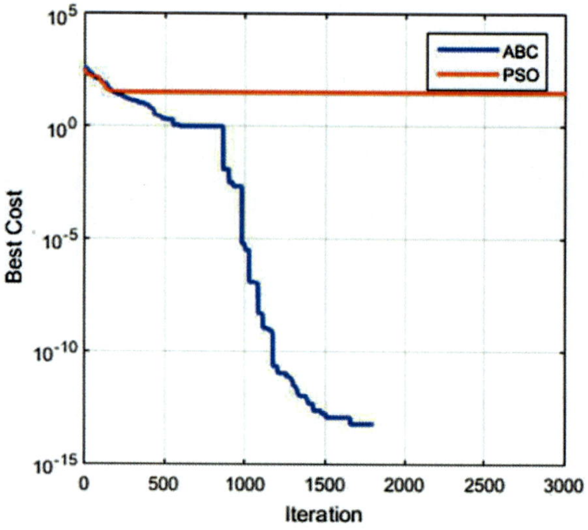

Fig. 9 Simulation result of Rastrigin for dimension = 30

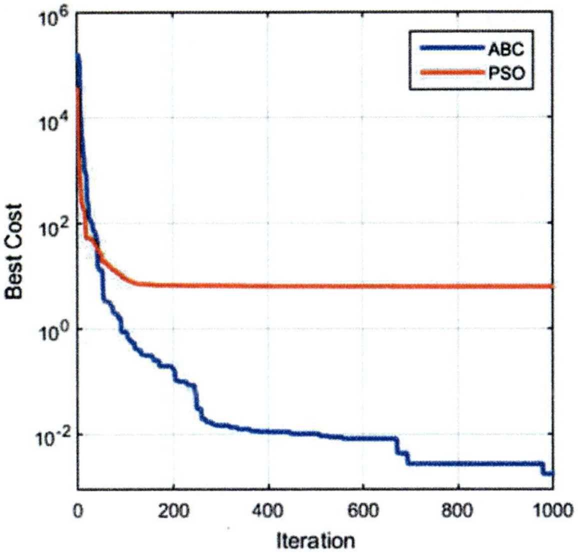

Fig. 10 Simulation result of Rosenbrock for dimension = 10

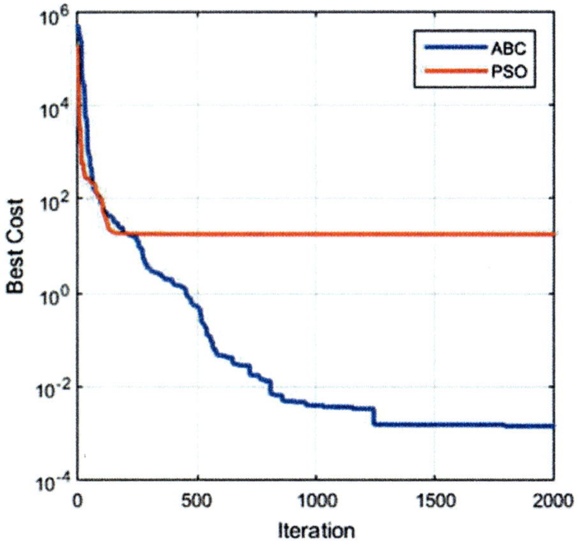

Fig. 11 Simulation result of Rosenbrock for dimension = 20

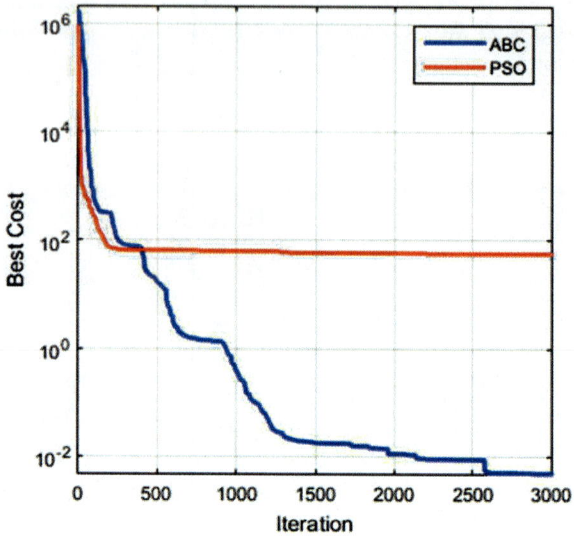

Fig. 12 Simulation result of Rosenbrock for dimension = 30

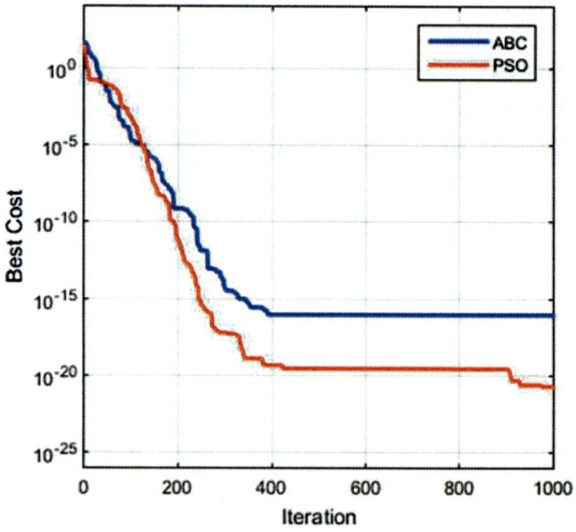

Fig. 13 Simulation result of Sphere for dimension = 10

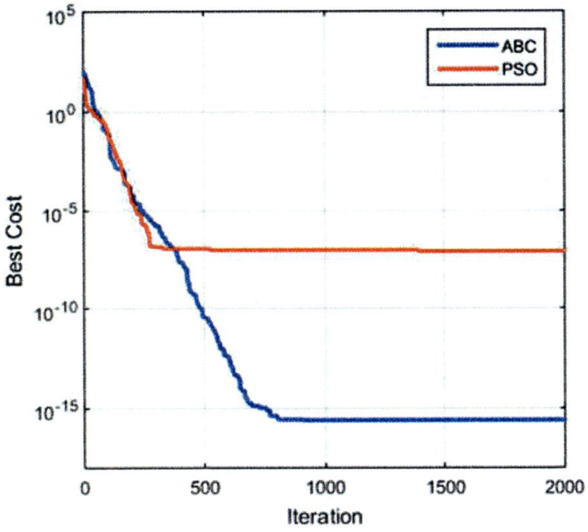

Fig. 14 Simulation result of Sphere for dimension = 20

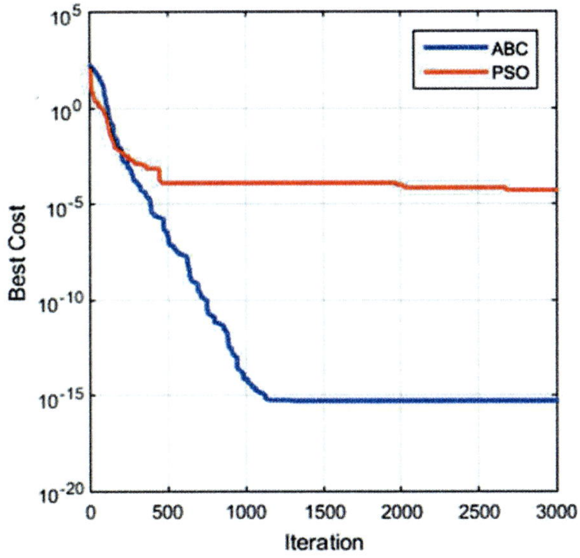

Fig. 15 Simulation result of Sphere for dimension = 30

6 Conclusion

This paper compares the performance of two-swarm intelligence and population-based algorithms ABC and PSO with five benchmark functions having multi-dimensionality property. The result obtained from the simulation shows that for both unimodal and multimodal function, ABC outperforms PSO, if the quality of the solutions is concerned. ABC algorithm efficiently comes out of the local minimum without staying there. Though ABC suffers from delayed convergence, it may not be used for time-sensitive applications. Otherwise, ABC can be used for a better result because its accuracy is better than that of PSO. To optimize the real-world problem, the work in this paper can be further extended. Further research can be done on ABC and PSO by introducing the hybridization of the algorithms with other methods of optimization.

References

1. Akbari, R., Mohammadi, A., Ziarati, K.: A novel bee swarm optimization algorithm for numerical function optimization. Commun. Nonlinear Sci. Numer. Simul. **15**(10), 3142–3155 (2010)
2. Bansal, J.C., Sharma, H., Jadon, S.S.: Artificial bee colony algorithm: a survey. Int. J. Adv. Intell. Paradig **5**(1–2), 123–159 (2013)
3. Dorigo, M., Birattari, M., Stutzle, T.: Ant colony optimization. IEEE Comput. Intell. Mag. **1** (4), 28–39 (2006)
4. Eberhart, R., Kennedy, J.: Particle swarm optimization. In: Proceedings of the IEEE International Conference on Neural Networks, Vol. 4. Citeseer (1995)
5. Jamil, M., Yang, X.-S.: A literature survey of benchmark functions for global optimisation problems. Int. J. Math. Modell. Numer. Optim. **4**(2), 150–194 (2013)
6. Karaboga, D., Basturk, B.: A powerful and efficient algorithm for numerical function optimization: artificial bee colony (ABC) algorithm. J. Global Optim. **39**(3), 459–471 (2007)
7. Karaboga, D.: Artificial bee colony algorithm. Scholarpedia **5**(3), 6915 (2010)
8. Karaboga, D.: An idea based on honey bee swarm for numerical optimization, Vol. 200. Technical report-tr06, Erciyes University, Engineering Faculty, Computer Engineering Department (2005)
9. Karaboga, D., et al.: A comprehensive survey: artificial bee colony (ABC) algorithm and applications. Artif. Intell. Rev. **42**(1), 21–57 (2014)
10. Kennedy, J., Eberhart, R.: Particle swarm optimization. In: Proceedings of ICNN'95-International Conference on Neural Networks. vol. 4. IEEE (1995)
11. Kothari, V., et al.: A survey on particle swarm optimization in feature selection. In: International Conference on Computing and Communication Systems. Springer, Berlin, Heidelberg (2011)
12. Sengupta, S., Basak, S., Peters, R.A.: Particle swarm optimization: a survey of historical and recent developments with hybridization perspectives. Mach. Learn. Knowl. Extr. **1**(1), 157–191 (2019)
13. Shahzad, F., Masood, S., Khan, N.K.: Probabilistic opposition-based particle swarm optimization with velocity clamping. Knowl. Inf. Syst. **39**(3), 703–737 (2014)
14. Shi, Y., Eberhart, R.C.: Empirical study of particle swarm optimization. In: Proceedings of the 1999 congress on evolutionary computation-CEC99 (Cat. No. 99TH8406), vol. 3. IEEE (1999)
15. Voges, K.E., Pope, N. (eds). Business Applications and Computational Intelligence. Igi Global (2006)
16. Wahab, A., Nadhir, M., Nefti-Meziani, S., Atyabi, A.: A comprehensive review of swarm optimization algorithms. PLoS ONE **10**(5), e0122827 (2015)
17. Zhang, Y., Wang, S., Ji, G.: A comprehensive survey on particle swarm optimization algorithm and its applications. Math. Probl. Eng. (2015)

A Comprehensive Study on Security in IoT and Resolving Security Threats Using Machine Learning (ML)

Mahendra Prasad Nath, Sushree Bibhuprada B. Priyadarshini, and Debahuti Mishra

Abstract With the dispatch of the "Amazon Echo family" and "Google applications," for example, "Chromecast and home," the use of Internet of Things (IoT) gadgets in the family unit is increasing exponentially this year. The next age in connectivity is the IoT. With the IoT, smart devices may be engaged to render and exchange knowledge both reliably and efficiently. Distinctive IoT advances center around robotizing different undertakings and trying to empower the working of the lifeless physical structures without human mediation. While convenience is at the bleeding edge of advancing the acknowledgment and utilization of these net assets and the focal point of training; security and protection are regularly post-thoroughly considered. As a consequence, there is a dangerous climate of potential for malicious attackers to hack compromised devices while sitting in homes. Recent history demonstrates that an assault on IoT devices will be both simple and also have obliterating effects. Current and future IoT technologies are very encouraging for the user to increase levels of comfort, performance, and automation. It needs high security, anonymity, authentication, and recovery of assault to be able to integrate such an environment in an ever-increasing manner. In the above context, to accomplish start to finish IoT conditions, it is imperative to make the vital enhancements in structure of the IoT framework. This chapter offers an in-depth analysis of the protection-related issues and vulnerability factors in the IoT systems. If the security issues are resolved, different current and emerging innovations are tackled and actions are initiated to achieve a high degree of confidence in IoT systems.

M. P. Nath · D. Mishra (✉)
Computer Science and Engineering, Siksha 'O' Anusandhan, Deemed to be University, Bhubaneswar, Odisha, India
e-mail: debahutimishra@soa.ac.in

M. P. Nath
e-mail: mahendranath272@gmail.com

S. B. B. Priyadarshini (✉)
Computer Science and Information Technology, Siksha 'O' Anusandhan, Deemed to be University, Bhubaneswar, Odisha, India
e-mail: bimalabibhuprada@gmail.com

S. Das and M. N. Mohanty (eds.), *Advances in Intelligent Computing and Communication*, Lecture Notes in Networks and Systems 202,
https://doi.org/10.1007/978-981-16-0695-3_51

Keywords AI · DNN · IoT · IoT security · ML · SVM

1 Introduction to Internet of Things (IoT)

Gadgets are getting progressively unavoidable trying to satisfy our necessities while affording accommodation and incentive in our regular exercises. Home appliances, mechanical mechanization, and shrewd city framework are totally associated with the Internet currently. The speed at which physical devices are connected to the Internet around us is growing quickly. There will be about "8.4 billion" linked devices worldwide in "2020" according to a recent "Gartner survey" [1–5]. Throughout all areas of the world, IoT innovations are commonly used. The most powerful moving nations are "North America," "Western Europe," and "China" [1, 3]. The amount of "machine-to-machine (M2M)" interactions is forecasted to expand from "5.6 billion" in "2016" to "27 billion" in "2024" [2, 4]. This statistical leap also labels IoT as significant developing markets that could act as a foundation for the expansion of the industrialized economy. The IoT business is relied upon to develop as far as income from "$892 billion" of every "2018" to "$4 trillion" by "2025" [5].

"M2M" systems spread a wide scope of utilizations, for example, brilliant atmosphere, shrewd urban areas, savvy matrices, keen retail, keen horticulture, and so on [6]. The gadgets can not only be linked later to the internet and other surrounding gadgets, but are often used to legally discuss the Internet with other gadgets. The possibility of social IoT (SIoT) is likewise developing beside the gadgets or things being related. The gadgets can not only be connected to the Internet and other gadgets in close proximity later, but also be used to discuss them legally with other gadgets on the Internet. The possibility of social IoT (SIoT) is likewise developing beside the gadgets or things being related [7]. The issue of security and protection accompanies this expansive range of IoT applications. The development of IoT technologies cannot satisfy a high demand and can destroy all latent capacities without a secure and interoperative IoT biological system. Notwithstanding the security constraints that internet, cell systems, and WSNs normally face, IoT additionally has its own remarkable security challenges, for example, protection issues, confirmation issues, the board issues, data stockpiling, etc. Due to these challenges and weaknesses, the IoT implementations establish a good base for numerous forms of software hazards [8]. Several IoT technologies previously spread throughout the world have contributed to vulnerability and privacy attacks. Figure 1 demonstrates the IoT's current and future architecture.

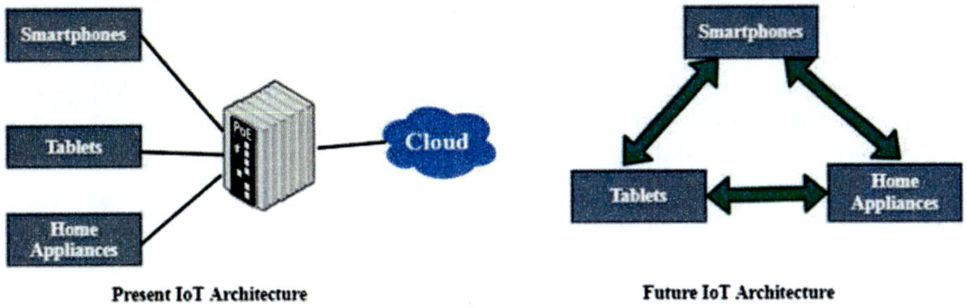

Fig. 1 Present and future architecture of IoT

2 Security in IoT Applications

Basically, for all the IoT applications which have either been sent or are in the sending stage, protection is crucial. Moreover, IoT applications extend quickly and enter the greater part of the present enterprises. While administrators utilize existing systems administration innovations to help these IoT applications, a large number of these applications need increasingly rigid security assurance from the advancements that they use. Different basic IoT insurance applications are shrouded in this section [9–12].

2.1 Security in Smart Home

Currently, home automation is among the IoT technologies that are most commonly used and deployed. It extends to such systems as those for remote control of energy-saving electrical devices, systems being mounted on shutters to track trespassers, etc. Vulnerability scanning is used to monitor use of fuel and energy supply, and consumers are encouraged to save on costs [13–16]. Researchers suggested the use of reasoning-based protection mechanisms to enhance the level of protection in homes. Infringements are identified by matching the client behaviors at the main home areas with the user's usual activity at those locations. However, perpetrators can obtain access to home IoT systems and attempt to hurt end-users. For example, since the introduction of many home automation systems, house burglaries have escalated rapidly [17, 18]. In the past, enemy people often try to investigate the composition and calculation of Internet traffic from and to the genius home and determine the actions of people and their surroundings.

2.2 Security in Smart Cities

Intelligent cities need the best use of modern electronic and networking technology to boost the overall quality of life for the general population [19, 20]. This includes intelligent home protection, intelligent traffic control, intelligent emergency prevention, smart utilities, etc. The urban areas worldwide are motivated to be more glossy by various opportunities to encourage their prosperity [21, 22]. While the use of smart applications is expected to enhance citizen's life quality, it does pose a challenge to citizens' confidentiality. Smart card systems aim to place citizens' card information and buying behavior at risk. Smart mobility applications can leak traces of users' locations. There are ways where parents can keep an eye on their child. If these frameworks are modified, then the child's protection may be at danger.

2.3 Security in Smart Surroundings

Smart ecosystems include various IoT technologies, such as detecting forest fires, monitoring snow levels in high-altitude areas, preventing landslides, and regulating pollution. Many of these IoT technologies are closely linked to human and animal life in those regions [23]. Government agencies will also rely on the expertise from certain IoT applications in these areas. In all the conditions associated with these IoT programmes, vulnerability and safety breaches can have serious consequences. In this sense, both false negative and false positive outcomes for these IoT applications will lead to catastrophic outcomes. For example, if the programme tries to wrongly predict earthquakes, it can create considerable financial damages for businesses and governments. At the other hand, if the system cannot predict the earthquake, then both property and life will be lost. Smart technology implementations therefore need to be extremely accurate, and violations of security and data tampering need to be avoided [24].

2.4 Security in Smart Grids and Smart Monitoring

Smart monitoring includes different calculation, tracking, and control applications. Smart monitoring is most commonly used on smart grids, where they calculate and track energy usage. Smart tracking may also be used to address the issue of energy theft [25]. Many uses of intelligent monitoring include tracking gas, oil, and water levels in tanks and cisterns for storage. Smart management also tracks and optimises the output of solar power plants by dynamically adjusting the angle of the solar panels in order to receive as much solar power as possible. "Smart meters" are regularly expected to perform works on account of the conventional vitality use [10]. All electrical equipment at home is "smart metres" connected to a smart area network, and load and cost tracking information collected from these systems is available [11]. Purposeful obstruction by the client or an enemy into these correspondence frameworks may change the data gathered, prompting financial misfortune for specialist organizations or customers.

2.5 Security in Safety and Emergency

This is another primary environment in which multiple IoT applications are submitted. It includes structures that tolerate, for example, only supported persons in protected environments. The capacity in this setting is the distinguishing proof of spillage of harmful gasses in assembling territories or close by concoction plants. Radiation levels can also be studied within the regions protected by nuclear power plants or cell base stations, and cautions can be developed if radiation levels are as strong. There are different systems that have characterized data in their frameworks or that house private products [22, 23]. Anyone can convey security applications to make sure about secret information and items. IoT gadgets that distinguish explicit fluids may likewise be utilized in these touchy structures to keep away from consumption and break downs. Security vulnerabilities may likewise have numerous genuine ramifications in these applications. For instance, by assaulting the vulnerabilities in these applications the hoodlums can attempt to get to the confined territories. False alerts about radiation fair and square can also have enormous, fast and long-distance implications. For starters, if young people face high radiation, it may cause important dangerous long-term diseases.

2.6 Security in Smart Marketing

IoT applications are broadly utilized in the retail division. Various software were developed to screen the goods' performance states as they pass gracefully down the chain. IoT is often used to monitor the different items in the store to make restocking preferably possible. Different sophisticated shopping applications are also built to help customers rely on their inclinations, propensities, bigotry for particular parts, and so on [19, 20]. Systems to give the experience of purchasing on the web to disconnected retailers utilizing expanded veracity techniques have likewise been created. In the sending and use of different IoT programmes, numerous retail firms have faced security problems like "Apple," "Home Depot," "JP Morgan Chase," and "Sony" [20]. Adversaries may try to manage IoT applications relating to the storage conditions of objects and may seek to provide purchasers with misdirected knowledge about products to improve market. In the event that security highlights are not executed in shrewd retail, assailants that may take from clients' charge and visa data, telephone numbers, email addresses, thereby bringing about money-related misfortunes to clients and retailers [16].

2.7 *Security in Intelligent Agriculture and Animal Farming*

Shrewd cultivation includes soil dampness following, smaller-scale atmosphere in the board, specific water system in dry territories, and mugginess and temperature of the board. The use of these powered highlights in agri-business can help produce outstanding returns and can save farmers from economic troubles. Temperature and humidity management levels in a variety of grain and vegetable creations can help deter microbes and other microbial pollutants. Weather recommendations can also help increase crop production and harvest efficiency. Similarly, as with crop following, there are IoT applications to follow livestock's practices and wellbeing status by adding sensors to the creatures while we are dealing with huge volume of big data [7, 8]. On-the-off chance that these applications are penetrated, it can bring about the burglary of the livestock and foes can likewise hurt the yields

3 Resolving IoT Security Threats Through Machine Learning (ML) Strategies

Lately, the field of machine learning (ML) has pulled in impressive intrigue in modern arena. Numerous spaces use ML for their creation, and its additionally utilized for IoT assurance. ML seems, by all accounts, to be a potential alternative for shielding IoT gadgets from digital dangers by having an alternate way to deal with guarding from dangers than other customary methodologies do.

3.1 *DoS Attack*

A significant issue is DoS assaults on IoT gadgets or those beginning from IoT gadgets. Another way to avoid these attacks is by using a “multilayer perceptron (MLP)” [13] convention, which guarantees DoS assault devices. The writers [17] have proposed an enhancement of molecule swarm and back spread calculation to prepare a MLP that improves the security of remote systems. ML procedures help increment the exactness of reasoning and stable IoT gadgets powerless against DoS attacks [19].

3.2 *Spoofing*

The “deep neural network (DNN)” model, “support vector machines (SVM),” Q-learning, Dyna-Q, “incremental aggregated gradient (IAG),” and “distributed Frank Wolfe (dFW)” procedures can be utilized to keep spoofers from assaulting.

The two strategies not just improve the exactness of ID and grouping yet in addition, they help to lessen the normal mistake rate and the bogus caution rate [3].

3.3 *Eavesdropping*

Attackers can spy messages while transmitting information. ML procedures, such as "Q-learning based offloading methodology" or "nonparametric Bayesian strategies," may be used to protect against such attacks. Plans, such as "Q-learning" and "Dyna-Q," are ML methodologies that can likewise be utilized to abstain from listening stealthily of applications. An assessment of these plans by investigations and support learning is portrayed in [5, 6].

3.4 *Privacy Leakage*

Gathering individual data, for example, wellbeing records, area or pictures, puts the protection of the client in danger. Logical calculations that secure protection (PPSC) should be utilized to dodge distribution of information. Another method proposed to build up IoT application trust is a "Product Integrity Detection Algorithm (CIDA)" that depends on the "Chinese Remaining Theorem (CRT)" [7, 8].

3.5 *Digital Fingerprinting*

Computerized fingerprinting is among the emerging and energizing solutions to ensure IoT frameworks, and it helps end-customers to build satisfactory confidence in application areas. Fingerprints are normally utilized for portable opening, installment acknowledgment, vehicle and home entryway opening, and so forth. Advanced fingerprinting is rising as a predominant biometric recognizable proof apparatus in light of its minimal effort, dependability, agreeableness and high-security level. Despite computerized fingerprinting upsides, the effectiveness of IoT's use of this technique is challenging to accomplish. For example, fingerprints should be sorted, pictures enhanced, highlights synchronized, etc.

Distinctive machine-based learning calculations have been created to give some non-customary ways to deal with these problems, some of which are talked about underneath.

- **Deep Neural Networks (DNN)**: DNN is one of the calculation algorithms most generally utilized in artificial intelligence (AI). This gives different points of interest, for example, adaptation to internal failure, incorporated learning, and speculation. In such technique, for utilizing ANN to carefully arrange fingerprints is carried out. For preparing purposes, the computerized estimations of different highlights in the unique finger impression, for example, particulars, edge completion, and bifurcation are added as the contribution to the neural system utilizing ANN back engendering calculation. Unique finger impression confirmation is performed based on past experiential qualities put away in the database [9, 10].

- **Support Vector Machine (SVM)**: SVM is a preparation calculation for non-linear and direct grouping, key part investigation, the arrangement of archives, recognizable proof of speakers, and relapse. This augments the separation between the limit of judgment and the preparation designs. The utilization of SVM in computerized fingerprinting was investigated inside and out by creators. They additionally contrasted it with another run of the mill models, as well. A capacity vector is manufactured dependent on the unique mark pixel esteems and is utilized to prepare the SVM. Different examples are analyzed behind the unique mark, and afterward, a finger impression coordinating is performed dependent on designs found [17, 18].

4 Conclusion

In IoT, the essential need is to secure all the frameworks and gadgets connected to the system. ML has the capacity to utilize and prepare calculations to recognize inconsistencies in IoT gadgets or to identify any unapproved conduct happening in the IoT framework to maintain a strategic distance from information misfortune or different issues. Therefore, ML offers a promising discussion for defeating the troubles in making sure about IoT applications. More commitments are required to help IoT's development around there.

References

1. Ganz, F., Puschmann, D., Barnaghi, P., Carrez, F.: A practical evaluation of information processing and abstraction techniques for the Internet of Things. IEEE Internet Things J. **2**(4), 340–354 (2015)
2. Anjomshoa, F., Aloqaily, M., Kantarci, B., Erol-Kantarci, M., Schuckers, S.: Social behaviometrics for personalized devices in the Internet of Things era. IEEE Acc. **5**, 12199–12213 (2017)
3. Dey, N., Ashour, A.S., Bhatt, C., Internet of things driven connected healthcare. In: Internet of Things and Big Data Technologies for Next Generation Healthcare, pp. 3–12. Springer, Cham (2017)
4. Frustaci, M., Pace, P., Aloi, G., Fortino, G.: Evaluating critical security issues of the IoT world: present and future challenges. IEEE Internet Things J. **5**(4), 2483–2495 (2018)

5. Fernández-Caramés, T.M., Fraga-Lamas, P.: A review on the use of blockchain for the Internet of Things. IEEE Acc. **6**, 32979–33001 (2018)
6. Nath, M.P., Goyal, K., Prasad, J., Kallur, B.: Chat Bot—an edge to customer insight. Int. J. Res. Sci. Innovation (IJRSI) **V**(V), 29–32 (2018)
7. Suthaharan, S.: Big data classification: problems and challenges in network intrusion prediction with machine learning. ACM Sigmetrics Perform. Eval. Rev. **41**(4), 70–73 (2014)
8. Nath, M.P., Sagnika, S., Das, M., Pandey, M.: Object recognition using cat swarm optimization. Int. J. Res. Sci. Innovation (IJRSI) **IV**(VIIS), July (2017)
9. Nath, M.P., Sagnika, S.: Capabilities of chatbots and its performance enhancements in machine learning. In: Machine Learning and Information Processing, Advances in Intelligent Systems and Computing, vol. 1101, pp. 183–192 (2020)
10. Elhayatmy, G., Dey, N., Ashour, A.S.: Internet of Things based wireless body area network in healthcare. In: Internet of Things and Big Data Analytics Toward Next-Generation Intelligence, pp. 3–20. Springer, Cham (2018)
11. Bhatt, C., Dey, N., Ashour, A.S. (eds.): Internet of Things and Big Data Technologies for Next Generation Healthcare (2017)
12. Meidan, Y., et al.: ProfilIoT: A machine learning approach for IoT device identification based on network traffic analysis. In: SAC 2017 32nd ACM Symposium on Applied Computing, pp. 506–509 (2017)
13. Nath, M., Muralikrishnan, J., Sundarrajan, K., Varadarajanna, M.: Continuous integration, delivery, and deployment: a revolutionary approach in software development. IJRSI **5**(VII), 185–190 (2018)
14. Mhetre, N.A., Deshpande, A.V., Mahalle, P.N.: Trust management model based on fuzzy approach for ubiquitous computing. Int. J. Ambient Comput. Intell. (IJACI) **7**(2), 33–46 (2016)
15. Nath, M.P., Pandey, P., Somu, K., Amalraj, P.: Artificial intelligence and machine learning: the emerging milestones in software development. IJRSI **5**(IX), 36–44 (2018)
16. Lin, J., Yu, W., Zhang, N., Yang, X., Zhang, H., Zhao, W.: A survey on Internet of Things: architecture, enabling technologies, security and privacy, and applications. IEEE Internet Things J. **4**(5), 1125–1142 (2017)
17. Chen, L., Thombre, S., Järvinen, K., Lohan, E.S., Alén-Savikko, A., Leppäkoski, H., Bhuiyan, M.Z.H., Bu-Pasha, S., Ferrara, G.N., Honkala, S., Lindqvist, J., Ruotsalainen, L., Korpisaari, P., Kuusniemi, H.: Robustness, security and privacy in location-based services for future IoT: a survey. IEEE Acc. **5**, 8956–8977 (2017)
18. Farris, I., Taleb, T., Khettab, Y., Song, J.: A survey on emerging SDN and NFV security mechanisms for IoT systems. IEEE Commun. Surv. Tuts. **21**(1), 812–837, 1st Quart. (2019)
19. Gharaibeh, A., Salahuddin, M.A., Hussini, S.J., Khreishah, A., Khalil, I., Guizani, M., Al-Fuqaha, A.: Smart cities: a survey on data management, security, and enabling technologies. IEEE Commun. Surveys Tuts. **19**(4), 2456–2501, 4th Quart. (2017)
20. Din, I.U., Guizani, M., Kim, B.-S., Hassan, S., Khan, M.K.: Trust management techniques for the Internet of Things: a survey. IEEE Acc. **7**, 29763–29787 (2019)
21. Mosenia, A., Jha, N.K.: A comprehensive study of security of Internet-of-Things. IEEE Trans. Emerg. Topics Comput. **5**(4), 586–602 (2017)
22. Nath, M.P., Sridharan, R., Bhargava, A., Mohammed, T.: Cloud computing: an overview, benefits, issues and research challenges. IJRSI **6**(II), 25–35 (2019)
23. Yang, Y., Wu, L., Yin, G., Li, L., Zhao, H.: A survey on security and privacy issues in Internet-of-Things. IEEE Internet Things J. **4**(5), 1250–1258 (2017)
24. Yu, W., Liang, F., He, X., Hatcher, W.G., Lu, C., Lin, J., Yang, X.: A survey on the edge computing for the Internet of Things. IEEE Acc. **6**, 6900–6919 (2018)
25. Priyadarshini, S.B.B., Bagjadab, A.B., Mishra, B.K.: The role of IoT and big data in modern technological arena. In: Internet of Things and Big data Analytics for Smart Generation, pp. 13–25. Springer (2018)

Optimal Energy Storage Allocation in Smart Distribution Systems: A Review

Vivekananda Pattanaik, Binaya Kumar Malika, Sarthak Mohanty, Pravat Kumar Rout, and Binod Kumar Sahu

Abstract The major reason for energy storage system (ESS) integration to the smart distribution system is to provide additional system security, reliability, stability, and flexibility in response to the changes due to disturbances. However, the establishment of ESS in the distribution network arises many issues and challenges that need to be focused on during distribution planning and design. The optimal planning of the ESS is very active and upcoming research and needs to find an optimal methodology considering diversified objective function, constraints, and variables. Even though various techniques are suggested in recent times, still it needs further improvement for an effective solution against all issues and challenges related to the planning and optimization of ESS. This manuscript is intended to present a review of all the suggestions and innovations of various authors and discuss the findings with possibilities of future developments particularly to placement, sizing, and economic value of ESS.

V. Pattanaik
Department of Electrical Engineering, Synergy Institute of Technology, Phulnakhara, Bhubaneswar 754001, India
e-mail: vivek7048@gmail.com

B. K. Malika
Department of Electrical Engineering, Einstein Academy of Technology and Management, Bhubaneswar 751015, India
e-mail: er.binaya@gmail.com

S. Mohanty (✉) · B. K. Sahu
Department of Electrical Engineering, Siksha 'O' Anusandhan University, Bhubaneswar 751030, India
e-mail: sarthak.mohanty1995@gmail.com

B. K. Sahu
e-mail: binoditer@gmail.com

P. K. Rout
Department of Electrical and Electronics Engineering, Siksha 'O' Anusandhan University, Bhubaneswar 751030, India
e-mail: pkrout_india@yahoo.com

S. Das and M. N. Mohanty (eds.), *Advances in Intelligent Computing and Communication*, Lecture Notes in Networks and Systems 202,
https://doi.org/10.1007/978-981-16-0695-3_52

Keywords Energy storage system · Optimal sizing and siting · Optimal ESS allocation · Analytical methods · Heuristic methods

1 Introduction

ESS is much needed for the optimal operation of the power system to support the distributed generation (DG) integration and enhanced performance. The general working of the ESS in terms of charging and discharging operation can be categorized into two types as the charging period and the discharging period. In the first operation during off-peak and low electricity price conditions, the ESSs store the energy from the network. This energy is discharged, and energy is in feed to the network during peak period and high electricity price condition. The ESSs are supplying energy in DC form and so its integration and control are more challenging as they are integrated through power electronic-based interfacing devices to the grid system. Apart from that, ESSs can be supplied in variety to microgrid systems in terms of technology, placement, capacity, requirement, and costs of investment. Among many factors related to its influence on the microgrid performance, location and sizing are one of the vital issues to be optimally set for ESSs in the microgrid structure. The multifunctionality of ESSs integration can be summarized as [1, 2]:

- Dependency on the additional need for peak generating capacity is reduced.
- Improve the optimal generation of existing generation units.
- Enhance the possibility and feasibility of integrating intermittent renewable sources.
- Offer ancillary services like load following, area regulation, and spinning reserve.
- Transmission congestion is reduced.
- Increase the flexibility and capacity of power transmission in both transmission and distribution sections.
- Sustain and improve demand-side response system dynamics.

This review is done with a primary intention to discuss the proposed approach suggested by various authors to optimally find the location and sizing the ESS for the microgrid.

The smart microgrid distribution system can operate in two modes either in grid-connected or in islanded mode. The switching operation from grid-connected to the islanded mode and reverse during reconnection with the synchronized operation is difficult to handle. Secondly, in islanded mode, the controllability of the microgrid gets reduced drastically due to the low inertia of the microgrid [3]. The voltage and frequency control poses an enormous challenge to operate in normal desirable conditions due to low damping capabilities. Under the steady-state operation also the microgrid needs to be planned to operate smoothly and efficiently with reduced distribution line loss, better voltage and frequency constraint handling capability, enhanced readability, increased transmission capability, and improved stability [4]. All these issues and challenges are more prominent in the case of islanded mode than the grid-connected mode. The placement of the ESS with proper rating considering many operational constraints is having an impact on the microgrid operation from many angles. This gives the motivation to present and discuss the proposed approaches with their advantages, limitations, and comparisons.

2 ESS Applications in a Smart Distribution System

The role of ESS on the distribution side can be summarized as follows [5, 6]:

- Time shifting of load can be possible to utilize the power during peak hours by storing the energy during up-peak periods. This helps the utility by increasing the utility factor with the peak loading reduction. In addition to that, it also helps the consumer in terms of optimally using energy at a lesser price.
- The supply of power by the distribution generators (DGs) can be smoothly regulated by the coordinated operation of ESS like PV and wind-based DGs. Apart from that, quick supply of power can regulate the voltage and frequency of the system for better power quality.
- The ESS can act as a voltage control support by providing reactive power to the network. In other words, the reactive power regulation and control for system stability can be handled by proper coordinated ESS operation.
- ESS can be a solution to the reverse power flow issues and challenges that may arise in case of improper placement of the DGs in the distribution network. Replacement of spinning reserve and peak shaving and provision of black starting in case of grid outages are other issues that can be resolved by the ESS integration.
- On the generator side, the ESS can avoid wind curtailment, minimize investment and operation cost, provide firm capacity, and minimize forecast error.
- For system operators, the ESS can provide energy arbitrage, minimize production cost, operating reserve, ramping, and replacement of network configuration, defer network expansion planning, enable higher DG integration, minimize investment and electricity cost, avoid system violations, island-reliability and island-dynamic security, provide firm capacity, minimize forecast error, and improve reliability.
- More responsive supply systems, replacement of insufficient capacity power plants, and improvements in transmission and distribution equipment are the major benefits as far as electric utilities are concerned.
- Managing electricity cost and reducing financial losses caused by outages are the benefits for the individual users by energy storage integration.

- Balancing regional loads and stabilizing transmission and distribution systems are the benefits of energy storage integration for the independent system operators.

At the end-user point, the importance of ESS includes [7]: (1) Power reliability is improved by the ESS integration at or nearer end users; (2) the ESS acts as a short backup during power outages; (3) the ESS may refine the fault ride-through capability of load for both commercial and domestic consumers by allowing the provision of an uninterrupted power supply (UPS) source; and (4) apart from that at the consumers' end, ESS can act as backup power, capacity contract savings/peak shaving, energy arbitrage, demand response, and smart home.

3 Categories and Types of ESS

Generally, ESS technologies are broadly classified into five types depending on the type of storage used [8]. They are shown in Fig. 1. They are as follows [9, 10]:

- **Battery-Based ESS**: These are generally electrochemical-based battery storage technologies with the most developed and used type being lead acid batteries. Other BESS technologies like sodium sulfur (NaS), metal–air, lithium-ion (Li-ion), and flow batteries like vanadium redox and zinc–bromine are also employed.
- **Thermal ESS**: Water storage tanks are employed as thermal ESS in case of load shifting functions during peak summertime for control and regulation of load, but are generally non-flexible.
- **Electromagnetic ESS**: These ESSs store energy in magnetic or electric forms (superconducting magnetic or supercapacitors). These have higher reliability and faster dynamic time of response but are compromised in lower energy storage densities and operating temperatures.

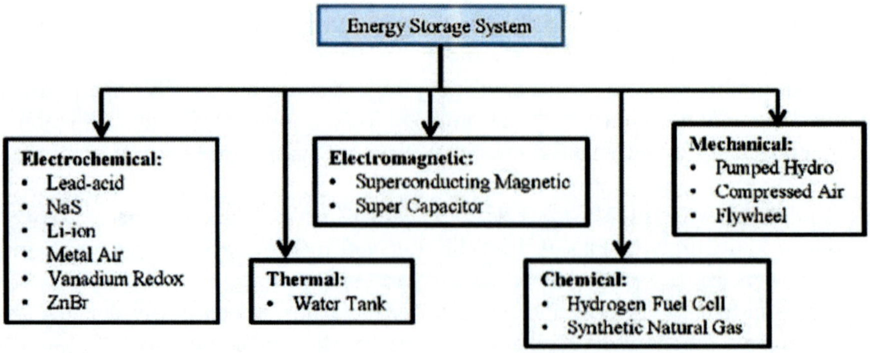

Fig. 1 Classification of ESS

- **Chemical ESS**: Hydrogen fuel cells and synthetic natural gas (SNG) ESS are excellent applications of chemical ESS. These are very proficient in long-term storage of energy, but the higher cost of hydrogen fuel cells and the conversion losses of SNG make these less desirable for ESS.
- **Mechanical ESS**: These ESSs inter-convert mechanical and electrical energy sources by conversion of energy using gravitational or potential energy. They are often classified into pumped-hydro, flywheel, and compressed-air ESS. Mechanical ESS has a high energy–power rating, reliability, and higher usage cycles but very slow dynamic response times.

4 Placement and Sizing of ESS

Financial indicators, technical indicators, and hybrid indicators are the major sizing criteria for ESSs devices. Major issues and challenges toward achieving organization and optimal ESS sizing are [11, 12]: (1) implementation of general policies for reduction of emissions; (2) selection of energy storage medium; (3) thorough analysis of cost-to-benefit ratio; (4) discharging depth capacity of the ESS; (5) connection mode/methodology to load or microgrid; (6) rate of optimal charging and discharging; (7) peak load shaving during heavy load demand; (8) power quality, reliability, and voltage control; (9) technological advancement of microgrid planning and optimization; (10) unpredictability of RES integration in microgrid; and (11) consequences of unpredictable nature in the addressed optimization algorithms for ESS size determination.

ESS technologies are distinguished owing to power handling capacity, energy storage capacity, placement, response granularity, response frequency/communication, ramp rate, and time of response and implementation requirements. In distribution planning and design, power rating, energy rating, and location play a vital role. The problem of optimizing the planning of ESS is a non-deterministic polynomial time (NP) hard problem. To date, much research has been looked into by various authors but does not arrive at a proper methodology for large-scale systems.

5 Approach for Problem-Solving

Figure 2 illustrates a general procedure for finding an optimal solution for the placement and sizing of the ESS irrespective of techniques applied [13, 14].

Objective function: The objective function is formulated based on the factors such as (1) minimization of cost of consumption and maximization of profit for utility, (2) minimization of the frequency deviation, (3) minimization of total storage capacity, (4) minimization of the area control error or forecast error,

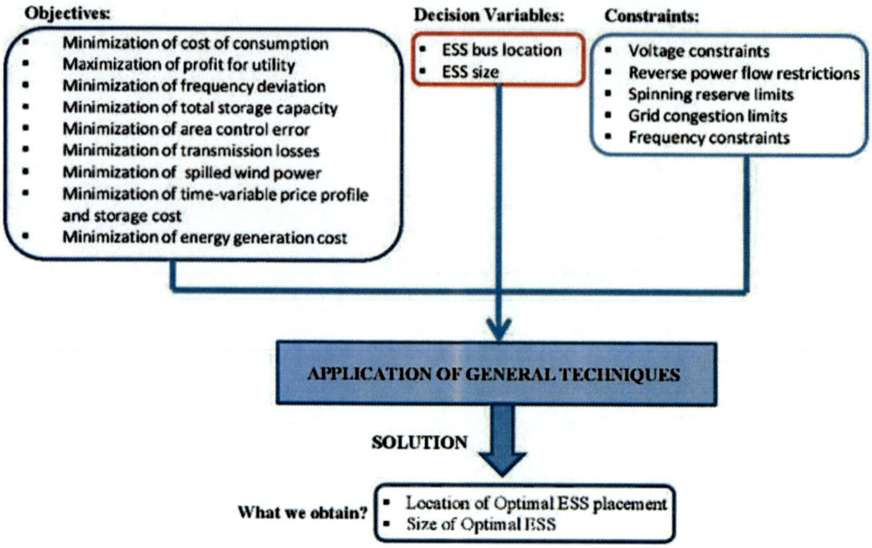

Fig. 2 General procedure for optimal solution of ESS placement and sizing

(5) minimization of transmission losses, (6) minimization of the spilled wind power, (7) minimization of the time-variable-dependent price profiling and storage cost, and (8) minimization of energy generation cost. A multi-objective study can be carried out by considering multiple combinations of objectives from the above simultaneously.

Constraints: Many real-time constraints are considered during the testing of the approaches for solving the planning of the ESS such as (1) voltage constraints, (2) grid congestion limits, (3) frequency constraints, (4) reverse power flow restrictions, and (5) spinning reserve limits.

Variables: Two major variables are considered for this problem such as (1) ESS bus location, (2) ESS size, and (3) combination of ESS location and size.

6 Algorithm for ESS Placement and Sizing

Figure 3 illustrates the general optimization algorithm for ESS placement and sizing [8, 15]. Here, the decision variables are taken into consideration with the constraints in place to formulate the objective function. Then, the objective function is evaluated for fitness using the proposed optimization algorithms (analytical or AI technique) to update the fitness value and optimal planning of the ESS. Finally, the test run of the power system is performed to evaluate the validation of the obtained optimal location and size of ESS.

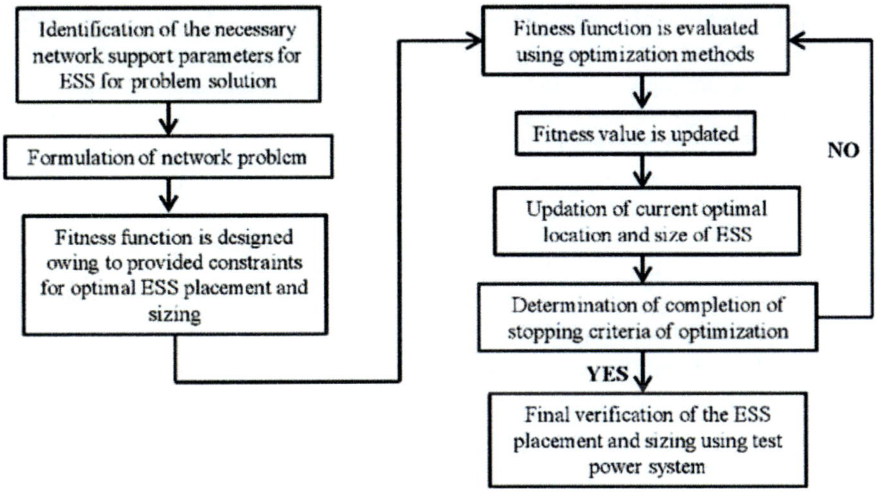

Fig. 3 General optimization algorithm for ESS placement and sizing

7 Techniques for ESS Placement and Sizing

There are many approaches proposed for the optimal placement and sizing of ESSs in the smart distribution system as illustrated in Fig. 4 and also described below.

Type-1: Analytical method (AM): In this approach, the optimal solution is determined by derivation and evaluating a combination of mathematical equations and algorithms. In this approach, the objective function and the system constraints associated with it are observed repeatedly with a distinct set of parameters and are compared repeatedly to find out the best solutions. However, the complexity of this approach increases with larger networks and system constraints. Secondly, no specific methodology is there for optimization applicable to all types of optimization problems, unlike other approaches.

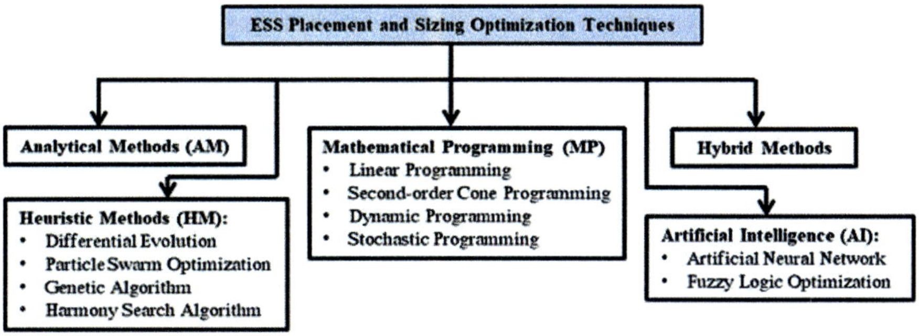

Fig. 4 ESS placement and optimization techniques

Type-2: Mathematical programming (MP): In this category, linear programming (LP), second-order cone programming (SOCP), dynamic programming (DP), and stochastic programming (SP) are the most prominent techniques applied extensively for the problem to optimally find out proper planning of location and capacity of the ESS in the smart distribution system. In this approach, numerical methods are applied to find out the optimal solutions. As this is a deterministic approach, the benefit is the attainment of an optimal solution, strong in finding the optimal solution, mature in power systems, and has good computational performance. However, major limitations lie in larger computational time, convergence difficulties, and explicit mathematical formulation required concerning complexity and larger dimension of the problems.

Type-3: Artificial Intelligence (AI): Artificial neural network (ANN) and fuzzy logic optimization algorithm are the major optimization techniques in this category. Unlike analytical and numerical methods, these approaches do not require any complex mathematical expressions and modeling. Apart from that in comparison, the computational time is less particularly under complex and larger dimension systems. The same code can be applied again and again even with the system parameter changes. However, the AI-based approaches do not guarantee the optimal solution.

Type-4: Heuristic methods (HM): These techniques are very popular and applied extensively in many engineering problems owing to their convenience and clarity in the application, non-derivative computation, not dependent on the exact system modeling and mathematical formulation, and less probability to trap in local minima. Many approaches are suggested in recent times under this category and are applied comprehensively in the optimal placement problem of ESS such as differential evolution (DE), particle swarm optimization (PSO), genetic algorithm (GA), and harmony search algorithm (HSA). Flexibility to solve all problems and avoid complex derivatives, and less computationally demanding are the major pros of this approach, while many convergences to the local optimum and less accuracy are the cons of this approach [15–17].

Type-5: Hybrid methods: Hybrid optimization techniques are more efficient in finding the optimum solution due to reducing the individual limitations by hybrid and coordinated searching approach. Fuzzy logic-based hybrid approaches are more popular due to the handling capability of system uncertainty and parameter changes. Combining the strengths of different methods, improving robustness, and ensuring global optimum solutions are the major advantages of these types of techniques. However, they are likely to increase the complexity, and many of them need high computational time than heuristic methods which are the major disadvantages of these types of techniques.

8 Discussions on Major Findings and Future Research

Although many research works have been done in recent times, many of the issues and challenges related to ESS placement, planning, and sizing have not been considered to date. The following findings on these limitations would be the guidelines for the researchers to develop optimal solutions in this regard.

- In the DG integration, the reversal of power flow in a microgrid is a very prominent issue that needs to be focused on during the planning and design stage. It has been found that the complex situation of a smart distribution system assuming the reverse power flow issue has not been investigated.
- The load variations of smart distribution systems are varying erratically such as duck curve shape variation in some cases. ESS with adequate control strategy and planning has a significant prospective in reduction and optimization of the aforesaid issues and challenges.
- ESS performance and cost optimization are needed to be balanced. The results provided by various authors to date have not considered both the factors in their objective. The maximization of performance also increases the cost of the establishment of ESS. So, a balanced investigation should be done to provide a solution that is more practically feasible.
- The profit and cost of the establishment are also considered simultaneously because ESS management leads to more profit to utility and consumers. Simultaneously, this maximization of profit is restricted with the constraints on the cost of establishment. Very rarely, authors considered this issue in the testing of their proposed approach for optimal placement and sizing of ESS.
- Considering all the practical constraints in every angle makes the convolution of the problem more. Again, the scenario in a large power system is quite different than the small distribution network. So, further enhancement is necessary to find a novel approach with complex and large smart distribution system considering all the real-time constraints in the problem-solving.
- An electric market study can be incorporated into this problem for providing a solution with a wider scope of application.
- Future research should be focused on model development that accurately describes the battery performance and lifetime, without sacrificing the computational time benefits. This will be beneficial to both short-term scheduling and network planning.
- Further research is essential on how to integrate optimally to provide multiple services from the same storage unit [18].
- Even if the decisions are made on optimal placement and sizing of ESSs, the advanced design of advanced circuit topologies and enhanced performance DC–AC converter design is essential for better functionality and improved power transfer mechanism of the ESSs.
- Adaptation of advance control strategies is necessary to reduce the impact of irregularity of renewable sources and load variations for the islanded and grid-connected environment.

- A simultaneous study is necessary to investigate comprised with energy management strategy with power quality, cost, reliability, and intermittency for proper fruitful implementation of ESS in the power system with proper location and sizing.
- Advanced simulation techniques are needed to be developed addressing the existing simulation program limitations. Consider all the factors like power quality, available power, voltage and frequency stability, power-sharing, peak shaving, cost, reliability, charging–discharging conditions, etc., while deciding the optimal location and sizing make the problem more complex.
- Future studies based on optimal sizing can be carried out to improve materials, reduce installations, operational, and maintenance costs, attenuation of transmission line losses, execution of easier control, provision of customer contentment, and achieving of affordable integration.
- Environmental and geographical constraints are required to be considered along with technical and economic constraints to arrive at a realistic solution for an ESS optimal placement.
- Very few investigations have been done considering the transient and dynamic conditions of the distribution and transmission system. Most of the studies have been done considering only the steady-state operation. The optimal strategy for placement and sizing needs to take the transient and dynamic factors in its formulation.

9 Conclusion

This paper contributes the following on the ESS optimal planning, location, and size problem review.

- Present the ESS role in the present and future smart distribution system.
- Provide the discussion on the objective function, constraints, and variables during the problem formulation.
- Discuss the general algorithm followed by almost all methods for optimal planning, formulation, location, and sizing of the ESS.
- Future research possibility on the problem to find out the optimal formulation of location and sizing of ESS are presented.

References

1. Wong, L.A., et al.: Review on the optimal placement, sizing and control of an energy storage system in the distribution network. J. Energy Storage **21**, 489–504 (2019)
2. Zhao, H., et al.: Review of energy storage system for wind power integration support. Appl. Energy **137**, 545–553 (2015)

3. Hoffman, M.G., et al.: Analysis tools for sizing and placement of energy storage in grid applications. In: ASME 2011 5th International Conference on Energy Sustainability, pp. 1563–1573 (2011)
4. Yang, Y., et al.: Battery energy storage system size determination in renewable energy systems: a review. Renew. Sustain. Energy Rev. **91**, 109–125 (2018)
5. Das, C.K., et al.: Overview of energy storage systems in distribution networks: placement, sizing, operation, and power quality. Renew. Sustain. Energy Rev. **91**, 1205–1230 (2018)
6. Hannan, M.A., et al.: Review of optimal methods and algorithms for sizing energy storage systems to achieve decarbonization in microgrid applications. Renew. Sustain. Energy Rev. **131**, 110022 (2020)
7. Hoffman, M., et al.: Analysis tools for sizing and placement of energy storage in grid applications. In: Energy Sustainability, vol. 54686 (2011)
8. Chang, L., et al.: Review on distributed energy storage systems for utility applications. CPSS Trans. Power Electron. Appl. **2**(4), 267–276 (2017)
9. Jannesar, M.R., et al.: Optimal placement, sizing, and daily charge/discharge of battery energy storage in low voltage distribution network with high photovoltaic penetration. Appl. Energy **226**, 957–966 (2018)
10. Sufyan, M., et al.: Sizing and applications of battery energy storage technologies in smart grid system: a review. J. Renew. Sustain. Energy **11**(1), 014105 (2019)
11. Ramírez, M., et al.: Placement and sizing of battery energy storage for primary frequency control in an isolated section of the Mexican power system. Electric Power Syst. Res. **160**, 142–150 (2018)
12. Motalleb, M., Reihani, E., Ghorbani, R.: Optimal placement and sizing of the storage supporting transmission and distribution networks. Renew. Energy **94**, 651–659 (2016)
13. Dvijotham, K., Backhaus, S., Chertkov, M.: Operations-based planning for placement and sizing of energy storage in a grid with a high penetration of renewables. arXiv preprint: arXiv:1107.1382 (2011)
14. Li, X., Wang, S.: A review on energy management, operation control and application methods for grid battery energy storage systems. CSEE J. Power Energy Syst. (2019)
15. Nayak, M.R., Dash, S.K., Rout, P.K.: Optimal placement and sizing of distributed generation in radial distribution system using differential evolution algorithm. In: International Conference on Swarm, Evolutionary, and Memetic Computing. Springer, Berlin, Heidelberg (2012)
16. Nayak, M.R., Kasturi, K., Rout, P.K.: Optimal placement and sizing of multiple capacitors in radial distribution systems using modified TLBO algorithm. Int. Rev. Autom. Control **6**(6), 701–713 (2013)
17. Nayak, M.R., Panigrahi, B.K., Rout, P.K.: Optimal reactive power dispatch based on the improved seeker optimisation algorithm and statistical analysis. Int. J. Power Energy Conv. **5** (3), 250–273 (2014)
18. Stecca, M., et al.: A comprehensive review of the integration of battery energy storage systems into distribution networks. IEEE Open J. Ind. Electron. Soc. **1**, 46–65 (2020)

OFDM-FSO Communication System Analysis

Prasant Kumar Sahu

Abstract In this paper, we have presented the performance analysis of a QAM-orthogonal frequency-division multiplexing (OFDM)-free space optics (FSO) (QAM-OFDM-FSO) transmission system, based on optical double sideband (ODSB), optical single sideband (OSSB), and optical vestigial sideband transmission scheme. Simulation models have been developed for optimizing the performance of the FSO system under different atmospheric conditions. The work is carried out using Optisystem 17.1.

Keywords FSO · QAM · OFDM · OSSB · ODSB · OVSB · Beam divergence

1 Introduction

The rapidly increased demand for high-speed connectivity services along with bandwidth hungry applications is a few of the driving forces behind the research and commercialization of next-generation communication technology. The main tasks emerging for designers and researchers for next-generation communication system are losses, selection of the modulation techniques, and nonlinear impairments [1–5]. Due to the inherent limitations of RF schemes, it may not be very attractive for very high-speed (several Gigabits per sec) next-generation wireless communication. On the other hand, fiber optic or copper cables requires significant maintenance costs and incurs deployment difficulties. Optical wireless communication (OWC) offers a license-free services along with a considerable advancement in terms of capacity, mobility, network reach, and number of supported users. Because of these advantages, OWC can be used as a complementary solution to RF. The research challenges toward the realization of optical wireless networks is multidisciplinary in nature. The performance of such system is affected by a number of factors [6–11]. Several system-related parameters and channel-related parameters

P. K. Sahu (✉)
School of Electrical Sciences, IIT Bhubaneswar, Bhubaneswar, India
e-mail: pks@iitbbs.ac.in

S. Das and M. N. Mohanty (eds.), *Advances in Intelligent Computing and Communication*, Lecture Notes in Networks and Systems 202,
https://doi.org/10.1007/978-981-16-0695-3_53

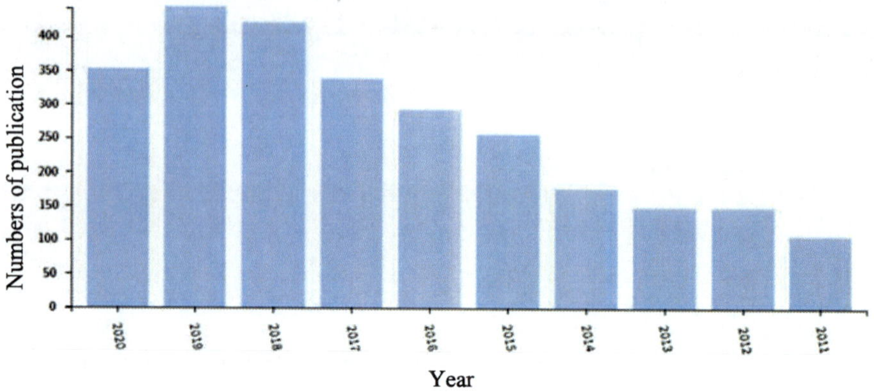

Fig. 1 Numbers of publication versus year

affect the system link performance, and hence the system needs to adopt to appropriate mitigation techniques to avoid unfavorable conditions. The transmission optics at the transmitter section have control over the light beam width; thus, a proper selection of light beam width compensates the inter-channel interference effect in FSO. The atmospheric conditions and compositions like rain, fog, snow, smoke, aerosol, dust, turbulence, and wind speed, etc., significantly impact the transmission rate and performance of the OWC system link. The transmitted light beam is absorbed and scattered because of the atmospheric particles. Optical signal intensity fluctuates due to the turbulence or scintillation effect. As well as multipath fading significantly affects the performance of FSO system. Due to the above-mentioned reasons, signal quality degrades, and bit error rate (BER) degrades, and many times, lead to link failure and signal fade [11–19]. Over the last one decade, a significant number of researchers are working in the filed of free space optical communication. Figure 1 is obtained through date collected from Web of Science which clearly indicates the number of publications in the areas of free space optical communication from 2011–2020 and signifies the importance of FSO research. In this work, we have analyzed the performance of a high bit rate QAM-OFDM-FSO system.

2 Free Space Optical Communication (FSO) System

The FSO system utilizes optical beams for data transmission through the atmosphere, or free space. The typical channel conditions vary and degrade because of different atmospheric attenuation, turbulence, fading, system loss like geometrical loss, and pointing error or misalignment errors which led to degrading the FSO system performance [8, 9, 14–19]. Atmospheric loss is one of the main contributors to optical signal power attenuation due to absorption and scattering. The particles present in the atmosphere which may be gaseous molecules, aerosol, and suspended particles etc., absorb the signal power. Let the absorption coefficient of the light beam with a specific wavelength (λ) is denoted by α_{abs}, which is related to the absorbing gas molecules and their concentration is given by [11].

$$\alpha_{\text{abs}} = \frac{4\pi K}{\lambda}\sigma_{\text{a}}N_{\text{a}} \tag{1}$$

where, K is the imaginary part of the refractive index of the absorbing particles, σ_{a} is related to the effective cross-section of the absorbing particles, and N_{a} is the concentration per unit volume of the particle. Attenuation due to the scattering is characterized by a widely used model known as Kruse and Kim model [12, 13] and the attenuation coefficient is presented as

$$\beta_{\text{scat}} = \frac{3.91}{V}\left(\frac{550}{\lambda}\right)^{i} \tag{2}$$

where, V is the visibility (km), λ is the operating wavelength (nm), and 'i' related to the particle size distribution, and its values are considered based on the visibility condition.

According to the Kim model [13]:

$$\begin{array}{ll} i = 1.6 & \text{for } V > 50\,\text{km};\ i = 1.3 \quad \text{for } 6\,\text{km} < V < 50\,\text{km} \\ i = 0.16V + 0.34 & \text{for } 1\,\text{km} < V < 6\,\text{km};\ i = V - 0.5 \quad \text{for } 0.5\,\text{km} < V < 1\,\text{km} \end{array}$$

According to the Kruse model [12]:

$$\begin{array}{ll} i = 1.6 & \text{for } V > 50\,\text{km};\ i = 1.3 \quad \text{for } 3\,\text{km} < V < 50\,\text{km} \\ i = 0.585V1/3 & \text{for } V < 6\,\text{km} \end{array}$$

The other major atmospheric factors which cause scattering and absorption are rain, fog, snow, cloud, etc. Fog is the most dominating factor affecting the performance of FSO communication under both scattering and absorption conditions. The thickness layer of the fog depends on the size of the particle, which varies the visibility of the atmosphere, and fog impacts severely in low altitude, which makes the model complicated. The light beam is significantly absorbed and scattered under thick fog condition, which leads to FSO communication link failure. The visibility distance values under rain and fog environments are presented in Table 1. Atmospheric turbulence which results in random variation of the air refractive index is mainly due to the variation of temperature and pressure in the atmosphere. It depends on a number of factors such as the wind speed, altitude, and time of the day, humidity, and atmospheric pressure. Besides the atmospheric effect, the pointing error which is due to the misalignment between the transmitter and receiver section has a significant effect on the FSO system performance and link availability [15–20].

Table 1 Weather conditions and visibility distance values [14]

Weather condition	Visibility distance (m)
Very clear	23,000–50,000
Clear/drizzle (R = 0.25 mm/hr)	18,000–20,000
Haze/rain (R = 12.5 mm/hr)	2800–4000
Thin fog/rain (R = 25 mm/hr)	1900–2000
Light fog/rain (R = 50 mm/hr)	770–1000
Medium fog	500
Thick fog	200

In literature, two statistical model are adopted for zero and nonzero boresight pointing error. The nonzero boresight pointing error is described by its PDF which is expressed as [17],

$$f_{I_p}(I_p) = \frac{\xi^2 \exp\left(\frac{-s^2}{2\sigma_s^2}\right)}{A_o^{\xi^2}} I_p^{\xi^2-1} I_o\left(\frac{r_s}{\sigma_s^2}\sqrt{\frac{-W_{eq}^2 \ln\frac{I_p}{A_o}}{2}}\right), \quad 0 \le I_p \le A_0 \tag{3}$$

where ξ is the pointing error parameter which is the ratio between equivalent beam width W_{eq} and jitter standard deviation (σ_s) at the receiver and it is presented as $\xi = W_{eq}/2\sigma_s$.I_o is the modified Bessel function of the first kind with zero order, I_p is the received irradiance due to pointing error, A_o= [erf(v)]^2 is the fraction of received power, $v = \sqrt{(\pi/2)}\ (r/W_z)$, related to the ratio between the aperture radius (r) and normalized beam waist (W_z), erf(.) is the standard error function and S is the boresight displacement. The PDF of zero boresight pointing error model can be obtained by using boresight displacement $s = 0$ in the above equation. Various mathematical channel models have been proposed in the literature such as Log-normal, Gamma-Gamma, Negative exponential, K-distribution, Double-Weibull distribution, I-K distribution, Double generalized gamma distribution and Exponential Weibull distribution [20]. A number of works is reported in the literature for realization of free space optics to address the issues of atmospheric effects [21–24]. In our work, we have considered the Log-normal channel model.

The setup used in the simulation study for analyzing the QAM-OFDM-FSO system performance comprises of the components as shown in the block diagram in Fig. 2.

In order to process the signal over a longer distance, both pre- and post-amplifications are realized using optical amplifier at the transmitter side and semiconductor optical amplifier at the receiver side. The simulation experiment was carried out using the Optisystem 17.1 software. Figure 3 shows the simulation system using Optisystem 17.1. In order to measure the atmospheric effects and other detrimental effects like geometrical loss, pointing error, and fading on the considered 4-QAM OFDM-FSO system, simulation is carried out under different modulation conditions. The system was also optimized for beam divergence using single parameter optimization method. The reason behind the selection of the OFDM-FSO system is to mitigate the multipath fading as well as two offers secure

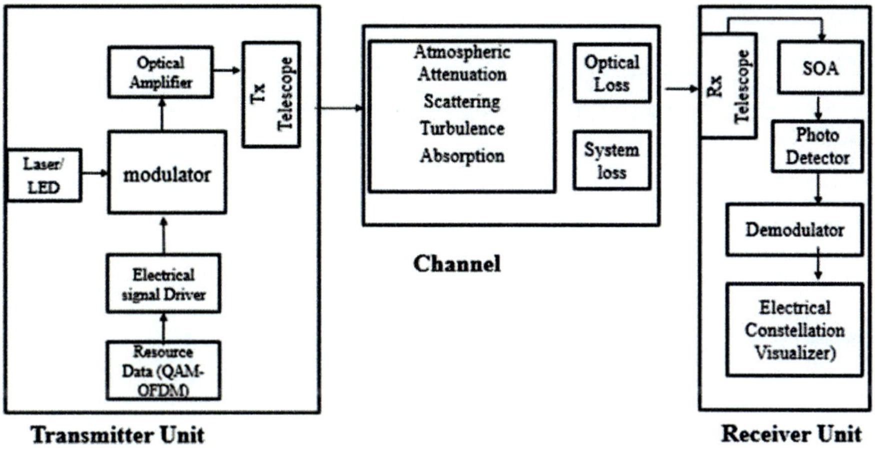

Fig. 2 Block diagram of FSO System

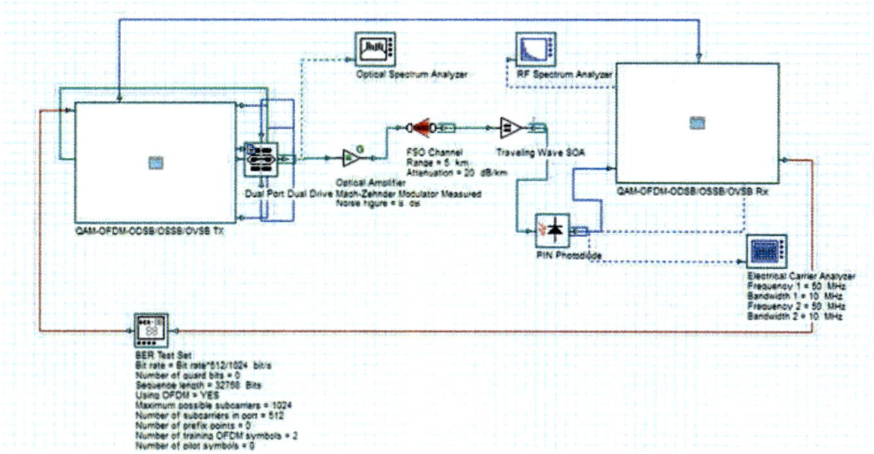

Fig. 3 Simulation experiment setup for QAM-OFDM using ODSB/OSSB/OVSB

higher data-rate transmission. The system performance was analyzed using optical single sideband (OSSB), optical double sideband (ODSB), and optical vestigial sideband (OVSB) transmission scheme.

The OVSB signal is generated using a phase shifter arrangement and the OSSB signal is generated using uniform FBG. The simulation parameters are listed in Table 2.

The resultant ODSB, OSSB, and OVSB signals are shown in Fig. 4. The ODSB generation can be easily implemented using an external MZM modulator. The OSSB signal was generated using a uniform FBG, whereas the OVSB signal was generated using an additional unit known as the 90° phase shifter block.

Table 2 Simulation parameters

Parameters	Values	Parameters	Values
Data rate	10 Gbps	Channel model	Log-normal
Wavelength	1550 nm	Turbulence conditions	No turbulence to strong
Tx LASER power	0 dBm	Photodetector	PIN
Optical amplifier	EDFA and SOA	Thermal power density	15e-24 (W/Hz)
Attenuation	0.2–100 dB/km	Responsivity	1A/w
Link range	0.5–200 km	Receiver dark current	10 nA
Pointing error and geometric loss	3 dB	Sequence length	65536 bits
Rx aperture diameter	20 cm	Bits per symbol	2
Beam divergence	0.2 mrad–2 mrad		
Tx aperture diameter	5 cm		

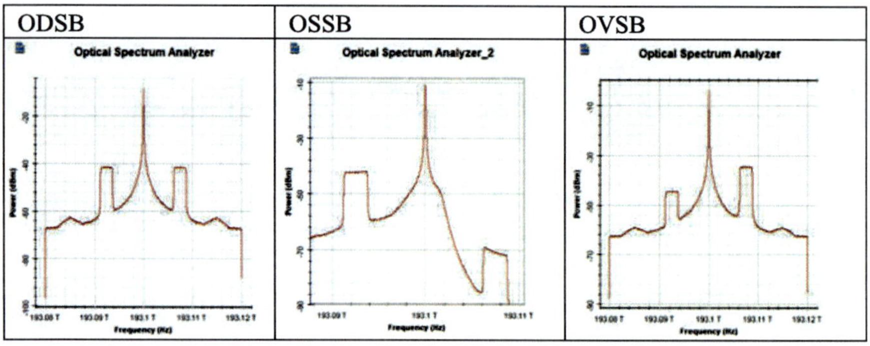

Fig. 4 Optical sideband generation

3 Results

The results obtained from above simulation setup for different atmospheric conditions varying from clear weather conditions to extreme weather conditions including turbulence, pointing error, and the scintillation have been modelled using the log-normal turbulence model presented below. Figure 5 presents the SNR variations with distance under bad weather conditions with an attenuation of 25 dB/km.

The constellation diagram for different weather conditions ranging from extremely bad weather to clear weather to conditions is reported in Fig. 6.

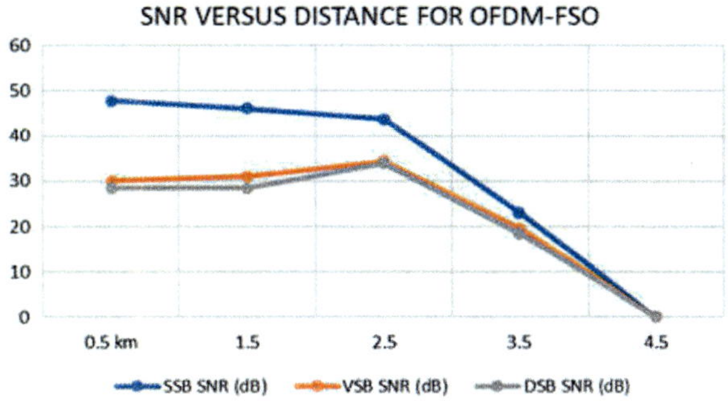

Fig. 5 SNR versus distance for the QAM-OFDM FSO system with atmospheric attenuation of 25 dB/km

From the above results one can conclude that a larger gain separates the symbols farther than a small gain and helps in combatting the noise generated in the receiver. Moreover, when the gain decreases, it becomes evident, that the constellation is not centered at (0, 0). Also, the results indicates that, when the symbols spacing becomes small enough, then the noise makes symbols unrecognizable. This happens mainly due to the nonlinear effects. If the extinction ratio of the MZM modulator is not properly selected, then it results in the off-center characteristic of the constellation. The beam divergence also plays a significant role in FSO transmission. The other important detrimental effect is the generation method of the DSB, SSB, and the VSB signal itself. A proper care must be taken for the generation of these signals, as it contribute a lot to the performance metrics of FSO. From our results, it is very much clear that the OSSB-based QAM-OFDM FSO system outperforms the ODSB and OVSB-based QAM-OFDM FSO system.

4 Conclusion

In this work, the utility of different modulation techniques was analyzed for a high bit rate (10 Gbps) QAM-OFDM-FSO system. If the spacing between subcarriers, transmitter-receiver aperture, and beam divergence can be optimized, then the overall performance of the OFDM-FSO link can be further improved. The ODSB modulation format requires higher transmission bandwidth and is more susceptible to the chromatic dispersion, as compared to OSSB. However, the OVSB scheme is an intermediate scheme which offers the performance similar to the ODSB scheme with a relatively lower bandwidth requirement. From the results, it is assumed that the operational range of FSO system can be increased significantly up to 200 km with an acceptable system performance using proper signal conditioning scheme. QAM-OFDM-FSO system is a potential candidate for fulfilling the high-speed data rates and for transmission of multi-type information propagation. Moreover, OFDM technique provides better transmission performance by efficiently utilizing spectral efficiency and minimizing the effect of dispersion. The QAM-OFDM-FSO system with OSSB performs better in diverse channel state at a higher data rate as compared to the other schemes like IMDD, ODSB, and OVSB.

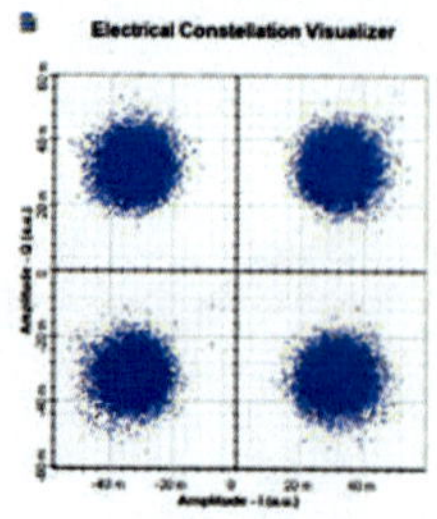

(A) OSSB over 0.5 km with 100 dB/km attenuation

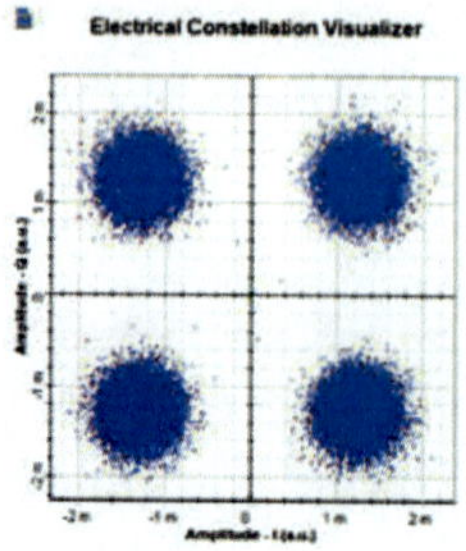

(B) OSSB over 1.0 km with 100 dB/km attenuation

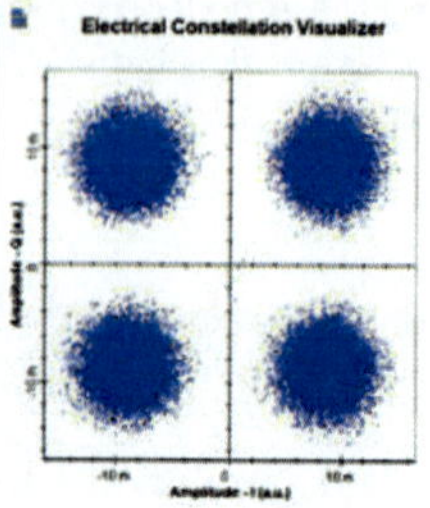

(C) OVSB over .5 km with 100 dB/ km attenuation

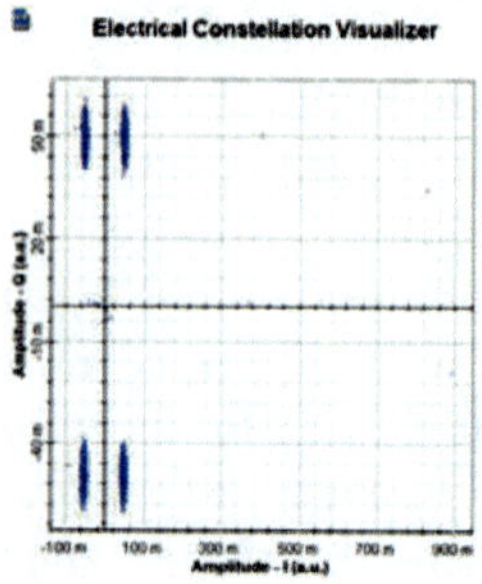

(D) ODSB over 1 km with 100 dB/ km attenuation

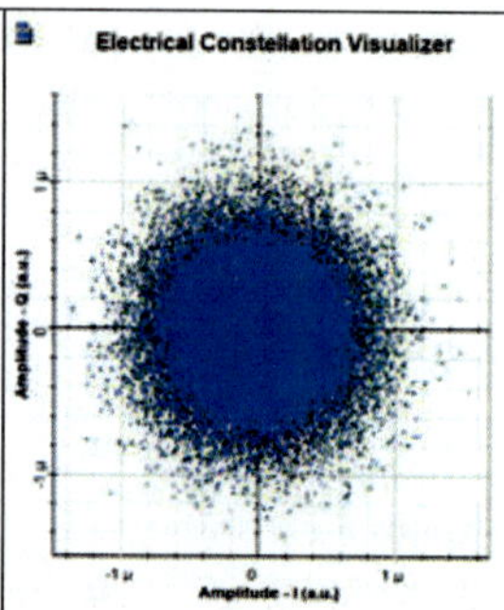

(E) ODSB over 5 km with 100 dB/km attenuation

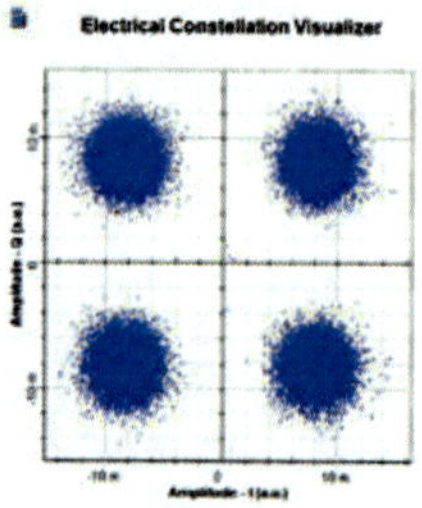

(F) OSSB over 200 km with .2 dB/km attenuation (Clear weather)

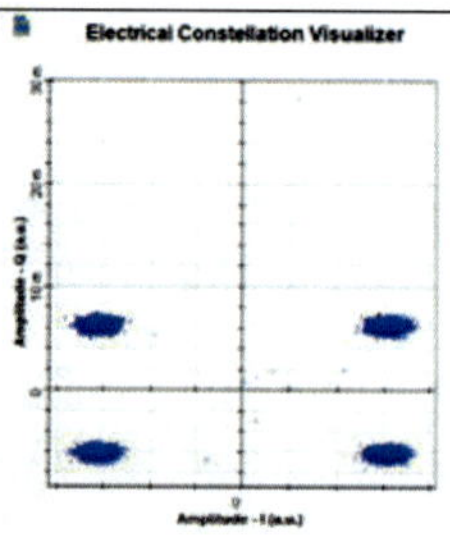

(G) ODSB over 200 km with .2 dB/km attenuation (Clear weather)

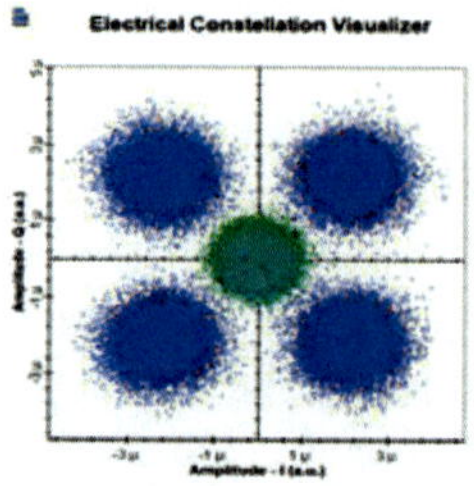

(H) OVSB over 200 km with .2 dB/km attenuation (Clear weather)

Fig. 6 Electrical constellation diagram of the received signal

Acknowledgements The author acknowledge the financial support by DST-FIST.

References

1. Muhammad, S.S., Leitgeb, E., Koudelkat, O.: Multilevel modulation and channel codes for terrestrial FSO links. In: IEEE 2nd International Symposium on Wireless Communication Systems, pp. 795–799 (2005)
2. Majumdar, A.K.: Advanced Free Space Optics (FSO): A Systems Approach. Springer, New York (2015) (ISBN: 978-1-4939-0918-6/978-1-4939-0917-9)
3. Fawaz, W., Abou-Rjeily, C., Assi, C.: UAV-aided cooperation for FSO communication systems. IEEE Commun. Mag. **56**(1), 70–75 (2018). https://doi.org/10.1109/MCOM.2017.1700320
4. Yadav, S., Kumar, V.: Optimal clustering in underwater wireless sensor networks: acoustic, EM and FSO communication compliant technique. IEEE Acc. **5**, 12761–12776 (2017). https://doi.org/10.1109/ACCESS.2017.2723506
5. Ghafary, B., Kashani, F.D., Kazemian, E.: Reliability analysis of FSO communication links using a berated divergent rectangular partially coherent flat-topped beam. Iran. J. Electr. Electron. Eng. **9**(1), 13–19 (2013)
6. Majumdar, A., Ghassemlooy, Z., Raj, A.: Principles and applications of free space optical (FSO) communication. IET, UK (2019)
7. Shen, H., Yu, L., Fan, C.: Temporal spectrum of atmospheric scintillation and the effects of aperture averaging and time averaging. Opt. Commun. **330**, 160–164 (2014)
8. Raj, A.A.B., Selvi, J.A.V., Durairaj, S.: Comparison of different models for ground-level atmospheric turbulence strength $[Cn]^2$ prediction with a new model according to local weather data for FSO applications. Appl. Opt. **54**, 802–815 (2015)
9. Ren, Y., Huang, H., Xie, G., et al.: Atmospheric turbulence effects on the performance of a free space optical link employing orbital angular momentum multiplexing. Opt. Lett. **38**(20), 4062–4065 (2013)
10. Kazemlou, S.: All-optical multihop free space optical communication systems. Master's thesis, McMaster University (2010)
11. Lightpointe company: How to design a reliable FSO system. http://www.lightpointe.com/images/LightPointe_How_to_Design_a_Reliable_FSO_System.pdf (2009)
12. Kruse, P.W., McGlauchlin, L.D., McQuistan, E.B.: Elements of Infrared Technology: Generation, Transmission and Detection. Wiley and Sons (1962)
13. Kim, I.I., McArthur, B., Korevaar, E.: Comparison of laser beam propagation at 785 nm and 1550 nm in fog and haze for optical wireless communications. In: Proceedings SPIE 4214, Boston, MA, USA (2001)
14. Popoola, W., Ghassemlooy, Z., Awan, M.S., Piteti, E.L.: Atmospheric channel effects on terrestrial free space optical communication link. In: International Conference on ECAI, 3rd ed., pp. 17–23 (2009)
15. Park, J., Lee, E., Chae, C., Yoon, G.: Impact of pointing errors on the performance of coherent free-space optical systems. IEEE Photon. Technol. Lett. **28**(2), 181–184 (2016)
16. Varotsos, G.K., Nistazakis, H.E., Volos, C.K., Tombras, G.S.: FSO links with diversity pointing errors and temporal broadening of the pulses over weak to strong atmospheric turbulence channels. Optik **127**(6), 3402–3409 (2016)
17. Yang, F., Cheng, J., Tsiftsis, T.A.: Free space optical communication with nonzero boresight pointing errors. IEEE Trans. Commun. **62**(2), 713–725 (2014)
18. Farid, A.A., Hranilovic, S.: Outage capacity optimization for free space optical links with pointing errors. IEEE J. Lightw. Technol. **25**, 1702–1710 (2007)
19. Andrews, L.C., Phillips, R.L., Hopen, C.Y.: Laser Beam Scintillation with Applications. SPIE Publications (2001)

20. Al-Habash, M., Andrews, L.C., Phillips, R.L.: Mathematical model for the irradiance probability density function of a laser beam propagating through turbulent media. Opt. Eng. **40**(8), 1554–1562 (2001)
21. Chaudhary, S., Amphawan, A., Nisar, K.: Realization of free space optics with OFDM under atmospheric turbulence. Optik **125**(18), 5196–5198 (2014)
22. Ismail, T., Leitgeb, E.: Performance analysis of SIM-DPSK FSO system over lognormal fading with pointing errors. In: IEEE 18th International Conference on Transparent Optical Networks (ICTON) (2016)
23. Malik, S., Sahu, P.K.: Parameter optimization of free space optical system using particle swarm optimization. Int. J. Commun. Syst. **33**(9) (2020)
24. Sandalidis, H.G., Tsiftsis, T.A., Karagiannidis, G.K., Uysal, M.: BER performance of FSO links over strong atmospheric turbulence channels with pointing errors. IEEE Commun. Lett. **12**(1), 44–46 (2008)

Control and Management of Fuel Cell Micro-grid Using Optimal Model Predictive Controller

Amba S. Nayak, Anshuman Satpathy, Lalit Mohan Satapathy, and Niranjan Nayak

Abstract In the present days, the renewable energy generation is increasing exponentially as per the demand. But in the distribution it is not economical and convenient to supply energy to a particular area separately. Thus, the renewable sources should be integrated with conventional grid, which is still a challenging task. It is because of the nonlinearity and fluctuating nature of the sustainable sources like PV and fuel cell. Here in this paper, a micro-grid (MG) is designed with PV and polymer exchange fuel cell as distributed energy resources (DER) and the MG is integrating with conventional grid with an online load. Since most of the loads are nonlinear, it can cause a power quality issue called harmonics distortion which may cause huge damage to the system. To address the above problem, an optimized model predictive controller is introduced in shunt active power filter to improve the robustness of controller. To validate the strength of the controller, the system is introduced with high reactive nonlinear load. All the results and model are simulated in MATLAB/Simulink environment.

Keywords Micro-grid · Fuel cell · Distributed energy resources · Renewable power resources

A. Satpathy · L. M. Satapathy (✉) · N. Nayak
Department of EEE, SOA Deemed to be University, Bhubaneswar, Odisha, India
e-mail: lalitsatapathy@soa.ac.in

A. Satpathy
e-mail: anshumanas1001@gmail.com

N. Nayak
e-mail: niranjannayak@soa.ac.in

A. S. Nayak
IIIT, Bhubaneswar, Odisha, India
e-mail: b318003@iiit-bh.ac.in

S. Das and M. N. Mohanty (eds.), *Advances in Intelligent Computing and Communication*, Lecture Notes in Networks and Systems 202,
https://doi.org/10.1007/978-981-16-0695-3_54

1 Introduction

The rapid global energy demand created an essentiality of renewable power resources (RPR) in the present energy market. In next two decades due to shortage of conventional energy sources, RES will dominate the entire energy market. The photovoltaic and fuel cells are two prominent RES due to their high efficiency, low emission, less noise and high modularity and other beneficial characteristics [1]. The solar cell is extremely used for alternative source due to its easy availability, visually and unobtrusive, small maintenance and pollution free generation. A Micro-grid can be defined as a combination of micro sources like PV fuel cell.

The MG technology is accepted worldwide as a auspicious solution of shortage of power in the system. Distribution energy resources, DERs like wind, PV fuel cell storage cell are connected parallel to the load.

Among renewable energy resources, photovoltaic (PV) and fuel cells are less significant and more accessible than the conventional power plants to which the MGs are integrated. Photovoltaic system (PV) is one of the recent technical processes to convert sunlight directly into electricity [2]. Many review articles on MG control scheme have been published. Mainly there are three control strategies like centralized, local, master–slave and distributed control schemes [3]. Many researches have been conducted on hybrid MG system, and cost effectiveness of the DERs has been analyzed with intermittent load conditions [4]. SOFC is a recent tool which produces clean and green energy in the range of few kilowatts [5].

A solar power plant of capacity 25.5 kW rating-based micro-grid model monitoring system is analyzed in [6]. A 25-km-long transmission and distribution system with SCADA and PLC are designed for a two-stage energy management integrated with a DC micro-grid is presented [7]. Besides this so many research works have been done on the optimized controllers for micro-grid management. In Ref. [8], a new water cycle is optimized ELM for MPPT of PV-based MG. The PV integration is done in two stages like DC–DC and DC–AC system. Here in this work DC–DC two-stage converters estimate the control reference. A SAPF designed to reduce the harmonic effects in a fuel cell connected micro-grid [9].

The authors in the paper [10] explained the elimination of harmonics in a solar MG integrated with SAPF. The solar cell is designed with MPPT controller in association with fuzzy controller based on particle swarm optimization (PSO). The research offered in Ref. [11] aims to show the priority-based internal power flow management system on the local power system. A 3 and 69 bus system consisting of seven hybrid micro-grids is studied by the use of a fast distributed controller design (FDCD) algorithm. It could effectively overwhelm several types of faults injected to the network. The converging rate is considerably inferior to some other competitor algorithms and faster than the other algorithms [12].

In this paper, a micro-grid is designed with PV and fuel cell as distributed energy resources. A robust controller known as model predictive controller based on competitive optimization algorithm (COA) is applied to the common boost converter. The controller improves the system stability and reduces harmonics below 5% as per the IEEE standard.

The remaining portion of the paper is presented as the system model is discussed in Sect. 2. The model predictive controller is derived in Sect. 3. The competitive optimization algorithm is discussed in Sect. 4. Result analysis is produced in Sect. 5, and finally the conclusions are drawn in Sect. 6.

2 System Modeling

Figure 1 indicates the design of the micro-grid which is the buildup of the one polymer exchange fuel cell and a photovoltaic cell. The fuel cell is supplied by 50 l hydrogen bottles, at 200 bar pressure. The energy supplied by the fuel cell is managed by the boost converter and one inverter, and the principle is same for PV cell also. The controllers associated with the MG manage the DERs for steady operation.

2.1 PEM Fuel Cell

A fuel cell structure is almost the same that of a battery. The components are anode and cathode as two electrodes, one electrolyte and a catalyst for chemical reaction in the electrolyte to generate electricity. The main difference between these two is fuel cell produces electricity from external electrolyte, whereas battery works within it. Here in the PEMFC a solid polymer electrolyte is used (Fig. 2).

The voltage generated by the fuel cell is given by

$$V = V_0 - V_{act} - V_{res} - V_{con} \tag{1}$$

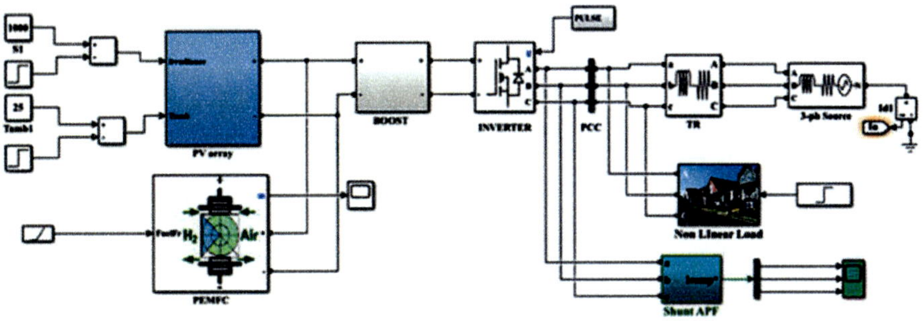

Fig. 1 Simulink model of MG

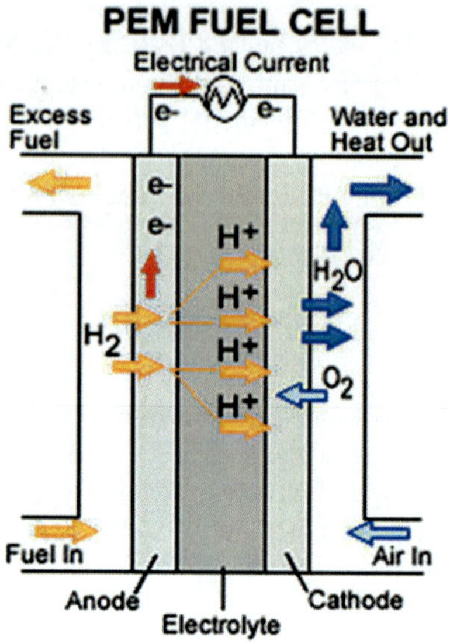

Fig. 2 Structure of PEM fuel cell

where V_0 is oc voltage, V_{act} is the activation drop, V_{res} is the resistance loss, and V_{con} is the concentration loss. The fuel cell stack voltage is as follows:

$$V_{stack} = nV \tag{2}$$

By using the standard state entropy change, the oc voltage for the PEMFC can be shown as follows:

$$V_0 = \frac{-\Delta g_f^{-0}}{2F} + \frac{RT}{2F}\ln\left[\frac{\left[aH_2 a_2^{\frac{1}{2}}\right]}{aH_2O}\right] \tag{3}$$

The generated voltage is

$$V = V_0 - A\ln\left(\frac{i + i_L}{i_0}\right) - ri \tag{4}$$

The data in this paper is taken as V_0=1.2 V, A = 0.06 V, i_0 = 0.04 mA cm^{-2}, and r = 0.003 Ω cm^2. The fuel cell stack comprises 16 cells (n = 16) which are connected in series. The error is

$$\text{er}(\%) = \frac{\text{experimental value} - \text{theoritical value}}{\text{experimental value}} \tag{5}$$

2.2 Photovoltaic System

In micro-grid, solar cell is one of the main DER in present scenario, due to its simple structure, can be installed anywhere and the easy availability of input resources. In order to reduce the operation and storage cost, the PV-based micro-grids are integrated with conventional grid (Fig. 3).

The intermittent natures of photovoltaic systems do not give any consistent operation. Thus, the maximum power point tracking is associated with the system to achieve steady operation. The PV output current is given by

$$I_c = I_{\text{ph}} - I_0\left\{e^{\frac{q(V+RIS)}{AKT}} - 1\right\} - \left(\frac{V+IRS}{R_{\text{sh}}}\right) \tag{6}$$

Boost converter provides the control of the voltage in MPPT in the system. PID controller is not a nonlinear controller, and it is a linear controller. So when we apply it in a nonlinear system, it fails to give the efficiency. So we apply optimization which will make linear PID controller to receive the nonlinearity of the system.

The PV current is expressed as

$$i_{\text{pv}} = I_{\text{L}} - I_{\text{s}}\left[e^{\alpha(V_{\text{pv}}+R_{\text{s}}i_{\text{pv}})-1}\right] - \frac{V_{\text{pv}}+i_{\text{PV}}}{R_{\text{sh}}} \tag{7}$$

$$I_L = \{I_{\text{sc}} + k_{\text{i}}(T_{\text{c}} - T_{\text{ref}})\}\frac{S}{1000} \tag{8}$$

The short-circuit current expression is given by

$$I_{\text{s}} = I_{\text{Rs}}\left[\frac{T_{\text{c}}}{T_{\text{ref}}}\right]^3 \text{e}^{\frac{qE_{\text{g}}}{Ak}\left(\frac{1}{T_{\text{ref}}}-\frac{1}{T_{\text{c}}}\right)} \tag{9}$$

$$I_{\text{s}} = I_{\text{R}}S\left[\frac{T_{\text{C}}}{T_{\text{r}}ef}\right]^3 \exp\left[\frac{(qE_{\text{g}})}{A_k}(1/T_{\text{r}}ef - 1/T_{\text{C}})\right] \tag{10}$$

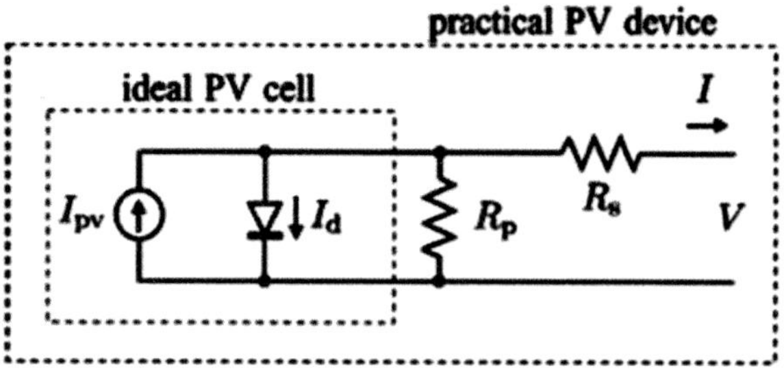

Fig. 3 Electrical equivalent of PV cell

2.3 Shunt Active Power Filter (SAPF)

SAPF consists of a DC link capacitor and a voltage source converter. They produce a current equal in magnitude of the harmonic current with 180° phase opposition. The device injects current in phase opposition at PCC in the MG. Thus, it cancels the harmonic current produced by the nonlinearity of the system (Fig. 4).

The values of L_F and V_C can be obtained by solving the following equation simultaneously

$$Q_c = 3V_s I_c = 3V_s \frac{V_c}{LF\omega}\left(1 - \frac{V_s}{V_c}\right) I_h = \frac{V_h}{mfLF\omega} \quad (11)$$

The DC capacitor of the filter is obtained from the above equation as

$$C_{dc} = \pi \times \frac{I_c^{\text{rated}}}{\sqrt{3}\omega V_{\text{ripples}}} \quad (12)$$

3 SAPF Controller

Here the model predictive controller is applied to the shunt active power filter which provides the control action along with several advantages. This reduces the uncertainties and reduces the cost function. In this control scheme, an open-loop scheme is designed to predict and select the best values.

This is only the first constituent of the best possible actuation of the sequence is applied, and every one of the optimization is calculated for each sampling. The

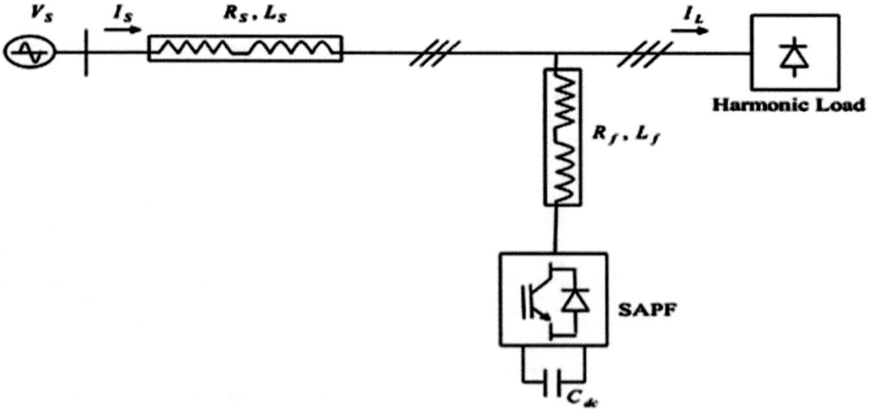

Fig. 4 Simple model of SAPF

block diagram of the proposed predictive control for a three-phase inverter with output LC filter is shown in Fig. 5. $V_c(k)$ and $I_f(k)$ are used to predict by the use of the following equation.

$$x(k+1) = A_q x(k) + B_q V_i(k) + B_{dq} i_o(k) \tag{13}$$

$$A_q = e^{AT_2},\ B_q = \int_0^{T_2} e^{AT_2} B dt,\ B_{dq} = \int_0^{T_2} e^{AT_2} B_d dt \tag{14}$$

The next sampling output voltage $V_c(k + 1)$ is calculated for all the probable voltage vectors that the inverter generates. The estimation of output current $i_O(k)$ is an essential parameter to predict $V_c(k + 1)$. The load current is measured by a load current observer given by the equation.

$$i_O(k-1) = i_f(k-1) - \frac{C}{T_3}\{V_c(k) - V_c(k-1)\} \tag{15}$$

The state space model of the observer is given by

$$A_{obse} = A - JC \tag{16}$$

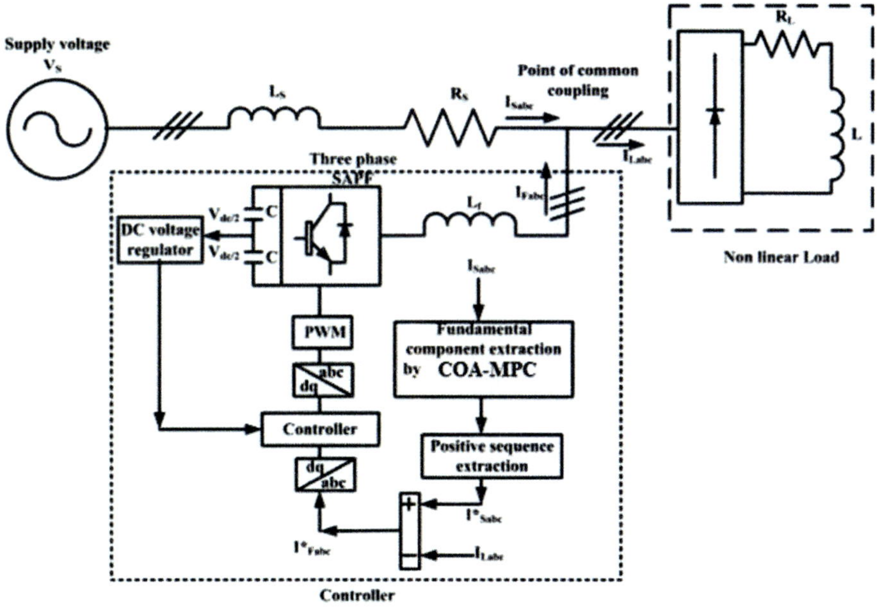

Fig. 5 Structure of SAPF controller

$$A = \begin{bmatrix} 0 & -\frac{1}{L} & 0 \\ \frac{1}{C} & 0 & -\frac{1}{C} \\ 0 & 0 & 0 \end{bmatrix}, \quad C = \begin{bmatrix} 1 & 0 & 0 \\ 0 & 1 & 0 \end{bmatrix} \tag{17}$$

To select $V_i(k)$ for the inverter, seven predictions obtained for $V_c(k + 1)$ are compared using a cost function g,

$$g = \left(V_{c\alpha}^{*} - V_{c\alpha}\right)^2 + \left(V_{c\beta}^{*} - V_{c\beta}\right)^2 \tag{18}$$

where $V_{c\alpha}^{*}$ and $V_{c\beta}^{*}$ are the real and imaginary component of inverter voltage and $V_{c\alpha}^{*}$ and $V_{c\beta}^{*}$ are the real and imaginary part of the predicted voltage $V_c(k+1)$. The SAPF control structure is shown in Fig. 5.

4 Imperialist Competitive Optimization Algorithm (ICOA)

It is a type of evolutionary metaheuristic algorithms, which works on animal behavior. Unlike other evolutionary algorithm, it also starts with initial population. There are two groups—imperialist country and imperialist colonies. The best countries belong to imperialist country. The colonies are divided into several groups and moves toward the imperialist colonies. The competition starts among empires. If an empire does not have its companion, then it may be removed. The algorithm is initialized as follows:

$$\begin{aligned} &\text{country} = [p_1, p_2, p_3, \ldots, p_N], \\ &\cos t = f(\text{country}) = f(p_1, p_2, p_3, \ldots, p_n), \\ &N_{\text{col}} = N_{\text{pop}} - N_{\text{imp}} \end{aligned} \tag{19}$$

$$\max_i\{C_i\} - c_n \tag{20}$$

$$p_n = \frac{C_n}{\sum_{i=1}^{N_{\text{imp}}} C_i} \tag{21}$$

$$NC_n = \text{Round}\{p_n N_{\text{col}}\} \tag{22}$$

where country = number of country, N = number of variables, N_{col} = number of colonies, c_n= imperialist cost, C_n=normalized cost, P_n= power of nth imperialist. The velocity and position equations are given by

$$\begin{aligned} x &\sim V(0, \mu d) \\ \theta &\sim V(-\lambda, +\lambda) \end{aligned} \tag{23}$$

Equation (22) shows how a colony moves toward the imperialist through a distance x, and position θ. μ is a constant between [1, 2], and 'd' is the distance between the colony and imperialist. The equation of full power of an imperialist is given by

$$TC_n = \mathrm{Cos}t(\mathrm{imp}) + \lambda \mathrm{mean}\{\cos t(\mathrm{col})\} \tag{24}$$

The parameters used in this work are number of countries = 200, Nnmber of imperialist = 10, revolution rate = 0.2, $\mu = 1.4$, $\lambda = 0.02$, and dimension = 30.

Pseudo Code

Step-1: Initial population is created for every empire.
Step-2: Accommodation and uprising of the colonies those move to the emperialist.
Step-3: Compare the imperialist and the best colony. If the cost of best colony is more, they are eligible to change their positions.
Step-4: The objective function of each empire is to be calculated.
Step-5: Find out the weakest colony and placed in the best empire.
Step-6: Failure of the weak empires.
Stage 7: When only one empire remains, termination reaches, otherwise returned to Step 2.

Objective Function
SAPF DC voltage error is taken as objective function.

$$\mathrm{err} = V_{\mathrm{fdcref}} - V_{\mathrm{fdc}} \tag{25}$$

f(err) is the objective function. The ICO algorithm is applied to model predictive controller to optimize $V_{\alpha c}$ and $V_{\beta C}$.

5 Result and Analysis

A MG with renewable energy system such as fuel cell and PV has been designed using MATLAB/Simulink platform. A single-phase bidirectional buck/boost converter is assembled for operation in two opposite directions being controlled by a PWM generated gain signal. A nonlinear load is connected to the MG. The system is simulated with the nonlinear load, and the PCC voltage and current of the fuel cell and PV are simulated and shown in Fig. 6. It is simulated with model predictive controller with imperialist competitive optimization algorithm.

Figures 6 and 7 represent the transient performance of conventional system optimal model predictive controller. It is depicted that the optimal MPC performs better and dominates all the current and voltage distortion.

In Figs. 8 and 9, the PCC voltage simulation with switching on a nonlinear load is shown. In Fig. 8, the phase voltages waveforms are not in proper phase sequence due to unbalanced condition of the system. But in Fig. 9, the voltage regains its original condition with the MPC and the ICOA optimization.

Harmonic Analysis

Here, in this work, a micro-grid is designed with PV and fuel cell as distributed energy resources DERs. Due to the intermittent generation of solar power and fuel cell, the harmonic distortion in this case is a predominate power quality issue. As we know the harmonic can cause various types of damages in the system. In order to suppress the harmonics, an optimized robust controller known as imperialist competitive optimization algorithm-based model predictive controller is applied to the shunt active power filter. It is seen that the ICOA-SAPF reduces the total harmonic distortion (THD) to a minimum value, as shown in Figs. 10, 11, 12 and 13. The PCC voltage THD was measured as 43.75% with SAPF but no controller.

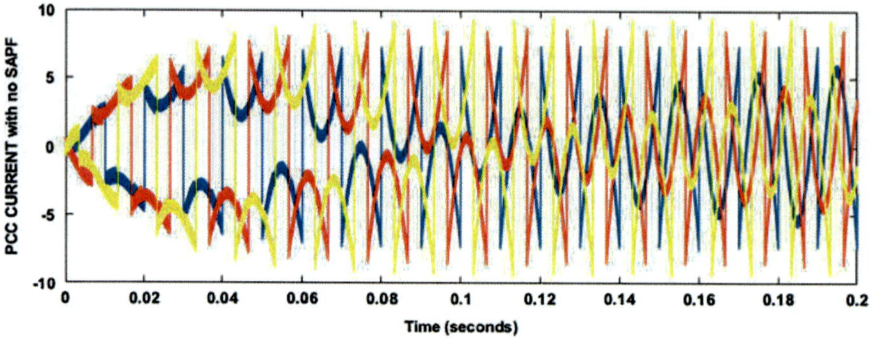

Fig. 6 PCC current with no controller

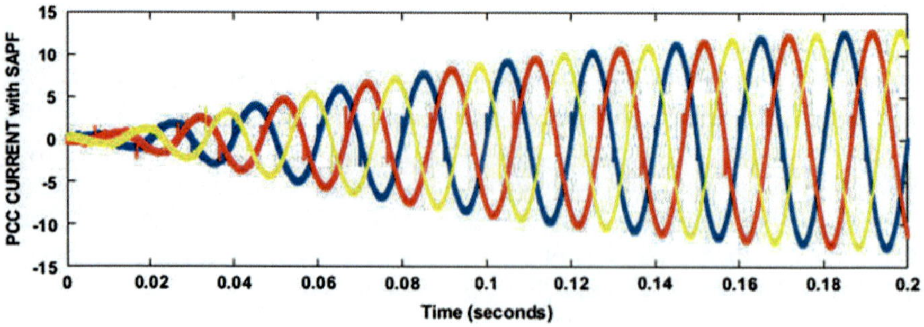

Fig. 7 PCC current with ICOA-MPC-SAPF

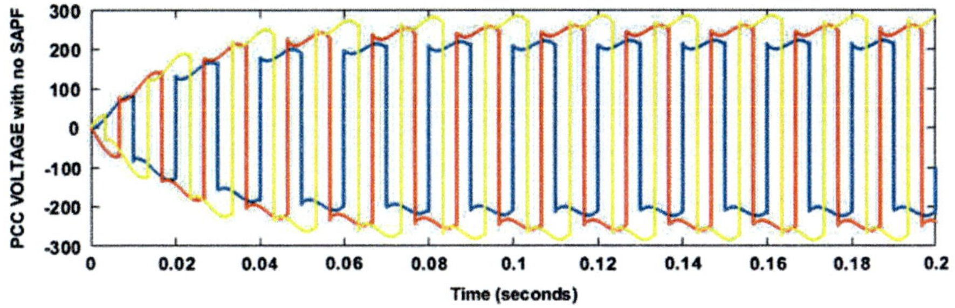

Fig. 8 PCC voltage of system with no controller

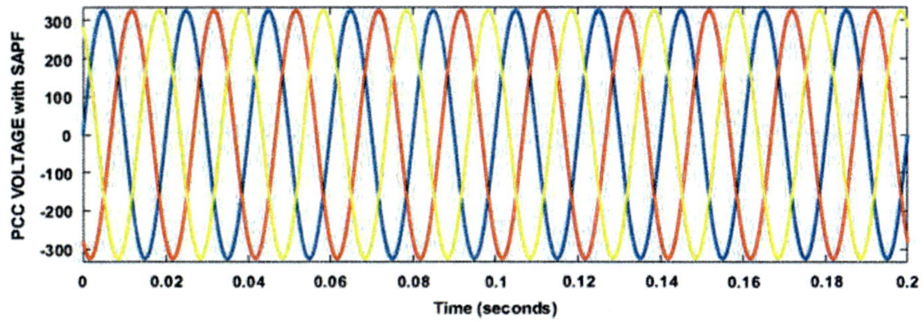

Fig. 9 PCC voltage of system with ICOA-MPC-SAPF

That has been reduced to 11.34% with MPC controller, and it is further reduced to 1.69% with ICOA-MPC-based SAPF. Similarly, the THD of PCC current appears as 50.20% with no controller. This has been reduced to 11.56% with model predictive controller only. Further with help of the optimized controller, it is lowered to 3.87% which meets the IEEE standard.

6 Conclusion

In this paper, a MG designed with a SAPF and two distributed energy resources such as solar cell and PEM fuel cell are shown in Fig. 1. The system is simulated with a nonlinear load and an optimized model predictive controller. An imperialist competitive optimization algorithm is applied to the SAPF with the DC voltage error as objective function. The grid voltage was distorted with SAPF only. The optimized SAPF cancels the disturbance and becomes perfect sinusoidal, as shown in Fig. 9. As per the harmonic analysis, the proposed optimized robust controller brings down the total harmonic distortion THD to below 5% as per the IEEE standard, as shown in Figs. 10–13.

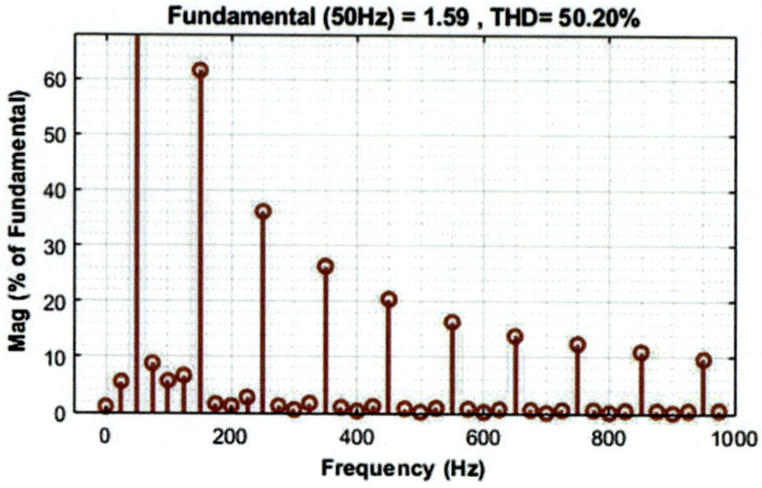

Fig. 10 THD of PCC current with SAPF

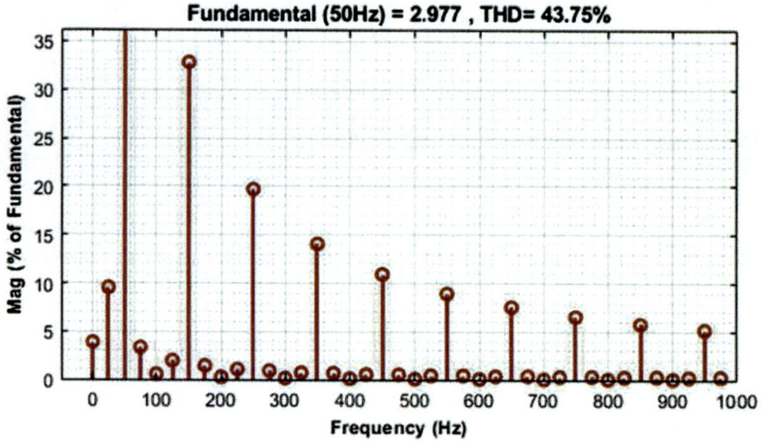

Fig. 11 THD of PCC voltage with SAPF

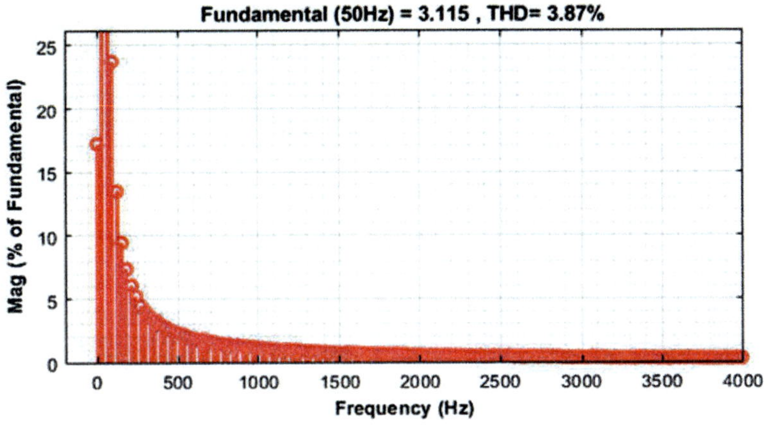

Fig. 12 THD of current with ICOA-based predictive controller

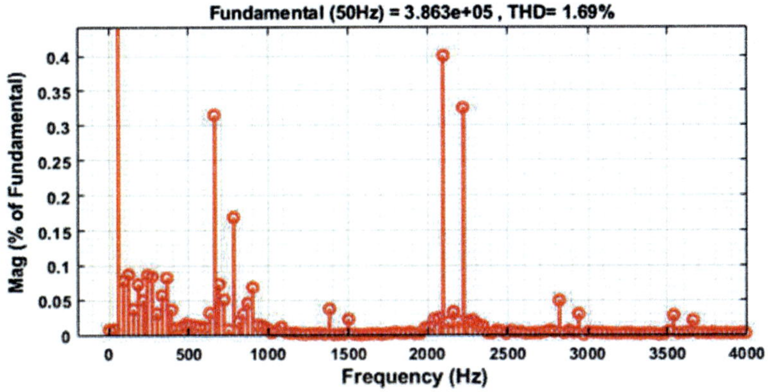

Fig. 13 THD of PCC voltage with ICOA-based model predictive controller model

References

1. Chiari, L., Zecca, A.: Constraints of fossil fuels depletionon global warming projections. Energy Police **39**, 5026–5034 (2011)
2. Kersten, F., Doll, R., Kux, A., Huljic, D.M., Gorig, M.A., Breyer, C., Müller, J.W., Wawer, P.: PV learning curves: past and future drivers of cost reduction. In: Proceedings of the 26th European Photovoltaic Solar Energy Conference, 5–9 September, Germany (2011)
3. Prodanovic, M., Green, T.C., Mansir, H.: Survey of control methods for three-phase inverters in parallel connection. In: Proceedings of the 8th IEEE Conference on Power Electronics and Variable Speed Drives, pp. 472–477 (2000)
4. Patterson, M., Macia, N.F., Kannan, A.M.: Hybrid microgrid model based on solar photovoltaic battery fuel cell system for intermittent load applications. IEEE Trans. Energy Convers. **30**(1), 359–366 (2015)

5. Roos, I., Soosaar, S., Volkova, A., Streimikene, D.: Green house gas emission reduction perspectives in the Baltic States in frames of EU energy and climate policy. Renew. Sustain. Energy Rev. **16**, 2133–2146 (2012)
6. Kabalci, Y., Kabalci, E.: Modeling and analysis of a smart grid monitoring system for renewable energy sources. Sol. Energy **153**, 262–275 (2017)
7. Sharma, R.K., Mishra, S.: Dynamic power management and control of a PV PEM fuel-cell-based standalone ac/dc micro-grid using hybrid energy storage. IEEE Trans. Ind. Appl. **54**(1):526–538 (2018)
8. Nayak, N., Pattnayak, S.K., Dhar, S., Bagarty, D.P., Viswavandya, M.: Accurate maximum power point tracking with new modified water cycle based ridge extreme learning machine for photovoltaic based micro-grids'. In: Advances in Electrical Control and Signal Systems: Select Proceedings of AECSS 2019, vol. 665, pp. 553–565, Springer Nature
9. Acharya, D.P., Nayak, N., Choudhury, S., Padhy, R.: Power quality improvement in a fuel-cell based micro-grid with shunt active power filter. Int. J. Renew. Energy Res. (IJRER) **10**(3):1071–1082
10. Acharya, D.P., Nayak, N., Choudhury, S.: Power quality enhancement of a photovoltaic based micro grid system using optimized fuzzy controller with SAPF. In: 2019 International Conference on Smart Systems and Inventive Technology (ICSSIT), Tirunelveli, India, pp. 67–72 (2019)
11. Spasova, B., Kawamoto, D., Takefuji, Y.: A study of fuel cell scheduling effect on local energy markets with heterogeneous renewable sources. Energies **12**, 854 (2019)
12. Nabi, B., Seyedtabaii, S.: Fast distributed control design for DC linked microgrids. Int. J. Electr. Power Energy Syst. **122** (2020)

Multilevel Thresholding Using Black Widow Optimization

Bibekananda Jena, Manoj Kumar Naik, Aneesh Wunnava, and Rutuparna Panda

Abstract The key objective of any multilevel thresholding method is to obtain the optimal thresholds on a target image for the region of interest. Traditional methods of multilevel thresholding suffer from poor accuracy and computationally expensive. Optimization algorithms are found suitable to answer these difficulties effectively. In this paper, we suggest a new multilevel thresholding algorithm using Black Widow Optimization (BWO). The paper uses the unique mating behavior concept of the black widow spider to optimize six different objective functions: Otsu's between-class variance, Tsallis entropy, Masi entropy, Kapur's entropy, Renyi's entropy, and fuzzy entropy for obtaining optimal threshold values. The performance of the proposed thresholding method is evaluated by generating the segmented images of a set of standard grayscale and color images using well-known performance metrics such as PSNR, SSIM, and FSIM. The experimental outcomes show a significant improvement in the performance of the proposed method in segmenting both grayscale and color images.

Keywords Nature-inspired algorithm · Black widow optimization · Entropy · Multilevel thresholding

---

B. Jena
Dept. of Electronics and Communication Engineering, Anil Neerukonda Institute of Technology & Science, Sangivalasa, Visakhapatnam 531162, Andhra Pradesh, India
e-mail: bibekananda.jena@gmail.com

M. K. Naik (✉) · A. Wunnava
Faculty of Engineering and Technology, Siksha O Anusandhan, Bhubaneswar 751030, Odisha, India
e-mail: naik.manoj.kumar@gmail.com

A. Wunnava
e-mail: aneeshwunnava@gmail.com

R. Panda
Dept. of Electronics and Telecommunication Engineering, Veer Surendra Sai University of Technology, Burla 768018, Odisha, India
e-mail: r_ppanda@yahoo.co.in

S. Das and M. N. Mohanty (eds.), *Advances in Intelligent Computing and Communication*, Lecture Notes in Networks and Systems 202,
https://doi.org/10.1007/978-981-16-0695-3_55

1 Introduction

Images are used for carrying useful information in the field of image processing and computer vision. In most of the applications such as remote sensing, medical diagnosis, and pattern recognition, the desired region extracted from the raw images is more used for further processing. Therefore, the task of extraction of the desired region from a test image is very important and considered as one of the important preprocessing steps in most of the applications. The process of dividing the raw images into meaningful regions of interest is called image segmentation. There are two different approaches to perform image segmentation: discontinuity-based approach and similarity-based approach. The similarity-based approach gains its popularity by its fast and convenient way of producing segmentation results. One of the similarity-based segmentation techniques is the thresholding method. Thresholding methods partition an image with the help of threshold values selected by analyzing the test image before segmentation. In multilevel thresholding, more than one threshold values are used for segmentation, which divides the target image into multiple segments used for a certain application.

The challenging task of the multilevel thresholding algorithm is the selection of appropriate threshold values by properly analyzing the image which decided the effectiveness of the methods. The two different approaches present in the literature for obtaining optimal threshold values from the test image are parametric and nonparametric methods. Due to the high computational complexity of the parametric method, it has been used rarely, whereas nonparametric methods are more convenient to implement than the parametric approach. In the nonparametric method, the optimal threshold values are obtained by optimizing some objective functions such as Tsallis entropy [1], Kapur's entropy [2], Otsu between-class variance [3], Masi entropy [4], Renyi's entropy [5], and fuzzy entropy [6]. Metaheuristic algorithms are extensively used in the nonparametric method of threshold selection by maximizing the above objective functions in a short period. Some of the well-known nature-inspired optimization algorithms which are successful are particle swarm optimization (PSO) [7], genetic algorithm (GA) [8], artificial bee colony (ABC) [9], gray wolf algorithm (GWO) [10], etc.

In this paper, a new multilevel thresholding algorithm using black widow optimization [11] is proposed. BWO is used here to obtain the optimal threshold by maximizing the six different well-known objective functions used extensively for multilevel thresholding techniques. The rest part of the paper is organized as follows: Sect. 2 deals with the multilevel thresholding method and description of different objective functions, the BWO algorithm is discussed in Sect. 3, the proposed multilevel thresholding using BWO is presented in Sect. 4, and result analysis is presented in Sect. 5. At last, the paper ends with a concluding remark in Sect. 6.

2 Image Segmentation Using Multilevel Thresholding

Multilevel thresholding is one of the convenient and effective methods for image segmentation and splits the image into several regions with the help of more than one threshold values obtained by optimizing a suitable objective function.

2.1 *The Mathematical Formulation of Multilevel Thresholding*

Let us consider an image of size $M \times N$ having L gray levels $\{0, 1, 2, \ldots, L-1\}$ divided into $k+1$ distinct regions depending upon their intensity level. For generating $k+1$ distinct regions k threshold values are required which can be shown by following the simple rule of thresholding, where p represents the intensity of the pixel, R_{i} represents the ith segmented region, and $\{T_1, T_2, \ldots, T_k\}$ represents the threshold values used for segmentation.

$$R_{\mathrm{i}} \leftarrow p, \text{ if } \quad T_{\mathrm{i}-1} \leq p \leq T_{\mathrm{i}} \quad \forall i = [0, (k+1)] \tag{1}$$

where $T_0 = 0$ and $T_{k+1} = L - 1$.

2.2 *Objective Functions*

The objective function (OF) in the nonparametric approach of threshold selection from images is needed to be maximized or minimized for obtaining the optimal threshold values as follows:

$$\left[T_1^*, T_2^*, T_3^*, \ldots, T_k^*\right] = \arg\max/\min(\mathrm{OF}) \tag{2}$$

where T_{i}^* represents the optimal threshold values.

In this paper, six commonly used OFs have been used which are discussed below. The objective functions used in this paper are given in Table 1

Table 1 Various objective functions (OF) used for multilevel thresholding

OF	Expression	Reference
Tsallis entropy	$OF_{Tsallis} = \sum_{i=0}^{k+1} S_q^{R^i} + (1-q)\prod_{i=0}^{k+1} S_q^{R^i}$ where, $S_q^{R_i} = \frac{1-\sum_i^{T_i+1}(pr_i/P_i)^q}{q-1}$	[1]
Kapur's entropy	$OF_{Kapur} = \sum_{i=1}^{k+1} H_i$ where $H_i = \sum_{T_j}^{T_{i+1}} \frac{Pr_{(i)}}{P_i} \ln\left(\frac{Pr_{(i)}}{P_i}\right)$	[2]
Otsu between-class variance	$OF_{Otsu} = \sum_{i=1}^{k+1} P_i(\mu_i - \mu)^2$ $\mu_i = \sum_{T_i}^{T_{i+1}} i.\left(\frac{Pr_{(i)}}{P_i}\right)$	[3]
Masi entropy	$OF_{Masi} = \sum_{i=1}^{k+1} S_r(R_i)$ where, $S_r(R_i) = \frac{1}{r-1}\left(\log\left(1-(1-r)\sum_{T_i}^{T_{i+1}}\left(\frac{Pr_{(i)}}{P_i}\right).\log\left(\frac{Pr_{(i)}}{P_i}\right)\right)\right)$	[4]
Renyi's entropy	$OF_{Renyi} = \sum_{i=1}^{k+1} Rn_\alpha^i$ Where, $Rn_\alpha^i = \frac{1}{1-\alpha}\ln\sum_{T_i}^{T_{i+1}}\left(\frac{Pr_{(i)}}{P_i}\right)^\alpha$	[5]
Fuzzy entropy	$OF_{Fuzzy} = \sum_{i=1}^{k+1} F_i$ where $F_i = -\sum_{j=0}^{L-1}\frac{Pr_{(j)} * \mu_i(j)}{Q_i}\ln\frac{Pr_{(j)} * \mu_i(j)}{Q_i}$	[6]

3 Black Widow Optimization

Black widow optimization (BWO) algorithm imitates the unique mating mechanism of the black widow spider for searching for a global optimum. There are three different steps generally found in the mating process of the black widow spider: *procreate*, *cannibalism*, and *mutation*. The mathematical model of all these steps is discussed below.

Let us consider the initial population of the spider as given by Eq. (3) where d denotes the number in a vector and NP represents the population size.

$$X = \begin{bmatrix} X_1^d & X_2^d & X_3^d & \ldots & X_{NP}^d \end{bmatrix} \tag{3}$$

Procreate
The mating process of spider is performed in procreate steps. Depending upon the procreate rate '*nr*', the best *nr* solutions are selected in place of the new population group *pop*1.Two random parents selected from the new group started mating to reproduce the new generation. The mathematical model of offspring generation from two independent parents' vectors X_i and X_j can be formulated as follows:

$$Y_i = \alpha.X_i + (1 - \alpha).X_j \tag{4}$$

$$Y_j = \alpha.X_j + (1 - \alpha).X_i \tag{5}$$

where Y_i and Y_j represent the offspring, and the process is repeated $d/2$ times.α is a random vector defined in the range [0, 1]. Finally, the selected mother and the set of children generated from the process are sorted as per their fitness value.

Cannibalism
There are two different types of cannibalism processes found in spiders' lifestyle: sexual cannibalism and sibling cannibalism. In the first one, the female spider eats her partner during or after mating, whereas in the later one, stronger spider eats their weaker sibling. In the algorithm, spiders having poor fitness values are rejected after the procreate steps depend upon the cannibalism rate. The remaining solutions are kept in a new population group *pop*2

Mutation
In this stage, some randomly selected individuals from *pop*1 depend upon mutation rate 'mt' which undergoes mutation by exchanging two elements in the vector.

The initial population is then updated by selecting the best candidate from *pop*1 and *pop*2 in each iteration until the termination criteria are met.

4 BWO-Based Multilevel Thresholding

In BWO-based multilevel thresholding for segmenting $k+1$ non-overlapping regions in the image, the initial population is formed by taking a set of individuals of dimension k. We have used Tsallis entropy, Kapur's entropy, Otsu between-class variance, Masi entropy, Renyi's entropy, and fuzzy entropy as the objective function for fitness computation in the optimization algorithm. In each iteration, the individual vectors in the population are updated by going through the three different stages of the mating mechanism of the black widow spider. In each iteration, the spiders with maximum fitness functions are survived, and the rest are removed from the populations. At the end of the search process, the best spider having the highest fitness value is declared as the optimal threshold for the segmentation.

5 Results and Discussion

The performance of the proposed multilevel thresholding algorithm is tested on a set of standard grayscale images like Lena, boat, peppers, and baboon as well as with a set of color images taken from the Berkeley segmentation dataset (BSDS 500) for 2, 3, and 4 thresholding levels. To obtain the optimal threshold values, the population size and the number of iterations are set to 30 and 100, respectively. The parameters of the BWO algorithm are taken as same in the original paper. For each image, the segmentation process is repeated 20 times, and then the mean and the best values of the performance quality matric such as peak signal-to-noise ratio (PSNR), structure similarity index (SSIM), and feature similarity index (FSIM) are noted down in Tables 2 and 3. From the result, it is clear that Otsu-based thresholding is superior to other thresholding methods for both grayscales and color images. Table 4 presents the optimal threshold values obtained from grayscale image Lena and the color image of number 480025 taken from the Berkeley database. For the color image, optimal threshold values are obtained from the histogram of the red (*R*), green (*G*), and blue (*B*) plane separately. Table 3 presents the optimal threshold values obtained from grayscale image Lena and the color image of number 480025 taken from the BSDS 500 dataset. For the color image, optimal threshold values are obtained from the histogram of the red (*R*), green (*G*), and blue (*B*) plane separately. Thresholded images obtained by BWO using various objective functions are shown in Figs. 1 and 2. It can be observed that BWO can effectively produce an optimal threshold by optimizing various objective functions.

Table 2 Performance measures the values of the Lena image (grayscale image)

Threshold	Measures	Tsallis	Kapur's	Otsu	Masi	Renyi's	Fuzzy
2	$PSNR_{mean}$	22.9759	23.0168	**23.1635**	23.0319	22.9700	21.4173
	$SSIM_{mean}$	0.2177	0.2161	**0.2291**	0.2169	0.2174	0.1946
	$FSIM_{mean}$	0.7484	0.7491	**0.7631**	0.7498	0.7484	0.7237
	$PSNR_{best}$	23.0080	23.0351	**23.1720**	23.0757	23.0080	22.5713
	OF_{mean}	0.8889	12.3210	**1.8837e+03**	5.4573	12.1234	13.6081
3	$PSNR_{mean}$	25.6116	25.9781	**26.0602**	25.9565	25.6638	24.3442
	$SSIM_{mean}$	0.2905	0.2967	**0.3061**	0.2961	0.2911	0.2678
	$FSIM_{mean}$	0.8165	0.8141	**0.8145**	0.8138	0.8173	0.7922
	$PSNR_{best}$	25.7702	26.0324	**26.0751**	26.0529	25.8607	25.7398
	OF_{mean}	1.2963	15.2704	**2.0364e+03**	6.9980	15.0006	17.2668
4	$PSNR_{mean}$	27.3925	27.7320	**28.186**	27.7328	27.4069	25.9284
	$SSIM_{mean}$	0.3432	0.3480	**0.3642**	0.3484	0.3438	0.3205
	$FSIM_{mean}$	0.8551	0.8590	**0.8631**	0.8591	0.8563	0.8294
	$PSNR_{best}$	27.4728	27.8337	**28.2524**	27.8363	27.4950	27.0398
	OF_{mean}	1.6543	17.9784	**2.0969e+03**	8.4757	17.7175	20.3949

Table 3 Performance measures values of the color image from BSDS 500 dataset for a subject 48025

Threshold	Measures	Tsallis	Kapur's	Otsu	Masi	Renyi's	Fuzzy
2	$PSNR_{mean}$	20.6354	21.0886	**21.2648**	21.0777	21.2118	21.1136
	$SSIM_{mean}$	**0.3418**	0.3175	0.2737	0.3203	0.3038	0.2583
	$FSIM_{mean}$	**0.6576**	0.6465	0.6328	0.6477	0.6425	0.6314
	$PSNR_{best}$	20.6404	21.1019	**21.2706**	21.0780	21.2168	21.2402
	OF_{mean}	0.8889	12.6567	**2.6485e+03**	5.5347	12.2084	13.8014
3	$PSNR_{mean}$	23.1947	24.0085	**24.6298**	24.0707	24.5394	23.1929
	$SSIM_{mean}$	**0.4544**	0.4345	0.3933	0.4299	0.3827	0.4039
	$FSIM_{mean}$	**0.7514**	0.7459	0.7304	0.7453	0.7248	0.7223
	$PSNR_{best}$	23.2563	24.0284	**24.6361**	24.1153	24.5874	24.4516
	OF_{mean}	1.2963	15.6544	**2.9085e+03**	7.0908	15.0449	17.3630
4	$PSNR_{mean}$	24.3628	25.5119	**26.4677**	25.6601	25.8714	25.1598
	$SSIM_{mean}$	0.5059	0.4859	**0.5247**	0.4866	0.4834	0.4669
	$FSIM_{mean}$	0.7819	0.7844	**0.8172**	0.7867	0.7975	0.7728
	$PSNR_{best}$	24.5675	25.8997	**26.5004**	26.1137	26.0496	25.7018
	OF_{mean}	1.6543	18.4770	**2.9876e+03**	8.5899	17.8600	20.3127

Table 4 Optimal threshold values obtained from various OF using BWO

OF	*k*	Color image (48025.jpg)			Grayscale image (Lena. jpg)
		R	*G*	*B*	
Tsallis	2	121 179	120 175	88 128	86 156
	3	118 173 215	110 148 192	77 107 137	67 118 170
	4	102 139 178 216	103 140 174 207	70 96 118 144	64 104 143 185
Kapur's	2	94 172	101 174	73 127	90 156
	3	85 134 185	90 134 182	71 123 172	71 114 161
	4	80 124 171 213	57 9 141 188	63 97 132 172	65 103 140 174
Otsu	2	88 179	91 179	62 124	87 147
	3	64 131 191	74 138 191	51 101 138	67 111 157
	4	56 112 147 197	67 117 149 191	45 87 114 142	60 97 128 165
Masi	2	94 172	101 173	73 127	91 154
	3	83 134 188	89 134 180	76 120 170	70 113 160
	4	86 127 168 205	46 89 132 180	64 99 134 172	67 103 139 171
Reniy's	2	106 173	109 174	78 128	86 156
	3	106 173 211	99 139 180	68 99 133	69 118 168
	4	90 130 173 210	45 89 134 176	70 100 132 172	64 104 144 184
Fuzzy	2	95 190	103 190	68 154	89 169
	3	71 116 183	112 158 213	69 103 166	64 117 169
	4	92 144 176 216	93 140 177 212	58 99 141 188	66 110 143 191

Fig. 1 Thresholded image of the subject 48025 from BSDS 500 dataset for $k = 4$

Fig. 2 Thresholded image of the Lena image for $k = 4$

6 Conclusion

This paper proposed BWO-based multilevel thresholding algorithms and observed its performance in color and gray images. For obtaining the optimal threshold values from the images, six different well-known objective functions have been used. The segmented images obtained from various objective functions using BWO are compared using different quality matrices. Though the selection of threshold values is a challenging task, BWO performs this task effectively and generates a thresholded image close to the accurate.

References

1. Portesde Albuquerque, M., Esquef, I.A., Gesualdi Mello, A.R.: Image thresholding using Tsallis entropy. Pattern Recognit. Letters **25**(9), 1059–1065 (2004)
2. Kapur, E.J.N., Sahoo, P.K., Wong, A.K.C.: A new method for grey-level picture thresholding using the entropy of the histogram. Comput. V. Gr. Image Process. **29**, 273–285 (1985)
3. Otsu, N.: A threshold selection method for grey level histograms. IEEE Trans. System, Man Cybern. SMC-9 (1), 62–66 (1979)
4. Khairuzzaman, A.K.M., Chaudhury, S.: Masi entropy based multilevel thresholding for image segmentation. Multimed. Tools Appl. **78**, 33573–33591 (2019)
5. Liu, W., Huang, Y., Ye, Z., Cai, W., Yang, S., Cheng, X., Frank, I.: Renyi's entropy based multilevel thresholding using a novel meta-heuristics algorithm. Appl. Sci. **10**, 3225 (2020). https://doi.org/10.3390/app10093225
6. Sarkar, S., Paul, S., Burman, R., Das, S., Chaudhuri, S.S.: A fuzzy entropy based multi-level image thresholding using differential evolution. In: SEMCCO 2014, LNCS 8947, pp. 386–395 (2015)
7. Eberhart, R., Kennedy, J.: A new optimizer using particle swarm theory. In: Proceedings of the Sixth International Symposium on Micro Machine and Human Science, pp. 39–43. IEEE (1995)
8. Holland, J.H.: Adaptation in Natural and Artificial Systems. University of Michigan Press, Ann Arbor, MI (1975)
9. Basturk, B., Karaboga, D.: An artificial bee colony (ABC) algorithm for numeric function optimization. In: IEEE Swarm Intell. Symp. **8**:687–697 (2006)
10. Mirjalili, S., Mirjalili, S.M., Lewis, A.: Grey wolf optimizer. Adv. Eng. Softw. **69**, 46–61 (2014)
11. Hayyolalam, V., Kazem, A.: Black Widow Optimization Algorithm: a novel meta-heuristic approach for solving engineering optimization problems. Eng. Appl. Artif. Intell. **87**, 103249 (2020)

Demonstration of a 4 × 3 × 10 Gbps WI-WDM Transmission Over MDM Link Using Ring-Core FMF

Bhagyalaxmi Behera, S. K. Varshney, and Mihir Narayan Mohanty

Abstract Few-mode fiber (FMF)-based spatial division multiplexing (SDM) and mode division multiplexing (MDM) are the most encouraging method for considerable growth of the spectral efficiency of current optical communication links. FMFs are the unique sort of fiber where every individual guiding mode is treated as a different information channel. This work presents the transmission of wavelength-interleaved (WI) wavelength division multiplexed (WDM) data channels over the MDM link. This wavelength-interleaved WDM-MDM transmission link is established to multiplex the three wavelength-interleaved WDM channels over four linearly polarized (LP) modes of a ring-core FMF over different link length to compare the link performance. A 4 × 3 × 10 Gbps wavelength-interleaved WDM-MDM link is established and verified through simulation over 100 km using intensity modulation and direct detection (IM-DD) with acceptable bit error rate (BER).

Keywords Few-mode fibers (FMFs) · Wavelength division multiplexing · Mode division multiplexing · Spatial division multiplexing · Wavelength-interleaved

1 Introduction

The immense network traffic can be taken care of with the single-mode fiber (SMF)-based optical communication networks, where the transmission capacity increments through WDM. But the fiber fuse phenomenon, nonlinear effects and Shanon's

---

B. Behera · M. N. Mohanty (✉)
ITER, Siksha 'O' Anusandhan (Deemed to Be University), Bhubaneswar, India
e-mail: researchmihir16@gmail.com

B. Behera
e-mail: bhagyaiter@gmail.com

S. K. Varshney
Department of Electronics and Electrical Communication Engineering, IIT, Kharagpur, India
e-mail: skvarshney.iitkgp@gmail.com

S. Das and M. N. Mohanty (eds.), *Advances in Intelligent Computing and Communication*, Lecture Notes in Networks and Systems 202,
https://doi.org/10.1007/978-981-16-0695-3_56

limit restrict this exponential growth in the data rate of long haul WDM networks up to 100 Tbit/s [1]. The fiber nonlinear effects are more prominent with the consistent market requirement for high-speed network services that causes the capacity crunch in current WDM optical networks [2]. However, in recent years, the usages of short-reach optical networks are again gaining their attention to handle the huge network traffic. These networks are mainly deployed for fifth-generation (5G) communication, data centers and virtual reality (VR) [3]. Currently, MDM [4–6] and SDM [7–9] are the two emerging techniques to establish high-speed short-reach optical networks with the help of few-mode fibers (FMFs) [9–11] and multi-core fibers (MCFs) [12, 13]. These are the special types of fiber where each of the guided mode or mode groups through a single core or core lattice is treated as a unique spatial data channel. Hence, MDM and SDM are emerged to enhance the transmission capacity further by increasing the number of spatial channels or number of modes in FMFs and MCFs [10]. Besides, this MDM or SDM transmission adequately decreases the fiber volume and transmission link complexity [14], optical fibers designed for short-reach applications are reused, and also facilitates the integration of optical transponders [15], and these are quite needed for upgrading optical networks.

However, the establishment of long haul MDM and SDM transmission is the biggest challenge for the researchers, because the success of these techniques is possible with precise coherent detection and multi-input multi-output (MIMO) signal processing. Here, the transmission capacity increases by increasing the number of modes or spatial channels at the cost of inter-channel crosstalk (ICC) among the modes due to mode coupling and pulse distortion due to large differential mode group delay (DMGD). Hence, to mitigate the effect of ICC and for simultaneous detection of each spatial channel at the receiver end, we need coherent detection with MIMO signal processing [16]. This, on the other hand, increases the complexity level and cost of MDM and SDM link and also not acceptable for large-scale implementation. However, the high-speed short-reach MDM and SDM system are possible through the traditional less complex intensity modulation and direct detection (IM-DD) techniques by designing weakly coupled FMFs [6, 17, 18].

Recently, most research works focus on the design of weakly coupled FMF for suppressing the mode coupling between the guided modes [19, 20]. However, wavelength interleaved is another promising technique [21] to reduce the modal crosstalk between the adjacent modes. In this work, we present the wavelength-interleaved approach to establish and compare WDM-MDM transmission using a weakly coupled IM-DD technique. This is one of the most efficient techniques to suppress the modal crosstalk between the neighboring modes and thus boost the performance of IM-DD weakly coupled MDM transmission. Here the wavelength-interleaved WDM-MDM transmission link is established to multiplex three wavelength-interleaved WDM channels over four modes of a ring-core FMF over different link length to compare the link performance. A $4 \times 3 \times 10$ Gbps wavelength-interleaved WDM-MDM link is established using IM-DD and verified through simulation over 100 km FMF with an acceptable bit error rate (BER).

This paper is organized as follows: Sect. 2 gives a brief theory of wavelength-interleaved techniques for WDM-MDM transmission using IM-DD. Section 3 shows the simulation-based link setup for wavelength-interleaved WDM-MDM transmission using IM-DD over ring-core FMF with the link performance. In Sect. 4, the results are discussed, and Sect. 5 describes the concluding remarks of this work.

2 Wavelength-Interleaved Technique in WDM-MDM System

Several research works show that the ICC in a weakly coupled FMF is robust between the neighboring spatial channels with less effective index difference (Δn_{eff}), whereas it is weaker between the non-adjacent channels with large Δn_{eff}. Hence, the acceptable BER performance for the adjacent spatial channels is not achieved through DD for weakly coupled MDM transmission. And the DD technique is not sufficient to mitigate the modal crosstalk between the adjacent spatial channels. Hence, wavelength interleaved is one of the emerged techniques to enhance the performance of weakly coupled MDM transmission through DD.

Currently, the researchers focus on the growth of the transmission capacity of an optical network by upgrading the link from SMF to FMF. Still, the conventional WDM channels are multiplexed over all the modes or spatial channels of an FMF through MDM and SDM by following the ITU-T standards. According to the ITU-T standard, all the WDM channels are maintaining a space of 100 or 50 GHz between the channels. This is referred to as a wavelength-aligned (WA) technique, as depicted in Fig. 1a, whereas the WDM channels, multiplexed over the neighboring spatial modes of an MDM system are interleaved by half of the WDM grid spacing in the wavelength-interleaved technique [21], as shown in Fig. 1b. In conventional MDM transmission using IM-DD at the receiver end, both the signal and modal crosstalk are fed to the photodetector (PD). Hence, the PD not only detects the signal but also detects the crosstalk-to-crosstalk beating interface (CCBI) and signal-to-crosstalk beating interface (SCBI). In a weakly coupled IM-DD MDM system, the crosstalk penalty is mainly dominated by SCBI than CCBI. Hence, the wavelength-interleaved technique was introduced by Tian et al., to suppress the modal crosstalk, i.e., SCBI between the adjacent modes of an MDM system. This SCBI is present out of the signal bandwidth and can easily be eliminated by using a low-pass filter after the PD's output as depicted in Fig. 2.

This work presents the use of the WI technique to transmit four spatial channels over 4-LP modes of an MDM link using a ring-core FMF. The fiber profile and parameters are given in [22]. Here the MDM transmission link is established by using the ring-core FMF that is designed to guide 4-LP modes (LP_{01}, LP_{11}, LP_{21}, and LP_{31}). Each of the modes is carrying three wavelength channels multiplexed through WDM and each wavelength channel is capable of transmitting 10 Gbps,

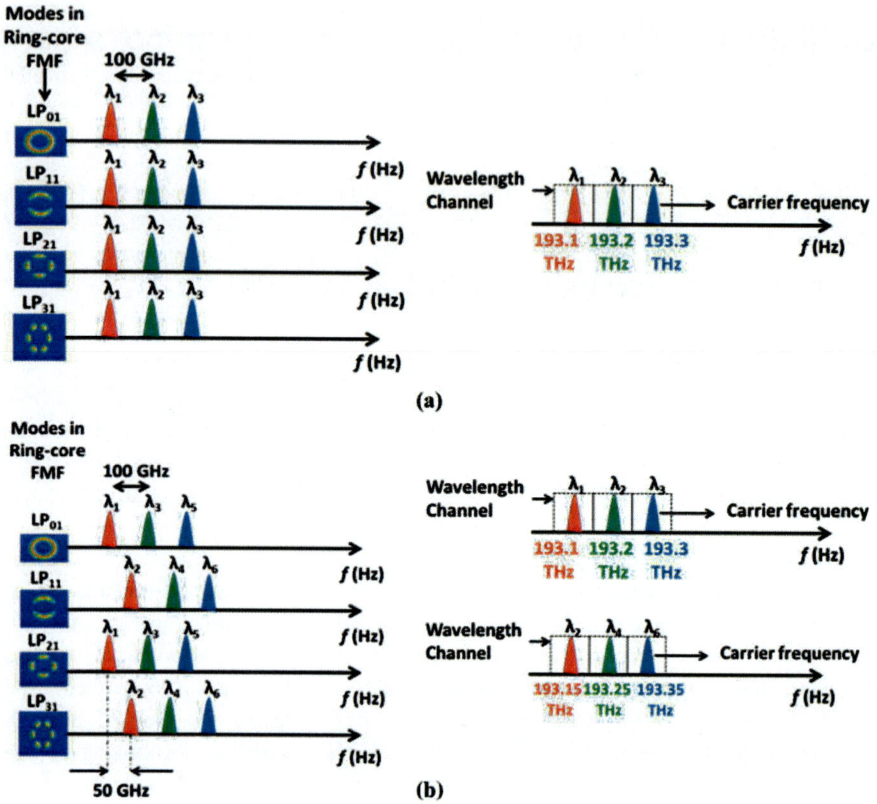

Fig. 1 Concept of **a** wavelength-aligned WDM, **b** wavelength-interleaved WDM channels over different modes of a ring-core FMF MDM links

thus establishing a $4 \times 3 \times 10$ Gbps WI-WDM transmission over the MDM link using ring-core FMF. The ring-core FMF is used to establish this MDM transmission link which exhibits a large effective index difference $\min|\Delta n_{\text{eff}}| \geq 1 \times 10^{-3}$ between the adjacent modes and ensures low mode coupling among the modes and suitable for MDM transmission.

3 Simulation-Based Experimental Link Setup

This work is carried out through the commercially available Optiwave platform. Optiwave is one of the emerging platforms for the development of advanced software tools. These tools are beneficial in the evolving fields of optoelectronics, photonic applications and optical networks. The products of Optiwave are OptiFiber [23], OptiGratting, OptiSPICE, OptiSystem [24] and OptiFDTD. These

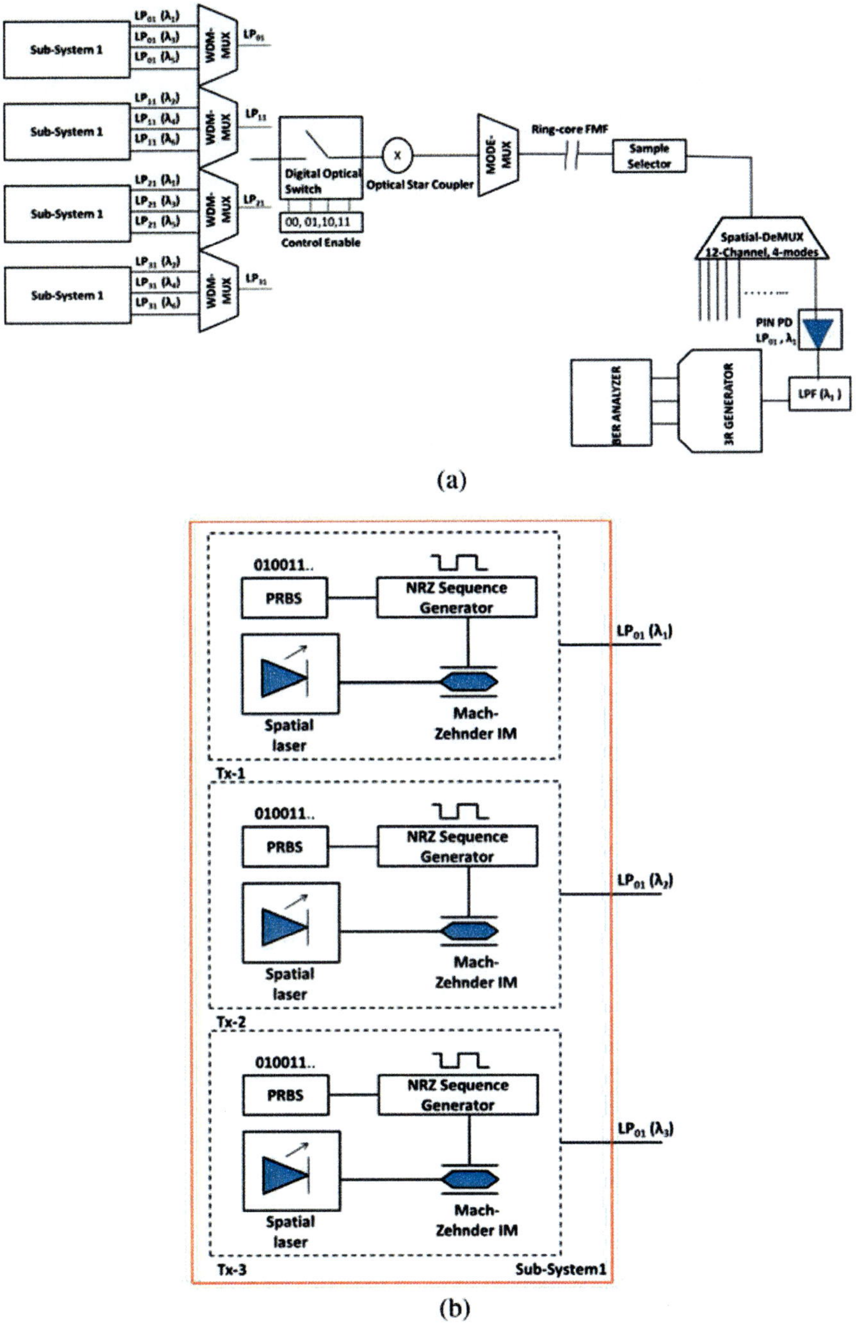

Fig. 2 Internal connection of the **a** sub-system at the transmitter end, **b** block diagram of the experimental link setup for the proposed 4 × 3 × 10 Gbps wavelength-interleaved WDM transmission over the MDM link using IM-DD and ring-core FMF

are useful for designing, optimization and simulating the optoelectronic components, optical communication systems, optical transmission links and optical networks. In this work, we have used the OptiFiber and Optisystem modules for design and verification. In the first phase of this work, the OptiFiber software package is utilized for modeling and design of the ring-core fiber. The finite difference method is chosen to investigate the modal properties of the proposed ring-core FMF [22]. In the second phase, the viability of the proposed fiber for the wavelength-interleaved WDM transmission over the MDM link is being accomplished through OptiSystem.

The block diagram of the experimental link setup for 4 × 3 × 10 Gbps WI-WDM transmission over the MDM link using ring-core FMF is shown in Fig. 2a. Weakly coupled MDM is the multiplexing technique that allows single or multiple (WDM) light channels to propagate through different modes of a few-mode fiber with minimum inter-channel interference. But for the adjacent mode with less Δn_{eff}, the crosstalk is more and difficult to detect each of the modes through DD. To suppress this crosstalk penalty due to SCBI in WDM-MDM transmission using IM-DD, wavelength interleave is one of the emerging technologies. Here each mode is treated as an individual data channel and can be detected at the receiver end through DD as long as the crosstalk between the adjacent mode is less. Hence, to suppress the modal crosstalk, instead of wavelength alignment technique ,the wavelength-interleaved WDM-MDM transmission is demonstrated here. We present a four-mode MDM link for the transmission of three wavelength-interleaved WDM channels using "Optisystem" as shown in Fig. 2. The performance of the proposed 4 × 3 × 10 Gbps wavelength-interleaved WDM-MDM link is being evaluated and compared over different link lengths. This transmission link is based on the concept of intensity modulation and direct detection to facilitate a MIMO less signal processing at the receiver end. The transmitter end consists of four distinct sub-systems as shown in Fig. 2b, each assigned with three spatial lasers to generate WDM channels with different wavelengths, and the channel spacing is maintained at 100 GHz as per the ITU-T standard. The spatial lasers of the non-adjacent modes are set to modulate with the carrier frequencies of 193.1, 193.2 and 193.3 Thz, whereas the wavelength channels for the adjacent modes are interleaved with 50 GHz with a channel spacing of 100 GHz. The spatial lasers for the adjacent modes are modulated with the carrier frequencies of 193.15, 193.25 and 193.35 Thz, respectively. Each of the spatial–optical sub-systems comprised of four modules like a spatial laser, pseudo-random binary sequence (PRBS) generator, NRZ pulse generator and a Mach–Zehnder intensity modulator. The PRBS data is encoded with the non-return-to-zero (NRZ) format and followed by an intensity modulator (IM) to modulate the optical source and generate three unique 10 Gbps optical on–off keying (OOK) signals in the following sequence LP_{01}, LP_{21}, at 193.1 Thz, 193.2 Thz and 193.3 Thz, whereas for LP_{11} and LP_{31} at 193.15 Thz, 193.25 Thz and 193.35 Thz, respectively, as given in Table 1. Each of the spatial lasers is set with 0 dB m input power. The wavelength channels of the respective modes are first multiplexed by a WDM multiplexer (MUX) to create four 3 × 10 Gbps WDM channels. Then these wavelength-interleaved WDM channels for a respective mode are selected by a

Table 1 Modes and WDM channels for the proposed 4 × 3 x 10 Gbps WDM-MDM transmission using the wavelength-interleaved technique

MODES over MDM link	Carrier frequency (THz) for WDM channels		
	Ch-1	Ch-2	Ch-3
LP_{01}	193.1	193.2	193.3
LP_{11}	193.15	193.25	193.35
LP_{21}	193.1	193.2	193.3
LP_{31}	193.15	193.25	193.35

digital optical switch. This optical switch is enabled to select any of the mode-specific WDM channels through the control input. To select four distinct modes, a group of two control bits is used. For combination 00, the optical switch is enabled to select LP_{01} mode, likewise, 01 selects LP_{11}, 10 sects LP_{21} and 11 selects LP_{31} mode, respectively. The mode MUX multiplexes these wavelengths interleaved WDM channels into weakly coupled MDM transmission link using ring-core FMF. The link is set up over different lengths without amplifiers and with a loss of 0.18 dB/km over the C-band. The MDM signals are selected in the form of samples and are de-multiplexed by the 12-channel 4-mode spatial de-multiplexer (DEMUX) to separate the 4 × 3-mode multiplexed WI-WDM channels. Each of the DEMUX output is followed by a PIN photodetector (PD) with a receiver sensitivity of −18 dB m for DD. The SCBI due to modal crosstalk can easily be separated from the signal when the PD output is followed and low-pass filter (LPF) with a cutoff frequency = 0.75 × symbol rate. The LPF output is followed by a 3R generator for reshaping, retiming and reamplification of the received bit sequence. The performance of each spatial channel is evaluated by the bit error rate (BER) analyzer connected to the respective 3R generator output. Here, the MUX/DEMUX is assumed to be lossless.

4 Results

Figure 3a shows the mode field distribution of four modes before and after the ring-core FMF transmission link. However, Fig. 3b depicts the WI-WDM channels over the specific modes. Figure 4 shows the eye diagrams of each spatial channel of the 4 × 3 × 10 Gbps wavelength-interleaved WDM transmission over the MDM link for 80 and 100 km. The eye diagram of each spatial channel describes the ring-core FMF can transmit effectively three WDM channels over four modes with acceptable minimum BER $\geq 10^{-9}$ for telecommunication applications using the WI method. The minimum BER, maximum Q-factor and received power for each spatial channel with 10 Gbps/channel over 80 km and 100 km fiber lengths without amplifiers are listed in Table 2. A good BER performance in the order of 10^{-100}–10^{-14} has been achieved for all the spatial channels over 80 km link length, whereas over 100 km link length the achieved BER is in the order of 10^{-50}–10^{-9}. However,

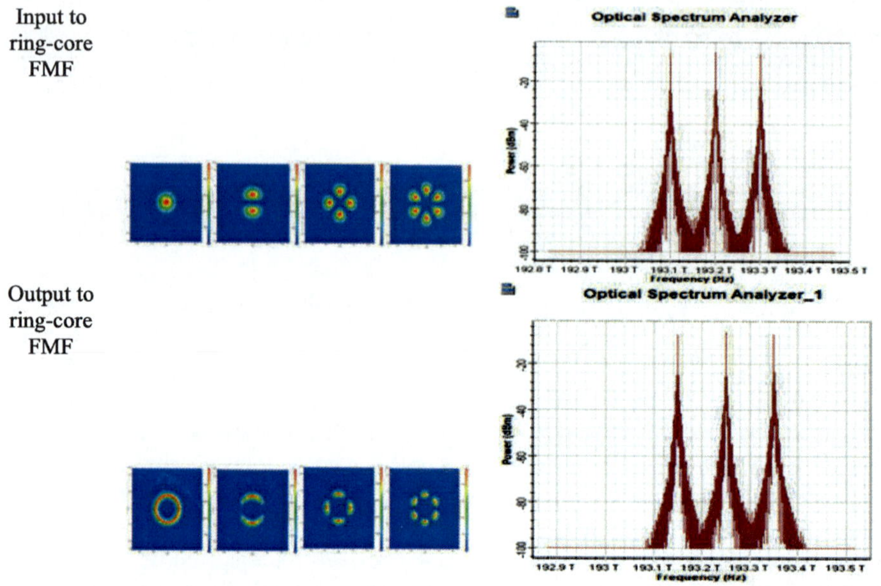

Fig. 3 **a** Mode field distribution of the modes (LP_{01}, LP_{11}, $LP_{21,}$ and LP_{31}) at the input and output of ring-core FMF based MDM link, **b** the wavelength-interleaved channels over the spatial channels

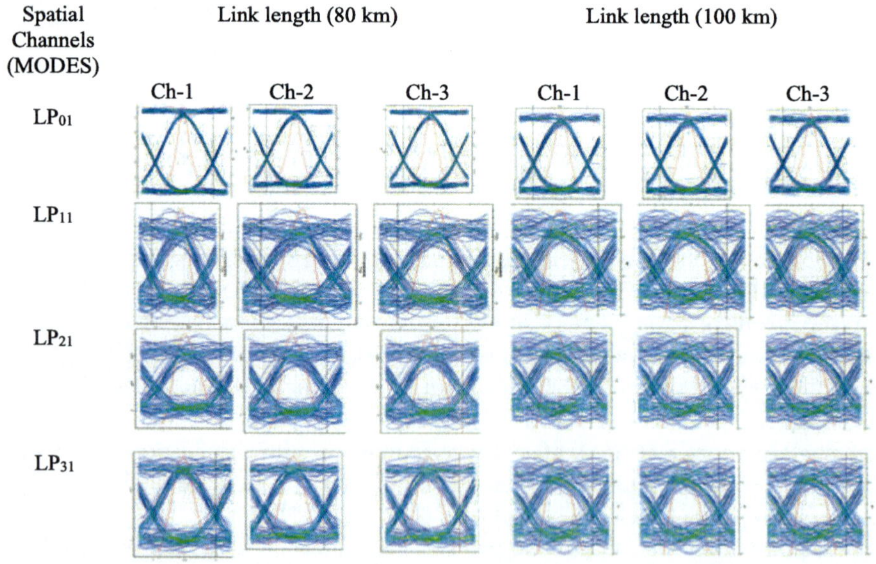

Fig. 4 Eye diagram of the wavelength-interleaved WDM channels over the spatial modes of the MDM link over 80 and 100 km length

Table 2 Performance of the spatial channels of the proposed 4 × 3 × 10 Gbps wavelength-interleaved WDM transmission over MDM link using IM-DD and ring-core FMF

Spatial channels (MODES)	Link length (80 km)				Link length (100 km)		
	Ch-1		Ch-2	Ch-3	Ch-1	Ch-2	Ch-3
LP_{01}	Q-factor	22.5	20.1	22.1	16.5	15.6	15.6
	BER	1×10^{-113}	1×10^{-90}	2×10^{-109}	1×10^{-61}	2×10^{-55}	2×10^{-57}
LP_{11}	Q-factor	8.1	7.2	6.3	5.7	5.8	6.0
	BER	2×10^{-16}	1×10^{-13}	1×10^{-12}	2×10^{-9}	1×10^{-9}	1×10^{-10}
LP_{21}	Q-factor	8.1	7.2	6.3	4.7	4.2	5.3
	BER	2×10^{-16}	1×10^{-13}	1×10^{-12}	2×10^{-8}	1×10^{-8}	1×10^{-9}
LP_{31}	Q-factor	8.1	7.5	7.5	4.7	4.2	5.3
	BER	2×10^{-16}	1×10^{-14}	1×10^{-14}	2×10^{-8}	1×10^{-8}	1×10^{-9}

the performance is slightly degraded for LP_{21} and LP_{31} mode at 100 km as both of these modes got attenuated due to mode coupling over a long distance. This indicates the wavelength-interleaved technique as one of the acceptable methods for DD of WDM channels over the weakly coupled MDM link.

5 Conclusion

This work adapted the wavelength-interleaved technique to establish a 4 × 3 × 10 Gbps WDM transmission over the MDM link using ring-core FMF. This work demonstrates the direct detection of each spatial channel with an acceptable BER over 100 km link length without amplifiers. This enhances the uses of weakly coupled MDM transmission using IM-DD for the establishment of short-reach high-speed optical networks.

References

1. Desurvire, E.B.: Capacity demand and technology challenges for lightwave systems in the next two decades. J. Lightwave Technol. **24**, 4697–4710 (2006)
2. Essiambre, R.-J., Kramer, G., Winzer, P.J., Foschini, G.J., Goebel, B.: Capacity limits of optical fiber networks. J. Lightwave Technol. **28**, 662–701 (2010)
3. Index, C.V.N.: Forecast and methodology, 2015–2020. White paper 1–41 (2016)
4. Berdagué, S., Facq, P.: Mode division multiplexing in optical fibers. Appl. Opt. **21**, 1950–1955 (1982)
5. Bigot-Astruc, M., Trinel, J.B., Maerten, H., Stralen, M.V., Milicevic, I., Bigot, L., Plus, S., Masselot, A., Habert, R., Simonneau, C., Benyahya, K., Labroille, G., Sillard, P.: Weakly-coupled 6-LP-mode fiber with low differential mode attenuation. In: 2019 Optical Fiber Communications Conference and Exhibition (OFC), pp. 1–3 (2019)

6. Ge, D., Li, J., Zhu, J., Shen, L., Gao, Y., Yu, J., Wu, Z., Li, Z., Chen, Z., He, Y.: Design of a weakly-coupled ring-core FMF and demonstration of 6-mode 10-km IM/DD transmission. In: 2018 Optical Fiber Communications Conference and Exposition (OFC), pp. 1–3 (2018)
7. Ryf, R., Randel, S., Gnauck, A.H., Bolle, C., Essiambre, R.-J., Winzer, P., Peckham, D.W., McCurdy, A., Lingle, R.: Space-division multiplexing over 10 km of three-mode fiber using coherent 6 × 6 MIMO processing. In: National Fiber Optic Engineers Conference, pp. PDPB10. Optical Society of America (2011)
8. Richardson, D., Fini, J., Nelson, L.E.: Space-division multiplexing in optical fibres. Nat. Photonics **7**, 354 (2013)
9. Sillard, P.: Few-mode fibers for space division multiplexing. In: Optical Fiber Communication Conference, pp. Th1J. 1. Optical Society of America (2016)
10. Sakamoto, T., Saitoh, K., Saito, S., Abe, Y., Takenaga, K., Urushibara, A., Wada, M., Matsui, T., Aikawa, K., Nakajima, K.: Spatial density and splicing characteristic optimized few-mode multi-core fiber. J. Lightwave Technol. 1–1 (2020)
11. Kitayama, K.-I., Diamantopoulos, N.-P.: Few-mode optical fibers: original motivation and recent progress. IEEE Commun. Mag. **55**, 163–169 (2017)
12. Sakamoto, T., Saitoh, K., Saitoh, S., Shibahara, K., Wada, M., Abe, Y., Urushibara, A., Takenaga, K., Mizuno, T., Matsui, T., Aikawa, K., Miyamoto, Y., Nakajima, K.: Six-mode seven-core fiber for repeated dense space-division multiplexing transmission. J. Lightwave Technol. **36**, 1226–1232 (2018)
13. Chang, J.H., Bae, S., Kim, H., Chung, Y.C.: Heterogeneous 12-core 4-LP-mode fiber based on trench-assisted graded-index profile. IEEE Photonics J. **9**, 1–10 (2017)
14. Fiorani, M., Tornatore, M., Chen, J., Wosinska, L., Mukherjee, B.: Optical spatial division multiplexing for ultra-high-capacity modular data centers. In: 2016 Optical Fiber Communications Conference and Exhibition (OFC), pp. 1–3. IEEE (2016)
15. Grieco, A., Porter, G., Fainman, Y.: Integrated space-division multiplexer for application to data center networks. IEEE J. Sel. Top. Quantum Electron. **22**, 1–6 (2015)
16. Ryf, R., Randel, S., Gnauck, A.H., Bolle, C., Sierra, A., Mumtaz, S., Esmaeelpour, M., Burrows, E.C., Essiambre, R.-J., Winzer, P.J.: Mode-division multiplexing over 96 km of few-mode fiber using coherent 6 × 6 MIMO Processing. J. Lightwave Technol. **30**, 521–531 (2011)
17. He, W., Yu, H., Sillard, P., Correa, R.A., Li, G.: Weakly-coupled few-mode fibers and their applications. In: 2017 European Conference on Optical Communication (ECOC), pp. 1–2 (2017)
18. Zhang, R., Tan, H., Zhang, J., Shen, L., Liu, J., Liu, Y., Zhang, L., Yu, S.: A novel ring-core fiber supporting MIMO-free 50 km Transmission Over High-Order OAM modes. In: 2019 Optical Fiber Communications Conference and Exhibition (OFC), pp. 1–3 (2019)
19. Rjeb, A., Seleem, H., Fathallah, H., Machhout, M.: Design of 12 OAM-Graded index few mode fibers for next generation short haul interconnect transmission. Optical Fiber Technology **55**, 102148 (2020)
20. Li, Y., Wang, X., Zheng, H., Li, X., Bai, C., Hu, W., Liu, Y., Dong, Q.: A novel six-core few-mode fiber with low loss and low crosstalk. Optical Fiber Technol. **57**, 102211 (2020)
21. Tian, Y., Li, J., Wu, Z., Chen, Y., Zhu, P., Tang, R., Mo, Q., He, Y., Chen, Z.: Wavelength-interleaved MDM-WDM transmission over weakly-coupled FMF. Opt. Express **25**, 16603–16617 (2017)
22. Behera, B., Mohanty, M.N.: Design of bend-limited large-mode area dispersion shifted few-mode fiber for fast communication. In: 2019 International Conference on Applied Machine Learning (ICAML), pp. 277–281 (2019)
23. Sabitu, R.I., Dong-Nhat, N., Malekmohammadi, A.: High dispersion four-mode fiber for mode-division multiplexing systems. Optik **181**, 1–12 (2019)
24. Ab-Rahman, M.S., Shaltami, F.M., Hwang, I.-S., Swedan, A.A.: Analysis of 10 Gbps SOA-based optical network unit with low seeding power that uses feedback seeding scheme. Optik **183**, 602–611 (2019)

Comparative Study of Different Orange Data Mining Tool-Based AI Techniques in Image Classification

Subhashree Mohapatra and Tripti Swarnkar

Abstract In the past few years, artificial intelligence (AI) has contributed significantly to the domain of image processing and image classification. There are many tools available for pre-processing of data, visualization of data and training of models. Orange is one such tool which has great data visualization, multiple pre-processing modules along with special add-on for image processing. The proposed work has applied this tool and presented a comparative analysis of different classifier models on cat and dog image dataset from Kaggle. Classification of the images is based on specific features which are extracted from the trained images with the help of a particular convolutional neural network model that is SqueezeNet. The result shows that the artificial neural network-based approach is found to be the best AI technique for classifying images with highest accuracy as compared to other techniques such as logistic regression, *k*-nearest neighbours, decision tree, random forests, Naïve Bayes and SVM.

Keywords Artificial intelligence · Image processing · Classification · Kaggle · Orange data mining tool

1 Introduction

Nowadays, image classification has taken the front position in different areas of research such as data mining, computer vision, medical image analysis, artificial intelligence and so on [1]. Figure 1 shows the rapid rise in the unstructured data

S. Mohapatra (✉)
Department of Computer Science and Engineering, Siksha 'O' Anusandhan Deemed to be University, Bhubaneswar 751030, India
e-mail: subhashreemohapatra68@gmail.com

T. Swarnkar
Department of Computer Application, Siksha 'O' Anusandhan Deemed to be University, Bhubaneswar 751030, India
e-mail: triptiswarnakar@soa.ac.in

S. Das and M. N. Mohanty (eds.), *Advances in Intelligent Computing and Communication*, Lecture Notes in Networks and Systems 202,
https://doi.org/10.1007/978-981-16-0695-3_57

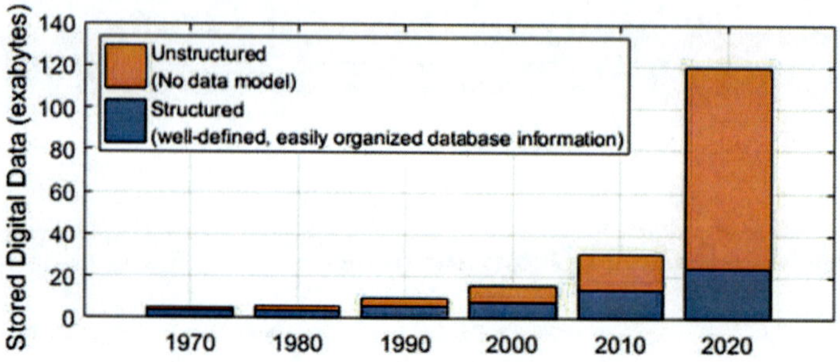

Fig. 1 Plot showing the drastic rise in unstructured data [2]

from the year 1970–2020 [2]. As images are mostly the unstructured data [3], they need to be transformed to a structured format so that it can be efficiently used for further processing. This is done by extracting relevant features from the images so that different AI techniques can be used for image classification [4]. There are billions and billions of images available in the Internet which can be easily collected and different data mining tools can be used to implement the AI techniques for proper classification of the images [1].

Deep learning and machine learning are the sub-branches of AI which involves the process of feature mining and use that features to learn specific form and characteristics of the given image data [5]. The basic difference between the two learning process is that in machine learning handcrafted features are used to train a model to classify images, whereas deep learning techniques automatically mines the features and use it to process the images [6]. As there are various models available for the classification of images, to find out which one is best suited for image classification, needs to be justified. These algorithms can be implemented in several platforms, may it be programming languages or using some tools. Orange data mining tool is one such open-source tool useful for pre-processing, data visualization and implementation of different classification and clustering models. Orange has the facility of extended modules for various other research applications such as bioinformatics, network, image analysis and many more [7].

The main motivation behind this study is the rise in demand for the application of image classification in various research fields and the selection of the best methodology to do so. Such a huge amount of data freely available needs to be handled efficiently so that it could benefit the researchers. In this work, authors have performed image classification through various AI techniques and compared them on basis of various parameters such as area under curve (AUC), classification accuracy (CA), F-1 score, precision and recall. Model comparison is done using the Orange data mining tool.

The remaining portion of the paper is divided into different sections as mentioned. Section 2 explains the methodology applied in the study followed by Sect. 3 which gives a critical analysis of the final result. Finally, the work is concluded in Sect. 4 along with a crisp insight into the future work.

2 Experimental Setup

In this section, the detailed methodology used for the comparison of the classification models is discussed. Figure 2 gives the work flow diagram of the steps involved in the study using the Orange data mining tool. Each process involved in the proposed work is explained in the following subsections.

2.1 Dataset

The comparative study is carried out using the cat and dog images available in the Kaggle repository [8]. The repository contained two directories as training and testing. In the training as well as testing folder, it had two sub-folders named as cat and dog. Each of the class contained 4000 images in training, whereas for testing each of the class had 1000 images. But for the image classification purpose, both the training and testing images were kept in single place. So, the final comparative work is carried out using total 5000 images of cat and 5000 images of dog.

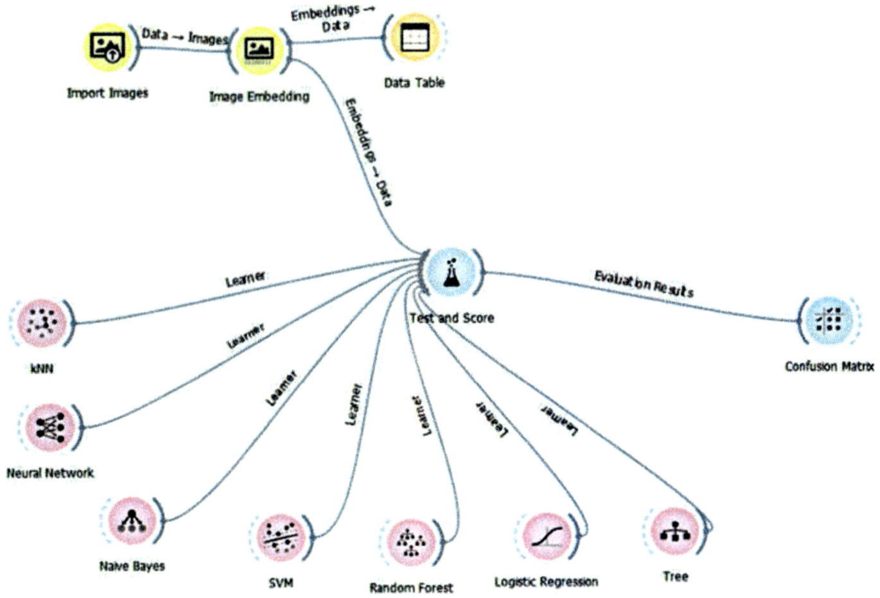

Fig. 2 Work flow diagram of the steps involved in the proposed study

2.2 Importing of Images

In the flow diagram, it can be observed that the first step is importing the images into the workspace. The image analysis add-on available in the Orange tool has a widget for importing the images. The widget is the 'Import Images' widget which helps to load the necessary images for the image processing or image classification purpose. Figure 3 shows an example of the cat and dog images used for the study.

2.3 Image Embedding

After the images are imported, the next step is the feature extraction from the images. This step is carried out by the 'Image Embedding' widget of the image analysis tool of Orange. The images are sent to the server for feature extraction. This widget uses several deep convolutional neural network (DCCN) models for feature extraction purpose. Different DCCN models such as SqueezeNet [9], VGG-16 [10], Inception V3 [11] and few others are available for feature mining purpose. For our work, SqueezeNet model is applied which extracts the features locally without sending the images to the server. At the end of the process, the features that are mined could be seen in a tabular form using the 'Data Table' widget of Orange. SqueezeNet generated 1000 features for individual images along with width, height and size as the meta-features of the images.

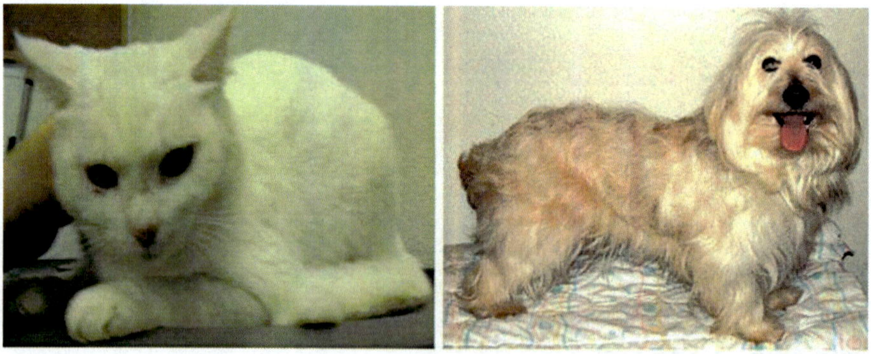

(a) (b)

Fig. 3 Sample **a** cat and **b** dog image used for the image classification

2.4 *Image Classification*

In this phase, seven different classifiers are applied for the classification of images based on the features that are extracted using the SqueezeNet model. The process is performed by using tenfold cross validation where 90% of the images are used for training and 10% images are used for testing of the models. Different classifiers that are trained are discussed below. All the models are trained and tested with various parameters. After testing with different parameters, the parameters giving the best results are mentioned along with the description of each models.

2.4.1 *K*-Nearest Neighbour (KNN)

KNN is a basic classification technique [12] which classifies an image by finding out its best '*K*' number of nearest images during the training process. The nearest neighbour of an item is calculated by using a similarity index. In this work, the index is calculated using the Euclidean distance and the number of neighbours is taken as 5 ($K = 5$).

2.4.2 Artificial Neural Network (ANN)

ANN is a classification algorithm [13] that recognizes the hidden relation between images through a process similar to the working of brain neurons. A simple neural network has an input layer which takes in the features as input, a hidden layer and the output layer which gives the class to which a particular image belongs to. In this work, the number of hidden layers is taken as 20, activation function used is the ReLU function, stochastic gradient descent is used as the optimizer and the number of iterations is taken as 1000.

2.4.3 Naïve Bayes

Naïve Bayes classifier is a basic and simple classifier [14] that classifies an image on basis of its probability of belonging to a particular class. This technique uses the Bayes theory having strong assumption of independence between the features.

2.4.4 Support Vector Machine (SVM)

SVM is one of the supervised learning techniques [15] that is used for both classification as well as regression purpose. It is one of the robust techniques used for prediction based on statistical learning system. In this work, the radial basis function kernel is used along with iteration limit as 100 and numerical tolerance as 0.0010.

2.4.5 Random Forest (RF)

RF is a classifier which creates a decision tree [16] such that the tree is a random subset from the training dataset. The output class is determined by calculating the mode of the classes from the individual trees. In this comparative study, the number of trees to be generated is specified to be 20.

2.4.6 Logistic Regression (LR)

LR uses a logistic function so as to design a binary dependent variable. Basically, it is a curve-fitting model such that the dependent value does change with reference to the independent value and the alignment of the data points are done with the smallest distance from the curve. In the experimental work, the regularization is done using ridge (L2) which minimizes the sum of squares of the weight [17].

2.4.7 Decision Tree (DT)

DT is a classification technique that uses some rules which is applied to divide the dataset into disjoint sets constructed on posterior probabilities [18]. It generates a tree-like structure such that each branch represents a group and each leaf represents the output of a group that gives the ultimate decision. In the particular experiment, the tree is binary induced with minimum number of leaf instance to be 2 and maximum tree depth is limited to 5.

2.5 Test and Scoring with Confusion Matrix

In the last step, all the different classifier models are given as input to the 'Test and Score' widget which gives a comparative value of all the classifiers on basis of CA, precision, recall, F-1 score and AUC. When this widget is connected to the 'Confusion Matrix' widget, it gives all the confusion matrix corresponding to each of the classifying models used. The detail analysis of the result is done in the next section.

3 Critical Analysis

All the experimental task is implemented using Orange version 3.27.0, on an Intel Core 2.60 GHz i7 CPU running Window 10 with 16 GB RAM. In this work, total 10,000 images of cat and dogs are taken for the comparative analysis, out of which the dataset is divided into 9:1 ratio for training and testing. Tenfold cross validation is used for training the models. Here, seven different classifiers are compared to check which one is performing the best.

The models are compared by using five parameters: CA, precision, recall, F-1 score and AUC. These indices are automatically calculated by the 'Test and Score' widget of the Orange tool. Classification accuracy is measured by the total number of correctly classified images. Precision generates the proportion of positive values which are actually positive, whereas recall gives the actual positive values that are identified correctly. F-1 score gives the harmonic mean of precision and recall. Likewise, area under curve gives the quality of the prediction done by the models that is trained. The parameter of interest for this comparative work is the accuracy of the classifiers.

Table 1 gives the values of the measuring indices of all the different classifiers used for the comparative work. In Table 1, it is observed that neural network has the highest classification accuracy of 96.6%, whereas tree has the lowest CA of 90.7%. Next to neural network is the accuracy of logistic regression as 95.3% and random forest as 95.1%. Figure 4 represents all the values corresponding to the evaluation parameter of Table 1.

Confusion matrix is useful in finding the error in classification problem. The diagonal elements of the matrix give the correct classification value, whereas the non-diagonal elements give the error while classifying the images. The confusion

Table 1 Values of the measuring indices of all the classifiers

Classifier	CA	Precision	Recall	F-1	AUC
KNN	0.947	0.948	0.948	0.948	0.982
Neural network	0.966	0.966	0.966	0.966	0.995
Naïve Bayes	0.923	0.923	0.923	0.923	0.964
SVM	0.908	0.908	0.908	0.908	0.967
RF	0.951	0.951	0.951	0.951	0.990
LR	0.953	0.953	0.953	0.953	0.991
Tree	0.907	0.908	0.907	0.907	0.929

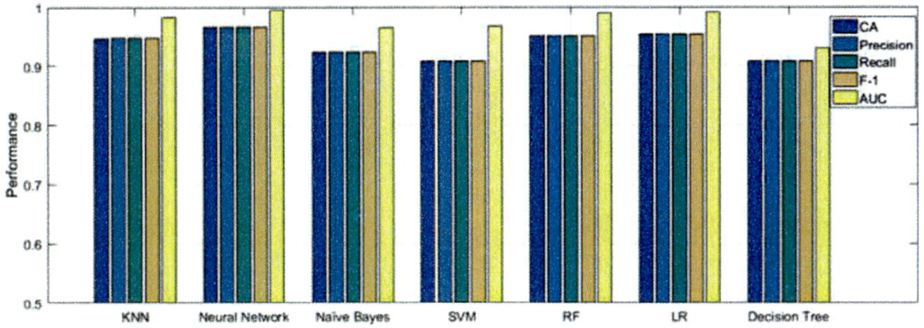

Fig. 4 Graph of the evaluation results of all the classifiers

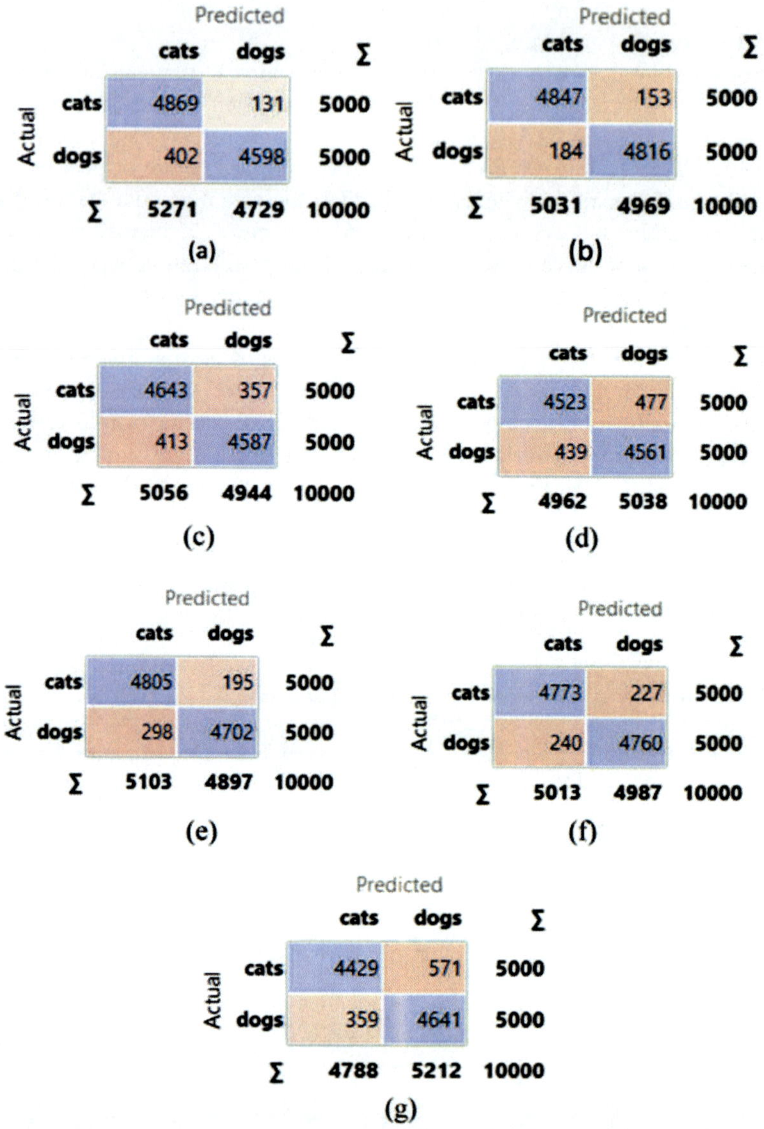

Fig. 5 Confusion matrix of **a** KNN, **b** neural network, **c** Naïve Bayes **d** SVM, **e** RF, **f** LR and **g** tree

matrix for each of the classifiers is generated by the 'Confusion Matrix' widget of the Orange tool. Figure 5 shows the confusion matrix for all the classification models used to classify the images.

4 Conclusion

In this work, the applicability of Orange data mining tool for automatic image classification has been tested on a publicly available cat and dog image dataset from Kaggle. In this tool, seven different AI-based intelligent classifiers such as KNN, ANN, RF, DT, SVM, LR and Naïve Bayes were trained and tested based on features extracted from the images. The feature extraction process was initially done through SqueezeNet-based image embedded model, which is also a part of Orange tool. The performance of each classifiers has been analysed through confusion matrix and several evaluation indices such as area under curve (AUC), classification accuracy (CA), F-1 score, precision and recall. Numerical output result shows that the artificial neural network-based classier outperform with accuracy 96.6%. From this work, we have found Orange data mining tool is one of the easier and handy tools for complex data analysis and processing. This work justifies the efficiency of neural network for image classification which gives a scope for future research work. In the future, authors will try to explore more into the neural network-based models for image classification taking into account a specific field of application.

References

1. Dash, S.K., Panda, M.: Image classification using data mining techniques. Adv. Comput. Sci. Inform. Technol. (ACSIT) **3**(3), 157–162 (2016)
2. What is structured data vs unstructured data?. https://www.m-files.com/blog/what-is-structured-data-vs-unstructured-data/. Accessed 11/10/2020
3. Mohapatra, S., Swarnkar, T.: Artificial intelligence for smart healthcare management: brief study. In: Intelligent and cloud computing, pp. 365–373. Springer, Singapore (2019)
4. Liu, H., Cocea, M., Ding, W.: Decision tree learning based feature evaluation and selection for image classification. In: 2017 International conference on machine learning and cybernetics (ICMLC), IEEE, vol. 2, pp. 569–574 (2017)
5. Vaishnav, D., Rao, B.R.: Comparison of machine learning algorithms and fruit classification using orange data mining tool. In: 2018 3rd International conference on inventive computation technologies (ICICT), IEEE, pp. 603–607 (2018)
6. Hoogenboom, S.A., Bagci, U., Wallace, M.B.: AI in gastroenterology. The current state of play and the potential. How will it affect our practice and when? Techniques in gastrointestinal endoscopy, p. 150634 (2019)
7. Demšar, J., Zupan, B.: Orange: data mining fruitful and fun-a historical perspective. Informatica **37**(1) (2013)
8. Dogs & Cats Images. https://www.kaggle.com/chetankv/dogs-cats-images?. Accessed 01/10/2020
9. Iandola, F.N., Han, S., Moskewicz, M.W., Ashraf, K., Dally, W.J., Keutzer, K.: SqueezeNet: AlexNet-level accuracy with 50x fewer parameters and <0.5 MB model size. arXiv preprint arXiv:1602.07360 (2016)
10. Simonyan, K., Zisserman, A.: Very deep convolutional networks for large-scale image recognition. arXiv preprint arXiv:1409.1556 (2014)
11. Szegedy, C., Vanhoucke, V., Ioffe, S., Shlens, J., Wojna, Z.: Rethinking the inception architecture for computer vision. In: Proceedings of the IEEE conference on computer vision and pattern recognition, pp. 2818–2826 (2016)

12. Chen, G.H., Shah, D.: Explaining the success of nearest neighbor methods in prediction. Trends® in Mach. Learn., vol. 10, no. 5–6, pp. 337–588 (2018)
13. Van Gerven, M., Bohte, S.: Artificial neural networks as models of neural information processing. Front. Comput. Neurosci. **11**, 114 (2017)
14. Rennie, J.D., Shih, L., Teevan, J., Karger, D.R.: Tackling the poor assumptions of naive bayes text classifiers. In: Proceedings of the 20th international conference on machine learning (ICML-03), pp. 616–623 (2003)
15. Smola, A.J., Schölkopf, B.: A tutorial on support vector regression. Stat. Comput. **14**(3), 199–222 (2004)
16. Davies, A., Ghahramani, Z.: The random forest kernel and other kernels for big data from random partitions. arXiv preprint arXiv:1402.4293 (2014)
17. Muthukrishnan, R., Rohini, R.: October. LASSO: a feature selection technique in predictive modeling for machine learning. In: 2016 IEEE international conference on advances in computer applications (ICACA), pp. 18–20 (2016)
18. Serban, R., Kupraszewicz, A., Hu, G.: Predicting the characteristics of people living in the South USA using logistic regression and decision tree. In: 2011 9th IEEE International conference on industrial informatics, pp. 688–693 (2011)

An Approach for Artificial Addition of Rain in Videos

Pankaj Prusty and Bibhu Prasad Mohanty

Abstract This paper contains a real-time novel approach for addition of rain synthetically in videos. The visual properties of the rain drops have been analyzed spatiotemporally. The darkening effects are considered in a more simplified way such that the appearance looks realistic. Also, the wind direction is considered along with blurring and darkening effect to make the video a real one. This rain simulation algorithm may be helpful for any rain removal algorithm used for automatic driver assistance systems as a ground truth. This can also be treated as game for rain rendering in videos.

Keywords Rain drops · Darkening · Blurring

1 Introduction

Many applications of rain removal algorithm require ground truth creation of rain affected frames. Ground truth creation requires rain and non-rain videos of same video sequences. To achieve that rain simulation in a video, it is highly required which is quiet interesting task. Generally, rain simulation is achieved by modeling the particle system [1–3]. In [4], it is analyzed that rain particles are dynamic in nature and difficult to identify rain pixels by just analyzing a single video frame. In [5], it has been explained that if static background rainy video frames are shuffled then the video looks as natural which explains the temporal independence property of rain. As shown [6], rain particles give a positive fluctuations in intensity values without affecting the color components. In our algorithm, all the works are being done in intensity plane and the chrominance components remain unaffected. As per

P. Prusty (✉) · B. P. Mohanty
Faculty of Engineering, Electronics and Communication Engineering,
Siksha 'O' Anusandhan (Deemed to be University), Bhubaneswar, India
e-mail: pankajprusty@soa.ac.in

B. P. Mohanty
e-mail: bibhumohanty@soa.ac.in

S. Das and M. N. Mohanty (eds.), *Advances in Intelligent Computing and Communication*, Lecture Notes in Networks and Systems 202,
https://doi.org/10.1007/978-981-16-0695-3_58

[5] due to first motion, the rain drops cannot be tracked by human perception. Many researchers are working on rain removal algorithm. As explained in [7], they compared their algorithm with the help of true positive and false positive. To do that individuals have to create their own ground truth. Using our method, the ground truth can be created for individual frames simply by subtracting the rain simulated frame from original frame.

This paper is organized as follows Sect. 2 explains all the details of the algorithm being used. Section 3 gives the results followed by conclusion and references.

2 Proposed Work

The entire algorithm is going to be discussed in this section. Artificially rain is being added to each frame of the video sequences in such a way that it appears to be real. Also, darkening and blurring effects are added to the video frames to give a realistic flavor to the rain simulated frames. Here, the videos can be static or dynamic and are being recorded in a sunny day environment to observe the rain simulation effects more effectively.

2.1 Gray Scaling of the Image

Initially, all the frames will be converted to gray scale by preserving the color components. The reason behind preserving the color components is to create a realistic colorful environment after rain simulation. All the darkening, blurring and rain rendering are done in intensity plane which makes the algorithm more easy and robust.

2.2 Temporal Modeling

It is required to know the dependencies of rain drops between consecutive frames. Many researchers have proposed many models regarding temporal dependencies of rain drops. If we are able to track the motion of the rain drops in each frame, then it will be quiet easy to analyze the temporal dependencies of rain drops. But generally rain moves very fast in consecutive frames and it becomes very difficult to track their path. Also, the drops which are even close to the camera are very fast in video due to their perspective projection. In past, many researchers [8, 9] have proposed rain detection algorithm with certain assumption that a rain drop lasts up to 2–3 frames and then it disappears. The location in which the drop is present increases the intensity value to a little high. We can say that rain drops give positive fluctuations over the intensity values.

2.3 Shape of Rain Drops

As per Beard and Chuang [10], small raindrops have spherical shape and when the size of raindrop increases it becomes flattened at the bottom. So they have suggested a parametric model to describe the shape of the raindrop as a weighted sum of tenth order of cosine.

$$d(a) = a\left(1 + \sum_{p=0}^{10} (C_p \cos(pa))\right)$$

Here, a represents the radius of the sphere in undistorted form, α represents the polar angle of elevation, whereas C_p represents the coefficients shape and is represented in Table 1 with α = 0. The shapes of the raindrops of various sizes [10] are shown in Fig. 1.

2.4 Weather Effect

To give more realistic effect to rain appearance, first, we need to blur the image so that it will make the image hazy. One important thing to consider here is to make the image darker because generally the scene of atmosphere becomes darker when the rain starts. The darkening effect can be observed by the contraction of histogram of the image. In this paper, we have adopted an approach to add darkening effect to individual frames of an image which is explained in Eq. 1.

$$y = kx \quad 0 < k < 1 \tag{1}$$

where y is the output intensity values and x is the input intensity values. The value of k needs to be decided empirically.

Table 1 Shape coefficients C_p

a (mm)	C_p for p ($C_p \times 10^4$)										
	0	1	2	3	4	5	6	7	8	9	10
1	−28	−30	−83	−22	−3	2	1	0	0	0	0
2	−131	−120	−376	−96	−4	15	5	0	−2	0	1
3	−282	−230	−779	−175	−21	46	11	−6	−7	0	3
4	−458	−335	−1211	−227	−83	89	12	−21	−13	1	8
5	−644	−416	−1629	−246	176	131	2	−44	−18	9	14
6	−840	−480	−2034	−237	297	166	−21	−72	−19	24	23

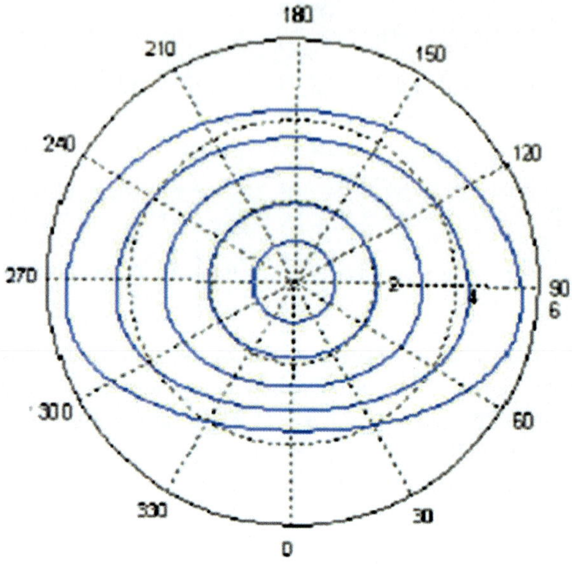

Fig. 1 Shape of rain drops of various sizes

2.5 Rain Mask Addition

At first, different rain densities and velocities are being assigned to different raindrops of different sizes. While assigning different sizes to rain drops, the shapes of rain drops and length of rain streaks are to be taken care of. To begin with a random matrix consisting of only ones and zeros is being created, where 1 denotes rain drops and o denotes non-rain objects. Then, diameters and intensities are being assigned to the rain drops. Then, the frames (masks) are gone through motion blur with some angle, which is computed according to wind speed. The mask is being normalized. Using this process, adequate number of rain masks are being generated. The exact steps of rain mask addition from beginning are explained below.

All the works are being done in intensity plane. To do that all RGB video frames are converted to YIQ plane, so that after all the processing in intensity plane, the color components can be added to it to give a realistic appearance. Then, the mean intensity value is calculated from the normalized intensity values. Darkening the frames is an important task to give a realistic effect for visualization of rain. To darken the image frames, Eq. (1) is used with suitably chosen values of k. Now, this darkened image is being blurred with a simple 3 × 3 average filter. Now, rain mask is being computed and added for each frame with the help of the following equation as explained in [8].

$$X = Y + Z \tag{2}$$

where X is the output intensity plane after rain addition, Y is the background, and Z is the rain mask for the frame which is computed as per the following equation.

$$Z = \alpha(K - \beta Y) \quad (3)$$

where K is the normalized, generated blurred mask. The value of $\beta \in (0, 0.039)$ and $\alpha \in (0, 0.5)$. Once mask is added to each frame, then image is scaled to [0, 255]. Then, preserved color component is being added to corresponding frames for visual appearance.

3 Experimental Results

The simulation has been done in MATLAB 2016a on Windows 10 operating system. Our algorithm was run on a laptop with Intel Corei3 3110 M CPU with 2.4 GHz clock speed and 4 GB RAM. Figure 2 shows two video frames from each of the two videos with darkening effect.

Figure 3 shows two videos with where rain is added to the image frames with effect of darkening.

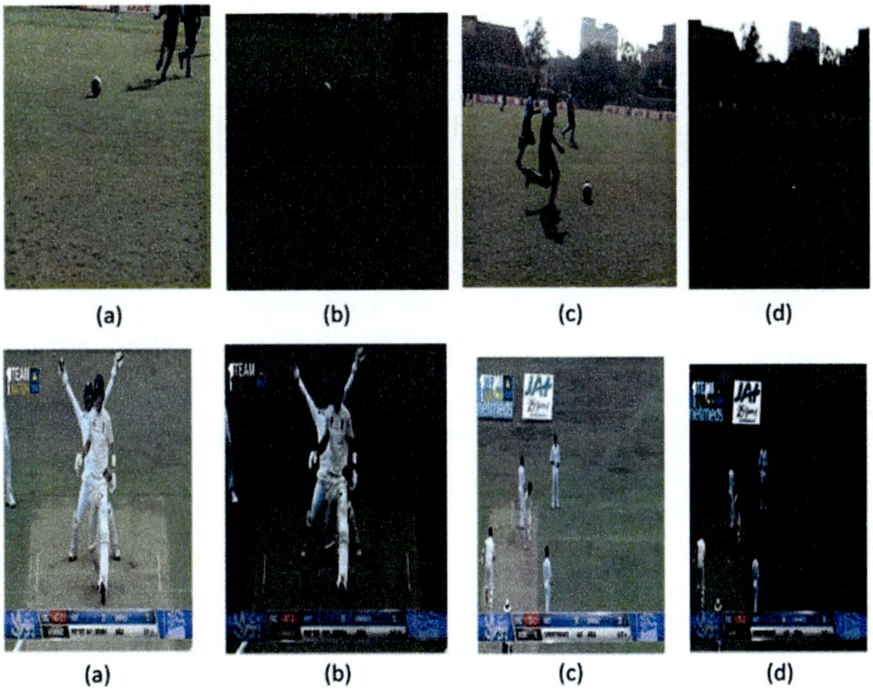

Fig. 2 Row 1: **a** one video frame from football video, **b** effect of darkening, **c** another video frame from same video, **d** effect of darkening. Row 2: **a** one video frame from cricket video, **b** effect of darkening, **c** another video frame from same video, **d** effect of darkening

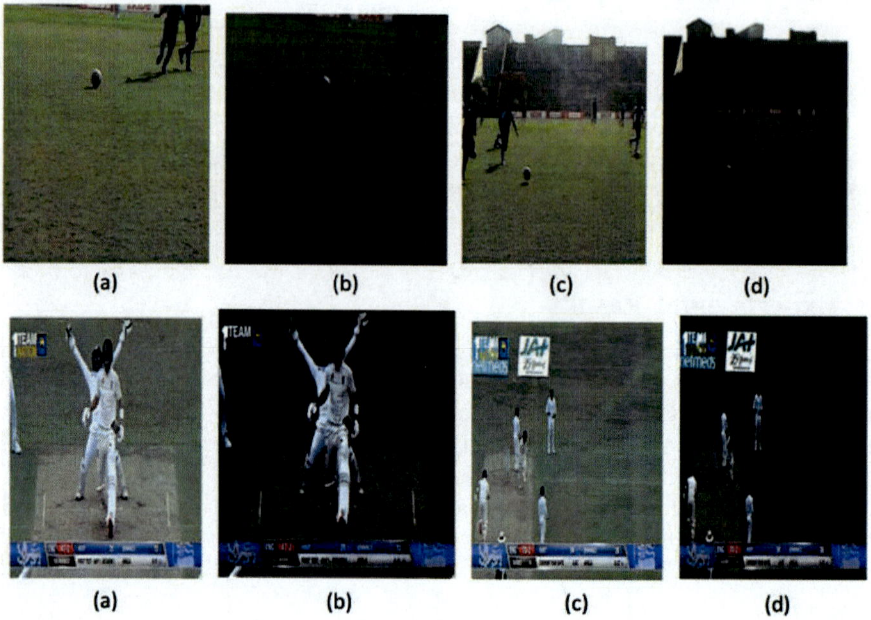

Fig. 3 Row 1: **a** one video frame from football video, **b** effect of rain addition, **c** another video frame from same video, **d** effect of rain addition. Row 2: **a** one video frame from cricket video, **b** effect of rain addition, **c** another video frame from same video, **d** effect of rain addition

4 Conclusion

In this paper, we have added rain particles to a general environment video. In addition to this, we have gone for darkening effect which gives the algorithm a realistic weather effect. Also, there is a provision in the work to give wind direction to the video frames. This work will be very helpful for the researchers who are currently working in the field of rain detection and removal for ground truth creation. Also, the simulation work can be used as a game for enjoyment purpose. All the works are being done in intensity plane for making the algorithm robust.

References

1. Fearing, P.: Computer modelling of falling snow. In: Computer graphics (SIGRAPH conference proceedings), pp. 37–46 (2000)
2. Reeves, W.T.: Particles system-a technique for modelling a class of fuzzy objects. ACM Trans. Graph. **2**, 91–108 (1983)
3. Zhang, G., Vivekanandan, J., Brandes, E.: A model for estimating rain rate and drop size distribution from polarimetric radar measurements. IEEE Trans. Geosci. Remote Sens. vol. 39, no. 4, pp. 830–841 (2011)

4. Narasimhan, S.G., Nayar, S.K.: Vision and the atmosphere. Int. J. Comput. Vision **48**(3), 233–254 (2002)
5. Starik, S., Werman, M.: Simulation of rain in videos. In: Proceedings of texture: the 3rd international workshop on texture analysis and synthesis, France, pp. 95–100 (2003)
6. Tripathi, A.K., Mukhopadhyay, S.: Rain rendering in videos. In: 4th international conference on computer applications in electrical engineering recent advances, India (2010 February)
7. Prusty, P., Nayak, J.: Detection and removal of rain from satic background videos. Int. J. Appl. Eng. Res. **10**(9), 22791–22797 (2015)
8. Garg, K., Nayar, S.K.: Detection and removal of rain from videos. In: IEEE computer society conference on computer vision and pattern recognition, vol. 1, pp. 528–535 (2004)
9. Tripathi, A.K., Mukhopadhyay,S.: A probabilistic approach for detection and removal of rain from videos. IETE J. Res. **57** (2011)
10. Beard, K.V., Chuang, C.: A new model for the equilibrium shape of rain drops. J. Atmos. Sci. **44**(11), 1509–1524 (1987)

Single Channel Speech Enhancement Using Fractional Wavelet Transform

Rashmirekha Ram, Saumendra Kumar Mohapatra, Prashant Kumar Nayak, and Mihir Narayan Mohanty

Abstract The occurrence of various noises attenuates the nature of the original speech signal while recording or transmitting. To enhance the quality and intelligibility of the signals, speech enhancement algorithms are still in demand. For this, different algorithms have been proposed since several years. In this work, fractional wavelet transforms (FrWT)-based speech enhancement method is proposed for quality improvement. Wavelet transform has its inherent property of noise minimization, and in fractional domain, noise filters out without overlapping with the desired signal. The enhanced speech signal is estimated by thresholding the fractional wavelet coefficients. For this, SURE method with hard thresholding is chosen. Simulation results show the efficacy of the proposed method compared to fractional fourier transform (FrFT) and fractional discrete cosine transform (FrDCT). Different subjective and objective results using NOIZEUS corpus show that the proposed method is better than other fractional methods.

Keywords Fractional discrete cosine transform · Fractional wavelet transform · Segmental signal-to-noise ratio · Short-time intelligibility measure · Mean opinion score · Speech enhancement

---

R. Ram · S. K. Mohapatra (✉) · M. N. Mohanty
Department of Electronics and Communication Engineering, ITER,
Siksha 'O' Anusandhan (Deemed to be University), Bhubaneswar, Odisha, India
e-mail: saumendra.mohapatra.27@gmail.com

R. Ram
e-mail: ram.rashmirekha14@gmail.com

M. N. Mohanty
e-mail: mihir.n.mohanty@gmail.com

P. K. Nayak
Department of Electronics and Telecommunication Engineering, Synergy Institute of Engineering and Technology, Bhubaneswar, Odisha, India

S. Das and M. N. Mohanty (eds.), *Advances in Intelligent Computing and Communication*, Lecture Notes in Networks and Systems 202,
https://doi.org/10.1007/978-981-16-0695-3_59

1 Introduction

In day-to-day life, we come across different sounds like speech, music, and unwanted sounds from different source. The focus is specifically on a single sound in a particular moment for further activity. For this, various methods have been focused with blind source separation, detection, recognition, identification, and noise elimination [1, 2]. Out of all these, speech enhancement by eliminating noise is focused more attention of researchers. The primary objective of enhancement algorithms is to reduce the noise levels and to improve the quality and intelligibility of signals. The demand grows more and more in real world situations where the clean signals are very much essential.

In most real-time applications, single channel enhancement techniques are more concentrated, e.g., speaker identification, voice recognition, hearing aids design, etc. In such cases, second channel is not available. Single channel enhancement devices are easy to build and less expensive than multiple channel devices. Many well-known enhancement methods are employed for single and multiple channels [3, 4]. Paliwal et al. proposed spectral subtraction (SS) using short-time Fourier transform (STFT) for single channel speech enhancement. The signal processing in modulation domain drastically reduced the musical noise. The modified SS method with minimum mean square error estimator (MMSE) improved the speech quality in STFT magnitude domain. But in this method, the quality of the original speech was degraded [5]. To overcome this limitation, Kalman filter was designed in modulation domain where the MMSE estimator used the statistics of both speech and noise magnitude spectrum. The temporal changes of the spectrum were highly beneficial for non-stationary noise, while the comparison was made with other time domain methods [6]. For multi-channel speech enhancement, authors used adaptive Wiener filter and SS for car environment. These methods with the combination of beamformer reduced the musical noise and overcome the problem of using several microphones [7]. While the speaker independent speech enhancement method was proposed by Leglaive et al. for multi-channel. They used the neural network model with non-negative matrix factorization (NMF). The major limitation of this method was the computational time [8]. Though multi-channel enhancement methods often result in better performance than single channel enhancement methods, the additional cost, computational complexities, power usage, and size demands occasionally limit their usability. On the other hand, single channel enhancement methods reduce the background noise coming from the same background where the speaker is present. Due to these reasons, we concentrate on single channel speech enhancement method.

Speech enhancement techniques are used in different domains. In time domain, the processing is implemented directly on the speech signal, and the signal is obtained from the time samples without signal transformation. Conversely, transform domain allows observing the signal components in different domain. Different transforms have been used for signal analysis such as Fourier transform (FT), discrete cosine transform (DCT), and wavelet transform (WT) [1]. FT provides the information about the spectral component of the signal, but it fails to give information about the spectral component at a particular time interval. In this case, STFT is well suited where the signal is processed block wise. But when the signals have low frequency in long duration and high frequency in short duration, STFT is not applicable. In such cases, WT is more beneficial and informative. It can simultaneously analyze the signal in both frequency and time domain. It can be able to estimate the local properties of the signal and analyze the non-stationary; transient signals as well [9–11].

Nowadays, researchers focus more on fractional transform such as fractional Fourier transform (FrFT) and fractional discrete cosine transform (FrDCT). FrFT avoids the limitation of FT and represents the signal in fractional Fourier domain [12]. FrFT and FrDCT were used by the authors in [9, 13] for speech enhancement purpose. FrDCT coefficients were used in the neural network model to increase the quality of speech [14]. However, to avoid the limitation of FrFT and FrDCT, fractional wavelet transform (FrWT) was proposed [15, 16]. For optical image encryption and medical image fusion, FRWT was used in [17, 18]. The fractional order was changed in the interval of 0 and 1, and the medical image was decomposed. The mode coefficients were enhanced by the process and by the use of inverse transform, and the fused images were obtained. Different application related to FrWT was proposed in [19].

As fractional spectrum of FrFT corresponds only the fractional domain frequency over the full length of the signal, it is difficult to know the exact frequency content. FRWT provides both the time and fractional domain frequency information of the speech signal over the entire duration. Moreover, it has more graceful mathematical properties than other transforms [16]. Motivated by this, a single channel speech enhancement method using FrWT is proposed and compared with FrFT and FrDCT filter through simulation. The paper is arranged as follows: In Sect. 2, brief idea about fractional wavelet transform is given, and in Sect. 2, enhancements using FrWT are presented. In Sect. 3, simulation results are discussed, and finally, this article concludes with remarks in Sect. 4.

2 Methodology

Though many fractional transforms have been used for enhancement purpose, FrWT has still not used for this purpose. The objective of the work is to design a technique based on FrWT and to compare with other fractional transforms.

A. **Preliminaries**

FrWT is derived based on the principle of FrFT and wavelet transform [15]. For the sake of better understanding, the concept of FrWT and WT is presented in this section. The βth order FrFT of a signal $y(t)$ is

$$Y_\beta(v) = F^\beta[y(t)] = \int_{-\infty}^{\infty} y(t)K_\beta(t,v)\mathrm{d}t \tag{1}$$

where $K_\beta(t,v)$ is the Kernel function of the fractional transform and is given by

$$K_\beta(t,v) = \begin{cases} A_\beta e^{j/2\left(t^2+v^2\right)\cot\beta - jtv\csc\beta}, & \beta \neq k\pi \\ \delta(t,v), & \beta = 2k\pi \\ \delta(t+v), & \beta = (2k+1)\pi \end{cases}$$

And A_β is given by

$$A_\beta = \sqrt{\frac{1-j\cot\beta}{2\pi}}$$

To get the signal in original form, inverse FrFT is applied. The inverse FrFT of the signal $y(t)$ is given by

$$y(t) = \int_{-\infty}^{+\infty} Y_\beta(u)K_\beta^*(t,v)\mathrm{d}v \tag{2}$$

The FrFT decomposes the signal into the chirp functions as $\left\{K_\beta^*(t,v)\right\}$. The argument v is the fractional frequency component. Similarly, the WT of a signal $y(t)$ is defined as

$$WT_y(a,b) = \frac{1}{\sqrt{a}}\int_{-\infty}^{\infty} y(t)\psi^*\left(\frac{t-b}{a}\right)\mathrm{d}t \tag{3}$$

where $\psi(t)$ is the mother wavelet. The scaling and shifting of $\psi(t)$ is defined as a and b, respectively. Both scaling and shifting parameters are continuous and real. Both time and frequency depend on the value of the scale parameter in WT. Depending on these two transforms, FrWT is defined and presented in the next section.

B. **Fractional Wavelet Transform (FrWT)**

The fractional WT of the signal is defined based on the principle of WT and FrFT. From the fundamental properties of these transforms, the βth order FrWT of a signal $y(t)$ is defined as

$$
\begin{aligned}
W_y^{\beta}(a,b) &= e^{-(-j/2)b^2\cot\beta}\left\{\left[y(t)e^{(-j/2)t^2\cot\beta}\right] * \left[\frac{1}{\sqrt{a}}\psi\left(-\frac{t}{a}\right)e^{(j/2)\left(\frac{t}{a}\right)^2\cot\beta}\right]^*\right\} \\
&= e^{(-j/2)b^2\cot\beta}\left\langle y(t)e^{(j/2)t^2\cot\beta}, \psi_{a,b}(t)e^{(j/2)\left(\frac{t-b}{a}\right)^2\cot\beta}\right\rangle \\
&= \int_{-\infty}^{\infty} y(t)\psi^*_{\beta,a,b}(t)\mathrm{d}t
\end{aligned}
\tag{4}
$$

When $\beta = 1$, FrWT coincides with basic WT. Based on these principles, FrWT for speech enhancement is proposed and is explained in the next section.

C. **Enhancement using FrWT**

FrWT decomposes the signal into approximation and details coefficients to examine the signal at various fractional frequency bands and with different scales. Assuming the noisy speech signal $y(t)$ consists of clean speech signal $s(t)$ and the noise signal $n(t)$. According to this, the noisy signal is expressed as

$$
y(t) = s(t) + n(t) \tag{5}
$$

FrWT of signal $y(t)$ can be represented as

$$
W_y^{\beta}(a,b) = \int_{-\infty}^{\infty} y(t)\psi^*_{\beta,a,b}(t)\mathrm{d}t \tag{6}
$$

where $\psi_{\beta,a,b}(t)$ is the Kernel function and can be expressed as

$$
\psi_{\beta,a,b}(t) = e^{(-j/2)\left(t^2 - b^2 - \left(\frac{t-b}{a}\right)^2\right)\cot\beta}\,\psi_{a,b}(t) \tag{7}
$$

The above equation signifies the FrWT of a signal $y(t)$ can also be expressed as the inner product of the signal with the βth order FrWT. It can be defined as the multiplication of wavelet function $\psi_{a,b}(t)$ and a chirp where $\psi_{a,b}(t)$ can be expressed as

$$
\psi_{a,b}(t) = \frac{1}{\sqrt{a}}\psi\left(\frac{t-b}{a}\right) \tag{8}
$$

The Kernel function of FrWT supports in the time domain. Similarly, the inner product of the signal $y(t)$ and $\psi_{\beta,a,b}(t)$ ensures that $W_y^{\beta}(a,b)$ provides the information in the time domain. As in WT, the approximation coefficients indicate the sampled signal and help to know the shape of the signal. Similarly, the details coefficients provide information about changes in the signal as it contains the derivatives of the signal. The coefficients of $W_y^{\beta}(a,b)$ contain the information of

$y(t)$ that lies near b. To get back the original signal, inverse FrWT is applied. The inverse FrWT can be expressed as

$$y(t) = \frac{1}{2\pi C_\psi} \int_{-\infty}^{\infty} W_y^\beta(a,b)\psi_{\beta,a,b}(t)\frac{da}{a^2}db \tag{9}$$

where C_ψ satisfies the condition [19].

$C_\psi = \int_{-\infty}^{\infty} \frac{Y(\omega)}{\omega} d\omega < \infty$ and $Y(\omega)$ is the FT of $y(t)$. Equation (5) can be represented as

$$Y_{WT} = S_{WT} + N_{WT} \tag{10}$$

where $Y_{WT} = W_y^\beta(a,b).y(t)$, $S_{WT} = W_y^\beta(a,b).s(t)$ and $N_{WT} = W_y^\beta(a,b).n(t)$. The enhanced speech signal $\hat{s}(t)$ is estimated by thresholding the fractional wavelet coefficients Y_{WT} as

$$\hat{s}(t) = \mathrm{Thr}(Y_{WT}, T) \tag{11}$$

T is the appropriate threshold value. The threshold value is chosen based on hard and soft threshold mentioned below.

$$\mathrm{Thr}_{\mathrm{Hard}}(Y_{WT}, T) = \begin{cases} Y_{WT} & |Y_{WT}| \geq T \\ 0 & |Y_{WT}| < T \end{cases} \tag{12}$$

$$\mathrm{Thr}_{\mathrm{Soft}}(Y_{WT}, T) = \begin{cases} \mathrm{sign}(|Y_{WT}|)(|Y_{WT}| - T) & |Y_{WT}| \geq T \\ 0 & |Y_{WT}| < T \end{cases} \tag{13}$$

Different methods are there to select the threshold value as SURE method, minimax method, and universal method, etc [20]. Various noisy signals are tested, and the results are presented in the results and discussion section.

3 Results and Discussion

In this section, the experimental results are shown to validate our proposed enhancement method. Moreover, FrWT is compared with FrFT and FrDCT. All experiments are performed using MATLAB. This section is again divided into (3.1) database used and (3.2) subjective and objective measures.

3.1 Database Used

The proposed speech enhancement method is evaluated with the NOIZEUS database. It includes 30 IEEE sentences where three male and three female speakers are recorded. Eight different real-world noises are there with SNR 0–15 dB (increase of 5 dB). The sampling frequency of all voice signals is 8 kHz. Each sound is of duration 2–3 s. To test the proposed method, train noise of different SNR level is considered. Figure 1 shows the clean signal '*He wrote down a long list of items*', and its corresponding spectrogram is shown in Fig. 1a. Similarly, Fig. 2a and b shows the noisy speech of 0 dB SNR both in time domain and frequency domain, respectively. FrWT is applied to the noisy signal, and its corresponding output is shown in Fig. 5. The noisy signal is decomposed into approximation and detail coefficients using FrWT. Due to the presence of noise, the details of the signal contain the randomness. Different values of fractional order (β) are being tested to remove the noise. In this experiment, 'dB6' wavelet is used as the mother wavelet, and β is chosen 0.5 empirically. As the details of FrWT are varied frequently, we chose thresholding to remove the noise levels. SURE method with hard thresholding provides the maximum SNR as 17.92 dB (Table 1). The modified coefficients are utilized in inverse FrWT to achieve the enhanced signal. The output of FrWT is shown in Fig. 5. From the output graph, it is clearly visible that the spectrogram of FrWT (Fig. 5b) is very much similar to the spectrogram of the clean signal. For comparison, the output of FrFT and FrDCT is shown in Figs. 3 and 4. Both time and frequency domain outputs are presented.

3.2 Subjective and Objective Measures

(i) *Mean Opinion Score (MOS)*

For subjective measure, we chose MOS for quality evaluation of the signals. It compares the original speech and the output speech by a group of people. In this, listeners rate the quality on a five-point scale based on the auditory perception. The scale rating is assigned with a score, and the details of the rating are presented in Table 1. The minimum score signifies very poor quality; however, the maximum score signifies excellent quality. The listening test is the most consistent method for evaluating the speech quality. In this test, 30 listeners of ITER College are participated between the age group of 17–30 years.

(ii) *Signal-to Noise Ratio (SNR)*

Different signals are tested to calculate the SNR where the input noise level is predefined and the output SNR is estimated using the following equation

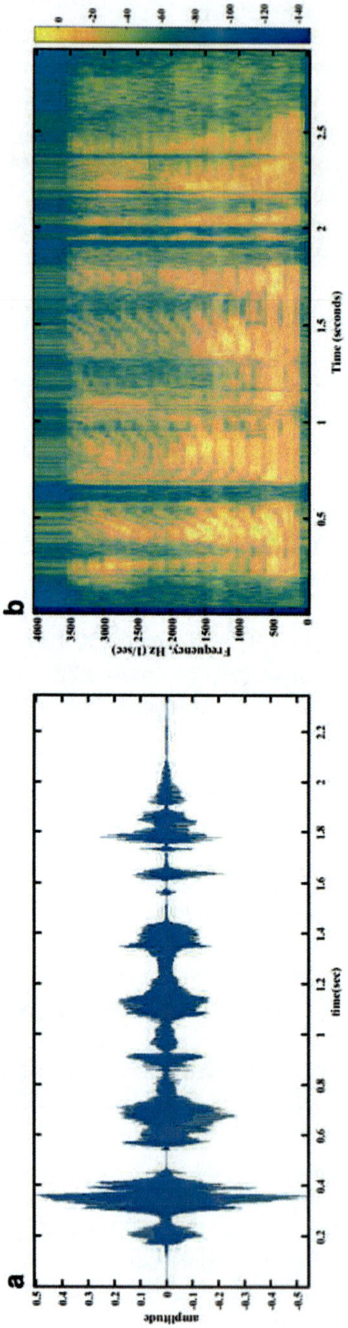

Fig. 1 **a** Clean speech signal '*He wrote down a long list of items*'. **b** Spectrogram of clean speech signal

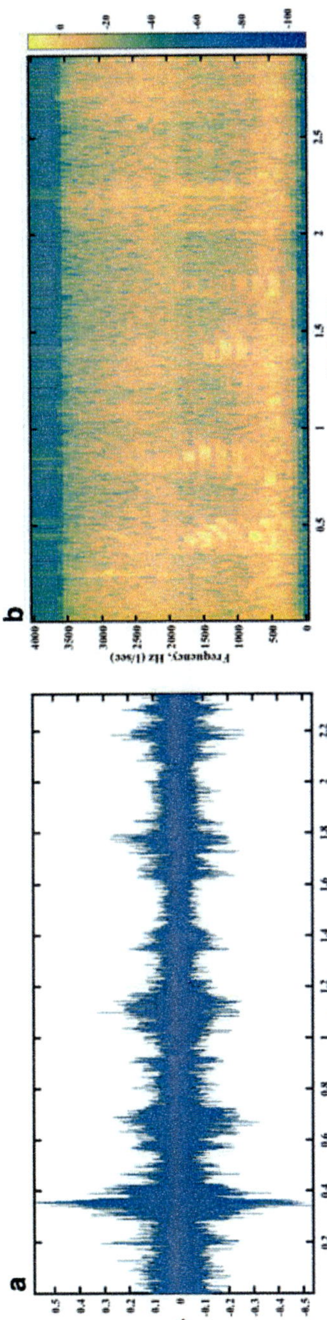

Fig. 2 **a** Noisy signal of train noise of SNR 0 dB. **b** Spectrogram of noisy signal of train noise of SNR 0 dB

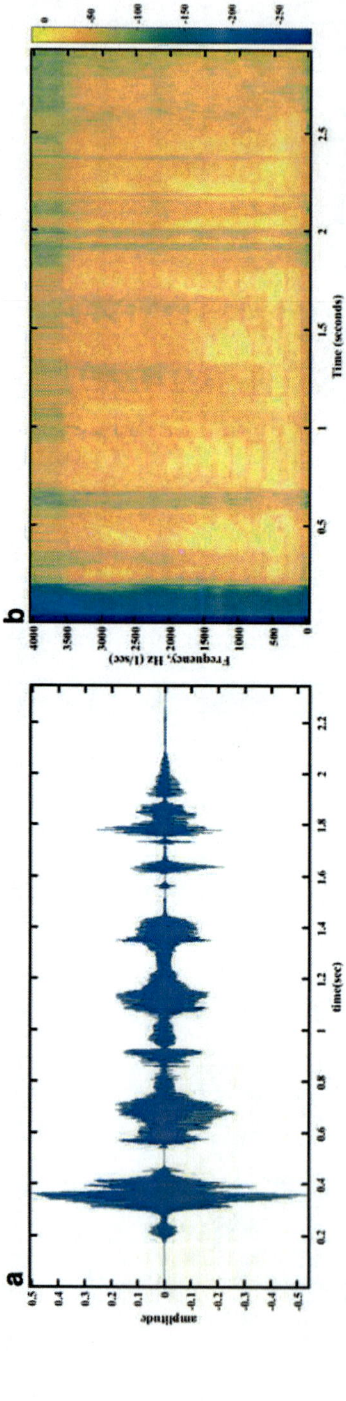

Fig. 3 **a** Output signal of FrFT. **b** Spectrogram of FrFT output

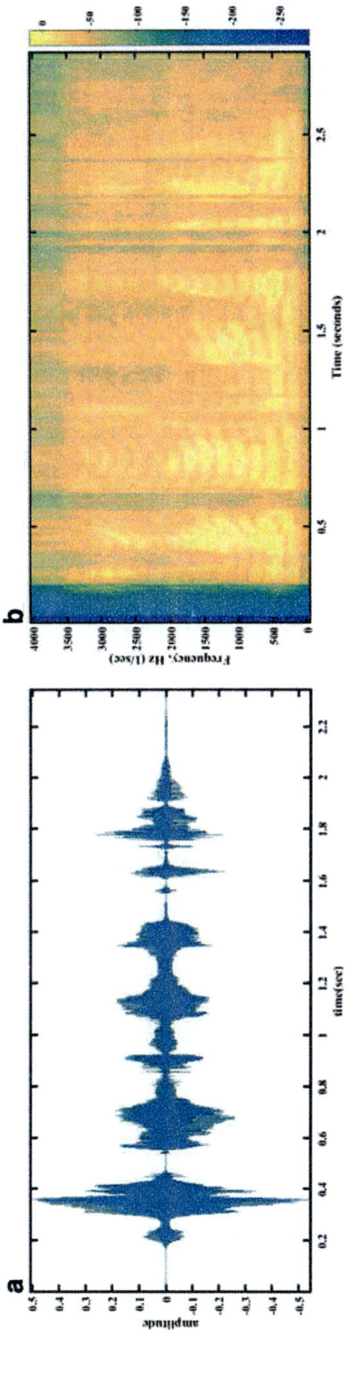

Fig. 4 **a** Output signal of FrDCT. **b** Spectrogram of FrDCT output

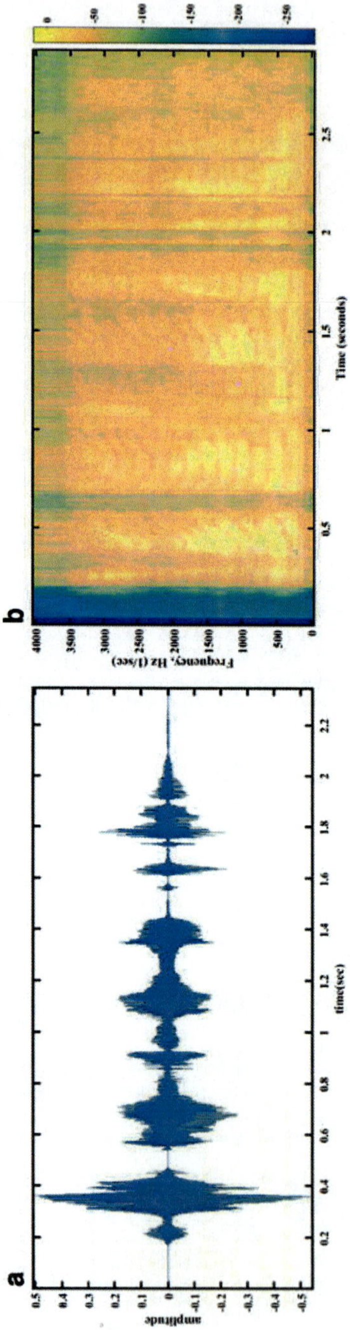

Fig. 5 **a** Output signal of FrWT. **b** Spectrogram of FrWT output

Table 1 MOS scale rating

Scores	Quality
1	Very Poor
2	Poor
3	Bad
4	Good
5	Excellent

$$(\mathrm{SNR})_{\mathrm{dB}} = 10\log_{10}\frac{\sum_{t=1}^{N}(y(t))^2}{\sum_{t=1}^{N}(y(t)-\hat{s}(t))^2} \quad (14)$$

where N is the length of the speech signal.

(iii) *Segmental Signal-to Noise Ratio (SegSNR)*

SegSNR is used to measure the speech quality. Signals are divided into frames to process and measure SegSNR. For a frame length of L, the SegSNR is defined as

$$(\mathrm{SegSNR})_{\mathrm{dB}} = \frac{1}{q}\sum_{m=0}^{q-1} 10\log_{10}\frac{\sum_{m=N_n}^{m_n+N\_1} y^2(t)}{\sum_{m=N_n}^{m_n+N\_1}[y(t)-\hat{s}(t)]^2} \quad (15)$$

where q = number of active speech frames in the signal. The comparison results are shown in Table 2. The maximum SegSNR is 3.49 dB obtained from 15 dB of train noise, and the maximum MOS is 3.25 obtained using FrWT.

Table 2 Comparison of FrWT method with FrFT and FrDCT

Train Noise (dB)	Methods	SNR (dB)	SegSNR (dB)	MOS
0	FrFT	1.99	2.90	2.65
	FrDCT	2.26	2.93	2.67
	FrWT	2.49	3.41	3.17
5	FrFT	6.12	2.94	2.66
	FrDCT	6.50	2.95	2.67
	FrWT	7.55	3.42	3.20
10	FrFT	11.32	3.02	2.70
	FrDCT	11.98	3.11	2.72
	FrWT	12.87	3.45	3.22
15	FrFT	15.67	3.10	2.73
	FrDCT	16.61	3.13	2.74
	FrWT	17.92	3.49	3.25

4 Conclusion

In this work, FrWT is proposed for single channel speech enhancement. For this, train noise is tested, and the results are shown in terms of SNR, SegSNR, and MOS. The proposed method is compared with FrFT and FrDCT. Due to FrFT and time-frequency distribution of the signal, FrWT provides better enhanced signals than other methods. As FrWT has multi-resolution property, it can efficiently reduce the noise in fractional domain. The empirically chosen 0.5th fractional order eliminates more noise from the noisy signal and increases the SNR. The hard thresholding with SURE provides a maximum SNR of 17.92 dB. The novel FrWT for speech enhancement provides better result than other fractional methods. Different thresholding may provide better enhanced signal with other fractional order.

References

1. Benesty, J., Makino, S., Chen, J.: Speech enhancement. Springer, Berlin (2005)
2. Loizou, P.C.: Speech enhancement: theory and practice. CRC press (2013)
3. Krawczyk, M., Gerkmann, T.: STFT phase reconstruction in voiced speech for an improved single-channel speech enhancement. IEEE/ACM Trans. Audio Speech Language Process. **22** (12), 1931–1940 (2014)
4. Mowlaee, P., Saeidi, R.: Iterative closed-loop phase-aware single-channel speech enhancement. IEEE Signal Process. Lett. **20**(12), 1235–1239 (2013)
5. Paliwal, K., Wójcicki, K., Schwerin, B.: Single-channel speech enhancement using spectral subtraction in the short-time modulation domain. Speech Commun. **52**(5), 450–475 (2010)
6. So, S., Paliwal, K.K.: Modulation-domain Kalman filtering for single-channel speech enhancement. Speech Commun. **53**(6), 818–829 (2011)
7. Meyer, J., Simmer, K.U.: Multi-channel speech enhancement in a car environment using Wiener filtering and spectral subtraction. In: 1997 IEEE international conference on acoustics, speech, and signal processing, vol. 2. IEEE, pp. 1167–1170 (1997 April)
8. Leglaive, S., Girin, L., Horaud, R.: Semi-supervised multichannel speech enhancement with variational autoencoders and non-negative matrix factorization. In: ICASSP 2019–2019 IEEE International Conference on Acoustics, Speech and Signal Processing (ICASSP). IEEE, pp. 101–105 (2019 May)
9. Ram, R., Palo, H.K., Mohanty, M.N.: Fractional segmental transform for speech enhancement. In: International conference on intelligent computing and applications. Springer, Singapore, pp. 135–142 (2019)
10. Almeida, L.B.: The fractional Fourier transform and time-frequency representations. IEEE Trans. Signal Process. **42**(11), 3084–3091 (1994)
11. Ram, R., Mohanty, M.N.: Use of fractional domain for speech enhancement. Int. J. Imaging Robotics **18**, 85–93 (2018)
12. Ram, R., Mohanty, M.N.: Design of fractional fourier transform based filter for speech enhancement. Int. J. Control Theory Appl. **10**(7), 235–243 (2017)
13. Ram, R., Mohanty, M.N.: Design of filter using fractional-DCT for speech enhancement. In: International conference on sustainable computing techniques in engineering, science and management (2017)

14. Ram, R., Mohanty, M.N.: Fractional DCT ADALINE method for speech enhancement. In: International conference on machine learning and computational intelligence (2017)
15. Mendlovic, D., Zalevsky, Z., Mas, D., García, J., Ferreira, C.: Fractional wavelet transform. Appl. Opt. **36**(20), 4801–4806 (1997)
16. Dai, H., Zheng, Z., Wang, W.: A new fractional wavelet transform. Commun. Nonlinear Sci. Numer. Simul. **44**, 19–36 (2017)
17. Chen, L., Zhao, D.: Optical image encryption based on fractional wavelet transform. Optics Commun. **254**(4–6), 361–367 (2005)
18. Xu, X., Wang, Y., Chen, S.: Medical image fusion using discrete fractional wavelet transform. Biomed. Signal Process. Control **27**, 103–111 (2016)
19. Shi, J., Zhang, N., Liu, X.: A novel fractional wavelet transform and its applications. Sci. China Inform. Sci. **55**(6), 1270–1279 (2012)
20. Wojtaszczyk, P.: A mathematical introduction to wavelets, vol. 37. Cambridge University Press (1997)

Random Forest Classification-Based Video Event Detection Utilizing Hand Crafted Features

A. Susmitha, Sanjay Jain, and Mihir Narayan Mohanty

Abstract Huge amount of user-generated video content is shared and downloaded by millions of users across the globe, every minute. The sudden raise of this web content made event detection in videos a major demanding field of research in recent days. Detecting events automatically in large user-generated video datasets (UGV) is a challenging task because of its poor quality and un structured content. In this work, random forest classifier is used to detect events in UCF 101 dataset by utilizing handcrafted features like HOG and Tamura. The performance is measured in terms of accuracy, specificity, sensitivity, false discovery rate, false omission rate, and error rate.

Keywords Multimedia · Video event detection · Random forest classifier · HOG · Tamura

1 Introduction

From the past few years, world is witnessing an impulsive increase in user-generated multimedia content on the web due to the growing attractiveness of the social networking sites. Each minute, 300 h of user-generated video are uploaded on YouTube by the users. Video content is a unique resource which provides both challenges and opportunities [1]. People may not have sufficient time and patience to watch all the details of the video content among a large number of videos. In

A. Susmitha
Department of ECE, CMR Institute of Technology, Bengaluru, India
e-mail: susmitha.academic@gmail.com

S. Jain
CMR Institute of Technology, Bengaluru, India
e-mail: principal@cmrit.ac.in

M. N. Mohanty (✉)
ITER, Siksha 'O' Anushandhan (Deemed to be University), Bhubaneswar, India
e-mail: mihir.n.mohanty@gmail.com

S. Das and M. N. Mohanty (eds.), *Advances in Intelligent Computing and Communication*, Lecture Notes in Networks and Systems 202,
https://doi.org/10.1007/978-981-16-0695-3_60

general user's interest levels are retained only for small periods of time. As events are unforgettable and significant, they can be easily recalled. Therefore, it is beneficial to identify unique interesting events from users. So, it has become increasingly important to find the solution to automatically figure out the events captured in this large collection of videos according to the user's interest. As it is highly challenging, extensive research has begun for developing efficient algorithms to identify the most exciting events and concepts from videos [2]. The need of automatic event detection in video content is obvious in many real-world applications such as natural disaster detection, abnormal detection, surveillance indexing, retrieval, and summarization.

Multimedia video content on the web provides both opportunities as well as challenges to the researchers. The first challenge that needs to be addressed is how to deal effectively with such huge Internet video content. The second challenge is that the videos shot by the users under unregulated circumstances are not structured, not edited, and suffer from poor quality. Therefore, it is difficult to search the videos for particular events important to the user.

Figure 1 shows frames from different videos containing an event "Feeding an animal." In this scenario, various kinds of animals provide a significant challenge in the detection of events.

Features are very important for any multimedia task. Therefore, in multimedia systems how to obtain positive features from video to boost the results of event detection is another considerable problem. The paper is arranged as follows: Some relevant works are given in Sect. 2. In Sect. 3, the proposed event detection framework is discussed. In Sect. 4, the experimental assessment of the proposed mechanism is discussed. Section 5 concludes the paper.

2 Related Work

Detecting high-level events in multimedia data is essential for many real-world applications. Due to the rapid rise in the amount of web video data, multimedia event detection has become a hot research topic in recent years. The goal of event detection is to automatically identify video clips based on real user related events. Event is an important occurrence which happens at a certain place at a certain time. Events can be of news events, sports events, social events, abnormal events, and every day events. A few examples of the above-mentioned events are earthquake, disasters, scoring a goal, concert, debates, exhibitions, suspicious activities, and interactions of people and objects such as grooming an animal and birthday party. In the last ten years, a variety of methodologies for event detection in videos have been suggested by a separate group of researchers. In [3], the authors used the Hidden Markov Model classifier to detect interesting clips in a soccer sports video by using low-level visual features.

Fig. 1 Feeding an animal event example video frames

Guo and Scott [4] summarized the novel properties of user-generated video and explored how to utilize the concept classifiers to detect complex events in UGV. In [5], authors proposed a framework for multimedia event detection. The framework recognizes dynamic events based on a wide variety of features from different modalities and classifies unseen videos by integrating different responses effectively. Mu et al. [6] extracted motion features as a vector and used support vector machines to identify suspicious activity in HD videos. Multiscale optical flow histogram feature vector and KNN classifier were used by Gnanavel et al. to detect abnormal events in crowded videos in [7]. In [8], Gaussian optical flow model and optical flow orientation histogram were computed by the authors for the detection of group violence in surveillance videos.

3 Proposed Framework

The framework consists of four main modules, namely (1) collection of dataset, (2) preprocessing, (3) extraction of features, and (4) classification using random forest classifier (Fig. 2).

3.1 *The UCF101*

There are 101 events in this dataset [9], ranging from basic sports to everyday life activities. There are 13,320 video clips such as playing violin, shaving beard, biking, applying eye makeup, etc.

3.2 *Data Preprocessing*

How to preprocess the unstructured data is the first step in each data analysis. We used histogram equalization for our video processing. The video is converted into frames. Histogram equalization was applied for contrast enhancement and smoothing of the collected frames. After the preprocessing, the selected frames are used for detecting events or concepts in videos.

3.3 *Feature Extraction*

Features play a significant task to understand the video events properly. High-quality features are likely to be strong next to changes so that under various capturing circumstances different videos of the same event type will be able to be appropriately detected. A combination of visual characteristics based on stationary frames is used by some video event detection systems [10]. HOG [11] developed by Dalal and scale-invariant feature transform (SIFT), developed by Lowe [12] were the most commonly used dominant features. Speeded up robust features (SURF) were proposed by Bay et al. [13]. In our proposed framework, histogram of oriented gradients (HOG) and Tamura's texture features [14] (coarseness, contrast, and directionality) are extracted from the preprocessed frames. The main reason to

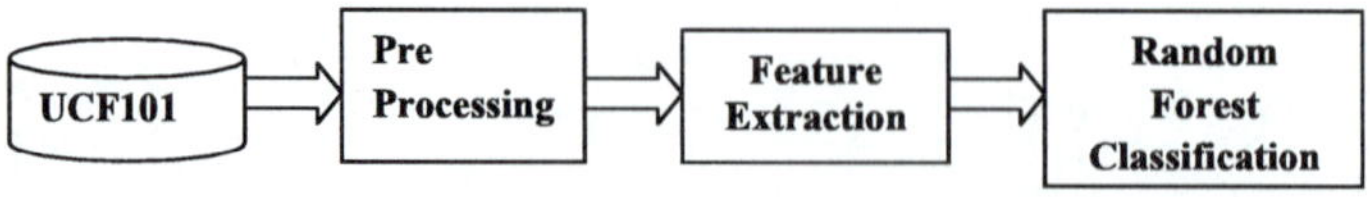

Fig. 2 Proposed Framework

extract HOG feature is that it is precise and prompt. HOG is a feature descriptor. By calculating gradients in horizontal and vertical directions, useful information of the image can be obtained.

3.4 *Random Forest Classifier*

It is a supervised learning algorithm which has number of trees with some sort of randomization used to grow each tree. Because of number of decision trees, it is highly accurate and robust. Use of multiple trees reduce the over fitting problem as it takes the average of all the predictions. Each random tree is passed down by the test image until it reaches a leaf node. Then, all the posterior probabilities are combined, and the average maximum is taken as the classification output.

4 Experimental Analysis

We evaluated our framework on UCF 101 video dataset. It consists of 101 action scenarios. Among 101 action scenarios, 20 random scenarios are considered for analysis (Table 1).

The performance analysis is measured in terms of accuracy, specificity, sensitivity, false discovery rate, and false omission rate.

$$\text{Acc.} = \frac{\text{TrP} + \text{TrN}}{\text{TrP} + \text{TrN} + \text{FaP} + \text{FaN}}$$

$$\text{Sensit.} = \frac{\text{TrP}}{\text{TrP} + \text{FaN}}$$

$$\text{Specificity} = \frac{\text{TrN}}{\text{TrN} + \text{FaP}}$$

$$\text{FOR} = \frac{\text{FaN}}{\text{FaN} + \text{TrN}}$$

$$\text{FDR} = \frac{\text{FaP}}{\text{FaP} + \text{TrP}}$$

where TrP is true positives, TrN is true negatives, Fap is false positives, and FaN is false negatives (Table 2 and Fig. 3).

Table 1 Random scenarios

S. no	Random scenarios
1	Horse race
2	Ice dancing
3	Knitting
4	Punch
5	Apply eye makeup
6	Haircut
7	Typing
8	Military parade
9	Surfing
10	Cricket shot
11	Baby crawling
12	Blowing candle
13	Basketball
14	Diving
15	Drumming
16	Writing on board
17	Ice dancing
18	Surfing
19	Biking
20	Rafting

Table 2 Performance evaluation of proposed frame work on UCF 101 database

Accuracy (%)	Sensitivity (%)	Specificity (%)	FOR (%)	FDR (%)
77.18	55.17	80	1.08	44.83

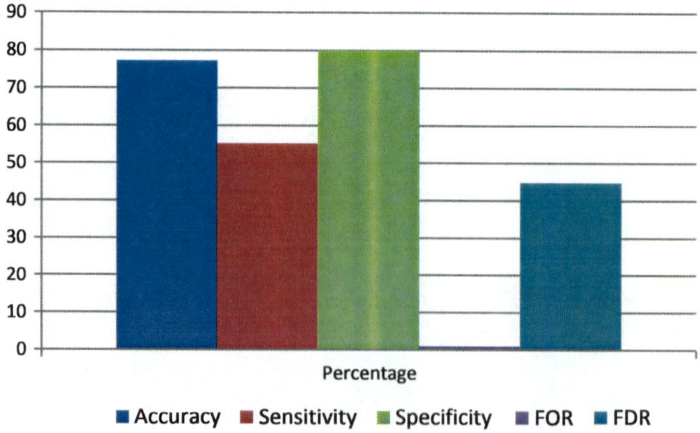

Fig. 3 Performance evaluation

5 Conclusion

In this article, we presented event detection structure to identify events in a UCF 101 dataset. Multiple features like HOG and Tamura's features are extracted from the preprocessing video frames. Random forest classifier was used to classify the events. Performance evaluation was considered in terms of accuracy, specificity, sensitivity, FOR, and FDR.

References

1. Shyu, M.-L., Xie, Z., Chen, M., Chen, S.-C.: Video semantic event/concept detection using a subspace-based multimedia data mining framework. IEEE Trans. Multimedia **10**(2), 252–259 (2008)
2. Lin, L., Shyu, M.-L.: Weighted association rule mining for video semantic detection. In: Methods and innovations for multimedia database content management, vol. 1, no. 1, pp. 37–54 (2012)
3. Nisha, J., Santanu, C., Roy, S., Mukherjee, P., Seal, K., Talluri, K.: A novel learning-based framework for detecting interesting events in soccer videos. In: Proceedings of the sixth indian conference on computer vision, graphics image processing (2008 Dec)
4. Guo, J., Scott, D., Hopfgartner, F., Gurrin, C.: Detecting complex events in user-generated video using concept classifiers. In: CBMI (2012)
5. Oh, S., McCloskey, S., IlseoKim: Multimedia event detection with multimodal feature fusion and temporal concept localization. Mach. Vis. Appl. **25**, 49–69 (2014)
6. Mu, C., Xie, J., Yan, W., Liu, T., Li, P.: A fast recognition algorithm for suspicious behavior in high definition videos. Multimedia Syst. **22**(3), 275–285 (2016)
7. Gnanavel, V., Srinivasan, A.: Abnormal event detection in crowded video scenes. In: Proceedings of the 3rd international conference on frontiers of intelligent computing: theory and applications (Ficta) 2014, pp. 441–448. Springer, Berlin (2015)
8. Zhang, T., Yang, Z., Jia, W., Yang, B., Yang, J., He, X.: A new method for violence detection in surveillance scenes. Multimedia Tools Appl. **75**(12), 7327–7349 (2016)
9. Soomro, K., Zamir, A.R., Shah, M.: UCF101: a dataset of 101 human actions classes from videos in the wild. CRCV-TR-12-01 (2012)
10. Jiang, Y.-G., Bhattacharya, S., Chang, S.-F., Shah, M.: High-level event recognition in unconstrained videos. In: Multimedia information retrieval (2013)
11. Dalal, N., Triggs, B.: Histograms of oriented gradients for human detection. In: IEEE computer society conference on computer vision and pattern recognition, pp. 886–893 (2005)
12. Lowe, D.G.: Distinctive image features from scale invariant keypoints. Int. J. Comput. Vis. **60** (2), 91–110 (2004)
13. Bay, H., Ess, A., Tuytelaars, T., Van Gool, L.: Speededup robust features (SURF). Comput. Vis. Image Underst. **110**(3), 346–359 (2008)
14. Humeau-Heurtier, A.: Texture feature extraction methods: a survey. IEEE Access **7**, 8975–9000 (2019)

Minimization of Collision Through Retransmission and Optimal Power Allocation in Wireless Sensor Networks (WSNs)

Jyotishree Bhanipati, Debabrata Singh, Anil Kumar Biswal, and Saroja Kumar Rout

Abstract The periodic data collection and energy consumption are the main issue for collision in wireless sensor networks, which generally rely on non-renewable energy sources. In recent days, the energy harvesting sources are based on solar, wind, and vibration for harvesting electric energy. The sensor nodes are unable to communicate with other nodes due to an unsteady supply of energy from energy harvesting sources, whereas a node cannot predict an energy level of nearer nodes that can receive a transmitted data packet from it. Sometimes the probability of losing or collision of data packets in the target node is more due to insufficient power supply. This proposed paper is designed to collect data by using protocols like probabilistic retransmission (PRT) and also probabilistic retransmission (PRT) with collision impact (PRT-CI), in addition to the supply of optimal power allocation source to sensor nodes which enhances the performance of the overall process. The number of data packets is retransmitted on the probability of reception using PRT as well as the collision impact (PRT-CI) evaluates the data packet with a response probability of each node.

Keywords Energy harvesting · Probabilistic retransmission (PRT) · PRT with collision impact (PRT-CI) · Wireless sensor networks (WSNs) · Data packet

J. Bhanipati · A. K. Biswal (✉)
Department of CSE, ITER, Siksha O Anusandhan University, BBSR, Odisha, India
e-mail: anil.biswal123@gmail.com

D. Singh
Department of CSIT, ITER, Siksha O Anusandhan University, BBSR, Odisha, India

S. K. Rout
Department of CSE, Gandhi Institute for Technology (GIFT), BBSR, Odisha, India

S. Das and M. N. Mohanty (eds.), *Advances in Intelligent Computing and Communication*, Lecture Notes in Networks and Systems 202,
https://doi.org/10.1007/978-981-16-0695-3_61

1 Introduction

In a recent situation, the most important part of WSNs is data transmission, as it is used for transferring the data from the sender to the receiver. Data transmission is an energy-consuming process for which it needs an optimal routing system for avoiding data loss [1]. Hence for this, the data transmission should be in a secured way but the main issue is that there is limited energy in the data nodes for their limited capacities of memory for which maintaining security is very much difficult. Some aspects of security are maintained to make secure data transmission while transmission. The main factor for which retransmission is performed is for reliability as it makes sure that there is no loss of data while wireless communication [2]. WSNs mostly use the retransmitting mechanism to make sure the reliability is the main priority. When a large number of data exchange occurs in various network nodes, which forms a high collision rate, due to that data packet lost [3, 4]. The cooperative diversity methods can be minimized with the transmission collision area by adding the high power around the data transmitting node. Through diversity, techniques provide cooperation to the nearer node to forward a data packet to the destination in the same signal quality, less power consumption with a large capacity, and optimum coverage [5, 6].

So, this diversity has been derived the robust and effective technique of the wireless communication channels [7].

Due to the excess energy consumption is causing a collision, that why to avoid the collision by retransmission of the packet. The WSNs are generally implemented for target detection or various fields of military applications, and the collision of a packet can create an issue like target missing and delays in transmission. Thus, the probability of collision needs to be reduced in WSNs with the help of cooperative diversity. The WSNs face the main issues like saving power, which is related to the desired quality of communications. Then, these issues can be minimized through two types of proposed techniques such as probabilistic retransmission (PRT) and also follow probabilistic retransmission (PRT) with packet collision impact (PRT-CI). Here, the main objective is to gather a data packet efficiently and reliably, in addition to supplied by energy harvesting method.

In this paper, the minimization of probability of collision in WSNs by using two protocols, cooperative diversity principle, and also optimal energy allocation, which can be formulated through Lagrange multiplier technique to calculate the optimal energy level of power supplier to target nodes.

2 Literature Review

Priyadarshini et al. [8] proposed a cross-layer co-operative routing algorithm in the WSNs for reducing the probability of the collision. By considering path collection for the election of nodes, allocating optimal energy allocation and cooperative transmission the author tries to reduce the collision. The author also allocates the power for helping the minimization of the collision probability.

Onsy et al. [9] adopted a methodology for calculating the throughput and energy consumed in a wireless sensor network environment under the effect of collision and retransmission. They also adopt a jamming node modeling interface in WSN. The author mainly focuses on dropped data packets due to interferences, retransmission, and collision inside the network. Finally, they find Wi-Fi under packet generation per second, which is possible by comparing ZigBee, Wi-Fi module with low energy. The result shows that the Wi-Fi LP unit and Wi-Fi module must be reflected in heavy traffic WSN applications. Tay and Balakrishna [10] proposed a protocol CSMA/*P**, which helps in motivating the event-driven workload and optimization techniques. The author approaches a non-persistent CSMA with no uniform probability *P** distribution, which selects randomly contention slots. Sudha et al. [11] describe the energy efficiency related to the collision. They lead to an effective algorithm to reduce the possibility of collision from the occurrence of congestion. By introducing a concept of count size of the source conflict window, the author tries to minimize the collision. Also, the author increases the source conflict window size and also the effect of conserving the power of data packets with delay in medium access.

Mansourkiail et al. [12] reviewed the probability of collision minimization with the help of optimal power allocation and co-operative transmissions. The author determines the relay nodes and the optimal power of the source, which minimizes the collision area. Their proposed technique takes equal time for both direct transmission and co-operative systems in the networks. Sadek et al. [13] discussed the technique to reduce energy consumption with the help of the co-operative diversity principle and also optimal energy allocation in wireless sensor networks.

The co-operative diversity system provides a method for total less consumed energy in the direct transmission process if the distance between the two neighbor nodes is small.

3 Proposed Model and Methodology

In our sensor network, we have several homogeneous static sensor nodes with sensing range S and a fixed transmission range T. All the sensor nodes are deployed over a 2D region in the networks. Example: In a WSNs with n homogenous sensor nodes $X = \{1, 2, 3 \ldots n\}$ and to use a sink node $(n + 1)$, deployed above the 2D region of the process. The topology of sensor network can be denoted through a graph $G\ (X, Y)$, where $Y = \{(a, b)|\ a, b \in X\}$, if the transmission range between the nodes a and b is $d\ (a, b)\ \leq\ T$. By choosing the shortest path from a to $n + 1$ the number of hops is our hop counts. The set of nodes $S_i \subseteq X$ is one-hop neighbor nodes of I, if $b \in S_i$ is neighboring to node a. The $G\ (X, Y)$ is a unidirectional graph as the links in $G\ (X, Y)$ are bidirectional.

3.1 Energy Model for Sensor Network

The individual sensor node requires energy for transmitting and receives. Each sensor node has also initial energy *E*. For sensing and computing, each node requires a negligible amount of energy, so we can neglect them. As in the network for load balancing, we must have considered the energy consumptions across the nodes. So, the first-order radio model is applied in a simple way to consumed power through a sensor node with l bit data packets receives as $R_x = E_{\text{ele.}}.l$. Here, the required power of sensor nodes of the transmitter or receiver circuit is denoted by $E_{\text{ele.}}$. To transmit all the l bit data packets, the energy consumed for transmission is represented in Eq. 1 as

$$T_x = R_x + E_{\text{amp}} * l * T^{\propto} \tag{1}$$

where $E_{\text{ele.}}$ is used to transmit a signal bit of transmitter amplifier, and the transmission radius denoted by T and also the component of path loss denoted by $\propto$ where $2 \leq \leq 6$. Our model contains homogeneous sensor nodes with packet length l, so the same amount of energy will be consumed for each packet transmission.

3.2 Data Collection Examples

The depicted Fig. 1a shows the data transmission from node $n3$ to the sink. Each node has a different charge level, i.e., charging state, receiving state, transmitting state, and sink to have stable power (always in receive state). Here, a data packet of node $n3$ is to be sent to sink. Now, we assume $n3$ is in a charged state and idle. It will broadcast the packet. Nodes $\{n5, n6\}$ are receiving, and $n4$ is charging. Nodes $\{n1, n2\}$ are physically far away from $n3$ to sink, so, it discards the receiving packet. The sensor node of the network like $\{n7, n8, n9, n10, n11\}$ lies out of the range level of node $n3$. So, nodes $\{n5, n6\}$ will receive the packet from $n3$. And in these manner nodes $\{n5, n6\}$ will broadcast, and the sink will receive the packet.

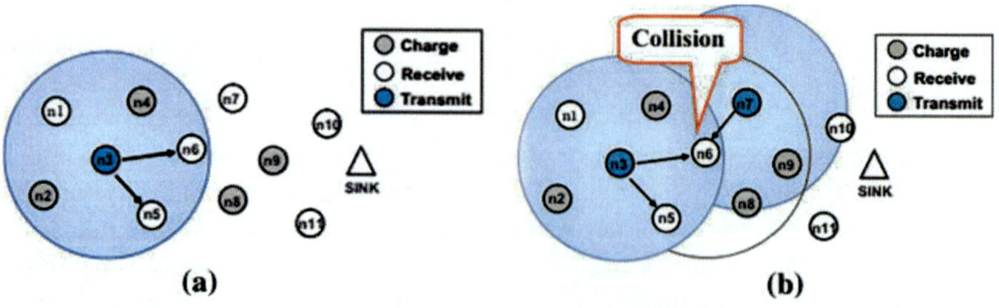

Fig. 1 **a** Data transmission scenario from source to sink node and **b** data transmission scenario from source to sink node with collision

In another condition, we will consider nodes $\{n5, n6\}$ are denoted in the charge state of sensor nodes as well as there is no node to receive the packet that will be lost. So, this proposed model accepts to communicate in one-way without forwarding ACK message or handshaking packets like request-to-send (RTS) and clear-to-send (CTS) due to that the sender does not have a record of its neighbor node is received or not. During the transmission of a data packet from one node to another, that causes packet loss, due to relying on other nodes as depicted in Fig. 1b.

3.3 *Technical Approach*

3.3.1 Probabilistic Retransmissions (PRT)

PRT is defined as retransmitting a data packet in large numbers which are based on the reception probability, and the various nodes are computed with an operating timer. Where the nodes Ndi denotes $\{n1, n2, n4, n5, n6\}$ which lies in the range of communication n_i, i.e., $n3$ and also SNi denotes $\{n4, n5, n6\}$ that is a subset of Ndi which are very nearer to sink. Here, T_{chr} denoted as total charging time, T_{rci} denoted as a total receiving time, and also T_{tri} demoted as a total transmitting time in $[\text{ti} - \pi, \text{ti}]$ period where the current time denoted as ti and $\pi = T_{\text{chr}} + T_{\text{tri}} + T_{\text{tri}}$ is kept by $n3$. Then, reception probability can be calculated by:

$$\text{pi} = \frac{T_{\text{rci}}}{T_{\text{chr}} + T_{\text{tri}} + T_{\text{tri}}} \cdot \frac{t_{\text{rx}} - t_{\text{tx}}}{t_{\text{rx}}} \tag{2}$$

Here, $\frac{t_{\text{rx}} - t_{\text{tx}}}{t_{\text{rx}}}$ refer to the time when sender node transmits the packet but receiver node does not receive the packet which have very less reception probability. It means the receiver node is charging. If the reception probability denotes pi = 0, then the sensor node is charging and cannot receive, and if pi = 1, it means the node is charged and in a receiving state. When a sensor node x_i receives a data packet from the sensor node y_j then node x_i sets the initial value p_j to 1. τ depends on the environment of the sensor network like a different type of rate of charging and uses of energy harvesting device. The value of τ would be considered as longer than the state of charge, to receive and transmit. But it becomes difficult to adjust to changing environments and hence become larger. When at least one node n_i will receive a broadcast packet is expressed by:

$$q_i = 1 - \prod_{j}^{S_i} (1 - p_j) \tag{3}$$

Here, if the reception probability is high or more no. of node is present in range of n_i, then probability of receiving by at least one node is more. If n_i start a transmission again, then probability of receiving by at least one node is denoted by:

$$r_{ai} = 1 - (1 - q_i)^a \tag{4}$$

Whereas this considered as reliability threshold in PRT on q_i, $r_{a'i} \geq T_h$. a' denote optimal no. of retransmission such that it remains above reliability threshold.

$$r_{a'i} \geq T_h,\ 1(1 - q_i)a_i \geq T_h,\ 1(1 - q_i)a_i \leq 1 - T_h,\ a' \geq \log(1 - q_i)(1 - T_h) \tag{5}$$

If T_h increases then a' and energy consumption also increases, so T_h is defined in different settings in it.

3.3.2 PRT with Collision Impact (PRT-CI)

In PRT-CI, each node process reception probability with collision consideration unlike in PRT, it processes with CI. In these Fig. 1b node $n3$ = $\{n1, n2, n4, n5, n6\}$ are in communication range and sensor nodes $n7$, $n8$, $n9$ are out of range of node $n3$ in communication, which will not able to receive packet sent by $n3$, and these nodes are known as a hidden terminal for $n3$ as depicted in Fig. 2. The reception probability of n_j can receive a packet from n_i is expressed as:

$$p'_{ji} = p_j \cdot \prod_{k}^{N_j \backslash N_i \backslash \{n_i\}} (1 - c_k) \tag{6}$$

where c_k € $N_j \backslash N_i \backslash \{n_i\}$ starts transmission of packet as soon as node n_i complete transmission is expressed by:

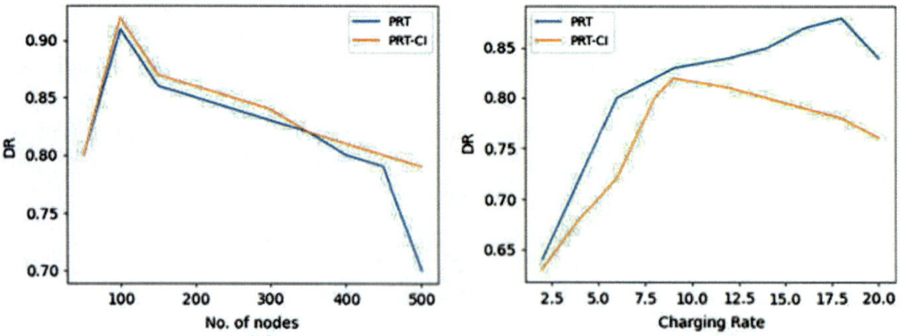

(a) (b)

Fig. 2 **a** Number of sensor nodes (n) versus delivery ratio comparison in PRT and PRT-CI and **b** charging rate (Ch) versus delivery ratio comparison in PRT and PRT-CI

$$c_k = \frac{T_{\mathrm{txk}}}{T_{\mathrm{chk}} + T_{\mathrm{rxk}} + T_{\mathrm{txk}}} \tag{7}$$

Here, equation implies that if the reception probability will reduce due to increases in the collision, at least one sensor node will accept a data packet once then probability is expressed by:

$$q_i^{'} = 1 - \prod_{k}^{S_i}\left(1 - p_{ki}^{'}\right) \tag{8}$$

Now calculating a' by reliability threshold with q'_i instead of q_i, a node will calculate retransmission time with collision consideration. The transmission of a data packet is to verify a' times, protocols like PRT and PRT-CI use additional duplicate detection technique of GR-DD. If a packet is received multiple times, then the latter packet is discarded.

3.4 Optimal Power Allocation in WSN

In WSN, the power allocation is the main subject where a number of distributed sensors transfer data through high traffic to the fusion center via different channels.

3.4.1 Routing and Flow Conservation of WSNs

Here, there requires to discuss about the rate of data transmission and data routing in network through flow conservation equation.

$$\sum_{j \in N_i}\left(r_{ij}(t) - r_{ji}(t)\right) = R_{ij}(t), \forall i \in N, j \in N_i \tag{9}$$

r_{ij} = information flow rate from sensor node.
R_{ij} = All sensor node information rate collected at source (sender) node (SR) i and to the sink node (SN)$j \in N_i$, where N = Sensor nodes set, i = Outgoing SR and j = Incoming SN.

3.4.2 Energy Cost Model

The lifetime of networks is dependable on the energy consumption from sensor node P_i and active time slot T_i of a node.

1. The energy communicated per time slot $E_{\mathrm{comm}}(t)$ consists $E_{\mathrm{RX}}(t)$ and $E_{\mathrm{TX}}(t)$.
2. The computed energy consists $E_{\mathrm{PR}}(t)$ and $E_{\mathrm{SN}}(t)$.

Assume, $E_B(t) \geq 0$ is our sum of residual energy, hence the power consumption for t is:

$$p_i(t) = \sum_{i \in N, j \in N_j} r_{ij}(t) E_{\mathrm{TX}}(t) + \sum_{i \in N, j \in N_i} r_{ij}(t) E_{\mathrm{RX}}(t) + \sum_{i \in N, j \in N_i} R_{ij}(t) E_{\mathrm{PR}}(t) + \sum_{i \in N, j \in N_i} R_{ij}(t) E_{\mathrm{SN}}(t) \tag{10}$$

$P_i(t)$ = Power consumption in (t),
$E_{\mathrm{TX}}(t)$ = Transmitted energy (t),
$E_{\mathrm{RX}}(t)$ = Receiver energy (t),
$E_{\mathrm{PR}}(t)$ = Processing energy (t),
$E_{\mathrm{SN}}(t)$ = Sensing energy (t).

For transmitting a single bit of data by transmitter energy from $i \in N$ to $j \in N_i$ over a distance (d) is

$$E_{\mathrm{TX}} = a_1 + a_2 * d_{ij}^{y} \tag{11}$$

where y = Path loss exponent, d = Distance of communication, a_1 and a_2 = Constants depending upon T_c.

3.4.3 Loss in Packet and Retransmission of Data

Here, the analysis of data packet error: (ARQ scheme), data packet is decrypted by a process known as cyclic redundancy check (CRC) by

$$P_{\mathrm{e}}^{\mathrm{ARQ}} = 1 - (1 - P_{\mathrm{b}})^{L_{\mathrm{p}}} \tag{12}$$

L_{p} = Packet length, P_{e} = Loss rate of packets, P_{b} = Bit error rate.

Analysis of packet error (ECC scheme), here the packets transmission is between the sinks node and sensor node. The sink node packet loss rate is:

$$P_{\mathrm{e}}^{\mathrm{ECC}} = 1 - \left(1 - \sum_{i=e+1}^{n} \binom{n}{i} P_{\mathrm{b}}^{i} (1 - P_{\mathrm{b}})^{n-i}\right)^{\frac{L_{\mathrm{p}}}{k}} \tag{13}$$

P_e = Probability of on event, here the no. of retransmission is done by

$$E(\mathrm{Tr}) = \frac{1}{(1 - P_e)} \tag{14}$$

where $E(\mathrm{Tr})$ = Expected no. of retransmission, P_e = Loss rate of packets and packet loss rate for h—hop scenario while taking each node transmission as independent.

$$E(\mathrm{Tr}, h) = \frac{h}{(1 - P_e)} \quad \text{where}\, h = \text{no. of hops.} \tag{15}$$

3.5 Collision Detection

The combination of carrier sensing and collision avoidance is the best way of dealing with packet collisions in the MAC layer in the wireless networks. But there is a con in this method because if the collision avoidance strategy failed, then the combination process will not work, which returns the data frames. But these are corrupted and are still transmitted, then the results in a loss of channel bandwidth. The arrival time of data frames expected end of the difference of transmission.

Hence to overcome this problem, a new wireless MAC protocol is proposed which is capable of collision detection. This MAC protocol has an increase in performance and requires no time synchronization.

3.6 Collision Avoidance

Multiple accesses with collision avoidance is a medium access control (MAC) layer protocol which is used in wireless networks to solve the hidden terminal problems. Collision avoidance is a basic task of the MAC protocol. Then, the number of senders may be sending data at the same time to the destination end that needs to avoid collisions through MAC protocol (802.11). In collision avoidance, it is required at the source end, where the rate of transmission data from the source is boosted up which in return the arrived period of the data packets is changed. Due to that, collisions will be avoided. Many medium access control (MAC) protocols are proposed for wireless networks which are based on collision avoidance between sender and receiver.

3.7 Packet Reception Probability

A sender node uses its neighbor node to data transmission and uses the packet reception probability for data retransmission to the other nodes. It also set a scheduled time to retransmission of message after exchanging the information probability among nodes. The reception probabilities of their neighbor nodes are updated. In case of WSNs, entire node should have sufficient energy to exchange the data for packet reception probability due to energy harvesting; we should have use a special time interval to satisfy the model or protocol [14]. Considering the delivery ratio (DR), delay have to consider for performance of the system. To find all the performance of the system, we have to simulate the PRT and PRT-CI Protocols and compare with them.

4 Simulation Setup and Result Analysis

Our simulation setup is consisting of 500 × 500 m area sizes with 300 nodes are deployed in the deployment area. The sink node has also located in the center of the deployment area. The sink node has sufficient power to supply and receive the information exchange. We assume if the collision occurs, there is no packet can be received and in the data packet loss is not allowed in the whole networks within the range of communication. We set the threshold value T_h (Reliability threshold) = 0.9 and need to set the depth of period to be 0 to Θ, where the sink node is placed at the center area. So, this sink can be ready to receive a packet which is powered by a steady power supply. In this simulation process, assume that no data packet loss within the range of transmission, but the no accurate data packet is a collision in the network means no correct packet can be received. Considering the power consumption for transmission Ptx = 76.2 Mw, power consumption for receiving Prx = 83.1 mW, transmission power = −3.0 dbm, receiving power = −85.0 dbm, and the simulation time = 1000 s. The data size is set to 800 bits, and the transmission rate is 250 kbps with the frequency f = 2.4 GHz. The simulation result depends on different factors as below.

Number of Sensor Nodes (*n*) versus Delivery Ratio: If the number of sensor nodes changes, then the delivery ratio will be changed. Each node changes its density by receiving their optimal number of data retransmission through PRT and PRT-CI. The depicted graph in Fig. 2a shows that PRT-CI gives the better result as compared with PRT; if the collision does not repeatedly occur, then the collision is not considered as an effective. However, the delivery ratio also decreases by increasing the packet size.

Delivery ratio versus charging rate (Ch) of nodes: The performance of a system is represented by the charging rate as depicted in Fig. 2b. When each protocol achieves a higher charging rate the delivery ratio also increases, but after some time,

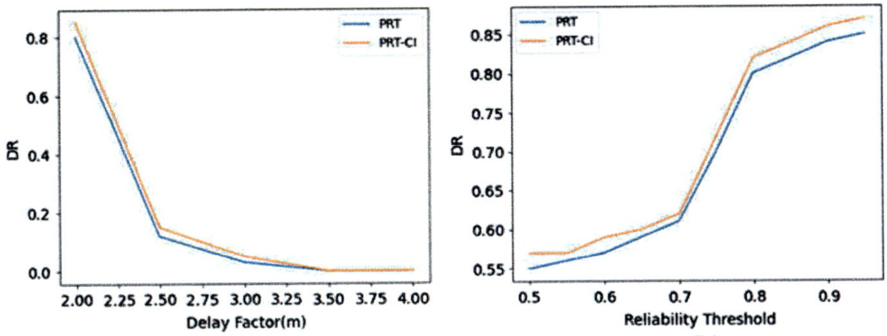

(a) **(b)**

Fig. 3 **a** Delay Factor (*m*) versus Delivery Ratio of nodes comparison in PRT and PRT-CI, and **b** Reliability threshold versus delivery ratio of nodes comparison in PRT and PRT-CI

the delivery ratio will decrease for the charging time. The sensor nodes charge quickly with higher duty cycle when the charging rate increases and delivers more packets to the sink.

Delivery Ratio versus Decay Factor (*m*): Each sensor node sets its information to their neighbor nodes within the communication range through PRT and PRT-CI as depicted in Fig. 3a. When the protocol gives a higher delay ratio, the delivery ratio also decreases. When the value of m increases, it is not certain to deliver the packets in the protocol.

Delivery Ratio versus Reliability Threshold (T_h): If the nodes reliability threshold increases, the data delivery ratio also increases in PRT and PRT-CI as depicted in Fig. 3b. When the T_h value increases, the node retransmits the same packet to achieve higher reliability.

Delay versus No. of Sensor nodes: Some sensor nodes require quickly gathering data from the WSNs if we put more sensor nodes in the deployment region. If the number of sensor nodes increases, the delay factor also increases.

In Fig. 4 as depicted, to evaluate the delay with various nodes. The delay decreases when the rate of charge increases in each protocol due to the small interval of charging time.

5 Conclusions

In this proposed paper, we analyzed the data packet loss in wireless sensor networks due to more energy consumption and collision in transmission. So that, this issue is to overcome by enhancing the data collection energy efficiency in between neighboring nodes, which proposed two data packet gathering protocols such as probabilistic retransmission (PRT) and probabilistic retransmission with collision interference (PRT-CI). Similarly, the data packets are not collided with each other

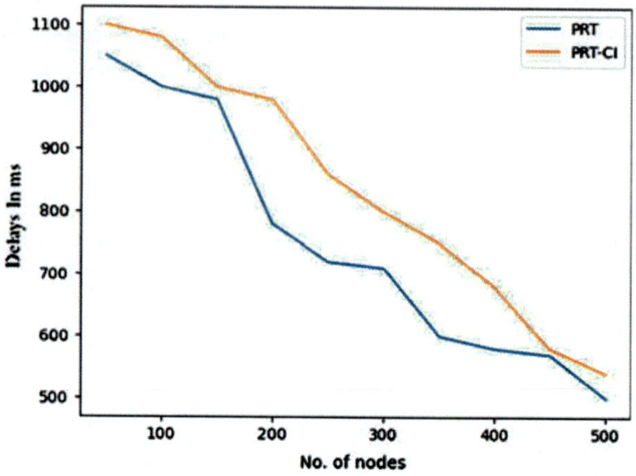

Fig. 4 No. of sensor nodes versus delay (ms) comparison in PRT and PRT-CI

due to the usages of optimal power allocation in the neighboring node. Here, the PRT and PRT-CI are evaluated in a sender node to find the probability of receiving a data packet from its own enable or active periods and also its archive higher delivery ratio through the simulation results of these two protocols. In the future analysis, it is based on how to define the best possible value of T_h and τ for various sensor network fields and needs to enhance the numerical analysis of the WSNs models and protocols.

References

1. Tse, D.N.C., Wornell, G.W.: Cooperative diversity in wireless networks: efficient protocols and outage behavior. IEEE Trans. Inform Theory **50**, 3062–3080 (2004)
2. Laneman, J.N., Sirimoungpirat, W.P., Han, Z., Liu, K.J.R.: Lifetime maximization by cooperative sensor and relay deployment in wireless sensor networks. IEEE J. Sel. Areas Commun. **25**(2), 306–317 (2007)
3. Bletsas, A., Khisti, A., Win, M.Z.: Opportunistic cooperative diversity with feedback and cheap radios. IEEE Trans. Wireless Commun. **7**(5), 1823–1827 (2008)
4. Liu, L., Ge, H.: Space-time coding for wireless sensor networks with cooperative routing diversity. In: Proceedings Asilomar conference on signals, systems & computers, pp. 1271–1275 (2004)
5. Gokturk, M., Gurbuz, O.: Cooperative MAC protocol with distributed relay actuation. In: Proceedings IEEE wirless communications and networking conference (WCNC'09) (2009)
6. Gokturk, M., Gurbuz, O.: Cooperation in wireless sensor networks: design and performance analysis of MAC protocol. In: Proceedings IEEE international communications conference (ICC'09), pp. 4284–4289 (2008)
7. El Batt, T.A., et al.: Power management for throughput enhancement in wireless Ad-Hoc networks. In: IEEE Int'l. conference communications (ICC'00), pp. 1506–1513 (2000)
8. Priyadarshini, S.B.B., Singh, D., Panda, M.: A comparative study of redundant data minimization and event coverage in wireless multimedia sensor networks (WMSNs). In: 2019 International conference on applied machine learning (ICAML), pp. 252–257. IEEE (2019)

9. Onsy, M., et al.: Performance of WSNs under the effect of collisions and interference. Wirel. Sens. Netw. **6**, 93–103 (2014)
10. Tay, Y.C., Jamieson, K., Balakrishnan, H.: Collision-minimizing CSMA and its applications to wireless sensor networks. IEEE J. Sel. Areas Commun. **22**(6), 1048–1057 (2004)
11. Sudha, M.N., Valarmathi, M.L., et al.: Minimization of collision in energy constrained wireless sensor network. Wirel. Sens. Netw. **2009**(1), 350–357 (2009)
12. Fatemeh, M., Hossam Ahmed, M., Gadallah, Y.: Minimizing the probability of collision in wireless sensor networks using cooperative diversity and optimal power allocation. In: Wireless communications and mobile computing conference (IWCMC-2013), pp. 120–124 (2013)
13. Sadek, A., Wei, Y., Liu, K.: When does cooperation have better performance in sensor networks? In: Proceedings IEEE the 3rd sensor and Ad Hoc communications and networks (SECON'06), pp. 188–197 (2006)
14. Singh, D., Kumar Pattanayak, B.: Markovian model analysis for energy harvesting nodes in a modified opportunistic routing protocol. Int. J. Electron. (2020). https://doi.org/10.1080/00207217.2020.1756444

Segmentation of Satellite Images Using Contractive Autoencoder (CAE) Aided Deep Learning Approach

Manami Barthakur and Kandarpa Kumar Sarma

Abstract The reliable extraction of contents from satellite images is a challenging problem as these images are very much useful in many real-life applications. In this paper, a deep learning-based method to extract different regions of interest (ROIs) is presented. In the work, at first, the high-resolution satellite images are processed with the contractive autoencoder (CAE) to remove the noises which are inherently present in the images during acquisition. The output of the CAE is taken as input to train a segmentation network which is a deep neural network. Finally, the output of the segmentation network is used to train both supervised and unsupervised classifiers to extract different ROIs from the satellite images. Experimental results show that the method is reliable for real-time application.

Keywords Denoising · Semantic segmentation · Deep learning · Satellite images

1 Introduction

The interpretation of satellite imagery datasets is a challenging problem due to the increase in the availability of these data in recent times. A good understanding and insights of the retrieved information from the images obtained from a satellite are necessary for many real-life applications such as weather prediction, flood prediction, deforestation, etc. [1]. To analyze a satellite image properly, the extraction of the different regions in the image is a crucial and important task. The automating of extraction of the different regions of these large-scale data has been addressed by the researchers in different ways over the years. In the last few years, machine learning (ML) and deep learning (DL)-based techniques have proven to be successful and strong prospects for this purpose. The state of the art, DL-aided

M. Barthakur (✉) · K. K. Sarma
Gauhati University, Guwahati, Assam 781014, India
e-mail: manamibarthakur@gmail.com

K. K. Sarma
e-mail: kandarpaks@gmail.com

S. Das and M. N. Mohanty (eds.), *Advances in Intelligent Computing and Communication*, Lecture Notes in Networks and Systems 202,
https://doi.org/10.1007/978-981-16-0695-3_62

approaches have been adopted for satellite image analysis during the last decade [2]. In [3], the semantic segmentation using the DeepLab v3 + model and VGG-16 [4] of farm areas in low-resolution satellite images was presented in [5]. Again, in [6] the encoder–decoder architecture for SegNet [7]-based network was used to identify buildings, vegetation, and buildings on high-resolution satellite images. Similarly, in [8], the U-net architecture was used to perform semantic segmentation of clouds in satellite images by utilizing the transfer learning method.

In this work, a deep learning-based method to extract different regions of interest (ROIs) is presented. In the work, at first, the high-resolution satellite images are processed with the contractive autoencoder (CAE) to remove the noises which are inherently present in the images during acquisition. The output of the CAE is taken as input to train a segmentation network which is a deep neural network. Finally, the output of the segmentation network is used to train both supervised and unsupervised classifiers to extract different ROIs from the satellite images. Here, the DeepGlobe Land Cover Classification Challenge [9] dataset is used to validate the results obtained. In the following sections, at first, the proposed method is discussed. Then in Sects. 3 and 4, the results obtained are discussed and finally, the conclusion is presented in Sect. 4.

2 The Proposed Method

In this section, the proposed method is discussed briefly. At first, the architecture of the proposed system and finally the methodology is discussed.

2.1 *Architecture of the Network*

In the proposed approach, the CAE is used for denoising the satellite images. For semantic segmentation, a segmentation network is designed and finally three classifiers are used to extract the required ROIs from the satellite images. The detailed descriptions of all these networks are presented here.

Contractive autoencoder: The contractive autoencoder (CAE) [10] is used to reduce the noise present in the satellite images during acquisition. It also makes the output less varying due to the variation in the input. Here, the input, I, is encoded by the CAE using function g, and to make the output identical to the input, the encoded values $g(I)$ are decoded with a function h. As the objective of the CAE is to obtain a vigorous representation of the learned pattern, therefore, to make the output less sensitive to the small variations in the input, a penalty term called the Frobenius norm of the Jacobian matrix (J_f) added to the loss function of the CAE. J_f is the sum of the square of all the elements, and it is calculated with respect to the inputs for the hidden layers. Therefore, the loss function of the CAE is given by,

$$l = |I - h(g(I))| + \lambda \|J_f(I)\|_f^2 \tag{1}$$

$$\|J_f(I)\|^2 = \sum_{ij} \left(\frac{\partial h_j(I)}{\partial I_i} \right)^2 \tag{2}$$

where λ is the coefficient of weight decay which controls the relative importance of regularization.

Segmentation network: For the semantic segmentation of the satellite images, the Deeplab-v2 [11] framework with ResNet-101 [12] model is used as the segmentation network. To match the output of the network with the input image size, the strides of the last two layers of the ResNet-101 [12] model are taken as 1. In the conv4 and conv5 layers of the network, to make the receptive field enlarge, dilated convolution [13] is applied with 2 and 4 strides, respectively. Again as the architecture of ResNet 101 [12] does not have an upsampling layer before the final classification layer, therefore, to match the size of the output of the segmentation network with the input image and to make the resolution better, the last classification layer of the ResNet-101 is removed and an upsampling layer with the softmax layer is applied as the final layer of the network. These variations also make the proposed method robust [14].

Classifier used: The classifiers used in this work provide better reliability in the segmentation process by accurately identifying the regions of the satellite images to be extracted. In this work, both supervised and unsupervised classifiers are used to check the robustness of the work. As an unsupervised classifier, the fuzzy c-means (FCM) [15] is used. Based on the application and data, different similarity measures such as intensity, connectivity, and distance are used in the clustering process to generate a decision cluster. Again, as the supervised classifier, multi-layer perceptron (MPL) [16] and softmax classifier [14] are used in this work. Both the MLP and softmax classifiers are trained with the backpropagation algorithm. The classifiers also add an accurate and strengthening mechanism to contribute to the effectiveness of the work. While dealing with a large set of images to be accurately segmented, these classifiers are effective and reliable in the proposed work.

2.2 *Methodology*

In this work, the input to the system is the dataset of high-resolution satellite images. The dataset is obtained from DeepGlobe land cover classification challenge [9]. The dataset consists of a total of 1146 number of images of size 20,448 × 20,448 which are high-resolution satellite images of rural areas. The label map annotation image for each of the satellite images are also there in the dataset. The images are divided into 803 (70% of total images) number of training images,

171 (15% of total images) number of images for validation, and 171 (15% of total images) number of test images. In the images, forest land, agriculture land, rangeland, urban land, barren land, water, and unknown region are the class distribution considered for segmentation in the proposed system. At first, as shown in the block diagram in Fig. 1, the input images are fed to the CAE. The CAE consists of one input layer, one hidden layer, and one output layer. The loss function used to train the CAE is given by Eq. 1. The output images of the CAE are the de-noised reconstructed images. Next, the de-noised images obtained from the CAE are taken as the input to the segmentation network. Here, the network is a Deeplab-v2 [11] framework with ResNet-101 [12]. The network is trained to obtain the semantic segmentation of the satellite images. The output images of the network are the maps of class probabilities with semantic categories. As explained earlier, the last layer of the segmentation network is the softmax layer, and therefore, the probability of the xth pixel i which belongs to the kth class and in cth batch is given by,

$$p_k(i) = \frac{\exp\left(\omega_k^T l(\alpha, i)\right)}{\sum_{j=1}^{k} \exp\left(\omega_j^T l(\alpha, i)\right)} \tag{3}$$

where w_k weight matrix for the kth class of the last convolutional layer, $l(\alpha, i)$ is the learned deep features of the network, and α are the other parameters related to the last convolutional layer of the segmentation network [14].

Thirdly, to train the MLP and the softmax classifier, the targets are fixed by selecting the particular ROIs from the labeled maps of ground truth images. The selection is done by keeping the pixel values of the ROIs as 1 and other pixel values as 0.

Subsequently, the semantic maps obtained from the segmentation network are taken as input to the classifiers. To extract a specific ROI from the output of the segmentation network, one classifier is used. That means when the unsupervised classifier is used, for the seven ROIs, seven FCM classifiers are used. Similarly, when supervised classifiers are used; at first, seven MLP classifiers are trained to obtain for the seven ROIs, and then, seven softmax classifiers are trained for each of the ROIs. In the block diagram in Fig. 1, only three classifiers are shown for three

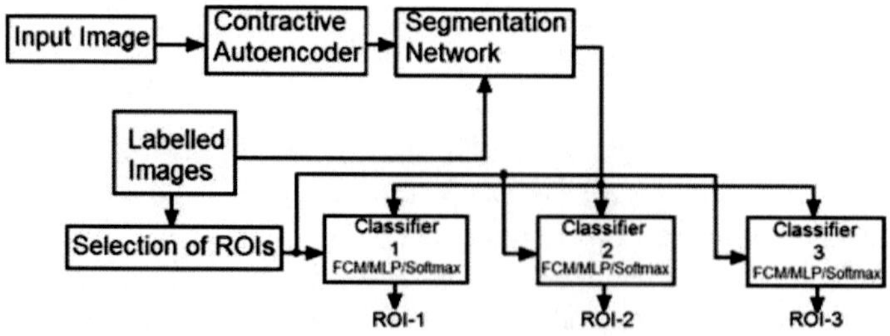

Fig. 1 The block diagram of the proposed method

ROIs for better understanding. For the FCM classifier, the output of the segmentation network is at first converted $L \times a \times b$ color space. This is done as the $L \times a \times b$ color space quantifies the visual difference of the colors in the images. Then, to cluster the required ROIs, the FCM algorithm is applied to the images. In the algorithm, the grouping of the pixel values of the ROIs in the images is done by minimizing the Euclidean distance between the cluster center and the pixel. The outputs of the algorithm are the index values to the clusters, and therefore, with these values, every pixel is labeled in the image. Thus, the FCM classifier outputs the required ROIs. Again, when the MLP classifier is used, it is trained with two hidden layers. The training is done with the BP algorithm. When the softmax classifier is used, it is trained with the BP algorithm and with stochastic gradient descent (SGD). The different classifiers are used to check which classifier is suitable for the proposed system and makes the system robust. Finally, the outputs obtained from the classifiers are the required ROIs of the satellite images.

3 Results and Discussion

In the proposed method, the CAE is trained with one hidden layer, and the number of the hidden neuron is taken as 75 as discussed earlier the loss is calculated with Eq. 1, and here, the value of the weight regularization parameter is taken as 0.005. The CAE is trained for a maximum of 5000 iterations. The peak signal-to-noise ratio (PSNR) value between the input satellite image, and the output reconstructed image is calculated after an interval of 1000 iteration. In Fig. 2, a graph between the PSNR value, and the iteration is shown.

The output of the CAE obtained after 5000 iterations is taken as the input to the segmentation network. At first, the segmentation network is trained with the input as the original satellite image. Then, the value of mean intersection over union (IOU), structural similarity index (SSIM), and pixel accuracy is calculated. Then, the network is trained with the input as the output of CAE. Then again the values mean IOU, SSIM, and pixel accuracy are calculated. The comparisons of these values are shown in Table 1. It is seen that when the segmentation network is

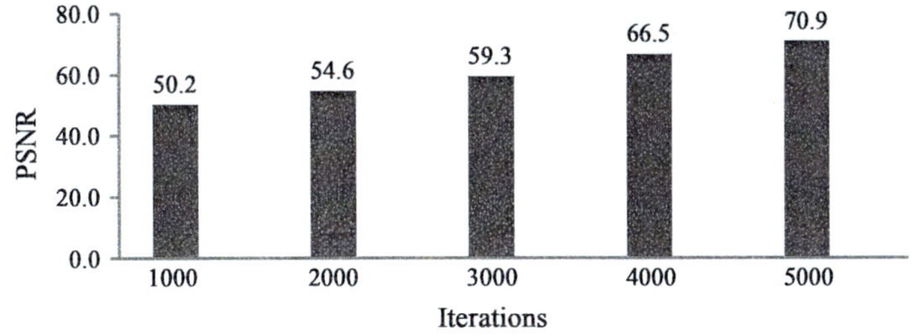

Fig. 2 Iteration versus PSNR values for CAE

Table 1 Comparison of the results obtained for output of the segmentation network with input as original image and output of CAE

Input to the segmentation network	Mean IOU	SSIM	Pixel accuracy
Original image	80.35	0.8458	81.73
Output of CAE	88.42	0.9035	89.78

trained with the output of the CAE, the mean IOU is 10.04%, SSIM is 6.8%, and pixel accuracy is 9.8% better than the output of the network when trained with original images. The network is trained with the SGD optimization method, and the weight decay is set to 10^{-5}. The momentum is taken to be 0.9, and the initial learning rate is set as 2.5×10^{-5} [14]. In Fig. 3, the comparison of the output of the segmentation network with input as the original images and the output of CAE is shown.

With the output of the segmentation network, the classifiers are trained. At first, the FCM classifier is used to obtain different ROIs from the output of the segmentation network. Then, the supervised classifiers, the MLP, and the softmax classifier are trained to obtain the required ROIs. The MLP is trained with a double hidden layer where the numbers of hidden neurons for the first, and second hidden layers are taken as 25 and 50, respectively. It is trained with the Levenberg–Marquardt (LM) training function. The softmax classifier is trained with the SGD method as discussed earlier. Both the MLP and the softmax classifier are trained for 1000 iterations. In Table 2, the comparison of mean square error (MSE) and SSIM values for different regions of the output of all the classifiers are shown. Table 2, it is seen that the softmax classifier gives a better result than the other classifiers. It is seen that the MSE, SSIM, and pixel accuracy of softmax is 6.7, 5.4, and 3.2%,

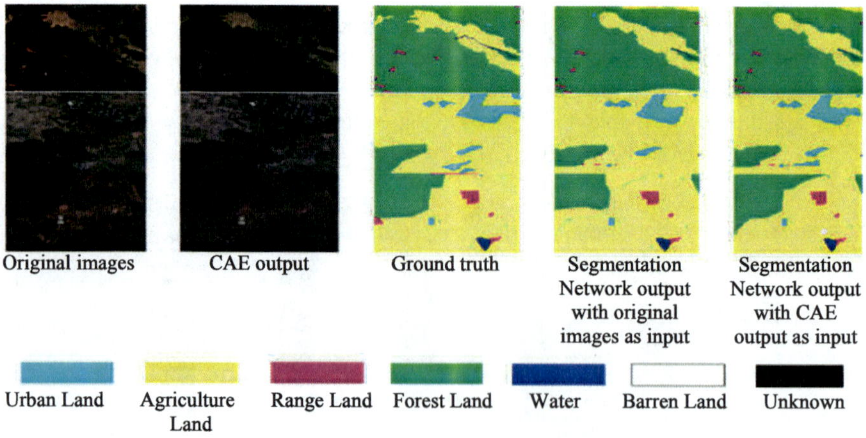

Fig. 3 Comparison of the output of the segmentation network with input as original image and output of CAE

Table 2 Comparison of MSE and SSIM values for different regions of the output of the classifier

Regions	FCM			MLP			Softmax		
	Time required in seconds	MSE	SSIM	Time required in seconds	MSE	SSIM	Time required in seconds	MSE	SSIM
Urban land	16.65	0.0955	0.8882	41.61	0.0053	0.8996	56.26	0.0043	0.9043
Agricultural land	21.25	0.0360	0.8846	51.29	0.0096	0.8986	46.41	0.070	0.9042
Range land	26.71	0.0666	0.8886	67.78	0.0074	0.8992	47.73	0.0054	0.9067
Forest land	29.33	0.0685	0.8844	75.37	0.0068	0.8984	63.42	0.0047	0.9059
Water	36.93	0.0943	0.8885	68.63	0.0074	0.8983	58.10	0.0051	0.9053
Barren land	46.04	0.0349	0.8884	96.14	0.0096	0.8991	102.23	0.0082	0.9036
Unknown region	35.31	0.0295	0.8881	85.31	0.0107	0.8994	91.23	0.0092	0.9029

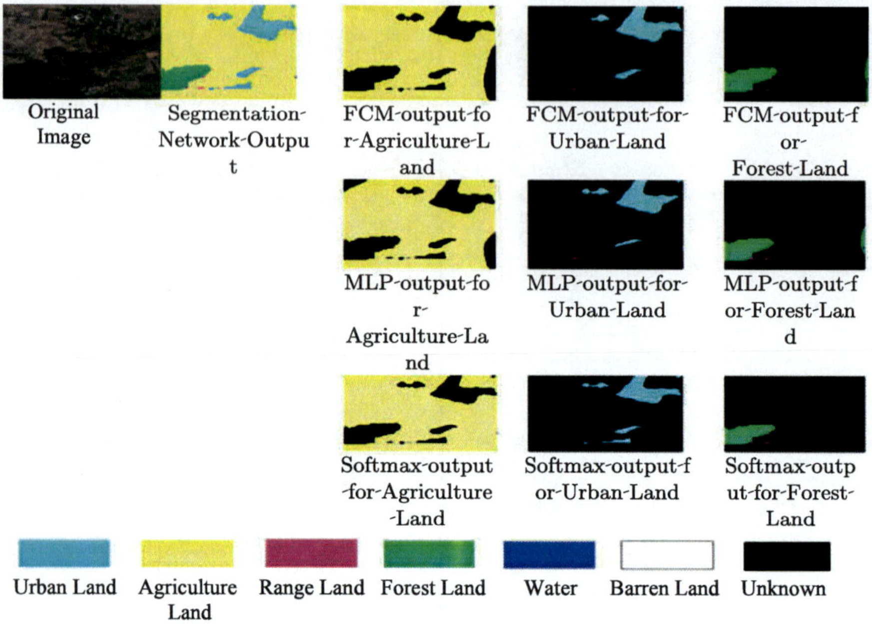

Fig. 4 Comparison of the output of the FCM, MLP and softmax classifier

respectively, better than the MLP and 8.4, 6.3, and 4.2%, respectively better than the FCM classifier. Figure 4 shows the comparison of the output of MLP, FCM, and softmax classifier.

4 Conclusion

A deep learning-based method for the extraction of different regions of satellite images is presented in this paper. At first, the high-resolution satellite images are processed with the contractive autoencoder (CAE) to remove the noise which is inherently present in the images during acquisition. The output of the CAE is taken as input to train a segmentation network which is a deep neural network. It is seen that the segmentation network when trained with input as the output of CAE gives a better result than with input as original images. Finally, the output of the segmentation network is used to train both supervised and unsupervised classifiers to extract different ROIs from the satellite images. A comparison of the results obtained from the classifiers is also presented. Among the classifiers, the softmax classifier gives better results. Experimental results show the method is reliable for real-time applications.

References

1. Saifi, M.Y., Singla, J., Nikita: Deep learning based framework for semantic segmentation of satellite images. In: Fourth international conference on computing methodologies and communication (ICCMC), Erode, India, pp. 369–374 (2020)
2. Barthakur, M., Sarma, K.K.: Deep learning based semantic segmentation applied to satellite image. In: Hemanth, J., Bhatia, M., Geman, O. (eds.) Data visualization and knowledge engineering. Lecture notes on data engineering and communications technologies, vol. 32. Spriger, Cham (2020)
3. Chen, L.C., Zhu, Y., Papandreou, G., Schroff, F., Adam, H.: Encoder-decoder with atrous separable convolution for semantic image segmentation. In: Ferrari, V., Hebert, M., Sminchisescu, C., Weiss, Y. (eds.) Computer vision—ECCV 2018. ECCV 2018. Lecture notes in computer science, vol. 11211. Springer, Cham (2018)
4. Simonyan, K., Zisserman, A.: Very deep convolutional networks for large-scale image recognition. arXiv:1409.1556. Available [online] https://arxiv.org/abs/1409.1556 (2014)
5. Sharifzadeh, S., Tata, J., Sharifzadeh, H., Tan, B.: Farm area segmentation in satellite images using deeplab v3 + neural networks. In: Hammoudi, S., Quix, C., Bernardino, J. (eds.) Data management technologies and applications. DATA 2019. Communications in computer and information science, vol. 1255. Springer, Cham (2020)
6. Younis, M.C., Keedwell, E.: Semantic segmentation on small datasets of satellite images using convolutional neural networks. J. Appl. Remote Sens. **13**(4), 046510 (2019)
7. Badrinarayanan, V., Kendall, A., Cipolla, R.: SegNet: a deep convolutional encoder-decoder architecture for image segmentation. IEEE Trans. Pattern Anal. Mach. Intell. **39**(12), 2481–2495 (2017)
8. Gonzales, C., Sakla, W.: Semantic segmentation of clouds in satellite imagery using deep pre-trained U-nets. In: IEEE applied imagery pattern recognition workshop (AIPR), Washington, DC, USA, pp. 1–7 (2019)
9. Demir, I., Koperski, K., Lindenbaum, D., Pang, G., Huang, J., Basu, S.: DeepGlobe 2018: a challenge to parse the earth through satellite images. In: Proceedings IEEE/CVF conference computer vision pattern recognition workshops (CVPRW), pp. 172–181 (2018)
10. Aamir, M., Mohd Nawi, N., Wahid, F., et al.: A deep contractive autoencoder for solving multiclass classification problems. Evol. Intel. (2020)
11. Chen, L.C., Papandreou, G., Kokkinos, I., Murphy, K., Yuille, A.L.: DeepLab: semantic image segmentation with deep convolutional nets atrous convolution and fully connected CRFs. IEEE Trans. Pattern Anal. Mach. Intell. **40**(4), 834–848 (2018)
12. He, K., Zhang, X., Ren, S., Sun, J.: Deep residual learning for image recognition. In: Proceedings IEEE conference computer vision pattern recognition (CVPR), pp. 770–778 (2016)
13. Yu, F., Koltun, V.: Multi-scale context aggregation by dilated convolutions. In: Proceedings of the international conference on learning representations (ICLR), p. 1 (2016)
14. Barthakur, M., Sarma, K.K., Mastorakis, N.: Modified semi-supervised adversarial deep network and classifier combination for segmentation of satellite images. IEEE Access **8**, 117972–117985 (2020)
15. Patra, D., Das, M.K., Pradhan, S.: Integration of FCM PCA and neural networks for classification of ECG arrhythmias. IAENG Int. J. Comput. Sci. **36**(3), 1–5 (2010)
16. Barthakur, M., Sarma, K.K., Mastorakis, N.: Neural network methods for image segmentation. In: Proceedings of the 2nd international conference on applied physics, system science computer (APSAC), pp. 176–182 (2017)

Role of Deep Learning in Screening and Tracking of COVID-19

Arya, Lambodar Jena, Rajanikanta Mohanty, and Ramakrushna Swain

Abstract As we head toward more than 34 million cases of Coronavirus disease of 2019 and a million deaths worldwide, there is an urgent need to control the spread and recover the patients as soon as possible. Technology has resolved many complex problems of human beings. Machine learning and deep learning have been intensely used in areas like medical image diagnosis, drug discovery, and manufacturing. It has achieved success in complex problems like the detection of cancerous tumors, skin cancer, and diabetic retinopathy. In this study, considering the power and ability of deep learning, we will study the possibilities and role of deep learning in detecting COVID-19. Several deep learning models can be used to monitor, detect, and predict the spread of the virus. The convolutional neural network has great potential to detect COVID-19 as it is currently used to detect pneumonia using chest X-ray images. This knowledge is used to build a model to detect the presence of COVID-19 which gives a high accuracy with minimum error percentage. The model is highly capable of detecting patients with COVID-19. Deep learning can be implemented widely to tackle this pandemic.

Keywords Deep neural network · Deep learning architectures · Convolutional neural network · Chest X-ray images · Max pooling

---

Arya
School of Electronics and Computer Science Engineering, Kalinga Institute of Industrial Technology (KIIT), Deemed to be University, Bhubaneswar, Odisha, India
e-mail: 1805559@kiit.ac.in

L. Jena (✉)
Department of Computer Science and Engineering, Siksha 'O' Anusandhan (Deemed to be) University, Bhubaneswar, Odisha, India
e-mail: lambodarjena@soa.ac.in

R. Mohanty
Department of Information Technology, PSIT, Kanpur, Uttar Pradesh, India
e-mail: rkm.bbs@gmail.com

R. Swain
Department of Computer Science and Engineering, Silicon Institute of Technology, Bhubaneswar, Odisha, India
e-mail: rkswain24@gmail.com

S. Das and M. N. Mohanty (eds.), *Advances in Intelligent Computing and Communication*, Lecture Notes in Networks and Systems 202,
https://doi.org/10.1007/978-981-16-0695-3_63

1 Introduction

In the current scenario, deep learning has emerged out as an integral part of the family of machine learning. It has been built based upon two concepts, namely artificial neural networks and visual representation of data. This process of learning can be classified under three categories: unsupervised, semi-supervised, and supervised [1]. Some major deep learning architecture comprises deep belief networks (DBN), deep neural networks (DNN) and convolutional neural networks (CNN) and recurrent neural networks (RNN). These are vastly used in essential fields such as vision recognition (machine vision, computer vision), speech recognition, machine translations, social networking filters, bio-informatics, design of drugs for medical purposes, analyzing images, inspection of materials, game programs, and many other [2]. The use of deep learning in these areas has proven out to be very efficient and has even surpassed human expert performances in many cases. In artificial neural network (ANN) when there are multiple layers introduced between input and output layers, it forms a deep neural network (DNN). The input and output could be related either linearly or nonlinearly. DNN's task is to find the correct mathematical manipulation techniques to extract output from the given input. As there are multiple layers in between, the DNN works on all the probabilities obtained after each stage or layer.

As mentioned earlier, there are several layers in between input and output. The term deep refers to this aspect of the neural networks. It signifies how deep or how many intermediate layers are present. Deep learning models also known as deep neural networks (which are used for such tasks mentioned here) as the name suggests are the combination of deep learning methods that use neural network architecture. One major advantage here is there is no necessity of manual feature extractions (which has higher probabilities of errors/missing values) as the model here studies and trains itself on its own from the huge sets of labeled data provided to it.

Raw data is fed as input to the system, and using multiple layers of deep learning network algorithm, the information or the pattern's decryption is done. The algorithms work in a manner such as identifying the patterns present in the raw data alike how a human brain when given data would have analyzed and decrypt it. When numerical patterns are observed by the algorithm, they use vectors to store them, and they are treated as inputs given in real time. These neural networks help us to re-arrange the data based upon similarities (if found any) between two subsets or more, extract the distinct attributes of the raw data provided, and also, data classification is performed when we have the identifications from already present data. The three major tasks performed by the algorithms are learning, classification, and regression. These are similar in a manner to how machine learning operates but only with added features to it. The hidden layers or intermediate layers are referring to the deep learning, and the algorithms used to traverse between these layers are known as neural networks. The feature which gives it superiority from machine learning is when we encounter unsupervised data. We may encounter mistakes or errors, less time-efficient, less cost-efficient methods if the tasks are performed by a human. Whereas on the other hand, deep learning overcomes these problems faced and is more optimal to use. Figure 1 depicts a sample deep neural network model.

2 Deep Learning in COVID-19

As we are all familiar with the ongoing pandemic situation caused due to the COVID-19, we would require every piece of technology which will help us beat this disease and ensure the safety and harmony of people. To defeat the pandemic, we first need to study and analyze the situation we are currently in and what better alternative there is other than the use of machine learning and artificial learning to study the situation. If we could find the pattern of how this deadly virus is spreading, we could probably break the chain of the transition of this virus from one to another at this rapid rate. Also, we plan to use our knowledge in these fields to perform the diagnosis of patients suffering from this and formulate ways to perform testing on people to identify who needs treatment. As we discuss, there are more than 30,000,000 cases all over the world out which there have been more than 1,000,000 reported dead already. It is high time we come up with a solution to this pandemic. Using deep learning in our project, we would make these tasks easier and results accurate. With the ability to learn on its own the multiple layers, deep learning framework provides mathematical models that are much expected from a human brain [3]. But being automated it reduces errors, does enormous amounts of data transfer tasks at a faster rate, and does not miss any valid point that we may overlook. The presented method used for diagnosing is RT-PCR, that is, reverse transcription-polymerase chain reaction. But it has several constraints such as its

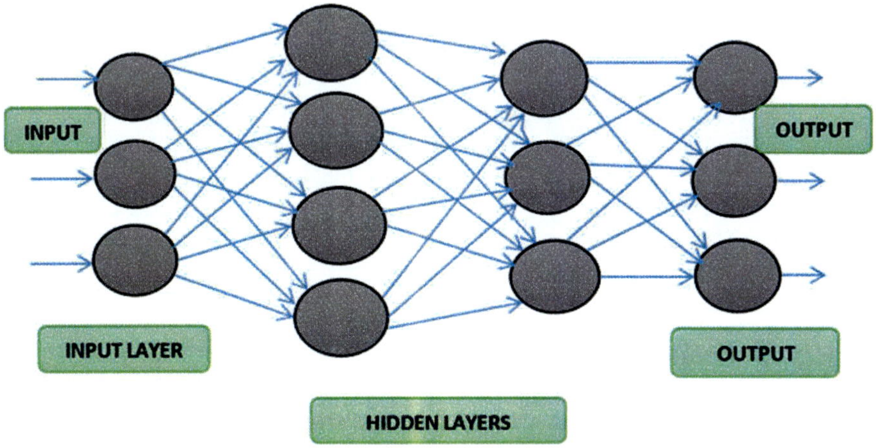

Fig. 1 Typical deep neural network

highly sensitive nature and unavailability everywhere (especially epidemic areas) due to a shortage of kits required for testing. Due to this, accurate evaluation of everyone is not taking place. Some other constraints are

- It is very costly, and the chemicals used are not available everywhere.
- Its inaccurate testing causes many patients to be left untreated.
- This leads to the spread of the virus at a rapid rate and increases the difficulty to control it.
- Does not tell the priority or the stage at which the patient is. If known, the treatment can be given accordingly. i.e., higher priority to patients who are more affected by this virus.

Here are some models that can help us break through this pandemic and resolve the issues which we discussed.

2.1 Convolutional Neural Network (CNN)

A convolutional neural network is built up of multiple layers within it for optimal performance. This uses perceptrons for supervised learning. It is also used to analyze the data which is fed into the system. CNN has its main application revolving around images or visual representations in a more obvious manner. Every image that is fed into the system is treated by CNN as a complex mixture of tensors, a two-dimensional matrix composed of numeric figures, and many other dimensions included in it.

To describe briefly what a tensor exactly is we can say tensor is when multiple (almost infinite times) number of arrays are nested among themselves. Tensors again have their own dimension. In the case of images, we refer to them as 4D tensors. To explain in-depth, consider a point as 0 dimensions, a line having one, 1 plain sheet having 2, and a box (made up of multiple such plain sheets) of 3 dimensions. Now, when we stack up all these 3D objects where each has unique features attached to it, we form a 4D tensor.

As CNNs have multiple layers, to name some of them, they are fully connected, pooling, convolutional, and normalization. When the system is fed an image, it assigns specific priorities to various aspects of the image and then sends it to multiple layers of preprocessing. As we go deep into layers, the features which they store or asses also get more complicated.

2.2 Deep Belief Network (DBN)

Deep belief network (DBN) is a deep learning network model where we depend on probabilities while considering algorithms. As the name suggests (deep belief), we presumably consider probabilities for a generative self-learning networked model. Here, we consider undirected and directed graphs to study the model. The pre-training comprises two parts. First a learning stage before processing and second a feed-forward network which has inputs of the outputs formed due to preprocessing. As mentioned, DBN consists of multiple such layers here, and each layer further has sub-units hidden in them. The algorithm which is followed works on a greedy approach starting from the bottom-most layer moving upwards. The main reason we use DBN models is that they are energy effective or energy-based, and they can even help us to process data which has not been labeled as we have a preprocessing stage.

2.3 *Long Short-Term Memory (LSTM)*

The long short-term memory algorithm or better known as LSTM algorithm helps us in cases where the network layers have to be processed recurrently. It comes under a type of recurrent neural network (RNN) that allows us to update weights of hidden layers (that are passed to subsequent function calls) without making the next recursive calls disrupted or malfunction. The patterns on which analysis or processing is done are available in the memory of the system for a longer duration, so we can recall the whole data or some subset of it. This makes reusability simpler. The algorithm can backtrack or use backpropagation to assess the data previously assessed. This makes use of memory blocks to store the data in a sequence which are present in the multiple intermediate layers. While propagating through the network, we perform multiple operations over the sequenced data. Examples, where we can use this model, are the prediction of the stock market or fields where we deal with a time-based platform. This algorithm is suited best in the classification and prediction of sequenced data related to a time factor. It handles the most complex problems in a very efficient manner and hence is vastly used in machine learning.

2.4 *Nonlinear Auto-Regression Neural Network (NARNN)*

Another frequently used approach in case of predictions or studies related to time-based patterns is NARNN. Nonlinear auto-regression neural network as the name suggests is an artificial neural network (ANN) that uses timestamps in its algorithm to produce the multiplier weights stored in different layers. If we were to represent a general formula for the algorithm, we would show it as

$$Y(t) = Y(t-1) + Y(t-2) + Y(t-3) + \cdots + Y(t-d+1) + Y(t-d)$$

where Y is a function of time t and d is a past number up to which we are considering the data to study or train our model. The value currently obtained is processed by looking into the historical data present in memory blocks. We calculate the values from the neural network model processing, and this is subtracted from the older values that we have stored with ourselves. If the mod of difference lies close to null or 0, it is said to be a good model or model with high performance else we try to improvise our model training for better performance.

3 Literature Survey

Recently, there has been a huge development in the field of medical data analysis with the help of present age technologies. Data scientists have been constantly applying the concepts of artificial intelligence, data visualization, machine learning, and deep learning in the medical field to improve data interpretation and analysis. It works with the help of feature extraction and the implementation of various models to give efficient and precise results. Images are classified as normal or disease affected with the help of many deep learning models. These concepts have been widely used for diseases like cancer, heart problems, and lung problems. Singh and Gupta [4] applied machine learning-based approaches to detect and classify lung cancers in human and got a successful detection rate of 85.55%. Rajpurkar et al. [5], looking at the present scenario, implemented a CNN-based model over chest X-ray images resembling the pneumonia detection approach. Alqudah et al. [6] suggested an automated method to diagnose COVID-19 by using machine learning classifiers such as KNN with CNN over chest X-ray images. By using soft-max, they got an accuracy of about 98%. Esteva et al. [7] used a deep neural network (single CNN here) with end to end direct training from images. Pixels and disease labels were given as inputs to the model. The model classifies critical cases: common cancer and deadly cancer and also provide some applications in the tuberculosis data set. Liu et al. [8] suggested a method to detect tuberculosis in the chest with X-ray images. The model used CNN and transfer learning to achieve a classification accuracy of 85.68% over large data set of TB images. Butt et al. [9] developed a deep learning system to screen COVID-19. CT chest scans provided better specificity than RT-PCR in COVID-19 detection, so they trained the model on CT chest scans. The model gained 86.7% accuracy by using ResNet23 and ResNet-18. Fanelli and Piazza [10] used data set collected by a University to analyze and forecast the spread of COVID-19 in various regions. Apostolopoulos and Mpesiana [11] have applied AI to detect COVID-19 cases with X-ray images by utilizing transfer learning and CNN. VGG19 gave them successful results at 98.75% accuracy.

4 Demonstration of Deep Learning in Screening of COVID-19

Deep learning and CNNs act similarly to the human visual processing system. They emulate their structure as well as their function. For this, it carries an out hierarchical layer of feature representation. This approach facilitates automatic image feature learning, allowing CNNs to surpass manually engineered features. Over the years, deep learning has shown remarkable success in several fields, one of them being medical image processing. The deep learning models use techniques of medical imaging like X-rays, endoscopy, MRIs by analyzing an image and signal data. The analysis could detect and diagnose maladies like cancer, rheumatoid arthritis, hepatitis, hearth, and liver-related diseases, and many more. Efficiently trained CNNs can give precise rapid and cost reduced medical imaging [12]. Segmentation map forecasting can be done by segmentation CNNs using 2D or 3D convolution kernels. Voxel information is used by 3D CNNs to carry out volumetric patch segmentation maps prediction. One by one dicing is performed on 2D maps for full volume generation. This interslicing context improves performance. Thus, the associated high computation is ignored as it is a sensitive case, and performance is vital here.

The study is carried out by designing a CNN-based model for detecting the Coronavirus disease of 2019 (COVID-19) which is caused by the novel Coronavirus SARS-CoV-2 [13]. It uses chest X-ray images to detect the disease. The correct classification and detection are highly important in a limited time and that is a major concern. The X-rays would first be distinguished by the Radiologists as COVID-19 infected X-rays or regular X-rays. Cases of pneumonia are specially observed. A different number of filters are used to detect local features and global features. This is done to detect what is being missed by other filters.

- There are five convolution layers and a fully connected layer in the model where the convolution layer deals with developing idea of the features that need to be extracted out of each image when given as input. It gets the feature extracted done.
- Next, batch normalization and ReLU are applied over the transformed images as well as to the succeeding two distinct forms of the pooling operation. The separate forms of pooling operations are being (1) the average pooling and (2) the maximum pooling.

Several pooling operations have been set up on purpose to reduce the dimensionality. There could be several redundant characteristics in the images which are unnecessary and should be avoided. A stride of 3×3 size is taken along with the pooling operations to decrease the dimensionality of the feature maps. Several particular information on the actual image detail can be present in the feature maps. The ReLU activated feature maps are sampled down. The negative values are reduced to zero, and other values remain intact. The max pooling (MaxP) layer helps in reducing the values to half. It works on the temporal pooling principle and selects the maximum value from the kernel matrix. The neural network observes very slow output change with respect to input change, and to observe these small changes with the input values, the sigmoid function is used. The function is applied to the point product of weights along with a bias value to give a resultant smooth output. The use of the sigmoid function makes the function cheaper computationally. Also, as it is a kind of activation function or more specifically a squashing function, it restricts the resultant output range in [0, 1].

COVID-19 affected chest has bilateral nodular and peripheral ground glass blurry patches and consolidation. It serves as an alert to radiologists that COVID-19 may be present in certain patients. Pneumonia affected chest X-rays have areas of lung inflammation designating the presence of infection.

The collection of good data set is essential for a properly working model (Fig. 2). Evaluation and testing of the CNN model are required. For the purpose, the data set has been collected from two publicly available sources (Kaggle data). The first source is the 1.15 GB data containing 5856 chest X-ray images collected by Paul Mooney. It has three sets namely test (234 normal files and 390 pneumonia files), train (1341 normal files and 3875 pneumonia files), and val (eight normal files and eight pneumonia files). The second source is the 85.24 MB data containing 188 chest X-ray images of confirmed COVID-19 viral infections with pneumonia collected by Wei Hao Khoong. It has two sets namely test (20 normal files and 20 COVID-19 files) and train (74 normal files and 74 COVID-19 files). The sample examples are shown in Fig. 3. Collected data is resized to fit the dimension of model design, and training parameters are set to batch size = 32 and epochs = 10. Training is carried out over the model followed by performance evaluation, which is done based on accuracy and loss metrics maintained in the training phase. The ability of the model to differentiate between normal, pneumonia, and COVID-19 X-ray is evaluated.

Figure 4 presents the classification accuracy rate when data is implemented using both training as well as testing data set. Total ten epochs were used for the

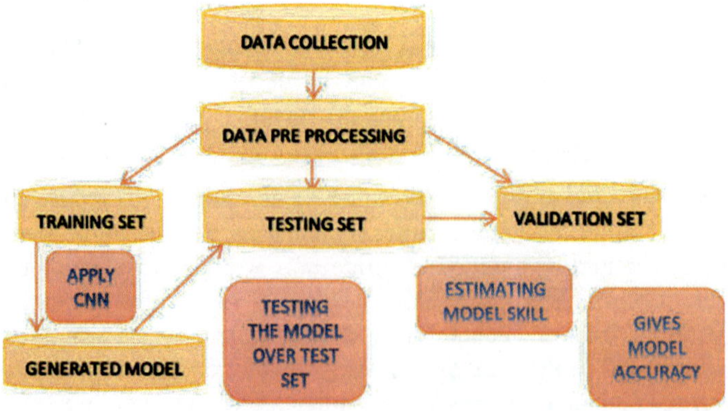

Fig. 2 Deep learning model for COVID-19 diagnosis using X-ray images

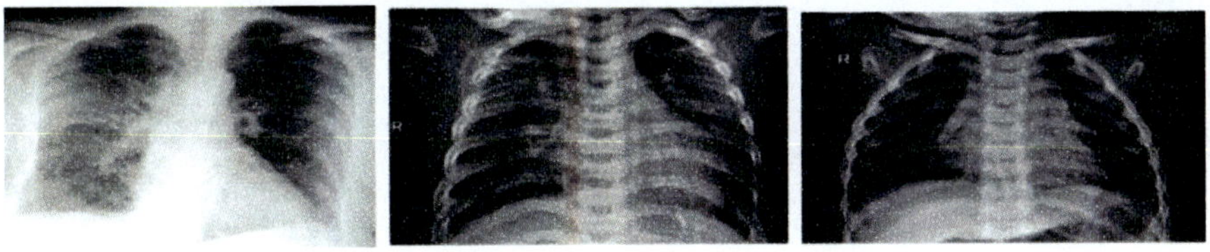

Fig. 3 Sample COVID-19 infected X-ray, bacterial pneumonia X-ray, and normal X-ray

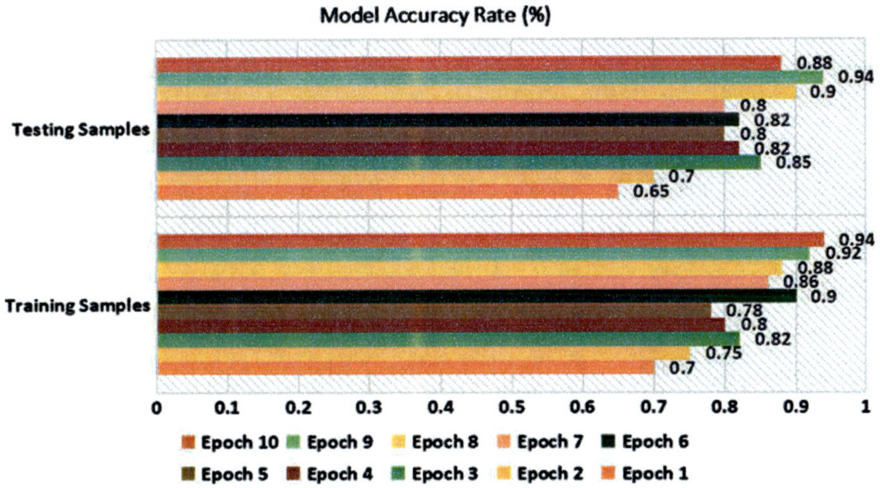

Fig. 4 Model accuracy rate analysis with training and testing data

purpose. During the training of the model, an optimum accuracy of 0.94 was observed during epoch 10, while the least accuracy of 0.7 was noted in epoch 1. While during testing phase, epoch 9 gave the highest accuracy of 0.94, and the least accuracy of 0.65 was provided by epoch 1.

Figure 5 plots the model loss during the implementation process. As shown, the X-axis represents the number of epochs, and the Y-axis represents the loss rate for training and validation. A minimum error rate of 0.675 was observed with epoch 4 during the training of data, while the least model loss error of 0.46 was noted in epoch 8 during testing or validation of data records.

The deep learning approach for the detection of the Coronavirus with the presented model significantly gives better results with various data sets during the different phases of the study. The model performed well enough with a normal chest X-ray, pneumonia X-ray, and COVID-19 X-rays. This work can be used for distinguishing and detecting the pandemic virus. It can serve the society in a much better way for the rapid detection and diagnosing of the virus disease.

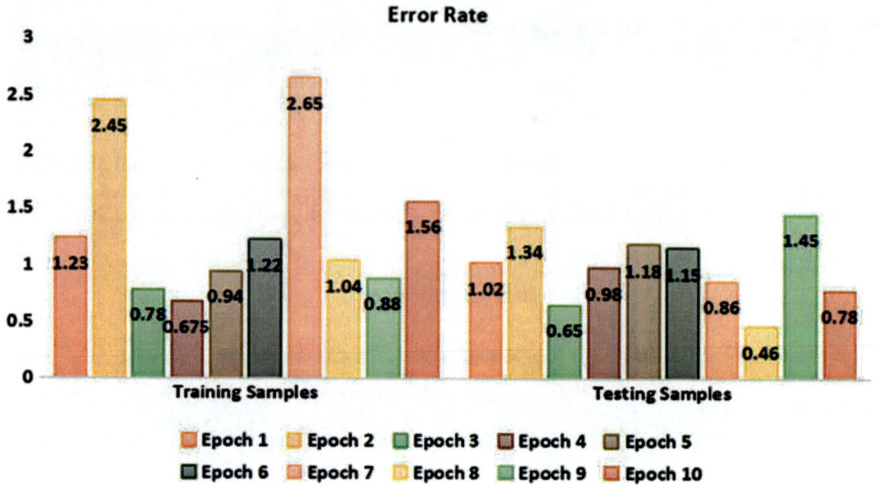

Fig. 5 Error rate comparison with respect to number of epochs

5 Conclusion

COVID-19 has caused huge disrupt in the lives of people, economy, and development. Scientists and medical professionals all over the world are working day and night to bring the situation back to normal. Various ML techniques could be used to tackle the situation. This study revolves around the concept of deep learning and its potential to monitor and detect COVID-19. Different deep learning models could be applied to solve different problems like the LSTM model could be applied to forecast the spread of the COVID-19 or the deep belief network (DBN) model could be applied to screen COVID-19. The CNN model studied here detects COVID-19 by classifying the chest X-rays as normal, pneumonia, or COVID-19 affected. It provides high classification accuracy and less error rate, but there is a scope of further enhancement of accuracy. The model could be further optimized with better parameters and hyperparameters to give better performance. Overall, deep learning hoards great capabilities to screen and tackle Coronavirus.

References

1. Mishra, S., Mallick, P.K., Jena, L., Chae, G.S.: Optimization of skewed data using sampling-based preprocessing approach. Front. Public Health **8**, 274 (2020). https://doi.org/10.3389/fpubh.2020.00274
2. Mishra, S., Tripathy, H.K., Mallick, P.K., Bhoi, A.K., Barsocchi, P.: EAGA-MLP—an enhanced and adaptive hybrid classification model for diabetes diagnosis. Sensors **20**(14), 4036 (2020)

3. Mishra, S., Tripathy, H.K., Mishra, B.K.: Implementation of biologically motivated optimisation approach for tumour categorisation. Int J. Comput. Aided Eng. Technol. **10** (3), 244–256 (2018)
4. Singh, G.A.P., Gupta, P.: Performance analysis of various machine learning-based approaches for detection and classification of lung cancer in humans. Neural Comput. Appl. **31**(10), 6863–6877 (2019)
5. Rajpurkar, P., Irvin, J., Zhu, K., Yang, B., Mehta, H., Duan, T., et al.: Chexnet: radiologist-level pneumonia detection on chest x-rays with deep learning. arXiv preprint arXiv:1711052252017
6. Alqudah, A., Qazan, S., Alqudah, A.: Automated systems for detection of covid-19 using chest x-ray images and lightweight convolutional neural networks (2020). https://doi.org/10.21203/rs.3.rs-26500/v1
7. Esteva, A., Kuprel, B., Novoa, R.A., Ko, J., Swetter, S.M., Blau, H.M., et al.: Dermatologist-level classification of skin cancer with deep neural networks. Nature **542** (7639), 115–118 (2017)
8. Liu, C., Cao, Y., Alcantara, M., Liu, B., Brunette, M., Peinado, J., et al.: Tx-cnn: detecting tuberculosis in chest x-ray images using convolutional neural network. In: 2017 IEEE international conference on image processing (ICIP). IEEE, pp. 2314–2318 (2017)
9. Butt, C., Gill, J., Chun, D., Babu, B.A.: Deep learning system to screen coronavirus disease 2019 pneumonia. Appl. Intell., 1 (2020)
10. Fanelli, D., Piazza, F.: Analysis and forecast of covid-19 spreading in China, Italy and France. Chaos, Solitons Fractals **134**, 109761 (2020)
11. Apostolopoulos, I.D., Mpesiana, T.A.: Covid-19: automatic detection from x-ray images utilizing transfer learning with convolutional neural networks. Phys. Eng. Sci. Med., 1 (2020)
12. Mallick, P.K., Mishra, S., Chae, G.S.: Digital media news categorization using Bernoulli document model for web content convergence. Pers. Ubiquit. Comput. (2020). https://doi.org/10.1007/s00779-020-01461-9
13. Mishra, S., Mishra, B.K., Tripathy, H.K.: Significance of biologically inspired optimization techniques in real-time applications. In: robotic systems: concepts, methodologies, tools, and applications, pp. 224–248. IGI Global (2020)

Emoji Analysis Using Deep Learning

Ambik Mitra, Lambodar Jena, and Soumya Sahoo

Abstract Emojis or avatars have been the ways for indicating nonverbal cues. These cues have become the most essential part of online chatting, product review, brand emotion, and so on. It has been leading to increasing data science research dedicated to emoji-driven storytelling. With advancements in computer vision and deep learning, it has been currently possible to detect human emotions from various images. In this deep learning project, we have classified human facial expressions to filter and mapped corresponding emojis or avatars. In 1999, Japanese artist Shigetaka Kurita created the first emoji. He worked with the people, who worked for a renowned mobile Internet platform, named 'I-mode' from DOCOMO which was Japan's operator company for the mobile phones. The original 176 emojis introduced by Kurita are currently among the persistent collection at Museum of Modern Art which is situated in New York, which are the confidential symbols over faces, because it was DOCOMO's objective and motive to search for better ways to communicate and convey ourselves. Emoji became a boon and sense of popularity in Japan very quickly. On the other hand, DOCOMO's idea was being taken by other rival mobile companies.

Keywords Computer vision · Transfer learning · Emojis · DOCOMO · Avatars

A. Mitra
School of Electronics and Computer Engineering, Kalinga Institute of Industrial Technology (KIIT) Deemed to be University, Bhubaneswar, Odisha, India
e-mail: 1830007@kiit.ac.in

L. Jena (✉)
Department of Computer Science and Engineering, Siksha 'O' Anusandhan (Deemed to be) University, Bhubaneswar, Odisha, India
e-mail: lambodarjena@soa.ac.in

S. Sahoo
Department of Computer Science and Engineering, C. V. Raman Global University, Bhubaneswar, Odisha, India
e-mail: soumya.sahoo685@gmail.com

S. Das and M. N. Mohanty (eds.), *Advances in Intelligent Computing and Communication*, Lecture Notes in Networks and Systems 202,
https://doi.org/10.1007/978-981-16-0695-3_64

1 Introduction

The document shows how an emoji is created from real face using the transfer learning method of machine learning [1]. So, we have built a convolution neural network to recognize facial emotions. We have trained our model on the FER dataset. Then, we are mapping those emotions with the corresponding emojis or avatars [2]. Using OpenCV's haar cascade xml, we are getting the bounding box of the faces in the live images. Then, we feed these boxes to the trained model for classification. We are also using other Python libraries like NumPy, OS, Tkinter and Keras in this domain. Emoji creation has various applications in today's world. Emojis have been very useful in the social media platform such as Facebook, Twitter, Snapchat, and Instagram and also used in crime and security issues [3, 4]. Emojis make the conversation and a normal sentence very understandable. Emojis help a person seem more mobile friendly and make a person up-to-date. Emoticons usage should be only for informal conversations, and more importantly, the usage should be very much limited in business communications as the sentences need to be specific and to the point. Emojis are also included in enterprise communication applications such as skype.

2 Literature Survey

In 2015 [5], Zhu believed that emojis were the facial expressions in the form of cartoons which helped in conveying specific emotions on the basis of text format, while the rest were similar to, but alike as compared to the emoticons, which diversified the thought process of the people of understanding the message in online interactions Eisenstein and Pavalanathan in 2015 [6], said that the launch of emojis became the most substantial shift in online writing. People used to send the appropriate predefined graphic icons on the replacement of typing or writing elaborate and huge texts. The people had the access to a vast amount of colourful and expressive emoji pictographs which resulted in attractive look of the texts and made it convenient to understand. It became very obvious for people to allow wayward orthographies for better communication in the social media. Kelly and Watts in 2015 [7] told that emoji might play an important role to make our conversions interesting, rather get aware about specific expressions of emotion, and could also come out as a huge role in limiting the threading in conversations or in supporting playful behaviour. Stark and Crawford in 2015 [8] told that the regular usage of the emojis were planned to normalize, capitalize, and stress on the impact it had created on the social relations of people through online mode. Gullberg in 2016 [9] told that except the fact that emojis were helpful for conveying emotions, they could also respectfully and kindly throw a reaction to a second person's message, without giving a lengthy reply. Emojis played an important role in maintaining relationships which included inside jokes, or be a sign of the enthusiasm of a specific person. Andral and Larroque in 2016 [10] found the fact that companies and marketing people became aware of sending appropriate emojis to the concerned person and also they had appropriate emojis that can be used for betterment of the brand logo of their company, to attract the consumers and to make people more connected to the brand. Kyle et al. in 2017 [11] told that emojis became a craze in online mode of communication. They also believed in some psychological theories such as miserable expression, nostalgic mimicry, judgement, pragmatics, and intention detection. Tian et al. in 2017 [12] had written that emojis acted as a signal to point of view understanding of tweets' result with good accuracy in comparison with using elaborate texts, where emojis showing positivity were preferred over the negative ones. Novak et al. in 2015 [13] had proven that emojis were equipments that reflected human emotions, which were observed at the period of creation of sentiment classification models, and were applied to different real-life situations, in the study of 1.6 million explained tweets over 13 different languages. Dürscheid and Siever in 2017 [14] said that the emojis were becoming a global language, but also believed that there is less probability, as emojis cannot be used in tense matters, and very rarely have a reference function. Except this, they spoke about specific conditions in which a sign must satisfy to be a Unicode emoji.

3 Objective

In this experiment, we had created various categories of emojis or emoticons from the live images stored in a database named FER dataset which is very easy to access. We used the Python libraries like Tkinter, OpenCV, OS, Keras and NumPy to perform this experiment. The method of deep learning used is transfer learning.

4 Proposed Model and Description

The proposed model is represented in Fig. 1.

4.1 Facial Detection and Tracking

Face detection and tracking is the technique of spotting faces in live images and video sequences. It is a way for identifying or verifying an unknown person with the help of database full of predefined facial expressions [15]. The Python library used to perform this task is **OpenCV**. It detects the faces in any live video or imagery.

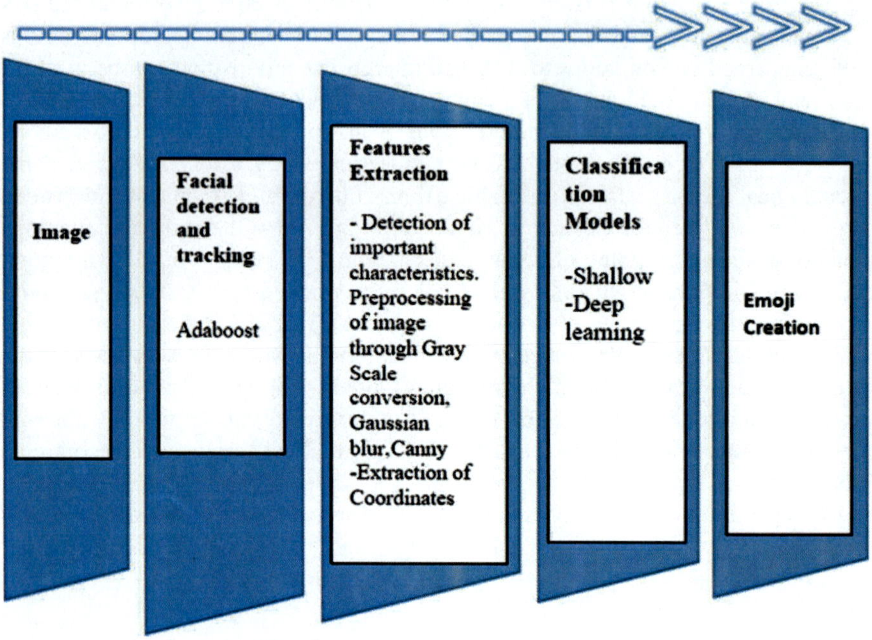

Fig. 1 Model for the experiment

4.2 Features Extraction

Pattern recognition as well as image processing includes feature extraction [16] which is an unusual form of decreasing of dimension. The only motive of this process is to procure the most apt information from the predefined data and present that information in a less space. **Keras** is the Python library used for this process in emoji creation. The Keras library extracts the various mood features from the face of a person in a live video or imagery.

4.3 Classification Models

A classification model is used in drawing some conclusions from the input values provided for training. It will predict the class labels or categories for the fresh data [17, 18]. **NumPy** is the Python library used for the model classification in the concept of emoji creation.

4.4 *Important Python Libraries*

4.4.1 Tkinter

Tkinter is a qualitative GUI library. Both Python and Tkinter provide a quick and efficient method for creating GUI applications. A power packed object-oriented interface is provided by Tkinter to the Tk GUI toolkit.

4.4.2 Os

The OS module consists of various functionality used in interaction with the operating system. OS module is one of the standard utility modules in Python. It shows a handy way of using dependent functionality in operating system. Both 'os' and 'os.path' modules involve various functionalities in interaction with the file system.

4.5 *Facial Emotion Recognition (FER) Dataset*

The facial expression recognition (FER) dataset consists of 48*48 pixel grayscale face images. The images are centred, and they occupy an equal amount of space. This dataset consists of facial emotions of following categories: angry, disgust, fear, happy, sad, surprise, and natural.

Basically, in this experiment, we have made two Python files, one is the training file (train.py). In this file, we had followed the following steps. Firstly, we imported all the modules useful. Secondly, we initialized the training and validation parameters. Thirdly, we built the convolution network architecture. Then, we compiled and trained the model. Then, we saved the model weights, and at last using OpenCV haar cascade xml, we detected the bounding boxes of face in the images and predicted the emotions.

Another file we had made, the GUI file (GUI.py), for what we had done is that we created a folder named emojis and saved the emojis corresponding to each of the seven emotions in the dataset. Then, we used the Tkinter, OpenCV, NumPy, and OS libraries and diplayed the emojis according to the different facial expressions.

Table 1 shows the emojis according to the categories present in the FER dataset. Each emoji has its own sentiment score.

Table 1 Sentiment scores of the emojis

Emoji	Description	Sentiment score
	Angry	−0.172
	Disgust	−0.154
	Fear	−0.141
	Happy	0.567
	Sad	−0.145
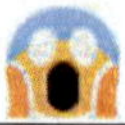	Surprise	0.191
	Natural	0.643

4.6 Sentiment Score

The position of an emoji is known by knowing its sentiment score. The range of the sentiment score is (−1, +1).

5 Result and Analysis

Table 2 shows the sentiment score of different moods under various categories like express emotions, clarify or disambiguate message, and light mood. The negative values of the shapes in an image are never considered in the facial detection process.

Table 2 Sentiment scores of different facial features

Features	Express emotions	Clarify/disambiguate message	Lighten up mood
Shyness	0.05	−0.071	−0.11
Neuroticism	0.06	0.11	0.031
Extraversion	−0.082	0.11	−0.011
Agreeableness	0.32	0.51	0.41
Openness	−0.11	−0.0031	−0.011
Conscientiousness	0.081	−0.082	−0.041

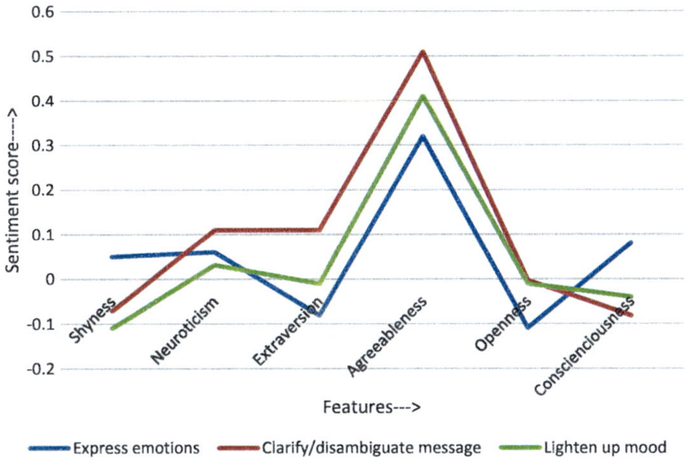

Fig. 2 Line graph for the variations in the sentiment score

Figure 2 shows the line graph which is obtained from Table 2. The variations in the sentiment scores are immense, and the highest sentiment score is observed for the 'agreeableness' facial feature.

5.1 Emoji Creations

Table 3 shows the emojis created from facial features present in the images taken from the FER dataset.

6 Conclusion

The images had been created appropriately according to the images, and a graph was created showing the variations in the sentiment score of each facial feature. The Python libraries like Tkinter, OS, NumPy, Keras, and OpenCV were very useful in accomplishing the experiment.

Table 3 Emoji creations

Live Image	Emoji
 (angry)	
 (disgust)	
 (fearful)	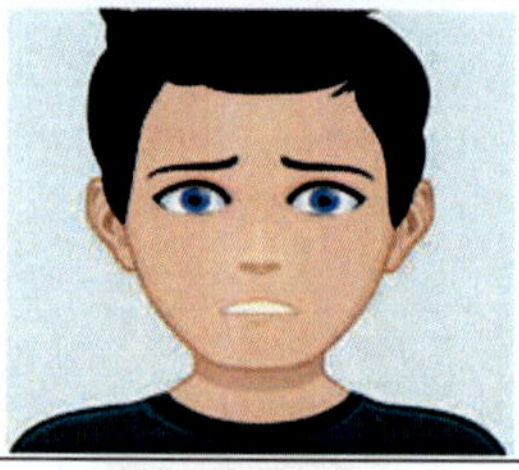
 (happy)	
 (neutral)	

(continued)

Table 3 (continued)

Live Image	Emoji
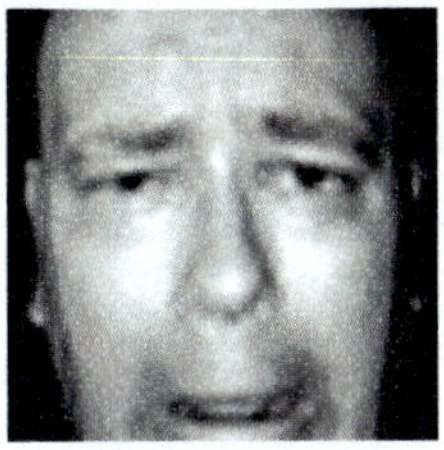 (sad)	
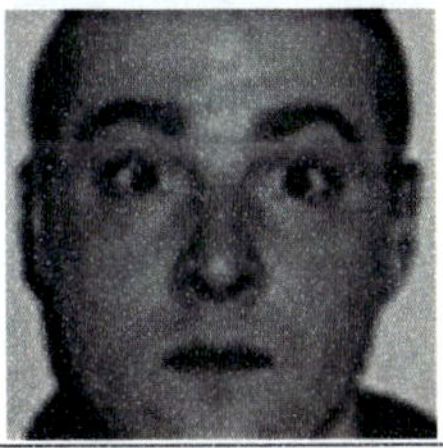 (surprised)	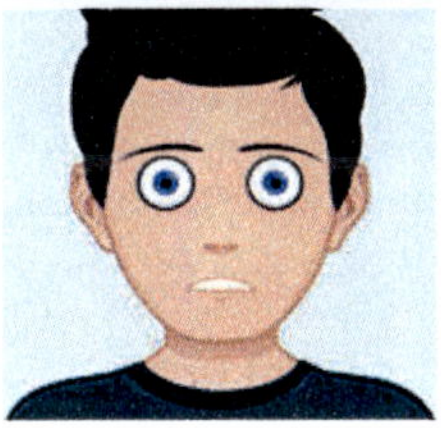

References

1. Mishra, S., Mallick, P.K., Jena, L., Chae, G.S.: Optimization of skewed data using sampling-based preprocessing approach. Front. Public Health **8**, 274 (2020). https://doi.org/10.3389/fpubh.2020.00274
2. Mishra, S., Tripathy, H.K., Mallick, P.K., Bhoi, A.K., Barsocchi, P.: EAGA-MLP—an enhanced and adaptive hybrid classification model for diabetes diagnosis. Sensors **20**(14), 4036 (2020)
3. Mishra, S., Tripathy, H.K., Mishra, B.K.: Implementation of biologically motivated optimisation approach for tumour categorisation. Int. J. Comput. Aided Eng. Technol. **10** (3), 244–256 (2018)
4. Mallick, P.K., Mishra, S., Chae, G.S.: Digital media news categorization using Bernoulli document model for web content convergence. Pers. Ubiquit. Comput. (2020). https://doi.org/10.1007/s00779-020-01461-9
5. Zhu, X.: A symbolism study of expression in text-based communication. Thesis and Dissertations, Iowa State University Graduate (2015)
6. Eisenstein, J., Pavalanathan, U.: Emoticons vs. emojis on twitter: a casual inference approach. CoRR abs/1510.08480 (2015)
7. Kelly, R., Watts, L.: Characterising the inventive appropriation of emoji as relationally meaningful in mediated close personal relationships. Experiences of technology appropriation: unanticipated users, usage, circumstances, and design (2015)
8. Stark, L., Crawford, K.: The conservatism of emoji: work, affect and communication. Social Media + Society 1.2: 2056305115604853 (2015)
9. Gullberg, K.: Laughing face with tears of joy: a study of the production and interpretation of emojis among Swedish University Students. Bachelor Thesis, Lund University (2016)
10. Andral, M., Larroque, A.: The emojis consumer perception in the online advertising. Halmstad University Bachelor's Valiation (2016)
11. Kyle, L.K., Malone, S.A., Wall, H.J.: Emojis: insights, affordances and possibilites for psychological science. Trends Cogn. Sci. **21**(2), 66–68 (2017)

12. Tian, Y., Galery, T., Dulcinati, G., Molimpakis, E., Sun, C.: Facebook sentiment: reactions and emojis. In: Proceedings of the fifth international workshop on natural language processing for social media, pp. 11–16 (2017). https://doi.org/10.18653/v1/w17-1102
13. Novak, P.K., Smailović, J., Sluban, B., Mozetiĉ, I.: Sentiment of emojis. CoRR abs/1509.07761 (2015)
14. Dürscheid, C., Siever, C.M.: Beyond the alphabet—communcataion of emojis. Kurzfassung eines (auf Deutsch) zur Publikation eingereichten Manuskripts (2017)
15. Mishra, S., Sahoo, S., Mishra, B.K.: Neuro-fuzzy models and applications. In: Emerging trends and applications in cognitive computing , pp. 78–98. IGI Global (2019)
16. Jena, L., Kamila, N.K.: Data extraction and web page categorization using text mining. IJAIEM, **2**(6). ISSN 2319-4847 (2013)
17. Mishra, S., Tadesse, Y., Dash, A., Jena, L., Ranjan, P.: Thyroid disorder analysis using random forest classifier. In: Intelligent and cloud computing, pp. 385–390. Springer, Singapore (2019)
18. Kamila, N.K., Jena, L., Bhuyan, H.K.: Pareto-based multi-objective optimization for classification in data mining. Cluster Comput. **19**, 1723–1745 (2016). https://doi.org/10.1007/s10586-016-0643-0

A Dual-Band Inset-Fed Octagonal Patch Antenna for Wearable Applications

Shaktijeet Mahapatra, Jibesh Mishra, and Madhumita Dey

Abstract In this paper, we propose an octagonal microstrip patch antenna optimized to use on human body. The antenna is fed by a microstrip line at an inset from the edge. The antenna has overall dimension of 22 × 22 × 1.6 mm and can be integrated with wearable devices easily. The antenna was found to resonate in two bands within 10.6 GHz in free space. The full ground plane reduces the back lobe radiations and limits the detuning effect of the human tissue as well. The antenna was found to resonate around 4.6 and 8.2 GHz with an average bandwidth of around 260 MHz. The observed specific absorption rate (SAR) values were below 1 W/kg for both the bands. The antenna can be used in wireless body area networks (WBAN) in ultra-wideband frequency region for wearable antennas.

Keywords Body area network · Dual-band · Inset-fed microstrip patch antenna · Polygonal antenna · Wideband antenna

1 Introduction

The wireless devices have become an inseparable part of everyone's life. Spurred by the growing need to collect more data remotely by the Internet of Things (IOT) and cyber-physical systems (CPS) and the miniaturization of the sensing devices, the development of wearable antennas has been picking up pace. Moreover, the challenges presented by the wearable antennas, like integration with the fabric or with the device, effect of proximity to living tissues, development of

S. Mahapatra (✉) · J. Mishra · M. Dey
Department of ECE, ITER, Siksha O' Anusandhan (Deemed to be University), Bhubaneswar, Odisha, India
e-mail: shaktijeetmahapatra@gmail.com

J. Mishra
e-mail: jibeshmishra86@gmail.com

M. Dey
e-mail: shreyamadhumita1@gmail.com

S. Das and M. N. Mohanty (eds.), *Advances in Intelligent Computing and Communication*, Lecture Notes in Networks and Systems 202,
https://doi.org/10.1007/978-981-16-0695-3_65

compact, conformal and flexible antennas, make them an interesting subject for research. The antennas for devices for the medical or military purposes may be implanted in the body, may be placed on the body, or may be handheld. This puts severe constraints on power consumed by the devices, the placement of the antennas, the substrate used for the antennas, the size and the radiation characteristics of the antennas, and the specific absorption rate (SAR) [1].

A monopole antenna with modified ground structure for wearable applications, fabricated on a denim substrate and operating in four bands, has been described in [2]. An inkjet-printed antenna system, fed by a coplanar waveguide structure and fabricated on polyethyl terephthalate (PET) film for operation at 2.45 and 5.8 GHz, is described in [3]. A microstrip-fed inverted-A monopole system for wireless body area networks (WBAN) presented in [4] resonated in 3.8 GHz band, 3.9–4.2 GHz band and 5.1–5.85 GHz band. An implantable antenna designed on FR4-epoxy is presented in [5] resonated at 4.8 GHz. An antenna proposed in [6] is a hexagonal-shaped ring with hexagonal slots in the annular region, with a bandwidth of 400 MHz and a SAR of 1.08 W/kg. For achieving low SAR values with compact antennas, electronic band gap (EBG) structures, frequency selective structures (FSS), or artificial magnetic conductor (AMC) structures have been designed in [7–10]. These structures help in reduction of back lobe radiations from the antennas to prevent excessive heating and the subsequent ionization by the EM radiations, and the SAR values were restricted to below 0.6 W/kg. In [11], three flexible antennas for wireless sensor node in shoe soles were designed, and the maximum SAR value observed was around 17 W/kg. A low profile, Koch fractal geometry-based antenna, designed on Rogers 5880, for on-body communications is proposed in [12], and SAR values observed were just below 1.6 W/kg. Conformal antenna designed on paper and working at 5.2 GHz, proposed in [13], is tested with various flexing angles. An all-textile antenna with a full ground plane and operating at 2.4 GHz, presented in [14], concluded that a full ground plane reduced the effects of proximity of antenna and human body. An H-I shaped, coaxial-fed microstrip patch antenna, presented in [15], exhibited a very low SAR of 5.6×10^{-6} W/kg.

The paper is further organized into the following sections: The design of the antenna is described in Sect. 2, the observations from the simulations are discussed in Sect. 3, and the findings of the work are concluded in the conclusion section. The referred literary works are listed in references section.

2 Antenna Design

The proposed antenna has been designed on 1.6-mm-thick FR4-epoxy substrate (relative dielectric permittivity, $\varepsilon r = 4.4$ and loss tangent, $\tan \delta = 0.02$). The substrate was chosen for its average dielectric permittivity, lightweight, moisture and temperature resistance, flexibility, wide availability, and low cost. As can be seen from Table 1, the overall dimension of the antenna is $22 \times 22 \times 1.6$ mm^3. The radiating patch is of octagonal shape, with each side being 7.46 mm. A full ground

Table 1 Design parameters

Parameters	Value (in mm)
Length of the substrate, Ls	22
Width of the substrate, Ws	22
Height/Thickness of the substrate	1.6
Side of the octagon, *a*	7.46
Width of inset slit, *b*	0.5
Distance of the slit from corner, *c*	1.76
Length of the feedline, *d*	7.9
Width of the feedline, *e*	3

plane is used so that back lobe radiations can be minimized, and the antenna becomes suitable for use on the human body. The octagonal shape increases the effective electrical path length. The side feeding was chosen to keep feed cable to one of the sides of the antenna and to avoid causing discomfort to the human user. The feed is given at an inset of 5.91 mm from the edge for impedance matching with a 50 Ω input port. The dimensions were finalized after running the parametric optimization (Fig. 1).

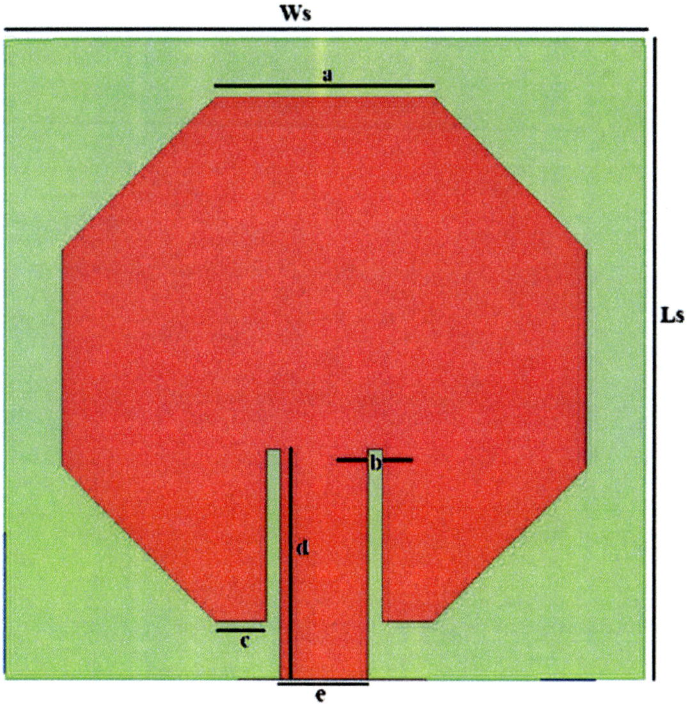

Fig. 1 Structure of the proposed inset-fed octagonal antenna

3 Results and Discussion

Return Loss: S11 represents the reflection loss from the port 1 if the input is from the port 1. This reflection loss is also known as return loss. So, lower the return loss, the more efficient is the radiation from the antenna. The acceptable value of return loss is less than −10 dB for the antenna to work efficiently. Considering the WBAN applications, the proposed antenna resonates in two bands, as can be seen in Fig. 2: 4.45–4.70 GHz with a resonant frequency of 4.58 GHz and 7.99–8.42 GHz with a resonant frequency of 8.19 GHz. Beyond 10.6 GHz, the antenna resonates in 15.07–16.17 GHz band with a resonant frequency of 15.57 GHz.
Gain: The simulated radiation pattern of the antenna in free space at 4.58 and 8.19 GHz is presented in Figs. 3 and 4, respectively. The antenna has broadside radiation pattern at 4.58 GHz with a highly directive nature. The maximum broadside gain obtained is 9.5 dB. The radiation efficiency was found to be 97.5%. The antenna has a gain of 2.41 dB in 30° direction with a radiation efficiency of 6.7%. Due to a full ground plane, the back lobes at 4.58 GHz are limited to −5.5 dB and back lobes at 8.19 GHz are limited to −1.1 dB, thus eliminating the need of any extra structure to limit the back lobe levels. As the back lobes were very efficiently limited, the effect of the back lobe radiation has been also very less.
Evaluation of SAR: The specific absorption rate (SAR) is a very important parameter in determining the extent of safety of usage of the antenna on a human body. For evaluation of SAR, right arm model of a human male phantom is used.

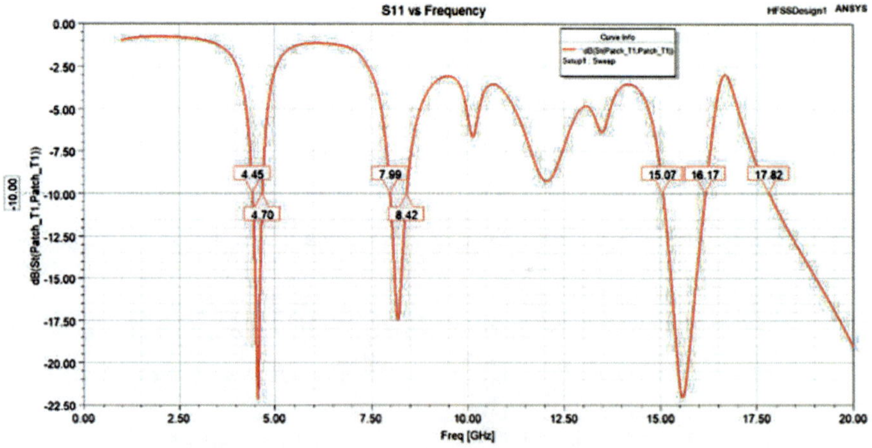

Fig. 2 S11 versus frequency of the proposed antenna in free space

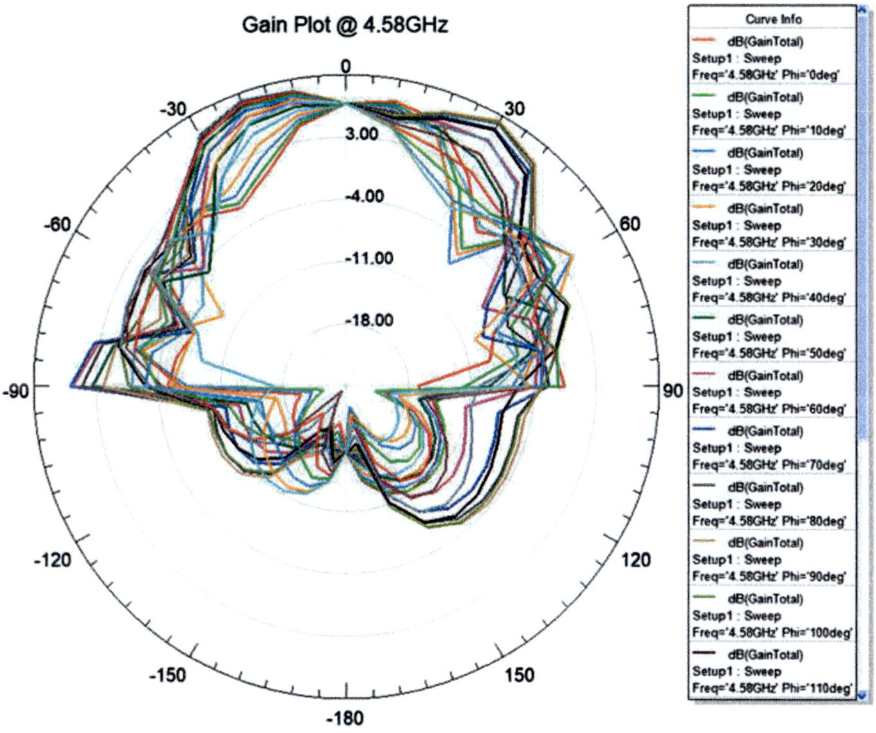

Fig. 3 Gain plot at 4.58 GHz

The model used is a four-layered 2/3rd homogeneous muscle model. The S11 vs frequency plot, in Fig. 5, of the antenna in free space (a) is compared with the antenna on the phantom (b) and antenna at 5 mm away from the phantom. It can be seen that even in the close proximity to human tissue, the antenna did not detune much, due to the presence of the full ground structure. The average SAR values observed at 4.6 and 8.2 GHz were 0.8014 and 0.9113 W/kg, respectively, when the antenna was placed on the phantom and 0.0084 and 0.5023 W/kg, respectively, when the antenna is located 5 mm away from the phantom. The observed SAR values were below the 1.6 W/kg cutoff [16]. The values of peak gain, peak directivity, and radiation efficiency are summarized in Table 2 (Fig. 1).

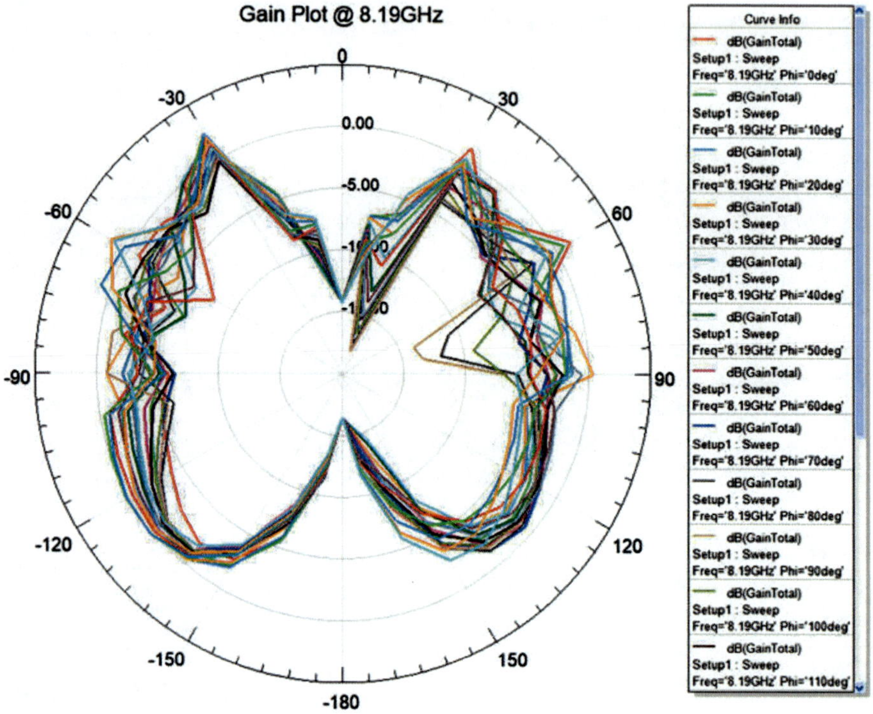

Fig. 4 Gain plot at 8.19 GHz

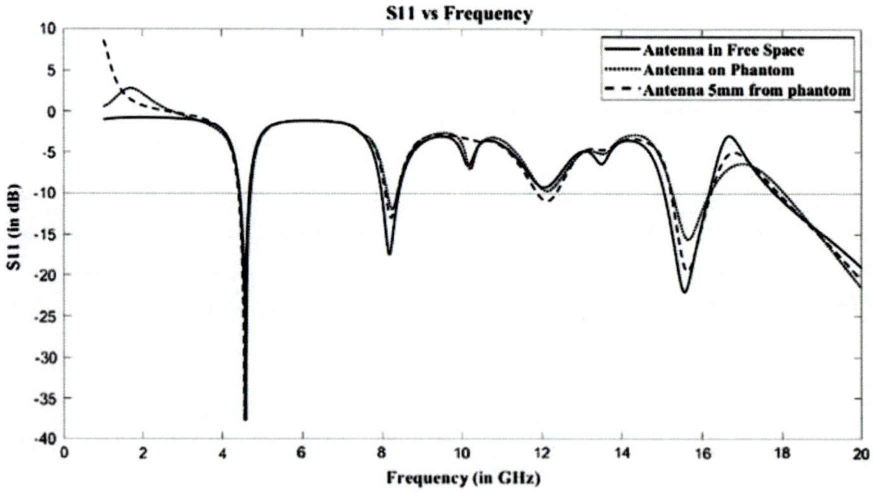

Fig. 5 S11 versus frequency of antenna in free space, antenna on phantom, and antenna at 5 mm from phantom

Table 2 Summary of results

Parameters	(a) Free Space		(b) Antenna on Phantom		(c) Antenna 5 mm from Phantom	
Resonant frequency (GHz)	4.58	8.19	4.6	8.2	4.6	8.2
S11 (dB)	−22.12	−17.46	−37.71	−11.90	−37.72	−12.96
Bandwidth (GHz)	0.25	0.41	0.27	0.24	0.28	0.29
Peak gain (dB)	9.36	2.41	8.73	1.99	5.44	3.36
Peak directivity (dB)	9.47	4.48	7.46	6.54	7.03	6.99
Radiation efficiency (%)	97.5	61.96	133.8	35.39	69.43	43.28
SAR (W/kg)	NA	NA	0.8014	0.9113	0.0084	0.5023

4 Conclusions

In this paper, we have presented an octagonal patch antenna fed by a microstrip line at an inset from the edge. The length of the inset was determined through parametric optimization to yield better matching point. The size of the overall antenna structure is kept small, i.e., 2.2 cm × 2.2 cm × 1.6 mm. The antenna was found to resonate in two bands, 4.45–4.70 GHz with a resonant frequency of 4.58 GHz and 7.99–8.42 GHz with a resonant frequency of 8.19 GHz. Various antenna parameters like bandwidth, peak gain, peak directivity, and radiation efficiency were determined for antenna in free space, antenna placed on the phantom, and antenna at 5 mm away from the phantom. The average SAR evaluated around 4.6 GHz was found to be around 0.8 W/kg when antenna is placed on the phantom and 0.0081 W/kg when antenna is placed 5 mm away from the phantom. The average SAR around 8.2 GHz was found to be around 0.9 W/kg when antenna is placed on the phantom and 0.5 W/kg when antenna is placed 5 mm away from the phantom. The SAR values were found to be below the cutoff value of 1.6 W/kg. The use of full ground plane served the purpose of reducing the effect of proximity of antenna with the human body by limiting the back lobe radiations and limiting the detuning effect posed by the human tissue. On the basis of the observations, the antenna was found to be a suitable candidate for wireless body area network applications.

References

1. Kiourti, A., Nikita, K.S.: A review of implantable patch antennas for biomedical telemetry: challenges and solutions [wireless corner]. IEEE Antennas Propag. Mag. **54**(3), 210–228 (2012)
2. Roopan, D., Samantaray, Bhattacharyya, S.: A multiband wearable antenna with defected ground structure. In: 2019 URSI Asia-Pacific radio science conference (AP-RASC), New Delhi, India, pp. 1–4 (2019)

3. Bait-Suwailam, M.M., Alomainy, A.: Flexible analytical curve-based dual-band antenna for wireless body area networks. Prog. Electromagn. Res. **84**, 73–84 (2019)
4. Suriya, I., Anbazhagan, R.: Inverted-A based UWB MIMO antenna with triple-band notch and improved isolation for WBAN applications. AEU-Int. J. Electron. Commun. **99**, 25–33 (2019)
5. Ramli, N., Ayob, M.: Design of a chip printed antenna at 4.8 GHZ for wireless implantable body area network applications. J. Telecommun. Electron. Comput. Eng. **10**(1–17), 27–31 (2018)
6. Gupta, V., Anupma, Narayan Mohanty, M.: Design of a microstrip antenna structure for WBAN devices at 2.45 GHz (ISM) band. Int. J. Eng. Technol. **7**(4.39), 222–226 (2018)
7. Gao, G.-P., et al.: Wearable circular ring slot antenna with EBG structure for wireless body area network. IEEE Antennas Wirel. Propag. Lett. **17**(3), 434–437 (2018)
8. Ashyap, A.Y., et al.: Highly efficient wearable CPW antenna enabled by EBG-FSS structure for medical body area network applications. IEEE Access **6**, 77529–77541 (2018)
9. Yin, B., et al.: A low SAR value wearable antenna for wireless body area network based on AMC structure. Prog. Electromagn. Res. **95**, 119–129 (2019)
10. Neebha, T.M., Nesasudha, M., Janapala, D.K.: A stable miniaturised AMC loaded flexible monopole antenna for ingestible applications. Comput. Biol. Med. **116**, 103578 (2020)
11. Wagih, M., Wei, Y., Beeby, S.: Flexible 2.4 GHz node for body area networks with a compact high-gain planar antenna. IEEE Antennas Wirel. Propag. Lett. **18**(1), 49–53 (2018)
12. Arif, A., et al.: A compact, low-profile fractal antenna for wearable on-body WBAN applications. IEEE Antennas Wirel. Propag. Lett. **18**(5), 981–985 (2019)
13. Singh, S., Kumari, S., Gupta, V.R.: Paper based conformal antenna. In: 2019 3rd international conference on trends in electronics and informatics (ICOEI). IEEE (2019)
14. Ashyap, A.Y.I., et al.: A wearable antenna based on fabric materials with circular polarization for body-centric wireless communications. Indones. J. Electr. Eng. Comput. Sci. **18**(1), 335–342 (2020)
15. Jumle, P., et al.: Shorted compact implantable patch antenna for ISM band-biomedical regime. In: 2019 9th international conference on emerging trends in engineering and technology-signal and information processing (ICETET-SIP-19). IEEE (2019)
16. IEEE Standard for Safety Levels with Respect to Human Exposure to Electric, Magnetic, and Electromagnetic Fields, 0 Hz to 300 GHz. IEEE Std C95.1™ (2019)

Feasibility Study of Hybrid Energy System for Indian Telecom Tower—A Case Study

Sonali Goel and Renu Sharma

Abstract Renewable energy can be used successfully to run the electrical equipment of telecom towers in remote areas where grid power is unavailable. It removes the possibility of interruptions of working of telecom equipment due to frequent power cut as electricity from renewable sources are most reliable one. This paper deals with an economic study by using solar–wind hybrid energy system in telecom towers in off-grid locations in a remote village in Odisha.

Keywords Telecom tower · Feasibility study · Hybrid energy system

1 Introduction

Our country India is one of the largest telecom customers in the world next to China [1]. Our country currently has a subscriber of over 1211.80 million and is expected to grow in coming years [2]. Majority of the telecom towers are installed in remote location where availability of grid power is a major concern. As most of these areas face frequent power cut, telecom companies should install backup power system to run those towers without interruption. Due to unreliable electric power grid, tower companies use various power management equipments such as diesel generator and batteries to ensure continuous power supply. Diesel generators are widely used in these areas to run the telecom equipment during power failure, and about 2.5 billion litres of diesel is burnt annually in these generators that cause environmental pollution by emitting 6.6 million tones of CO_2 [3]. The Ministry of New and Renewable Energy (MNRE), Government of India, instructed the telecom companies to opt for the renewable energy sources to run their towers and reduce the

S. Goel (✉) · R. Sharma
Department of Electrical Engineering, ITER, SOA (Deemed to be University), Bhubaneswar, India
e-mail: sonali19881@gmail.com

R. Sharma
e-mail: renusharma@soa.ac.in

S. Das and M. N. Mohanty (eds.), *Advances in Intelligent Computing and Communication*, Lecture Notes in Networks and Systems 202,
https://doi.org/10.1007/978-981-16-0695-3_66

use of diesel fuel. So it is a high time to focus on renewable energy technologies solutions like solar photovoltaic system, wind, biomass and fuel cells.

By using solar and other alternate sources of energy instead of diesel generators can reduce about 5 million tons of CO_2 emissions and savings of huge amount of operating expenditure of telecom tower companies [4].

As Odisha has a vast potential of solar energy as well as other renewable sources, telecom infrastructure companies should exploit these and thereby reduce the dependency on grid or DG and help the nation by reducing GHG emissions [5]. There is a vast scope for the researchers and academicians to make the study and analyse the feasibility of application of renewable energy for powering telecom towers.

Hence, the present study aims at a study of application of renewable energy for supplying power to telecom tower equipment in remote, rural off-grid location of Odisha.

2 Power Problem in Telecom Towers

As supply of grid electricity is uncertain in many rural areas, most of the sites face power outage of about 12 h a day. Hence, about 90% of telecom towers use backup power by incorporating a diesel generator and battery unit to run the equipment during this period. Figure 1 shows the availability of grid power in different states in India to run the telecom towers.

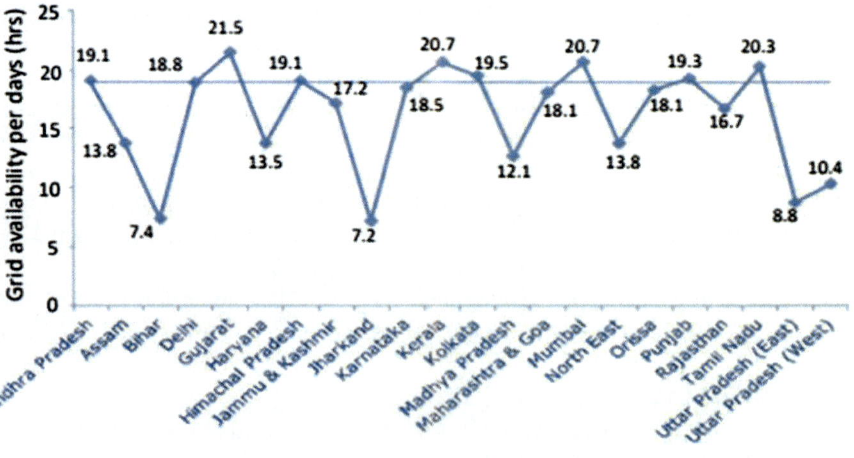

Fig. 1 Grid power availability per day in telecom towers in various Indian states. *Source* Greenomics, 2018 [3]

3 Hybrid Energy System

Solar–wind hybrid system can successfully be utilised to power the telecom tower in rural areas. During the bright sunshine hours of the day, the solar panel produces electricity, charges the batteries and stores enough energy reserve to be utilised during night hours, while the wind turbine produces electricity during monsoon period when solar power generation is minimum due to cloudy weather. Hence, solar–wind hybrid energy system is the ideal option for sustainable power supply to telecom towers in rural areas [6]. Various renewable energy technologies options for telecom tower are solar photovoltaic, wind, biomass, fuel cell and their hybrid combinations.

4 Methodology

In Odisha, there is one telecom tower installed by BSNL in coastal island of Paradip port (20.330 N latitude and 86.730 E longitude) that is at present running with solar–wind–DG hybrid system for supplying electricity to the tower equipment. This tower is powered by 10 kWp solar system and 5.1 kWp wind turbine with a battery unit of 48 numbers, each of 2 V, 1250 Ah and a 20 KVA diesel generator for uninterrupted power supply to the tower. The load data power generated by the solar–wind hybrid system at the site and the renewable recourses like solar radiation and wind speed data were analysed, and the system configuration is optimized by using HOMER software. Different possible models with their economics were analysed.

5 Study Location

The study location is mentioned in Table 1.

The single line diagram of solar–wind hybrid system is presented in Fig. 2.

The solar–wind hybrid system installed at village Barakolikhola in Paradip is mentioned in Fig. 3.

6 System Description

The location of the site is at a remote village, namely Barakolikhola of Mahakalpada block in Kendrapara district and is at 20.330 N latitude and 86.730 E longitude, at a distance of 10 km from Paradip port. The tower in the site has one BTS and some ancillary equipment that bring the power requirement of

Table 1 Location and specification of telecom tower

Sl. No.	Parameters	Location/specifications
1	Site location of tower	Village Barakolikhola (An island near Paradip Port) P.O.—Barakolikhola Dist.—Kendrapara, Odisha, India
2	Geographical location	Latitude: 20.33° N Longitude: 86.73° E
3	Customer	BSNL
4	Year of commissioning	2011
5	Installed capacity	15.1 kW = 10 kWp (Solar) + 5.1 kW (Wind)
6	Module type	ES-250M60B
7	Glass type	Albarino P
8	Installation type	Hybrid (Solar + Wind)
9	Orientation	True south and 20° to horizontal
10	Applications	Telecom Solutions-(BTS Sites)

approximately 22.7 kWh/day. The tower has only one service provider, i.e. Airtel. There is no grid power supply available for powering the tower equipment. A 20 KVA diesel generator set is installed for power backup during night/bad weather condition. The average clearness index is 0.63, and the average daily radiation of the site is 4.71 kWh/m^2/day.

Solar PV panels of 10 kW and a horizontal wind turbine of a 5.1 kW were installed at this site to supply the power requirement of site. Forty nos. of PV panel each of 250 Wp are used to get a total power of kWp. They are connected in series–parallel matrix. The BTS is typically backed by 48 nos. of batteries, 24 nos. in each rack. Each battery is of 2 V, 1250 Ah. The battery is operated at 80% depth of discharge (DoD). When fully charged, it can power the BTS for 88.8 h (by two battery racks).

6.1 System Equipment and Cost for Hybrid Power System

The cost of different equipment installed at the site along with their lifetimes is given in Table 2, and the different sizes taken for analysis are presented in Table 3.

6.2 Load

The actual load data of the site for the period of 1 year were collected from the BSNL office as well as from the site. The load data of a particular day are presented in Table 4.

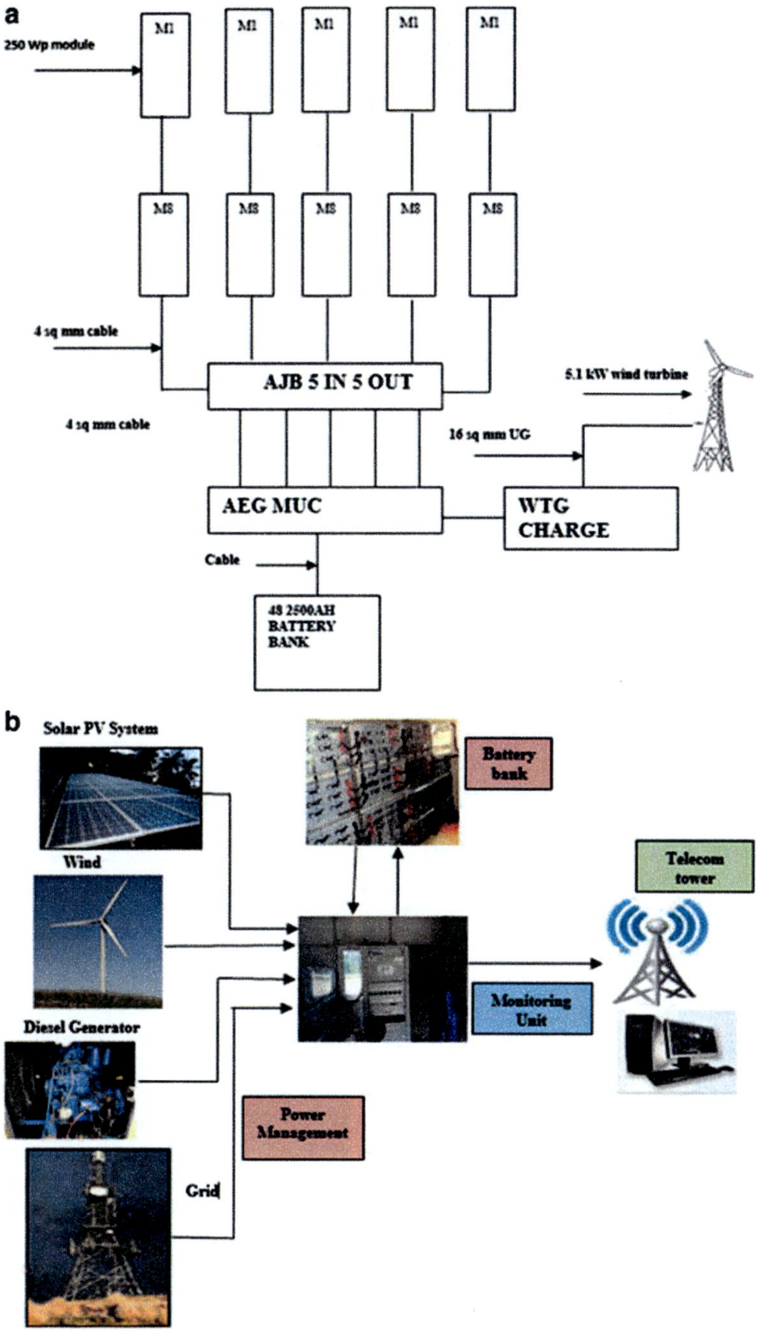

Fig. 2 **a** Single line diagram of solar–wind hybrid system at the site. **b** Layout of solar–wind hybrid system at the site

Fig. 3 Solar–wind hybrid system installed at village Barakolikhola

Table 2 System equipment and cost

Element	Initial cost ($/kW)	Replacement cost ($/kW)	O&M ($/kW)	Life time (years)
Solar system	4363	3927	0	20
Wind turbine	1641	1477	82	20
Battery	268 (per unit)	268 (per unit)	13 (per unit)	8
Converter	1000	1000	100	15
DG	453	408	0.02 (per hour)	10 (15,000 h)

Source Goel [7]

Table 3 Sizes of equipments and resources considered for analysis

Equipments	Sizes taken for analysis
Solar PV	0, 10 kW
Wind turbine	0, 5.1 kW
DG	0, 8, 12, 16 kW
Converter	6, 12, 15 kW
Resources	
Solar insolation, kWh/m^2/d	4.7, 5.0, 6.0
Wind velocity, m/s	4.36, 3.0, 5.0, 5.5, 6.0, 6.5, 7.0
Diesel price, $/L	0.9, 1.0, 1.1, 1.2

The annual average energy consumption of the site was found to be 22.7 kWh/d (say 23 kWh/d) with a peak load of 2 kW. From Table 5, it is seen that in a particular day, during 7.00–8.00 h, the power consumption of the tower is 1.138 kW/h say 1.14 kW/h which is maximum in a day. Keeping in view that in future, if another two BTS will be installed, then load of 55 and 83 kW/d will be consumed (Table 5).

Table 4 Power production by solar–wind hybrid system and load profile in telecom tower at different hours of a particular day

Time (hours)	Power generated (W)			Electric load (W)
	Solar PV	Wind Turbine	Total	
0.00–1.00	0	228	228	978
1.00–2.00	0	274	274	810
2.00–3.00	0	112	112	814
3.00–4.00	0	72	72	814
4.00–5.00	0	246	246	918
5.00–6.00	0	120	120	860
6.00–7.00	702	102	804	1032
7.00–8.00	790	124	914	1138
8.00–9.00	1620	94	1714	1120
9.00–10.00	1740	0	1740	1038
10.00–11.00	2426	72	2498	910
11.00–12.00	1104	114	1218	1048
12.00–13.00	1084	198	1282	858
13.00–14.00	404	394	798	976
14.00–15.00	662	412	1074	1048
15.00–16.00	528	188	716	886
16.00–17.00	0	184	184	960
17.00–18.00	0	232	232	980
18.00–19.00	0	104	104	1020
19.00–20.00	0	194	194	980
20.00–21.00	0	158	158	1126
21.00–22.00	0	292	292	888
22.00–23.00	0	104	104	882
23.00–00.00	0	98	98	812
Total	11,060	4116	15,176	22,896

Table 5 Present and projected load at telecom site

No of BTS	Power consumption/h (kW/h)	Power consumption/day (kW)
1	1.14	27.36 (Actual av. 22.7 kW)
2	2.28	55.0
3	3.42	83

Hence, considering future increase in demand, loads of 55 and 83 kWh/d were considered along with present demand of 22.7 kWh/d for analysis.

6.3 *Feasibility Study of the Hybrid Power System*

The low-cost model for supplying energy to the remote telecom tower (located at village Barakolikhola, Latitude 20033′ N and Longitude 86073′ E) has been optimised by using HOMER (Hybrid Optimisation Model for Electric Renewable) software (2.68 version) developed by National Renewable Energy Laboratory (NREL), Colorado, USA.

In the present analysis, different variables taken are as follows:

Equipment sizes considered:

Solar PV: 2 (0, 10 kW).
Wind turbine: 2 (0, 5.1 kW).
DG: 4 (0, 8, 12, 16 kW).
Converter: 3 (6, 12, 15 kW).

Resource sizes considered:

Solar radiation: 3 (4.7, 5.0, 6.0 kWh/m^2/d).
Wind speed: 7 (4.36, 3.0, 5.0, 5.5, 6.0, 6.5, 7.0) m/s.
Diesel price: 4 (0.9, 1.0, 1.1, 1.2 $/L).
Load: 3 (22.7, 55, 83 kWh/d).

In all the simulations models, 5% annual maintenance cost (except solar PV), salvage value of 10% (except battery) and annual nominal interest (actual interest—inflation rate) of 6% were taken. The net present cost (NPC), cost of energy (COE) and operating expenses for each system were determined and compared with that of conventional grid–DG system.

Various models considered for optimisation analysis are:

Off-grid condition

Model-1 Solar–wind-DG system
Model-2 Wind–DG system
Model-3 Solar–DG system
Model-4 Solar–wind system
Model-5 DG only system.

On-grid condition

Model-6 Grid DG system
Model 7 Grid only system.

Table 6 Hybrid systems having different combinations (Figures are in $/kW)

Off-Grid Systems	Load (kWh/d)		
	22.7	**55**	**83**
Solar–wind–DG	0.839	**0.552**	**0.506**
Wind–DG	**0.787**	0.590	0.536
Solar–DG	0.851	0.609	0.539
Solar–wind	0.963	–	–
DG only	0.966	0.648	0.568
Grid connected systems			
Grid–DG	0.519	0.269	0.211
Grid only	0.409	0.226	0.182

6.3.1 Cost of Energy of Different Models

The cost of energy of seven different models was simulated, and the results are calculated by HOMER software. For off-grid remote areas, the hybrid systems having different combinations were analysed and mentioned in Table 6.

7 Conclusion

Many telecom towers in India are now utilising solar–wind hybrid power system for powering their telecom equipment. The hybrid systems with possible combinations of energy resources were analysed and optimised by using HOMER software. Three levels of loads of 22.7, 55 and 83 kWh/d were considered for analysis with different other variables like diesel price, wind speed and solar radiation.

The lowest cost of energy was obtained with PV–wind–DG hybrid system at higher load of 55 ($0.552/kW) and 83 kWh/d ($0.506/kW). But at lower load of 22.7 kWh/d, wind–DG system was found to be most optimum cleaner energy solution with a lowest cost of energy ($0.787/kW). The highest cost of energy was found with only DG system at all three levels of loads. Though conventional grid power is most economical for the telecom tower, it should be avoided to reduce environmental pollution and should be moved towards cleaner form of energy as per the availability and site locations.

References

1. Anonymus. Consultation Paper on Green Telecommunications, Consultation Paper No 3/2011, Telecom Regulatory Authority of India, Mahanagar Doorsanchar Bhawan, Jawahar Lal Nehru Marg, New Delhi-110002 (2011)
2. Anonymus. Telephone statistics India, 2018. Economic Research Unit-Statistics, Department of Telecommunications, Ministry of Communication, Govt. of India (2018)

3. Anonymous. Telecom sector: A paradigm shift towards cleaner energy, Greenomics World, Deep Research, Bhubaneswar, India (2018)
4. Goel, S., Majid Ali, S.: Hybrid energy systems for off-grid remote telecom tower in Odisha, India. Int. J. Ambient Energy **36**(3), 116–122 (2015)
5. Goel, S., Ali, S.M.: Feasibility study of hybrid energy systems for remote area electrification in Odisha, India by using HOMER. Int. J. Renew. Energy Res. (IJRER) **3**(3), 666–672 (2013)
6. Fabbri, G., Boccaletti, C., Marques Cardoso, A.J.: A hybrid stand alone power system for telecom applications in Minor Islands. IEEE (2010)
7. Goel, S.: Studies on application of solar photovoltaic-wind hybrid power systems for telecom towers. Un-published M. Tech thesis, KIIT University, Bhubaneswar (2013)

A Survey Paper on 5G Suitable Waveform Candidates

Mrinalini, Kamlesh Kumar Singh, and Himanshu Katiyar

Abstract Fifth-generation wireless transmission cellular networks require waveforms which have higher data rates, low latency and maximum spectral efficiency. OFDM and advanced LTE is utilized in 4G but due to higher-frequency range of 24.25 GHz and above with latency less than 30 ms for 5G, OFDM cannot be implemented further. Generalized frequency division multiplexing is best suitable multicarrier modulation technique as well as waveform suitable for high-frequency network. Nowadays, in the present scenario, there is a high demand of fast data rates and wider frequency range with efficient latency. This paper deals with various methods and analyses that led to the evolution of GFDM. Comparison of GFDM is done with different types of modulation techniques, and conclusion is drawn that proves its dynamic and advantageous characteristics. GFDM transreceiver can be designed by using MATLAB and Python software; in the paper, second one is taken into consideration. Bit error rate is analyzed that depicts the importance of GFDM in terms spectral efficiency and minimal out of band (OOB) emissions.

Keywords GFDM · OFDM · UFMC · FBMC

1 Introduction

Nowadays, formidable growth in wireless communication with increasing data rate demands needs a waveform that has wide range of bandwidth and supports data rate signals low latency transmissions. Orthogonal frequency division multiplexing (OFDM) was the technique behind 3G and 4G cellular networks. But due to its high

Mrinalini (✉) · K. K. Singh
ASET, Amity University, Lucknow Campus, Lucknow, Uttar Pradesh, India
e-mail: mrinalini.prep@gmail.com

K. K. Singh
e-mail: kksingh@lko.amity.edu

H. Katiyar
Department of Electronics Engineering, Rajkiya Engineering College, Sonbhadra, India

S. Das and M. N. Mohanty (eds.), *Advances in Intelligent Computing and Communication*, Lecture Notes in Networks and Systems 202,
https://doi.org/10.1007/978-981-16-0695-3_67

value PAPR, OOB and low bandwidth efficiency, it is not considered to be good option for future cellular networks. There is an urgent need of a waveform that comprises all positive characteristics of OFDM like robustness, orthogonality but at the same time is suitable for higher-frequency networks. So there came evolution of filter bank multicarrier (FBMC), universal filter multicarrier (UFMC) and generalized frequency division multiplexing (GFDM). In order to make future cellular networks robust, highly efficient and versatile in energy as well as spectrum dimensions, these multicarrier modulation techniques came into existence. The fifth-generation (5G) cellular systems networks support future applications like tactile communications, machine-to-machine (M2M) communications coordinated multi-point transmission, Internet of things (IOT) and many more [1]. Recently, 4G is using LTE-OFDM, but this technique has maximum delay spread that results in long cyclic prefix (CP) further in interference. Major drawbacks of OFDM have showed non-supportive characteristics toward future 5G networks communication. Various research works had been carried out to meet the demand of high cellular networks out of which FBMC, UFMC and GFDM are be short-listed as better options. This paper consists of four sections which comprise descriptions and advantages of modulation techniques [2]. Section 2 is about OFDM, its applications, advantages and disadvantages. Sections 3 and 4 are the elaboration of FBMC and UFMC with the help of block diagrams. Section 5 consists of discussion and analysis of 5G wireless system with the implementation of GFDM. Section 6 consists of comparison analysis on the basis of graphical representation and tabular values of BER for each modulation methods.

2 Orthogonal Frequency Division Multiplexing

OFDM is a multicarrier modulation scheme in which multicarrier transmission scheme, the present channel bandwidth, is divided into several parallel subcarriers. Distance between subcarriers in OFDM is $\Delta = 1/$, which creates minimal cross-talk that is known as orthogonality.

In Fig. 1, basic building block of OFDM is explained that consists of serial to parallel convertor that turns incoming data into N streams that is further converted into several carrier frequencies with values $f0$, $f1$, $f2$, etc. The central carrier is valued to zero. Pilot insertion is done in order to minimize interference caused due to orthogonality of OFDSM system. Inverse fast Fourier transform (IFFT) is carried out to convert frequency domain signal into time spectrum [3]. For enhancement of robustness to reduce interference resulted due to multipath propagation, cyclic prefix is added at the beginning to each symbol that has turned out to be one of the major disadvantages of OFDM in terms of degradation of bandwidth efficiency.

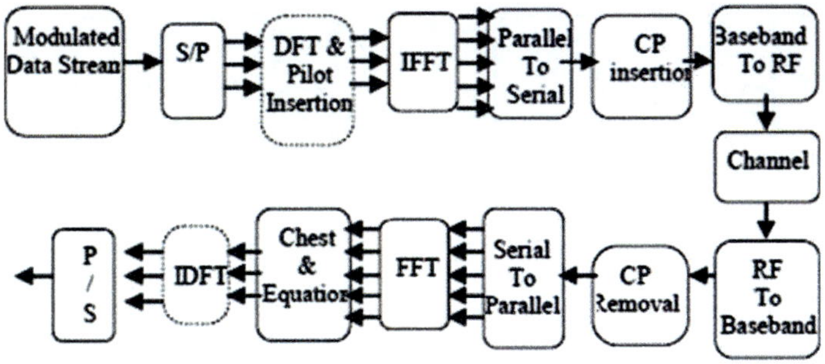

Fig. 1 OFDM transreceiver [3]

2.1 *Limitations of OFDM*

Cyclic Prefix overhead: CP overhead means same bandwidth being utilized in the beginning and end of the symbol. It is expressed as $ß_{overhead} = T_{CP}/(T_{cp} + T_{symbol})$. This results in wastage of bandwidth that lowers performance of cellular network in terms of bandwidth efficiency [4].

Frequency and Timing Offsets Sensitivity: It is believed that transmitter and receiver have similar reference frequency so there is a chance of losing orthogonality that results in leakage known as intercarrier interference (ICI).

High Peak-To-Average-Power Ratio (PAPR): This phenomenon occurs due to summation of several power components of subcarriers that led to further crest factor.

Out of Band (OOB) Emissions: Various subcarriers in OFDM systems cause emission in the outer area of the necessary bandwidth that makes a system to fall into category of low latency transmission which is not acceptable during this formidable growth of wireless networks [5].

3 Filter Bank Multicarrier (FBMC)

FBMC is a modulation method that is defined on the basis of filter banks present on transmitting receiving ends. Various aspects and implementations of FMBC are discussed that comes under staggered modulated multitone (SMT), cosine modulated multitone (CMT) and filtered multitone but mostly used one is SMT FBMC [6]. FBMC has been proved to be a successful method to sort out the problem of spectral containment present in OFDM by using prototype filter in transreceiver [6] (Fig. 2).

Polyphase network (PPN) FBMC system consists of bank of FIR filters those are prototype in nature. They depend on RRC that has value of roll-off factor 0.1.

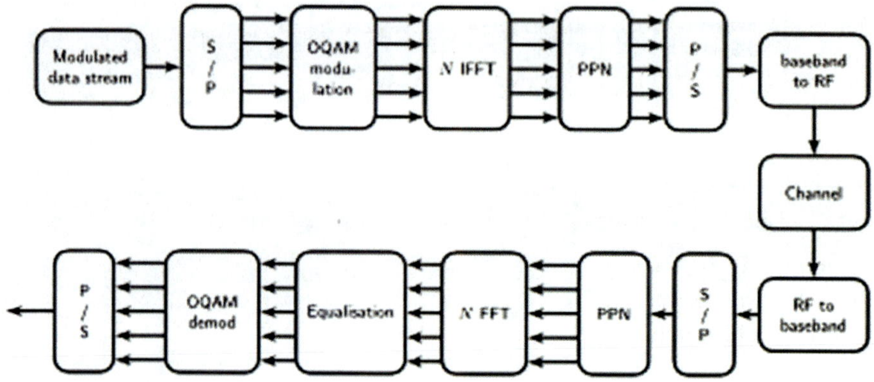

Fig. 2 FBMC transreceiver

Like OFDM, here also IFFT is implemented for windowing operations. With utilization of and long trail of filter banks, this method is not suggested for short burst communication that is the main demand of present high-frequency cellular system.

4 Universal Filter MultiCarrier (UFMC)

UFMC is also known as generalization form of FBMC and filtered OFDM. In this multicarrier modulation technique, subcarriers are filtered in groups that lowers filter length which prevents aliasing that makes UFMC capable of 5G waveform requirements [7]. This method is suggested for ISI and ICI robustness systems.

We can see in Fig. 3 that time overlap in between two consecutive UFMC symbols is not present. Value of symbol duration is equal to $N + L - 1$ where N is size FFT and L is assumed to be length of the filter. As in FBMC, UFMC also faces problem of complexity due to large size of FFT window size [3]. UFMC may or may not have insertion guard interval because addition of cyclic prefix can be optional. UFMC has better spectral efficiency and lesser cyclic overhead as compared to FBMC. It is very well suited for short burst communications and low latency transmission. As the orthogonality is lost due to the presence of filter banks, it is not recommendable for high data rates networks.

5 Generalized Frequency Division Multiplexing (GFDM)

GFDM is a flexible and robust multicarrier method that is considered to be best approach for 5G higher-frequency networks. This method of modulation contains CP of low bandwidth that results in building GFDM more spectral efficient. It has a

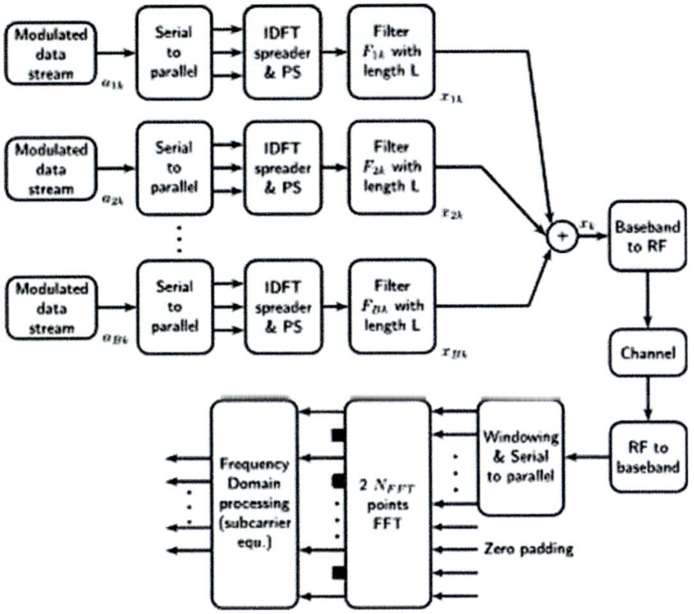

Fig. 3 UFMC transreceiver [3]

block structure in the form of M which consists of subcarriers and sub-channels. OOB is minimized with the implementation circular convolution method on sub-carriers [8]. One major advantage of GFDM is the concept to provide the present spectrum to individual user into various spectral segments with each segment of less bandwidth. In comparison with FBMC, GFDM is a platform that provides with additional characteristics of introduction of CP done after multiplexing or before filtering. Subcarrier can have a different bandwidth. Tail biting is carried out on counted symbols with implementation of circular convolution.

In Fig. 4, there is an explanation for the block structure of GFDM that has $g_{k,m(n)}$ representing a filter having N samples with k, m and n are assumed to be subcarrier, subsymbol and time indices. Every block is given a cyclic prefix with various several subsymbols. This results in making GFDM more flexible for TTI length helping in carrying data for specific user making it best waveform for low latency applications [9]. From the above diagram, it can be seen that GFDM has adjustable filters present at transmitter so it has minimal OOB emissions. It provides with scheduling many users in frequency as well as time domain [10, 11]. This system is free from high PAPR values, and it provides utilization in white space aggregation.

GFDM signal is given by the following expression:

$$g_{k,m(n)} = g(n - mk) \bmod (N) \exp(j2\Pi(kn/N)) \quad (1)$$

where n is sampling index.

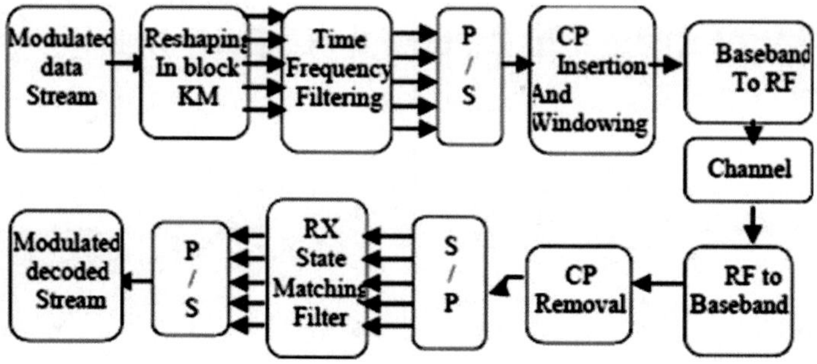

Fig. 4 GFDM transreceiver [3]

6 Comparison Analysis of 5G Waveforms

6.1 Parameterization

FBMC	UFMC	GFDM
RRC Filter	Dolph Chebyshev	Dirichlet filter
CP Length = 144	144 samples	144 samples
–	Filter length = 145	–
Mapping 16QAM	Mapping 16QAM	Mapping 16QAM
$K = 4$, PPN = 3	$N = 64$	$N = 4$, $M_a = 2$
–	No. of subbands = 40	–
Subcarriers = 2048	2048	2048
Subcarrier spacing = 15 kHz	15 kHz	15 kHz

7 Conclusion

It has been found that bit error rate has been controlled that means decreased for GFDM systems which means performance is upgraded. OFDM systems suffers lot of problems such as interferences, OOB and high PAPR so the value of BER is more in that case. Other techniques have higher values and again result in low spectral efficient and not as much suited for 5G wireless communication. It is seen in the graph that GFDM can be concluded as most suitable candidate for higher-frequency networks. In this paper, there is an extensive comparison of prominent waveforms that find its application in the 5G air interface. With certain same values of subcarriers, the analysis is done for UFMC, FBMC and GFDM [12]. Therefore, GFDM is best spectrally efficient waveform for short bursts and

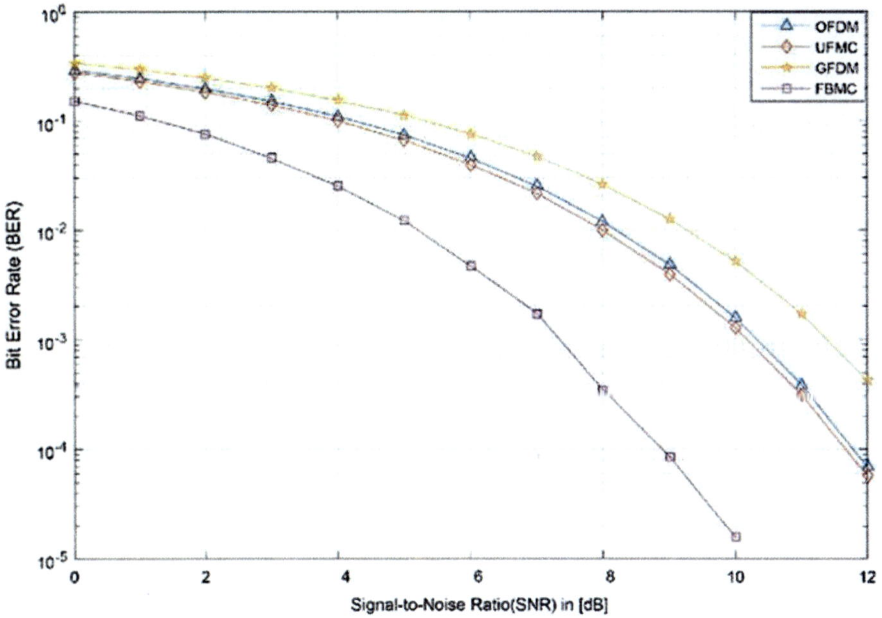

Fig. 5 BER of different modulation schemes [1]

upcoming low latency transmissions. Hence, better performance of system is seen under GFDM so it can extensively be applicable solution for 5G. Hardware setup can be developed on a flexible field programmable gate array (FPGA) which acts as a platform of experimental research. GFDM transreceiver can be designed by using MATLAB and Python software. It is easy to implement in Python as coding is easier through this software (Fig. 5).

References

1. 5G Waveform Candidate Selection D3.2 5GNOW_D3.2_v1.3.docx
2. Michailow, N., Matthé, M., Gaspar, I.S., et al.: Generalized frequency division multiplexing for 5th generation cellular networks. IEEE Trans. Commun. **62**(9), 3045–3061 (2014)
3. Mrinalini, Kumar Singh, K.: Survey paper on multicarrier modulation techniques. In: 5th IEEE international conference on electrical, electronics and computer engineering (2018)
4. Mirahmadi, M., Al-Dweik, A., Shami, A.: BER reduction of OFDM based broadband communication systems over multipath channels with impulsive noise. IEEE Trans. Commun. **61**(11), 4602–4615 (2013)
5. Maximilian, M., Luciano, L.M., Nicola, M., et al.: Widely linear estimation for space-timecoded GFDM in low-latency applications. IEEE Trans. Commun. **63**(11), 4501–4509 (2015)
6. Na, D., Choi, K.: Low PAPR FBMC. IEEE Trans. Wirel. Commun. **17**(1) (2017)

7. Nadal, J., Abdel Nour, C., Baghdadi, A.: Design and evaluation of a novel short prototype filter for FBMC/OQAM modulation. IEEE Access **6**(23) (2018)
8. Michailow, N., Matthé, M., Simões Gaspar, I., Navarro Caldevilla, A., Leonel Mendes, L., Festag, A.: Generalized frequency division multiplexing for 5th generation cellular networks. IEEE Trans. Commun. **62**(9) (2014)
9. El Gamal, H., Damen, M.O.: An algebraic number theoretic framework for space-time coding. In: Proceedings IEEE international symposium on information theory, p. 132. IEEE (2002)
10. Mohammadian, A., Mohammadi, A., Abdipour, A., Baghani, M.: Spectral analysis of GFDM modulated signal under nonlinear behavior of power amplifier. Electr. Eng. Syst. Sci. Sig. Process. **6** (2018)
11. Maximilian, M., Luciano, L.M., Ivan, G., et al.: Multi-user time-reversal STC-GFDMA for future wireless networks. J. Wirel. Commun. Netw. **2015**, 1–8 (2015)
12. Kim, J., Lee, J., Kim, J., Yun, J.: M2M service platforms: survey, issues, and enabling technologies. IEEE Commun. Surveys Tuts. **16**(1), 61–76 (2014)

Author Index

S. Das and M. N. Mohanty (eds.), *Advances in Intelligent Computing and Communication*, Lecture Notes in Networks and Systems 202,
https://doi.org/10.1007/978-981-16-0695-3